Areas of the Standard Normal Distribution

D0076013

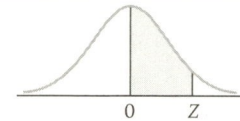

The entries in this table are the probabilities that a standard normal random variable is ~~~~
and Z (the shaded area).

	SECOND DECIMAL PLACE IN Z								
Z	0.00	0.01	0.02	0.03	0.04	0.05	0.06	0.07	0.08	0.09
0.0	.0000	.0040	.0080	.0120	.0160	.0199	.0239	.0279	.0319	.0359
0.1	.0398	.0438	.0478	.0517	.0557	.0596	.0636	.0675	.0714	.0753
0.2	.0793	.0832	.0871	.0910	.0948	.0987	.1026	.1064	.1103	.1141
0.3	.1179	.1217	.1255	.1293	.1331	.1368	.1406	.1443	.1480	.1517
0.4	.1554	.1591	.1628	.1664	.1700	.1736	.1772	.1808	.1844	.1879
0.5	.1915	.1950	.1985	.2019	.2054	.2088	.2123	.2157	.2190	.2224
0.6	.2257	.2291	.2324	.2357	.2389	.2422	.2454	.2486	.2517	.2549
0.7	.2580	.2611	.2642	.2673	.2704	.2734	.2764	.2794	.2823	.2852
0.8	.2881	.2910	.2939	.2967	.2995	.3023	.3051	.3078	.3106	.3133
0.9	.3159	.3186	.3212	.3238	.3264	.3289	.3315	.3340	.3365	.3389
1.0	.3413	.3438	.3461	.3485	.3508	.3531	.3554	.3577	.3599	.3621
1.1	.3643	.3665	.3686	.3708	.3729	.3749	.3770	.3790	.3810	.3830
1.2	.3849	.3869	.3888	.3907	.3925	.3944	.3962	.3980	.3997	.4015
1.3	.4032	.4049	.4066	.4082	.4099	.4115	.4131	.4147	.4162	.4177
1.4	.4192	.4207	.4222	.4236	.4251	.4265	.4279	.4292	.4306	.4319
1.5	.4332	.4345	.4357	.4370	.4382	.4394	.4406	.4418	.4429	.4441
1.6	.4452	.4463	.4474	.4484	.4495	.4505	.4515	.4525	.4535	.4545
1.7	.4554	.4564	.4573	.4582	.4591	.4599	.4608	.4616	.4625	.4633
1.8	.4641	.4649	.4656	.4664	.4671	.4678	.4686	.4693	.4699	.4706
1.9	.4713	.4719	.4726	.4732	.4738	.4744	.4750	.4756	.4761	.4767
2.0	.4772	.4778	.4783	.4788	.4793	.4798	.4803	.4808	.4812	.4817
2.1	.4821	.4826	.4830	.4834	.4838	.4842	.4846	.4850	.4854	.4857
2.2	.4861	.4864	.4868	.4871	.4875	.4878	.4881	.4884	.4887	.4890
2.3	.4893	.4896	.4898	.4901	.4904	.4906	.4909	.4911	.4913	.4916
2.4	.4918	.4920	.4922	.4925	.4927	.4929	.4931	.4932	.4934	.4936
2.5	.4938	.4940	.4941	.4943	.4945	.4946	.4948	.4949	.4951	.4952
2.6	.4953	.4955	.4956	.4957	.4959	.4960	.4961	.4962	.4963	.4964
2.7	.4965	.4966	.4967	.4968	.4969	.4970	.4971	.4972	.4973	.4974
2.8	.4974	.4975	.4976	.4977	.4977	.4978	.4979	.4979	.4980	.4981
2.9	.4981	.4982	.4982	.4983	.4984	.4984	.4985	.4985	.4986	.4986
3.0	.4987	.4987	.4987	.4988	.4988	.4989	.4989	.4989	.4990	.4990
3.1	.4990	.4991	.4991	.4991	.4992	.4992	.4992	.4992	.4993	.4993
3.2	.4993	.4993	.4994	.4994	.4994	.4994	.4994	.4995	.4995	.4995
3.3	.4995	.4995	.4995	.4996	.4996	.4996	.4996	.4996	.4996	.4997
3.4	.4997	.4997	.4997	.4997	.4997	.4997	.4997	.4997	.4997	.4998
3.5	.4998									
4.0	.49997									
4.5	.499997									
5.0	.4999997									
6.0	.499999983									

Business and Economic Statistics
Using Microsoft® Excel

Ken Black

University of Houston—Clear Lake

David L. Eldredge

Murray State University

SOUTH-WESTERN
™
THOMSON LEARNING

Australia · Canada · Mexico · Singapore · Spain · United Kingdom · United States

Business and Economic Statistics Using Microsoft® Excel, by Ken Black, David L. Eldredge

Vice President/Team Director: Melissa Acuna
Senior Acquisitions Editor: Charles McCormick Jr.
Senior Developmental Editor: Alice C. Denny
Senior Marketing Manager: Joseph A. Sabatino
Production Editor: Robert Dreas
Manufacturing Coordinator: Diane Lohman
Media Development Editor: Christine Wittmer
Media Production Editor: Robin Browning
Media Technology Editor: Diane Van Bakel
Internal Design: Sandy Kent, Kent and Co., Cincinnati
Internal Design Revisions: Rik Moore
Cover Design/Photo Illustration/Photography Manager: Rik Moore
Cover Photo Source: PhotoDisc, Inc.
Production House: Shepherd Incorporated
Printer: R.R. Donnelley & Sons Company - Willard Manufacturing Division

Printed in the United States of America
1 2 3 4 5 04 03 02 01

For more information contact South-Western, 5101 Madison Road, Cincinnati, Ohio, 45227 or find us on the Internet at http://www.swcollege.com

For permission to use material from this text or product, contact us by
• **telephone: 1-800-730-2214**
• **fax: 1-800-730-2215**
• **web: http://www.thomsonrights.com**

Library of Congress Cataloging-in-Publication Data

Black, Ken.
  Business and economic statistics using Microsoft Excel / Ken Black,
David Eldredge.
      p. cm.
Includes bibliographical references and index.
  ISBN 0-324-01726-X
  1. Commercial statistics—Computer programs. 2. Microsoft Excel
(Computer file) I. Eldredge, David L. II. Title.
  HF1017 .B568 2001
  519.5'0285'5369—dc21
                                2001042825

Contents in Brief

Contents

Preface

With pleasure and great pride, we present our new text, *Business and Economic Statistics Using Microsoft® Excel.* The text, geared for both introductory undergraduate and MBA-level statistics courses, contains 15 chapters of material focusing on the essentials of applied business statistics. Virtually all key business statistics topics are included in this well-targeted and fast-moving text. We believe that *Business and Economic Statistics Using Microsoft® Excel,* a perfect blend of practical business statistics and Microsoft Excel, is the text that instructors have sought for themselves and their students.

The text is targeted at the large number of business and economics students who will become decision-makers rather than statisticians. Concepts are illustrated through the use of real business and economic data and examples. Along with the major objective of providing a clear, concise presentation of applied statistical techniques, a second objective of this text is to demonstrate and explain how Microsoft's Excel can be used to analyze business and economic data. Thus, users will find a balance between statistical concept/technique presentation and understanding the statistical computer output contained in Excel. We assume that the student has a background in college algebra, but calculus is neither used nor needed.

One of the major assets of this text is the synergy gained by the joint efforts of its authors, Ken Black and David L. Eldredge. Ken Black's experience includes two successful business statistics texts, regarded for their excellent examples and problems. David L. Eldredge is an Excel expert who is well known for his popular *Microsoft® Excel Companion to Business Statistics* that explains and demonstrates how to use Excel to analyze data with statistics.

Excel is seamlessly woven into the presentation of business statistics techniques in this text. After setting the stage in Chapter One by giving an overview of Excel's statistical and graphical capabilities, each Excel statistical technique appears within the relevant section in the text. In other words, Excel techniques are presented *at the point of impact* rather than at the end of the chapter or in an appendix. The explanation of each Excel technique includes its input dialog box along with an example of output so that the student knows exactly how to input data, and how to recognize and interpret the output. Virtually all problems include raw data so that the student will be able to analyze the data using Excel. All data used in problems is included on the CD-ROM that accompanies this text.

There are several statistical topics for which no Microsoft Excel techniques exist. To meet this need, we created FAST ⦚ STAT™, a package of add-ins, which look and act like Excel and can be used for statistical analyses where Excel does not have a technique. Between Excel and FAST ⦚ STAT, most of the statistics presented in this text can be computerized.

FAST ⦚ STAT™

The FAST ⦚ STAT add-ins feature written by David L. Eldredge includes 15 macro procedures that augment and extend the capabilities of the statistical features of Excel. FAST ⦚ STAT is easy to install and is available along with documentation and installation information on the CD-ROM that accompanies this text. This software was designed and tested with Excel 2000 and Excel 97 under the Windows 98 and Windows 95 operating systems.

FAST ⦚ STAT is seamless with Excel and, when installed, becomes another selection on your Excel menu bar. It has a pull-down menu that makes it easy to select the macro you want to use. The macros are written so that both the input dialog boxes and the output very closely resemble Excel. However, while data analysis in Excel is static,

FAST ⑄ STAT's macro output is dynamic. What this means is that if the user of the macro goes back to the original data that was entered and changes any of the input numbers, FAST ⑄ STAT will recompute and update the output for the new data.

The 15 FAST ⑄ STAT macros include: Box and Whiskers Plot (chapter 3), Probabilities (chapter 4), Uniform Distribution (chapter 6), One Mean Using the Z Distribution (chapters 8 & 9), One Mean Using the t Distribution (chapters 8 & 9), Sample Size for One Mean (chapter 8), One Proportion C.I. and H.T. (chapters 8 & 9), Sample Size One Proportion (chapter 8), Two Proportions Hyp. Test (chapter 10), Chi-Square Indep. Test (chapter 11), Regression C.I. and P.I. (chapter 12), Durbin Watson Test (chapter 14), X bar and R charts (chapter 15), P Chart (chapter 15), and c Chart (chapter 15).

FEATURES OF THE TEXTBOOK

Exclusively Microsoft Excel

Today's students have access to Excel at home, school, or work, and they want to use it. Excel is the software featured exclusively in this text. Excel explanations are included in most sections of all chapters of the book. The authors have gone to great lengths to make the Excel discussion seamless with the statistical presentation so that Excel is easy and convenient to use and interpret, but does not distract from the statistical material. There are sixty computer input dialog boxes displayed in the text and more than 200 statistical computer outputs. As discussed earlier, to augment Excel for situations where Excel does not have a statistical feature, the authors created FAST ⑄ STAT.

Topical Coverage

Topical coverage, while traditional, parallels the types of analyses contained in Excel. Designed for a single term course, the text includes the usual treatment of graphical depiction, descriptive statistics, probability, discrete and continuous distributions, sampling distributions, hypothesis testing, and estimation. There is a chapter on one-way ANOVA, randomized block designs, and two-way ANOVA. Included are the chi-square test of goodness-of-fit and the chi-square test of independence. Because business decision-makers often use regression and forecasting tools, separate chapters are dedicated to multiple regression analysis and time-series forecasting. Finally, a chapter on statistical quality control is included.

Demonstration Problems

Virtually every section of every chapter contains demonstration problems. A demonstration problem is an additional example problem that contains a full solution and is used to augment the statistical analysis being discussed. This pedagogical tool supplements explanations and reinforces the material for the student.

Databases

This text contains seven databases, all of which are available in Excel format on the CD-ROM. A manufacturing database, a financial database, a stock market database, an international employment database, an energy database, a healthcare database, and an agri-business database provide more than 8,350 observations and 56 variables. Data are from reliable and recognizable sources, including the U.S. Bureau of Labor Statistics, the New York Stock Exchange, the U.S. Department of Agriculture, Moody's Handbook of Common Stocks, the American Hospital Association, and the U.S. Bureau of Census.

Four of the seven databases have time-series data; one contains 168 months of time-series data ideal for demonstrating and analyzing forecasting techniques.

Problems

More than 600 problems are presented in this book. Many of the problems are taken from actual business and economic situations with *real* companies. Most of the problems contain raw data so that they may be analyzed using Excel or FAST ＼ STAT.

Analyzing the Databases

This feature, located at the end of each chapter, contains several questions/problems that require the application of the techniques from the chapter to data in the variables of the databases. Solving these questions/problems will provide solid computer experience.

Cases

Each chapter ends with a case that is based on a real company. These cases give the student an opportunity to use statistical concepts and techniques presented in the chapter to solve a business dilemma. Some cases feature very large companies, such as Shell Oil, Coca-Cola, or Colgate-Palmolive. Others pertain to small businesses—such as Thermatrix, Robotron, or Fletcher-Terry—that have overcome obstacles to survive and thrive. Most cases include raw data for analysis and questions that encourage the student to use several of the techniques presented in the chapter. In most instances, the student must analyze software output in order to reach conclusions or make decisions.

CD-ROM

Each copy of this text comes with a CD-ROM that contains valuable resources to assist both the student and the instructor. The CD-ROM contains:

- FAST ＼ STAT software and operating manual
- Data files in Excel for all the problems and cases in the text
- All seven databases in Excel format
- For students with no prior experience with Microsoft® Windows and Excel, an introduction to these, including a number of worksheet practices to make statistical worksheets more effective
- A guide to Excel's DATA ANALYSIS TOOLS
- A guide to Excel's STATISTICAL FUNCTIONS including how they differ from DATA ANALYSIS TOOLS
- Additional information and data, and some specialized Excel templates for time series forecasting (Chapter 14)

Ancillary Teaching and Learning Materials

Two ancillaries are available to students either through their bookstore or for direct purchase through the online catalog at *http://www.swcollege.com*.

- Prepared by the textbook authors Ken Black and David L. Eldredge, **WebTutor**™ is a brand-new electronic ancillary. There are two main formats for this product:

  WebTutor is used by an entire class under the direction of the instructor. It provides web-based learning resources (including review materials, study questions with answers, and solutions to the odd-numbered text problems) to students as well as powerful communication and other course management tools, including course calendar, chat, and e-mail for instructors. WebTutor is available on WebCT and Blackboard. See *http://webtutor.thomsonlearning.com* for more information.

Personal WebTutor provides the learning resources for individual students to purchase and use for study and review. See *http://pwt.swcollege.com* for more information about this product.

■ *A Microsoft® Excel Companion for Business Statistics* (ISBN: 0-324-06898-0) by David L. Eldredge of Murray State University, provides step-by-step instructions for using Excel to solve many of the problems included in an introductory business and economics statistics course.

In addition, the *Instructor's Resource CD* (ISBN: 0-324-01729-4) provides all instructor ancillaries. Included in this convenient format are:

■ *Instructor's Manual*—The Instructor's Manual, prepared by the text authors, contains chapter outlines, teaching strategies, and full solutions to all problems and cases presented in the text.

■ *PowerPoint™ Presentation Slides*—The presentation slides contain graphics to help instructors create stimulating lectures. The slides may be adapted using PowerPoint software to facilitate classroom use.

■ *Test Bank* and *ExamView™*—The Test Bank includes multiple choice questions for each chapter. The Test Bank is provided in Microsoft® Word format on the IRCD. The ExamView computerized testing software allows instructors to create print and on-line exams.

Acknowledgments

We join South-Western in thanking the reviewers and advisors who cared enough to provide us with their excellent insights and advice in the development of this text. These colleagues include:

Pricilla Chaffe-Stengel, California State University, Fresno
Alexander Holmes, University of Oklahoma
June Lapidus, Roosevelt University
Valarie P. Muehsam, Sam Houston State University
Handanhal (H.V.) Ravinder, University of New Mexico
Farhad Saboori, Albright College

We would like to thank our families for their unwavering support of this project and for the sacrifices that they have made to allow us to bring this project to fruition. These special women are Carolyn, Judy, Kristi, Caycee, Lynn, Tracey, and Wendi. We would also like to express our appreciation for the personal interest and support shown to us by colleagues at our universities. Included at the University of Houston–Clear Lake are President William Staples and Ted Cummings, dean of the School of Business and Public Administration. Assistant Professor George Rice and Dannie E. Harrison, Dean of the College of Business and Public Affairs from Murray State University, merit special note.

There are many people within the South-Western/Thomson Learning™ publishing group that we would like to thank for their invaluable assistance on this project. In particular, we would like to cite: Charles McCormick, Jr., senior acquisitions editor, who had the foresight to envision this project and who gave us encouragement from the inception; Joe Sabatino, senior marketing manager, who provided market insights and believed in us and the project from the beginning; Alice Denny, senior development editor, who orchestrated the project plan, kept the text and ancillaries on schedule, and provided insight and helpful advice; Rick Moore, designer, who prepared the layout and photos; and Bob Dreas, production editor, who provided professional assistance in producing the finished text.

Ken Black and David L. Eldredge

About the Authors

Ken Black is currently Professor of Decision Sciences in the School of Business and Public Administration at the University of Houston–Clear Lake. Born in Cambridge, Massachusettes and raised in Missouri, he earned a Bachelor's degree in mathematics from Graceland College, a Master's degree in math education from the University of Texas at El Paso, a Ph.D. in business administration in management science, and a Ph.D. in educational research from the University of North Texas.

Since joining the faculty in 1979, Professor Black has taught all levels of statistics courses, forecasting, management science, market research, and production/operations management. He has published fifteen journal articles and more than twenty professional papers, as well as two textbooks, *Business Statistics: An Introductory Course* and *Business Statistics: Contemporary Decision Making.* He has consulted with many different companies, including Aetna, City of Houston, NYLCare, AT&T, Johnson Space Center, Southwest Information Resources, Connect Corporation, and Eagle Engineering.

Ken Black and his wife Carolyn have two daughters, Caycee and Wendi. His hobbies include playing the guitar, reading, coaching girls' softball, and participating in Master's track and field as a long jumper and a triple jumper.

David L. Eldredge is Professor of Computer Science in the College of Business and Public Policy at Murray State University. He earned a Bachelor of Science degree in aerospace engineering from Iowa State University, and an M.S. and a Ph.D. in operations research from The Ohio State University.

Professor Eldredge's current teaching responsibilities include management science for Murray State University's MBA program and business statistics for the undergraduate business administration program. He served as dean of the College of Business for eight years. Dave Eldredge has had more than fifteen books and papers published, primarily in the areas of simulation and gaming, and the use of Microsoft® Excel for data analysis. He also has ten years of industrial experience with four different Fortune 500 firms.

Dave Eldredge and his wife Judy are parents of three daughters—Kristi, Lynn and Tracey—and have seven grandchildren. The Eldredges share a love for camping and travel; they have visited all 50 states and many other countries.

1

Introduction to Business Statistics with Excel

Learning Objectives

The primary objective of Chapter 1 is to introduce you to the world of statistics, thereby enabling you to:

1. Define statistics.

2. Become aware of a wide range of applications of statistics in business.

3. Differentiate between descriptive and inferential statistics.

4. Classify numbers by level and type of data and understand why doing so is important.

5. Become aware of the statistical analysis capabilities of Excel.

Every minute of the working day, decisions are made in businesses and industries around the world that determine if companies are profitable and growing or if they are stagnating and failing. Most of these decisions are made with the assistance of information gathered about the marketplace, the economic and financial environment, the work force, the competition, and other relevant factors. Such information usually comes in the form of data or is accompanied by data. Business statistics is the means through which such data are collected, analyzed, summarized, and presented to facilitate the decision-making process. Thus, in the twenty-first century, business statistics plays an important role in decision making within the dynamic world of business.

Business statistics entails working with large amounts of data that usually require the data to be computerized. Oftentimes the data are presented in the form of computer spreadsheets, since spreadsheet software is the most powerful and most available general-purpose managerial software in business and industry. Currently, the most widely used spreadsheet program is Microsoft® **Excel.** Business and industrial managers and analysts have used Excel throughout their organizations for their computational, charting, and data management needs for years. Beyond these three uses, current versions of Excel include a number of features that provide the capability for easily conducting many statistical analyses. Within this book, we use Excel to computerize most of our statistical procedures.

Excel
Computerized spreadsheet software for analyzing, charting, and managing data.

We begin in the next section of this chapter by citing a number of real-world examples of the use of business statistics. In the following section, we discuss some statistical concepts that are basic to understanding and communicating about business statistics. We continue with a discussion of data, the essential element of statistics, in the third section. Finally, the last six sections of this chapter introduce the use of Excel features for conducting statistical analyses.

1.1
Business Statistics Applications

Virtually every area of business uses statistics in decision making. A survey of 302 directors and vice presidents of marketing and marketing communications conducted by Pitney Bowes at midsize and large U.S. companies revealed that almost 35 percent said that direct mail or catalogues was the most cost-effective way to reach customers. An additional 11 percent felt that the Internet was most cost-effective. The study also showed that over 25 percent said that the best way to increase brand identity was through direct mail or catalogues. These and other statistics gathered and summarized in this study can help decision makers to solve the dilemma of finding cost-effective vehicles for their products.

If decision makers are looking for ways to reduce healthcare expenses among their work force, then they might do well to learn from a study of some 46,000 employees conducted by the Health Enhancement Research Organization in which it was discovered that depression and stress seem to have a greater impact on higher medical expenses than do high blood sugar, obesity, or smoking. The study showed that depressed workers had medical expenses that were 70 percent higher than nondepressed workers and that workers who said they were under constant stress had expenditures that were 46 percent higher than stress-free peers. On the other hand, the medical expenses for people who suffer from high blood pressure were just 11 percent higher than for those who do not. This information, along with other statistics reported in this study, can help decision makers to develop a strategy for reducing medical expenses among workers.

Yankelovich Partners have identified a special subgroup of Generation Xers which they call "Friend-zied Xers." What does this group look like? Business statistics gathered on this group show that they possess a median household income of $40,000 and a median personal income of $19,000. Twenty percent hold at least one college degree, 63 percent have never been married, and 55 percent are under 25 years old. These types of statistics can help decision makers to size up the market potential of "Friend-zied Xers" and guide marketers and product designers in meeting the needs of this target group.

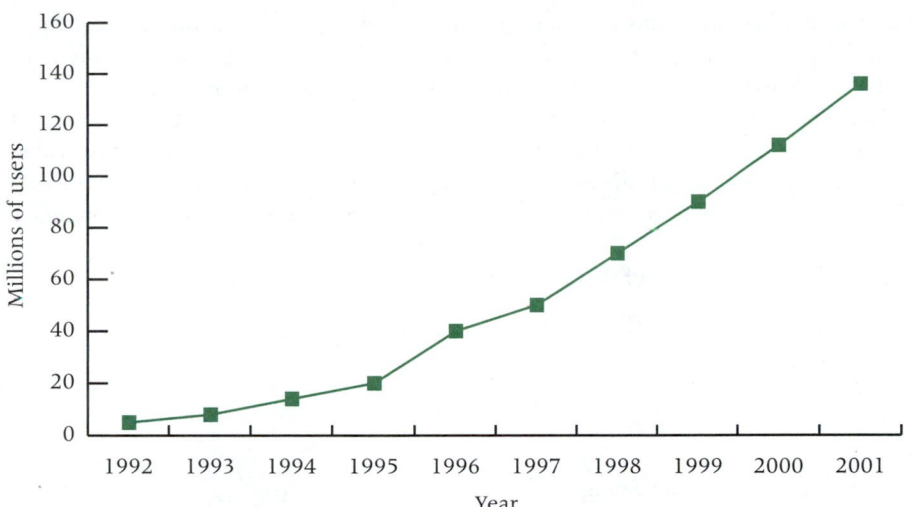

Figure 1.1

Excel graph of consumer use of e-mail in the United States

How is the economy doing? One *Wall Street Journal* report published statistics to help investors and other decision makers to track the state of the economy. Some of the business statistics included the number of new home sales, an index of consumer confidence, the percentage increase in the gross domestic product, the number of initial jobless claims, and the unemployment rate. These statistics and others can serve as indicators of economic and financial states to come and can be used by forecasters as they attempt to predict future business climates.

Consumer use of E-mail has grown exponentially as shown by the Excel graph in Figure 1.1. The data for this graph, gathered by Forrester Research, Inc., indicate that consumer use of E-mail in 1999 was almost twenty times what it was in 1992 with expectations that it will nearly double again in the next five years. These business statistics can be used by managers who are hoping to find better ways for employees to communicate and by communications companies in planning new market opportunities.

In this text we will examine several types of graphs for depicting data and study ways to rearrange and/or restructure data into forms that are both meaningful and useful to decision makers. We will learn about techniques for sampling from a population in such a way that studies of the business world can be conducted more inexpensively and more timely. We will explore various ways to forecast future values and examine techniques for predicting trends. This text also includes many statistical tools for testing hypotheses and for estimating population values. These and many other exciting statistics and statistical techniques await us on this journey through business statistics. Let us begin.

1.2
Basic Statistical Concepts

Business statistics, like any area of study, requires an introduction of some basic concepts in order to understand and communicate about the subject. We begin with a discussion of the word *statistics*. It has many different meanings in our culture. *Webster's Third New International Dictionary* gives a comprehensive definition of **statistics** as *a science dealing with the collection, analysis, interpretation, and presentation of numerical data.* Viewed from this perspective, statistics includes all the topics presented in this text. Statistics also is a branch of mathematics, and most of the science of statistics is based on mathematical thought and derivation. Many academic areas, including business, offer statistics courses within their own disciplines. However, statistics has become a course of study in its own right.

Statistics
A science dealing with the collection, analysis, interpretation, and presentation of numerical data.

People often use the word *statistics* to refer to a group of data. They may say, for example, that they gathered statistics from their business operation. What they are referring to is measured facts and figures. The media and others also use the word *statistic* to refer to a death. Becoming a statistic in this sense of the word is obviously undesirable.

The expression *statistics* is used in at least two other important ways. First, statistics can be descriptive measures computed from a sample and used to make determinations about a population. This usage is discussed later. Second, statistics can be the distributions used in the analysis of data. For example, a researcher using the t distribution to analyze data might refer to use of the t statistic in analyzing the data.

The following are some of the common uses of the word *statistics*.

1. Science of gathering, analyzing, interpreting, and presenting data
2. Branch of mathematics
3. Course of study
4. Facts and figures
5. A death
6. Measurement taken on a sample
7. Type of distribution used to analyze data

Population
A collection of persons, objects, or items of interest.

Census
A process of gathering data from the whole population for a given measurement of interest.

Sample
A portion of the whole.

Descriptive statistics
Statistics that have been gathered on a group to describe or reach conclusions about that same group.

The study of statistics can be organized in a variety of ways. One of the main ways is to subdivide statistics into two branches: descriptive statistics and inferential statistics. To understand the difference between descriptive and inferential statistics, definitions of population and sample are helpful. *Webster's Third New International Dictionary* defines **population** as *a collection of persons, objects, or items of interest.* The population can be a widely defined category, such as "all automobiles" or it can be narrowly defined such as "all Ford Mustang cars produced from 1997 to 1999." A population can be a group of people, such as "all workers presently employed by Microsoft," or it can be set of objects, such as "all dishwashers produced on February 3, 2000 by the General Electric Company at the Louisville plant." The researcher defines the population to be whatever he or she is studying. When a researcher *gathers data from the whole population for a given measurement of interest,* it is called a **census.** Most people are familiar with the U.S. Census. It is an attempt made every 10 years to measure all persons living in this country. If a researcher is interested in ascertaining the Scholastic Aptitude Test (SAT) scores for all students at the University of Arizona, one way to do so is to conduct a census of all students currently enrolled at that university.

A **sample** is *a portion of the whole* and, if property taken, is representative of the whole. For various reasons (explained in Chapter 7), researchers often prefer to work with a sample of the population instead of the entire population. For example, in conducting quality control experiments to determine the average life of light bulbs, a light bulb manufacturer might randomly sample only 75 light bulbs during a production run. Because of time and money limitations, a human resources manager might take a random sample of 40 employees instead of using a census to measure company morale.

If a business analyst is *using data gathered on a group to describe or reach conclusions about that same group,* the statistics are called **descriptive statistics.** For example, if an instructor produces statistics to summarize a class's examination effort and uses those statistics to reach conclusions about that class only, the statistics are descriptive. The instructor can use these statistics to discuss class average, talk about the range of class scores, or present any other data measurements for the class based on the test.

Most athletic statistics, such as batting average, rebounds, and first downs, are descriptive statistics because they are used to describe an individual or team effort. Many of the statistical data generated by businesses are descriptive. They might include number of employees on vacation during June, average salary at the Denver office, corporate sales for 2001, average managerial satisfaction score on a companywide census of employee attitudes and average return on investment for the Lofton Company for the years 1988 through 2001.

Another type of statistics is called **inferential statistics.** If a business analyst *gathers data from a sample and uses the statistics generated to reach conclusions about the population from which the sample was taken,* the statistics are inferential statistics. The data gathered are used to *infer* something about a larger group. Inferential statistics are sometimes referred to as *inductive statistics.* The use and importance of inferential statistics are ever growing.

One application of inferential statistics is in pharmaceutical research. Some new drugs are very expensive to produce, and therefore tests must be limited to small samples of patients. Utilizing inferential statistics, researchers can design experiments with small randomly selected samples of patients and attempt to reach conclusions and make inferences about the population.

Market researchers use inferential statistics to study the impact of advertising on various market segments. Suppose a soft drink company creates an advertisement depicting a dispensing machine that talks to the buyer, and market researchers want to measure the impact of the new advertisements on various age groups. The researcher could stratify the population into age categories ranging from young to old, randomly sample each stratum, and use inferential statistics to determine the effectiveness of the advertisement for the various age groups in the population. The advantage of using inferential statistics is that they enable the researcher to study effectively a wide range of phenomena without having to conduct a census. Most of the topics discussed in this text pertain to inferential statistics.

A *descriptive measure of the population* is called a **parameter.** Parameters are usually denoted by Greek letters. Examples of parameters are population mean (μ), population variance (σ^2), and population standard deviation (σ). A *descriptive measure of a sample* is called a **statistic.** Statistics are usually denoted by Roman letters. Examples of statistics are sample mean ($\bar{X}$), sample variance (S^2) and sample standard deviation (S).

Differentiation between the terms *parameter* and *statistic* is important only in the use of inferential statistics. A statistician often wants to estimate the value of a parameter or conduct tests about the parameter. However, the calculation of parameters is usually either impossible or infeasible because of the amount of time and money required to take a census. In such cases, the statistician can take a random sample of the population, calculate a statistic on the sample, and infer by estimation the value of the parameter. The basis for inferential statistics, then, is the ability to make decisions about parameters without having to complete a census of the population.

For example, a manufacturer of washing machines would probably want to determine the average number of loads that a new machine can wash before it has to be repaired. The parameter is the population mean or average number of washes per machine before repair. A company statistician takes a sample of machines, computes the number of washes before repair for each machine, averages the numbers, and estimates the population value or parameter by using the statistic, which in this case is the sample average. Figure 1.2 demonstrates this process.

Inferential statistics
Statistics that have been gathered from a sample and used to reach conclusions about the population from which the sample was taken.

Parameter
A descriptive measure of the population.

Statistic
A descriptive measure of a sample.

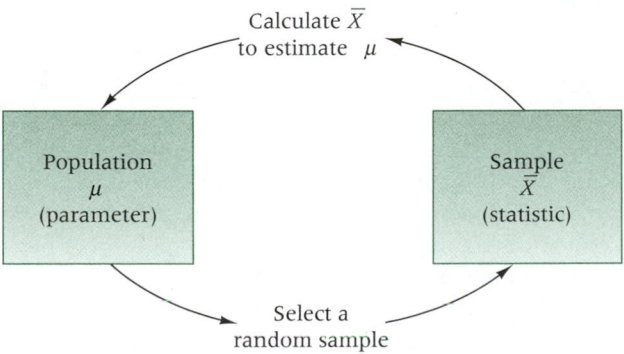

Figure 1.2

Process of inferential statistics to estimate a population mean (μ)

Inferences about parameters are made under uncertainty. Unless parameters are computed directly from the population, the statistician never knows with certainty whether the estimates or inferences made from samples are true or not. In an effort to estimate the level of confidence in the result of the process, business analysts use probability statements. Therefore, part of this text is devoted to probability (Chapter 4).

1.3
Data

Millions of pieces of data are gathered in businesses every day, representing countless items. For example, numbers represent dollar cost of items produced, geographical locations of retail outlets, weights of shipments, and rankings of subordinates at yearly appraisal reviews. All such data should not be analyzed the same way statistically because the entities represented by the numbers are different. For this reason, the business analysts need to be able to identify the *level of data measurement* used to collect the data being analyzed.

The disparate use of numbers can be illustrated by the numbers 40 and 80, which could represent the weights of two objects being shipped, the ratings received on a consumer test by two different products, or the football jersey numbers of a fullback and a wide receiver. Although 80 lb is twice as much as 40 lb, the wide receiver is probably not twice as big as the fullback. Averaging the two weights seems reasonable, but averaging the football jersey numbers makes no sense. The appropriateness of the data analysis depends on the level of measurement of the data gathered. The phenomenon represented by the numbers determines the level of data measurement. Four common levels of data measurement follow.

1. Nominal
2. Ordinal
3. Interval
4. Ratio

Nominal Level

Nominal level data
The lowest level of data measurement; used only to classify or catagorize.

The *lowest level of data measurement* is the nominal level. Numbers representing **nominal level data** (the word *level* often is omitted) can be *used only to classify or categorize*. Employee identification numbers are an example of nominal data. The numbers are used only to differentiate employees and not to make a value statement about them. Many demographic questions in surveys result in data that are nominal because the questions are used for classification only. An example of such a question is:

Which of the following employment classifications best describes your area of work?

A. Educator   **D.** Lawyer
B. Construction worker   **E.** Doctor
C. Manufacturing worker   **F.** Other

Suppose that, for computing purposes, an educator is assigned a 1, a construction worker is assigned a 2, a manufacturing worker is assigned a 3, and so on. These numbers should be used only to classify respondents. The number 1 does not denote the top classification. It is used only to differentiate an educator (1) from a lawyer (4).

Some other types of variables that often produce nominal level data are gender, religion, ethnicity, geographic location, and place of birth. Social security numbers, telephone numbers, employee ID numbers, and ZIP code numbers are further examples of nominal data. Statistical techniques that are appropriate for analyzing nominal data are limited. However, some of the more widely used statistics, such as the chi-square statistic, can be applied to nominal data, often producing useful information.

Ordinal Level

Ordinal level data measurement is higher than the nominal level. In addition to the nominal level capabilities, ordinal level measurement can be used to rank or order objects. For example, using ordinal data, a supervisor can evaluate three employees by ranking their productivity with the numbers 1 through 3. The supervisor could identify one employee as the most productive, one as the least productive, and one as middling by using ordinal data. However, the supervisor could not use ordinal data to establish that the intervals between the employees ranked 1 and 2 and between the employees ranked 2 and 3 are equal. That is, she could not say that the differences in the amount of productivity between workers ranked 1, 2, and 3 are necessary the same. With ordinal data, the distances or spacing represented by consecutive numbers are not always equal.

Some questionnaire Likert-type scales are considered by many researchers to be ordinal in level. The following is an example of one such scale:

This computer tutorial is _____   _____   _____   _____   _____

	not helpful	somewhat helpful	moderately helpful	very helpful	extremely helpful
	1	2	3	4	5

When this survey question is coded for the computer, only the numbers 1 through 5 will remain, not the adjectives. Virtually everyone would agree that a 5 is higher than a 4 on this scale and that ranking responses is possible. However, most respondents would not consider the differences between not helpful, somewhat helpful; moderately helpful, very helpful, and extremely helpful to be equal.

Mutual funds as investments are sometimes rated in terms of risk by using measures of default risk, currency risk, and interest rate risk. These three measures are applied to investments by rating them as having high, medium, and low risk. Suppose high risk is assigned a 3, medium risk a 2, and low risk a 1. If a fund is awarded a 3 rather than a 2, it carries more risk, and so on. However, the differences in risk between categories 1, 2, and 3 are not necessarily equal. Thus, these measurements of risk are only ordinal level measurements. Another example of the use of ordinal numbers in business is the ranking of the top 50 most admired companies in *Fortune* magazine. The numbers ranking the companies are only ordinal in measurement. Certain statistical techniques are specifically suited to ordinal data, but many other techniques are not appropriate for use on ordinal data.

Ordinal level data
Next-higher level of data from nominal level data; can be used to order or rank items, objects, or people.

Interval Level

Interval level data measurement is the *next to the highest level of data*. Interval measurements *have all the properties of ordinal level data, but in addition the distances between consecutive numbers have meaning* and the data are always numerical. The distances represented by the differences between consecutive numbers are equal. That is, interval data have *equal intervals*. An example of interval measurement is Fahrenheit temperature. With Fahrenheit temperature numbers, the temperatures can be ranked, and the amounts of heat between consecutive readings, such as 20°, 21°, and 22°, are the same.

In addition, with interval level data, the zero point is a matter of convention or convenience and not a natural or fixed zero point. That is, zero is just another point on the scale and does not mean the absence of the phenomenon. For example, zero degrees Fahrenheit is not the lowest possible temperature. Some other examples of interval level data are the percentage change in employment, the percentage return on a stock, and the dollar change in stock price.

Interval level data
Next to highest level of data. These data have all the properties of ordinal level data, but in addition, intervals between consecutive numbers have meaning.

With interval level data, converting the units from one measurement to another involves multiplying by some factor, *a,* and adding another factor, *b,* such that $y = b + ax$. As an example, converting from centigrade temperature to Fahrenheit temperature involves the relationship.

$$\text{Fahrenheit} = 32 + \frac{9}{5}\,\text{centigrade}$$

Ratio Level

Ratio level data
Highest level of data measurement; contains the same properties as interval level data, with the additional property that zero has meaning and represents the absence of the phenomenon being measured.

Ratio level data measurement is the *highest level of data measurement.* Ratio data *have the same properties as interval data,* but ratio data have an *absolute zero* and the ratio of two numbers is meaningful. The notion of absolute zero means that zero is fixed and there *is a zero value in the data that represents the absence of the characteristic being studied.* The value of zero cannot be arbitrarily assigned because it represents a fixed point. This definition enables the business analyst to create *ratios* with the data.

Examples of ratio data are height, weight, time, volume, and Kelvin temperature. With ratio data, a researcher can state that 180 pounds of weight is twice as much as 90 pounds or, in other words, make a ratio of 180:90. Many of the data gathered by machines in industry are ratio data.

Other examples in the business world that are ratio level in measurement are production cycle time, work measurement time, passenger miles, number of trucks sold, complaints per 10,000 fliers, and number of employees. With ratio level data, there is no *b* factor in converting units from one measurement to another, that is, $y = ax$. As an example, in converting height from yards to feet: feet = 3 · yards.

Comparison of the Four Levels of Data

Figure 1.3 shows the relationships of the usage potential among the four levels of data measurement. The concentric squares denote that each higher level of data can be analyzed by any of the techniques used on lower levels of data but, in addition, can be used in other statistical techniques. That is, ratio data can be analyzed by any statistical technique applicable to the other three levels of data plus some others.

Nominal data are the most limited data in terms of the types of statistical analysis that can be used with them. Ordinal data allow the business analyst to perform any analysis that can be done with nominal data and some additional analyses. With ratio data, a business analyst can make ratio comparisons and appropriately do any analysis that can be

---

Figure 1.3

Usage potential of various levels of data

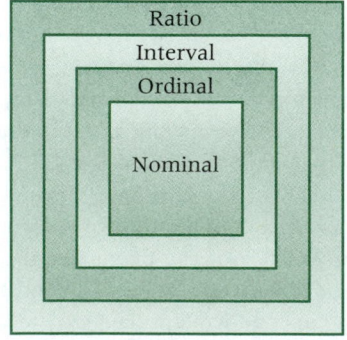

performed on nominal, ordinal, or interval data. Some statistical techniques require ratio data and cannot be used to analyze other levels of data.

Qualitative versus Quantitative Data

Data are collected using one of the four levels of data measurement discussed above. These four levels can be combined into two *types of data.* The data type determines the statistical analysis that is appropriate.

Because nominal and ordinal level data are derived from imprecise measurements such as demographic questions, the categorization of people or objects, or the ranking of items, *nominal* and *ordinal data* are *nonmetric data.* They are referred to as *categorical* or **qualitative data.** Qualitative data essentially consist of labels or categories that are either unordered—as for nominal data, or ordered—as for ordinal data. They can be either nonnumeric or numeric. For example you could refer to your college classification by the nonnumeric category of freshman, sophomore, junior, or senior, or by the corresponding numeric category of 1, 2, 3, or 4.

On the other hand, because interval and ratio level data are usually gathered by precise instruments that often are used in production and engineering processes, in national standardized testing, or in standardized accounting procedures, *interval* and *ratio* data are *metric data.* They are referred to as *numerical* or **quantitative data.** Quantitative data consist of values that are naturally numerical. Thus, quantitative data are always numeric.

Some statistical analysis procedures are appropriate for qualitative data and some for quantitative data. *Qualitative data* are usually summarized by counting the number of items in each category or by computing the proportion of the items in each category. The tabular and graphical procedures of Chapter 2 are appropriate descriptive procedures for qualitative data. Inferential procedures for qualitative data include the sections of Chapters 7, 8, 9, 10, and 15 that treat proportions plus the sections of Chapter 11 that utilize the chi-square statistic.

Quantitative data can be summarized by counting the number of values within a specified interval or by computing the proportion of items in each interval. Thus, the tabular and graphical procedures of Chapter 2 are appropriate descriptive procedures for quantitative data. However, you can also perform the usual analysis procedures, such as computing average values and measures of data variability. Accordingly, the numerical procedures of Chapter 3 are appropriate descriptive procedures for quantitative data. Inferential procedures for quantitative data include the sections of Chapters 7, 8, 9, 10, and 15 that treat averages (means) plus Chapters 4, 5, 6, 11, 12, 13, and 14.

In closing this section, we note one further classification for data. Numeric data are either *discrete* or *continuous* depending on the values that are possible. If the values can only come *from a list of specific possible values,* the data are **discrete.** For example, the number of students in a statistics class is a discrete data item. Discrete data are numerical values resulting from a counting process.

On the other hand, if you *cannot list all the possible values for a data item,* it is **continuous.** For example, the weight of a student may be said to be 135 pounds. However, the actual weight might be 135.089 or 134.97 or an infinite number of possible values. Thus, the student's weight is a continuous data item. Continuous data are numerical values resulting from a measuring process.

Qualitative data that are presented numerically are always discrete. However, quantitative data can either be discrete or continuous. A summary of all the data classifications we have considered in this section is given in Figure 1.4. The dichotomy of discrete versus continuous data is important to our later consideration of describing the possible values for data items through the use of probability distributions. In Chapter 5 we discuss discrete distributions, and in Chapter 6 we discuss continuous distributions.

Qualitative data
Data of the nominal or ordinal level that classifies by a nonnumeric or numeric label or category.

Quantitative data
Data of the interval or ratio level that measures on a naturally occurring numeric scale.

Discrete data
Numeric data in which the set of all possible values is at most a finite or a countably infinite number of possible values.

Continuous data
Numeric data that take on values at every point over a given interval.

Figure 1.4   Summary of data classifications

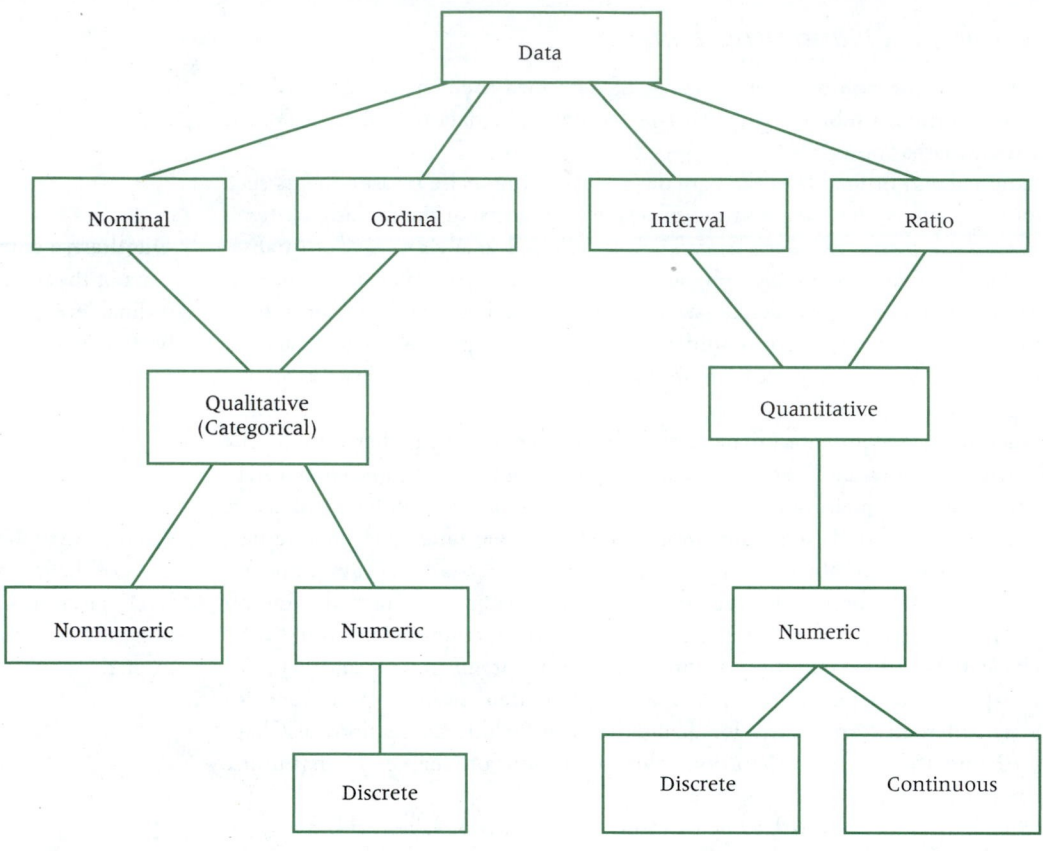

Many changes have occurred in the healthcare industry. Because there is more competition for patients among providers and to determine how they can better serve their clientele, hospital administrators sometimes mail a quality satisfaction survey to their patients after the patient is released. Below are types of questions that are sometimes asked on such a survey. These questions will result in what level of data measurement?

1. How long ago were you released from the hospital?
2. Which type of unit were you in for most of your stay?

    _____ Coronary care   _____ Medical unit
    _____ Intensive care   _____ Pediatric/children's unit
    _____ Maternity care   _____ Surgical unit

3. In choosing a hospital, how important was the hospital's location?

    (Circle One)

    Very   Somewhat   Not Very   Not at All
    Important   Important   Important   Important

4. How serious was your condition when you were first admitted to the hospital?

    _____ Critical   _____ Serious   _____ Moderate   _____ Minor

5. Rate the skill of your doctor:

    _____ Excellent   _____ Very Good   _____ Good   _____ Fair   _____ Poor

6. On the scale from one to seven shown below, rate the nursing care:

    Poor   1   2   3   4   5   6   7   Excellent

SOLUTION

Question 1 is a time measurement with an absolute zero and is therefore ratio level measurement. A person who has been out of the hospital for two weeks has been out twice as long as someone who has been out of the hospital for one week.

Question 2 yields nominal data because the patient is asked only to categorize the type of unit he or she was in. There is no hierarchy or ranking of type of unit in this question. Questions 3, 4, and 5 are likely to result in ordinal level data. Suppose a number is assigned the descriptors in each of these three questions. For question 3, "very important" might be assigned a 4, "somewhat important" a 3, "not very important" a 2, and "not at all important" a 1. Certainly, the higher the number, the more important is the hospital's location. Thus, these responses can be ranked by selection. However, the increases in importance from 1 to 2 to 3 to 4 are not necessarily equal. This same logic applies to the numeric values assigned in questions 4 and 5.

Question 6 displays seven numeric choices with equal distances between the numbers shown on the scale and no adjective descriptors assigned to the numbers. Many researchers would declare this to be interval level measurement because of the equal distance between numbers and the fact that there is no true zero on this scale. Some researchers might argue that because of the imprecision of the scale and the vagueness of selecting values between "poor" and "excellent" the measurement is only ordinal in level.

1.4
Statistical Analysis Using Excel

We use Microsoft Excel to computerize most of our statistical procedures within this textbook. Excel includes a number of features that facilitate the computation and charting requirements for statistical analyses. Primary among these are the 19 **Data Analysis Tools** and the 80 **Statistical Functions.** Both of these features are used throughout this textbook. Two additional Excel features that support specific aspects of statistical analysis within this textbook include the **Chart Wizard** and the **Trendline** feature for charts.

Although these four Excel features form a powerful foundation for statistical analyses, there are some additional needed analyses that are not treated by them. For many of these needs, we have created an Excel macro procedure to perform the analysis. A *macro procedure* is sequence of Excel commands that automatically performs a specific analysis task once the procedure has been initiated. We have developed 15 macros and have put them together in an Excel add-in application called FAST＼STAT. *Excel add-in applications* facilitate the distribution of custom features such as macro procedures. An add-in application acts as if it is built into Excel itself.

These five features provide the basis for the statistical analysis of this textbook. We discuss each of them in the following five sections. As you encounter the use of each of these features in Chapters 2 through 15, you may wish to refer back to the following appropriate section to review the introduction to the feature.

1.5
Data Analysis Tools

Excel includes 19 **Data Analysis Tools.** One of those provides the engineering application *Fourier Analysis* and the remaining 18 provide statistical analyses. We demonstrate the use of all but two of these 18 within the following chapters. In addition, the CD-ROM that accompanies this text presents a complete listing and description of all 18 statistical data analysis tools. Utilizing a question-and-answer format, the discussion on the CD-ROM addresses the following questions.

1. What are the data analysis tools?
2. Where can I find the data analysis tools?
3. Are the data analysis tools available on the computer I am using?

Data analysis tools
A statistical add-in feature that augments Excel's data analysis capabilities.

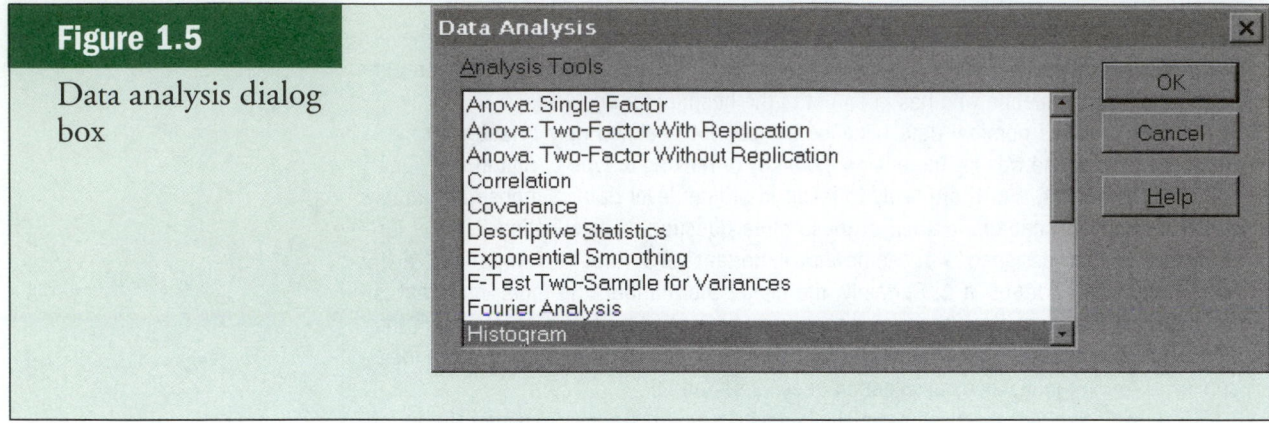

Figure 1.5

Data analysis dialog box

4. How do I use the data analysis tools?
5. For what analyses are the data analysis tools used?

Refer to the CD-ROM discussion for the answer to any of these questions.

To access the data analysis tools, you should click on **Tools** on the menu bar, and then click on **Data Analysis** on the subsequent pull-down menu. The result will be the dialog box shown in Figure 1.5. It may happen that the pull-down menu for the computer you are using does not include the entry **Data Analysis.** If so, you will need to do some preparation before accessing the **Data Analysis** dialog box. The details of the necessary preparation are found in the discussion on the CD-ROM of *Question 3* in the above list.

From the dialog box of Figure 1.5, click on the tool you wish to use. If the one you wish to use is not displayed in the list box, click on the vertical scroll bar arrow to scroll to it. After you have selected the tool you wish to use, click the **OK** button. The result will be a second dialog box allowing you to enter the specific cell ranges and values for your data analysis problem. You may wish to try selecting a tool and viewing its dialog box. For example, refer to Figure 2.1 in Chapter 2 to see the dialog box for the **Histogram** data analysis tool.

The discussion on the accompanying CD-ROM of *Question 5* in the prior list gives a short description of the use of each of the statistical tools. In addition, it classifies the tools into six categories: *Descriptive Statistics; Sampling from Distributions; Hypothesis Testing; Analysis of Variance; Regression and Correlation; and Time Series Forecasting.* The chapter(s) within this textbook that demonstrates each statistical tool is also noted.

You may wish to explore the purpose and use of some of these tools at this time through the *Help System.* First select the analysis tool of interest from the **Data Analysis** dialog box (Figure 1.5). Then in the subsequent dialog box click on the **Help** command button.

1.6
Statistical Functions

There are hundreds of built-in functions (predefined formulas) in Excel. Eighty are classified as **Statistical Functions** beginning with **AVEDEV** and ending with **ZTEST.** We use many of these within this textbook. In addition, a discussion on the accompanying CD-ROM presents a complete listing and description of all 80 of the statistical functions. Again we use a question-and-answer format and address the following questions on the CD-ROM.

Statistical functions
Predefined Excel formulas for performing statistical analyses.

1. What are the statistical functions?
2. How do the statistical functions and the data analysis tools differ?
3. How do I use the statistical functions?
4. For what analyses are the statistical functions used?

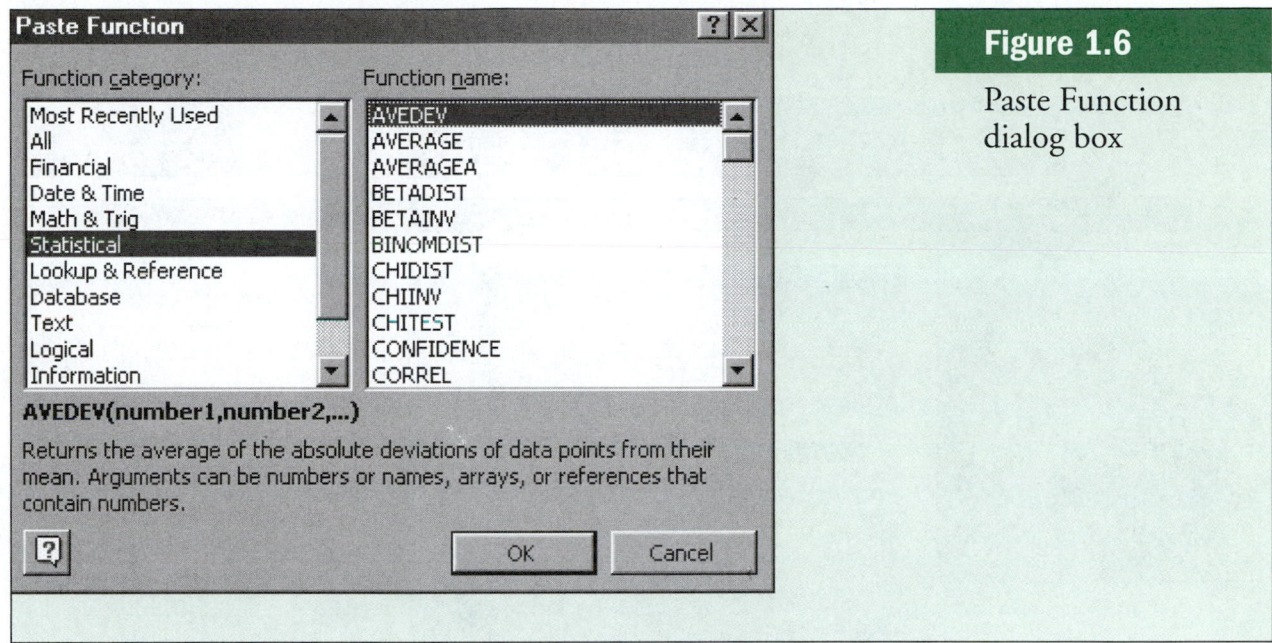

Figure 1.6

Paste Function
dialog box

Refer to the CD-ROM discussion for the answer to any of these questions.

The Statistical Functions both supplement and duplicate the analysis capabilities of the Data Analysis Tools. However there are a number of differences that are presented in the discussion on the CD-ROM of *Question 2* of above list. The primary difference is that the results from the *Tools* usually are numbers and the results from the *Functions* are formulas.

Access to the statistical functions is facilitated through the **Paste Function.** It has an icon on the standard toolbar labeled with the symbol *fx.* Click on the icon and the dialog box of Figure 1.6 will be displayed. Next click on the category **Statistical** in the list box on the left. As a result the right list box will present the 80 statistical functions. You can scroll through the list to find the function you wish to use. Next click on the function you wish to use and a second dialog box will be shown. For example, refer to Figure 3.4 of Chapter 3 to see the dialog box for the function **AVERAGE,** which computes the average or mean of a data set.

The discussion on the CD-ROM of *Question 4* in the above list gives a short description of the use of each of the statistical functions. In addition, we have classified the functions into fifteen categories beginning with *Descriptive Statistics—Measures of Location* and ending with *Regression and Correlation—Exponential Regression Analysis.*

You will note in Figure 1.6 that a brief description of the function that is highlighted in the right list is given below the lists. Much greater detail about the selected function can be obtained by clicking on the **Help** button in the lower left corner of the dialog box. This will activate the **Office Assistant** feature of Excel. Click on the option labeled as **Help with this feature** within the help balloon. This action will result in a second help balloon. Click on the option **Help on selected function.** The result will be the presentation of a detailed help window. You may wish to explore the help facility for one or more functions of interest to you at this time.

Graphs and charts are effective in summarizing and visually presenting statistical data. Excel's **Chart Wizard** makes the development of graphs and charts somewhat easy. An Excel Wizard is a step-by-step process with a sequential series of dialog boxes you complete to perform a task such as creating a chart or a table. The **Chart Wizard** provides

1.7

The Chart Wizard

| Figure 1.7 | Chart Wizard Step 1: **Chart type** |

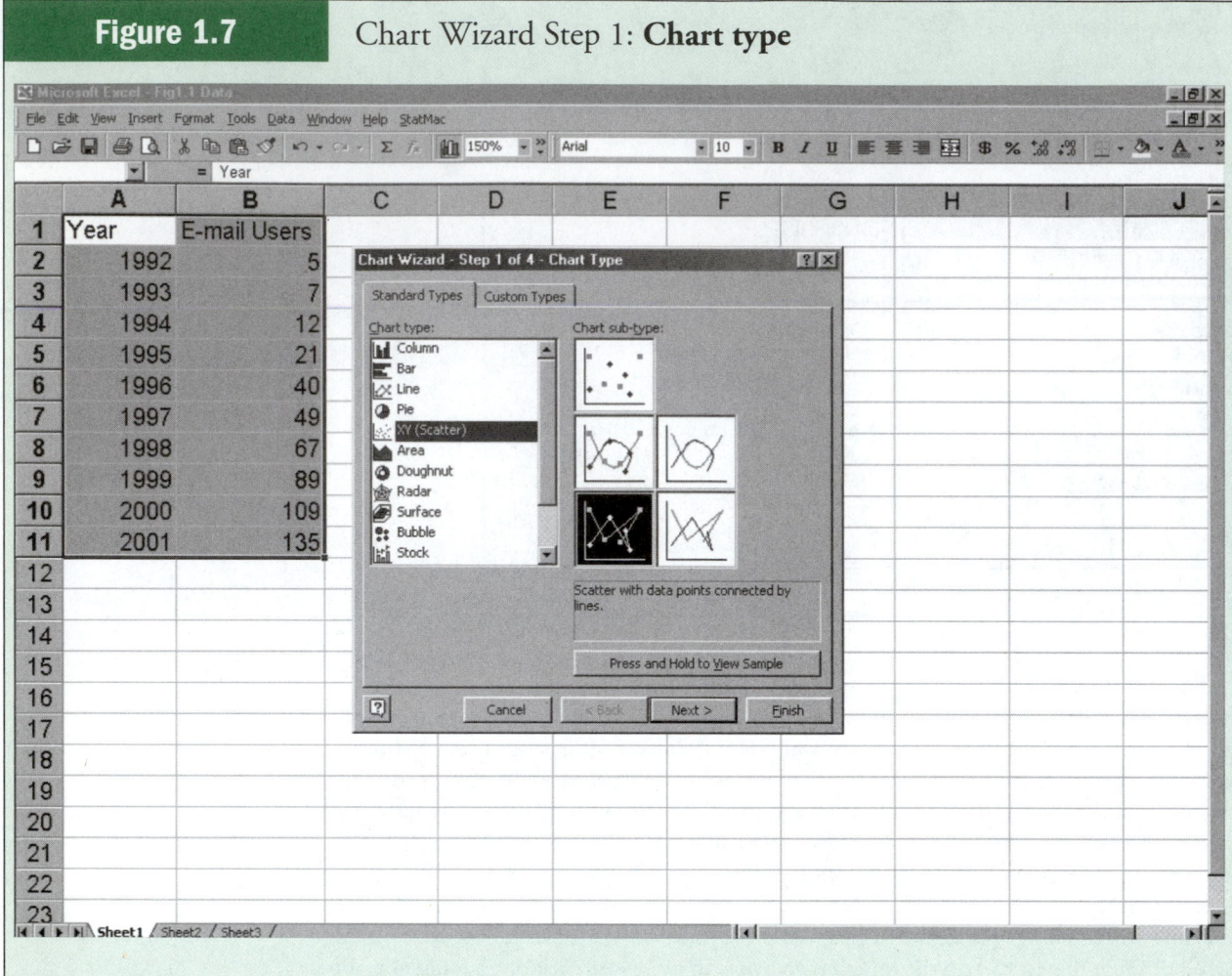

Chart Wizard

A step-by-step series of Excel dialog boxes for creating a chart/graph.

14 *standard chart types.* Furthermore, each of these standard chart types has from two to seven *chart sub-types* to provide variations in the presentation of the chart type. In addition, Excel is distributed with 20 *custom chart types,* and it is possible for the Excel user to add to the list of custom chart types.

The **Chart Wizard** is made up of four dialog boxes. We will take this initial look at these four dialog boxes using the data for Figure 1.1 as an example. Begin by entering the data and labels in columns A and B as shown in Figure 1.7. Next highlight these cells by dragging through them.

The **Chart Wizard** is next accessed with a mouse click on the **Chart Wizard** icon on the standard toolbar. Its icon depicts a bar/column chart. A click on the icon will result in the first of four dialog boxes that make up the **Chart Wizard.** The Step 1 dialog box is shown in Figure 1.7. You may wish to try this on your computer. If the **Standard Types** tab is not in front as shown in Figure 1.7, click on it.

Figure 1.7 lists 11 of the 14 chart types in the scrolling list on the left. The remaining three are at the bottom of the scrolling list and are hidden from view. These three are the

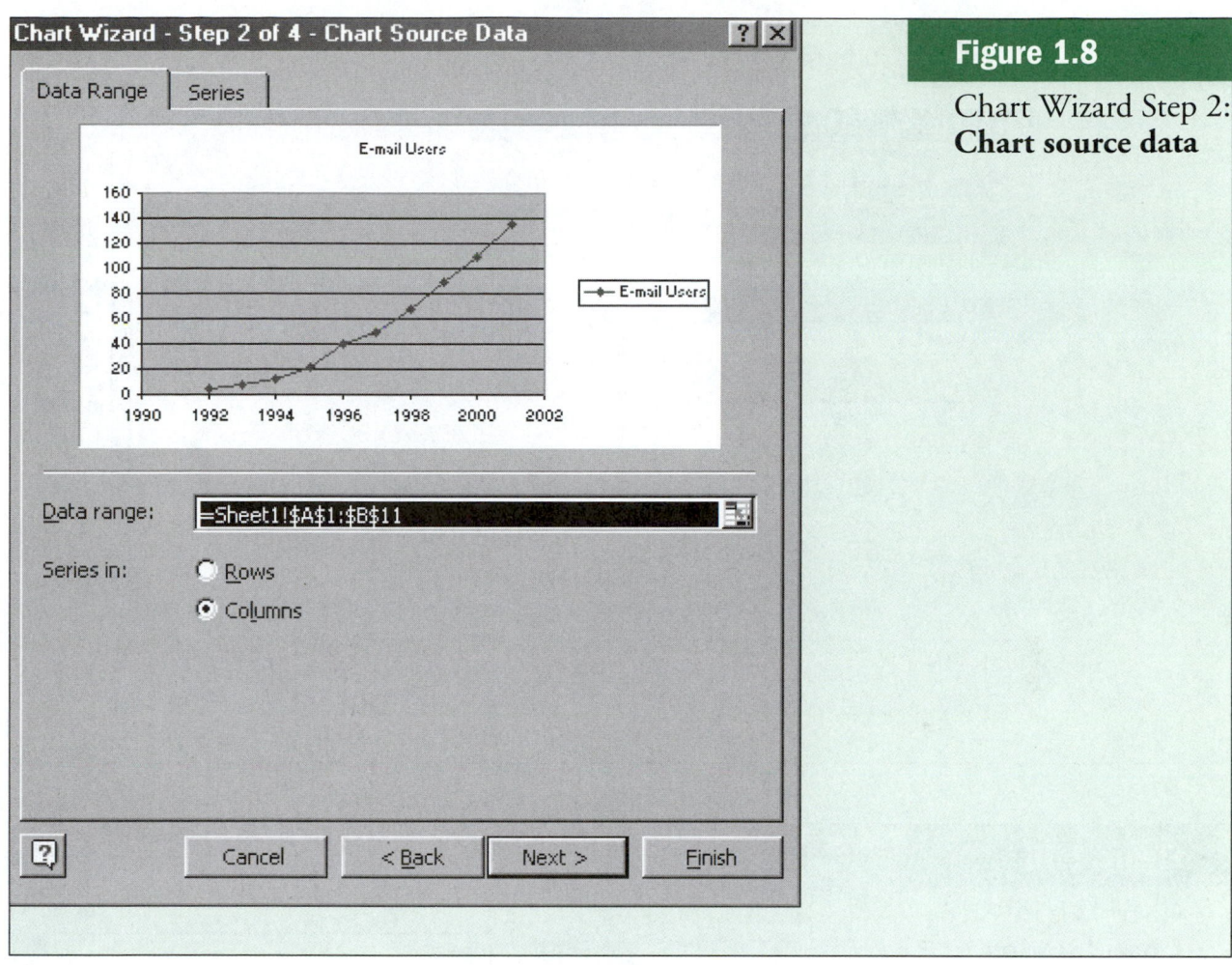

Figure 1.8

Chart Wizard Step 2:
Chart source data

Cylinder, the *Cone,* and the *Pyramid* charts. All three of these resemble a three-dimensional bar/column chart with the bars either in the form of a cylinder, cone, or pyramid.

Within this dialog box, first select a **Chart Type** with a mouse click and then a **Chart sub-type** with a click. As indicated in Figure 1.7 for this example, select the **XY (Scatter)** chart type and the sub-type in the lower left corner.

A click on the **Next** command button will result in the second **Chart Wizard** dialog box as shown in Figure 1.8. If the **Data Range** tab is not in front as shown in Figure 1.8, click on it. For this dialog box check to make sure the *Data range* has the correct entry and that **Columns** is selected for the *Series in* option. A click on the **Next** button will produce the third **Chart Wizard** dialog box as shown in Figure 1.9.

Figure 1.9 displays five tabs. These allow you to select options for the chart's **Titles, Axes, Gridlines, Legend** and **Data Labels.** For this example we have entered the titles for the chart, the X axis and the Y axis as shown in Figure 1.9 for the **Titles** tab. Other entries we have made in order to obtain the chart of Figure 1.9 include deselecting all four of the options on the **Gridlines** tab and deselecting *Show legend* on the **Legend** tab. No entries were made on the **Axes** and **Data Labels** tabs.

Once you have specified all the options for a chart, click on **Next** in order to view the final dialog box as shown in Figure 1.10. This box allows you to display the chart either in the worksheet that contains the data or in a new worksheet. After making this selection,

Figure 1.9   Chart Wizard Step 3: **Chart options**

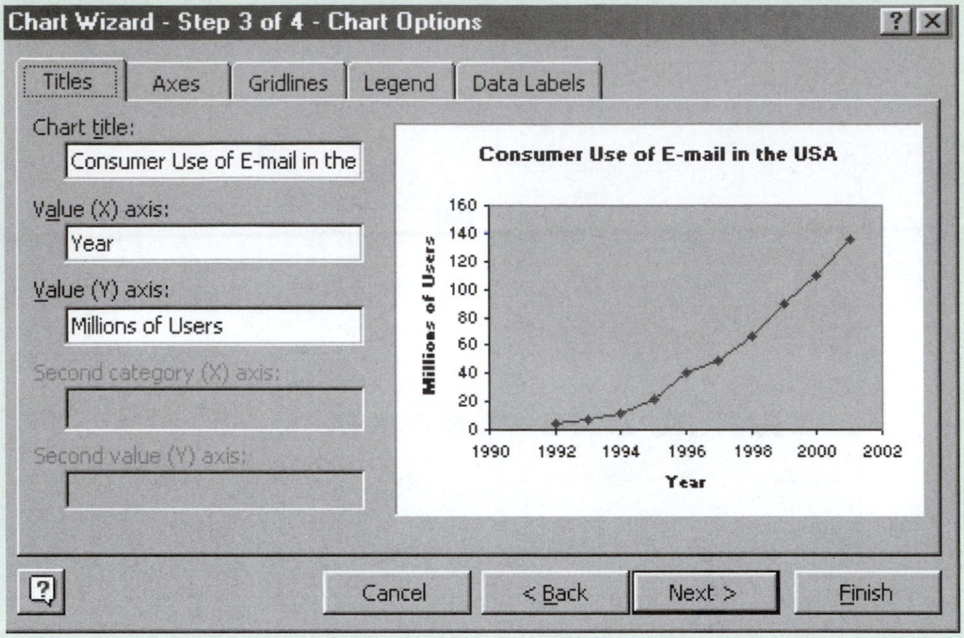

Figure 1.10

Chart Wizard Step 4:
Chart location

click on the **Finish** button to complete the chart. For this example we have selected *As an object in* option. The resulting chart is shown in Figure 1.11.

Once you have constructed a chart, you may wish to revise it. This can be accomplished easily in one of two ways. The first is to click within the chart to make it the active object and then click on the **Chart Wizard** icon. The result will be the Step 1 dialog box as shown in Figure 1.7. Step through the remainder of the **Chart Wizard** steps, Figures 1.8 to 1.10, and make whatever revisions you wish.

A second method is to right mouse click inside the chart. The result will be a shortcut menu that will allow you to access whichever of the four steps of the Chart Wizard that you wish to revise. For our example, this process will yield the results of Figure 1.12. Note that the shortcut menu provides a number of editing features in addition to allowing you to revise each of the four steps of the **Chart Wizard.**

Figure 1.11   Chart Wizard final results

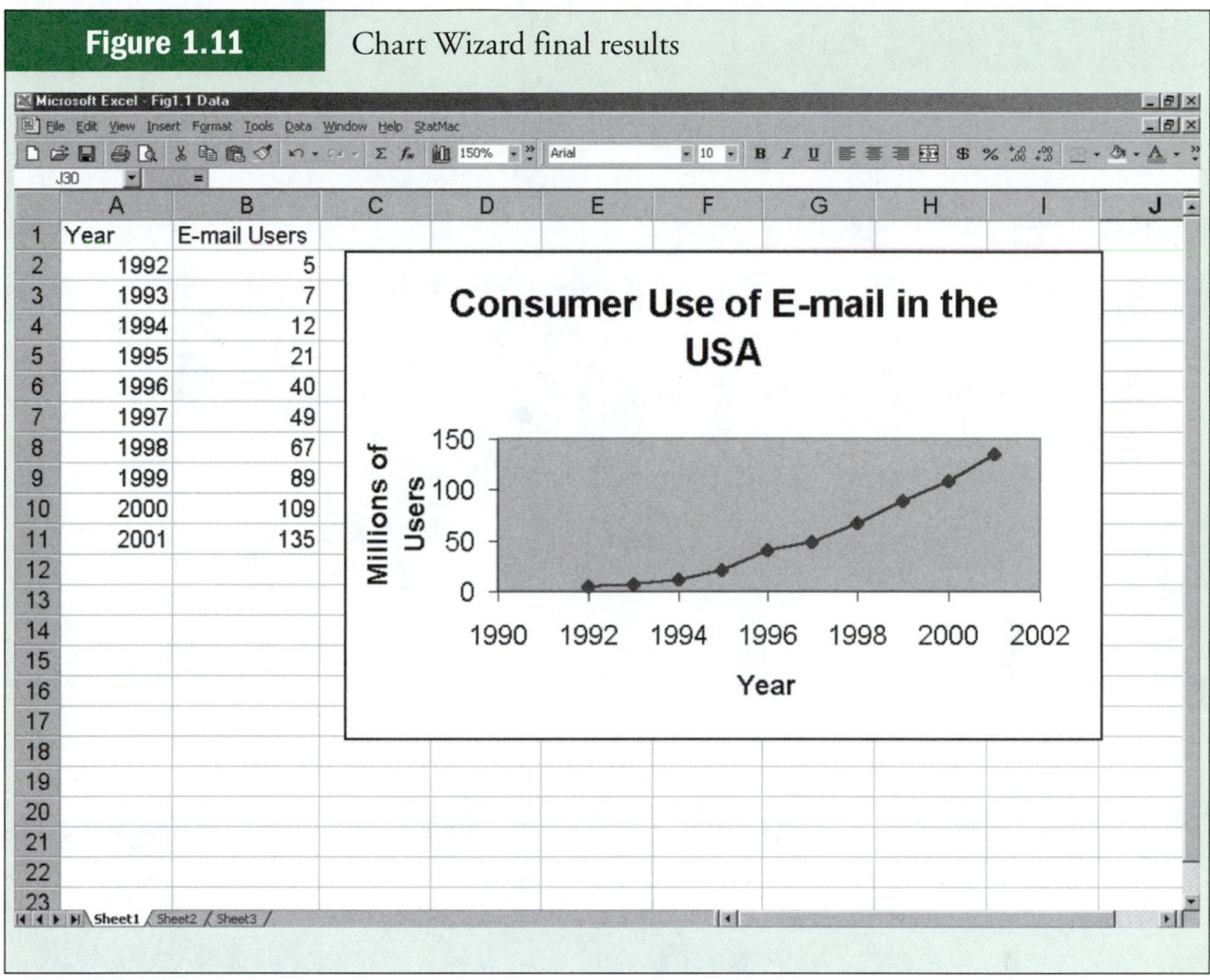

Regression analysis may well be the most widely used statistical procedure in business and industry. It is a statistical procedure for studying the relationship between two or more variables. Usually the objective of the analysis is to arrive at a mathematical relationship that will predict values for one variable based on values of the remaining variables. Since business, industrial, governmental, and social organizations need to forecast values for such items as demand, interest rates, inflation rates, prices, costs and so on, there is great interest in the use of regression analysis.

As a result of the widespread use of regression analysis, Excel includes a number of features for performing the required computations and charting. These include three data analysis tools, twelve statistical functions, the **Chart Wizard** for plotting relationships and **Trendline** for fitting relationships to data that have been charted. Although **Trendline** is a part of the charting feature of Excel, its ease of use and additional capabilities merits it being considered separately here.

Trendlines can be added to seven of the fourteen standard types of charts available from the **Chart Wizard** (refer to Figure 1.7). These include the unstacked, two-dimensional sub-type versions of the area, bar, bubble, column, line, stock and XY (Scatter) charts. For

1.8

The Trendline Feature

Trendline feature
An Excel feature for fitting mathematical relationships to charted data.

Figure 1.12   Chart revision menu

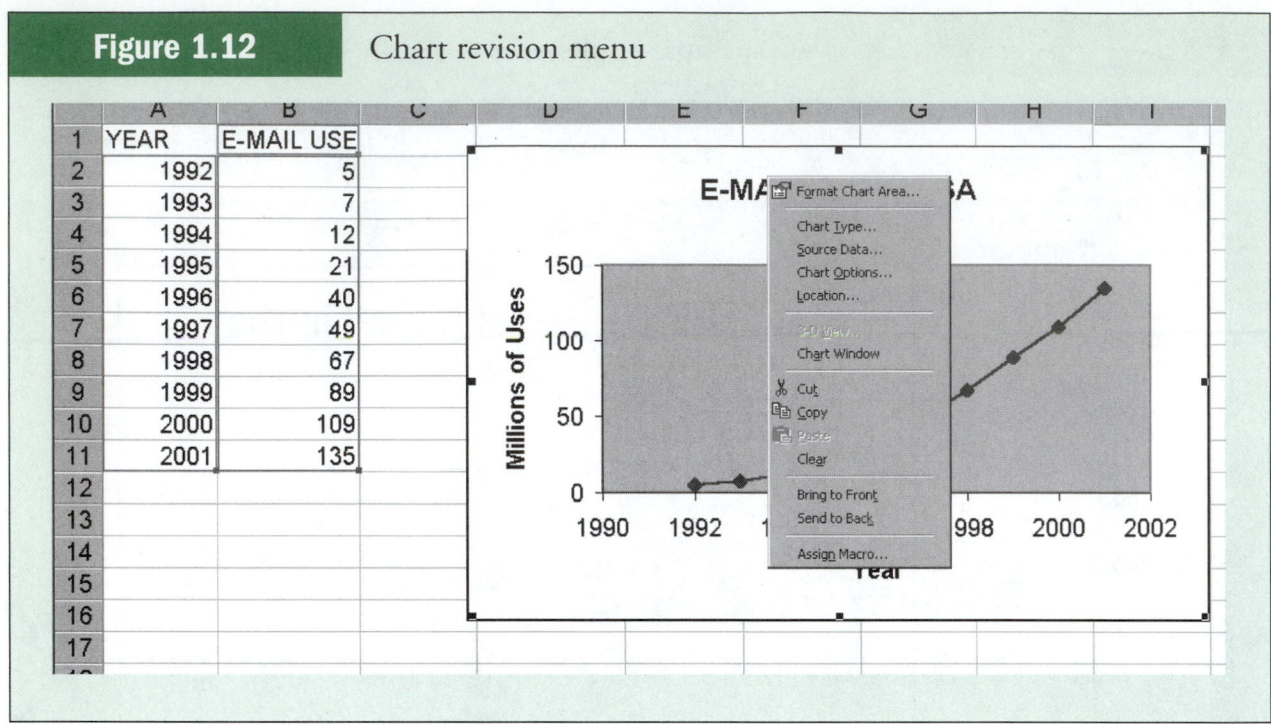

Figure 1.13

Add Trendline
dialog box

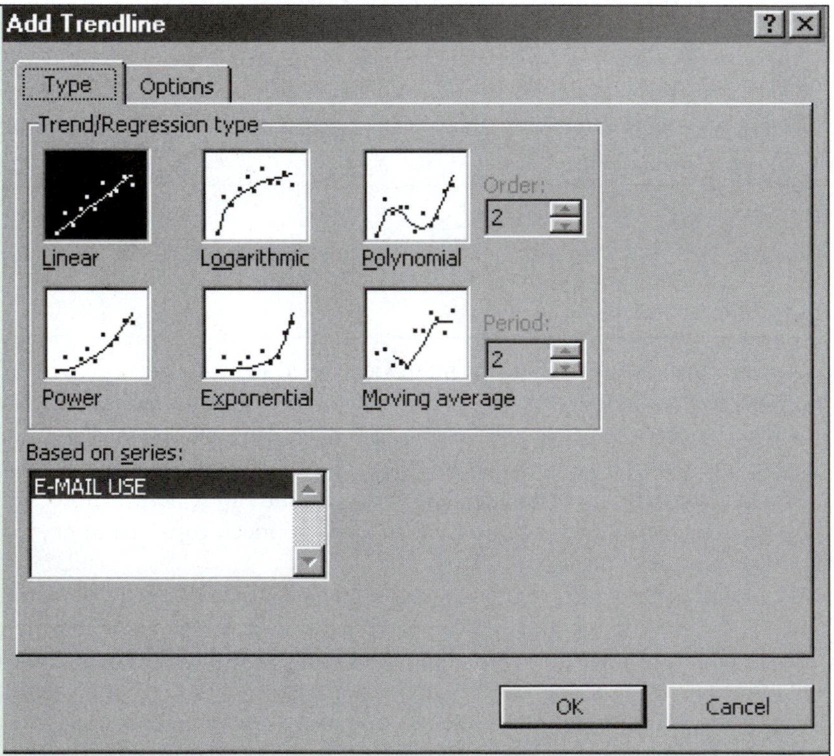

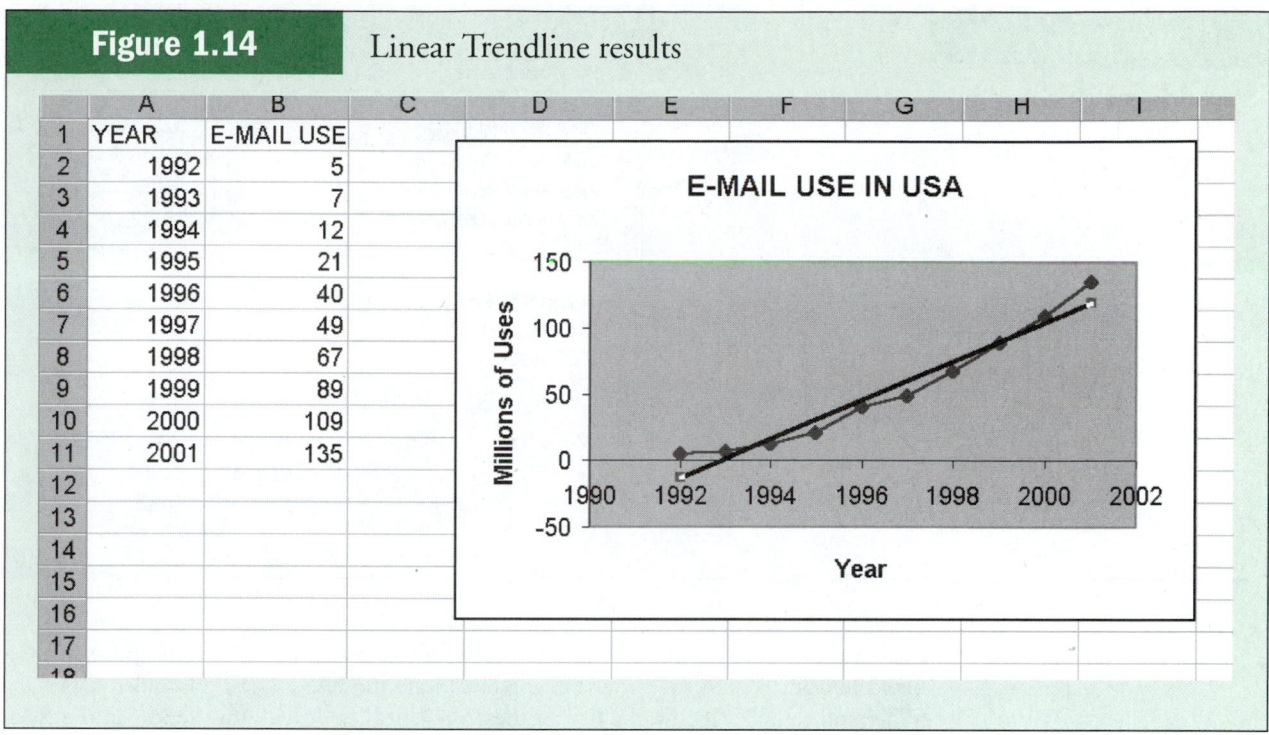

Figure 1.14 Linear Trendline results

instance, consider the example shown in Figure 1.12. If we wished to fit a trendline to these data we would begin by mouse clicking inside the chart. This action results in the **Chart** menu being added to the menu bar. Next select the **Chart** menu and the **Add Trendline** selection from the subsequent pull-down menu. The result will be as shown in Figure 1.13. Alternatively you could right click the plotted data line and choose **Add Trendline** from the subsequent shortcut menu to obtain the dialog box of Figure 1.13.

As shown, there are six *Trend/Regression types* that can be used to fit the data. If we select **Linear** and then click on **OK,** the result will be as shown in Figure 1.14.

The chart now includes the original plotted data values plus a linear trendline fit to the data. You will learn of the additional capabilities of **Trendline** in the later regression analysis and time series forecasting chapters.

The FAST \ STAT add-in feature is distributed on the CD-ROM accompanying this textbook.The CD-ROM includes the FAST \ STAT software and a 62-page manual with detailed instructions for both its installation and operation.

In brief, the installation of the FAST \ STAT add-in on your computer involves two major steps. First, you will need to copy the FAST \ STAT files to your hard drive. Begin by creating a folder for FAST \ STAT, for example, *C:***FAST** \ **STAT** (assuming your hard drive is designated as the *C* drive). Copy all the FAST \ STAT files from the CD-ROM to this new folder on your hard drive.

Second, you will activate FAST \ STAT as an Excel add-in. Begin by starting Excel. Next select the **Tools** from the menu bar and then **Add-Ins** from the resulting pull-down menu. The result will be the **Add-Ins** dialog box as shown in Figure 1.15.

This dialog box lists 15 or so add-ins that are available. The add-ins that are checked are those currently loaded in Excel. It does not list FAST \ STAT.

1.9
FAST \ STAT Add-in Macros

FAST \ STAT
A statistical add-in feature that augments Excel's statistical capabilities beyond those of the data analysis tools.

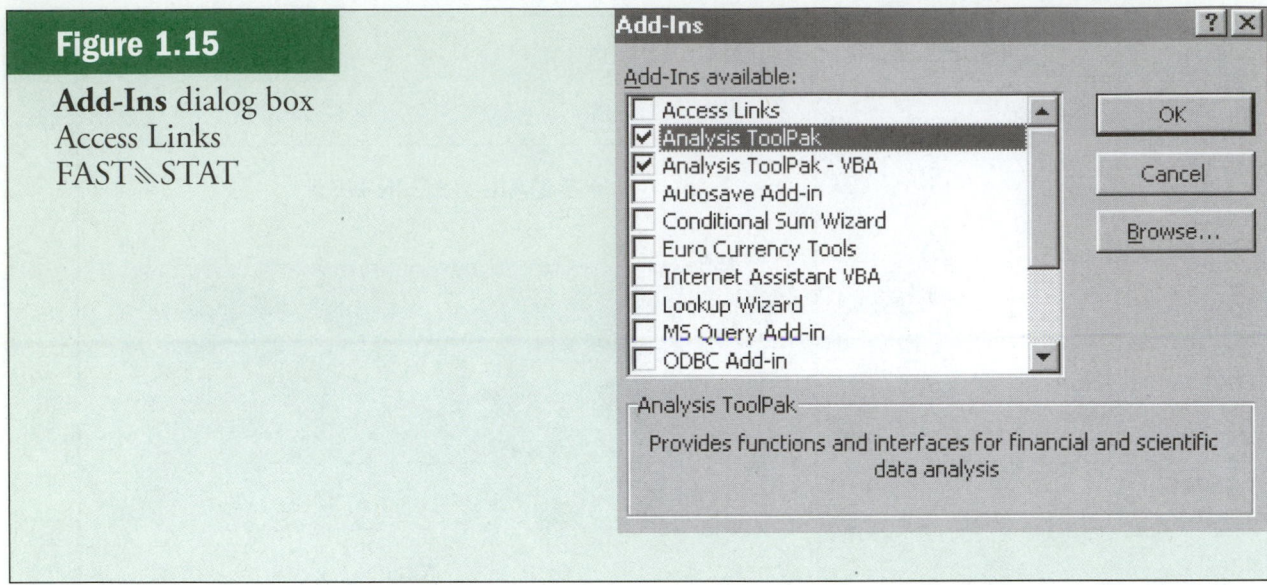

Figure 1.15

Add-Ins dialog box
Access Links
FAST\STAT

To begin the process of adding FAST\STAT to this list, click on the **Browse** command button. Within the Browse dialog box locate the FAST\STAT folder. Open the folder and double click on the first of the 16 macro files. It will now appear in the **Add-Ins** dialog box. Repeat for the remaining 15 FAST\STAT files and click **OK.** To activate the FAST\STAT menu, simultaneously press the **Shift, Ctrl,** and **R** keys. A new item labeled **FAST\STAT** will be added to the menu bar just to the left of the *Help* menu (see page 6 of the CD-ROM manual if the new menu doesn't appear). A click on **FAST\STAT** will result in a pull-down menu for FAST\STAT that lists its 15 macro procedures as shown in Figure 1.16.

A summary of the macro procedures included with FAST\STAT is given in Table 1.1. In addition to the macro name and statistical function each performs, the table indicates in which chapter(s) each is discussed.

A click on one of the FAST\STAT menu entries will result in an input dialog box for that macro. The inputs are generally one of two types of inputs. One type is specific input values needed to perform the required computations. The second type is the location of input values within the worksheet. This second type of input is usually the sample values or input values that are the results from one of the other statistical features of Excel.

For example, Figure 1.17 shows the dialog box for the FAST\STAT macro procedure *Statistical Inference for One Mean with t Distribution.* The upper text box asks for the *Cell Range for the Sample Values.* The lower three text boxes ask for the *Confidence Level,* the *Hypothesized Mean* and *Significance Level.* Once these entries are made for the macro, a click on the **OK** command button initiates the procedure. Once initiated, the macro will automatically perform its specific analysis task.

Finally, some of the FAST\STAT macro procedures have limitations upon their input data. Five of them have the restriction that no more than 400 observations are allowed. These five include the macros entitled **Box and Whisker Plot, 1 Mean Using Z Dist., 1 Mean Using t Dist., Regression C.I. and P.I.,** and **Durbin-Watson Test.** The **Probabilities** macro is restricted to six or fewer categories for both of the variables, and the **Chi Square Indep. Test** macro to ten or fewer categories for both of its variables. The three control chart macros also have data limitations. The **X-bar and R Charts** macro treats up to 100 samples with up to eight items in each sample. The **P Chart** macro treats up to 100 samples and the **c Chart** up to 100 items. The remaining five macros do not have data restrictions.

Figure 1.16 FAST\\STAT Menu

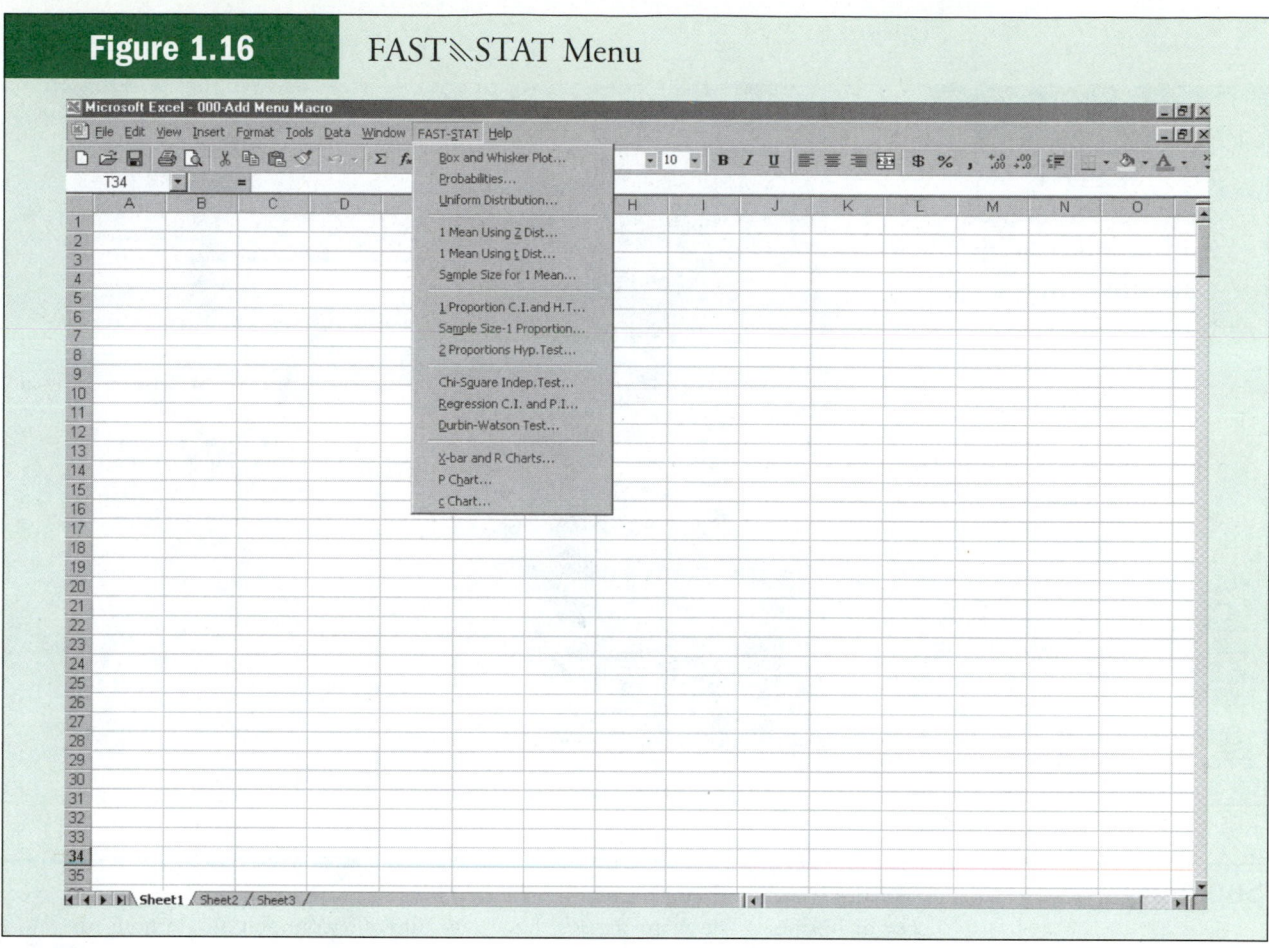

TABLE 1.1 FAST \\ STAT Macro Procedures

MACRO NAME	MACRO FUNCTION	USED IN TEXTBOOK
Box and Whisker Plot	Constructs a box and whisker plot	Chapter 3
Probabilities	Computes joint, marginal, union, and conditional, probabilities	Chapter 4
Uniform Distribution	Computes the cumulative probability between two points for a **continuous uniform** distribution	Chapter 6
1 Mean using Z Dist.	Computes point estimate, confidence interval, and hypothesis test for the **mean of a single population with the Z distribution**	Chapters 8 and 9
1 Mean using t Dist	Computes point estimate, confidence interval, and hypothesis test for the **mean of a single population with the t distribution**	Chapters 8 and 9
Sample Size for 1 Mean	Computes estimated sample size necessary for estimating the **mean** of a single population	Chapter 8
1 Proportion C.I. and H.T.	Computes point estimate, confidence interval, and hypothesis test for the **proportion of a single population with the Z distribution**	Chapters 8 and 9
Sample Size-1 Proportion	Computes estimated sample size necessary for estimating the **proportion** of a single population	Chapter 8
2 Proportions Hyp. Test	Computes a hypothesis test for determining the difference in **proportion of two populations with the Z distribution**	Chapter 10
Chi-Square Indep. Test	Computes a chi-square test of independence for a contingency table	Chapter 11
Regression CI and PI	Computes the confidence interval and prediction interval for simple linear regression	Chapter 12
Durbin-Watson Test	Computes Durbin-Watson test for autocorrelation	Chapter 14
X-bar and R Charts	Constructs an X-bar and R control charts	Chapter 15
P Chart	Constructs an P control chart	Chapter 15
c Chart	Constructs a c control chart	Chapter 15

Figure 1.17

FAST⧹STAT statistical inference for one mean with *t* distribution dialog box

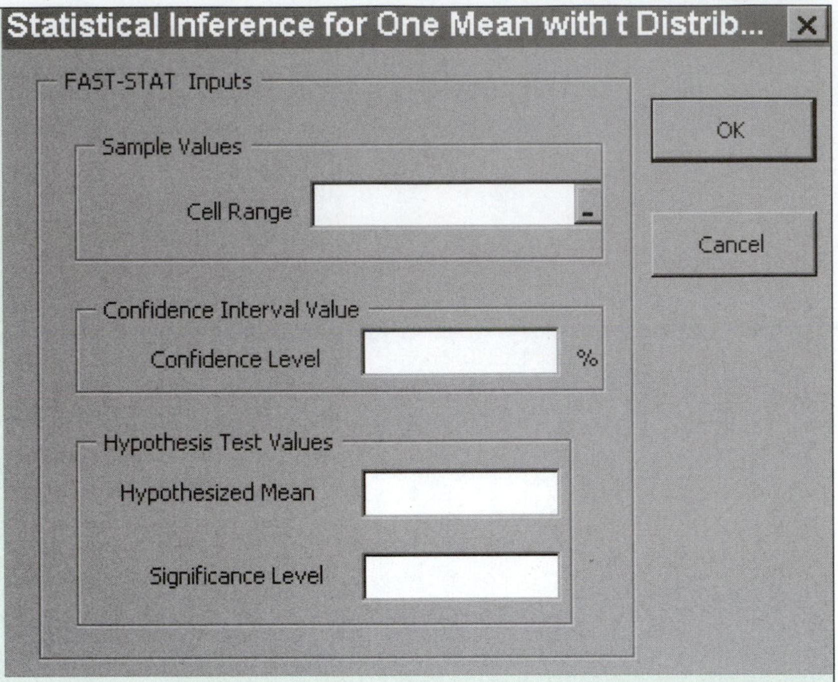

Statistical Inference for One Mean with t Distrib... ✕

FAST-STAT Inputs

Sample Values

Cell Range [   ] ▪

OK

Cancel

Confidence Interval Value

Confidence Level [   ] %

Hypothesis Test Values

Hypothesized Mean [   ]

Significance Level [   ]

Summary

Statistics is an important decision-making tool in business and is used in virtually every area of business. The word *statistics* has many different connotations. Among the more common meanings of the word are (1) the science of gathering, analyzing, interpreting, and presenting data, (2) a branch of mathematics, (3) a course of study, (4) facts and figures, (5) a death, (6) sample measurement, and (7) type of distribution used to analyze data. Statistics are broadly used in business, including the disciplines of accounting, decision sciences, economics, finance, management, management information systems, marketing, and production.

The study of statistics can be subdivided into two main areas: *descriptive statistics* and *inferential statistics*. Descriptive statistics result from gathering data from a body, group, or population and reaching conclusions only about that group. Inferential statistics are generated from the process of gathering sample data from a group, body, or population and reaching conclusions about the larger group from which the sample was drawn.

The type of statistical analysis that is appropriate depends on the level of data measurement, which can be either (1) *nominal,* (2) *ordinal,* (3) *interval,* or (4) *ratio.* Nominal is the lowest level, representing classification of only such data as geographic location, gender, or social security number. The next level is ordinal, which provides rank ordering measurements in which the intervals between consecutive numbers do not necessarily represent equal distances. Interval is the next to the highest level of data measurement, in which the distances represented by consecutive numbers are equal. The highest level of data measurement is ratio, which has all the qualities of interval measurement, but ratio data contain an absolute zero and ratios between numbers are meaningful. Interval and ratio data sometimes are called *metric* or *quantitative* data. Nominal and ordinal data sometimes are called *nonmetric* or *qualitative* data. Qualitative data can either be nonnumeric categories or numeric categories. If numeric, the data will be discrete numbers. Quantitative data are always numeric and can either be discrete or continuous numbers.

Statistical analysis with Excel is facilitated through the use of four features that are included with the software. These include (1) the *data analysis tools,* (2) the *statistical functions,* (3) the *Chart Wizard* and (4) the *trendline* feature for charts. To extend the statistical analysis capabilities of Excel, the CD-ROM accompanying this textbook includes the FAST ⦚ STAT add-in macro procedures.

Key Terms

census
Chart Wizard
continuous data
data analysis tools
descriptive statistics
discrete data
Excel
FAST ⦚ STAT
inferential statistics
interval level data
nominal level data

ordinal level data
parameter
population
qualitative data
quantitative data
ratio level data
sample
statistic
statistical functions
statistics
trendline feature

SUPPLEMENTARY PROBLEMS

1.1 Classify each of the following as nominal, ordinal, interval, or ratio level data.
   a. The time required to produce each tire on an assembly line
   b. The number of quarts of milk a family drinks in a month
   c. The ranking of four machines in your plant after they have been designated as excellent, good, satisfactory, or poor
   d. The telephone area code of clients in the United States
   e. The age of each of your employees
   f. The dollar sales at the local pizza house each month
   g. An employee's ID number
   h. The response time of an emergency unit

1.2 Classify each of the following as nominal, ordinal, interval, or ratio level data.
   a. The ranking of a company by *Fortune 500*
   b. The number of tickets sold at a movie theater on any given night
   c. The identification number on a questionnaire
   d. Per capita income
   e. The trade balance in dollars
   f. Socioeconomic class (low, middle, upper)
   g. Profit loss in dollars
   h. A company's tax ID

   i. The Standard & Poor's bond ratings of cities based on the following scales

RATING	GRADE
Highest quality	AAA
High quality	AA
Upper medium quality	A
Medium quality	BBB
Somewhat speculative	BB
Low quality, speculative	B
Low grade, default possible	CCC
Low grade, partial recovery possible	CC
Default, recovery unlikely	C

1.3 The Rathburn Manufacturing Company makes electric wiring, which it sells to contractors in the construction industry. Approximately 900 electric contractors purchase wire from Rathburn annually. Rathburn's director of marketing wants to determine electric contractors' satisfaction with Rathburn's wire. He developed a questionnaire that yields a satisfaction score of between 10 and 50 for participant responses. A random sample of 35 of the 900 contractors is asked to complete a satisfaction survey. The satisfaction scores for the 35 participants are averaged to produce a mean satisfaction score.
   a. What is the population for this study?
   b. What is the sample for this study?
   c. What is the statistic for this study?
   d. What would be a parameter for this study?

ANALYZING THE DATABASES

Seven major databases have been constructed for this text that can be used for applying the techniques presented in this course. These databases are found on the CD-ROM that accompanies this text, and each of these databases is available in Excel format for your convenience. These seven databases represent a wide variety of business areas, such as the stock market, manufacturing, international labor, finance, energy, healthcare, and agri-business. Altogether, there are 56 variables and 8,350 observations in these databases. The data are gathered from such reliable sources as the U.S. government's Bureau of Labor, the New York Stock Exchange, the U.S. Department of Agriculture, *Moody's Handbook of Common Stocks,* the American Hospital Association, and the U.S. Census Bureau. Four of the seven databases contain time-series data that can be especially useful in forecasting and regression analysis. Here is a description of each database along with information that may help you to interpret outcomes.

STOCK MARKET DATABASE

The stock market database contains eight variables on the New York Stock Exchange. There are three observations per month for nine years yielding a total of 324 observations per variable. The variables are: Composite Index, Industrial Index, Transportation Index, Utility Index, Stock Volume, Reported Trades, Dollar Value, and Warrants Volume. Dollar value is reported in units of millions of dollars. Recognizing that time of the month may make a difference in the value of an observation, each variable contains an observation from on or near to the tenth of the month denoted in the database as 1 under the variable Part of the Month, an observation from on or near to the twentieth of the month denoted as 2, and an observation from on or near to the thirtieth of the month denoted as 3. This database was constructed from data displayed on the Internet by the New York Stock Exchange. The original data can be accessed at www.nyse.com and searching for "NYSE Statistics Archive."

MANUFACTURING DATABASE

This database contains eight variables taken from 20 industries and 140 subindustries in the United States. The source of the database is the *1996 Annual Survey of Manufactures,* which is published by the Bureau of the Census of the U.S., Department of Commerce. Some of the industries are food products, textile mill products, furniture, chemicals, rubber products, primary metals, industrial machinery, and transportation equipment. The eight variables are Number of Employees, Number of Production Work-

ers, Value Added by Manufacture, Cost of Materials, Value of Industry Shipments, New Capital Expenditures, End-of-Year Inventories, and Industry Group, Two variables: Number of Employees and Number of Production Workers, are in units of 1000. Four variables: Value Added by Manufacture, Cost of Materials, New Capital Expenditures, and End-of-Year Inventories, are in million-dollar units. The Industry Group variable consists of numbers from 1 to 20 to denote the industry group to which the particular subindustry belongs Value of Industry Shipments has been recoded to the following 1 to 4 scale.

1 = $0 to $4.9 billion
2 = $5 billion to $13.9 billion
3 = $14 billion to $28.9 billion
4 = $29 billion or more

INTERNATIONAL LABOR DATABASE

This time-series database contains the civilian unemployment rates in percent from seven countries presented yearly from 1959 through 1998. The data are published by the Bureau of Labor Statistics of the U.S. Department of Labor. The countries are the United States, Canada, Australia, Japan, France, Germany, and Italy.

FINANCIAL DATABASE

The financial database contains observations on eight variables for 100 companies. The variables are Type of Industry, Total Revenues ($ millions), Total Assets ($ millions), Return on Equity (%), Earnings per Share ($), Average Yield (%), Dividends per Share ($); and Average Price per Earnings (P/E) ratio. The data were gathered from *Monday's Handbook of Common Stocks,* summer 1998 edition. The companies represent seven different types of industries. The variable "Type" displays a company's industry type as:

1 = apparel
2 = chemical
3 = electric power
4 = grocery
5 = healthcare products
6 = insurance
7 = petroleum

ENERGY DATABASE

The energy database consists of data on seven energy variables over a period of 26 years. The database is adopted from *Monthly Energy Review,* February 1999 (Office of Energy Markets and End Use, Energy Information Administration, U.S. Department of Energy). The seven variables are World Crude Oil Production (million barrels per day), U.S. Energy Consumption (quadrillion, BTUs per year), U.S. Nuclear Electricity Gross Generation (billion kilowatt-hours), U.S. Coal Production (million short tons), U.S. Total Dry Gas Production (million cubic feet), U.S. Fuel Rate for Automobiles (miles per gallon), and Cost of Unleaded (regular) Gasoline (U.S. city average).

HOSPITAL DATABASE

This database contains observations for 11 variables on U.S. hospitals. These variables are: Geographic Region, Control, Service, Number of Beds, Number of Admissions, Census, Number of Outpatients, Number of Births, Total Expenditures, Payroll Expenditures, and Personnel. Information for these databases is taken from the *American Hospital Association Guide to the Health Care Field*, 1998–99 edition, published in Chicago, Illinois.

The region variable is coded from 1 to 7, and the numbers represent the following regions.

    1 = South
    2 = Northeast
    3 = Midwest
    4 = Southwest
    5 = Rocky Mountain
    6 = California
    7 = Northwest

Control is a type of ownership. Four categories of control are included in the database.

    1 = government, nonfederal
    2 = nongovernment, not-for-profit
    3 = for-profit
    4 = federal government

Service is the type of hospital. There are only two types of hospitals in this database:

    1 = general medical
    2 = psychiatric

The total expenditures and payroll variables are in units of $1000.

AGRI-BUSINESS TIME-SERIES DATABASE

The agri-business time-series database contains the monthly weight (in 1000 lbs) or cold storage holdings for six different vegetables and for total frozen vegetables over a 14-year period. Each of the seven variables represents 168 months of data from 1984 to 1997. The six vegetables are green beans, broccoli, carrots, sweet corn, onions, and green peas. The data are published by the National Agricultural Statistics Service of the U.S. Department of Agriculture.

Use the databases to answer the following questions.

1. In the manufacturing database, what is the level of data for each of the following variables?
   a. Number of production workers
   b. Cost of materials
   c. Value of industry shipments
   d. Industry group
2. In the hospital database, what is the level of data for each of the following variables?
   a. Region
   b. Control
   c. Number of beds
   d. Personnel
3. In the financial database, what is the level of data for each of the following variables?
   a. Type of industry
   b. Total assets
   c. P/E ratio

CASE

DIGIORNO PIZZA: INTRODUCING A FROZEN PIZZA TO COMPETE WITH CARRY-OUT

In 1996 Kraft's DiGiorno Pizza hit the market. DiGiorno Pizza was a booming success with sales of $120 million the first year followed by $200 million the next year. It was neither luck nor coincidence that DiGiorno Pizza was an instant success. Kraft conducted extensive research about the product and the marketplace before introducing this product to the public. Many questions had to be answered before Kraft began production. For example, why do people eat pizza? When do they eat pizza? Do consumers believe that carry-out pizza is always more tasty?

SMI-Alcott conducted a research study for Kraft in which they sent out 1000 surveys to pizza lovers. The results indicated that people ate pizza during fun social occasions or at home when no one wanted to cook. People used frozen pizza mostly for convenience but selected carry-out pizza for a variety of other reasons, including quality and the avoidance of cooking. The Loran Marketing Group conducted focus groups for Kraft with women ages 25 to 54. Their findings showed that consumers used frozen pizza for convenience but wanted carry-out pizza taste. To satisfy these seemingly divergent goals (convenience and taste), Kraft developed DiGiorno Pizza, which rises in the oven as it cooks. This impressed focus group members, and in a series of blind taste tests conducted by Product Dynamics, DiGiorno Pizza beat out all frozen pizza and finished second overall behind one carry-out brand.

Through advertising Kraft was able to overcome two concerns that were raised by marketing research: people had trouble pronouncing "DiGiorno" and people needed to be convinced that the frozen pizza actually tasted good. Kraft had the name DiGiorno repeated several times in advertisements to make certain that consumers could pronounce the name. As a by-product, the ads also generated strong brand identification. In addition, the ads emphasized "fresh-baked taste" and the rising dough aspect of the product, which helped convince people of DiGiorno's higher quality of taste.

DiGiorno Pizza now has a 13% market share of the U.S. $2.3 billion frozen pizza category. It is the fastest Kraft product ever to break the $200 million barrier.

DISCUSSION

Think about the market research that was conducted by Kraft and the fact that they used several companies. If you were in charge of conducting this research to help launch such a new product, what decisions would you make about whom to survey, where and when to survey, and what to measure?

1. What are some of the populations that Kraft might have been interested in measuring for these studies? Did Kraft actually attempt to contact entire populations? What samples were taken? In light of these two questions, how was the inferential process used by Kraft in their market research? Can you think of any descriptive statistics that might have been used by Kraft in their decision-making process?

2. In the various market research efforts made by Kraft for DiGiorno, some of the measurements that might have been made are listed below. Categorize these by level of data. Think of some other measurements that Kraft researchers might have made that could have helped them in this research effort and categorize them by level of data.

   a. Number of pizzas consumed per week per household
   b. Age of pizza purchaser
   c. Zip code of the survey respondent
   d. Dollars spent per month on pizza per person
   e. Time in between purchases of pizza
   f. Rating of taste of a given pizza brand on a scale from 1 to 10 where 1 is very poor tasting and 10 is excellent taste
   g. Ranking of the taste of four pizza brands on a taste test
   h. Number representing the geographic location of the survey respondent
   i. Quality rating of a pizza brand as excellent, good, average, below average, poor
   j. Number representing the pizza brand being evaluated.
   k. Gender of survey respondent

SOURCE: Adapted from "Upper Crust," *American Demographics:* March 1999, p. 58; Marketwatch—News That Matters websites, "What's in a Name? Brand Extension Potential" and "DiGiorno Rising Crust Delivers $200 Million" at http://www.foodexplorer.com/BUSINESS/Products/MarketAnalysis/ PF02896b.htm (and PF022827a.htm).

2

Descriptive Charts and Graphs

Learning Objectives

The overall objective of Chapter 2 is for you to master several techniques for summarizing and depicting data, thereby enabling you to:

1. Construct a frequency distribution.

2. Determine the class midpoints, relative frequencies, and cumulative frequencies of a frequency distribution.

3. Construct a histogram, a frequency polygon, an ogive, and a pie chart.

Ungrouped data
Raw data, or data that have not been summarized in any way.

Frequency distribution
A summary of data presented in the form of class intervals and frequencies.

Grouped data
Data that have been organized into a frequency distribution.

In Chapters 2 and 3 many techniques are presented for reformatting or reducing data so that the data are more manageable and can be used to assist decision makers more effectively. One technique for grouping data is the frequency distribution presented in this chapter. In addition, Chapter 2 discusses and displays several graphical tools for summarizing and presenting data, including histogram, frequency polygon, ogive, and pie chart. By using these and other techniques, decision makers can begin to "get a handle" on information contained in data and begin to use the data to enhance the decision-making process.

Raw data, or data that have not been summarized in any way, are sometimes referred to as **ungrouped data.** Table 2.1 contains raw data that are the unemployment rates for France over 40 years. *Data that have been organized into a* **frequency distribution** *are called* **grouped data.** Table 2.2 presents a frequency distribution for the data displayed in Table 2.1.

2.1
Frequency Distributions

Range
The difference between the largest and the smallest values in a set of numbers.

Frequency distributions are relatively easy to construct. Although there are some guidelines for their construction, distributions vary in final shape and design, even when the original raw data are identical. In a sense, frequency distributions are constructed according to individual researchers' taste.

When constructing a frequency distribution, the researcher should first determine the range of the raw data. The **range** often is defined as *the difference between the largest and smallest numbers.* The range for the data in Table 2.1 is 11.3 (12.5–1.2).

The second step in constructing a frequency distribution is to determine how many classes it will have. One rule of thumb is to select between 5 and 15 classes. If the frequency distribution has too few classes, the data summary may be too general to be useful. Too many classes may result in a frequency distribution that does not aggregate the data enough to be helpful. The final number of classes is arbitrary. The researcher arrives at a number by examining the range and determining a number of classes that will span the range adequately and also be meaningful to the user. The data in Table 2.1 were grouped into six classes for Table 2.2.

After selecting the number of classes, the researcher must determine the width of the class interval. An approximation of the class width can be calculated by dividing the range by the number of classes. For the data in Table 2.1, this approximation would be (11.3)/6, or 1.9.

Normally, the number is rounded up to the next whole number, which in this case is 2. The frequency distribution must start at a value equal to or lower than the lowest number

TABLE 2.1
Unemployment Rates for France over 40 Years (Ungrouped Data)

1.6	2.1	4.2	8.6	9.6
1.5	2.7	4.6	10.0	10.4
1.2	2.3	5.2	10.5	11.8
1.4	2.5	5.4	10.6	12.3
1.6	2.8	6.1	10.8	11.8
1.2	2.9	6.5	10.3	12.5
1.6	2.8	7.6	9.6	12.4
1.6	2.9	8.3	9.1	11.8

TABLE 2.2
Frequency Distribution of the Unemployment Rates of France (Grouped Data)

CLASS INTERVAL	FREQUENCY
1-under 3	16
3-under 5	2
5-under 7	4
7-under 9	3
9-under 11	9
11-under 13	6

of the ungrouped data and end at a value equal to or higher than the highest number. The lowest unemployment rate is 1.2 and the highest is 12.5, so the researcher starts the frequency distribution at 1 and ends it at 13. Table 2.2 contains the completed frequency distribution for the data in Table 2.1. Class endpoints are selected so that no value of the data can fit into more than one class. The class interval expression, -under, in the distribution of Table 2.2 avoids such a problem.

Class Midpoint

The *midpoint of each class interval* is called the **class midpoint** and is sometimes referred to as the **class mark.** It is *the value halfway across the class interval* and can be calculated as *the average of the two class endpoints.* For example, in the distribution of Table 2.2, the midpoint of the class interval 3-under 5 is 4, or (3 + 5)/2. A second way to obtain the class midpoint is to calculate one-half the distance across the class interval (half the class width) and add it to the class beginning point, as for the unemployment rates distribution:

Class Beginning Point = 3

Class Width = 2

Class Midpoint = $3 + \frac{1}{2}(2) = 4$

The class midpoint is important, because it becomes the representative value for each class in most group statistics calculations. The third column in Table 2.3 contains the class midpoints for all classes of the data from Table 2.2.

Relative Frequency

Relative frequency is *the proportion of the total frequency that is in any given class interval in a frequency distribution.* Relative frequency is the individual class frequency divided by the total frequency. For example, from Table 2.3, the relative frequency for the class interval 5-under 7 is 4/40, or .10. Consideration of the relative frequency is preparatory to the study of probability in Chapter 4. Indeed, if values were selected randomly from the data in Table 2.1, the probability of drawing a number that is "5-under 7" would be .10, the relative frequency for that class interval. The fourth column of Table 2.3 lists the relative frequencies for the frequency distribution of Table 2.2.

Cumulative Frequency

The **cumulative frequency** is a *running total of frequencies through the class of a frequency distribution.* The cumulative frequency for each class interval is the frequency for that class interval added to the preceding cumulative total. In Table 2.3 the cumulative frequency for the first class is the same as the class frequency: 16. The cumulative frequency

Class midpoint
For any given class interval of a frequency distribution, the value halfway across the class interval; the average of the two class endpoints.

Class mark
Another name for class midpoint; the midpoint of each class interval in grouped data.

Relative frequency
The proportion of the total frequencies that fall into any given class interval in a frequency distribution.

Cumulative frequency
A running total of frequencies through the classes of a frequency distribution.

TABLE 2.3
Class Midpoints, Relative Frequencies, and Cumulative Frequencies for Unemployment Data

CLASS INTERVAL	FREQUENCY	MIDPOINT	RELATIVE FREQUENCY	CUMULATIVE FREQUENCY
1-under 3	16	2	.400	16
3-under 5	2	4	.050	18
5-under 7	4	6	.100	22
7-under 9	3	8	.075	25
9-under 11	9	10	.225	34
11-under 13	6	12	.150	40
	40		1.000	

for the second class interval is the frequency of that interval (2) plus the frequency of the first interval (16), which yields a new cumulative frequency of 18. This process continues through the last interval, at which point the cumulative total equals the sum of the frequencies (40). The concept of cumulative frequency is used in many areas, including sales cumulated over a fiscal year, sports scores during a contest (cumulated points), years of service, points earned in a course, and costs of doing business over a period of time. Table 2.3 gives cumulative frequencies for the data in Table 2.2.

DEMONSTRATION PROBLEM 2.1

The following data are the average weekly mortgage interest rates for a 60-week period.

7.29	7.03	7.14	6.77	6.35
6.69	7.02	7.40	7.16	6.96
6.98	7.56	6.75	6.87	7.11
7.39	7.28	6.97	6.90	6.57
7.11	6.95	7.23	7.31	7.00
7.30	7.17	6.96	6.78	7.30
7.16	6.78	6.79	7.07	7.03
6.87	6.80	7.10	7.13	6.95
7.08	7.24	7.34	7.47	7.31
6.96	6.70	6.57	6.88	6.84
7.02	7.40	7.12	7.16	7.16
6.99	6.94	7.29	7.05	6.84

Construct a frequency distribution for these data. Calculate and display the class midpoints, relative frequencies, and cumulative frequencies for this frequency distribution.

SOLUTION

How many classes should this frequency distribution contain? The range of the data is 1.21 (7.56–6.35). If 13 classes are used, each class width is approximately:

$$\text{Class width} = \frac{\text{Range}}{\text{Number of Classes}} = \frac{1.21}{13} = 0.093.$$

If a class width of .10 is used, a frequency distribution can be constructed with endpoints that are more uniform looking and allow presentation of the information in categories more familiar to mortgage interest rate users.

The first class endpoint must be 6.35 or lower to include the smallest value; the last endpoint must be 7.56 or higher to include the largest value. In this case the frequency distribution begins at 6.30 and ends at 7.60. The resulting frequency distribution, class midpoints, relative frequencies, and cumulative frequencies are listed in the following table.

CLASS INTERVAL	FREQUENCY	CLASS MIDPOINT	RELATIVE FREQUENCY	CUMULATIVE FREQUENCY
6.30-under 6.40	1	6.35	.0167	1
6.40-under 6.50	0	6.45	.0000	1
6.50-under 6.60	2	6.55	.0333	3
6.60-under 6.70	1	6.65	.0167	4
6.70-under 6.80	6	6.75	.1000	10
6.80-under 6.90	6	6.85	.1000	16
6.90-under 7.00	10	6.95	.1667	26
7.00-under 7.10	8	7.05	.1333	34
7.10-under 7.20	11	7.15	.1833	45
7.20-under 7.30	5	7.25	.0833	50
7.30-under 7.40	6	7.35	.1000	56
7.40-under 7.50	3	7.45	.0500	59
7.50-under 7.60	1	7.55	.0167	60
Totals	60		1.0000	

The frequencies and relative frequencies of these data reveal the mortgage interest rate classes that are likely to occur during the period. Most of the mortgage interest rates (52 of the 60) are in the classes starting with (6.70-under 6.80) and going through (7.30-under 7.40). The rates with the greatest frequency, 11, are in the (7.10-under 7.20) class.

Analysis Using Excel

To construct a frequency distribution from raw data using Excel, begin by selecting the **Data Analysis** feature of **Tools.** From the left-side options provided by the pull-down menu of **Data Analysis,** select **Histogram.** The **Histogram** feature has the capability of constructing a frequency distribution from raw data along with some types of graphical depictions that will be discussed in section 2.2. The dialog box for **Histogram** is displayed in Figure 2.1.

To construct a frequency distribution, enter the location of the raw data into the space provided beside *Input* **Range.** If you click **OK,** then Excel will group the data by determining its own endpoints. If you want to specify class endpoints for the frequency distribution, then list the upper endpoints of each class on your spreadsheet, along with the raw data, and then enter the location of the class endpoints in the space provided with *Bin* **Range.** Click **OK** and Excel will produce a frequency distribution from the data using the upper class endpoints provided. Figure 2.2 is Excel output for the frequency distribution from Demonstration Problem 2.1.

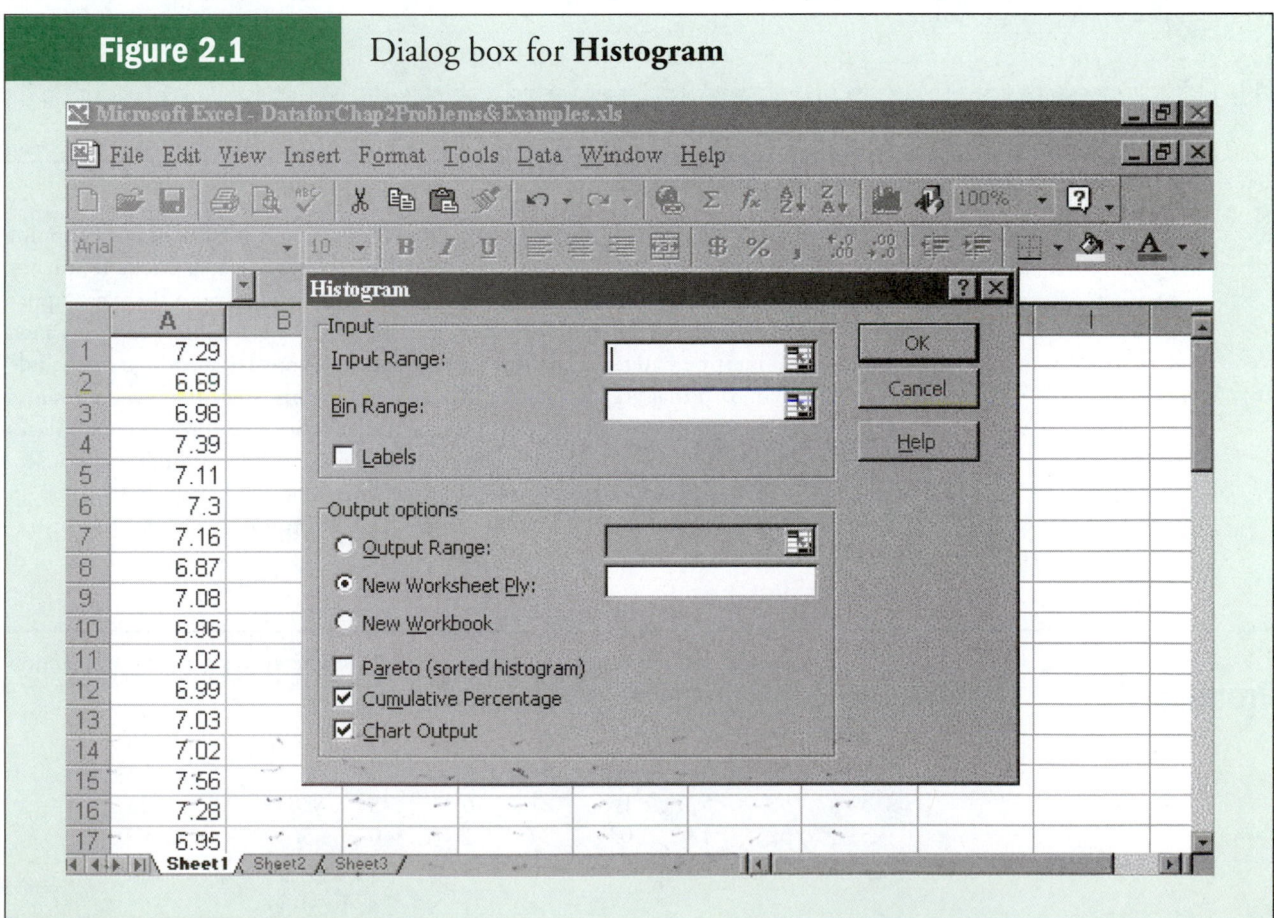

Figure 2.1   Dialog box for **Histogram**

Figure 2.2	Excel-produced frequency distribution for the data in Demonstration Problem 2.1

Microsoft Excel - DataforChap2Problems&Examples.xls

File Edit View Insert Format Tools Data Window Help

A1   =   Bin

	A	B	C	D	E	F	G	H	I
1	Bin	Frequency							
2	6.399	1							
3	6.499	0							
4	6.599	2							
5	6.699	1							
6	6.799	6							
7	6.899	6							
8	6.999	10							
9	7.099	8							
10	7.199	11							
11	7.299	5							
12	7.399	6							
13	7.499	3							
14	7.599	1							
15	More	0							
16									
17									

Sheet4 / Sheet5 / Sheet6 \ **Sheet7** / Sheet1 / Sheet2 / Sheet3 /

In constructing a frequency distribution, Excel includes values for the upper endpoint of each class interval in the class for which it is specified. In order to construct a frequency distribution like the one shown in Demonstration Problem 2.1 (where the class upper endpoint is not included in the class for which it is specified), class upper endpoints such as "under 6.70" must be entered using a value less than 6.70 but large enough to include all values less than 6.70 that are in that class. To accomplish this, since the raw data values of Demonstration Problem 2.1 include two decimal places, we use three decimal places (one more than the raw data) in the upper endpoint and enter 6.699 into the Excel analysis rather than 6.70. This correction allows us to get the same frequency distribution in the Excel output shown in Figure 2.2 as the frequency distribution shown in Demonstration Problem 2.1.

2.1 Problems

2.1 The following data represent the afternoon high temperatures for 50 construction days during a year in St. Louis.

42	70	64	47	66	55	85	10	24	45
16	40	81	15	35	38	79	35	36	23
31	38	52	16	81	69	73	38	48	25
31	62	47	63	84	17	40	36	44	17
64	75	53	31	60	12	61	43	30	33

   a. Construct a frequency distribution for the data using five class intervals.

   b. Construct a frequency distribution for the data using ten class intervals.

   c. Examine the results of (a) and (b) and comment on the usefulness of the frequency distribution in terms of temperature summarization capability.

2.2   A packaging process is supposed to fill small boxes of raisins with approximately 50 raisins so that each box will weigh the same. However, the number of raisins in each box will vary. Suppose 100 boxes of raisins are randomly sampled, the raisins counted, and the following data are obtained.

57	51	53	52	50	60	51	51	52	52
44	53	45	57	39	53	58	47	51	48
49	49	44	54	46	52	55	54	47	53
49	52	49	54	57	52	52	53	49	47
51	48	55	53	55	47	53	43	48	46
54	46	51	48	53	56	48	47	49	57
55	53	50	47	57	49	43	58	52	44
46	59	57	47	61	60	49	53	41	48
59	53	45	45	56	40	46	49	50	57
47	52	48	50	45	56	47	47	48	46

Construct a frequency distribution for these data. What does the frequency distribution reveal about the box fills?

2.3   The owner of a fast-food restaurant ascertains the ages of a sample of her customers. From these data, she constructs the frequency distribution shown. For each class interval of the frequency distribution, determine the class midpoint, the relative frequency, and the cumulative frequency.

CLASS INTERVAL	FREQUENCY
0-under 5	6
5-under 10	8
10-under 15	17
15-under 20	23
20-under 25	18
25-under 30	10
30-under 35	4

What does the relative frequency tell the fast food restaurant owner about customer ages?

2.4   The human resources manager for a large company commissions a study in which the employment records of 500 company employees are examined for absenteeism during the past year. The researcher conducting the study organizes the data into a frequency distribution to assist the human resources manager in analyzing the data. The frequency distribution is shown. For each class of the frequency distribution, determine the class midpoint, the relative frequency, and the cumulative frequency.

CLASS INTERVAL	FREQUENCY
0-under 2	218
2-under 4	207
4-under 6	56
6-under 8	11
8-under 10	8

2.2
Graphical Depiction of Data

One of the most effective mechanisms for presenting data in a form meaningful to decision makers is graphical depiction. Through graphs and charts, the decision maker can often get an overall picture of the data and reach some useful conclusions merely by studying the chart or graph. Converting data to graphics can be creative and artful. Often the most difficult step in this process is to reduce important and sometimes expensive data to a graphic picture that is both clear and concise and yet consistent with the message of the original data. One of the most important uses of graphical depiction in statistics is to help the researcher determine the shape of a distribution. Four types of graphic depiction are presented here: (1) histogram, (2) frequency polygon, (3) ogive, and (4) pie chart.

Histograms

Histogram

A type of vertical bar chart constructed by graphing line segments for the frequencies of classes across the class intervals and connecting each to the *X* axis to form a series of rectangles.

A **histogram** is *a type of vertical bar chart* that is used to depict a frequency distribution. Construction involves labeling the *X* axis (abscissa) with the class endpoints and the *Y* axis (ordinate) with the frequencies; drawing a horizontal line segment from class endpoint to class endpoint at each frequency value; and connecting each line segment vertically from the frequency value to the *X* axis, forming a series of rectangles. Figure 2.3 is an Excel histogram of the frequency distribution in Table 2.2.

A histogram is a useful tool for differentiating the frequencies of class intervals. A quick glance at a histogram reveals which class intervals produce the highest frequency totals.

Figure 2.3

Excel histogram of French unemployment data

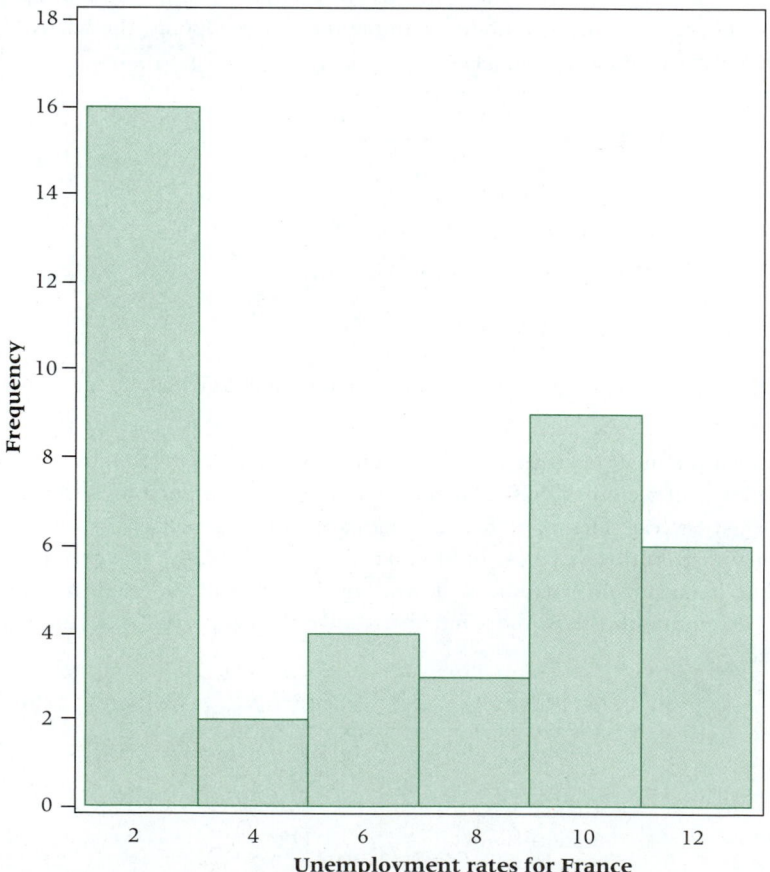

Figure 2.3 clearly shows that the class interval 1-under 3 yields by far the highest frequency count (16). Examination of the histogram reveals where large increases or decreases occur between classes, such as from the 1-under 3 class to the 3-under 5 class, a decrease of 14, and from the 7-under 9 class to the 9-under 11 class, an increase of 6.

Note that the scales used along the X and Y axes for the histogram in Figure 2.3 are almost identical. However, because ranges of meaningful numbers for the two variables being graphed often differ considerably, the graph may have different scales on the two axes. Figure 2.4 shows what the histogram of unemployment rates would look like if the scale on the Y axis were more compressed than that on the X axis. Notice that there appears to be less difference in the length of the rectangles representing the frequencies in Figure 2.4. It is important that the user of the graph has a clear understanding of the scales used for the axes of a histogram. Otherwise, a graph's creator can "lie with statistics" by stretching or compressing a graph to make a point.

USING HISTOGRAMS TO GET AN INITIAL OVERVIEW OF THE DATA   Because of the widespread availability of computers and statistical software packages to researchers and decision makers, the histogram has become more important. Sometimes decision makers are presented with a large database of information and do not know where to begin in attempting to understand what the data mean. Histogram analysis of such data can yield initial information about the shape of the distribution of the data, the amount of variability of data, the central location of the data, and outlier data. While most of these concepts are presented in Chapter 3, the notion of histogram as an initial tool to access these data characteristics is presented here.

For example, one of the variables in the Stock Market database (displayed on the CD-ROM) is "stock volume." There are 324 stock volume observations in this database. Suppose a financial decision maker wants to use these data to reach some conclusions about the stock market. Figure 2.5 shows a histogram of these data. What can we learn from this histogram? Virtually all stock market volumes fall between zero and a billion shares. The distribution takes on a shape that is high on the left end and tapered to the right. In Chapter 3 we will learn that the shape of this distribution is skewed toward the right end. In

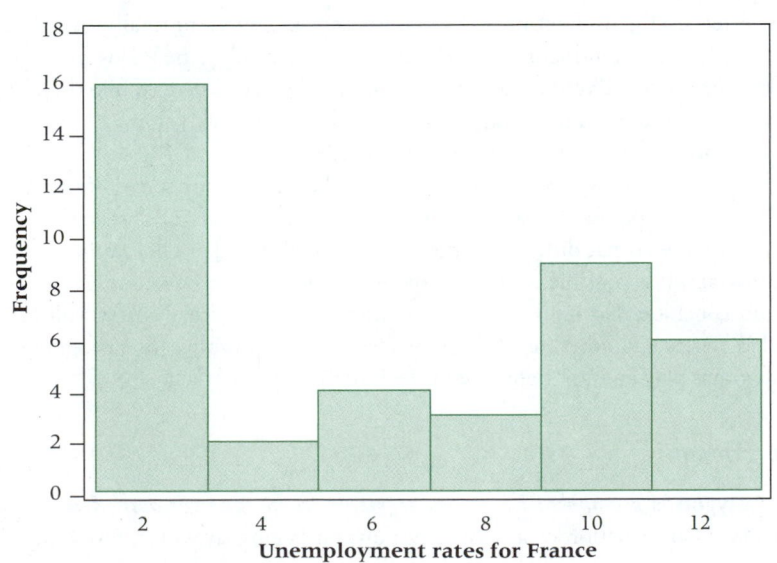

Figure 2.4

Excel histogram of French unemployment data (Y axis compressed)

Figure 2.5

Histogram of stock
volumes, 1990–1998

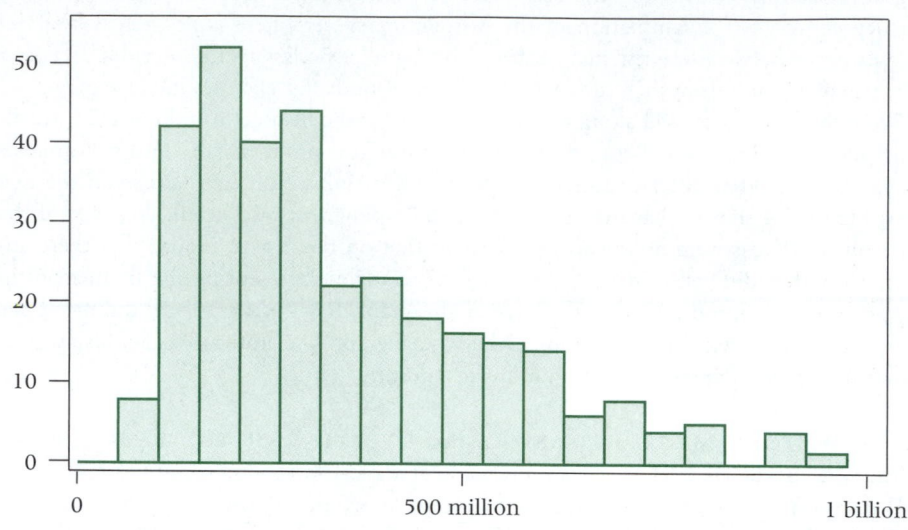

Figure 2.6

Normal distribution

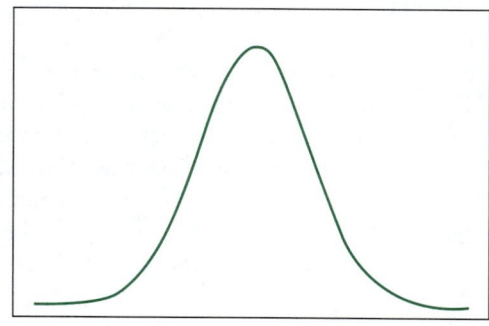

statistics, it is often useful to determine if data are approximately normally distributed (bell-curve shaped) as shown in Figure 2.6. We can see by examining the histogram in Figure 2.5 that the stock market volume data are not normally distributed. While the center of the histogram is around 500 million shares, a large portion of stock volume observations falls in the lower end of the data somewhere between 100 million and 400 million shares. In addition, the histogram shows that there are some outliers in the upper end of the distribution. Outliers are data points that appear outside of the main body of observations and may represent phenomena that differ from those represented by other data points. By observing the histogram, it is possible to see that there are a few data observations near 1 billion. One could conclude that on a few stock market days an unusually large volume of shares are traded. These and other insights can be gleaned by examining the histogram and show that histograms play an important role in the initial analysis of data.

Frequency polygon
A graph constructed by plotting a dot for the frequencies at the class midpoints and connecting the dots.

Frequency Polygons

A **frequency polygon** is *a graph in which line segments "connecting the dots" depict a frequency distribution.* Construction of a frequency polygon begins, as with a histogram, by scaling class endpoints along the X axis and the frequency values along the Y axis. A dot is plotted for the frequency value at the midpoint of each class interval (class midpoint).

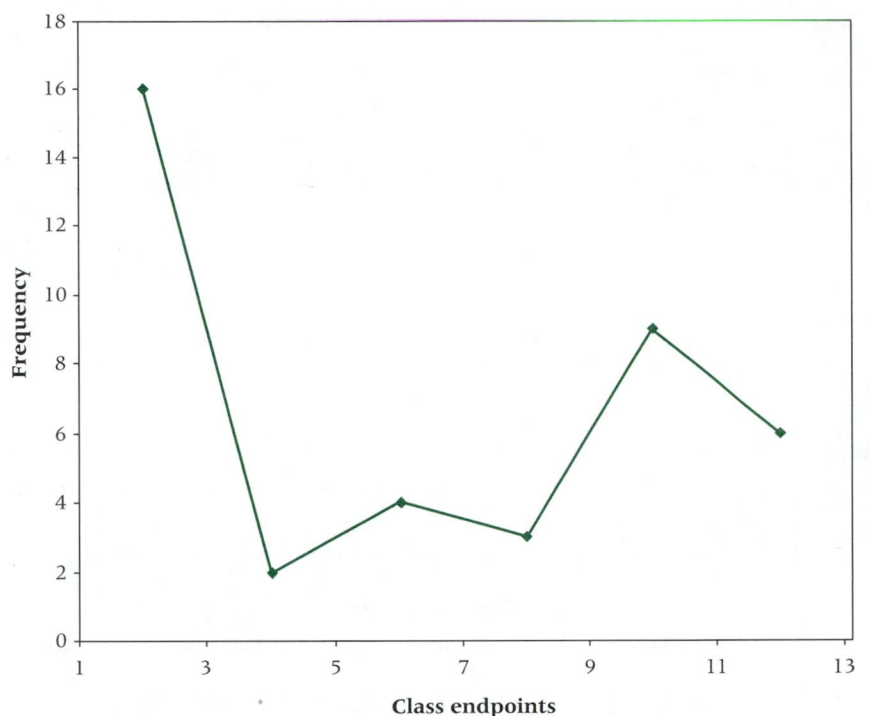

Figure 2.7

Excel-produced
frequency polygon
of the
unemployment data

Connecting these midpoint dots completes the graph. Figure 2.7 is a frequency polygon of the distribution data from Table 2.2 produced by using Excel. The information gleaned from frequency polygons and histograms is about the same. As with the histogram, changing the axes' scales can compress or stretch a frequency polygon, affecting the user's impression of what the graph represents.

Ogives

An **ogive** (o-jive) is a *cumulative frequency polygon.* Again, construction begins by labeling the X axis with the class endpoints and the Y axis with the frequencies. However, the use of cumulative frequency values requires that the scale along the Y axis be great enough to include the frequency total. A dot of zero frequency is plotted at the beginning of the first class and construction proceeds by marking a dot at the *end* of each class interval for the cumulative value. Connecting the dots then completes the ogive. Figure 2.8 is an ogive produced by using Excel for the data in Table 2.2.

Ogives are most useful when the decision maker wants to see *running totals.* For example, if a comptroller is interested in controlling costs, an ogive could depict cumulative costs over a fiscal year.

Steep slopes in an ogive can be used to identify sharp increases in frequencies. In Figure 2.8 steep slopes occur in the 1-under 3 class and the 9-under 11 class, signifying large class frequency totals.

Pie Charts

A **pie chart** is *a circular depiction of data where the area of the whole pie represents 100% of the data being studied and slices represent a percentage breakdown of the sublevels.* Pie charts show the relative magnitudes of parts to a whole. They are widely used in business, particularly to depict such things as budget categories, market share, and time and

Ogive
A cumulative frequency polygon; plotted by graphing a dot at each class endpoint for the cumulative or decumulative frequency value and connecting the dots.

Pie chart
A circular depiction of data where the area of the whole pie represents 100% of the data being studied and slices represent a percentage breakdown of the sublevels.

Figure 2.8

Excel ogive of the unemployment data

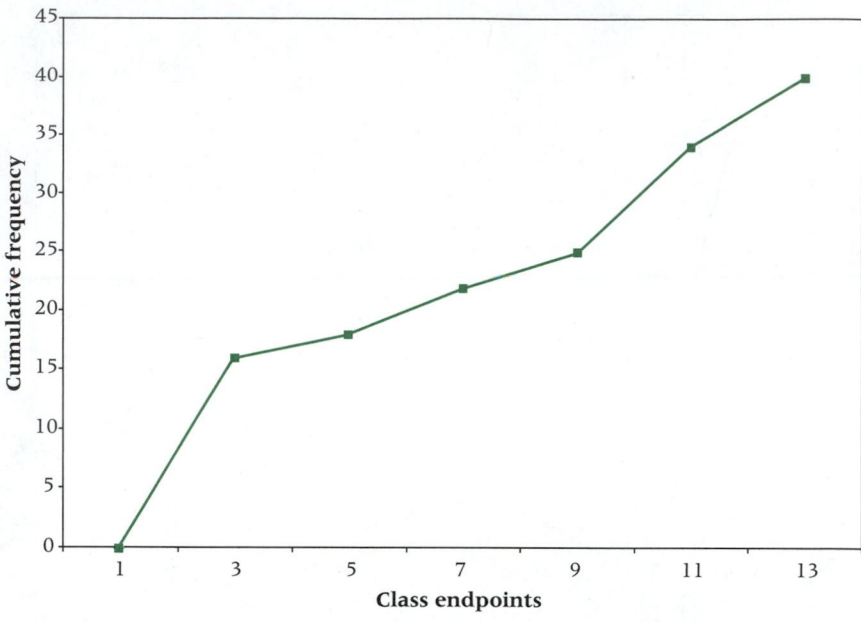

TABLE 2.4

Sales of Toothpaste for Top Ten Brands

BRAND	SALES	PROPORTION	DEGREES
Crest	$370,437,000	.2745	98.82
Colgate	321,084,000	.2380	85.68
Aquafresh	177,989,000	.1319	47.48
Mentadent	170,630,000	.1265	45.55
Arm & Hammer	109,512,000	.0812	29.23
Rembrandt	52,067,000	.0386	13.90
Sensodyn	50,133,000	.0372	13.39
Listerine	40,107,000	.0297	10.69
Closeup	32,009,000	.0237	8.53
Ultrabrite	25,358,000	.0187	6.73
Totals	$1,349,326,000	1.0000	360.00

resource allocations. However, the use of pie charts is minimized in the sciences and technology because pie charts can lead to less accurate judgments than are possible with other types of graphs.* Generally, it is more difficult for the viewer to interpret the relative size of angles in a pie chart than to judge the length of rectangles in a histogram or the relative distance of a frequency polygon dot from the X axis.

Construction of the pie chart begins by determining the proportion of the subunit to the whole. Table 2.4 contains sales figures generated by Information Resources, Inc. for the top 10 toothpaste brands for a recent year. First, the whole-number sales figures are converted to proportions by dividing each sales figure by the total sales figure. This proportion is analogous to the relative frequency computed for frequency distributions. Because there are 360° in a circle, each proportion is multiplied by 360 to obtain the correct number of degrees to represent each item. For example, Aquafresh sales of $177,989,000 represent .1319 proportion of the total sales (177,989,000/1,349,326,000 = .1319). Mul-

---

*William S. Cleveland, *The Elements of Graphing Data,* Monterey, CA: Wadsworth Advanced Books and Software, 1985.

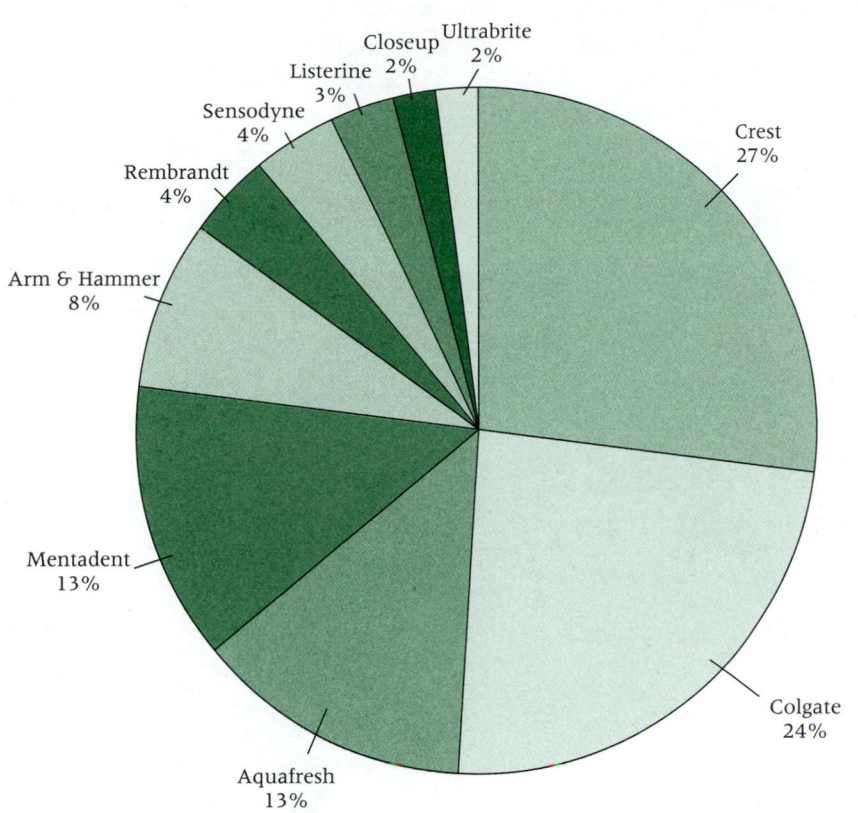

Figure 2.9

Excel pie chart
of toothpaste sales
by brand

tiplying this value by 360° results in 47.48°. Aquafresh sales will account for 47.48° of the
pie. The pie chart is then completed by using a compass to lay out the slices. The pie
chart in Figure 2.9, constructed by using Excel, depicts the data from Table 2.4.

According to the National Retail Federation and Center for Retailing Education at the
University of Florida, there are four main sources of inventory shrinkage: employee theft,
shoplifting, administrative error, and vendor fraud. The estimated annual dollar amount
in shrinkage ($ millions) associated with each of these sources follows:

Employee theft	$17,918.6
Shoplifting	15,191.9
Administrative error	7,617.6
Vendor fraud	2,553.6
Total	$43,281.7

Construct a pie chart to depict these data.

SOLUTION
Convert each raw dollar amount to a proportion by dividing each individual amount by
the total.

Employee theft	17,918.6/43,281.7 =	.414
Shoplifting	15,191.9/43,281.7 =	.351
Administrative error	7,617.6/43,281.7 =	.176
Vendor fraud	2,553.6/43,281.7 =	.059
Total		1.000

**DEMONSTRATION
PROBLEM 2.2**

Convert each proportion to degrees by multiplying each proportion by 360°.

Employee theft	.414 × 360° = 149.0°
Shoplifting	.351 × 360° = 126.4°
Administrative error	.176 × 360° = 63.4°
Vendor fraud	.059 × 360° = 21.2°
Total	360.0°

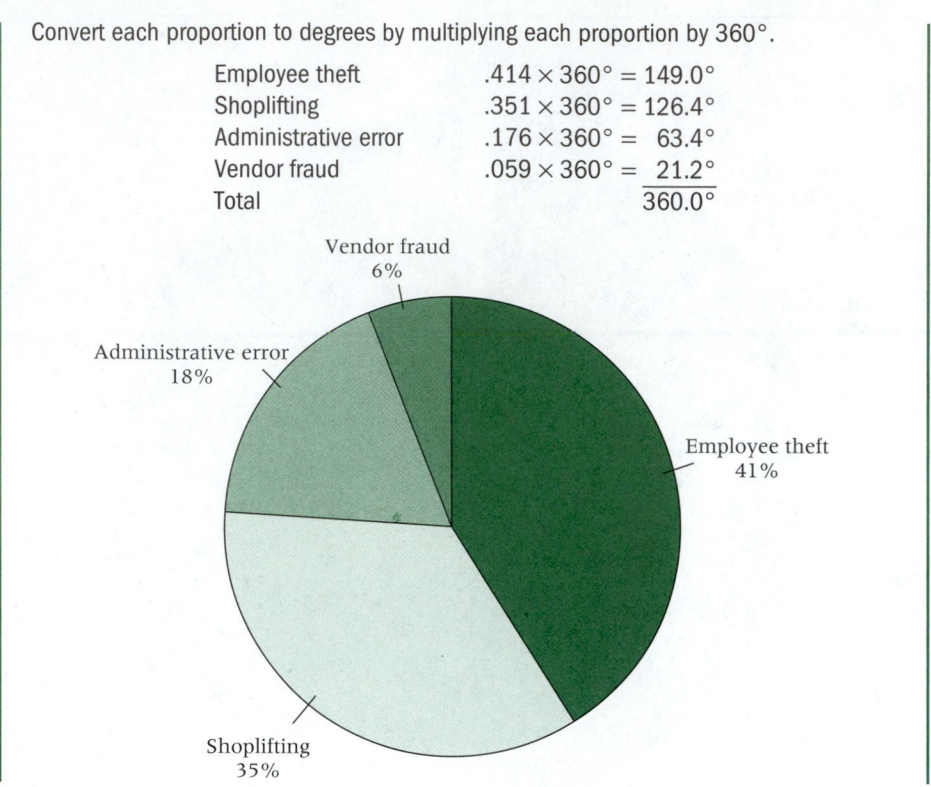

Analysis Using Excel

Histograms, frequency polygons, ogives, and pie charts can be constructed using the **Chart Wizard** (discussed in Chapter 1). In Step 1 of the **Chart Wizard** (**Chart Type**) select **Column** to construct a histogram, **Line** to construct a frequency polygon or an ogive, and **Pie** to construct a pie chart. The other three steps of the **Chart Wizard** are similar for each of the types of graphs and are discussed in Chapter 1.

Histograms and ogives can also be constructed using the **Histogram** feature of *Data Analysis.* In the **Histogram** dialog box presented in Figure 2.1, note the two spaces at the bottom of the dialog box, **Cumulative Percentage** and *Chart Output.* If you check **Cumulative Percentage** when you are using the **Histogram** feature to create a frequency distribution, you will also get an ogive as part of the output produced using percentages. If you check *Chart Output* when you are using the **Histogram** feature, you will get a histogram graph along with the frequency distribution. Both of these can be checked at the same time producing one chart with the two types of graphs overlaid.

2.2

Problems

2.5   Construct a histogram and a frequency polygon for the following data.

CLASS INTERVAL	FREQUENCY
30–32	5
32.1–34	7
34.1–36	15
36.1–38	21
38.1–40	34
40.1–42	24
42.1–44	17
44.1–46	8

2.6 Construct a histogram and a frequency polygon for the following data.

CLASS INTERVAL	FREQUENCY
10–20	9
20.1–30	7
30.1–40	10
40.1–50	6
50.1–60	13
60.1–70	18
70.1–80	15

2.7 Construct an ogive for the following data.

CLASS INTERVAL	FREQUENCY
3–6	2
6.1–9	5
9.1–12	10
12.1–15	11
15.1–18	17
18.1–21	5

2.8 A list of the largest accounting firms in the United States along with their net revenue figures for 1997 ($ millions) according to the Public Accounting Report follows.

FIRM	REVENUE
Andersen Worldwide	$5445
Ernst & Young	4416
Deloitte & Touche	3600
KPMG Peat Marwick	2698
Coopers & Lybrand	2504
PriceWaterhouse	2344
Grant Thornton	289
McGladrey & Pullen	270
BDO Seidman	240

Construct a pie chart to represent these data. Label the slices with the appropriate percentages. Comment on the effectiveness of using a pie chart to display the revenue of these top accounting firms.

2.9 According to the Air Transport Association of America, Delta Airlines led all U.S. carriers in the number of passengers flown in a recent year. The top five airlines were Delta, United, American, US Airways, and Southwest. The number of passengers flown (in thousands) by each of these airlines follow:

AIRLINE	PASSENGERS
Delta	103,133
United	84,203
American	81,083
US Airways	58,659
Southwest	55,946

Construct a pie chart to depict this information.

2.10 Information Resources, Inc., reports that in a recent year, Huggies was the top-selling diaper brand in the United States with 41.3% of the market share. Other leading brands included Pampers with 25.6%, Luvs with 12.1%, Drypers with 3.3%, Fitti with 0.9%, and private labels with 15.8%. Use this information to display the diaper market shares using a pie chart.

Summary

There are two types of data, grouped and ungrouped. Most statistical analysis is performed on ungrouped, or raw, data. Grouped data are data that have been organized into a frequency distribution.

Constructing a frequency distribution involves several steps. The first step is to determine the range of the data, which is the difference between the largest value and the smallest value. Next, the number of classes is determined, which is an arbitrary choice of the researcher. However, too few classes overaggregate the data into meaningless categories and too many classes do not summarize the data enough to be useful. The third step in constructing the frequency distribution is to determine the width of the class interval. Dividing the range of values by the number of classes yields the approximate width of the class interval.

The class midpoint is the midpoint of a class interval. It is the average of the class endpoints and represents the halfway point of the class interval. Relative frequency is a value computed by dividing an individual frequency by the sum of the frequencies. Relative frequency represents the proportion of total values that is in a given class interval, and it is analogous to the probability of randomly drawing a value from a given class interval out of all values. The cumulative frequency is a running total frequency tally that starts with the first frequency value and adds each ensuing frequency to the total.

The types of graphic depictions presented in this chapter are histograms, frequency polygons, ogives, and pie charts. Graphical depiction of data is especially useful in helping statisticians to determine the shape of distributions. A histogram is a vertical bar chart in which a line segment connects class endpoints at the value of the frequency. Two vertical lines connect this line segment down to the X axis, forming a rectangle. Histograms are taking on a growing importance as an initial analysis tool. The statistician can learn much about the shape of the distribution and other important characteristics of the data by examining a histogram of the data. A frequency polygon is constructed by plotting a dot at the midpoint of each class interval for the value of each frequency and then connecting the dots. Ogives are cumulative frequency polygons. Points on an ogive are plotted at the class endpoints. The ogive graph starts at the beginning of the first class interval with a value of zero and continues through the values of the cumulative frequencies to the class endpoints.

A pie chart is a circular depiction of data. The amount of each category is represented as a slice of the pie proportionate to the total. The slices are determined by multiplying the proportion of each category by 360° to compute the number of degrees of the circle allotted to each category. The researcher is cautioned in using pie charts because it is sometimes difficult to differentiate the relative sizes of the slices.

Key Terms

class mark	histogram
class midpoint	ogive
cumulative frequency	pie chart
frequency distribution	range
frequency polygon	relative frequency
grouped data	ungrouped data

SUPPLEMENTARY PROBLEMS

2.11 For the following data, construct a frequency distribution with six classes.

57	23	35	18	21
26	51	47	29	21
46	43	29	23	39
50	41	19	36	28
31	42	52	29	18
28	46	33	28	20

2.12 For each class interval of the frequency distribution given, determine the class midpoint, the relative frequency, and the cumulative frequency.

CLASS INTERVAL	FREQUENCY
20–25	17
25.1–30	20
30.1–35	16
35.1–40	15
40.1–45	8
45.1–50	6

2.13 Construct a histogram, a frequency polygon, and an ogive for the following frequency distribution.

CLASS INTERVAL	FREQUENCY
50–60	13
61–70	27
71–80	43
81–90	31
91–100	9

2.14 Construct a pie chart from the following data.

LABEL	VALUE
A	55
B	121
C	83
D	46

2.15 The Whitcomb Company manufactures a metal ring for industrial engines that usually weighs about 50 oz. A random sample of 50 of these metal rings produced the following weights (in ounces).

51	53	56	50	44	47
53	53	42	57	46	55
41	44	52	56	50	57
44	46	41	52	69	53
57	51	54	63	42	47
47	52	53	46	36	58
51	38	49	50	62	39
44	55	43	52	43	42
57	49				

Construct a frequency distribution for these data using eight classes. What can you observe from the frequency distribution about the data?

2.16 A northwestern distribution company surveyed 53 of its midlevel managers. The survey obtained the ages of these managers, which later were organized into the frequency distribution shown. Determine the class midpoint, relative frequency, and cumulative frequency for these data.

CLASS INTERVAL	FREQUENCY
20-under 25	8
25-under 30	6
30-under 35	5
35-under 40	12
40-under 45	15
45-under 50	7

2.17 The following data are shaped roughly like a normal distribution (discussed in Chapter 6).

61.4	27.3	26.4	37.4	30.4	47.5
63.9	46.8	67.9	19.1	81.6	47.9
73.4	54.6	65.1	53.3	71.6	58.6
57.3	87.8	71.1	74.1	48.9	60.2
54.8	60.5	32.5	61.7	55.1	48.2
56.8	60.1	52.9	60.5	55.6	38.1
76.4	46.8	19.9	27.3	77.4	58.1
32.1	54.9	32.7	40.1	52.7	32.5
35.3	39.1				

Construct a frequency distribution starting with 10 as the lowest class beginning point and use a class width of 10. Construct a histogram and a frequency polygon for this frequency distribution and observe the shape of a normal distribution. On the basis of your results from these graphs, what does a normal distribution look like?

2.18 In a medium-size southern city, 86 houses are for sale, each having about 2000 ft^2 of floor space. The asking prices vary. The frequency distribution shown contains the price categories for the 86 houses. Construct a histogram, a frequency polygon, and an ogive from these data.

ASKING PRICE	FREQUENCY
$ 60,000–$ 70,000	21
70,001– 80,000	27
80,001– 90,000	18
90,001– 100,000	11
100,001– 110,000	6
110,001– 120,000	3

2.19 Good, relatively inexpensive prenatal care often can prevent a lifetime of expense owing to complications resulting from a baby's low birth weight. A survey of a random sample of 57 new mothers asked them to estimate how much they spent on prenatal care. The researcher tallied the results and presented them in the frequency distribution shown. Use these data to construct a histogram, a frequency polygon, and an ogive.

AMOUNT SPENT ON PRENATAL CARE	FREQUENCY OF NEW MOTHERS
$ 0–$100	3
101– 200	6
201– 300	12
301– 400	19
401– 500	11
501– 600	6

2.20 A consumer group surveyed food prices at 87 stores on the East Coast. Among the food prices being measured was that of sugar. From the data collected, the group constructed the frequency distribution of the prices of 5 lbs of Domino's sugar in the stores surveyed. Compute a histogram, a frequency polygon, and an ogive for the following data.

PRICE	FREQUENCY
$1.75–$1.90	9
1.91– 2.05	14
2.06– 2.20	17
2.21– 2.35	16
2.36– 2.50	18
2.51– 2.65	8
2.66– 2.80	5

2.21 The top music genres according to SoundScan for a recent year are R&B, Alternative (Rock) Music, Rap, and Country. These and other music genres along with the number of albums sold in each (in millions) are shown.

GENRE	ALBUMS SOLD
R&B	146.4
Alternative	102.6
Rap	73.7
Country	64.5
Soundtrack	56.4
Metal	26.6
Classical	14.8
Latin	14.5

Construct a pie chart for these data displaying the percentage of the whole that each of these genre represents.

2.22 Below is a list of the industries with the largest total release of toxic chemicals in a recent year according to the U.S. Environmental Protection Agency. Construct a pie chart to depict this information.

INDUSTRY	TOTAL RELEASE (POUNDS)
Chemicals	785,178,163
Primary metals	564,535,183
Paper	227,563,372
Plastics	116,409,291
Transportation equipment	111,352,769
Fabricated metals	90,254,367
Food	83,303,395
Petroleum	68,887,258
Electrical equipment	41,765,377

ANALYZING THE DATABASES

1. Using the manufacturing database, construct a frequency distribution for the variable Number of Production Workers Across all Industries. What does the frequency distribution reveal about the number of production workers?
2. Using the stock market database, construct a histogram for the variable Reported Trades. How is the histogram shaped? Is it high in the middle or near one or both of the endpoints? Is it relatively constant in size across the classes (uniform) or does it appear to have no shape? Does it appear to be "normally" distributed?
3. Construct an ogive for the variable, type, in the financial data base. There are 100 companies in this database, each of which are categorized into one of seven types of companies. These types are listed at the end of Chapter 1. Construct a pie chart of these types and discuss the output. For example, which type is most prevalent in the database and which is the least?

SOAP COMPANIES DO BATTLE

Procter & Gamble has been the leading soap manufacturer in the United States since 1879, when it introduced Ivory soap. However, late in 1991 its major rival, Lever Bros. (Unilever), overtook it by grabbing 31.5% of the $1.6 billion personal soap market, of which Procter & Gamble had a 30.5% share. Lever Bros. had always trailed Procter & Gamble since it entered the soap market with Lifebuoy in 1895. In 1990 Lever Bros. entered a new soap, Lever 2000, into its product mix as a soap for the entire family. A niche for such a soap had been created because of the segmentation of the soap market into specialty soaps for children, women, and men. Lever Bros. felt that it could sell a soap for everyone in the family. Consumer response was strong; Lever 2000 rolled up $113,000,000 in sales in 1991, putting Lever Bros. ahead of Procter & Gamble for the first time in the personal-soap revenue contest. Procter & Gamble still sells more soap, but Lever's brands cost more, thereby resulting in greater overall sales.

Needless to say, Procter & Gamble was quick to search for a response to the success of Lever 2000. Procter & Gamble looked at several possible strategies, including repositioning Safeguard, which has been seen as a male soap.

Ultimately, Procter & Gamble responded to the challenge by introducing its Oil of Olay Moisturizing Bath Bar. In its first year of national distribution, this product was backed by a $24 million media effort. The new bath bar was quite successful and helped Procter & Gamble regain market share.

The top 10 personal soaps in the United States in 1999 are shown with their respective sales figures. Each of these soaps is produced by one of four soap manufacturers: Unilever, Procter & Gamble, Dial, or Colgate-Palmolive.

SOAP	MANUFACTURER	SALES ($ MILLIONS)
Dove	Unilever	271
Dial	Dial	193
Lever 2000	Unilever	138
Irish Spring	Colgate-Palmolive	121
Zest	Procter & Gamble	115
Ivory	Procter & Gamble	94
Caress	Unilever	93
Olay	Procter & Gamble	69
Safeguard	Procter & Gamble	48
Coast	Procter & Gamble	44

In 1983 the market shares for soap were Procter & Gamble with 37.1%, Lever Bros. (Unilever) with 24%, Dial with 15%, Colgate-Palmolive with 6.5%, and all others with 17.4%. By 1991 the market shares for soap were Lever Bros. (Unilever) with 31.5%, Procter & Gamble with 30.5%, Dial with 19%, Colgate-Palmolive with 8%, and all others with 11%.

DISCUSSION

1. Suppose you are making a report for Procter & Gamble displaying their share of the market along with the share of other companies for the years 1983, 1991, and 1999. Using Excel, produce graphs for the market shares of personal soap for each of these years. For the 1999 data, assume that the "all others" total is about $119 million. What do you observe about the market shares of the various companies by studying the graphs? In particular, how is Procter & Gamble doing relative to previous years?

2. Suppose Procter & Gamble sells about 20 million bars of soap per week, but the demand is not constant and production management would like to get a better handle on how sales are distributed over the year. Let the sales figures given below in units of million bars represent the sales of bars per week over one year. Construct a histogram to represent these data. What do you see in the graph that might be helpful to the production (and sales) people?

17.1	19.6	15.4	17.4	15.0	18.5	20.6	18.4	20.0
20.9	19.3	18.2	14.7	17.1	12.2	19.9	18.7	20.4
20.3	15.5	16.8	19.1	20.4	15.4	20.3	17.5	17.0
18.3	13.6	39.8	20.7	21.3	22.5	21.4	23.4	23.1
22.8	21.4	24.0	25.2	26.3	23.9	30.6	25.2	26.2
26.9	32.8	26.3	26.6	24.3	26.2	23.8		

SOURCE: Adapted from Valerie Reitman, "Buoyant Sales of Lever 2000 Soap Bring Sinking Sensation to Procter & Gamble," *Wall Street Journal,* March 19, 1992, B1. Reprinted by permission of *The Wall Street Journal* © 1992, Dow Jones & Company, Inc. All Rights Reserved Worldwide. And from Pam Weisz, "$40 M Extends Lever 2000 Family," *Brandweek,* vol. 36, no. 32, August 21, 1995, p. 6; Laurie Freeman, "P&G Pushes Back against Unilever in Soap," *Advertising Age,* vol. 65, no. 41, September 28, 1994, p. 21; Jeanne Whalen and Pat Sloan, "Intros Help Boost Soap Coupons," *Advertising Age,* vol. 65, no. 19, May 2, 1994, p. 30, and "P&G Places Coast Soap Up for Sale," *The Post,* World Wide Web Edition of *The Cincinnati Post,* February 2, 1999, at http://www.cincypost.com.business/pg022599.html.

3

Descriptive Statistics

Learning Objectives

The focus of Chapter 3 is the use of statistical techniques to describe data, thereby enabling you to:

1. Distinguish between measures of central tendency, measures of variability, and measures of shape.

2. Understand the meanings of mean, median, mode, quartile, and range.

3. Compute mean, median, mode, quartile, range, variance, standard deviation, and mean absolute deviation.

4. Differentiate between sample and population variance and standard deviation.

5. Understand the meaning of standard deviation as it is applied by using the empirical rule.

6. Understand box and whisker plots, skewness, and kurtosis.

Chapter 2 described graphical techniques for organizing and presenting data. While these graphs allow the researcher to make some general observations about the shape and spread of the data, a fuller understanding of the data can be attained by summarizing the data numerically using statistics. This chapter presents such statistical measures, including measures of central tendency, measures of variability, and measures of shape.

3.1 Measures of Central Tendency

Measure of central tendency
One type of measure that is used to yield information about the center of a group of numbers.

Mode
The most frequently occurring value in a set of data.

One type of measure that is used to describe a set of data is the **measure of central tendency.** Measures of central tendency *yield information about the center, or middle part, of a group of numbers.* Displayed in Table 3.1 are the offer price for the 20 largest U.S. initial public offerings in a recent year according to the Securities Data Co. For these data, measures of central tendency can yield such information as the average offer price, the middle offer price, and the most frequently occurring offer price. Measures of central tendency do not focus on the span of the data set or how far values are from the middle numbers. The measures of central tendency presented here for ungrouped data are the mode, the median, the mean, and quartiles.

Mode

The **mode** is *the most frequently occurring value* in a set of data. For the data in Table 3.1 the mode is $19.00 because the offer price that recurred the most times (4) was $19.00. Organizing the data into an *ordered array* (an ordering of the numbers from smallest to largest) helps to locate the mode. The following is an ordered array of the values from Table 3.1.

7.00	11.00	14.25	15.00	15.00	15.50	19.00	19.00	19.00	19.00
21.00	22.00	23.00	24.00	25.00	27.00	27.00	28.00	34.22	43.25

This grouping makes it easier to see that 19.00 is the most frequently occurring number.

If there is a tie for the most frequently occurring value, there are two modes. In that case the data are said to be **bimodal.** If a set of data is not exactly bimodal but contains two values that are more dominant than others, some researchers take the liberty of referring to the data set as bimodal even though there is not an exact tie for the mode. Data sets with more than two modes are referred to as **multimodal.**

Bimodal
Data sets that have two modes.

Multimodal
Data sets that contain more than two modes.

In the world of business, the concept of mode is often used in determining sizes. For example, shoe manufacturers might produce inexpensive shoes in three widths only: small, medium, and large. Each width size represents a modal width of feet. By reducing the number of sizes to a few modal sizes, companies can reduce total product costs by limiting machine setup costs. Similarly, the garment industry produces shirts, dresses, suits, and many other clothing products in modal sizes. For example, all size M shirts in a given lot are produced in the same size. This size is some modal size for medium-size men.

The mode is an appropriate measure of central tendency for nominal level data. The mode can be used to determine which category occurs most frequently.

TABLE 3.1
Offer Prices for the Twenty Largest U.S. Initial Public Offerings in a Recent Year ($)

14.25	19.00	11.00	28.00
24.00	23.00	43.25	19.00
27.00	25.00	15.00	7.00
34.22	15.50	15.00	22.00
19.00	19.00	27.00	21.00

Median

The **median** is *the middle value in an ordered array of numbers.* If there is an odd number of terms in the array, the median is the middle number. If there is an even number of terms, the median is the average of the two middle numbers. The following steps are used to determine the median.

STEP **1.**   Arrange the observations in an ordered data array.
STEP **2.**   If there is an odd number of terms, find the middle term of the ordered array. It is the median.
STEP **3.**   If there is an even number of terms, find the average of the middle two terms. This average is the median.

Suppose a business analyst wants to determine the median for the following numbers.

15   11   14   3   21   17   22   16   19   16   5   7   19   8   9   20   4

He or she arranges the numbers in an ordered array.

3   4   5   7   8   9   11   14   15   16   16   17   19   19   20   21   22

There are 17 terms (an odd number of terms), so the median is the middle number, or 15.
   If the number 22 is eliminated from the list, there are only 16 terms.

3   4   5   7   8   9   11   14   15   16   16   17   19   19   20   21

Now there is an even number of terms, and the business analyst determines the median by averaging the two middle values, 14 and 15. The resulting median value is 14.5.
   Another way to locate the median is by finding the $(n + 1)/2$ term in an ordered array. For example, if a data set contains 77 terms, the median is the 39th term. That is,

$$\frac{n+1}{2} = \frac{77+1}{2} = \frac{78}{2} = 39\text{th term.}$$

This formula is helpful when a large number of terms must be manipulated.
   Consider the offer price data in Table 3.1. Because there are 20 values and therefore $n = 20$, the median for these data is located at the $(20 + 1)/2$ term, or the 10.5th term. This indicates that the median is located halfway between the 10th and 11th term or the average of 19.00 and 21.00. Thus, the median offer price for the largest twenty U.S. initial public offerings is $20.00.
   The median is unaffected by the magnitude of extreme values. This characteristic is an advantage, because large and small values do not inordinately influence the median. For this reason, the median is often the best measure of location to use in the analysis of variables such as house costs, income, and age. Suppose, for example, that a real estate broker wants to determine the median selling price of 10 houses listed at the following prices.

$67,000	$105,000	$148,000	$5,250,000
91,000	116,000	167,000	
95,000	122,000	189,000	

The median is the average of the two middle terms, $116,000 and $122,000, or $119,000. This price is a reasonable representation of the prices of the 10 houses. Note that the house priced at $5,250,000 did not enter into the analysis other than to count as one of the 10 houses. If the price of the tenth house were $200,000, the results would be

Median
The middle value in an ordered array of numbers.

the same. However, if all the house prices were averaged, the resulting average price of the original 10 houses would be \$635,000, higher than nine of the 10 individual prices.

A disadvantage of the median is that not all the information from the numbers is used. That is, information about the specific asking price of the most expensive house does not really enter into the computation of the median. The level of data measurement must be at least ordinal for a median to be meaningful.

Mean

Arithmetic mean
The average of a group of numbers.

The **arithmetic mean** is synonymous with the *average of a group of numbers* and is computed by summing all numbers and dividing by the number of numbers. Because the arithmetic mean is so widely used, most statisticians refer to it simply as the *mean*.

The population mean is represented by the Greek letter mu (μ). The sample mean is represented by $\overline{X}$. The formulas for computing the population mean and the sample mean are given in the boxes that follow.

POPULATION MEAN
$$\mu = \frac{\Sigma X}{N} = \frac{X_1 + X_2 + X_3 + \cdots + X_N}{N}$$

SAMPLE MEAN
$$\overline{X} = \frac{\Sigma X}{n} = \frac{X_1 + X_2 + X_3 + \cdots + X_n}{n}$$

The capital Greek letter sigma (Σ) is commonly used in mathematics to represent a summation of all the numbers in a grouping.* Also, N is the number of terms in the population, and n is the number of terms in the sample. The algorithm for computing a mean is to sum all the numbers in the population or sample and divide by the number of terms.

A more formal definition of the mean is

$$\mu = \frac{\sum_{i=1}^{N} X_i}{N}.$$

However, for the purposes of this text,

$$\Sigma X \text{ denotes} \sum_{i=1}^{N} X_i.$$

It is inappropriate to use the mean to analyze data that are not at least interval level in measurement.

Suppose a company has five departments with 24, 13, 19, 26, and 11 workers each. The *population mean* number of workers in each department is 18.6 workers. The computations follow.

$$
\begin{array}{r}
24 \\
13 \\
19 \\
26 \\
\underline{11} \\
\Sigma X = 93
\end{array}
$$

---

*The mathematics of summations is not discussed here. A more detailed explanation is given on the CD-ROM.

and

$$\mu = \frac{\Sigma X}{N} = \frac{93}{5} = 18.6.$$

The calculation of a sample mean uses the same algorithm as for a population mean and will produce the same answer if computed on the same data. However, it is inappropriate to compute a sample mean for a population or a population mean for a sample. Since both populations and samples are important in statistics, a separate symbol is necessary for the population mean and for the sample mean.

DEMONSTRATION PROBLEM 3.1

The number of U.S. cars in service by top car rental companies in a recent year according to *Auto Rental News* follows.

COMPANY	NUMBER OF CARS IN SERVICE
Enterprise	355,000
Hertz	250,000
Avis	200,000
National	145,000
Alamo	130,000
Budget	125,000
Dollar	62,000
FRCS (Ford)	53,150
Thrifty	34,000
Republic Replacement	32,000
DRAC (Chrysler)	27,000
U-Save	12,000
Rent-a-Wreck	12,000
Payless	12,000
Advantage	9,000

Compute the mode, the median, and the mean.

SOLUTION

Mode: 12,000

Median: There are 15 different companies in this group, so $n = 15$. The median is located at the $(15 + 1)/2 = $ 8th position. Since the data are already ordered, the 8th term is 53,150, which is the median.

Mean: The total number of cars in service is $1,458,150 = \Sigma X$

$$\mu = \frac{\Sigma X}{n} = \frac{1,458,150}{15} = 97,210$$

The mean is affected by each and every value, which is an advantage. The mean uses all the data and each data item influences the mean. It is also a disadvantage, because extremely large or small values can cause the mean to be pulled toward the extreme value. Recall the preceding discussion of the 10 house prices. If the mean is computed on the 10 houses, the mean price is higher than the prices of nine of the houses because the $5,250,000 house is included in the calculation. The total price of the 10 houses is $6,350,000, and the mean price is

$$\bar{X} = \frac{\Sigma X}{n} = \frac{\$6,350,000}{10} = \$635,000.$$

The mean is the most commonly used measure of location because it uses each data item in its computation, it is a familiar measure, and it has mathematical properties that make it attractive to use in inferential statistics analysis.

Quartiles

Quartiles
Measures of central tendency that divide a group of data into four subgroups or parts.

Quartiles are *measures of central tendency that divide a group of data into four subgroups or parts.* There are three quartiles, denoted as Q_1, Q_2, and Q_3. The first quartile, Q_1, separates the first, or lowest, one-fourth of the data from the upper three-fourths. The second quartile, Q_2, separates the second quarter of the data from the third quarter and equals the median of the data. The third quartile, Q_3, divides the first three-quarters of the data from the last quarter. These three quartiles are shown in Figure 3.1.

Shown next is a summary of the steps used in determining the location of a quartile.

STEPS IN DETERMINING THE LOCATION OF A QUARTILE

1. Organize the numbers into an ascending-order array.
2. Calculate the quartile location (i) by:

$$i = \frac{Q}{4}(n)$$

where:
   Q = the quartile of interest,
   i = quartile location, and
   n = number in the data set.

3. Determine the location by either (a) or (b).
   a. If i *is* a whole number, quartile Q is the average of the value at the ith location and the value at the $(i + 1)$st location.
   b. If i *is not* a whole number, quartile Q value is located at the whole number part of $i + 1$.

Suppose we want to determine the values of Q_1, Q_2, and Q_3 for the following numbers.

106   109   114   116   121   122   125   129

The value of Q_1 is found by

$$\text{For } n = 8, i = \frac{1}{4}(8) = 2$$

Figure 3.1

Quartiles

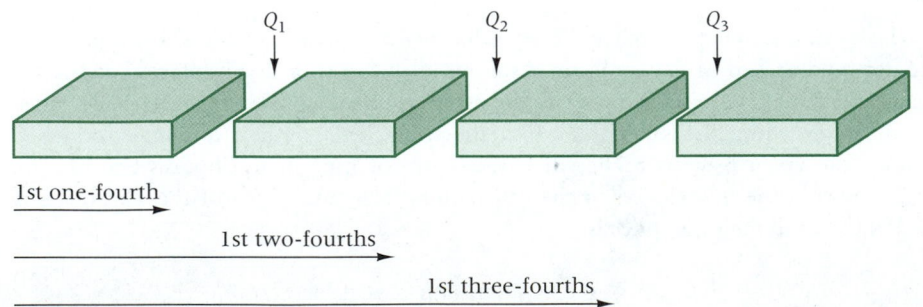

1st one-fourth

1st two-fourths

1st three-fourths

Because i is a whole number, Q_1 is found as the average of the second and third numbers.

$$Q_1 = \frac{(109 + 114)}{2} = 111.5$$

The value of Q_1 is 111.5. Notice that one-fourth, or two, of the values (106 and 109) are less than 111.5.

The value of Q_2 is equal to the median. As there is an even number of terms, the median is the average of the two middle terms.

$$Q_2 = \text{median} = \frac{(116 + 121)}{2} = 118.5$$

Notice that exactly half of the terms are less than Q_2 and half are greater than Q_2.

The value of Q_3 is determined as follows.

$$i = \frac{3}{4}(8) = 6$$

Because i is a whole number, Q_3 is the average of the sixth and the seventh numbers.

$$Q_3 = \frac{(122 + 125)}{2} = 123.5$$

The value of Q_3 is 123.5. Notice that three-fourths, or six, of the values are less than 123.5 and two of the values are greater than 123.5.

The following shows revenues for the world's top 20 advertising organizations according to *Advertising Age*, Crain Communications, Inc. Determine the first, the second, and the third quartiles for these data.

DEMONSTRATION PROBLEM 3.2

AD ORGANIZATION	HEADQUARTERS	WORLDWIDE GROSS INCOME ($ MILLIONS)
Omnicom Group	New York	4154
WPP Group	London	3647
Interpublic Group of Cos.	New York	3385
Dentsu	Tokyo	1988
Young & Rubicam	New York	1498
True North Communications	Chicago	1212
Grey Advertising	New York	1143
Havas Advertising	Paris	1033
Leo Burnett Co.	Chicago	878
Hakuhodo	Tokyo	848
MacManus Group	New York	843
Saatchi & Saatchi	London	657
Publicis Communication	Paris	625
Cordiant Communications Group	London	597
Carlson Marketing Group	Minneapolis	285
TMP Worldwide	New York	274
Asatsu	Tokyo	263
Tokyu Agency	Tokyo	205
Daiko Advertising	Tokyo	204
Abbott Mead Vickers	London	187

SOLUTION

There are 20 advertising organizations, $n = 20$. Q_1 is found by

$$i = \frac{1}{4}(20) = 5$$

Because i is a whole number, Q_1 is found to be the average of the fifth and sixth values from the bottom.

$$Q_1 = \frac{274 + 285}{2} = 279.5$$

Q_2 = median; as there are 20 terms, the median is the average of the tenth and eleventh terms.

$$Q_2 = \frac{843 + 848}{2} = 845.5$$

Q_3 is solved by

$$i = \frac{3}{4}(20) = 15$$

Q_3 is found by averaging the fifteenth and sixteenth terms.

$$Q_3 = \frac{1212 + 1498}{2} = 1355$$

Analysis Using Excel

Excel can compute a mode, a median, a mean, and quartiles. Each of these statistics is accessed using the paste function, f_x. Select **Statistical** from the options presented on the left side of the paste function dialog box, and a long list of statistical options are displayed on the right side. Among the options shown on the right side are **MODE, MEDIAN, AVERAGE** (used to compute means), and **QUARTILE.** The Excel dialog boxes for these four statistics are displayed in Figures 3.2 through 3.5.

To compute a mode, a median, or a mean (average), enter the location of the data in the first box of the dialog box labeled **Number1.** The answer will be displayed on the dialog box and will be shown on the spreadsheet after clicking **OK.** The quartile dialog box also requires that the location of the data be entered in the first box, but this box is labeled **Array** for quartile computation. In the second box of the quartile dialog box labeled **Quart,** insert the number 1 to compute the first quartile, the number 2 to compute the second quartile, and the number 3 to compute the third quartile.

Figure 3.6 displays the Excel output of the mean, median, mode, Q_1, Q_2, and Q_3 for Demonstration Problem 3.1. The answers obtained for the mode, median, mean, and Q_2 are the same as those computed manually in this text. However, Excel defines the first quartile, Q_1, as the $\left[\frac{n+3}{4}\right]^{th}$ item and the third quartile, Q_3, as the $\left[\frac{3n+1}{4}\right]^{th}$ item. Thus, the answers for Q_1 and Q_3 will either be the same or will differ by 1 at the most from the values obtained using methods presented in this chapter.

Figure 3.2   Dialog box for **MODE**

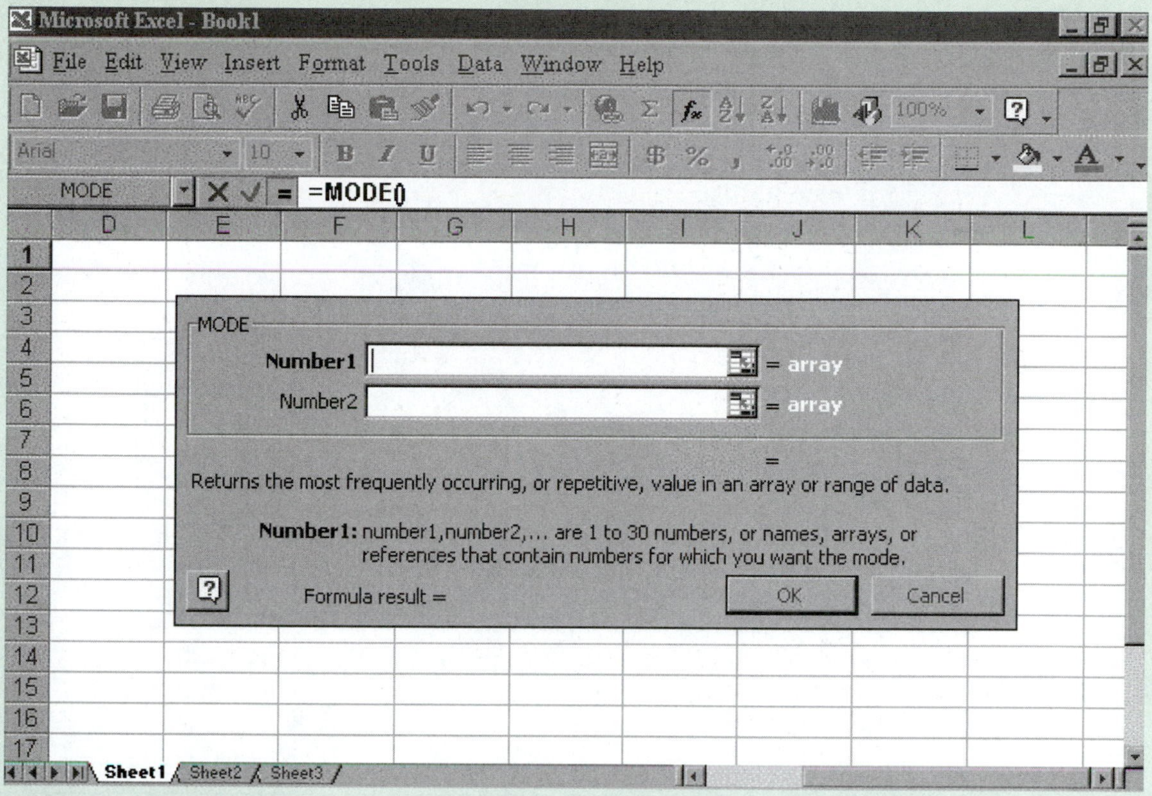

Figure 3.3   Dialog box for **MEDIAN**

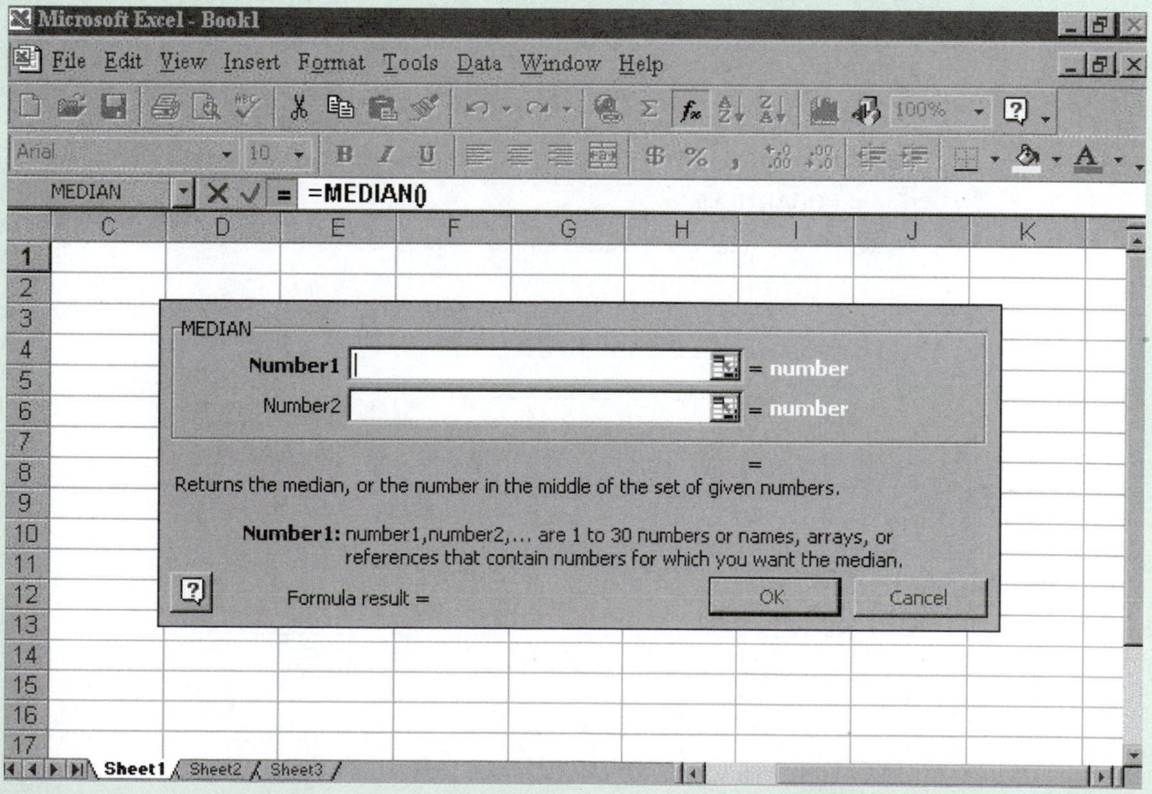

Figure 3.4 Dialog box for **MEAN**

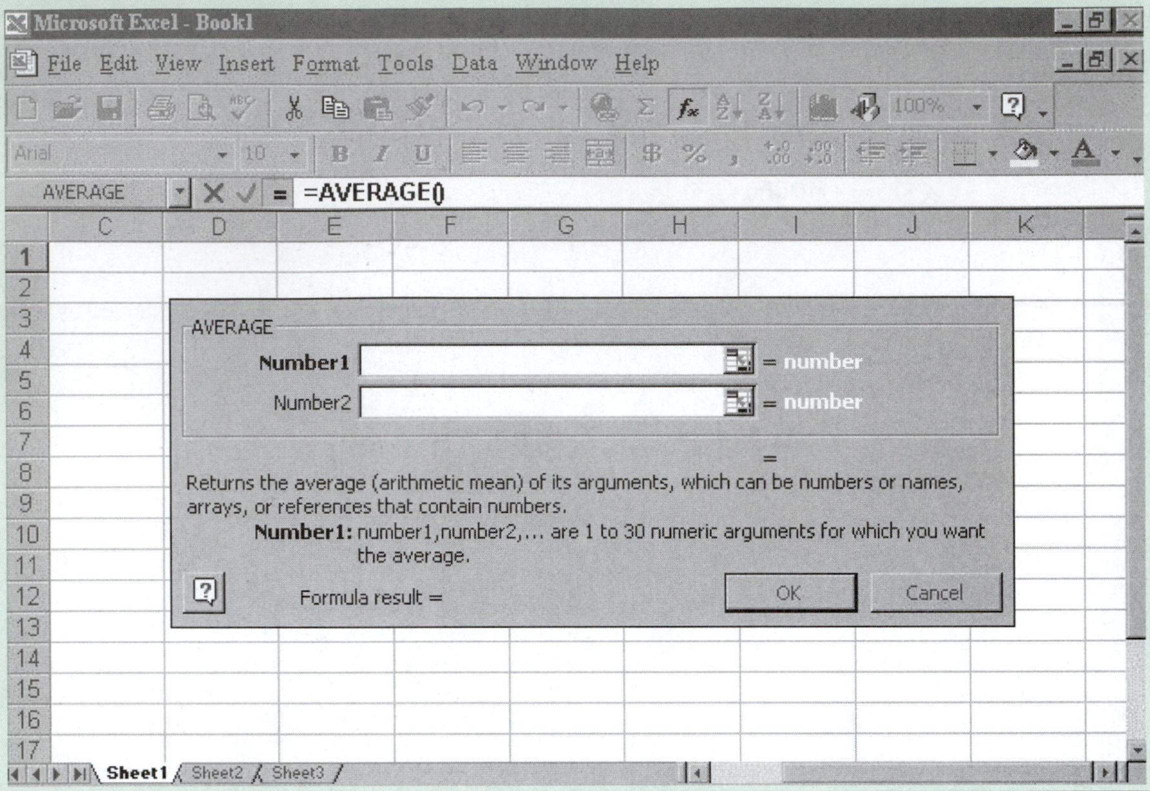

Figure 3.5 Dialog box for **QUARTILES**

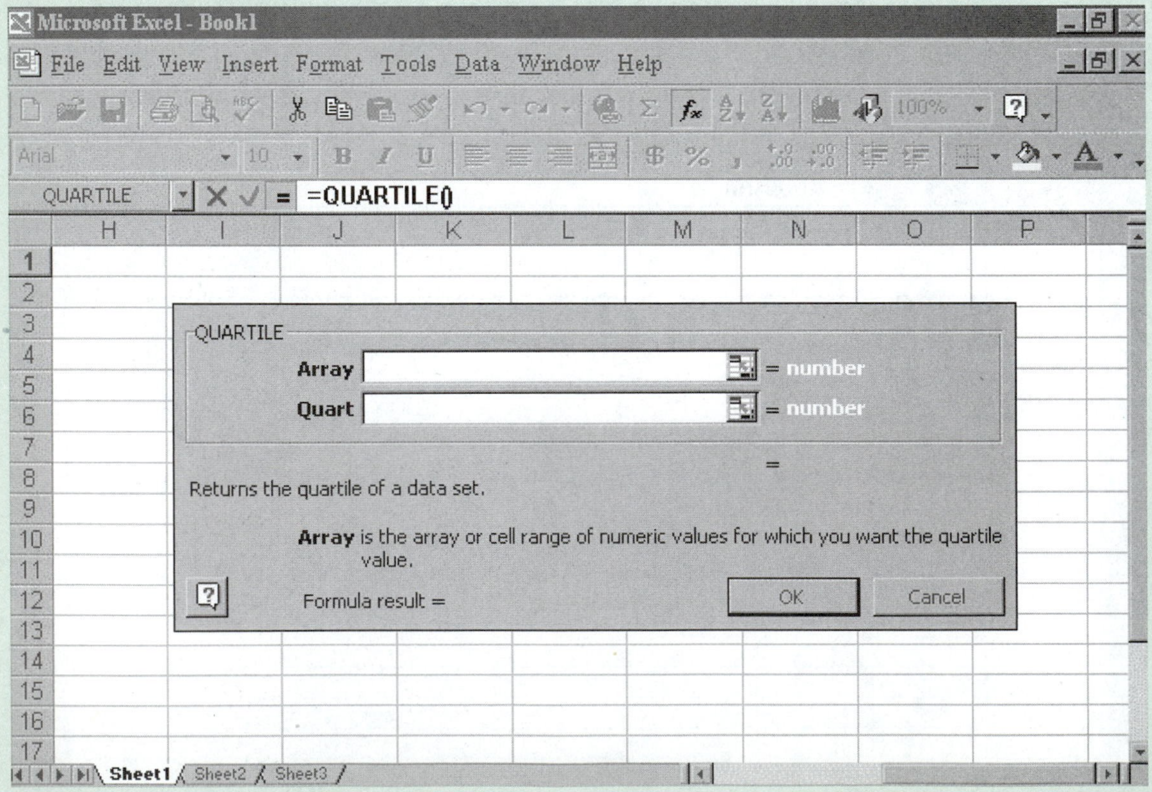

Figure 3.6	Excel output for Demonstration Problem 3.1

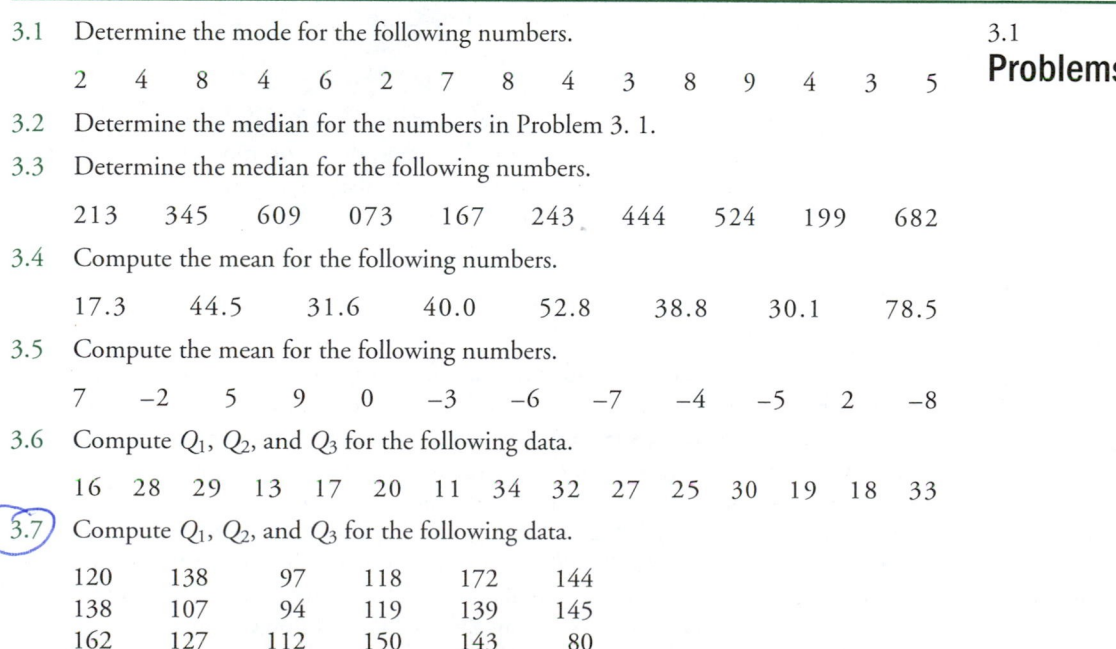

	A	B	C	D	E	F	G	H	I
1	9,000		The mode is:			12,000			
2	12,000								
3	12,000		The median is:			53,150			
4	12,000								
5	27,000		The mean is:			97,210			
6	32,000								
7	34,000		The first quartile is:			19,500			
8	53,150								
9	62,000		The second quartile is:			53,150			
10	125,000								
11	130,000		The third quartile is:			137,500			
12	145,000								
13	200,000								
14	250,000								
15	355,000								
16									
17									

Problems

3.1 Determine the mode for the following numbers.

2   4   8   4   6   2   7   8   4   3   8   9   4   3   5

3.2 Determine the median for the numbers in Problem 3. 1.

3.3 Determine the median for the following numbers.

213   345   609   073   167   243   444   524   199   682

3.4 Compute the mean for the following numbers.

17.3   44.5   31.6   40.0   52.8   38.8   30.1   78.5

3.5 Compute the mean for the following numbers.

7   −2   5   9   0   −3   −6   −7   −4   −5   2   −8

3.6 Compute Q_1, Q_2, and Q_3 for the following data.

16   28   29   13   17   20   11   34   32   27   25   30   19   18   33

3.7 Compute Q_1, Q_2, and Q_3 for the following data.

120	138	97	118	172	144
138	107	94	119	139	145
162	127	112	150	143	80
105	116	142	128	116	171

3.8   Shown here are the projected number of cars and light trucks for the year 2000 for the largest automakers in the world, as reported by AutoFacts, a unit of Coopers & Lybrand Consulting. Compute the mean and median. Which of these two measures do you think is most appropriate for summarizing these data and why? What is the value of Q_1, Q_2, and Q_3?

AUTOMAKER	PRODUCTION (THOUSANDS)
General Motors	7880
Ford Motors	6359
Toyota	4580
Volkswagen	4161
Chrysler	2968
Nissan	2646
Honda	2436
Fiat	2264
Peugeot	1767
Renault	1567
Mitsubishi	1535
Hyundai	1434
BMW	1341
Daimler-Benz	1227
Daewoo	898

3.9   The following lists the biggest banks in the world ranked by assets according to *The Banker,* bank reports. Compute the median Q_1 and Q_3.

BANK	ASSETS
BNP-SG-Paribas	$1096
Deutsche-Bank-BT	800
UBS	751
Citigroup	701
Bank of Tokyo-Mitsubishi	653
BankAmerica	595
Credit Suisse	516
Industrial and Commercial Bank of China	489
HSBC	483
Sumitomo Bank	468

3.10 The following lists the number of fatal accidents by scheduled commercial airlines over a 17-year period according to the Air Transport Association of America. Using these data, compute the mean, median, and mode. What is the value of the third quartile?

4   4   4   1   4   2   4   3   8   6   4   4   1   4   2   3   3

3.2
Measures
of Variability

Measures of variability
Statistics that describe the spread or dispersion of a set of data.

Measures of central tendency yield information about particular points of a data set. However, researchers can use another group of analytic tools to describe a set of data. These tools are **measures or variability,** which *describe the spread or the dispersion of a set of data.* Using measures of variability in conjunction with measures of central tendency makes possible a more complete numerical description of the data.

For example, a company has 25 salespeople in the field, and the median annual sales figure for these people is $1,200,000. Are the salespeople being successful as a *group* or not? The median provides information about the sales of the person in the middle, but what about the other salespeople? Are all of them selling $ 1,200,000 annually, or do the

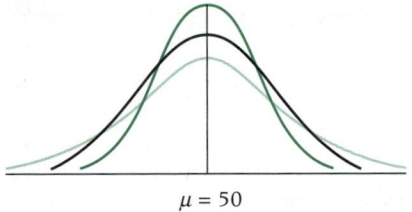

$\mu = 50$

Figure 3.7

Three distributions with the same mean but different dispersions

sales figures vary widely, with one person selling $5,000,000 annually and another selling only $150,000 annually? Measures of variability provide the additional information necessary to answer that question.

Figure 3.7 shows three distributions in which the mean of each distribution is the same ($\mu = 50$) but the variabilities differ. Observation of these distributions shows that a measure of variability is necessary to complement the mean value in describing the data. This section focuses on seven measures of variability: range, interquartile range, mean absolute deviation, variance, standard deviation, Z scores, and coefficient of variation.

Range

The **range** is *the difference between the largest value of a data set and the smallest value*. Although it is usually a single numeric value, some researchers define the range as *the ordered pair of smallest and largest numbers (smallest, largest)*. It is a crude measure of variability, describing the distance to the outer bounds of the data set. It reflects those extreme values because it is constructed from them. An advantage of the range is its ease of computation. One important use of the range is in quality assurance, where the range is used to construct control charts. A disadvantage of the range is that because it is computed with the values that are on the extremes of the data it is affected by extreme values and therefore its application as a measure of variability is limited.

The data in Table 3.1 represent the offer prices for the 20 largest U.S. initial public offerings in a recent year. The lowest offer price was $7.00 and the highest price was $43.25. The range of the offer prices can be computed as the difference of the highest and lowest values:

$$\text{Range} = \text{Highest} - \text{Lowest} = \$43.25 - \$7.00 = \$36.25$$

Range
The difference between the largest and the smallest values in a set of numbers.

Interquartile Range

Another measure of variability is the **interquartile range.** The interquartile range is *the range of values between the first and third quartile*. Essentially, it is the range of the middle 50% of the data, and it is determined by computing the value of $Q_3 - Q_1$. The interquartile range is especially useful in situations where data users are more interested in values toward the middle and less interested in extremes. In describing a real estate housing market, realtors might use the interquartile range as a measure of housing prices when describing the middle half of the market when buyers are interested in houses in the midrange. In addition, the interquartile range is used in the construction of box and whisker plots.

Interquartile range
The range of values between the first and the third quartile.

$$Q_3 - Q_1 \qquad \text{INTERQUARTILE RANGE}$$

The following lists the top 15 trading partners of the United States by U.S. exports to the country in a recent year according to the U.S. Census Bureau.

COUNTRY	EXPORTS ($ BILLIONS)
Canada	$151.8
Mexico	71.4
Japan	65.5
United Kingdom	36.4
South Korea	25.0
Germany	24.5
Taiwan	20.4
Netherlands	19.8
Singapore	17.7
France	16.0
Brazil	15.9
Hong Kong	15.1
Belgium	13.4
China	12.9
Australia	12.1

What is the interquartile range for these data? The process begins by computing the first and third quartiles as follows.

Solving for Q_1 when $n = 15$:

$$i = \frac{1}{4}(15) = 3.75$$

Since i is not a whole number, Q_1 is found as the 4th term from the bottom.

$$Q_1 = 15.1$$

Solving for Q_3:

$$i = \frac{3}{4}(15) = 11.25$$

Since i is not a whole number, Q_3 is found as the 12th term from the bottom.

$$Q_3 = 36.4$$

The interquartile range is:

$$Q_3 - Q_1 = 36.4 - 15.1 = 21.3$$

The middle 50% of the exports for the top 15 United States trading partners spans a range of 21.3 ($ billions).

Mean Absolute Deviation, Variance, and Standard Deviation

Three other measures of variability are the variance, the standard deviation, and the mean absolute deviation. They are obtained through similar processes and are therefore presented together. These measures are not meaningful unless the data are at least interval-level data. The variance and standard deviation are widely used in statistics. Although the standard deviation has some stand-alone potential, the importance of variance and standard deviation lies mainly in their role as tools used in conjunction with other statistical devices.

Suppose a small company has started a production line to build computers. During the first five weeks of production, the output is 5, 9, 16, 17, and 18 computers, respectively. Which descriptive statistics could the owner use to measure the early progress of production? In an attempt to summarize these figures, he could compute a mean.

$$\frac{X}{5}$$
$$9$$
$$16$$
$$17$$
$$\underline{18}$$

$$\Sigma X = 65 \qquad \mu = \frac{\Sigma X}{N} = \frac{65}{5} = 13$$

What is the variability in these five weeks of data? One way for the owner to begin to look at the spread of the data is to subtract the mean from each data value. *Subtracting the mean from each value of data* yields the **deviation from the mean** $(X - \mu)$. Table 3.2 shows these deviations for the computer company production. Note that some deviations from the mean are positive and some are negative. Figure 3.8 shows that geometrically the negative deviations represent values that are below (to the left of) the mean and positive deviations represent values that are above (to the right of) the mean.

An examination of deviations from the mean can reveal information about the variability of data. However, the deviations are used mostly as a tool to compute other measures of variability. Note that in both Table 3.2 and Figure 3.8 these deviations total zero. This phenomenon applies to all cases. For a given set of data, the sum of all deviations from the arithmetic mean is always zero.

Deviation from the mean
The difference between a number and the average of the set of numbers of which the number is a part.

$$\Sigma(X - \mu) = 0$$

SUM OF DEVIATIONS FROM THE ARITHMETIC MEAN IS ALWAYS ZERO

NUMBER (X)	DEVIATIONS FROM THE MEAN $(X - \mu)$
5	$5 - 13 = -8$
9	$9 - 13 = -4$
16	$16 - 13 = +3$
17	$17 - 13 = +4$
$\underline{18}$	$18 - 13 = \underline{+5}$
$\Sigma X = 65$	$\Sigma(X - \mu) = 0$

TABLE 3.2
Deviations from the Mean for Computer Production

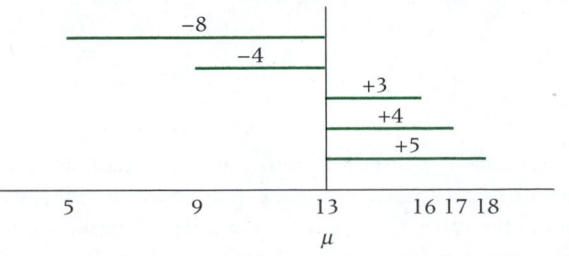

Figure 3.8

Geometric distances from the mean (from Table 3.2)

This property requires considering alternative ways to obtain measures of variability.

One obvious way to force the sum of deviations to have a nonzero total is to take the absolute value of each deviation around the mean. Utilizing the absolute value of the deviations about the mean makes solving for the mean absolute deviation possible.

Mean Absolute Deviation

The **mean absolute deviation (MAD)** is *the average of the absolute values of the deviations around the mean for a set of numbers.*

Mean absolute deviation (MAD)

The average of the absolute values of the deviations around the mean for a set of numbers.

<table>
<tr><td>MEAN ABSOLUTE DEVIATION</td><td>$$\text{MAD} = \frac{\Sigma |X - \mu|}{N}$$</td></tr>
</table>

Using the data from Table 3.2, the computer company owner can compute a mean absolute deviation by taking the absolute values of the deviations and averaging them, as shown in Table 3.3. The mean absolute deviation for the computer production data is 4.8.

Because it is computed by using absolute values, the mean absolute deviation is less useful in statistics than other measures of dispersion. However, in the field of forecasting, it is used occasionally as a measure of error.

Variance

Because absolute values are not conducive to easy manipulation, mathematicians developed an alternative mechanism for overcoming the zero-sum property of deviations from the mean. This approach utilizes the square of the deviations from the mean. The result is the variance, an important measure of variability.

The **variance** is *the average of the squared deviations about the arithmetic mean for a set of numbers.* The population variance is denoted by σ^2.

Variance

The average of the squared deviations about the arithmetic mean for a set of numbers.

<table>
<tr><td>POPULATION VARIANCE</td><td>$$\sigma^2 = \frac{\Sigma(X - \mu)^2}{N}$$</td></tr>
</table>

Table 3.4 shows the original production numbers for the computer company, the deviations from the mean, and the squared deviations from the mean.

The sum of the squared deviations about the mean of a set of values—called the **sum of squares of X** and sometimes abbreviated as SS_X—is used throughout statistics. For the computer company, this value is 130. Dividing it by the number of data values (5 wk) yields the variance for computer production.

Sum of squares of X

The sum of the squared deviations about the mean of a set of values.

$$\sigma^2 = \frac{130}{5} = 26.0$$

Because the variance is computed from squared deviations, the final result is expressed in terms of squared units of measurement. Statistics measured in squared units are problematic to interpret. Consider, for example, Mattel Toys attempting to interpret production costs in terms of squared dollars or Troy-Built measuring production

X	$X - \mu$	$|X - \mu|$
5	−8	+8
9	−4	+4
16	+3	+3
17	+4	+4
18	+5	+5
$\Sigma X = 65$	$\Sigma(X - \mu) = 0$	$\Sigma|X - \mu| = 24$

$$\text{MAD} = \frac{\Sigma|X - \mu|}{N} = \frac{24}{5} = 4.8$$

TABLE 3.3
MAD for Computer Production Data

X	$X - \mu$	$(X - \mu)^2$
5	−8	64
9	−4	16
16	+3	9
17	+4	16
18	+5	25
$\Sigma X = 65$	$\Sigma(X - \mu) = 0$	$\Sigma(X - \mu)^2 = 130$

$$SS_X = \Sigma(X - \mu)^2 = 130$$

$$\text{Variance} = \sigma^2 = \frac{SS_X}{N} = \frac{\Sigma(X - \mu)^2}{N} = \frac{130}{5} = 26.0$$

$$\text{Standard deviation} = \sigma = \sqrt{\frac{\Sigma(X - \mu)^2}{N}} = \sqrt{\frac{130}{5}} = 5.1$$

TABLE 3.4
Computing a Variance and a Standard Deviation from the Computer Production Data

output variation in terms of squared lawn mowers. Therefore, when used as a descriptive measure, variance can be considered as an intermediate calculation in the process of obtaining the sample standard deviation.

Standard Deviation

The standard deviation is a popular measure of variability. It is used both as a separate entity and as a part of other analyses, such as computing confidence intervals and in hypothesis testing (see Chapters 8, 9, and 10).

$$\sigma = \sqrt{\frac{\Sigma(X - \mu)^2}{N}}$$

POPULATION STANDARD DEVIATION

The **standard deviation** is *the square root of the variance.* The population standard deviation is denoted by σ.

Like the variance, the standard deviation utilizes the sum of the squared deviations about the mean (SS_X). It is computed by averaging these squared deviations (SS_X/N) and taking the square root of that average. One feature of the standard deviation that distinguishes it from a variance is that the standard deviation is expressed in the same units as the raw data, whereas the variance is expressed in those units squared. Table 3.4 shows the standard deviation for the computer production company: $\sqrt{26}$, or 5.1.

Standard deviation
The square root of the variance.

What does a standard deviation of 5.1 mean? The meaning of standard deviation is more readily understood from its use, which is explored in the next section. Although the standard deviation and the variance are closely related and can be computed from each other, differentiating between them is important, because both are widely used in statistics.

Meaning of Standard Deviation

What is a standard deviation? What does it do, and what does it mean? There is no precise way of defining a standard deviation other than reciting the formula used to compute it. However, insight into the concept of standard deviation can be gleaned by viewing the manner in which it is applied. One way of applying the standard deviation is the empirical rule.

Empirical rule
A guideline that states the approximate percentage of values that fall within a given number of standard deviations of a mean of a set of data that are normally distributed.

EMPIRICAL RULE   The **empirical rule** is a very important rule of thumb that is used to state the approximate percentage of values that lie within a given number of standard deviations from the mean of a set of data if the data are normally distributed.

The empirical rule is used only for three numbers of standard deviations: 1σ, 2σ, and 3σ. More detailed analysis of other numbers of σ values is presented in Chapter 6. Also discussed in further detail in Chapter 6 is the normal distribution, a unimodal, symmetrical distribution that is bell (or mound) shaped. The requirement that the data be normally distributed contains some tolerance, and the empirical rule generally applies so long as the data are approximately mound shaped.

Empirical Rule*	DISTANCE FROM THE MEAN	VALUES WITHIN DISTANCE
	$\mu \pm 1\sigma$	68%
	$\mu \pm 2\sigma$	95%
	$\mu \pm 3\sigma$	99.7%

*Based on the assumption that the data are approximately normally distributed.

If a set of data is normally distributed, or bell shaped, approximately 68% of the data values are within one standard deviation of the mean, 95% are within two standard deviations, and almost 100% are within three standard deviations.

For example, suppose a recent report states that for California the average statewide price of a gallon of regular gasoline is $1.52. Suppose regular gasoline prices vary across the state with a standard deviation of $0.08 and are normally distributed. According to the empirical rule, approximately 68% of the prices should fall within $\mu \pm 1\sigma$, or $1.52 $\pm$ 1 ($0.08). Approximately 68% of the prices would be between $1.44 and $1.60, as shown in Figure 3.9A. Approximately 95% should fall within $\mu \pm 2\sigma$ or $1.52 $\pm$ 2 ($0.08) = $1.52 $\pm$ $0.16, or between $1.36 and $1.68, as shown in Figure 3.9B. Nearly all regular gasoline prices (99.7%) should fall between $1.28 and $1.76 ($\mu \pm 3\sigma$).

Note that since 68% of the gasoline prices lie within one standard deviation of the mean, approximately 32% are outside this range. Since the normal distribution is symmetrical, the 32% can be split in half such that 16% lie in each tail of the distribution. Thus, approximately 16% of the gasoline prices should be less than $1.44 and approximately 16% of the prices should be greater than $1.60.

Because many phenomena are distributed approximately in a bell shape, including most human characteristics, such as height and weight, the empirical rule applies in many situations and is widely used.

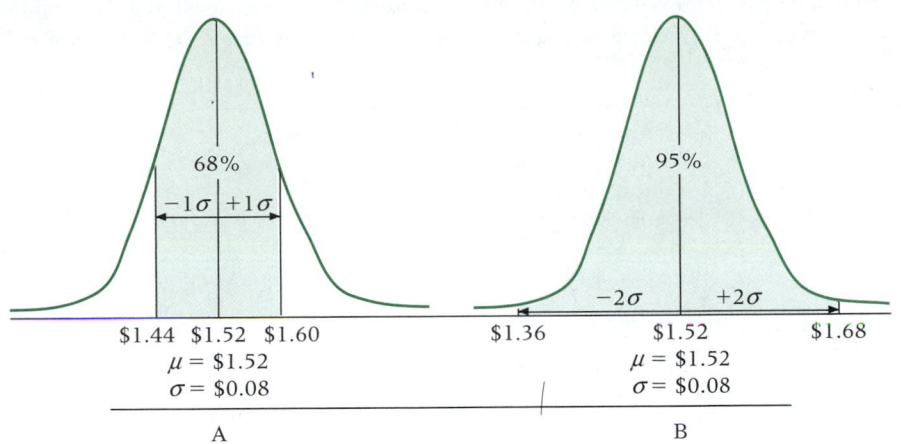

Figure 3.9

Empirical rule for one and two standard deviations of gasoline prices

A	B
68%	95%
$\$1.44 \quad \$1.52 \quad \$1.60$	$\$1.36 \qquad \$1.52 \qquad \$1.68$
$\mu = \$1.52$	$\mu = \$1.52$
$\sigma = \$0.08$	$\sigma = \$0.08$

A company produces a lightweight valve that is specified to weigh 1365 g. Unfortunately, because of imperfections in the manufacturing process not all of the valves produced weigh exactly 1365 grams. In fact, the weights of the valves produced are normally distributed with a mean weight of 1365 grams and a standard deviation of 294 grams. Within what range of weights would approximately 95% of the valve weights fall? Approximately 16% of the weights would be more than what value? Approximately 0.15% of the weights would be less than what value?

SOLUTION

Since the valve weights are normally distributed, the empirical rule applies. According to the empirical rule, approximately 95% of the weights should fall within $\mu \pm 2\sigma = 1365 \pm 2(294) = 1365 \pm 588$. Thus, approximately 95% should fall between 777 and 1953. Approximately 68% of the weights should fall within $\mu \pm 1\sigma$ and 32% should fall outside this interval. Because the normal distribution is symmetrical, approximately 16% should lie above $\mu + 1\sigma = 1365 + 294 = 1659$. Approximately 99.7% of the weights should fall within $\mu \pm 3\sigma$ and .3% should fall outside this interval. Half of these or .15% should lie below $\mu - 3\sigma = 1365 - 3(294) = 1365 - 882 = 483$.

Population versus Sample Variance and Standard Deviation

The sample variance is denoted by S^2 and the sample standard deviation by S. Computation of the sample variance and standard deviation differs slightly from computation of the population variance and standard deviation. The main use for sample variances and standard deviations is as estimators of population variances and standard deviations. Using $n - 1$ in the denominator of a sample variance or standard deviation, rather than n, results in a better estimate of the population values.

$$S^2 = \frac{\Sigma(X - \bar{X})^2}{n - 1}$$	SAMPLE VARIANCE

$$S = \sqrt{S^2}$$	SAMPLE STANDARD DEVIATION

Shown here is a sample of six of the largest accounting firms in the United States and the number of partners associated with each firm as reported by the *Public Accounting Report.*

FIRM	NUMBER OF PARTNERS
Price Waterhouse	1062
McGladrey & Pullen	381
Deloitte & Touche	1719
Andersen Worldwide	1673
Coopers & Lybrand	1277
BDO Seidman	217

The sample variance and sample standard deviation can be computed by:

X	$(X-\bar{X})^2$
1062	51.41
381	454,046.87
1719	441,121.79
1673	382,134.15
1277	49,359.51
217	701,959.11
$\Sigma X = 6329$	$SS_X = \Sigma(X-\bar{X})^2 = 2,028,672.84$

$$\bar{X} = \frac{6329}{6} = 1054.83$$

$$S^2 = \frac{\Sigma(X-\bar{X})^2}{n-1} = \frac{2,028,627.84}{5} = 405,734.57$$

$$S = \sqrt{S^2} = \sqrt{405,734.57} = 636.97$$

The sample variance is 405,734.57 and the sample standard deviation is 636.97.

Computational Formulas for Variance and Standard Deviation

An alternative method of computing variance and standard deviation, sometimes referred to as the computational method or shortcut method, is available. Algebraically,

$$\Sigma(X - \mu)^2 = \Sigma X^2 - \frac{(\Sigma X)^2}{N}$$

and

$$\Sigma(X - \bar{X})^2 = \Sigma X^2 - \frac{(\Sigma X)^2}{n}.$$

Substituting these equivalent expressions into the original formulas for variance and standard deviation yields the following computational formulas.

COMPUTATIONAL FORMULA FOR POPULATION VARIANCE AND STANDARD DEVIATION	$\sigma^2 = \dfrac{\Sigma X^2 - \dfrac{(\Sigma X)^2}{N}}{N}$ $\sigma = \sqrt{\sigma^2}$

TABLE 3.5
Computational Formula Calculations of Variance and Standard Deviation for Computer Production Data

X	X^2
5	25
9	81
16	256
17	289
18	324
$\Sigma X = 65$	$\Sigma X^2 = 975$

$$\sigma^2 = \frac{975 - \dfrac{(65)^2}{5}}{5} = \frac{975 - 845}{5} = \frac{130}{5} = 26$$

$$\sigma = \sqrt{26} = 5.1$$

COMPUTATIONAL FORMULA FOR SAMPLE VARIANCE AND STANDARD DEVIATION

$$S^2 = \frac{\Sigma X^2 - \dfrac{(\Sigma X)^2}{n}}{n - 1}$$

$$S = \sqrt{S^2}$$

These computational formulas utilize the sum of the X values and the sum of the X^2 values instead of the difference between the mean and each value and computed deviations. In the pre-calculator/computer era, this method usually was faster and easier than using the original formulas.

For situations in which the mean is already computed or is given, alternative forms of these formulas are

$$\sigma^2 = \frac{\Sigma X^2 - N\mu^2}{N}$$

$$S^2 = \frac{\Sigma X^2 - n(\overline{X})^2}{n - 1}$$

Using the computational method, the owner of the start-up computer production company can compute a population variance and standard deviation for the production data, as shown in Table 3.5. (Compare these results with those in Table 3.4.)

DEMONSTRATION PROBLEM 3.4

The effectiveness of district attorneys can be measured by several variables, including the number of convictions per month, the number of cases handled per month, and the total number of years of conviction per month. A researcher uses a sample of five district attorneys in a city. She determines the total number of years of conviction that each attorney won against defendants during the past month, as reported in the first column in the following tabulations. Compute the mean absolute deviation, the variance, and the standard deviation for these figures.

SOLUTION
The researcher computes the mean absolute deviation, the variance, and the standard deviation for these data in the following manner.

X	$\lvert X - \bar{X} \rvert$	$(X - \bar{X})^2$
55	41	1,681
100	4	16
125	29	841
140	44	1,936
60	36	1,296
$\Sigma X = 480$	$\Sigma \lvert X - \bar{X} \rvert = 154$	$SS_X = 5,770$

$$\bar{X} = \frac{\Sigma X}{n} = \frac{480}{5} = 96$$

$$\text{MAD} = \frac{154}{5} = 30.8$$

$$S^2 = \frac{5,770}{4} = 1,442.5 \quad \text{and} \quad S = \sqrt{S^2} = 37.98$$

She then uses computational formulas to solve for S^2 and S and compares the results.

X	X^2
55	3,025
100	10,000
125	15,625
140	19,600
60	3,600
$\Sigma X = 480$	$\Sigma X^2 = 51,850$

$$S^2 = \frac{51,850 - \dfrac{(480)^2}{5}}{4} = \frac{51,850 - 46,080}{4} = \frac{5,770}{4} = 1,442.5$$

$$S = \sqrt{1,442.5} = 37.98$$

The results are the same. The sample standard deviation obtained by both methods is 37.98, or 38, years.

Z Scores

A **Z score** represents *the number of standard deviations a value (X) is above or below the mean of a set of numbers when the data are normally distributed.* Using Z scores allows translation of a value's raw distance from the mean into units of standard deviations.

Z score
The number of standard deviations a value (X) is above or below the mean of a set of numbers when the data are normally distributed.

Z SCORE

$$Z = \frac{X - \mu}{\sigma}$$

For samples,

$$Z = \frac{X - \bar{X}}{S}.$$

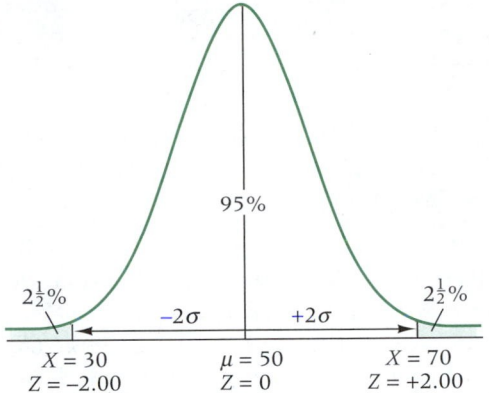

Figure 3.10

Percentage breakdown of scores two standard deviations from the mean

If a Z score is negative, the raw value (X) is below the mean. If the Z score is positive, the raw value (X) is above the mean.

For example, for a data set that is normally distributed with a mean of 50 and a standard deviation of 10, suppose a statistician wants to determine the Z score for a value of 70. This value ($X = 70$) is 20 units above the mean, so the Z value is

$$Z = \frac{70 - 50}{10} = +\frac{20}{10} = +2.00$$

This Z score signifies that the raw score of 70 is two standard deviations above the mean. How is this Z score interpreted? The empirical rule states that 95% of all values are within two standard deviations of the mean if the data are approximately normally distributed. Figure 3.10 shows that because the value of 70 is two standard deviations above the mean ($Z = +2.00$), 95% of the values are between 70 and the value ($X = 30$), that is two standard deviations below the mean ($Z = \frac{30-50}{10} = -2.00$). As 5% of the values are outside the range of two standard deviations from the mean and the normal distribution is symmetrical, 2½% (½ of the 5%) are below the value of 30. Thus 97½% of the values are below the value of 70. Because a Z score is the number of standard deviations an individual data value is from the mean, the empirical rule can be restated in terms of Z scores.

Between $Z = -1.00$ and $Z = +1.00$ are approximately 68% of the values.

Between $Z = -2.00$ and $Z = +2.00$ are approximately 95% of the values.

Between $Z = -3.00$ and $Z = +3.00$ are approximately 99.7% of the values.

The topic of Z scores is discussed more extensively in Chapter 6.

Coefficient of Variation

The **coefficient of variation** is a statistic that is *the ratio of the standard deviation to the mean expressed in percentage* and is denoted CV.

Coefficient of variation (CV)
The ratio of the standard deviation to the mean, expressed as a percentage.

COEFFICIENT OF VARIATION	$$CV = \frac{\sigma}{\mu}(100)$$

For sample data, $CV = \frac{S}{\overline{X}}(100)$.

The coefficient of variation essentially is a relative comparison of a standard deviation to its mean. The coefficient of variation can be useful in comparing standard deviations that have been computed from data with different means.

Suppose five weeks of average prices for stock A are 57, 68, 64, 71, and 62. To compute a coefficient of variation for these prices, first determine the mean and standard deviation: $\mu = 64.40$ and $\sigma = 4.84$. The coefficient of variation is:

$$CV_A = \frac{\sigma_A}{\mu_A}(100) = \frac{4.84}{64.40}(100) = .075 = 7.5\%$$

The standard deviation is 7.5% of the mean.

Sometimes financial investors use the coefficient of variation or the standard deviation or both as measures of risk. Imagine a stock with a price that never changes. There is no risk of losing money from the price going down because there is no variability to the price. Suppose, in contrast, that the price of the stock fluctuates wildly. An investor who buys at a low price and sells for a high price can make a nice profit. However, if the price drops below what the investor buys it for, there is a potential for loss. The greater the variability, the more the potential for loss. Hence, investors use measures of variability such as standard deviation or coefficient of variation to determine the risk of a stock. What does the coefficient of variation tell us about the risk of a stock that the standard deviation does not?

Suppose the average prices for a second stock, B, over these same five weeks are 12, 17, 8, 15, and 13. The mean for stock B is 13.00 with a standard deviation of 3.03. The coefficient of variation can be computed for stock B as:

$$CV_B = \frac{\sigma_B}{\mu_B}(100) = \frac{3.03}{13}(100) = .233 = 23.3\%$$

The standard deviation for stock B is 23.3% of the mean.

With the standard deviation as the measure of risk, stock A is more risky over this period of time because it has a larger standard deviation. However, the average price of stock A is almost five times as much as that of stock B. Relative to the amount invested in stock A, the standard deviation of $4.84 may not represent as much risk as the standard deviation of $3.03 for stock B, which has an average price of only $13.00. The coefficient of variation reveals the risk of a stock in terms of the size of standard deviation relative to the size of the mean (in percentage).

Stock B has a coefficient of variation that is nearly three times as much as the coefficient of variation for stock A. Using coefficient of variation as a measure of risk indicates that stock B is riskier.

The choice of whether to use a coefficient of variation or raw standard deviations to compare multiple standard deviations is a matter of preference. The coefficient of variation also provides an optional method of interpreting the value of a standard deviation.

Analysis Using Excel

Excel can compute the variance and the standard deviation for both a population and a sample. The range is computed as part of *Summary Statistics,* which are discussed later in Section 3.4. To compute the variance and standard deviation, begin with the paste function *fx.* Select **Statistical** from the left side of the paste function dialog box. Included in the menu on the right side of this dialog box is **STDEV,** which computes the sample standard deviation, **STDEVP,** which computes the population standard deviation, **VAR,** which computes the sample variance, and **VARP,** which computes the population variance. The dialog boxes for each of these functions are shown in Figures 3.11, 3.12, 3.13, and 3.14. In each of these dialog boxes, place the location of the data to be analyzed in the line labeled **Number1.** The resulting answer will be displayed on the dialog box; after clicking **OK,** the answer will be displayed on the worksheet.

Figure 3.15 displays the sample standard deviation and sample variance for the attorney data presented in Demonstration Problem 3.4. In addition, Figure 3.15 contains the population standard deviation and population variance for the computer production data presented at the beginning of the section. Note that the answers obtained from Excel are the same as those computed manually in the book.

Figure 3.11   Dialog box for **STDEV**

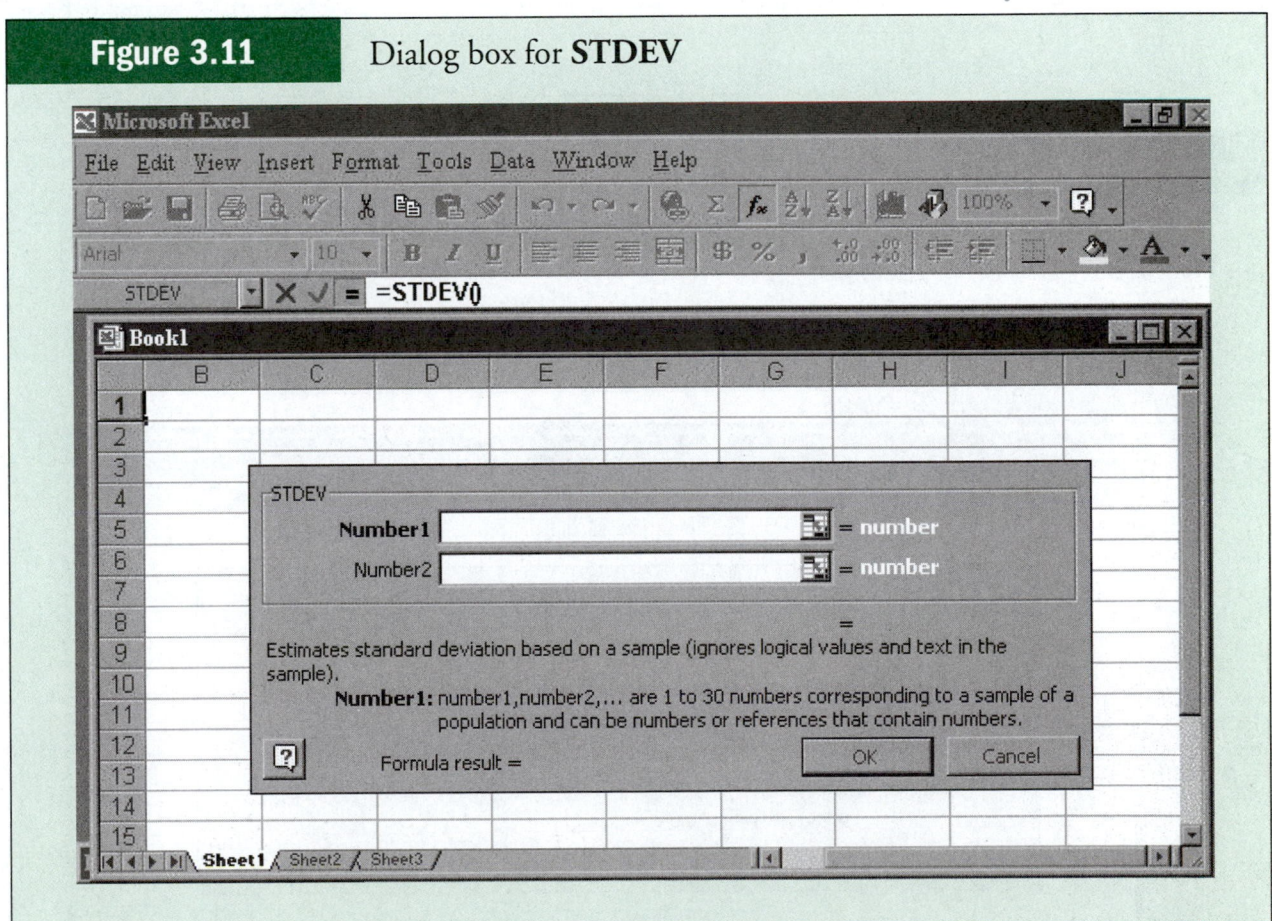

Figure 3.12   Dialog box for **STDEVP**

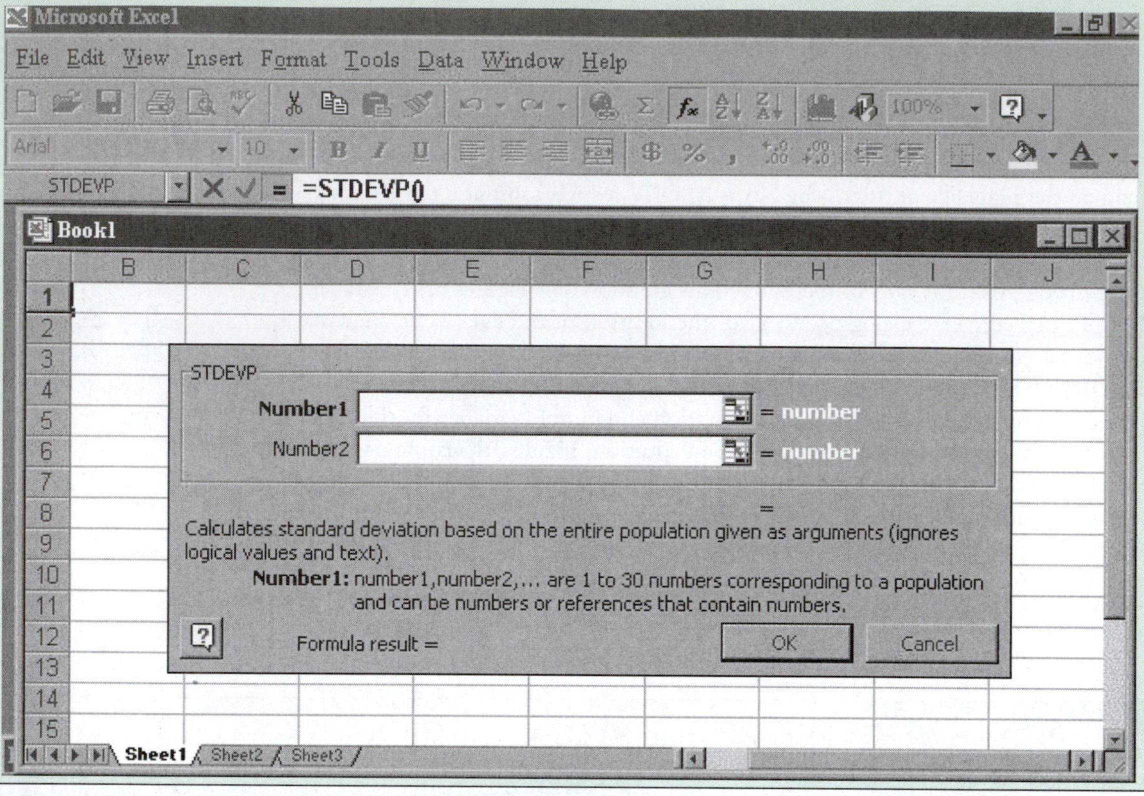

Figure 3.13   Dialog box for **VAR**

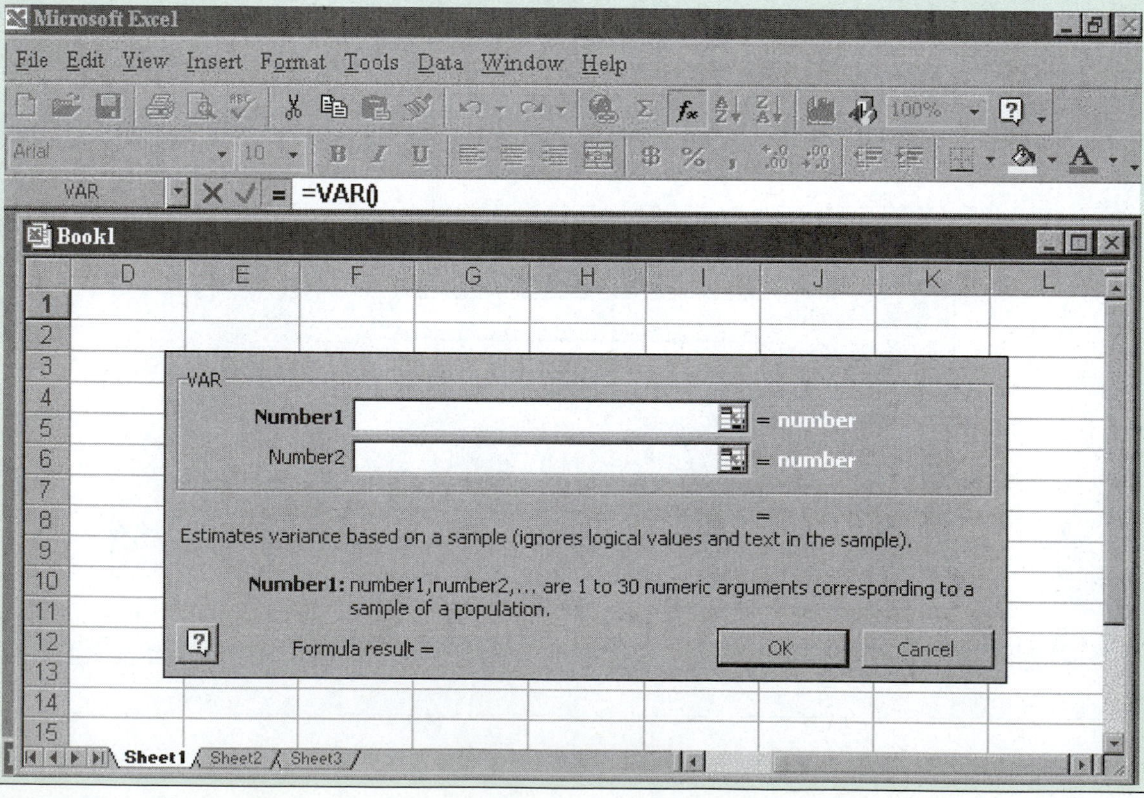

Figure 3.14 — Dialog box for **VARP**

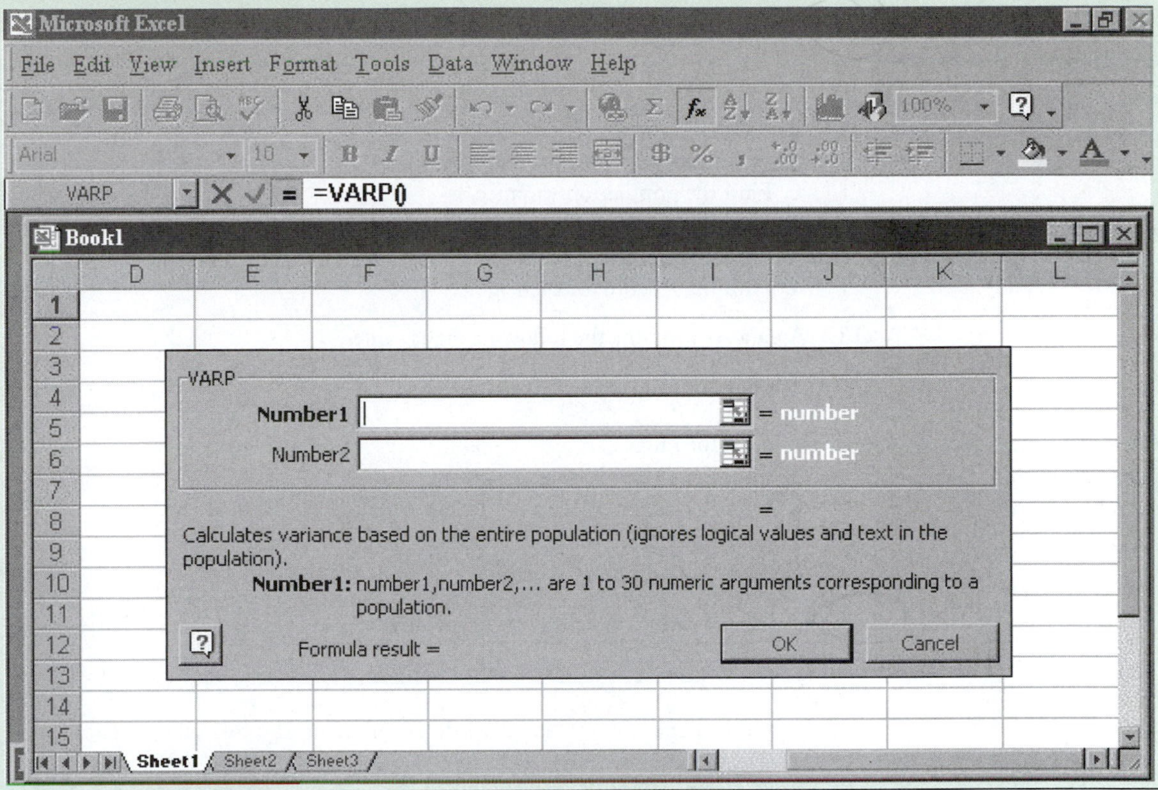

Figure 3.15 — Excel standard deviation and variance output

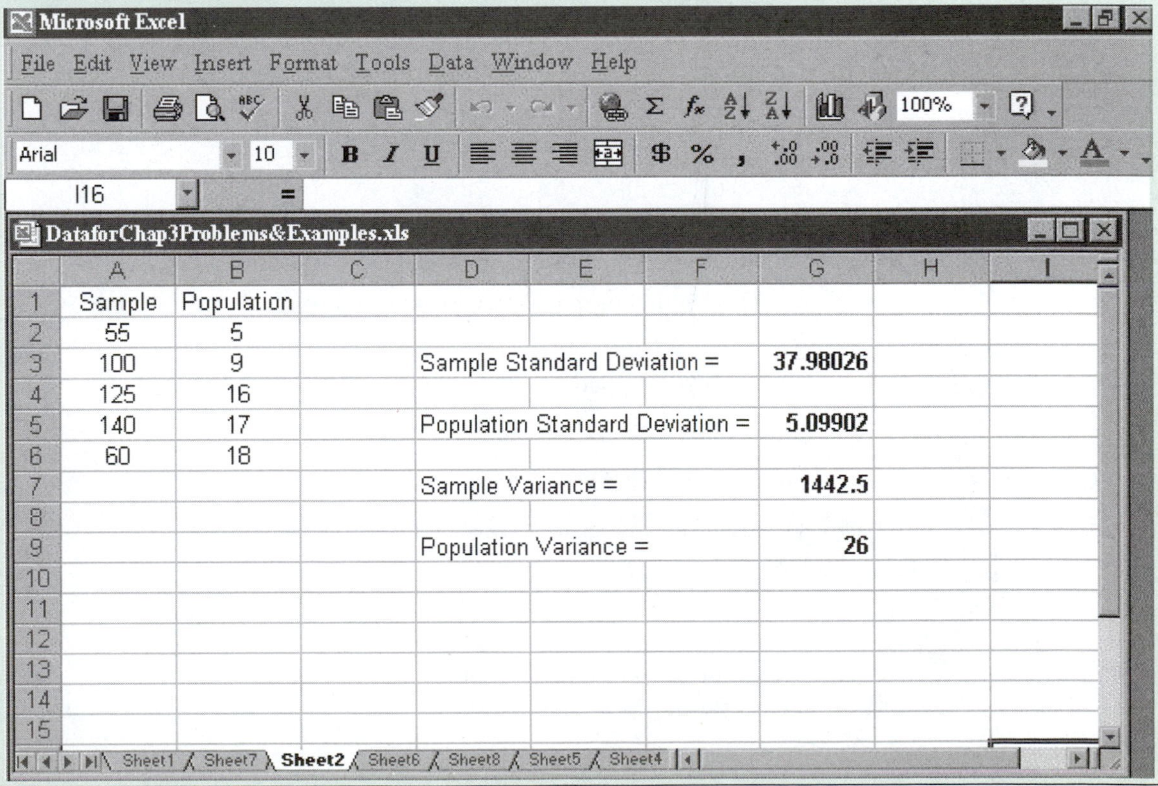

	A	B	C	D	E	F	G	H	I
1	Sample	Population							
2	55	5							
3	100	9		Sample Standard Deviation =			37.98026		
4	125	16							
5	140	17		Population Standard Deviation =			5.09902		
6	60	18							
7				Sample Variance =			1442.5		
8									
9				Population Variance =			26		

3.2

Problems

3.11 A data set contains the following seven values.

6   2   4   9   1   3   5

a. Find the range.
b. Find the mean absolute deviation.
c. Find the population variance.
d. Find the population standard deviation.
e. Find the interquartile range.
f. Find the Z score for each value.

3.12 A data set contains the following eight values.

4   3   0   5   2   9   4   5

a. Find the range.
b. Find the mean absolute deviation.
c. Find the sample variance.
d. Find the sample standard deviation.
e. Find the interquartile range.

3.13 A data set contains the following six values.

12   23   19   26   24   23

a. Find the population standard deviation using the formula containing the mean (the original formula).
b. Find the population standard deviation using the computational formula.
c. Compare the results. Which formula was faster to use? Which formula do you prefer? Why do you think the computational formula is sometimes referred to as the "shortcut" formula?

3.14 Use Excel to find the sample variance and sample standard deviation for the following data.

57	88	68	43	93
63	51	37	77	83
66	60	38	52	28
34	52	60	57	29
92	37	38	17	67

3.15 Use Excel to find the population variance and population standard deviation for the following data.

123	090	546	378
392	280	179	601
572	953	749	075
303	468	531	646

3.16 Determine the interquartile range on the following data.

44	18	39	40	59
46	59	37	15	73
23	19	90	58	35
82	14	38	27	24
71	25	39	84	70

3.17 Compare the variability of the following two sets of data by using both the standard deviation and the coefficient of variation.

DATA SET 1	DATA SET 2
49	159
82	121
77	138
54	152

3.18 A sample of 12 small accounting firms reveals the following numbers of professionals per office.

7	10	9	14	11	8
5	12	8	3	13	6

   a. Determine the mean absolute deviation.
   b. Determine the variance.
   c. Determine the standard deviation.
   d. Determine the interquartile range.
   e. What is the Z score for the firm that has six professionals?
   f. What is the coefficient of variation for this sample?

3.19 The following is a list supplied by Marketing Intelligence Service, Ltd., of the companies with the most new products in a recent year.

COMPANY	NUMBER OF NEW PRODUCTS
Avon Products, Inc.	768
L'Oreal	429
Unilever U.S. Inc.	323
Revlon, Inc.	306
Garden Botanika	286
Philip Morris, Inc.	262
Procter & Gamble Co.	215
Nestlé	172
Paradiso Ltd.	162
Tsumura International, Inc.	148
Grand Metropolitan, Inc.	145

   a. Find the range.
   b. Find the mean absolute deviation.
   c. Find the population variance.
   d. Find the population standard deviation.
   e. Find the interquartile range.
   f. Find the Z score for Nestlé.
   g. Find the coefficient of variation.

3.20 A distribution of numbers is approximately bell-shaped. If the mean of the numbers is 125 and the standard deviation is 12, between what two numbers would approximately 68% of the values be? Between what two numbers would 95% of the values be? Between what two values would 99.7% of the values be?

3.21 The time needed to assemble a particular piece of furniture with experience is normally distributed with a mean time of 43 minutes. If 68% of the assembly times are between 40 and 46 minutes, what is the value of the standard deviation? Suppose 99.7% of the assembly times are between 35 and 51 minutes and the mean is still 43 minutes. What would the value of the standard deviation be now?

3.22 Environmentalists are concerned about emissions of sulfur dioxide into the air. The average number of days per year in which sulfur dioxide levels exceed 150 mg/per cubic meter in Milan, Italy, is 29. The number of days per year in which emission limits are exceeded is normally distributed with a standard deviation of 4.0 days. What percentage of the years would average between 21 and 37 days of excess emissions of sulfur dioxide? What percentage of the years would exceed 37 days? What percentage of the years would exceed 41 days? In what percentage of the years would there be fewer than 25 days with excess sulfur dioxide emissions?

3.23 The Runzheimer Guide publishes a list of the most inexpensive cities in the world for the business traveler. Listed are the 10 most inexpensive cities with their respective per diem costs. Use this list to calculate the Z scores for Bordeaux, Montreal, Edmonton, and Hamilton. Treat this list as a sample.

CITY	PER DIEM ($)
Hamilton, Ontario	97
London, Ontario	109
Edmonton, Alberta	111
Jakarta, Indonesia	118
Ottawa	120
Montreal	130
Halifax, Nova Scotia	132
Winnipeg, Manitoba	133
Bordeaux, France	137
Bangkok, Thailand	137

3.3
Measures of Shape

Measures of shape

Tools that can be used to describe the shape of a distribution of data.

Skewness

The lack of symmetry of a distribution of values.

Measures of shape are *tools that can be used to describe the shape of a distribution of data.* In this section, we examine two measures of shape—skewness and kurtosis. We also look at box and whisker plots.

Skewness

A distribution of data in which the right half is a mirror image of the left half is said to be *symmetrical.* One example of a symmetrical distribution is the normal distribution, or *bell curve,* which is presented in more detail in Chapter 6.

Skewness occurs when a distribution is asymmetrical or lacks symmetry. The distribution in Figure 3.16 has no skewness because it is symmetric. Figure 3.17 shows a distribution that is skewed left, or negatively skewed, and Figure 3.18 shows a distribution that is skewed right, or positively skewed.

The skewed portion is the long, thin part of the curve. Many researchers use *skewed distribution* to mean that the data are sparse at one end of the distribution and piled up at the other end. Instructors sometimes refer to a grade distribution as *skewed,* meaning that few students scored at one end of the grading scale, and many students scored at the other end.

SKEWNESS AND THE RELATIONSHIP OF THE MEAN, MEDIAN, AND MODE The concept of skewness helps to understand the relationship of the mean, median, and mode. In a unimodal distribution (distribution with a single peak or mode) that is skewed, the mode is the apex (high point) of the curve and the median is the middle value. The mean tends to be located toward the tail of the distribution, because the mean is affected by all values, including the extreme ones. Because a bell-shaped or normal distribution has no skewness, the mean, median, and mode all are at the center of the distribution. Figure 3.19 displays the relationship of the mean, median, and mode for different types of skewness.

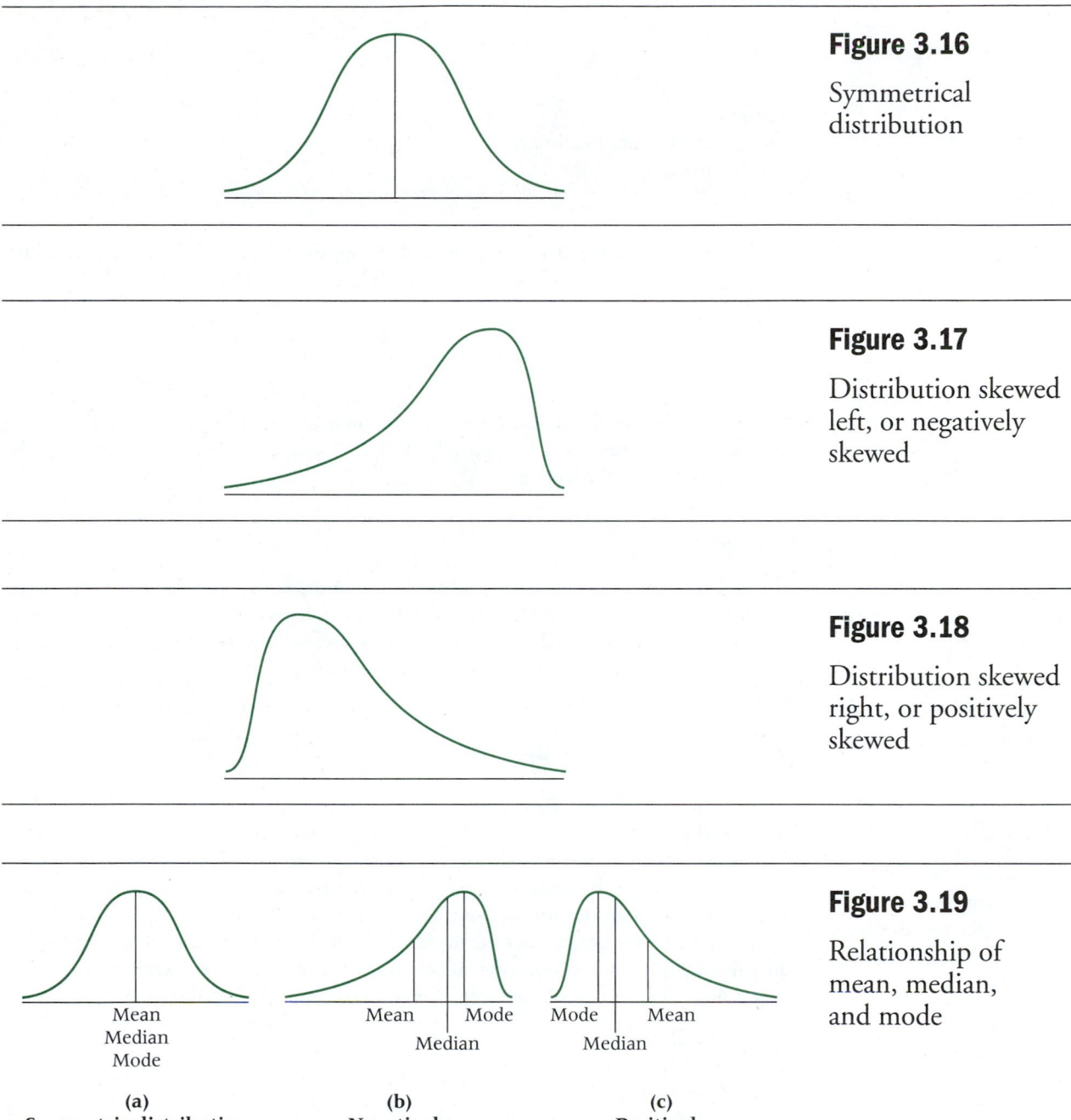

Figure 3.16

Symmetrical distribution

Figure 3.17

Distribution skewed left, or negatively skewed

Figure 3.18

Distribution skewed right, or positively skewed

Figure 3.19

Relationship of mean, median, and mode

Coefficient of skewness
A measure of the degree of skewness that exists in a distribution of numbers; compares the mean and the median in light of the magnitude of the standard deviation.

COEFFICIENT OF SKEWNESS   Statistician Karl Pearson is credited with developing at least two coefficients of skewness that can be used to determine the degree of skewness in a distribution. We present one of these coefficients here, referred to as a *Pearsonian* **coefficient of skewness.** This coefficient *compares the mean and median in light of the magnitude of the standard deviation.* Note that if the distribution is symmetrical, the mean and median are the same value and hence the coefficient of skewness is equal to zero.

COEFFICIENT OF SKEWNESS	$$S_k = \frac{3(\mu - M_d)}{\sigma}$$

where:

S_k = coefficient of skewness
M_d = median

Suppose, for example, that a distribution has a mean of 29, a median of 26, and a standard deviation of 12.3. The coefficient of skewness is computed as

$$S_k = \frac{3(29 - 26)}{12.3} = +0.73.$$

Because the value of S_k is positive, the distribution is positively skewed. If the value of S_k is negative, the distribution is negatively skewed. The greater the magnitude of S_k, the more skewed is the distribution.

Kurtosis

Kurtosis
The amount of peakedness of a distribution.

Leptokurtic
Distributions that are high and thin.

Platykurtic
Distributions that are flat and spread out.

Mesokurtic
Distributions that are normal in shape—that is, not too high or too flat.

Box and whisker plot
A diagram that utilizes the upper and lower quartiles along with the median and the two most extreme values to depict a distribution graphically; sometimes called a box plot.

Kurtosis describes the *amount of peakedness of a distribution. Distributions that are high and thin* are referred to as **leptokurtic** distributions. *Distributions that are flat and spread out* are referred to as **platykurtic** distributions. Between these two types are *distributions that are more "normal" in shape,* referred to as **mesokurtic** distributions. These three types of kurtosis are illustrated in Figure 3.20.

Box and Whisker Plots

Another way to describe a distribution of data is by using a box and whisker plot. A **box and whisker plot,** sometimes called a *box plot,* is *a diagram that utilizes the upper and lower quartiles along with the median and the two most extreme values to depict a distribution graphically.* The plot is constructed by using a box to enclose the median. This *box* is extended outward from the median along a continuum to the lower and upper quartiles, enclosing not only the median but the middle 50% of the data. From the lower and upper quartiles, lines referred to as *whiskers* are extended out from the box toward the outermost data values. The box and whisker plot is determined from five specific numbers.

1. The median (Q_2).
2. The lower quartile (Q_1).
3. The upper quartile (Q_3).
4. The smallest value in the distribution.
5. The largest value in the distribution.

The box of the plot is determined by locating the median and the lower and upper quartiles on a continuum. A box is drawn around the median with the lower and upper quartiles (Q_1 and Q_3) as the box endpoints. These box endpoints (Q_1 and Q_3) are referred to as the *hinges* of the box.

Next the value of the interquartile range (IQR) is computed by $Q_3 - Q_1$. The interquartile range includes the middle 50% of the data and should equal the length of the box. However, here the interquartile range is used outside of the box also. At a distance of $1.5 \cdot$ IQR outward from the lower and upper quartiles are what are referred to as *inner*

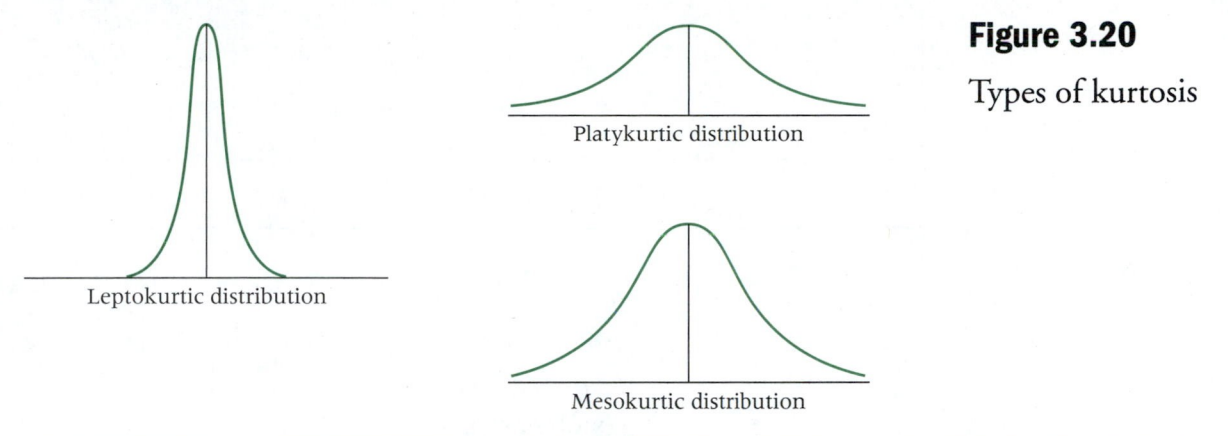

Figure 3.20

Types of kurtosis

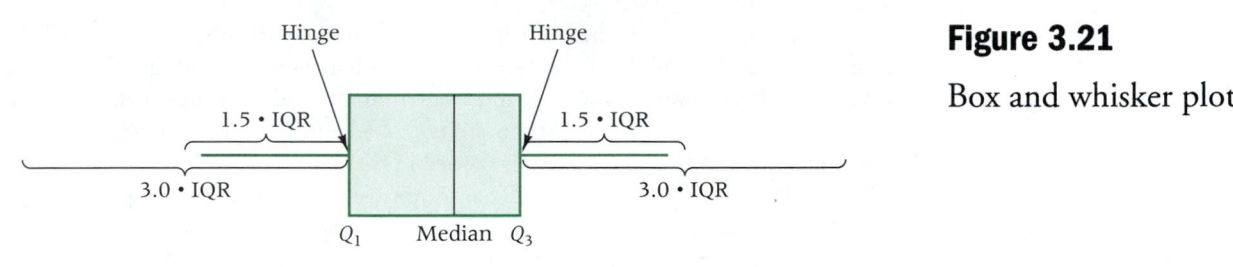

Figure 3.21

Box and whisker plot

fences. A *whisker,* a line segment, is drawn from the lower hinge of the box outward to the smallest data value. A second whisker is drawn from the upper hinge of the box outward to the largest data value. The inner fences are established as follows.

$$Q_1 - 1.5 \cdot IQR$$
$$Q_3 + 1.5 \cdot IQR$$

If there are data beyond the inner fences, then *outer fences* can be constructed:

$$Q_1 - 3.0 \cdot IQR$$
$$Q_3 + 3.0 \cdot IQR$$

Figure 3.21 shows the features of a box and whisker plot.

Data values that are outside the mainstream of values in a distribution are viewed as *outliers.* Outliers can be merely the more extreme values of a data set. However, sometimes outliers are due to measurement or recording errors. Other times they are values that are so unlike the other values that they should not be considered in the same analysis as the rest of the distribution. Values in the data distribution that are outside the inner fences but within the outer fences are referred to as *mild outliers.* Values that are outside the outer fences are called *extreme outliers.* Thus, one of the main uses of a box and whisker plot is to identify outliers.

TABLE 3.6
Data for Box and Whisker Plot

71	87	82	64	72	75	81	69
76	79	65	68	80	73	85	71
70	79	63	62	81	84	77	73
82	74	74	73	84	72	81	65
74	62	64	68	73	82	69	71

TABLE 3.7
Data in Ordered Array with Quartiles and Median

87	85	84	84	82	82	82	81	81	81
80	79	79	77	76	75	74	74	74	73
73	73	73	72	72	71	71	71	70	69
69	68	68	65	65	64	64	63	62	62

$$Q_1 = 69$$
$$Q_2 = \text{median} = 73$$
$$Q_3 = 80.5$$
$$IQR = Q_3 - Q_1 = 80.5 - 69 = 11.5$$

Another use of box and whisker plots is to determine if a distribution is skewed. If the median falls in the middle of the box, then there is no skewness. If the distribution is skewed, it will be skewed in the direction away from the median. If the median falls in the upper half of the box, then the distribution is skewed left. If the median falls in the lower half of the box, then the distribution is skewed to the right.

We shall use the data given in Table 3.6 to construct a box and whisker plot.

After organizing the data into an ordered array, as shown in Table 3.7, it is relatively easy to determine the values of the lower quartile (Q_1), the median, and the upper quartile (Q_3). From these, the value of the interquartile range can be computed.

The hinges of the box are located at the lower and upper quartiles, 69 and 80.5. The median is located within the box at distances of 4 from the lower quartile and 6.5 from the upper quartile. The distribution is skewed right, because the median is nearer to the lower or left hinge. The inner fence is constructed by

$$Q_1 - 1.5 \cdot IQR = 69 - 1.5 \cdot 11.5 = 69 - 17.25 = 51.75$$

and

$$Q_3 + 1.5 \cdot IQR = 80.5 + 1.5 \cdot 11.5 = 80.5 + 17.25 = 97.75.$$

The whiskers are constructed by drawing a line segment from the lower hinge outward to the smallest data value and a line segment from the upper hinge outward to the largest data value. An examination of the data reveals that there are no data values in this set of numbers that are outside the inner fence. The whiskers are constructed outward to the lowest value, which is 62, and to the highest value, which is 87.

To construct an outer fence, we calculate $Q_1 - 3 \cdot IQR$ and $Q_3 + 3 \cdot IQR$, as follows.

$$Q_1 - 3 \cdot IQR = 69 - 3 \cdot 11.5 = 69 - 34.5 = 34.5$$
$$Q_3 + 3 \cdot IQR = 80.5 + 3 \cdot 11.5 = 80.5 + 34.5 = 115.0$$

Analysis Using Excel

Computation of the previously presented Pearsonian coefficient of skewness is accomplished through the use of the three Excel functions, **AVERAGE, MEDIAN** and **STDEVP.**

They can be combined utilizing the Excel formula

= 3 * (AVERAGE (data range) – MEDIAN (data range)) / STDEVP (data range)

to yield the previous manually computed coefficient of skewness.

In addition, Excel has a statistical function, **SKEW,** that computes another accepted form of the coefficient of skewness. This coefficient is computed as a function of the third power of the deviations about the mean. It is accessed through the Paste Function, f_x, using **Statistical** on the left side of the dialog box and **SKEW** on the right side. Figure 3.22 displays the dialog box for **SKEW.** To use **SKEW,** insert the location of the data in the first line labeled **Number1.** The answer will appear on the dialog box; after clicking on **OK** it will appear on the spreadsheet.

Figure 3.23 displays the Excel computed value of skewness for the data from Table 3.6.

Excel cannot produce Box and Whisker Plots, but **FAST ⦥ STAT** has the capability. Figure 3.24 displays the **Box and Whisker Plot** dialog box for **FAST ⦥ STAT**. Note that the only entry requirement of this feature is the location of the data. The **FAST ⦥ STAT** Box and Whisker results consist of four general items. First, the input data values are repeated in Column A of the worksheet. Second, three output items are given. These include the box and whisker plot, the five-number summary values (smallest value, largest value, and the three quartiles) to the left of the plot, and five values on the right that are used for constructing the plot.

Shown in Figure 3.25 are two of the outputs for the data in Table 3.6. These include the five-number summary and the box and whisker plot.

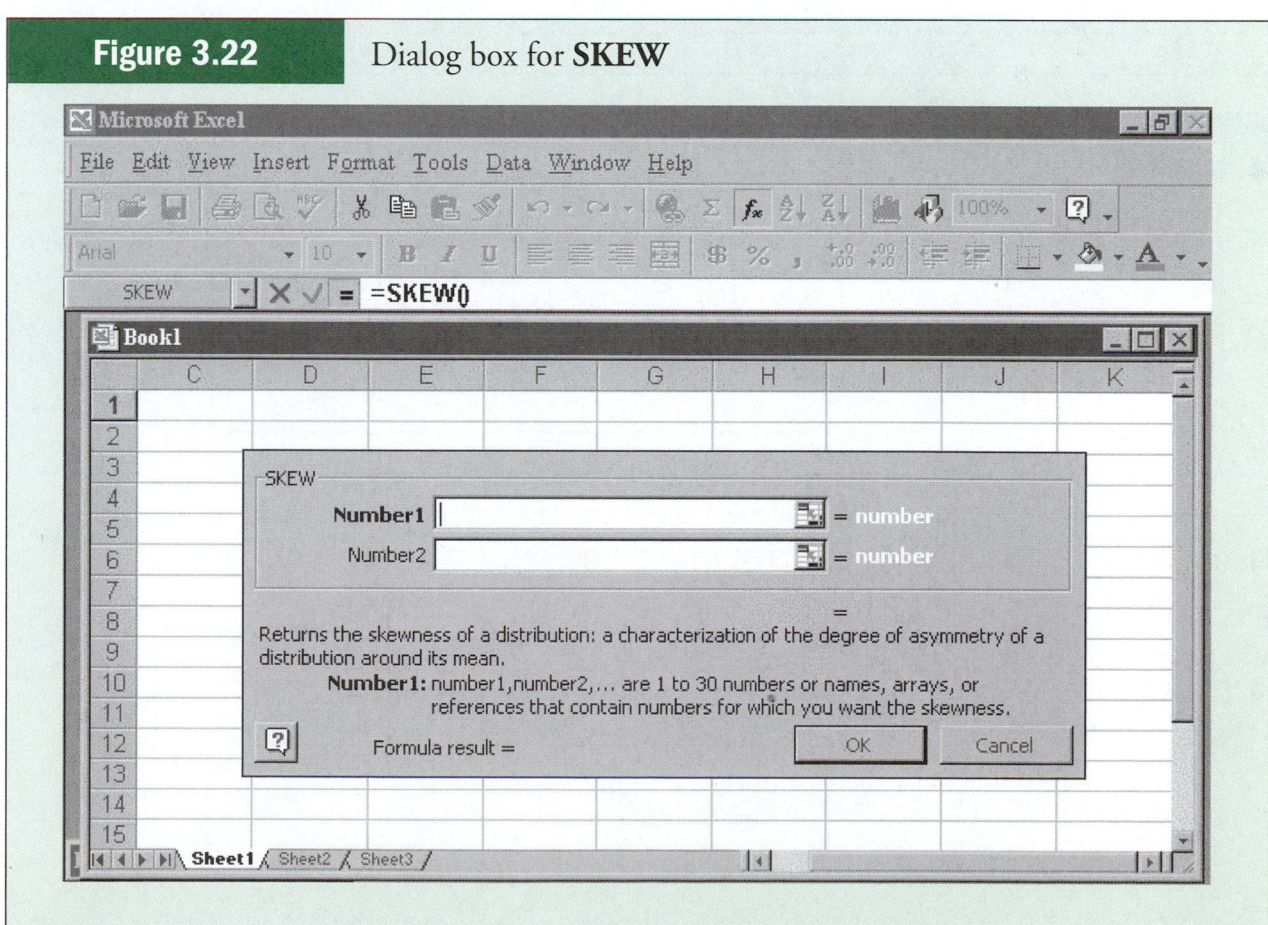

Figure 3.22   Dialog box for **SKEW**

Figure 3.23   Excel skewness output for the data from Table 3.6

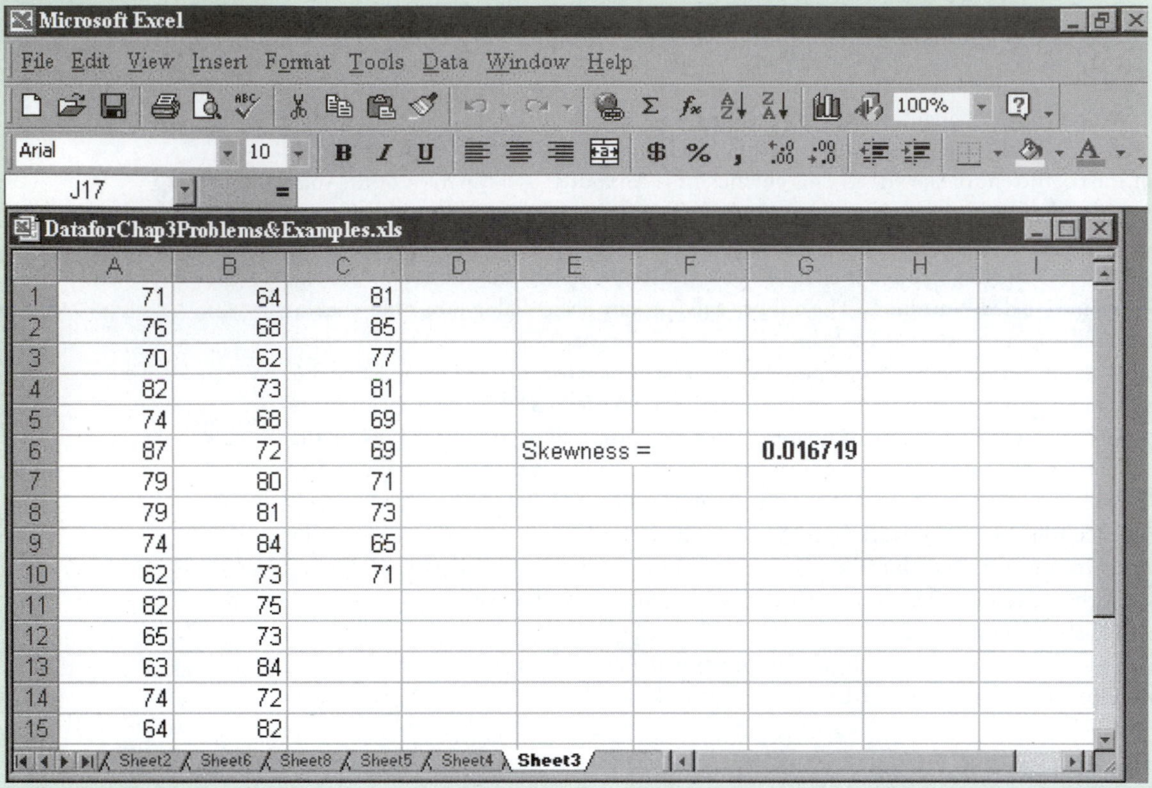

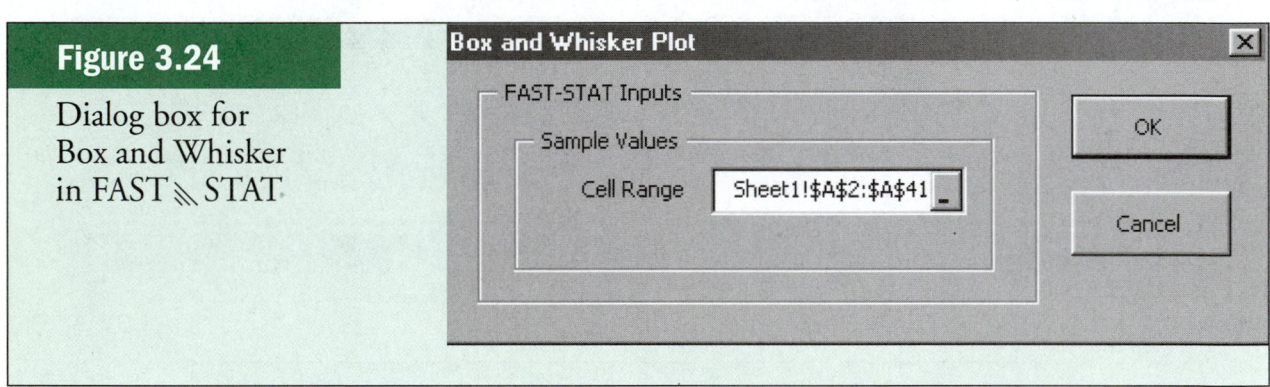

	A	B	C	D	E	F	G	H	I
1	71	64	81						
2	76	68	85						
3	70	62	77						
4	82	73	81						
5	74	68	69						
6	87	72	69		Skewness =		**0.016719**		
7	79	80	71						
8	79	81	73						
9	74	84	65						
10	62	73	71						
11	82	75							
12	65	73							
13	63	84							
14	74	72							
15	64	82							

Figure 3.24

Dialog box for
Box and Whisker
in FAST⟍STAT

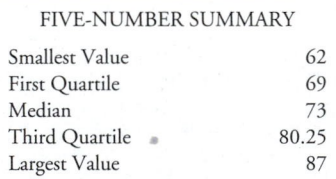

Figure 3.25

FAST⟍STAT box
and whisker analysis
of Table 3.6 data

FIVE-NUMBER SUMMARY	
Smallest Value	62
First Quartile	69
Median	73
Third Quartile	80.25
Largest Value	87

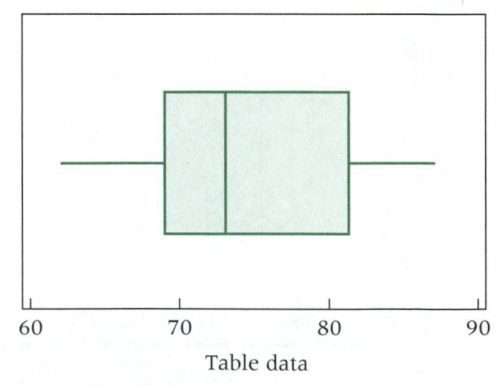

Table data

3.24 On a certain day the average closing price of a group of stocks on the New York Stock Exchange is $35 (to the nearest dollar). If the median value is $33 and the mode is $21, is the distribution of these stock prices skewed? if so, how?

3.25 A local hotel offers ballroom dancing on Friday nights. A researcher observes the customers and estimates their ages. Discuss the skewness of the distribution of ages, if the mean age is 51, the median age is 54, and the modal age is 59.

3.26 The sales volumes for the top real estate brokerage firms in the United States for a recent year were analyzed using descriptive statistics. The mean annual dollar volume for these firms was $5.51 billion, the median was $3.19 billion, and the standard deviation was $9.59 billion. Compute the value of the Pearsonian coefficient of skewness and discuss the meaning of it. Is the distribution skewed? If so, to what extent?

3.27 Suppose the data below are the ages of Internet users obtained from a sample. Use these data to compute a Pearsonian coefficient of skewness. What is the meaning of the coefficient?

41	15	31	25	24
23	21	22	22	18
30	20	19	19	16
23	27	38	34	24
19	20	29	17	23

3.28 Construct a box and whisker plot on the following data. Are there any outliers? Is the distribution of data skewed?

540	690	503	558	490	609
379	601	559	495	562	580
510	623	477	574	588	497
527	570	495	590	602	541

3.29 Suppose a consumer group asked 18 consumers to keep a yearly log of their shopping practices and that the following data represent the number of coupons used by each consumer over the yearly period. Use the data to construct a box and whisker plot. List the median, Q_1, Q_3, the endpoints for the inner fences, and the endpoints for the outer fences. Discuss the skewness of the distribution of these data and point out any outliers.

81	68	70	100	94	47	66	70	82
110	105	60	21	70	66	90	78	85

In this chapter we have introduced many descriptive statistics techniques that are useful in analyzing data. To this point we have taken an "a la carte", one-at-a-time, Excel-approach in presenting them. However, Excel has one tool that can perform many of these functions at once. This tool is the **Descriptive Statistics** tool, which is accessed as one of the options under **Data Analysis.** The **Descriptive Statistics** dialog box is displayed in Figure 3.26.

The location of the data is inserted into the **Input Range** blank on the first line of the dialog box. Check the **Summary Statistics** box to calculate the several descriptive measures at once. The output contains the mean, median, mode, sample standard deviation, sample variance, range, skewness, and a measure of kurtosis. Applying this Excel feature to the data in Table 3.6 results in the output shown in Figure 3.27.

Figure 3.26 | **Descriptive Statistics** dialog box

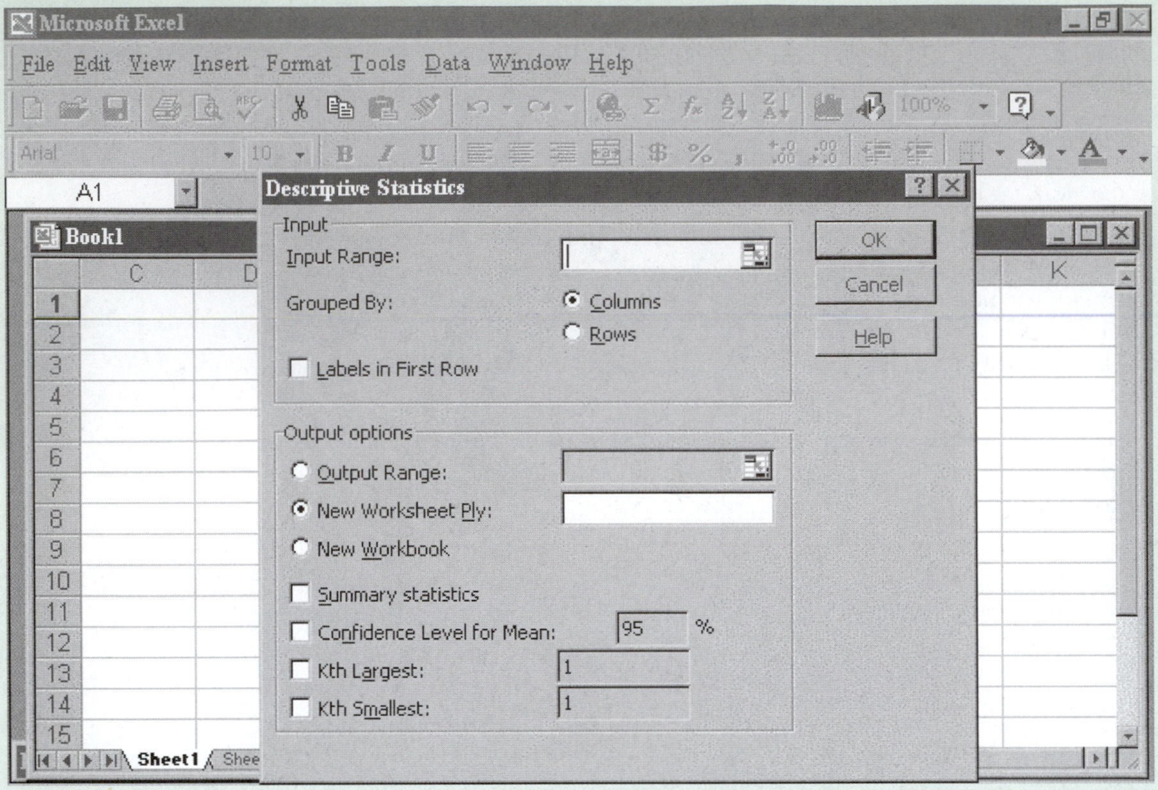

Microsoft Excel

File Edit View Insert Format Tools Data Window Help

A1

Book1

Descriptive Statistics

Input

Input Range:

Grouped By: ● Columns
               ○ Rows

☐ Labels in First Row

Output options

○ Output Range:

● New Worksheet Ply:

○ New Workbook

☐ Summary statistics

☐ Confidence Level for Mean: 95   %

☐ Kth Largest: 1

☐ Kth Smallest: 1

OK

Cancel

Help

Sheet1 Shee

Figure 3.27 | Excel **Descriptive Summary Statistics** for the Table 3.6 data

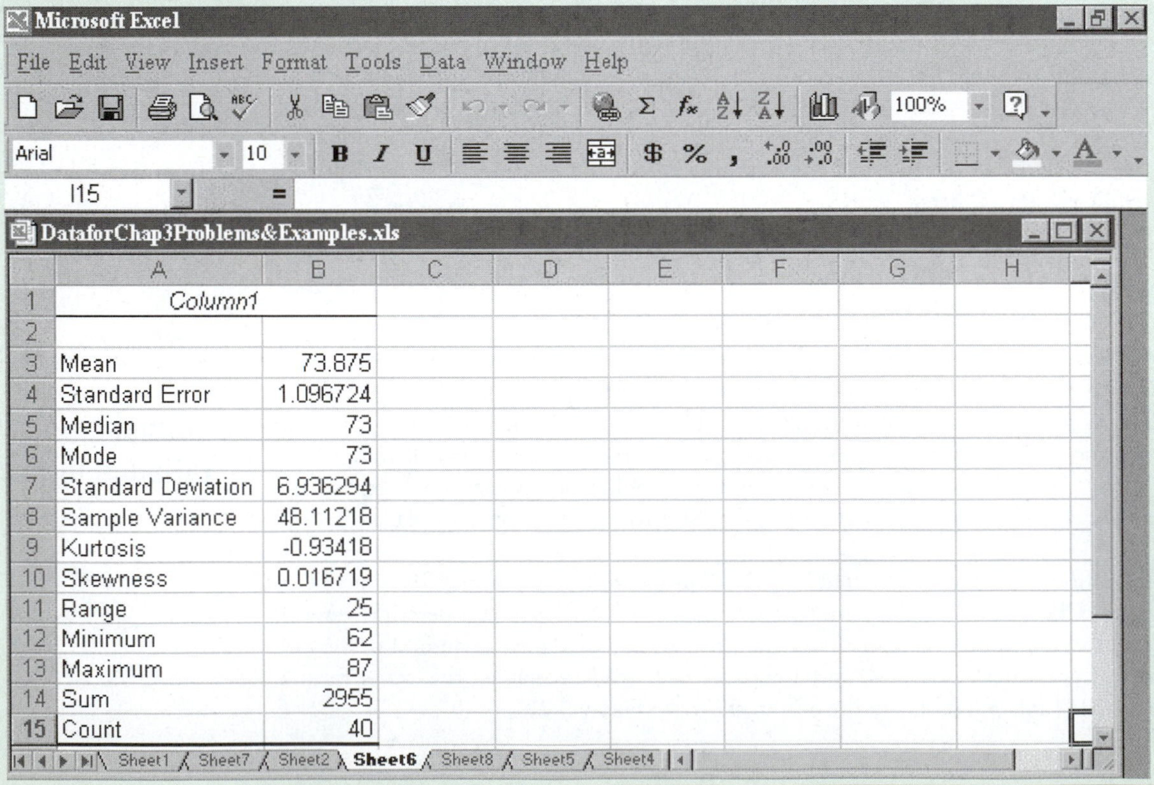

Microsoft Excel

File Edit View Insert Format Tools Data Window Help

I15   =

DataforChap3Problems&Examples.xls

	A	B	C	D	E	F	G	H
1	*Column1*							
2								
3	Mean	73.875						
4	Standard Error	1.096724						
5	Median	73						
6	Mode	73						
7	Standard Deviation	6.936294						
8	Sample Variance	48.11218						
9	Kurtosis	-0.93418						
10	Skewness	0.016719						
11	Range	25						
12	Minimum	62						
13	Maximum	87						
14	Sum	2955						
15	Count	40						

Sheet1 Sheet7 Sheet2 **Sheet6** Sheet8 Sheet5 Sheet4

Summary

Statistical descriptive measures include measures of central tendency, measures of variability, and measures of shape. Measures of central tendency are useful in describing data because they communicate information about the more central portions of the data. The most common measures of central tendency are the three m's: mode, median, and mean. In addition, in this text, quartiles are presented as measures of central tendency.

The mode is the most frequently occurring value in a set of data. If two values tie for the mode, the data are bimodal. Data sets can be multimodal. Among other things, the mode is used in business for determining sizes.

The median is the middle term in an ordered array of numbers if there is an odd number of terms. If there is an even number of terms, the median is the average of the two middle terms in an ordered array. The formula $(n + 1)/2$ specifies the location of the median. A median is unaffected by the magnitude of extreme values. This characteristic makes the median a most useful and appropriate measure of location in reporting such things as income, age, and prices of houses.

The arithmetic mean is widely used and is usually what researchers are referring to when they use the word *mean*. The arithmetic mean is the average. The population mean and the sample mean are computed in the same way but are denoted by different symbols. The arithmetic mean is affected by every value and can be inordinately influenced by extreme values.

Quartiles divide data into four groups. There are three quartiles: Q_1, which is the lower quartile; Q_2, which is the middle quartile and equals the median; and Q_3, which is the upper quartile.

Measures of variability are statistical tools used in combination with measures of central tendency to describe data. Measures of variability provide a description of data that measures of central tendency cannot give—information about the spread of the data values. These measures include the range, mean absolute deviation, variance, standard deviation, interquartile range, and coefficient of variation.

One of the most elementary measures of variability is the range. It is the difference between the largest and smallest values. Although the range is easy to compute, it has limited usefulness. The interquartile range is the difference between the third and first quartile. It equals the range of the middle 50% of the data.

The mean absolute deviation (MAD) is computed by averaging the absolute values of the deviations from the mean. The mean absolute deviation provides the magnitude of the average deviation but without specifying its direction. The mean absolute deviation has limited usage in statistics, but interest is growing for the use of MAD in the field of forecasting.

Variance is widely used as a tool in statistics but is little used as a stand-alone measure of variability. The variance is the average of the squared deviations about the mean.

The square root of the variance is the standard deviation. It also is a widely used tool in statistics. It is used more often than the variance as a stand-alone measure. The standard deviation is best understood by examining its applications in determining where data are in relation to the mean. The empirical rule contains statements about the proportions of data values that are within various numbers of standard deviations from the mean.

The empirical rule reveals the percentage of values that are within one, two, or three standard deviations of the mean for a set of data. The empirical rule applies only if the data are in a bell-shaped distribution. According to the empirical rule, approximately 68% of all values of a normal distribution are within plus or minus one standard deviation of the mean. Ninety-five percent of all values are within two standard deviations either side of the mean, and virtually all values are within three standard deviations of the mean. The Z score represents the number of standard deviations a value is from the mean for normally distributed data.

The coefficient of variation is a ratio of a standard deviation to its mean, given as a percentage. It is especially useful in comparing standard deviations or variances that represent data with different means.

Two measures of shape are skewness and kurtosis. Skewness is the lack of symmetry in a distribution. If a distribution is skewed, it is stretched in one direction or the other. The skewed part of a graph is its long, thin portion. One measure of skewness is the Pearsonian coefficient of skewness.

Kurtosis is the degree of peakedness of a distribution. A tall, thin distribution is referred to as leptokurtic. A flat distribution is platykurtic, and a distribution with a more normal peakedness is said to be mesokurtic.

A box and whisker plot is a graphical depiction of a distribution. The plot is constructed by using the median, the lower quartile, and the upper quartile. It can yield information about skewness and outliers.

Key Terms

arithmetic mean	measures of variability
bimodal	median
box and whisker plot	mesokurtic
coefficient of skewness	mode
coefficient of variation (CV)	multimodal
deviation from the mean	platykurtic
empirical rule	quartiles
interquartile range	range
kurtosis	skewness
leptokurtic	standard deviation
mean absolute deviation (MAD)	sum of squares of X
measures of central tendency	variance
measures of shape	Z score

SUPPLEMENTARY PROBLEMS

3.30 The 2000 U.S. Census asks every household to report information on each person living there. Suppose a sample of 30 households is selected and the number of persons living in each is reported as follows.

2 3 1 2 6 4 2 1 5 3 2 3 1 2 2
1 3 1 2 2 4 2 1 2 8 3 2 1 1 3

Compute the mean, median, mode, range, lower and upper quartiles, and interquartile range for these data.

3.31 The 2000 U.S. Census also asks for each person's age. Suppose that a sample of 40 households is taken from the census data and the age of the first person recorded on the census form is given as follows.

42	29	31	38	55	27	28
33	49	70	25	21	38	47
63	22	38	52	50	41	19
22	29	81	52	26	35	38
29	31	48	26	33	42	58
40	32	24	34	25		

Compute Q_1, Q_3, the interquartile range, and the range for these data.

3.32 According to the National Association of Investment Clubs, PepsiCo, Inc., is the most popular stock with investment clubs with 11,388 clubs holding PepsiCo stock. The Intel Corp. is a close second, followed by Motorola, Inc. We show a list of the most popular stocks with investment clubs. Compute the mean, median, Q_1, Q_3, range, and interquartile range for these figures.

COMPANY	NUMBER OF CLUBS HOLDING STOCK
PepsiCo, Inc.	11388
Intel Corp.	11019
Motorola, Inc.	9863
Tricon Global Restaurants	9168
Merck & Co., Inc.	8687
AFLAC Inc.	6796
Diebold, Inc.	6552

3.32 *continued*

McDonald's Corp.	6498
Coca-Cola Co.	6101
Lucent Technologies	5563
Home Depot, Inc.	5414
Clayton Homes, Inc.	5390
RPM, Inc.	5033
Cisco Systems, Inc.	4541
General Electric Co.	4507
Johnson & Johnson	4464
Microsoft Corp.	4152
Wendy's International, Inc.	4150
Walt Disney Co.	3999
AT&T Corp.	3619

3.33 *Editor & Publisher International Yearbook* published a listing of the top 10 daily newspapers in the United States, as shown here. Use these population data to compute a mean and a standard deviation. The figures are given in average daily circulation from Monday through Friday. Because the numbers are large, it may save you some effort to recode the data. One way to recode these data is to move the decimal point six places to the left (e.g., 1,774,880 becomes 1.77488). If you recode the data this way, the resulting mean and standard deviation will be correct for the recoded data. To rewrite the answers so that they are correct for the original data, move the decimal point back to the right six places in the answers.

NEWSPAPER	AVERAGE DAILY CIRCULATION
Wall Street Journal	1,774,880
USA Today	1,629,665
New York Times	1,074,741
Los Angeles Times	1,050,176
Washington Post	775,894
(N.Y.) *Daily News*	721,256
Chicago Tribune	653,554
Newsday	568,914
Houston Chronicle	549,101
Chicago Sun-Times	484,379

3.34 We show the companies with the largest oil refining capacity in the world according to the *Petroleum Intelligence Weekly*. Use these population data and answer the questions.

COMPANY	CAPACITY (1000s BARRELS PER DAY)
Exxon	4273
Royal Dutch/Shell	3791
China Petrochemical Corp.	2867
Petroleos de Venezuela	2437

3.34 *continued*

Saudi Arabian Oil Co.	1970
British Petroleum	1965
Chevron	1661
Petrobras	1540
Texaco	1532
Petroleos Mexicanos (Pemex)	1520
National Iranian Oil Co.	1092

a. What are the values of the mean and the median? Compare the answers and state which you prefer as a measure of location for these data and why.
b. What are the values of the range and interquartile range? How do they differ?
c. What are the values of variance and standard deviation for these data?
d. What is the Z score for Texaco? What is the Z score for Mobil? Interpret these Z scores.
e. Calculate the Pearsonian coefficient of skewness and comment on the skewness of this distribution.

3.35 The U.S. Department of the Interior's Bureau of Mines releases figures on mineral production. Following are the 10 leading states in nonfuel mineral production in terms of the percentage of the U.S. total.

STATE	PERCENT OF U.S. TOTAL
Arizona	8.91
Nevada	7.69
California	7.13
Georgia	4.49
Utah	4.46
Florida	4.42
Texas	4.31
Minnesota	4.06
Michigan	3.96
Missouri	3.34

SOURCE: Bureau of Mines, U.S. Department of the Interior (*1999 World Almanac*)

a. Calculate the mean, median, and mode.
b. Calculate the range, interquartile range, mean absolute deviation, sample variance, and sample standard deviation.
c. Compute the Pearsonian coefficient of skewness for these data.
d. Sketch a box and whisker plot.

3.36 Financial analysts like to use the standard deviation as a measure of risk for a stock. The greater the deviation in a stock price over time, the more risky it is to invest in the stock. However, the average prices of some stocks are considerably higher than the average price of others, allowing for the potential of a greater standard deviation of price. For example, a standard deviation of $5.00 on a $10.00 stock is considerably different from a $5.00 standard deviation on a $40.00 stock. In this situation, a coefficient of variation might provide insight into risk. Suppose stock X costs an average of $32.00 per share and has had a standard deviation of $3.45 for the past 60 days. Suppose stock Y costs an average of $84.00 per share and has had a standard deviation of $5.40 for the past 60 days. Use the coefficient of variation to determine the variability for each stock.

3.37 The Polk Company reported that the average age of a car on U.S. roads in a recent year was 7.5 years. Suppose the distribution of ages of cars on U.S. roads is approximately bell-shaped. If 99.7% of the ages are between 1 year and 14 years, what is the standard deviation of car age? Suppose the standard deviation is 1.7 years and the mean is 7.5 years. What two values would 95% of the car ages be between?

3.38 According to a *Human Resources Report,* a worker in the industrial countries spends on average 419 minutes a day on the job. Suppose the standard deviation of time spent on the job is 27 minutes.
   a. If the distribution of time spent on the job is approximately bell-shaped, between what two times would 68% of the figures be? 95%? 99.7%?
   b. Suppose a worker spent 400 minutes on the job. What would that worker's Z score be and what would it tell the researcher?

3.39 During the 1990s, businesses were expected to show a lot of interest in Central and Eastern European countries. As new markets begin to open, American business people need to gain a better understanding of the market potential there. The following are the per capita GNP figures for eight of these European countries published by the *World Almanac.*

COUNTRY	PER CAPITA INCOME (U.S. $)
Albania	1290
Bulgaria	4630
Croatia	4300
Germany	20400
Hungary	7500
Poland	6400
Romania	5200
Bosnia/Herzegovina	600

   a. Compute the mean and standard deviation for Albania, Bulgaria, Croatia, and Germany.
   b. Compute the mean and standard deviation for Hungary, Poland, Romania, and Bosnia/Herzegovina.

   c. Use a coefficient of variation to compare the two standard deviations. Treat the data as population data.

3.40 According to the Bureau of Labor Statistics, the average annual salary of a worker in Detroit, Michigan, is $35,748. Suppose the median annual salary for a worker in this group is $31,369 and the mode is $29,500. Is the distribution of salaries for this group skewed? If so, how and why? Which of these measures of central tendency would you use to describe these data? Why?

3.41 According to the U.S. Army Corps of Engineers, the top 20 U.S. ports, ranked by total tonnage (in million tons), were as follows.

PORT	TOTAL TONNAGE
Port of South Louisiana, LA	189.8
Houston, TX	148.2
New York, NY	131.6
New Orleans, LA	83.7
Baton Rouge, LA	81.0
Corpus Christi, TX	80.5
Valdez Harbor, AK	77.1
Port of Plaguemines, LA	66.9
Long Beach, CA	58.4
Texas City, TX	56.4
Mobile, AL	50.9
Pittsburgh, PA	50.9
Norfolk Harbor, VA	49.3
Tampa Harbor, FL	49.3
Lake Charles, LA	49.1
Los Angeles, CA	45.7
Baltimore Harbor, MD	43.6
Philadelphia, PA	41.9
Duluth-Superior, MN	41.4
Port Arthur, TX	37.2

   a. Construct a box and whisker plot for these data.
   b. Discuss the shape of the distribution from the plot.
   c. Are there outliers?
   d. What are they and why do you think they are outliers?

3.42 Runzheimer International publishes data on overseas business travel costs. They report that the average per diem total for a business traveler in Paris, France, is $349. Suppose the per diem costs of a business traveler to Paris are normally distributed, and 99.7% of the per diem figures are between $317 and $381. What is the value of the standard deviation? The average per diem total for a business traveler in Moscow is $415. If the shape of the distribution of per diem costs of a business traveler in Moscow is normal and if 95% of the per diem costs in Moscow lie between $371 and $459, what is the standard deviation?

ANALYZING THE DATABASES

1. Use the manufacturing database. The original data from the variable, Value of Industry Shipments, has been recoded in this database so that there are only four categories. What is the modal category? What is the mean amount of New Capital Expenditures? What is the median amount of New Capital Expenditures? What does the comparison of the mean and the median tell you about the data?

2. For the stock market database "describe" the Dollar Value variable. Include measures of central tendency, variability, and skewness. What did you find?

3. Using the financial database study Earnings per Share for Type 2 and Type 7 (chemical companies and petrochemical companies). Compute a coefficient of variability for Type 2 and for Type 7. Compare the two coefficients and comment.

4. Use the hospital database. Construct a box and whisker plot for Number of Births. Thinking about hospitals and birthing facilities, comment on why the box and whisker plot may look the way it does.

CASE

COCA-COLA GOES SMALL IN RUSSIA

The Coca-Cola Company is the number-one seller of soft drinks in the world. Every day an average of more than one billion servings of Coca-Cola, Diet Coke, Sprite, Fanta, and other products of Coca-Cola are enjoyed around the world. The company has the world's largest production and distribution system for soft drinks and sells more than twice as many soft drinks as its nearest competitor. Coca-Cola products are sold in more than 200 countries around the globe.

For several reasons, the company believes it will continue to grow internationally. One reason is that disposable income is rising. Another is that outside the United States and Europe, the world is getting younger. In addition, reaching world markets is becoming easier as political barriers fall and transportation difficulties are overcome. Still another reason is that the sharing of ideas, cultures, and news around the world creates market opportunities. Part of the company mission is for Coca-Cola to maintain the world's most powerful trademark and effectively utilize the world's most effective and pervasive distribution system.

In June 1999 Coca-Cola Russia introduced a 200-ml (about 6.8 oz) Coke bottle in Volgograd, Russia, in a campaign to market Coke to its poorest customers. This strategy was successful for Coca-Cola in other countries, such as India. The bottle sells for 12 cents, making it affordable to almost everyone.

DISCUSSION

1. Because of the variability of bottling machinery, it is likely that every bottle does not contain exactly 200 ml of fluid. Some bottles may contain more fluid and others less. Since 200-ml bottle fills are somewhat unusual, a production engineer wants to test some of the bottles from the first production runs to determine how close they are to the 200-ml specification. Suppose the following data are the fill measurements from a random sample of 50 bottles. Use the techniques presented in this chapter to describe the sample. Consider measures of central tendency, variability, and skewness. Based on this analysis, how is the bottling process working?

12.1	11.9	12.2	12.2	12.0	12.1	12.9	12.1	12.3	12.5
11.7	12.4	12.3	11.8	11.3	12.1	11.4	11.6	11.2	12.2
12.4	11.8	11.9	12.2	11.6	11.6	12.4	12.4	12.6	12.6
12.1	12.8	11.9	12.0	11.9	12.3	12.5	11.9	13.1	11.7
12.2	12.5	12.2	11.7	12.9	12.2	11.5	12.6	12.3	11.8

Suppose that at another plant Coca-Cola is filling bottles with the more traditional 20 oz of fluid. A lab randomly samples 150 bottles and tests the bottles for fill volume. The descriptive statistics are given in Excel computer output. Write a brief report to supervisors summarizing what this output is saying about the process.

Excel Output

BOTTLE FILLS

Mean	20.0085
Standard error	0.0023
Median	20.0092
Mode	20.0169
Standard deviation	0.0279
Sample variance	0.0008
Kurtosis	0.6028
Skewness	−0.1063
Range	0.1759
Minimum	19.9287
Maximum	20.1046
Sum	3001.2707
Count	150

SOURCE: Adapted from "Coke, Avis Adjust in Russia," *Advertising Age,* July 5, 1999, p. 25, and The Coca-Cola company's Web site at http://www.coca-cola.com/home.html.

4
Probability

Learning Objectives

The main objective of Chapter 4 is to help you understand the basic principles of probability, thereby enabling you to:

1. Comprehend the different ways of assigning probability.

2. Understand and apply marginal, union, joint, and conditional probabilities.

3. Select the appropriate law of probability to use in solving problems.

4. Solve problems by using the laws of probability, including the law of addition, the law of multiplication, and the law of conditional probability.

5. Revise probabilities by using Bayes' rule.

In business, most decision making involves uncertainty. For example, an operations manager does not know definitely whether a valve in the plant is going to malfunction or continue to function—or, if it continues, for how long. Should it be replaced? What is the chance that the valve will malfunction within the next week? In the banking industry, what are the new vice president's prospects for successfully turning a department around? The answers to these questions are uncertain.

In the case of a high-rise building, what are the chances that a fire-extinguishing system will work when needed if redundancies are built in? Business people must address these and thousands of similar questions daily. Because most such questions do not have definite answers, the decision making is based on uncertainty. In many of these situations, a probability can be assigned to the likelihood of an outcome. This chapter is about learning how to determine or assign probabilities.

4.1
Introduction to Probability

Chapter 1 discussed the difference between descriptive and inferential statistics. Much statistical analysis is inferential, and probability is the basis for inferential statistics. Recall that inferential statistics involves taking a sample from a population, computing a statistic on the sample, and inferring from the statistic the value of the corresponding parameter of the population. The reason for doing so is that the value of the parameter is unknown. Because it is unknown, the analyst conducts the inferential process under uncertainty. However, by applying rules and laws, the analyst can often assign a probability of obtaining the results. Figure 4.1 depicts this process.

Suppose a quality control inspector selects a random sample of 40 light bulbs from a population of brand X bulbs and computes the average number of hours of lumination for the sample bulbs. By using techniques discussed later in this text, the specialist estimates the average number of lumination hours for the *population* of brand X light bulbs from this *sample* information. Because the light bulbs being analyzed are only a sample of the population, the average number of lumination hours for the 40 bulbs may or may not accurately estimate the average for all bulbs in the population. The results are uncertain. By applying the laws presented in this chapter, the inspector can assign a value of probability to this estimate.

In addition, probabilities are used directly in certain industries and industry applications. For example, the insurance industry uses probabilities in actuarial tables to determine the likelihood of certain outcomes in determining rates and coverages. The gaming industry uses probability values to establish charges and payoffs. One way to determine whether a company's hiring practices meet the government's EEOC guidelines is to compare various proportional breakdowns of their employees (by ethnicity, gender, age, etc.)

Figure 4.1

Probability in the process of inferential statistics

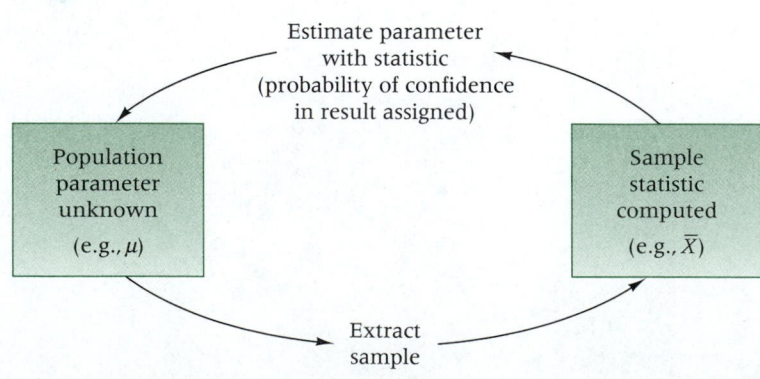

to the proportions in the general population from which the employees are hired. In comparing the company figures with those of the general population, the courts could study the probabilities of a company randomly hiring a certain profile of employees from a given population. In other industries, such as manufacturing and aerospace, it is important to know the life of a mechanized part and the probability that it will malfunction at any given length of time in order to protect the firm from major breakdowns.

The three general methods of assigning probabilities are (1) the classical method, (2) the relative frequency of occurrence method, and (3) subjective probabilities.

Classical Method of Assigning Probabilities

When *probabilities are assigned based on laws and rules,* the method is referred to as the **classical method of assigning probabilities.** This method involves an **experiment,** which is *a process that produces outcomes,* and an **event,** which is *an outcome of an experiment.*

$$P(E) = \frac{n_e}{N}$$

where:

N = total possible number of outcomes of an experiment

n_e = the number of outcomes in which the event occurs out of N outcomes

CLASSICAL METHOD OF ASSIGNING PROBABILITIES

4.2
Methods of Assigning Probabilities

Classical method of assigning probabilities
Probabilities assigned based on rules and laws.

For example, in a particular plant a product is made by three machines. Machine A always produces 40% of the total number of this product. Ten percent of the items produced by machine A are defective. If the finished products are well mixed with regard to which machine produced them and if one of these products is randomly selected, the classical method of assigning probabilities tells us that the probability that the part was produced by machine A *and* is defective is .04. This probability can be determined even before the part is sampled because with the classical method, the probabilities can be determined *a priori;* that is, they can be *determined prior to the experiment.*

When we assign probabilities using the classical method, the probability of an individual event occurring is determined as the ratio of the number of items in a population containing the event (n_e) to the total number of items in the population (N). That is, $P(E) = n_e/N$. For example, if a company has 200 workers and 70 are female, the probability of randomly selecting a female from this company is 70/200 = .35.

As n_e can never be greater than N (no more than N outcomes in the population could possibly have attribute e), the highest value of any probability is 1. If the probability of an outcome occurring is 1, the event is certain to occur. The smallest possible probability is 0. If none of the outcomes of the N possibilities has the desired characteristic, $e,$ the probability is $0/N = 0$, and the event is certain not to occur.

Experiment
A process that produces outcomes.

Event
An outcome of an experiment.

a priori
Determined before, or prior to, an experiment.

$$0 \le P(E) \le 1$$

RANGE OF POSSIBLE PROBABILITIES

Thus, probabilities are nonnegative proper fractions or nonnegative decimal values less than or equal to 1.

Probability values can be converted to percentages by multiplying by 100. Meteorologists often report weather probabilities in percentage form. For example, when they forecast a 60% chance of rain for tomorrow, they are saying that the probability of rain tomorrow is .60.

Relative Frequency of Occurrence

Relative frequency of occurrence
Assigning probability based on cumulated historical data.

The **relative frequency of occurrence** method of assigning probabilities is *based on cumulated historical data*. With this method, the probability of an event occurring is equal to the number of times the event has occurred in the past divided by the total number of opportunities for the event to have occurred.

$$\text{Probability by Relative Frequency of Occurrence} = \frac{\text{Number of Times an Event Occurred}}{\text{Total Number of Opportunitues for the Event to Occur}}$$

Relative frequency of occurrence is not based on rules or laws but on what has occurred in the past. For example, a company wants to determine the probability that its inspectors are going to reject the next batch of raw materials from a supplier. Data gathered from company record books show that the supplier had sent the company 90 batches in the past, and inspectors had rejected 10 of them. By the method of relative frequency of occurrence, the probability of the inspectors rejecting the next batch is 10/90, or .11. If the next batch is rejected, the relative frequency of occurrence probability for the subsequent shipment would change to 11/91 = .12.

Subjective Probability

Subjective probability
A probability assigned based on the intuition or reasoning of the person determining the probability.

The subjective method of assigning probability is based on the feelings or insights of the person determining the probability. **Subjective probability** comes from *the person's intuition or reasoning*. Although not a scientific approach to probability, the subjective method often is based on the accumulation of knowledge, understanding, and experience stored and processed in the human mind. At times it is merely a guess. At other times, subjective probability can potentially yield accurate probabilities. Subjective probability can be used to capitalize on the background of experienced workers and managers in decision making.

Suppose a director of transportation for an oil company is asked the probability of getting a shipment of oil out of Saudi Arabia to the United States within three weeks. A director who has scheduled many such shipments, has a knowledge of Saudi politics, and has an awareness of current climatological and economic conditions may be able to give an accurate probability that the shipment can be made on time.

Subjective probability also can be a potentially useful way of tapping a person's experience, knowledge, and insight and using them to forecast the occurrence of some event. An experienced airline mechanic can usually assign a meaningful probability that a particular plane will have a certain type of mechanical difficulty. Physicians sometimes assign subjective probabilities to the life expectancy of people who have certain diseases.

4.3
Structure
of Probability

In the study of probability, developing a language of terms and symbols is helpful. The structure of probability provides a common framework within which the topics of probability can be explored.

Experiment

As previously stated, an experiment is a process that produces outcomes. Examples of business-oriented experiments with outcomes that can be statistically analyzed might include the following.

- Interviewing 20 randomly selected consumers and asking them which brand of appliance they prefer
- Sampling every 200th bottle of ketchup from an assembly line and weighing the contents
- Testing new pharmaceutical drugs on samples of cancer patients and measuring the patient's improvement
- Auditing every 10th account to detect any errors
- Recording the Dow Jones Industrial Average on the first Monday of every month for 10 years

Event

Because an event is an outcome of an experiment, the experiment defines the possibilities of the event. If the experiment is to sample five bottles coming off a production line, an event could be to get one defective and four good bottles. In an experiment to roll a die, one event could be to roll an even number and another event could be to roll a number greater than two. Events are denoted by uppercase letters; italic capital letters (e.g., A and E_1, E_2, . . .) represent the general or abstract case, and roman capital letters (e.g., H and T for heads and tails) denote specific things and people.

Elementary Events

Events that cannot be decomposed or broken down into other events are called **elementary events.** Elementary events are denoted by lowercase letters (e.g., e_1, e_2, e_3, . . .). Suppose the experiment is to roll a die. The elementary events for this experiment are to roll a 1 or roll a 2 or roll a 3, and so on. Rolling an even number is an event, but it is not an elementary event because the even number can be broken down further into events 2, 4, and 6.

In the experiment of rolling a die, there are six elementary events {1, 2, 3, 4, 5, 6}. Rolling a pair of dice results in 36 possible elementary events (outcomes). For each of the six elementary events possible on the roll of one die, there are six possible elementary events on the roll of the second die, as depicted in the tree diagram in Figure 4.2. Table 4.1 contains a list of these 36 outcomes.

In the experiment of rolling a pair of dice, other events could include outcomes such as two even numbers, a sum of 10, a sum greater than five, and others. However, none of these events is an elementary event because each can be broken down into several of the elementary events displayed in Table 4.1.

Elementary events
Events that cannot be decomposed or broken down into other events.

Sample Space

A **sample space** is *a complete roster or listing of all elementary events for an experiment.* Table 4.1 is the sample space for the roll of a pair of dice. The sample space for the roll of a single die is {1, 2, 3, 4, 5, 6}.

Sample space can aid in finding probabilities. Suppose an experiment is to roll a pair of dice. What is the probability that the dice will sum to 7? An examination of the sample space shown in Table 4.1 reveals that there are six outcomes in which the dice sum to 7— {(1,6), (2,5), (3,4), (4,3), (5,2), (6,1)}—and there are 36 elementary events in the sample space. Using this information, we can conclude that the probability of rolling a pair of

Sample space
A complete roster or listing of all elementary events for an experiment.

Figure 4.2

Possible outcomes
for the roll of a pair
of dice

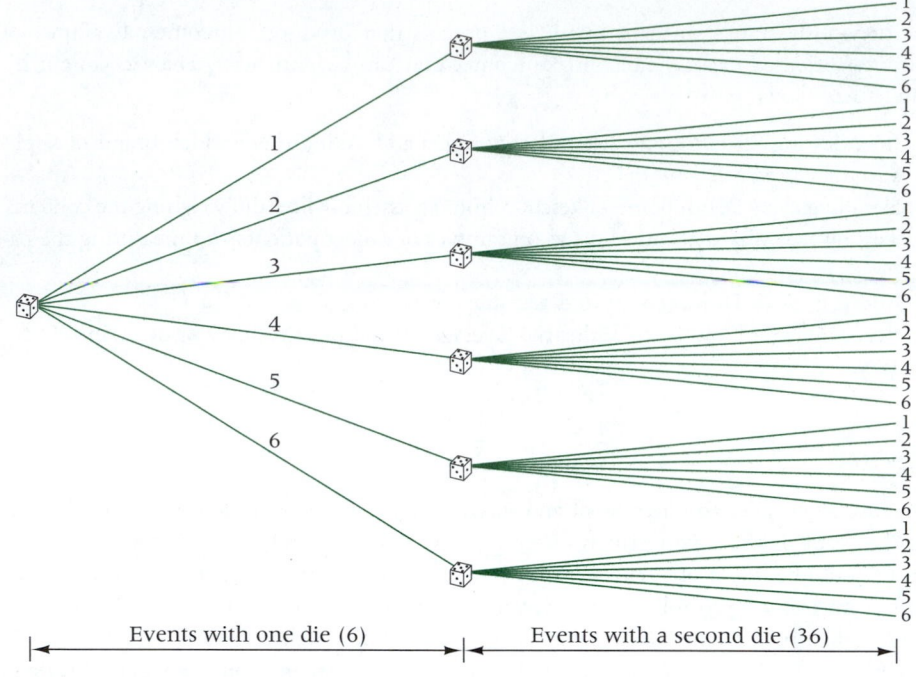

Events with one die (6)   Events with a second die (36)

TABLE 4.1
All Possible Elementary Events
in the Roll of a Pair of Dice
(Sample Space)

(1,1)	(2,1)	(3,1)	(4,1)	(5,1)	(6,1)
(1,2)	(2,2)	(3,2)	(4,2)	(5,2)	(6,2)
(1,3)	(2,3)	(3,3)	(4,3)	(5,3)	(6,3)
(1,4)	(2,4)	(3,4)	(4,4)	(5,4)	(6,4)
(1,5)	(2,5)	(3,5)	(4,5)	(5,5)	(6,5)
(1,6)	(2,6)	(3,6)	(4,6)	(5,6)	(6,6)

dice that sum to 7 is 6/36 or .1667. However, using the sample space to determine probabilities is unwieldy and cumbersome when the sample space is large. Hence statisticians usually use other, more effective methods of determining probability.

Unions and Intersections

Set notation
The use of braces to group
numbers that have some
specified characteristic.

Set notation, *the use of braces to group numbers,* is used as a symbolic tool for unions and intersections in this chapter. The **union** of X, Y formed by combining elements of both sets, is denoted $X \cup Y$. An element qualifies for the union of X, Y if it is in either X or Y or in both X and Y. The union expression $X \cup Y$ can be translated to X *or* Y. For example, if

Union
A new set of elements
formed by combining the
elements of two or more
other sets.

$$X = \{1, 4, 7, 9\} \quad \text{and} \quad Y = \{2, 3, 4, 5, 6\},$$
$$X \cup Y = \{1, 2, 3, 4, 5, 6, 7, 9\}.$$

Intersection
The portion of the
population that contains
elements that lie in both or
all groups of interest.

Note that all the values of X and all the values of Y qualify for the union. However, none of the values is listed more than once in the union. In Figure 4.3, the shaded region of the Venn diagram denotes the union.

An **intersection** is denoted $X \cap Y$. To qualify for intersection, an element must be in *both X and Y.* The intersection *contains the elements common to both sets.* Thus the intersec-

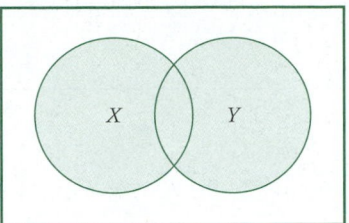

Figure 4.3

A union

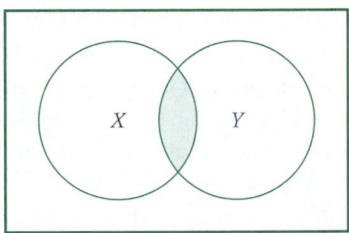

Figure 4.4

An intersection

tion symbol, ∩, is often read as *and*. The intersection of *X, Y* is referred to as *X and Y*. For example, if

$$X = \{1, 4, 7, 9\} \quad \text{and} \quad Y = \{2, 3, 4, 5, 6\},$$
$$X \cap Y = \{4\}.$$

Note that only the value 4 is common to both sets *X* and *Y*. The intersection is more exclusive than and hence equal to or (usually) smaller than the union. That is, elements must be characteristic of both *X* and *Y* to qualify. In Figure 4.4, the shaded region denotes the intersection.

Mutually Exclusive Events

Two or more events are **mutually exclusive events** if *the occurrence of one event precludes the occurrence of the other event(s)*. This characteristic means that mutually exclusive events cannot occur simultaneously and therefore can have no intersection.

The variable "gender" presents two mutually exclusive outcomes, male and female: An employee randomly selected to be part of a study is either male or female but cannot be both. A manufactured part is either defective or okay: The part cannot be both okay and defective at the same time because "okay" and "defective" are mutually exclusive categories. In a sample of the manufactured products, the event of selecting a defective part is mutually exclusive with the event of selecting a nondefective part. Suppose an office building is for sale and two different potential buyers have placed bids on the building. It is not possible for both buyers to purchase the building, therefore, the event of buyer A purchasing the building is mutually exclusive with the event of buyer B purchasing the building. In the toss of a single coin, heads and tails are mutually exclusive events. The person tossing the coin gets either a head or a tail but never both. On a toss of a pair of dice, the event (6, 6), "boxcars," is mutually exclusive with the event (1, 1), "snake eyes." Getting both boxcars and snake eyes on the same roll of the dice is impossible.

The probability of two mutually exclusive events occurring at the same time is zero.

Mutually exclusive events
Events such that the occurrence of one precludes the occurrence of the other.

MUTUALLY EXCLUSIVE EVENTS X AND Y	$P(X \cap Y) = 0$

Independent Events

Independent events
Events such that the occurrence or nonoccurrence of one has no effect on the occurrence of the others.

Two or more events are **independent events** if *the occurrence or nonoccurrence of one of the events does not affect the occurrence or nonoccurrence of the other event(s)*. Certain experiments, such as rolling dice, yield independent events; each die is independent of the other. Whether a 6 is rolled on the first die has no influence on whether a 6 is rolled on the second die. Coin tosses always are independent of each other. The event of getting a head on the first toss of a coin is independent of getting a head on the second toss. It is generally believed that certain human characteristics are independent of other events. For example, left-handedness is probably independent of the possession of a credit card. Whether a person wears glasses or not is probably independent of the brand of milk preferred.

Many experiments using random selection can produce either independent or nonindependent events. In these experiments, the outcomes are independent if sampling is done *with replacement.* That is, after each item is selected and the outcome is determined, the item is restored to the population and the population is shuffled. This way, each draw becomes independent of the previous draw. Suppose an inspector is randomly selecting bolts from a bin that contains 5% defects. If the inspector samples a defective bolt and returns it to the bin, on the second draw there are still 5% defects in the bin regardless of the fact that the first outcome was a defect. If the inspector does *not* replace the first draw, the second draw is not independent of the first; in this case, fewer than 5% defects remain in the population. Thus the probability of the second outcome is dependent on the first outcome.

If X and Y are *independent,* the following symbolic notation is used.

INDEPENDENT EVENTS X AND Y	$P(X \mid Y) = P(X)$ and $P(Y \mid X) = P(Y)$

$P(X \mid Y)$ denotes the probability of X occurring given that Y has occurred. If X and Y are independent, then the probability of X occurring given that Y has occurred is just the probability of X occurring. Knowledge that Y has occurred does not impact the probability of X occurring because X and Y are independent. For example, P(prefers Pepsi | person is right-handed) = P(prefers Pepsi) because a person's handedness is independent of brand preference.

Collectively Exhaustive Events

Collectively exhaustive events
A list containing all possible elementary events for an experiment.

A list of **collectively exhaustive events** contains *all possible elementary events for an experiment.* Thus, all sample spaces are collectively exhaustive lists. The list of possible outcomes for the tossing of a pair of dice contained in Table 4.1 is a collectively exhaustive list. The sample space for an experiment can be described as a list of events that are mutually exclusive and collectively exhaustive. Sample space events do not overlap or intersect, and the list is complete.

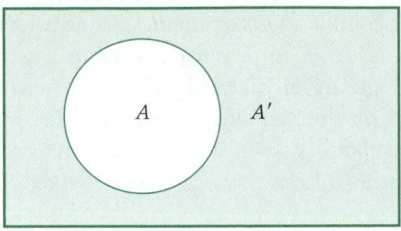

Figure 4.5

The complement of event *A*

Complementary Events

The **complement** of event *A* is denoted *A'*, pronounced "not A." *All the elementary events of an experiment not in A* comprise its complement. For example, if in rolling one die, event *A* is getting an even number, the complement of *A* is getting an odd number. If event *A* is getting a 5 on the roll of a die, the complement of *A* is getting a 1, 2, 3, 4, or 6. The complement of event *A* contains whatever portion of the sample space that event *A* does not contain, as the Venn diagram in Figure 4.5 shows.

Using the complement of an event sometimes can be helpful in solving for probabilities because of the following rule.

Complementary events
Two events, one of which comprises all the elementary events of an experiment that are not in the other event.

$$P(A') = 1 - P(A)$$

PROBABILITY OF THE COMPLEMENT OF *A*

Suppose 32% of the employees of a company have a college degree. If an employee is randomly selected from the company, the probability that the person does not have a college degree is 1 − .32 = .68. Suppose 42% of all parts produced in a plant are molded by machine A and 31% are molded by machine B. If a part is randomly selected, the probability that it was molded by neither machine A nor machine B is 1 − .73 = .27. (Assume that a part is only molded on one machine.)

Counting the Possibilities

In statistics, there is a collection of techniques and rules for counting the number of outcomes that can occur for a particular experiment. Some of these rules and techniques can be used to delineate the size of the sample space. Presented here are three of these counting methods.

1. THE *mn* COUNTING RULE   Suppose a customer has decided to buy a certain brand of new car. The car has as options two different engines, five different paint colors, and three interior packages. If each of these options is available with each of the others, how many different cars could the customer choose from? To determine this number, we can use the ***mn* counting rule.**

***mn* counting rule**
A rule used in probability to count the number of ways two operations can occur if the first operation has *m* possibilities and the second operation has *n* possibilities.

If an operation can be done *m* ways and a second operation can be done *n* ways, then there are *mn* ways for the two operations to occur in order. This rule can be extended to cases with three or more operations.

THE *mn* COUNTING RULE

Using the *mn* counting rule, we can determine that the automobile customer has (2)(5)(3) = 30 different car combinations of engines, paint colors, and interiors available.

Suppose a scientist wants to set up a research design to study the effects of gender (M, F), marital status (single never married, divorced, married), and economic class (lower, middle, and upper) on the frequency of airline ticket purchases per year. The researcher would have to set up a design in which 18 different samples are taken to represent all possible groups generated from these customer characteristics.

$$\text{Number of Groups} = (\text{Gender})(\text{Marital Status})(\text{Economic Class})$$

$$= (2)(3)(3) = 18 \text{ Groups}$$

2. SAMPLING FROM A POPULATION WITH REPLACEMENT   In sampling *n* items from a population of size *N* with replacement, there are

$$(N)^n \text{ possibilities}$$

where:

$$N = \text{population size, and}$$

$$n = \text{sample size.}$$

For example, there are six sides to a die. Each time a die is rolled, the outcomes are independent (with replacement) of the previous roll. If a die is rolled three times in succession, how many different outcomes can occur? That is, what is the size of the sample space for this experiment? The size of the population, *N,* is six, the six sides of the die. We are sampling three dice rolls, *n* = 3. The sample space is

$$(N)^n = (6)^3 = 216.$$

Suppose in a lottery six numbers are drawn from the digits 0 through 9, with replacement (digits can be reused). How many different groupings of six numbers can be drawn? *N* is the population of 10 numbers (0 through 9) and *n* is the sample size, six numbers. As a result,

$$(N)^n = (10)^6 = 1,000,000.$$

That is, a million six-digit numbers are available!

Combinations
Used to determine the number of possible ways *n* things can happen from *N* total possibilities when sampling without replacement.

3. COMBINATIONS: SAMPLING FROM A POPULATION WITHOUT REPLACEMENT
In sampling *n* items from a population of size *N* without replacement, there are

$$_NC_n = \binom{N}{n} = \frac{N!}{n!(N-n)!}$$

possibilities.

For example, suppose a small law firm has 16 employees and three are to be selected randomly to represent the company at the annual meeting of the American Bar Association. How many different combinations of lawyers could be sent to the meeting? This is not sampling with replacement because it is assumed that three *different* lawyers will be selected to go. This problem is solved by using combinations. *N* = 16 and *n* = 3, so

$$_NC_n = {}_{16}C_3 = \frac{16!}{3!13!} = 560.$$

There are 560 combinations of three lawyers who could represent the firm.

4.1 A supplier shipped a lot of six parts to a company. The lot contained three defective parts. Suppose the customer decided to randomly select two parts and test them for defects. How large a sample space is the customer potentially working with? List the sample space. Using the sample space list, determine the probability that the customer will select a sample with exactly one defect.

4.2 Given $X = \{1, 3, 5, 7, 8, 9\}$, $Y = \{2, 4, 7, 9\}$, and $Z = \{1, 2, 3, 4, 7\}$, solve the following.
  a. $X \cup Z = $ ___
  b. $X \cap Y = $ ___
  c. $X \cap Z = $ ___
  d. $X \cup Y \cup Z = $ ___
  e. $X \cap Y \cap Z = $ ___
  f. $(X \cup Y) \cap Z = $ ___
  g. $(Y \cap Z) \cup (X \cap Y) = $ ___
  h. X or $Y = $ ___
  i. Y and $X = $ ___

4.3 If a population consists of the positive even numbers through 30 and if $A = \{2, 6, 12, 24\}$, what is A'?

4.4 A company's customer service 800 telephone line is set up so that the caller has six options. Each of these six options has a menu with four options. For each of these four options, there are three more options. For each of these three options, there are three more options. If a person calls the 800 line for assistance, how many total options would be available?

4.5 A bin contains six parts. Two of the parts are defective and four are acceptable. If three of the six parts are selected from the bin, how large is the sample space? Which counting rule did you use and why? For this sample space, what is the probability that exactly one of the three sampled parts is defective?

4.6 A company places a seven-digit serial number on each part that is made. Each digit of the serial number can be any number from 0 through 9. Digits can be repeated in the serial number. How many different serial numbers are possible?

4.7 A small company has 20 employees. Six of these employees will be selected randomly to be interviewed as part of an employee satisfaction program. How many different groups of six can be selected?

Four particular types of probability are presented in this chapter. The first type is **marginal probability.** Marginal probability is denoted $P(E)$, where E is some event. A marginal probability is usually *computed by dividing some subtotal by the whole.* An example of marginal probability is the probability that a person owns a Ford car. This probability is computed by dividing the number of Ford owners by the total number of car owners. The probability of a person wearing glasses is also a marginal probability. This probability is computed by dividing the number of people wearing glasses by the total number of people.

A second type of probability is the union of two events. **Union probability** is denoted $P(E_1 \cup E_2)$, where E_1 and E_2 are two events. $P(E_1 \cup E_2)$ is *the probability that* E_1 *will occur or that* E_2 *will occur or that both* E_1 *and* E_2 *will occur.* An example of union probability is the probability that a person owns a Ford or a Chevrolet. To qualify for the union, the person only has to have at least one of these cars. Another example is the probability of a person wearing glasses or having red hair. All people wearing glasses are included in the union, along with all redheads and all redheads who wear glasses. In a company, the probability that a person is male *or* a clerical worker is a union probability. A person qualifies for the union by being male or by being a clerical worker or by being both (a male clerical worker).

4.4
Marginal, Union, Joint, and Conditional Probabilities

Marginal probability
A probability computed by dividing a subtotal of the population by the total of the population.

Union probability
The probability of one event occurring or the other event occurring or both occurring.

Figure 4.6

Marginal, union, joint, and conditional probabilities

Marginal	Union	Joint	Conditional
$P(X)$	$P(X \cup Y)$	$P(X \cap Y)$	$P(X \mid Y)$
The probability of X occurring	The probability of X or Y occurring	The probability of X and Y occurring	The probability of X occurring given that Y has occurred
Uses total possible outcomes in denominator	Uses total possible outcomes in denominator	Uses total possible outcomes in denominator	Uses subtotal of the possible outcomes in denominator

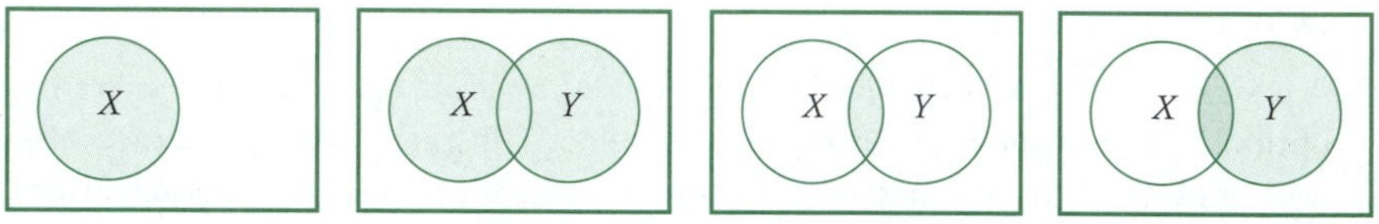

Joint probability
The probability of the intersection occurring, or the probability of two or more events happening at once.

Conditional probability
The probability of the occurrence of one event given that another event has occurred.

A third type of probability is *the intersection two events,* or **joint probability.** The joint probability of events E_1 and E_2 occurring is denoted $P(E_1 \cap E_2)$. Sometimes $P(E_1 \cap E_2)$ is read as the probability of E_1 and E_2. To qualify for the intersection, both events *must* occur. An example of joint probability is the probability of a person owning both a Ford and a Chevrolet. Owning one type of car is not sufficient. A second example of joint probability is the probability that a person is a redhead and wears glasses.

The fourth type is **conditional probability.** Conditional probability is denoted $P(E_1 \mid E_2)$. This expression is read: *the probability that E_1 will occur given that E_2 is known to have occurred.* Conditional probabilities involve knowledge of some prior information. The information that is known or given is written to the right of the vertical line in the probability statement. An example of conditional probability is the probability that a person owns a Chevrolet given that she owns a Ford. This conditional probability is only a measure of the proportion of Ford owners who have a Chevrolet—not the proportion of total car owners who own a Chevrolet. Conditional probabilities are computed by determining the number of items that have an outcome out of some subtotal of the population. In the car owner example, the possibilities are reduced to Ford owners, and then the number of Chevrolet owners out of those Ford owners is determined. Another example of a conditional probability is the probability that a worker in a company is a professional given that he is male. Of the four probability types, only conditional probability does not have the population total as its denominator. Conditional probabilities have a population subtotal in the denominator. Figure 4.6 summarizes these four types of probability.

4.5
Law of Addition

Several tools are available for use in solving probability problems. These tools include sample space, tree diagrams, the laws of probability, probability matrices, and insight. Because of the individuality and variety of probability problems, some techniques apply more readily in certain situations than in others. There is no best method for solving all probability problems. In some instances, the probability matrix lays out a problem in a readily solvable manner. In other cases, setting up the probability matrix is more difficult than solving the problem in another way. The probability laws almost always can be used to solve probability problems. However, for some problems the solution can be determined without formally applying the laws.

One of the tools already presented is sample space; others include the laws of probability. Four general laws of probability are presented in this chapter: The law of addition, the law of multiplication, the conditional law, and Bayes' rule. The law of addition and the law of multiplication each have a general law and a special law. The general law of addition is used to find the probability of the union of two events, $P(X \cup Y)$. The expression $P(X \cup Y)$ denotes the probability of X occurring or Y occurring or both X and Y occurring.

$$P(X \cup Y) = P(X) + P(Y) - P(X \cap Y)$$

GENERAL LAW OF ADDITION

where X, Y are events and $(X \cap Y)$ is the intersection of X and Y.

Yankelovich Partners conducted a survey for the American Society of Interior Designers in which workers were asked which changes in office design would increase productivity. Respondents were allowed to answer more than one type of design change. The number one change that 70% of the workers said would increase productivity was reducing noise. In second place was more storage/filing space, selected by 67%. If one of the survey respondents was randomly selected and asked what office design changes would increase worker productivity, what is the probability that this person would select reducing noise *or* more storage/filing space?

Let N represent the event "reducing noise." Let S represent the event "more storage/filing space." The probability of a person responding with N *or* S can be symbolized statistically as a union probability by using the law of addition.

$$P(\text{N} \cup \text{S})$$

To successfully satisfy the search for a person who responds with reducing noise *or* more storage/filing space, we need only find someone who wants *at least one* of those two events. As 70% of the surveyed people responded that reducing noise would create more productivity, $P(\text{N}) = .70$. In addition, as 67% responded that increased storage space would improve productivity, $P(\text{S}) = .67$. Either of these would satisfy the requirement of the union. Thus, the solution to the problem seems to be

$$P(\text{N} \cup \text{S}) = P(\text{N}) + P(\text{S}) = .70 + .67 = 1.37.$$

However, we have already established that probabilities cannot be more than 1.00. What is the problem here? Notice that all people who responded that *both* reducing noise *and* increasing storage space would improve productivity are included in *each* of the marginal probabilities $P(\text{N})$ and $P(\text{S})$. Certainly a respondent who recommends both of these improvements should be included as favoring at least one. However, because they are included in the $P(\text{N})$ *and* the $P(\text{S})$, the people who recommended both improvements are *double counted*. That is why the general law of addition subtracts the intersection probability, $P(\text{N} \cap \text{S})$.

In Figure 4.7 are Venn diagrams illustrating this discussion. Notice that the intersection area of N and S is double shaded in diagram A, indicating that it has been counted twice. In diagram B, the shading is consistent throughout N and S because the intersection area has been subtracted out. Thus diagram B illustrates the proper application of the general law of addition.

Figure 4.7

Solving for the union
in the office
productivity problem

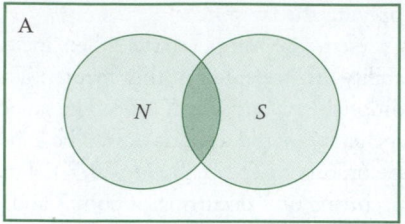

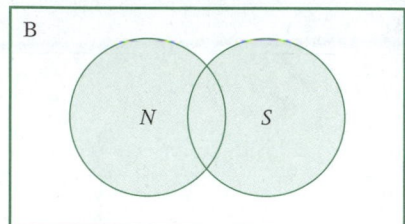

So what is the answer to Yankelovich Partners' union probability question? Suppose 56% of all respondents to the survey had said that *both* noise reduction *and* increased storage/filing space would improve productivity: $P(N \cap S) = .56$. Then we could use the general law of addition to solve for the probability that a person responds that *either* noise reduction *or* increased storage space would improve productivity.

$$P(N \cup S) = P(N) + P(S) - P(N \cap S) = .70 + .67 - .56 = .81$$

Hence, 81% of the workers surveyed responded that *either* noise reduction *or* increased storage space would improve productivity.

Probability Matrices

Probability matrix

A two-dimensional table that displays the marginal and intersection probabilities of a given problem.

In addition to the formulas, another useful tool in solving probability problems is a probability matrix. A **probability matrix** *displays the marginal probabilities and the intersection probabilities of a given problem.* Union probabilities and conditional probabilities must be *computed* from the matrix. Generally, a probability matrix is constructed as a two-dimensional table with one variable on each side of the table. For example, in the office design problem, noise reduction would be on one side of the table and increased storage space on the other. In this problem, a Yes row and a No row would be created for one variable and a Yes column and a No column would be created for the other variable, as shown in Table 4.2.

Once the matrix has been created, we can enter the marginal probabilities. $P(N) = .70$ is the marginal probability that a person responds *yes* to noise reduction. This value is placed in the "margin" in the row of Yes to noise reduction, as shown in Table 4.3. Since $P(N)$ is .70, this means that 30% of the people surveyed did not think that noise reduction would increase productivity. Thus, $P(\text{not } N) = 1 - .70 = .30$. This value, also a marginal probability, goes in the row indicated by No under noise reduction. In the column under Yes for increased storage space, the marginal probability $P(S) = .67$ is recorded. Finally, the marginal probability of No for increased storage space, $P(\text{not } S) = 1 - .67 = .33$, is placed in the No column.

In this probability matrix, all four marginal probabilities are given or can be computed simply by using the probability of a complement rule, $P(\text{not } S) = 1 - P(S)$. The intersection of noise reduction *and* increased storage space is given as $P(N \cap S) = .56$. This value is entered into the probability matrix in the cell under Yes Yes, as shown in Table 4.3. The rest of the matrix can be determined by subtracting the cell values from the marginal probabili-

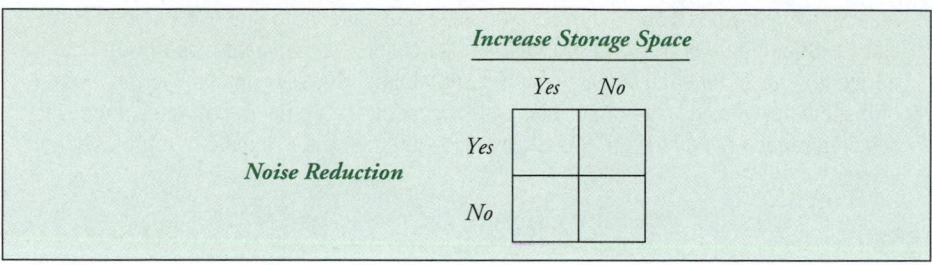

TABLE 4.2
Probability Matrix for the
Office Design Problem

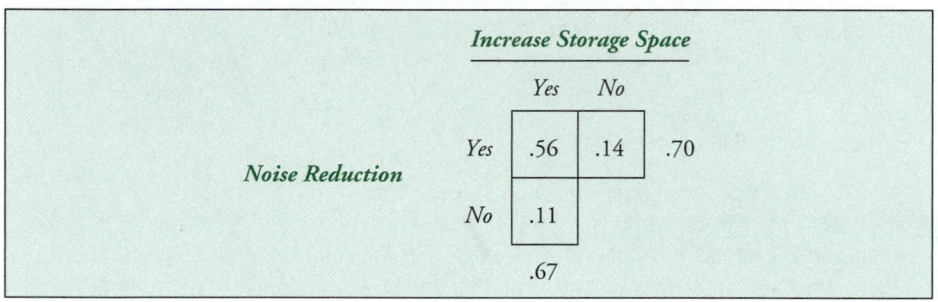

TABLE 4.3
Probability Matrix for the
Office Design Problem

Increase Storage Space

		Yes	*No*	
	Yes	.56	.14	.70
Noise Reduction				
	No	.11		
		.67		

TABLE 4.4
Yes Row and Column for
Probability Matrix of the
Office Design Problem

ties. For example, subtracting .56 from .70 and getting .14 yields the value for the cell under Yes for noise reduction and No for increased storage space. That is, 14% of all respondents said that noise reduction would improve productivity but increased storage space would not. Filling out the rest of the matrix results in the probabilities shown in Table 4.3.

Now we can solve the union probability, $P(N \cup S)$, in at least two different ways using the probability matrix. The focus is on the Yes row for noise reduction and the Yes column for increase storage space, as displayed in Table 4.4. The probability of a person suggesting noise reduction *or* increased storage space as a solution for improving productivity, $P(N \cup S)$, can be determined from the probability matrix by adding the marginal probabilities of Yes for noise reduction and Yes for increased storage space and then subtracting the Yes Yes cell, following the pattern of the general law of probabilities.

$P(N \cup S) = .70$ (from Yes row) $+ .67$ (from Yes column) $- .56$ (from Yes Yes cell) $= .81$

Another way to solve for the union probability from the information displayed in the probability matrix is to sum all cells in any of the Yes rows or columns. Observe the following from Table 4.4.

$P(N \cup S) = .56$ (from Yes Yes cell)

$+ .14$ (from Yes on noise reduction and No on increase storage space)

$+ .11$ (from No on noise reduction and Yes on increase storage space)

$= .81$

DEMONSTRATION PROBLEM 4.1

A small company has 155 employees who have been hired into one of four types of positions. Shown here is a *raw values matrix* (also called a *contingency table*) with actual frequency counts for each category and for subtotals and totals containing a breakdown of these employees by type of position and by gender. If an employee of the company is selected randomly, what is the probability that the employee is female or a professional worker?

COMPANY HUMAN RESOURCE DATA

		Gender		
		Male	*Female*	
Type of Position	*Managerial*	8	3	11
	Professional	31	13	44
	Technical	52	17	69
	Clerical	9	22	31
		100	55	155

SOLUTION

Let F denote the event of female and P denote the event of professional worker. The question is:

$$P(\text{F} \cup \text{P}) = ?$$

By the general law of addition,

$$P(\text{F} \cup \text{P}) = P(\text{F}) + P(\text{P}) - P(\text{F} \cap \text{P}).$$

There are 55 women in a total of 155 employees. Therefore, $P(\text{F}) = 55/155 = .355$. There are 44 professionals in a total of 155 employees. Therefore, $P(\text{P}) = 44/155 = .284$. Because 13 employees are both female and professional, $P(\text{F} \cap \text{P}) = 13/155 = .084$. The union probability is solved as

$$P(\text{F} \cup \text{P}) = .355 + .284 - .084 = .555.$$

To solve this probability using a matrix, you can either use the raw values matrix shown above or convert the raw values matrix to a probability matrix by dividing every value in the matrix by the value of N, 155. The raw value matrix is used in a manner similar to that of the probability matrix. To compute the union probability of selecting a person who is either female or a professional worker from the raw value matrix, add the number of people in the Female column (55) to the number of people in the Professional row (44), then subtract the number of people in the intersection cell of Female and Professional (13). This step yields the value $55 + 44 - 13 = 86$. Dividing this value (86) by the value of N (155) produces the union probability,

$$P(\text{F} \cup \text{P}) = 86/155 = .555.$$

A second way to produce the answer from the raw value matrix is to add all the cells one time that are in either the Female column or the Professional row,

$$3 + 13 + 17 + 22 + 31 = 86,$$

and then divide by the total number of employees, $N = 155$, which gives

$$P(\text{F} \cup \text{P}) = 86/155 = .555.$$

Shown here are the raw values matrix and corresponding probability matrix for the results of a national survey of 200 executives who were asked to identify the geographic locale of their company and their company's industry type. The executives were only allowed to select one locale and one industry type.

RAW VALUES MATRIX

		Geographic Location				
		Northeast D	*Southeast* E	*Midwest* F	*West* G	
Industry Type	Finance A	24	10	8	14	56
	Manufacturing B	30	6	22	12	70
	Communications C	28	18	12	16	74
		82	34	42	42	200

PROBABILITY MATRIX

		Geographic Location				
		Northeast D	*Southeast* E	*Midwest* F	*West* G	
Industry Type	Finance A	.12	.05	.04	.07	.28
	Manufacturing B	.15	.03	.11	.06	.35
	Communications C	.14	.09	.06	.08	.37
		.41	.17	.21	.21	1.00

Suppose a respondent is selected randomly from these data.

a. What is the probability that the respondent is from the Midwest (F)?
b. What is the probability that the respondent is from the communications industry (C) or from the Northeast (D)?
c. What is the probability that the respondent is from the Southeast (E) or from the finance industry (A)?

SOLUTION

a. $P(\text{Midwest}) = P(F) = .21$
b. $P(C \cup D) = P(C) + P(D) - P(C \cap D) = .37 + .41 - .14 = .64$
c. $P(E \cup A) = P(E) + P(A) - P(E \cap A) = .17 + .28 - .05 = .40$

In computing the union by using the general law of addition, the intersection probability is subtracted because it is already included in both marginal probabilities. This adjusted probability leaves a union probability that properly includes both marginal values and the intersection value. If the intersection probability is subtracted out a second time, the intersection is removed, leaving the probability of *X or Y* but *not both*.

$$P(X \text{ or } Y \text{ but not both}) = P(X) + P(Y) - P(X \cap Y) - P(X \cap Y) = P(X \cup Y) - P(X \cap Y)$$

Figure 4.8 is the Venn diagram for this probability.

Figure 4.8

The *X* or *Y* but not both case

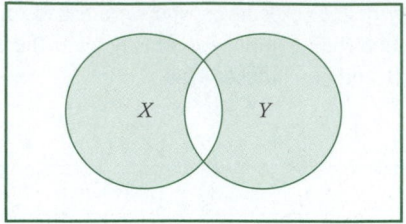

Figure 4.9

The complement of a union: The neither/nor region

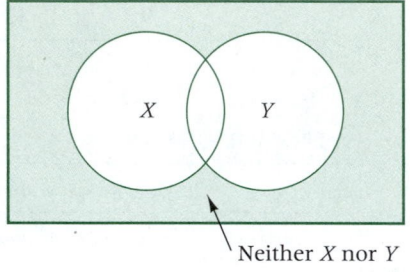

Neither *X* nor *Y*

Complement of a union
The only possible case other than the union of sets *X* and *Y*; the probability that neither *X* nor *Y* is in the outcome.

COMPLEMENT OF A UNION The probability of the union of two events *X* and *Y* represents the probability that the outcome is *either X or* it is *Y or* it is both *X and Y.* The union includes everything *except* the possibility that it is *neither* (*X* or *Y*). Another way to state this is as *neither X nor Y,* which can symbolically be represented as $P(\text{not } X \cap \text{not } Y)$. Because this is the only possible case other than the union of *X* or *Y,* it is the **complement of the union.** Stated more formally,

$$P(\text{neither } X \text{ nor } Y) = P(\text{not } X \cap \text{not } Y) = 1 - P(X \cup Y).$$

Examine the Venn diagram in Figure 4.9. Note that the complement of the union of *X, Y* is the shaded area outside the circles. This area represents the neither *X* nor *Y* region.

In the survey about increasing worker productivity by changing the office design discussed earlier, the probability that a randomly selected worker would respond with noise reduction *or* increased storage space was determined to be

$$P(\text{N} \cup \text{S}) = P(\text{N}) + P(\text{S}) - P(\text{N} \cap \text{S}) = .70 + .67 - .56 = .81.$$

The probability that a worker would respond with *neither* noise reduction *nor* increased storage space is calculated as the complement of this union.

$$P(\text{neither N nor S}) = P(\text{not N} \cap \text{not S}) = 1 - P(\text{N} \cup \text{S}) = 1 - .81 = .19$$

Thus 19% of the workers selected neither noise reduction nor increased storage space as solutions to increasing productivity. In Table 4.3, this *neither/nor* probability is found in the No No cell of the matrix, .19.

Special Law of Addition

If two events are mutually exclusive, the probability of the union of the two events is the probability of the first event plus the probability of the second event. Because mutually exclusive events do not intersect, nothing has to be subtracted.

If X, Y are mutually exclusive,

$$P(X \cup Y) = P(X) + P(Y).$$

SPECIAL LAW
OF ADDITION

The special law of addition is a special case of the general law of addition. In a sense, the general law fits all cases. However, when the events are mutually exclusive, a zero is inserted into the general law formula for the intersection, resulting in the special law formula.

In the survey about improving productivity by changing office design, the respondents were allowed to choose more than one possible office design change. Therefore, it is most likely that virtually none of the change choices were mutually exclusive, and the special law of addition would not apply to that example.

In another survey, however, respondents were allowed to select only one option for their answer, which made the possible options mutually exclusive. In this survey, conducted by Yankelovich Partners for William M. Mercer, Inc., workers were asked what most hinders their productivity and were given only the following selections from which to choose only one answer.

lack of direction

lack of support

too much work

inefficient process

not enough equipment/supplies

low pay/chance to advance

Lack of direction was cited by the most workers (20%), followed by lack of support (18%), too much work (18%), inefficient process (8%), not enough equipment/supplies (7%), low pay/chance to advance (7%), and a variety of other factors added by respondents. If a worker who responded to this survey is selected (or if the survey actually reflects the views of the working public and a worker in general is selected) and that worker is asked which of the given selections most hinders his or her productivity, what is the probability that the worker will respond that it is *either* too much work *or* inefficient process?

Let M denote the event "too much work" and I denote the event "inefficient process." The question is:

$$P(M \cup I) = ?$$

Because 18% of the survey respondents said "too much work,"

$$P(M) = .18.$$

Because 8% of the survey respondents said "inefficient process,"

$$P(I) = .08.$$

Because it was not possible to select more than one answer,

$$P(M \cap I) = .00.$$

Implementing the special law of addition gives

$$P(M \cup I) = P(M) + P(I) = .18 + .08 = .26.$$

DEMONSTRATION PROBLEM 4.3

If a worker is randomly selected from the company described in Demonstration Problem 4.1, what is the probability that the worker is either technical or clerical? What is the probability that the worker is either a professional or a clerical?

SOLUTION

Examine the raw value matrix of the company's human resources data shown in Demonstration Problem 4.1. In many raw value and probability matrices like this one, the rows are nonoverlapping or mutually exclusive, as are the columns. In this matrix, a worker can be classified as being in only *one* type of position and as either male or female but not both. Thus, the categories of type of position are mutually exclusive, as are the categories of gender, and the special law of addition can be applied to the human resource data to determine the union probabilities.

   Let T denote technical, C denote clerical, and P denote professional. The probability that a worker is either technical or clerical is

$$P(T \cup C) = P(T) + P(C) = 69/155 + 31/155 = 100/155 = .645.$$

The probability that a worker is either professional or clerical is

$$P(P \cup C) = P(P) + P(C) = 44/155 + 31/155 = 75/155 = .484.$$

DEMONSTRATION PROBLEM 4.4

Use the data from the matrices in Demonstration Problem 4.2. What is the probability that a randomly selected respondent is from the Southeast or the West? That is,

$$P(E \cup G) = ?$$

SOLUTION

As geographic location is mutually exclusive (the work location is either in the Southeast or in the West but not in both),

$$P(E \cup G) = P(E) + P(G) = .17 + .21 = .38.$$

4.5
Problems

4.8   Given $P(A) = .10$, $P(B) = .12$, $P(C) = .21$, $P(A \cap C) = .05$, and $P(B \cap C) = .03$, solve the following.
    a. $P(A \cup C) =$ ___
    b. $P(B \cup C) =$ ___
    c. If A and B are mutually exclusive, $P(A \cup B) =$ ___.

4.9   Use the values in the matrix to solve the equations given.

	D	E	F
A	5	8	12
B	10	6	4
C	8	2	5

a. $P(A \cup D) =$ ___
b. $P(E \cup B) =$ ___
c. $P(D \cup E) =$ ___
d. $P(C \cup F) =$ ___

4.10   Use the values in the matrix to solve the equations given.

	E	F
A	.10	.03
B	.04	.12
C	.27	.06
D	.31	.07

a. $P(A \cup F) =$ ___
b. $P(E \cup B) =$ ___
c. $P(B \cup C) =$ ___
d. $P(E \cup F) =$ ___

4.11   Suppose that 47% of all Americans have flown in an airplane at least once and that 28% of all Americans have ridden on a train at least once. What is the probability that a randomly selected American has either ridden on a train or flown in an airplane? Can this problem be solved? Under what conditions can it be solved? If the problem cannot be solved, what information is needed to make it solvable?

4.12   According to the U.S. Bureau of Labor Statistics, 75% of the women 25 through 49 years of age participate in the labor force. Suppose 78% of the women in that age group are married. Suppose also that 61% of all women 25 through 49 years of age are married and are participating in the labor force.
   a. What is the probability that a randomly selected woman in that age group is married or is participating in the labor force?
   b. What is the probability that a randomly selected woman in that age group is married or is participating in the labor force but not both?
   c. What is the probability that a randomly selected woman in that age group is neither married nor participating in the labor force?

4.13   According to Nielsen Media Research, approximately 67% of all U.S. households with television have cable TV. Seventy-four percent of all U.S. households with television have two or more TV sets. Suppose 55% of all U.S. households with television have cable TV *and* two or more TV sets. A U.S. household with television is randomly selected.
   a. What is the probability that the household has cable TV *or* two or more TV sets?
   b. What is the probability that the household has cable TV *or* two or more TV sets but not both?
   c. What is the probability that the household has neither cable TV nor two or more TV sets?
   d. Why does the special law of addition not apply to this problem?

4.14   A survey conducted by the Northwestern University Lindquist-Endicott Report asked 320 companies about the procedures they use in hiring. Only 54% of the responding companies review the applicant's college transcript as part of the hiring process, and only 44% consider faculty references. Assume that these percentages are true for the population of companies in the United States and that

35% of all companies use both the applicant's college transcript and faculty references.

a. What is the probability that a randomly selected company uses either faculty references or college transcript as part of the hiring process?

b. What is the probability that a randomly selected company uses either faculty references or college transcript but not both as part of the hiring process?

c. What is the probability that a randomly selected company uses neither faculty references nor college transcript as part of the hiring process?

d. Construct a probability matrix for this problem and indicate the locations of your answers for parts (a), (b), and (c) on the matrix.

4.6 **Law of** **Multiplication**	As stated in Section 4.4, the probability of the intersection of two events $(X \cap Y)$ is called the joint probability. The general law of multiplication is used to find the joint probability.

GENERAL LAW OF MULTIPLICATION	$$P(X \cap Y) = P(X) \cdot P(Y \mid X) = P(Y) \cdot P(X \mid Y)$$

The notation $X \cap Y$ means that both X *and* Y must happen. The general law of multiplication gives the probability that both event X and event Y will occur at the same time.

According to the U.S. Bureau of Labor Statistics, 46% of the U.S. labor force is female. In addition, 25% of the women in the labor force work part time. What is the probability that a randomly selected member of the U.S. labor force is a woman *and* works part time? This question is one of joint probability, and the general law of multiplication can be applied to answer it.

Let W denote the event that the member of the labor force is a woman. Let T denote the event that the member is a part-time worker. The question is:

$$P(W \cap T) = ?$$

According to the general law of multiplication, this can be solved by

$$P(W \cap T) = P(W) \cdot P(T \mid W).$$

As 46% of the labor force is women, $P(W) = .46$. $P(T \mid W)$ is a conditional probability that can be stated as the probability that a worker is a part-time worker given that the worker is a woman. This is what was given in the statement that 25% *of the women in the labor force* work part time. Hence, $P(T \mid W) = .25$. From this it follows that

$$P(W \cap T) = P(W) \cdot P(T \mid W) = (.46)(.25) = .115.$$

It can be stated that 11.5% of the U.S. labor force is women *and* works part time. The Venn diagram in Figure 4.10 shows these relationships and the joint probability.

Determining joint probabilities from raw value and/or probability matrices is easy because every cell of these matrices is a joint probability. In fact, some statisticians refer to a probability matrix as a *joint probability table.*

For example, suppose the raw value matrix of the company human resource data from Demonstration Problem 4.1 is converted to a probability matrix by dividing by the total

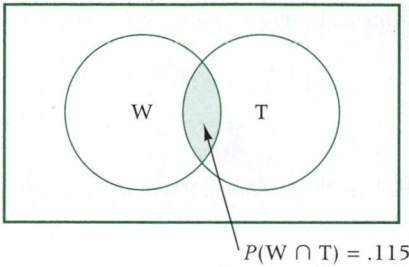

$P(W \cap T) = .115$

Figure 4.10

Joint probability that
a woman is in the
labor force and is a
part-time worker

	Gender		
	Male	*Female*	
Managerial	.052	.019	.071
Professional	.200	.084	.284
Technical	.335	.110	.445
Clerical	.058	.142	.200
	.645	.355	1.000

Type of Position

TABLE 4.5
Probability Matrix
of Company Human
Resource Data

number of employees ($N = 155$), resulting in Table 4.5. Each value in the cell of
Table 4.5 is an intersection, and the table contains all possible intersections (joint probabilities) for the events of gender and type of position. For example, the probability that a
randomly selected worker is male *and* a technical worker, $P(M \cap T)$, is .335. The probability that a randomly selected worker is female *and* a professional worker, $P(F \cap P)$, is
.084. Once a probability matrix has been constructed on a problem, usually the easiest
way to solve for the joint probability is to find the appropriate cell in the matrix and select
the answer. However, sometimes because of what is given in a problem, using the formula
is easier than constructing the matrix.

A company has 140 employees, of which 30 are supervisors. Eighty of the employees
are married, and 20% of the married employees are supervisors. If a company employee is randomly selected, what is the probability that the employee is married *and* is
a supervisor?

**DEMONSTRATION
PROBLEM 4.5**

SOLUTION
Let M denote married and S denote supervisor. The question is:

$$P(M \cap S) = ?$$

First, calculate the marginal probability.

$$P(M) = \frac{80}{140}$$
$$= .5714$$

Then, note that 20% of the married employees are supervisors, which is the conditional probability, $P(S \mid M) = .20$. Finally, applying the general law of multiplication gives

$$P(M \cap S) = P(M) \cdot P(S \mid M)$$
$$= (.5714)(.20) = .1143.$$

Hence, 11.43% of the 140 employees are married *and* are supervisors.

DEMONSTRATION PROBLEM 4.6

From the data obtained from the interviews of 200 executives in Demonstration Problem 4.2, find:

a. $P(B \cap E)$.
b. $P(G \cap A)$.
c. $P(B \cap C)$.

RAW VALUES MATRIX

		Geographic Location				
		Northeast D	Southeast E	Midwest F	West G	
	Finance A	24	10	8	14	56
Industry Type	Manufacturing B	30	6	22	12	70
	Communications C	28	18	12	16	74
		82	34	42	42	200

PROBABILITY MATRIX

		Geographic Location				
		Northeast D	Southeast E	Midwest F	West G	
	Finance A	.12	.05	.04	.07	.28
Industry Type	Manufacturing B	.15	.03	.11	.06	.35
	Communications C	.14	.09	.06	.08	.37
		.41	.17	.21	.21	1.00

SOLUTION

a. From the cell of the probability matrix, $P(B \cap E) = 6/200 = .03$. To solve by the formula, $P(B \cap E) = P(B) \cdot P(E \mid B)$, first find $P(B)$:

$$P(B) = \frac{70}{200} = .35.$$

The probability of E occurring given that B has occurred, $P(E \mid B)$, can be determined from the probability matrix. How many Bs are there? The probability matrix shows

.35 of B. As B is given, $P(E|B) = E/.35$. What are the Es in the B row? .03. Thus $P(E|B) = .03/35$. Therefore,

$$P(B \cap E) = P(B) \cdot P(E|B) = (.35)\left(\frac{.03}{.35}\right) = .03.$$

Although the formula works, finding the joint probability in the cell of the probability matrix is faster than using the formula.

An alternative formula is $P(B \cap E) = P(E) \cdot P(B|E)$, but $P(E) = .17$. Then $P(B|E)$ means the probability of B if E is given. There are .17 Es in the probability matrix and .03 Bs in these Es. Hence,

$$P(E|B) = \frac{.03}{.17} \quad \text{and} \quad P(B \cap E) = P(E) \cdot P(B|E) = (.17)\left(\frac{.03}{.17}\right) = .03.$$

b. To obtain $P(G \cap A)$, find the intersecting cell of G and A in the probability matrix, .07, or use one of the following formulas:

$$P(G \cap A) = P(G) \cdot P(A|G) = (.21)\left(\frac{.07}{.21}\right) = .07$$

or

$$P(G \cap A) = P(A) \cdot P(G|A) = (.28)\left(\frac{.07}{.28}\right) = .07.$$

c. The probability $P(B \cap C)$ means that one respondent would have to work both in the manufacturing industry and the communications industry. The survey used to gather data from the 200 executives, however, requested that each respondent specify only one industry type for his or her company. The matrix shows no intersection for these two events. Thus B and C are mutually exclusive. None of the respondents is in both manufacturing *and* communications. Hence,

$$P(B \cap C) = .00.$$

Special Law of Multiplication

If events X and Y are independent, a special law of multiplication can be used to find the intersection of X and Y. This special law utilizes the fact that when two events X, Y are independent, $P(X|Y) = P(X)$ and $P(Y|X) = P(Y)$. Thus, the general law of multiplication, $P(X \cap Y) = P(X) \cdot P(Y|X)$, becomes $P(X \cap Y) = P(X) \cdot P(Y)$ when X and Y are independent.

If X, Y are independent,

$$P(X \cap Y) = P(X) \cdot P(Y).$$

SPECIAL LAW
OF MULTIPLICATION

A study released by Bruskin-Goldring Research for SEIKO found that 28% of U.S. adults believe that the automated teller has had a most significant impact on everyday life. Another study by David Michaelson & Associates for Dale Carnegie & Associates examined employee views on team spirit in the workplace and discovered that 72% of all employees believe that working as a part of a team lowers stress. Are people's views on automated tellers independent of their views on team spirit in the workplace? If they are independent, then the probability of a person being randomly selected who believes that the automated teller has had a most significant impact on everyday life *and* that working as part of a team lowers stress is found as follows. Let A denote automated teller and S denote teamwork lowers stress.

$$P(A) = .28$$

$$P(S) = .72$$

$$P(A \cap S) = P(A) \cdot P(S) = (.28)(.72) = .2016$$

That is, 20.16% of the population believes that the automated teller has had a most significant impact on everyday life *and* that working as part of a team lowers stress.

DEMONSTRATION PROBLEM 4.7

A manufacturing firm produces pads of bound paper. Three percent of all paper pads produced are improperly bound. An inspector randomly samples two pads of paper, one at a time. Because a large number of pads are being produced during the inspection, the sampling being done, in essence, is *with* replacement. What is the probability that the two pads selected are both improperly bound?

SOLUTION
Let I denote improperly bound. The problem is to determine

$$P(I_1 \cap I_2) = ?$$

The probability of $I = .03$, or 3% are improperly bound. As the sampling is done *with* replacement, the two events are independent. Hence,

$$P(I_1 \cap I_2) = P(I_1) \cdot P(I_2) = (.03)(.03) = .0009.$$

Most probability matrices contain variables that are not independent. If a probability matrix contains independent events, the special law of multiplication can be applied. If not, the special law cannot be used. In Section 4.7 we explore a technique for determining whether events are independent. Table 4.6 contains data from independent events.

TABLE 4.6
Contingency Table of Data from Independent Events

	D	E	
A	8	12	20
B	20	30	50
C	6	9	15
	34	51	85

Use the data from Table 4.6 and the special law of multiplication to find $P(B \cap D)$.

SOLUTION

$$P(B \cap D) = P(B) \cdot P(D) = \frac{50}{85} \cdot \frac{34}{85} = .2353$$

This approach works *only* for contingency tables and probability matrices in which the variable along one side of the matrix is *independent* of the variable along the other side of the matrix. Note that the answer obtained by using the formula is the same as the answer obtained by using the cell information from Table 4.6.

$$P(B \cap D) = \frac{20}{85} = .2353$$

4.15 Use the values in the contingency table to solve the equations given.

	C	D	E	F
A	5	11	16	8
B	2	3	5	7

a. $P(A \cap E) = \_\_\_$
b. $P(D \cap B) = \_\_\_$
c. $P(D \cap E) = \_\_\_$
d. $P(A \cap B) = \_\_\_$

4.16 Use the values in the probability matrix to solve the equations given.

	D	E	F
A	.12	.13	.08
B	.18	.09	.04
C	.06	.24	.06

a. $P(E \cap B) = \_\_\_$
b. $P(C \cap F) = \_\_\_$
c. $P(E \cap D) = \_\_\_$

4.17 **a.** A batch of 50 parts contains six defects. If two parts are drawn randomly one at a time without replacement, what is the probability that both parts are defective?
b. If this experiment is repeated, *with* replacement, what is the probability that both parts are defective?

4.18 According to the nonprofit group Zero Population Growth, 78% of the U.S. population now lives in urban areas. Scientists at Princeton University and the University of Wisconsin report that about 15% of all U.S. adults care for ill relatives. Suppose that 11% of adults living in urban areas care for ill relatives.
a. Use the general law of multiplication to determine the probability of randomly selecting an adult from the U.S. population who lives in an urban area *and* is caring for an ill relative.
b. What is the probability of randomly selecting an adult from the U.S. population who lives in an urban area and does not care for an ill relative?
c. Construct a probability matrix and show where the answer to this problem lies in the matrix.
d. From the probability matrix, determine the probability that an adult lives in a nonurban area and cares for an ill relative.

4.19 A study by Peter D. Hart Research Associates for the Nasdaq Stock Market revealed that 43% of all U.S. adults are stockholders. In addition, the study determined that 75% of all U.S. adult stockholders have some college education.

Suppose 37% of all U.S. adults have some college education. A U.S. adult is randomly selected.

a. What is the probability that the adult does not own stock?

b. What is the probability that the adult owns stock *and* has some college education?

c. What is the probability that the adult owns stock *or* has some college education?

d. What is the probability that the adult has neither some college education nor owns stock?

e. What is the probability that the adult does not own stock or has no college education?

f. What is the probability that the adult has some college education and owns no stock?

4.20 According to the Consumer Electronics Manufacturers Association, 10% of all U.S. households have a fax machine and 52% have a personal computer. Suppose 91% of all U.S. households having a fax machine have a personal computer. A U.S. household is randomly selected.

a. What is the probability that the household has a fax machine *and* a personal computer?

b. What is the probability that the household has a fax machine *or* a personal computer?

c. What is the probability that the household has a fax machine and does not have a personal computer?

d. What is the probability that the household has neither a fax machine nor a personal computer?

e. What is the probability that the household does not have a fax machine and does have a personal computer?

4.21 A study by Becker Associates, a San Diego travel consultant, found that 30% of the traveling public said that their flight selections are influenced by perceptions of airline safety. Thirty-nine percent of the traveling public want to know the age of the aircraft. Suppose 87% of the traveling public who say that their flight selections are influenced by perceptions of airline safety want to know the age of the aircraft.

a. What is the probability of randomly selecting a member of the traveling public and finding out that she says that flight selection is influenced by perceptions of airline safety and that she does not want to know the age of the aircraft?

b. What is the probability of randomly selecting a member of the traveling public and finding out that she says that flight selection is neither influenced by perceptions of airline safety nor does she want to know the age of the aircraft?

c. What is the probability of randomly selecting a member of the traveling public and finding out that he says that flight selection is not influenced by perceptions of airline safety and he wants to know the age of the aircraft?

4.22 The U.S. Energy Department states that 60% of all U.S. households have ceiling fans. In addition, 29% of all U.S. households have an outdoor grill. Suppose 13% of all U.S. households have both a ceiling fan and an outdoor grill. A U.S. household is randomly selected.

a. What is the probability that the household has a ceiling fan or an outdoor grill?

b. What is the probability that the household has neither a ceiling fan nor an outdoor grill?

c. What is the probability that the household does not have a ceiling fan and does have an outdoor grill?

d. What is the probability that the household does have a ceiling fan and does not have an outdoor grill?

Conditional probabilities are based on knowledge of one of the variables. If X, Y are two events, the conditional probability of X occurring given that Y is known or has occurred is expressed as $P(X \mid Y)$ and is given in the *law of conditional probability*.

$$P(X \mid Y) = \frac{P(X \cap Y)}{P(Y)} = \frac{P(X) \cdot P(X \mid Y)}{P(Y)}$$

LAW OF
CONDITIONAL
PROBABILITY

The conditional probability of $(X \mid Y)$ is the probability that X will occur given Y. The formula for conditional probability is derived by dividing both sides of the general law of multiplication by $P(Y)$.

In the study by Yankelovich Partners to determine what changes in office design would improve productivity, 70% of the respondents believed noise reduction would improve productivity and 67% said increased storage space would improve productivity. In addition, suppose 56% of the respondents believed both noise reduction and increased storage space would improve productivity. A worker is selected randomly and asked about changes in office design. This worker believes that noise reduction would improve productivity. What is the probability that this worker believes increased storage space would improve productivity? That is, what is the probability that a randomly selected person would believe storage space would improve productivity *given that* he or she believes noise reduction improves productivity? In symbols, the question is:

$$P(S \mid N) = ?$$

Note that the given information is listed to the right of the vertical line in the conditional probability. The formula solution is:

$$P(S \mid N) = \frac{P(S \cap N)}{P(N)}$$

$$P(N) = .70 \quad \text{and} \quad P(S \cap N) = .56$$

$$P(S \mid N) = \frac{P(S \cap N)}{P(N)} = \frac{.56}{.70} = .80.$$

Eighty percent of workers who believe noise reduction would improve productivity believe increased storage space would improve productivity.

Note in Figure 4.11 that the area for N in the Venn diagram is completely shaded because it is given that the worker believes noise reduction will improve productivity. Also notice that the intersection of N and S is more heavily shaded. This is the portion of noise reduction that includes increased storage space. This is the only part of increased storage space that is in noise reduction and because the person is known to favor noise reduction, it is the only area of interest that includes increased storage space.

Examine the probability matrix in Table 4.7 for the office design problem. None of the figures given in the matrix are conditional probabilities. To reiterate what has been previously stated, a probability matrix contains only two types of probabilities, marginal and joint. The cell values are all joint probabilities and the subtotals in the margins are marginal probabilities. How are conditional probabilities determined from a probability

Figure 4.11

Conditional
probability of
increased storage
space given noise
reduction

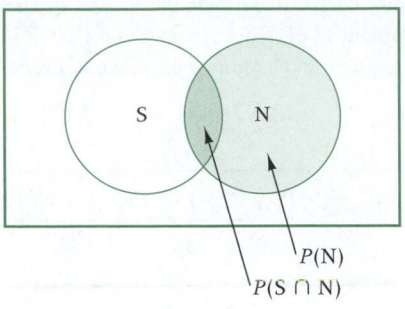

$P(N)$
$P(S \cap N)$

TABLE 4.7
Office Design Problem
Probability Matrix

		Increase Storage Space		
		Yes	*No*	
Noise Reduction	*Yes*	.56	.14	.70
	No	.11	.19	.30
		.67	.33	1.00

matrix? The law of conditional probabilities shows that a conditional probability is computed by dividing the joint probability by the marginal probability. Thus, the probability matrix has all the necessary information to solve for a conditional probability.

What is the probability that a randomly selected worker believes noise reduction would not improve productivity given that the worker does believe increased storage space would improve productivity? That is,

$$P(\text{not } N \mid S) = ?$$

The law of conditional probability states that

$$P(\text{not } N \mid S) = \frac{P(\text{not } N \cap S)}{P(S)}.$$

Notice that because S is given, we are interested only in the column that is shaded in Table 4.7, which is the Yes column for increase storage space. The marginal probability, $P(S)$, is the total of this column and is found in the margin at the bottom of the table as .67. $P(\text{not } N \cap S)$ is found as the intersection of No for noise and Yes for storage. This value is .11. Hence, $P(\text{not } N \cap S)$ is .11. Therefore,

$$P(\text{not } N \mid S) = \frac{P(\text{not } N \cap S)}{P(S)} = \frac{.11}{.67} = .164.$$

The second version of the conditional probability law formula is

$$P(X \mid Y) = \frac{P(X) \cdot P(Y \mid X)}{P(Y)}$$

This version is more complex than the first version, $[P(X \cap Y)/P(Y)]$. However, sometimes the second version must be used because of the information given in the problem—for example, when solving for $P(X \mid Y)$ but $P(Y \mid X)$ is given. The second version of the formula is obtained from the first version by substituting the formula for $P(X \cap Y) = P(X) \cdot P(Y \mid X)$ into the first version.

As an example, in Section 4.6 data relating to women in the U.S. labor force were presented. Included in this information was the fact that 46% of U.S. workers are women and that 25% of all female U.S. laborers are part-time workers. In addition, 17.4% of all U.S. laborers are known to be part-time workers. What is the probability that a randomly selected U.S. worker is a woman if that person is known to be a part-time worker? Let W denote the event of selecting a woman and T denote the event of selecting a part-time worker. In symbols, the question to be answered is:

$$P(W \mid T) = ?$$

The first form of the law of conditional probabilities is

$$P(W \mid T) = \frac{P(W \cap T)}{P(T)}.$$

Note that this version of the law of conditional probabilities requires knowledge of the joint probability, $P(W \cap T)$, which is not given here. We therefore try the second version of the law of conditional probabilities, which is

$$P(W \mid T) = \frac{P(W) \cdot P(T \mid W)}{P(T)}.$$

For this version, everything is given in the problem.

$$P(W) = .46$$
$$P(T) = .174$$
$$P(T \mid W) = .25$$

The probability of a laborer being a woman given that the person works part time can now be computed.

$$P(W \mid T) = \frac{P(W) \cdot P(T \mid W)}{P(T)} = \frac{(.46)(.25)}{(.174)} = .661$$

Hence, 66.1% of the part-time workers are women.

In general, this second version of the law of conditional probabilities is likely to be used when $P(X \cap Y)$ is unknown but $P(Y \mid X)$ is known.

The data from the executive interviews given in Demonstration Problem 4.2 are repeated here. Use these data to find:

a. $P(B \mid F)$.
b. $P(G \mid C)$.
c. $P(D \mid F)$.

RAW VALUES MATRIX

			Geographic Location			
		Northeast D	Southeast E	Midwest F	West G	
	Finance A	24	10	8	14	56
Industry Type	Manufacturing B	30	6	22	12	70
	Communications C	28	18	12	16	74
		82	34	42	42	200

PROBABILITY MATRIX

			Geographic Location			
		Northeast D	Southeast E	Midwest F	West G	
	Finance A	.12	.05	.04	.07	.28
Industry Type	Manufacturing B	.15	.03	.11	.06	.35
	Communications C	.14	.09	.06	.08	.37
		.41	.17	.21	.21	1.00

SOLUTION

a.
$$P(B \mid F) = \frac{P(B \cap F)}{P(F)} = \frac{.11}{.21} = .524$$

Determining conditional probabilities from a probability matrix by using the formula is a relatively painless process. In this case, the joint probability, $P(B \cap F)$, appears in a cell of the matrix (.11); the marginal probability, $P(F)$, appears in a margin (.21). Bringing these two probabilities together by formula produces the answer, $.11/.21 = .524$. This answer means that 52.4% of the Midwest executives (the F values) are in manufacturing (the B values).

b.
$$P(G \mid C) = \frac{P(G \cap C)}{P(C)} = \frac{.08}{.37} = .216$$

This result means that 21.6% of the responding communications industry executives (C) are from the West (G).

c.
$$P(D \mid F) = \frac{P(D \cap F)}{P(F)} = \frac{.00}{.21} = .00$$

Because D and F are mutually exclusive, $P(D \cap F)$ is zero and so is $P(D \mid F)$. The rationale behind $P(D \mid F) = .00$ is that, if F is given (the respondent is known to be located in the Midwest), the respondent could not be located in D (the Northeast).

Independent Events

> To determine whether *X* and *Y* are independent events, the following definition can be used.
>
> $$P(X \mid Y) = P(X) \quad \text{and} \quad P(Y \mid X) = P(Y)$$

It does not matter that *X* or *Y* is given in either case, because *X* and *Y* are *independent*. In these cases, the conditional probability is solved as a marginal probability.

Sometimes, it is important to test a contingency table of raw data to determine whether events are independent. If *any* combination of two events from the different sides of the matrix fail the test, $P(X \mid Y) = P(X)$, the matrix does *not* contain independent events.

Test the matrix for the 200 executive responses to determine whether industry type is independent of geographic location.

**DEMONSTRATION
PROBLEM 4.10**

RAW VALUES MATRIX

		Geographic Location				
		Northeast *D*	Southeast *E*	Midwest *F*	West *G*	
	Finance *A*	24	10	8	14	56
Industry Type	Manufacturing *B*	30	6	22	12	70
	Communications *C*	28	18	12	16	74
		82	34	42	42	200

SOLUTION

Select one industry and one geographic location (say, A-Finance and G-West). Does $P(A \mid G) = P(A)$?

$$P(A \mid G) = \frac{14}{42} \quad \text{and} \quad P(A) = \frac{56}{200}$$

Does 14/42 = 56/200? No: .33 ≠ .28. Industry and geographic location are not independent because there is at least one exception to the test.

The contingency table of data shown in Table 4.6 is repeated here. Determine whether the table contains independent events.

**DEMONSTRATION
PROBLEM 4.11**

	D	*E*	
A	8	12	20
B	20	30	50
C	6	9	15
	34	51	85

SOLUTION
Check the first cell in the matrix to find whether $P(A \mid D) = P(A)$.

$$P(A \mid D) = \frac{8}{34} = .2353$$

$$P(A) = \frac{20}{85} = .2353$$

The checking process must continue until all the events are determined to be independent. In this matrix, all the possibilities check out. Thus, Table 4.6 contains independent events.

Analysis Using Excel

Excel doesn't have a statistical function or data analysis tool for computing probabilities directly. However, FAST ⟍ STAT has a macro procedure called **Probabilities** that will. It computes all four types of the probabilities discussed in Sections 4.4 through 4.7: marginal, joint, union, and conditional probabilities for two variables. The conditional probabilities can be computed for both the probability of a first variable given a second variable, $P(X \mid Y)$, and the probability of the second variable given the first, $P(Y \mid X)$.

The first step for using **Probabilities** is to enter into an Excel worksheet the table of values together with the label or identifier for each row and each column. The macro will accept either frequency values or probability values. Thus for Demonstration Problem 4.9, enter either the raw values of the first matrix or the probability values of the second matrix. Also, enter the labels for each of the three rows and each of the four columns.

The next step is to click on FAST ⟍ STAT on the menu bar. The result will be the FAST ⟍ STAT pull-down menu as previously shown in Figure 1.16. Click on the entry labeled **Probabilities** and the dialog box of Figure 4.12 will be displayed.

We have entered the frequency values for Demonstration Problem 4.1 into the worksheet. We have labels for each row for the first variable, Type of Position, and for each column for the second variable, Gender. We have dragged through these labels and values to enter the cell range into the dialog box. In addition, we have checked that we wish to compute all five of the possible types of probabilities.

Since we have asked for all five of the types of probabilities, the output will be quite extensive. In your use, you will usually be interested in only one or two of the types of probabilities so the output will be more manageable. The output for all five types for Demonstration Problem 4.1 is given in Figures 4.13 and 4.14.

Figure 4.13 first presents the input data. This allows you to determine if all of it has been entered correctly. The first output section gives three summary measures for the problem. The marginal probability values are given in the second output section. In Figure 4.13, the marginal probability results include the symbol *P*(). These represent categories that are not used. The **Probabilities** macro is set up for problems with up to six categories for the first variable and up to six for the second variable. However, Demonstration Problem 4.1 only has four categories for the first variable and two categories for the second variable. Accordingly, there are two *P*() symbols for the first variable and four for the second.

The third output section of Figure 4.13 provides the joint probability values. As shown in Figure 4.14, the remainder of the outputs includes a section with the union probability values and two sections with the conditional probability values.

Figure 4.12 FAST\STAT dialog box for Probabilities

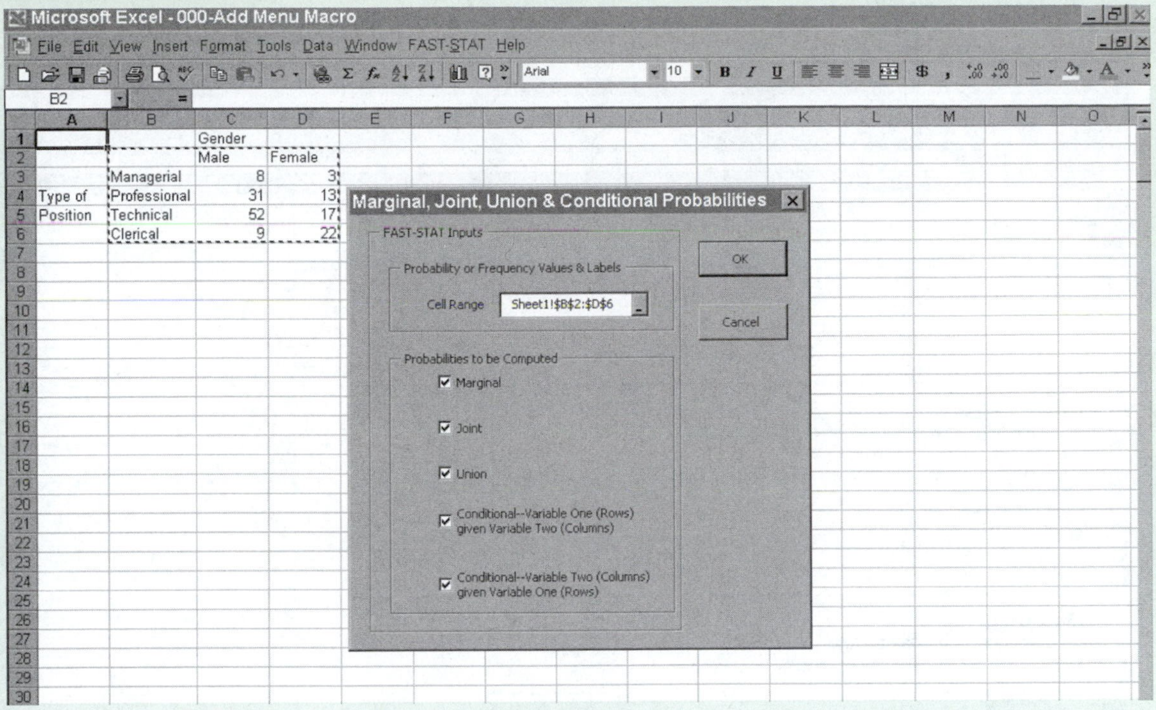

Figure 4.13 FAST\STAT output for demonstration Problem 4.1—Part 1

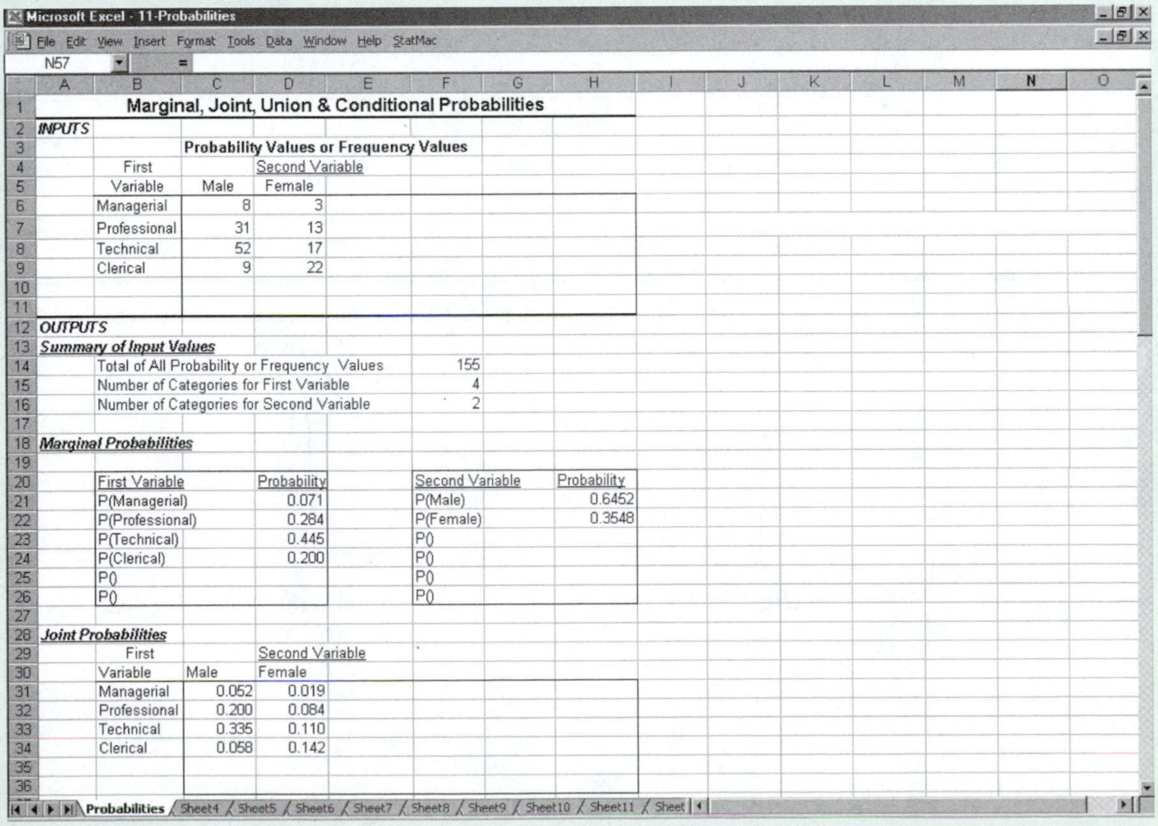

Marginal, Joint, Union & Conditional Probabilities

INPUTS

Probability Values or Frequency Values

First Variable	Second Variable: Male	Female
Managerial	8	3
Professional	31	13
Technical	52	17
Clerical	9	22

OUTPUTS

Summary of Input Values

Total of All Probability or Frequency Values	155
Number of Categories for First Variable	4
Number of Categories for Second Variable	2

Marginal Probabilities

First Variable	Probability
P(Managerial)	0.071
P(Professional)	0.284
P(Technical)	0.445
P(Clerical)	0.200
P()	
P()	

Second Variable	Probability
P(Male)	0.6452
P(Female)	0.3548
P()	
P()	
P()	
P()	

Joint Probabilities

First Variable	Second Variable: Male	Female
Managerial	0.052	0.019
Professional	0.200	0.084
Technical	0.335	0.110
Clerical	0.058	0.142

Figure 4.14	FAST STAT output for Demonstration Problem 4.1—Part 2

```
Microsoft Excel - 11-Probabilities
 File  Edit  View  Insert  Format  Tools  Data  Window  Help  StatMac
 L31              =
```

	A	B	C	D	E	F	G	H	I	J	K	L	M	N	O
37															
51	*Union Probabilities*														
52		First		Second Variable											
53		Variable	Male	Female											
54		Managerial	0.665	0.406											
55		Professional	0.729	0.555											
56		Technical	0.755	0.690											
57		Clerical	0.787	0.413											
58															
59															
60															
61	*Conditional Probabilities*														
62		*Probability of FIRST Variable given SECOND Variable*													
63		First		Second Variable											
64		Variable	Male	Female											
65		Managerial	0.080	0.055											
66		Professional	0.310	0.236											
67		Technical	0.520	0.309											
68		Clerical	0.090	0.400											
69															
70															
71															
72	*Conditional Probabilities*														
73		*Probability of SECOND Variable given FIRST Variable*													
74		First		Second Variable											
75		Variable	Male	Female											
76		Managerial	0.727	0.273											
77		Professional	0.705	0.295											
78		Technical	0.754	0.246											
79		Clerical	0.290	0.710											
80															
81															
82															
83															
84															

Probabilities / Sheet4 / Sheet5 / Sheet6 / Sheet7 / Sheet8 / Sheet9 / Sheet10 / Sheet11 / Sheet

4.7
Problems

4.23 Use the values in the contingency table to solve the equations given.

	E	F	G
A	15	12	8
B	11	17	19
C	21	32	27
D	18	13	12

a. $P(G \mid A) =$ ___
b. $P(B \mid F) =$ ___
c. $P(C \mid E) =$ ___
d. $P(E \mid G) =$ ___

4.24 Use the values in the probability matrix to solve the equations given.

	C	D
A	.36	.44
B	.11	.09

a. $P(C \mid A) =$ ___
b. $P(B \mid D) =$ ___
c. $P(A \mid B) =$ ___

4.25   The results of a survey asking, "Do you have a calculator and/or a computer in your home?" follow.

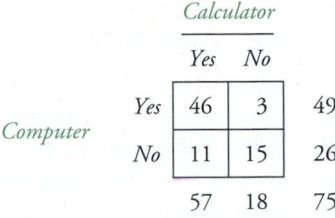

Is the variable "calculator" independent of the variable "computer"? Why or why not?

4.26   In 1997, there were 83,384 business failures in the United States, according to Dun & Bradstreet. The construction industry accounted for 10,867 of these. The South Atlantic states accounted for 8010 of the business failures. Suppose that 1258 of all business failures were construction businesses located in the South Atlantic states. A failed business from 1997 is randomly sampled.
  a. What is the probability that the business is located in the South Atlantic states?
  b. What is the probability that the business is in the construction industry or located in the South Atlantic states?
  c. What is the probability that the business is in the construction industry if it is known that the business is located in the South Atlantic states?
  d. What is the probability that the business is located in the South Atlantic states if it is known that the business is a construction business?
  e. What is the probability that the business is not located in the South Atlantic states if it is known that the business is not a construction business?
  f. Given that the business is a construction business, what is the probability that the business is not located in the South Atlantic states?

4.27   Arthur Andersen Enterprise Group/National Small Business United, Washington, conducted a national survey of small-business owners to determine the challenges for growth for their businesses. The top challenge, selected by 46% of the small-business owners, was the economy. A close second was finding qualified workers (37%). Suppose 15% of the small-business owners selected both the economy and finding qualified workers as challenges for growth. A small-business owner is randomly selected.
  a. What is the probability that the owner believes the economy is a challenge for growth if he believes that finding qualified workers is a challenge for growth?
  b. What is the probability that the owner believes that finding qualified workers is a challenge for growth if he believes that the economy is a challenge for growth?
  c. Given that the owner does not select the economy as a challenge for growth, what is the probability that he believes that finding qualified workers is a challenge for growth?
  d. What is the probability that the owner believes neither that the economy is a challenge for growth nor that finding qualified workers is a challenge for growth?

4.28   Late in 1998, a study of online users was conducted by Jupiter Communications to determine for which type of purchase a consumer prefers live customer service. Forty-seven percent of the users replied that when purchasing airline tickets, they prefer live customer service. Suppose that of those who prefer live customer service for purchasing airline tickets, 81% prefer live customer service for transacting loans. If an online user is randomly selected, what is the probability that she:
  a. prefers live customer service for both purchasing airline tickets and for transacting loans.

b. does not prefer live customer service for transacting loans given that she does prefer live customer service for purchasing airline tickets.

c. does not prefer live customer service for transacting loans and does prefer live customer service for purchasing airline tickets.

4.29 *Accounting Today* reported that 37% of accountants purchase their computer hardware by mail order direct and that 54% purchase their computer software by mail order direct. Suppose that 92% of the accountants who purchase their computer hardware by mail order direct purchase their computer software by mail order direct. If an accountant is randomly selected, what is the probability that he:

a. does not purchase his computer software by mail order direct given that he does purchase his computer hardware by mail order direct?

b. does purchase his computer software by mail order direct given that he does not purchase his computer hardware by mail order direct?

c. does not purchase his computer hardware by mail order direct if it is known that he does purchase his computer software by mail order direct?

d. does not purchase his computer hardware by mail order direct if it is known that he does not purchase his computer software by mail order direct?

4.8
Revision of Probabilities: Bayes' Rule

An extension to the conditional law of probabilities is Bayes' rule, which was developed by and named for Thomas Bayes (1702–1761). **Bayes' rule** is a formula that *extends the use of the law of conditional probabilities to allow revision of original probabilities* with new information.

BAYES' RULE	
	$$P(X_i|Y) = \frac{P(X_i) \cdot P(Y|X_i)}{P(X_1) \cdot P(Y|X_1) + P(X_2) \cdot P(Y|X_2) + \cdots + P(X_n) \cdot P(Y|X_n)}$$

Bayes' rule
An extension of the conditional law of probabilities discovered by Thomas Bayes that can be used to revise probabilities.

Recall that the law of conditional probability for

$$P(X_i|Y)$$

is

$$P(X_i|Y) = \frac{P(X_i) \cdot P(Y|X_i)}{P(Y)}.$$

Compare Bayes' rule to this law of conditional probability. The numerators of Bayes' rule and the law of conditional probability are the same, the intersection of X_i and Y shown in the form of the general rule of multiplication. The new feature that Bayes' rule uses is found in the denominator of the rule:

$$P(X_1) \cdot P(Y|X_1) + P(X_2) \cdot P(Y|X_2) + \cdots + P(X_n) \cdot P(Y|X_n).$$

The denominator of Bayes' rule includes a product expression (intersection) for every partition in the sample space, Y, including the event (X_i) itself. The denominator is thus a collective exhaustive listing of mutually exclusive outcomes of Y. This denominator is sometimes referred to as the "total probability formula." It represents a weighted average of the conditional probabilities, with the weights being the prior probabilities of the corresponding event.

By expressing the law of conditional probabilities in this new way, Bayes' rule enables the statistician to make new and different applications using conditional probabilities. In particular, statisticians use Bayes' rule to "revise" probabilities in light of new information.

A particular type of printer ribbon is produced by only two companies, Alamo Ribbon Company and South Jersey Products. Suppose Alamo produces 65% of the ribbons and that South Jersey produces 35%. Eight percent of the ribbons produced by Alamo are defective and 12% of the South Jersey ribbons are defective. A customer purchases a new ribbon. What is the probability that Alamo produced the ribbon? What is the probability that South Jersey produced the ribbon? The ribbon is tested, and it is defective. Now what is the probability that Alamo produced the ribbon? That South Jersey produced the ribbon?

The probability was .65 that the ribbon came from Alamo and .35 that it came from South Jersey. These are called *prior probabilities* because they are based on the original information.

The new information that the ribbon is defective changes the probabilities because one company produces a higher percentage of defective ribbons than the other company does. How can this information be used to update or revise the original probabilities? Bayes' rule allows such updating. One way to lay out a revision of probabilities problem is to use a table. Table 4.8 shows the analysis for the ribbon problem.

The process begins with the prior probabilities: .65 Alamo and .35 South Jersey. These prior probabilities appear in the second column of Table 4.8. Because the product is found to be defective, the conditional probabilities, $P(\text{defective} \mid \text{Alamo})$ and $P(\text{defective} \mid \text{South Jersey})$ should be used. Eight percent of Alamo's ribbons are defective: $P(\text{defective} \mid \text{Alamo}) = .08$. Twelve percent of South Jersey's ribbons are defective: $P(\text{defective} \mid \text{South Jersey}) = .12$. These two conditional probabilities appear in the third column. Eight percent of Alamo's 65% of the ribbons are defective: $(.08)(.65) = .052$, or 5.2% of the total. This figure appears in the fourth column of Table 4.8; it is the joint probability of getting a ribbon that was made by Alamo *and* is defective. Because the purchased ribbon is defective, these are the only Alamo ribbons of interest. Twelve percent of South Jersey's 35% of the ribbons are defective. Multiplying these two percentages yields the joint probability of getting a South Jersey ribbon that is defective. This figure also appears in the fourth column of Table 4.8: $(.12)(.35) = .042$. That is, 4.2% of all ribbons are made by South Jersey *and* are defective. These are the only South Jersey ribbons of interest because the ribbon purchased is defective.

Column 4 is totaled to get .094, indicating that 9.4% of all ribbons are defective (Alamo *and* defective = .052 + South Jersey *and* defective = .042). The other 90.6% of the ribbons, which are acceptable, are not of interest because the ribbon purchased is defective. To compute the fifth column, the *posterior or revised probabilities,* involves dividing each value in column 4 by the total of column 4. That is, for Alamo, .052 of the total ribbons are Alamo *and* defective out of the total of .094 that are defective. Dividing .052 by .094 yields .553 as a revised probability that the purchased ribbon was made by

EVENT	PRIOR PROBABILITY $P(E_i)$	CONDITIONAL PROBABILITY $P(d \mid E_i)$	JOINT PROBABILITY $P(E_i \cap d)$	POSTERIOR OR REVISED PROBABILITY
Alamo	.65	.08	.052	$\frac{.052}{.094} = .553$
South Jersey	.35	.12	.042	$\frac{.042}{.094} = .447$
			$P(\text{defective}) = .094$	

TABLE 4.8
Bayesian Table for Revision of Ribbon Problem Probabilities

Figure 4.15

Tree diagram for ribbon problem probabilities

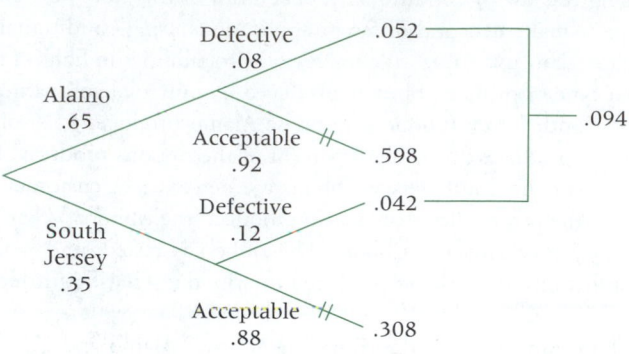

Alamo. This probability is lower than the prior or original probability of .65 because fewer of Alamo's ribbons (as a percentage) are defective than those produced by South Jersey. The defective ribbon is now less likely to have come from Alamo than before the knowledge of the defective ribbon. South Jersey's probability is revised by dividing the .042 joint probability of the ribbon being made by South Jersey *and* defective by the total probability of the ribbon being defective (.094). The result is .042/.094 = .447. The probability that the defective ribbon is from South Jersey has increased because a higher percentage of South Jersey ribbons are defective.

Tree diagrams are another common way to solve Bayes' rule problems. Figure 4.15 shows the solution for the ribbon problem. Note that the tree diagram contains all possibilities, including both defective and acceptable ribbons. When new information is given, only the pertinent branches are selected and used. The joint probability values at the end of the appropriate branches are used to revise and compute the posterior possibilities. Using the total number of defective ribbons = .052 + .042 = .094, the calculation is as follows.

$$\text{Revised Probability: Alamo} = \frac{.052}{.094} = .553$$

$$\text{Revised Probability: South Jersey} = \frac{.042}{.094} = .447$$

DEMONSTRATION PROBLEM 4.12

Machines A, B, and C all produce the same two parts, X and Y. Of all the parts produced, machine A produces 60%, machine B produces 30%, and machine C produces 10%. In addition,

40% of the parts made by machine A are part X.

50% of the parts made by machine B are part X.

70% of the parts made by machine C are part X.

A part produced by this company is randomly sampled and is determined to be an X part. With the knowledge that it is an X part, revise the probabilities that the part came from machine A, B, or C.

SOLUTION

The prior probability of the part coming from machine A is .60, because machine A produces 60% of all parts. The prior probability is .30 that the part came from B and .10 that it came from C. These prior probabilities are more pertinent if nothing is known about the part. However, the part is known to be an X part. The conditional probabilities show that different machines produce different proportions of X parts. For example, .40 of the parts made by machine A are X parts, but .50 of the parts made by machine B and .70 of the parts made by machine C are X parts. It makes sense that the probability of the part coming from machine C would increase and that the probability that the part was made on machine A would decrease because the part is an X part.

The following table shows how the prior probabilities, conditional probabilities, joint probabilities, and marginal probability, $P(X)$, can be used to revise the prior probabilities to obtain posterior probabilities.

EVENT	PRIOR $P(E_i)$	CONDITIONAL $P(E_i)$ $P(E_i)$	JOINT $P(X \cap E_i)$	POSTERIOR
A	.60	.40	$(.60)(.40) = .24$	$\dfrac{.24}{.46} = .52$
B	.30	.50	.15	$\dfrac{.15}{.46} = .33$
C	.10	.70	.07	$\dfrac{.07}{.46} = .15$
			$\overline{P(X) = .46}$	

After the probabilities have been revised, it is apparent that the probability of the part being made at machine A has decreased and that the probabilities that the part was made at machines B and C have increased. A tree diagram presents another view of this problem.

Revised probabilities: Machine A $\dfrac{.24}{.46} = .52$

Machine B $\dfrac{.15}{.46} = .33$

Machine C $\dfrac{.07}{.46} = .15$

4.8
Problems

4.30 In a manufacturing plant, machine A produces 10% of a certain product, machine B produces 40% of this product, and machine C produces 50% of this product. Five percent of machine A products are defective, 12% of machine B products are defective, and 8% of machine C products are defective. The company inspector has just sampled a product from this plant and has found it to be defective. Determine the revised probabilities that the sampled product was produced by machine A, machine B, or machine C.

4.31 Alex, Alicia, and Juan fill orders in a fast-food restaurant. Alex incorrectly fills 20% of the orders he takes. Alicia incorrectly fills 12% of the orders she takes. Juan incorrectly fills 5% of the orders he takes. Alex fills 30% of all orders, Alicia fills 45% of all orders, and Juan fills 25% of all orders. An order has just been filled.

a. What is the probability that Alicia filled the order?

b. If the order was filled by Juan, what is the probability that it was filled correctly?

c. Who filled the order is unknown, but the order was filled incorrectly. What are the revised probabilities that Alex, Alicia, or Juan filled the order?

d. Who filled the order is unknown, but the order was filled correctly. What are the revised probabilities that Alex, Alicia, or Juan filled the order?

4.32 In a small town, two lawn companies fertilize lawns during the summer. Tri-State Lawn Service has 72% of the market. Thirty percent of the lawns fertilized by Tri-State could be rated as very healthy one month after service. Greenchem has the other 28% of the market. Twenty percent of the lawns fertilized by Greenchem could be rated as very healthy one month after service. A lawn that has been treated with fertilizer by one of these companies within the last month is selected randomly. If the lawn is rated as very healthy one month after service, what are the revised probabilities that Tri- State or Greenchem treated the lawn?

4.33 Companies have many different reasons for giving training to employees. Among them are employee loyalty, employee retention, and quality of employee work. Suppose 65% of all companies give some training to their employees but that this figure varies by company size. Suppose further that 18% of all companies using training are small companies and that 75% of all companies that do not use training are small companies. A company is randomly sampled without regard to size. What is the probability that the company uses training? Suppose it is determined that the selected company is not a small company. What is the revised probability that the company uses training? What proportion of all companies is not small?

Summary

The study of probability addresses ways of assigning probabilities, types of probabilities, and laws of probabilities. Probabilities undergird the notion of inferential statistics. Using sample data to estimate and test hypotheses about population parameters is done with uncertainty. If samples are taken at random, probabilities can be assigned to outcomes of the inferential process.

Three methods of assigning probabilities are (1) the classical method, (2) the relative frequency of occurrence method, and (3) subjective probabilities. The classical method can assign probabilities *a priori,* or before the experiment takes place. It relies on the laws and rules of probability. The relative frequency of occurrence method assigns probabilities based on historical data or empirically derived data. Subjective probabilities are based on the feelings, knowledge, and experience of the person determining the probability.

Certain special types of events necessitate amendments to some of the laws of probability: mutually exclusive events and independent events. Mutually exclusive events are events that cannot occur at the same time, so the probability of their intersection is zero. In determining the union of two mutually exclusive events, the law of addition is amended by the deletion of the intersection. Independent events are events by which the occurrence of one has no impact or influence on the occurrence of the other. Certain experiments, such as those involving coins or dice, naturally produce independent events. Other experiments produce independent events when the experiment is conducted *with replacement.* If events are independent, the joint probability is computed by multiplying the individual probabilities, which is a special case of the law of multiplication.

Three techniques for counting the possibilities in an experiment are the *mn* counting rule, the N^n possibilities, and combinations. The *mn* counting rule is used to determine how many total possible ways an experiment can occur when there is a series of sequential operations. The N^n formula is applied when sampling is being done with replacement or events are independent. Combinations are used to determine the possibilities when sampling is being done without replacement.

Four types of probability are marginal probability, joint probability, conditional probability, and union probability. The general law of addition is used to compute the probability of a union. The general law of multiplication is used to compute joint probabilities. The conditional law is used to compute conditional probabilities.

Bayes' rule is a method that can be used to revise probabilities when new information becomes available; it is a variation of the conditional law. Bayes' rule takes prior probabilities of events occurring and adjusts or revises those probabilities on the basis of information about what subsequently occurs.

Key Terms

a priori
Bayes' rule
classical method of assigning probabilities
collectively exhaustive events
combinations
complement of a union
complementary events
conditional probability
elementary events
event
experiment
independent events
intersection
joint probability
marginal probability
mn counting rule
mutually exclusive events
probability matrix
relative frequency of occurrence
sample space
set notation
subjective probability
union
union probability

SUPPLEMENTARY PROBLEMS

4.34 Use the values in the contingency table to solve the equations given.

		Variable 1	
		D	E
Variable 2	A	10	20
	B	15	5
	C	30	15

a. $P(E) = $ ___
b. $P(B \cup D) = $ ___
c. $P(A \cap E) = $ ___
d. $P(B \mid E) = $ ___
e. $P(A \cup B) = $ ___
f. $P(B \cap C) = $ ___
g. $P(D \mid C) = $ ___
h. $P(A \mid B) = $ ___
i. Are variables 1 and 2 independent? Why or why not?

4.35 Use the values in the contingency table to solve the equations given.

	D	E	F	G
A	3	9	7	12
B	8	4	6	4
C	10	5	3	7

a. $P(F \cap A) = $ ___
b. $P(A \mid B) = $ ___
c. $P(B) = $ ___
d. $P(E \cap F) = $ ___
e. $P(D \mid B) = $ ___
f. $P(B \mid D) = $ ___
g. $P(D \cup C) = $ ___
h. $P(F) = $ ___

4.36 The following probability matrix contains a breakdown on the age and gender of U.S. physicians in a recent year, as reported by the American Medical Association.

U.S. PHYSICIANS IN A RECENT YEAR

		Age (years)					
		<35	35–44	45–54	55–64	>65	
Gender	Male	.11	.20	.19	.12	.16	.78
	Female	.07	.08	.04	.02	.01	.22
		.18	.28	.23	.14	.17	1.00

a. What is the probability that one randomly selected physician is 35–44 years old?
b. What is the probability that one randomly selected physician is both a woman and 45–54 years old?
c. What is the probability that one randomly selected physician is a man or is 35–44 years old?

d. What is the probability that one randomly selected physician is less than 35 years old or 55–64 years old?

e. What is the probability that one randomly selected physician is a woman if she is 45–54 years old?

f. What is the probability that a randomly selected physician is neither a woman nor 55–64 years old?

4.37 *Purchasing Survey* asked purchasing professionals what sales traits impressed them the most in a sales representative. Seventy-eight percent selected "thoroughness." Forty percent responded "knowledge of your own product." The purchasing professionals were allowed to list more than one trait. Suppose 27% of the purchasing professionals listed both "thoroughness" and "knowledge of your own product" as sales traits that impressed them the most. A purchasing professional is randomly sampled.

a. What is the probability that the professional selected "thoroughness" or "knowledge of your own product"?

b. What is the probability that the professional selected neither "thoroughness" nor "knowledge of your own product"?

c. If it is known that the professional selected "thoroughness," what is the probability that the professional selected "knowledge of your own product"?

d. What is the probability that the professional did not select "thoroughness" and did select "knowledge of your own product"?

4.38 The U.S. Bureau of Labor Statistics publishes data on the benefits offered by small companies to their employees. Only 42% offer retirement plans, while 61% offer life insurance. Suppose 33% offer both retirement plans and life insurance as benefits. If a small company is randomly selected, what is the probability that:

a. The company offers a retirement plan given that they offer life insurance?

b. The company offers life insurance given that they offer a retirement plan?

c. The company offers life insurance or a retirement plan?

d. The company offers a retirement plan and does not offer life insurance?

e. The company does not offer life insurance if it is known that they offer a retirement plan?

4.39 According to Link Resources, 16% of the U.S. population is technology-driven. However, these figures vary by region. For example, in the West the figure is 20% and in the Northeast the figure is 17%. Twenty-one percent of the U.S. population in general is in the West and 20% of the U.S. population is in the Northeast. Suppose an American is chosen randomly.

a. What is the probability that the person lives in the West and is a technology-driven person?

b. What is the probability that the person lives in the Northeast and is a technology-driven person?

c. Suppose the chosen person is known to be technology-driven. What is the probability that the person lives in the West?

d. Suppose the chosen person is known *not* to be technology-driven. What is the probability that the person lives in the Northeast?

e. Suppose the chosen person is known to be technology-driven. What is the probability that the person lives in neither the West nor the Northeast?

4.40 In a certain city, 30% of the families have a MasterCard, 20% have an American Express card, and 25% have a Visa card. Eight percent of the families have both a MasterCard and an American Express card. Twelve percent have both a Visa card and a MasterCard. Six percent have both an American Express card and a Visa card.

a. What is the probability of selecting a family that has either a Visa card or an American Express card?

b. If a family has a MasterCard, what is the probability that it has a Visa card?

c. If a family has a Visa card, what is the probability that it has a MasterCard?

d. Is possession of a Visa card independent of possession of a MasterCard? Why or why not?

e. Is possession of an American Express card mutually exclusive of possession of a Visa card?

4.41 A few years ago, a survey commissioned by *The World Almanac* and Maturity News Service reported that 51% of the respondents did not believe the Social Security system will be secure in 20 years. Of the respondents who were age 45 or older, 70% believed the system will be secure in 20 years. Of the people surveyed, 57% were under age 45. One respondent is selected randomly.

a. What is the probability that the person is age 45 or older?

b. What is the probability that the person is younger than age 45 and believes that the Social Security system will be secure in 20 years?

c. If the person selected believes the Social Security system will be secure in 20 years, what is the probability that the person is 45 years old or older?

d. What is the probability that the person is younger than age 45 or believes the Social Security system will not be secure in 20 years?

4.42 A telephone survey conducted by the Maritz Marketing Research company found that 43% of Americans expect to save more money next year than they saved last year. Forty-five percent of those surveyed plan to reduce debt next year. Of those who expect to save more money next year, 81% plan to reduce debt next year. An American is selected randomly.

a. What is the probability that this person expects to save more money next year and plans to reduce debt next year?

b. What is the probability that this person expects to save more money next year or plans to reduce debt next year?

c. What is the probability that this person neither expects to save more money next year nor plans to reduce debt next year?

d. What is the probability that this person expects to save more money next year and does not plan to reduce debt next year?

4.43 The Steelcase Workplace Index studied the types of work-related activities that Americans did while on vacation in the summer. Among other things, 40% read work-related material. Thirty-four percent checked in with the boss. Respondents to the study were allowed to select more than one activity. Suppose that of those who read work-related material, 78% checked in with the boss. One of these survey respondents is selected randomly. What is the probability that while on vacation this respondent:

a. Checked in with the boss and read work-related material?

b. Neither read work-related material nor checked in with the boss?

c. Read work-related material given that he checked in with the boss?

d. Did not check in with the boss given that he read work-related material?

e. Did not check in with the boss given that he did not read work-related material?

f. Construct a probability matrix for this problem.

4.44 Health Rights Hotline published the results of a survey of 2400 people in Northern California in which consumers were asked to share their complaints about managed care. The number one complaint was denial of care, with 17% of the participating consumers selecting it. Several other complaints were noted including inappropriate care (14%), customer service (14%), payment disputes (11%), specialty care (10%), delays in getting care (8%), and prescription drugs (7%). These complaint categories are mutually exclusive. Assume that the results of this survey can be inferred to all managed care consumers. If a managed care consumer is randomly selected, what is the probability that:

a. The consumer complains about payment disputes or specialty care?

b. The consumer complains about prescription drugs and customer service?

c. The consumer complains about inappropriate care given that she complains about specialty care?

d. The consumer does not complain about delays in getting care nor does she complain about payment disputes?

4.45 Companies use employee training for various reasons including employee loyalty, certification, quality, and process improvement. In a national survey of companies, BI Learning Systems reported that 56% percent of the responding companies named employee retention as a top reason for training. Suppose 36% of the companies replied that they use training for process improvement and for employee retention. In addition, suppose that of the companies that use training for process improvement, 90% use training for employee retention. A company that uses training is randomly selected.

a. What is the probability that the company uses training for employee retention and not for process improvement?

b. If it is known that the company uses training for employee retention, what is the probability that it uses training for process improvement?

c. What is the probability that the company uses training for process improvement?

d. What is the probability that the company uses training for employee retention or process improvement?

e. What is the probability that the company neither uses training for employee retention nor uses training for process improvement?

f. Suppose it is known that the company does not use training for process improvement. What is the probability that the company does use training for employee retention?

4.46 Pitney Bowes surveyed 302 directors and vice presidents of marketing at large and midsize U.S. companies to determine what they believe is the best vehicle for educating decision makers on complex issues in selling products and services. The highest percentage of companies chose direct mail/catalogs, followed by direct sales/sales rep. Direct mail/catalogs was selected by 38% of the companies. None of the companies selected both direct mail/catalogs and direct sales/sales rep. Suppose also that 41% selected neither direct mail/catalogs nor direct sales/sales rep. If one of these companies is selected randomly and their top marketing person interviewed about this matter, what is the probability that she:

  a. Selected direct mail/catalogs and did not select direct sales/sales rep?
  b. Selected direct sales/sales rep?
  c. Selected direct sales/sales rep given that she selected direct mail/catalogs?
  d. Did not select direct mail/catalogs given that she did not select direct sales/sales rep?

4.47 A small independent physicians' practice has three doctors. Doctor Sarabia sees 41% of the patients, Doctor Tran sees 32%, and Doctor Jackson sees the rest. Doctor Sarabia requests blood tests on 5% of her patients, Doctor Tran requests blood tests on 8% of his patients, and Doctor Jackson requests blood tests on 6% of her patients. An auditor randomly selects a patient from the past week and discovers that the patient has had a blood test as a result of his physician visit. Knowing this, what is the probability that the patient saw Doctor Sarabia? For what percentage of all patients at this practice are blood tests requested?

4.48 A survey by the Arthur Anderson Enterprise Group/National Small Business United attempted to determine what the leading challenges are for the growth and survival of small businesses. While the economy and finding qualified workers were the leading challenges, there were several others listed in the results of the study. Among those are regulations, listed by 30% of the companies, and the tax burden, listed by 35%. Suppose that 71% of the companies listing regulations as a challenge listed the tax burden as a challenge. Assume these percentages hold for all small businesses. If a small business is randomly selected, what is the probability that it:

  a. Lists both the tax burden and regulations as a challenge?
  b. Lists either the tax burden or regulations as a challenge?
  c. Lists either the tax burden or regulations but not both as a challenge?
  d. Lists regulations as a challenge given that it lists the tax burden as a challenge?
  e. Does not list regulations as a challenge given that it lists the tax burden as a challenge?
  f. Does not list regulations as a challenge given that it does not list the tax burden as a challenge?

4.49 According to the Public Voice for Food and Health Policy, approximately 27% of all soup products in a recent year did not carry nutritional labeling. Approximately 83% of breakfast meats and about 59% of hot dog products did not have nutritional labeling. Assume that if these three groups of foods were combined, 60% would be soup products, 35% would be breakfast meats, and 5% would be hot dogs. A researcher is blindly given a food product from one of these three groups and is told that the product *does* have nutritional labeling. Revise the probabilities that the product is a soup product, a breakfast meat, and a hot dog product.

4.50 A survey conducted for Lifetime's daily half-hour series "The Great American TV Poll" asked Americans what they consider to be the most important thing in their lives. Twenty-nine percent said "good health," 21% responded "a happy marriage," and 40% replied "faith in God." Because they were asked which of these things is *the most* important thing, a respondent could not select more than one answer.

  a. What is the probability that a person replied "a happy marriage," or "faith in God"?
  b. What is the probability that a person replied "a happy marriage" or "faith in God" or "good health"?
  c. What is the probability that a person replied "faith in God" and "good health"?
  d. What is the probability that a person replied neither "faith in God" nor "good health" nor "a happy marriage"?

ANALYZING THE DATABASES

1. In the manufacturing database, what is the probability that a randomly selected SIC Code industry is in industry group 13? What is the probability that a randomly selected SIC code industry has a value of industry shipments of 4? What is the probability that a randomly selected SIC Code industry is in industry group 13 and has a value of industry shipments of 2? What is the probability that a randomly selected SIC Code industry is in industry group 13 or has a value of industry shipments of 2? What is the probability that a randomly selected SIC code industry neither is in industry group 13 nor has a value of industry shipments of 2?

2. Use the hospital database. Construct a raw values matrix for region and for type of control. You should have a 7 × 4 matrix. Using this matrix, answer the following questions. (Refer to Chapter 1 for category members.) What is the probability that a randomly selected hospital is in the Midwest if the hospital is known to be for-profit? If the hospital is known to be in the South, what is the probability that it is a government, nonfederal hospital? What is the probability that a hospital is in the Rocky Mountain region or a not-for-profit, non-government hospital? What is the probability that a hospital is a for-profit hospital located in California?

COLGATE-PALMOLIVE MAKES A "TOTAL" EFFORT

In the mid-1990s, Colgate-Palmolive had developed a new toothpaste for the U.S. market, Colgate Total, with an antibacterial ingredient that was already being successfully sold overseas. However, the word *antibacterial* was not allowed for such products by the Food & Drug Administration rules. So Colgate-Palmolive had to come up with another way of marketing this and other features of their new toothpaste to U.S. consumers. Market researchers told Colgate-Palmolive that consumers were weary of trying to discern among the different advantages of various toothpaste brands and wanted simplification in their shopping lives. In response, the name "Total" was given to the product in the United States: The one word would convey that the toothpaste is the "total" package of various benefits.

Young & Rubicam developed several commercials illustrating Total's benefits and tested the commercials with focus groups. One commercial touting Total's long-lasting benefits was particularly successful. Meanwhile, in 1997, Colgate-Palmolive received FDA approval for Total, five years after the company had applied for it. The product was launched in the United States in January of 1998 using commercials that were designed from the more successful ideas of the focus group tests. A print campaign followed.

Within three months, Colgate-Palmolive grabbed the number one market share for toothpaste. Ten months later, 21% of all U.S. households had purchased Total for the first time. During this same time period, 43% of those who initially tried Total purchased it again. Colgate Total had been successfully introduced into the U.S. market.

DISCUSSION

1. What probabilities are given in this case? Use these probabilities and the probability laws to determine what percentage of U.S. households purchased Total at least twice in the first 10 months of its release.

2. Is age category independent of willingness to try new products? According to the U.S. Bureau of the Census, approximately 20% of all Americans are in the 45–64 age category. Suppose 24% of the consumers who purchased Total for the first time during the initial 10-month period were in the 45–64 age category. Use this information to determine whether age is independent of the initial purchase of Total during the introductory time period. Explain your answer.

3. Using the probabilities given in Question 2, calculate the probability that a randomly selected U.S. consumer is either in the 45–64 age category or purchased Total during the initial 10-month period. What is the probability that a randomly selected person purchased Total in the first 10 months given that they are in the 45–64 age category?

4. Suppose 32% of all toothpaste consumers in the United States saw the Total commercials. Of those who saw the commercials, 40% purchased Total at least once in the first 10 months of its introduction. Of those who did not see the commercials, 12.06% purchased Total at least once in the first 10 months of its introduction. Suppose a toothpaste consumer is randomly selected and it is learned that they purchased Total during the first 10 months of its introduction. Revise the probability that this person saw the Total commercials and the probability that they did not see the Total commercials.

5
Discrete Probability Distributions

Learning Objectives

The overall learning objective of Chapter 5 is to help you understand a category of probability distributions that produces only discrete outcomes, thereby enabling you to:

1. Distinguish between discrete random variables and continuous random variables.

2. Identify the type of statistical experiments that can be described by the binomial distribution and know how to work such problems.

3. Decide when to use the Poisson distribution in analyzing statistical experiments and know how to work such problems.

4. Decide when to use the hypergeometric distribution and know how to work such problems.

Statistical experiments produce outcomes. In experiments of chance, outcomes occur randomly. As an example of such an experiment, suppose a battery manufacturer randomly selects three batteries from a large batch of batteries to be tested for quality. Each battery selected is to be rated as good or defective. The batteries are numbered from 1 to 3, a good battery is designated with a G, a defective battery is designated with a D. All possible outcomes are shown in Table 5.1. The expression, $G_1 D_2 G_3$, denotes one particular outcome in which the first and third batteries are good, and the second battery is defective. In this chapter, we examine the probabilities of various outcomes that can occur with particular types of experiments.

5.1
Discrete versus Continuous Distributions

Random variable
A variable that contains the outcomes of a chance experiment.

A **random variable** is *a variable that contains the outcomes of a chance experiment.* For example, suppose an experiment is to measure the arrivals of automobiles at a turnpike toll booth during a 30-second period. The possible outcomes are: 0 cars, 1 car, 2 cars, . . . , n cars. These numbers $(0, 1, 2, . . . , n)$ are the values of a random variable. Suppose another experiment is to measure the time between the completion of two tasks in a production line. The values will range from 0 seconds to n seconds. These time measurements are the values of another random variable. There are two categories of random variables: (1) discrete random variables and (2) continuous random variables.

Discrete random variables
Random variables in which the set of all possible values is at most a finite or a countably infinite number of possible values.

A random variable is a **discrete random variable** *if the set of all possible values is at most a finite or a countably infinite number of possible values.* In most statistical situations, discrete random variables produce values that are nonnegative whole numbers. For example, if six people are randomly selected from a population and how many of the six are left-handed is to be determined, the random variable produced is discrete. The only possible numbers of left-handed people in the sample of six are 0, 1, 2, 3, 4, 5, and 6. There cannot be 2.75 left-handed people in a group of six people; obtaining non-whole number values is impossible. Other examples of experiments that yield discrete random variables include the following.

1. Randomly selecting 25 people who consume soft drinks and determining how many people prefer diet soft drinks
2. Determining the number of defects in a batch of 50 items
3. Counting the number of people who arrive at a store during a 5-minute period
4. Sampling 100 registered voters and determining how many voted for the president in the last election

The battery experiment described at the beginning of the chapter produces a distribution that has discrete outcomes. In any one trial of the experiment, there will be either 0, 1, 2, or 3 defective batteries. It is not possible to get 1.58 defective batteries. It could be said that discrete random variables are usually generated from experiments in which things are "counted" not "measured."

Continuous random variables
Variables that take on values at every point over a given interval.

Continuous random variables *take on values at every point over a given interval.* Thus continuous random variables have no gaps or unassumed values. It could be said that con-

TABLE 5.1
All Possible Outcomes for the Battery Experiment

$G_1 G_2 G_3$
$D_1 G_2 G_3$
$G_1 D_2 G_3$
$G_1 G_2 D_3$
$D_1 D_2 G_3$
$D_1 G_2 D_3$
$G_1 D_2 D_3$
$D_1 D_2 D_3$

tinuous random variables are generated from experiments in which things are "measured" not "counted." For example, if a person is assembling a product component, the time it takes to accomplish that feat could be any value within a reasonable range such as 3 minutes 36.4218 seconds or 5 minutes 17.5169 seconds. A noninclusive list of measures for which continuous random variables might be generated is time, height, weight, and volume. Other examples of experiments that yield continuous random variables include the following.

1. Sampling the volume of liquid nitrogen in a storage tank
2. Measuring the time between customer arrivals at a retail outlet
3. Measuring the lengths of newly designed automobiles
4. Measuring the weight of grain in a grain elevator at different points of time

Once continuous data are measured and recorded, they become discrete data because the data are rounded off to a discrete number. Thus in actual practice, virtually all business data are discrete. However, for practical reasons, data analysis is facilitated greatly by using continuous distributions on data that were continuous originally.

The outcomes for random variables and their associated probabilities can be organized into distributions. The two types of distributions are **discrete distributions,** *constructed from discrete random variables,* and **continuous distributions,** *based on continuous random variables.* Discrete distributions include the binomial distribution, Poisson distribution, and hypergeometric distribution. Continuous distributions include the normal distribution, uniform distribution, exponential distribution, *t* distribution, chi-square distribution, and *F* distribution. In this chapter, we will explore discrete distributions. Chapter 6 addresses continuous distributions.

Discrete distributions
Distributions constructed from discrete random variables.

Continuous distributions
Distributions constructed from continuous random variables.

Perhaps the most widely known of all discrete distributions is the **binomial distribution.** The binomial distribution has been used for hundreds of years. Several assumptions underlie the use of the binomial distribution:

5.2
Binomial Distribution

The experiment involves *n* identical trials.

Each trial has only two possible outcomes denoted as success or as failure.

Each trial is independent of the previous trials.

The terms *p* and *q* remain constant throughout the experiment, where the term *p* is the probability of getting a success on any one trial and the term $q = (1 - p)$ is the probability of getting a failure on any one trial.

ASSUMPTIONS OF THE BINOMIAL DISTRIBUTION

As the word *bi*nomial indicates, any single trial of a binomial experiment contains only two possible outcomes. These two outcomes are labeled *success* or *failure.* Usually the outcome of interest to the researcher is labeled a success. For example, if a quality control analyst is looking for defective products, he would consider finding a defective product a success even though the company would not consider a defective product a success. If researchers are studying left-handedness, the outcome of getting a left-handed person in a trial of an experiment is a success. The other possible outcome of a trial in a binomial experiment is called a failure. The word *failure* is used only in opposition to success. In the preceding experiments, a failure could be to get an acceptable part (as opposed to a defective part) or to get a right-handed person (as opposed to a left-handed person). In a binomial distribution experiment there are only two possible, mutually exclusive outcomes to any one trial (right-handed/left-handed, defective/good, male/female, etc.).

Binomial distribution
Widely known discrete distribution constructed by determining the probabilities of *X* successes in *n* trials.

The binomial distribution is a discrete distribution. In n trials, only X successes are possible, where X is a whole number between 0 and n. For example, if five parts are randomly selected from a batch of parts, only 0, 1, 2, 3, 4, or 5 defective parts are possible in that sample. In a sample of five parts, getting 2.714 defective parts is not possible, nor is getting eight defective parts possible.

In a binomial experiment, the trials must be independent. This constraint means that either the experiment is by nature one that produces independent trials (such as tossing coins or rolling dice) or the experiment is conducted with replacement. The effect of the independent trial requirement is that p, the probability of getting a success on one trial, remains constant from trial to trial. For example, suppose 5% of all parts in a bin are defective. The probability of drawing a defective part on the first draw is $p = .05$. If the first part drawn is not replaced, the second draw is not independent of the first, and the p value will change for the next draw. The binomial distribution does not allow for p to change from trial to trial within an experiment. However, if the population is large in comparison with the sample size, the effect of sampling without replacement is minimal, and the independence assumption essentially is met. That is, p remains relatively constant.

Generally, if the sample size, n, is less than 5% of the population, the independence assumption is not of great concern. Therefore the acceptable sample size for using the binomial distribution with samples taken *without* replacement is

$$n < 5\% \ N$$

where:

n = sample size, and
N = population size.

For example, suppose 10% of the population of the world is left-handed and that a sample of 20 people is selected randomly from the world's population. If the first person selected is left-handed—and the sampling is conducted without replacement—the value of $p = .10$ is virtually unaffected because the population of the world is so large. In addition, with many experiments the population is continually being replenished even as the sampling is being done. This condition often is the case with quality control sampling of products from large production runs. Some examples of binomial distribution problems follow.

1. Suppose a machine producing computer chips has a 6% defective rate. If a company purchases 30 of these chips, what is the probability that none is defective?
2. One ethics study suggested that 84% of U.S. companies have an ethics code. From a random sample of 15 companies, what is the probability that at least 10 have an ethics code?
3. Suppose brand X car battery has a 35% market share. If 70 cars are selected at random, what is the probability that at least 30 cars have a brand X battery?
4. A survey found that nearly 67% of company buyers stated that their company had programs for preferred buyers. If a random sample of 50 company buyers is taken, what is the probability that 40 or more have companies with programs for preferred buyers?

Solving a Binomial Problem

A survey of relocation administrators by Runzheimer International revealed several reasons why workers reject relocation offers. Included in the list were family considerations, financial reasons, and others. Four percent of the respondents said they rejected relocation offers because they received too little relocation help. Suppose five workers who have just rejected relocation offers are randomly selected and interviewed. Assuming the 4% figure holds for all workers being offered relocation, what is the probability that the first worker interviewed rejected the offer because of too little relocation help and the next four workers rejected the offer for other reasons?

Let T represent too little relocation help and R represent other reasons. The sequence of interviews for this problem is:

T_1, R_2, R_3, R_4, R_5.

The probability of getting this sequence of workers is calculated by using the special rule of multiplication for independent events (assuming the workers are independently selected and there is a large population of workers). If 4% of the workers rejecting relocation offers do so for too little relocation help, the probability of one person being randomly selected from workers rejecting relocation offers who does so for that reason is .04, which is the value of p. The other 96% of the workers who reject relocation offers do so for other reasons. Thus the probability of randomly selecting a worker from those who reject relocation offers who does so for other reasons is $1 - .04 = .96$, which is the value for q. The probability of obtaining this sequence of five workers who have rejected relocation offers is:

$$P(T_1 \cap R_2 \cap R_3 \cap R_4 \cap R_5) = (.04)(.96)(.96)(.96)(.96) = .03397.$$

Obviously, in the random selection of workers who have rejected relocation offers, the worker who did so because of too little relocation help could have been the second worker or the third or the fourth or the fifth. All the possible sequences of getting one worker who rejected relocation because of too little help and four workers who did so for other reasons follow.

T_1, R_2, R_3, R_4, R_5
R_1, T_2, R_3, R_4, R_5
R_1, R_2, T_3, R_4, R_5
R_1, R_2, R_3, T_4, R_5
R_1, R_2, R_3, R_4, T_5

The probability of each of these sequences occurring is calculated as follows.

$(.04)(.96)(.96)(.96)(.96) = .03397$
$(.96)(.04)(.96)(.96)(.96) = .03397$
$(.96)(.96)(.04)(.96)(.96) = .03397$
$(.96)(.96)(.96)(.04)(.96) = .03397$
$(.96)(.96)(.96)(.96)(.04) = .03397$

Note that in each case the probability is the same. Each of the five sequences contains the product of .04 and four .96s. The commutative property of multiplication allows for the reordering of the five individual probabilities in any one sequence. The probabilities in each of the five sequences may be reordered and summarized as $(.04)^1(.96)^4$. Each sequence contains the same five probabilities, so there is no point in recomputing the probability of each sequence. What *is* important is to determine how many different ways the sequences can be formed and multiply that figure by the probability of one sequence occurring. For the five sequences of this problem, the total probability of getting exactly one worker who rejected relocation because of too little relocation help in a random sample of five workers who rejected relocation offers is

$$5(.04)^1(.96)^4 = .16985.$$

An easier way to determine the number of sequences than listing all possibilities is to use *combinations* to calculate them. (The concept of combinations was introduced in Chapter 4.) Five workers are being sampled, so $n = 5$, and the problem is to get one worker who rejected a relocation offer because of too little relocation help, $X = 1$. Hence $_nC_X$ will yield the number of possible ways to get X successes in n trials. For this problem, $_5C_1$ tells the number of sequences of possibilities.

$$_5C_1 = \frac{5!}{1!(5-1)!} = 5$$

Weighting the probability of one sequence with the combination yields

$$_5C_1(.04)^1(.96)^4 = .16985.$$

Using combinations simplifies the determination of how many sequences of possibilities there are for a given value of X in a binomial distribution.

Now suppose 70% of all Americans believe cleaning up the environment is an important issue. What is the probability of randomly sampling four Americans and having exactly two of them say that they believe cleaning up the environment is an important issue? Let E represent the success of getting a person who believes cleaning up the environment is an important issue. For this example, $p = .70$. Let N represent the failure of not getting a person who believes cleaning up is an important issue (N denotes not important). The probability of getting one of these persons is $q = .30$.

The various sequences of getting two Es in a sample of four follow.

E_1, E_2, N_3, N_4
E_1, N_2, E_3, N_4
E_1, N_2, N_3, E_4
N_1, E_2, E_3, N_4
N_1, E_2, N_3, E_4
N_1, N_2, E_3, E_4

Two successes in a sample of four can occur six ways. Using combinations, the number of sequences is

$$_4C_2 = 6 \text{ ways.}$$

The probability of selecting any individual sequence is

$$(.70)^2(.30)^2 = .0441.$$

Thus the overall probability of getting exactly two people who believe cleaning up the environment is important out of four randomly selected people, when 70% of Americans believe cleaning up the environment is important, is

$$_4C_2(.70)^2(.30)^2 = .2646.$$

Generalizing from these two examples yields the binomial formula, which can be used to solve binomial problems.

$$P(X) = {}_nC_x \cdot p^X \cdot q^{n-X} = \frac{n!}{X!(n-X)!} \cdot p^X \cdot q^{n-X}$$

where:
$n =$ the number of trials (or the number being sampled)
$X =$ the number of successes desired
$p =$ the probability of getting a success in one trial
$q = 1 - p =$ the probability of getting a failure in one trial

The binomial formula summarizes the steps presented so far to solve binomial problems. The formula allows the solution of these problems quickly and efficiently.

A study conducted by The Gallup Organization found that 65% of all financial consumers are very satisfied with their primary financial institution. Suppose 40 financial consumers are sampled randomly. What is the probability that exactly 23 of the 40 are very satisfied with their primary financial institution?

DEMONSTRATION PROBLEM 5.1

SOLUTION
The value of p is .65 (very satisfied), the value of $q = 1 - p = 1 - .65 = .35$ (*not* very satisfied), $n = 40$, and $X = 23$. The binomial formula yields the final answer.

$${}_{40}C_{23}(.65)^{23}(.35)^{17} = (88732378800)(.000049775)(.000000018) = .0784$$

If 65% of the financial consumers are very satisfied, about 7.84% of the time the researcher would get exactly 23 out of 40 financial consumers who are very satisfied with their financial institution. The odds are against getting 23 out of 40 financial consumers by chance who are very satisfied with their financial institution. How many very satisfied financial consumers would one expect to get in 40 randomly selected financial consumers? If 65% of the financial consumers are very satisfied with their primary financial institution, one would expect to get about 65% of 40 or $(.65)(40) = 26$ very satisfied financial consumers. In any individual sample of 40 financial consumers, the number who are very satisfied is likely to differ from 26. On average, the expected number is 26. A researcher who gets 23 very satisfied financial consumers out of 40 can view this number in light of the 26 that would be expected.

According to the U.S. Bureau of the Census, approximately 6% of all workers in Jackson, Mississippi, are unemployed. In conducting a random telephone survey in Jackson, what is the probability of getting two or fewer unemployed workers in a sample of 20?

DEMONSTRATION PROBLEM 5.2

SOLUTION
This problem must be worked as the union of three problems: (1) zero unemployed, $X = 0$; (2) one unemployed, $X = 1$; and (3) two unemployed, $X = 2$. In each problem, $p = .06$, $q = .94$, and $n = 20$. The binomial formula gives the following result.

$$\begin{array}{ccc} X = 0 & X = 1 & X = 2 \\ {}_{20}C_0(.06)^0(.94)^{20} + & {}_{20}C_1(.06)^1(.94)^{19} + & {}_{20}C_2(.06)^2(.94)^{18} = \\ .2901 \quad + & .3703 \quad + & .2246 \quad = .8850 \end{array}$$

If 6% of the workers in Jackson, Mississippi, are unemployed, the telephone surveyor would get zero, one, or two unemployed workers 88.5% of the time in a random sample of 20 workers. The requirement of getting two or fewer is satisfied by getting zero, one, or two unemployed workers. Thus this problem is the union of three probabilities. Whenever the binomial formula is used to solve for cumulative success (not an exact number), the probability of each *X* value must be solved and the probabilities summed. If an actual survey produced such a result, it would serve to validate the census figures.

Using the Binomial Table

Anyone who works enough binomial problems will begin to recognize that the probability of getting $X = 5$ successes from a sample size of $n = 30$ when $p = .10$ is the same no matter whether the five successes are left-handed people, defective parts, brand X purchasers, or any other variable. Whether the sample involves people, parts, or products does not matter in terms of the final probabilities. The essence of the problem is the same: $n = 30$, $X = 5$, and $p = .10$. Recognizing this fact, mathematicians constructed a set of binomial tables containing presolved probabilities.

Two parameters, n and p, describe or characterize a binomial distribution. Binomial distributions actually are a family of distributions. Every different value of n and/or every different value of p gives a different binomial distribution, and tables are available for various combinations of n and p values. Because of space limitations, the binomial tables presented in this text are limited. Table A.2 in Appendix A contains binomial tables. Each table is headed by a value of n. Nine values of p are presented in each table of size n. In the column below each value of p is the binomial distribution for that combination of n and p. Table 5.2 contains a segment of Table A.2; the binomial probabilities for $n = 20$.

TABLE 5.2
Excerpt from Table A.2, Appendix A

$n = 20$				PROBABILITY					
X	.1	.2	.3	.4	.5	.6	.7	.8	.9
0	.122	.012	.001	.000	.000	.000	.000	.000	.000
1	.270	.058	.007	.000	.000	.000	.000	.000	.000
2	.285	.137	.028	.003	.000	.000	.000	.000	.000
3	.190	.205	.072	.012	.001	.000	.000	.000	.000
4	.090	.218	.130	.035	.005	.000	.000	.000	.000
5	.032	.175	.179	.075	.015	.001	.000	.000	.000
6	.009	.109	.192	.124	.037	.005	.000	.000	.000
7	.002	.055	.164	.166	.074	.015	.001	.000	.000
8	.000	.022	.114	.180	.120	.035	.004	.000	.000
9	.000	.007	.065	.160	.160	.071	.012	.000	.000
10	.000	.002	.031	.117	.176	.117	.031	.002	.000
11	.000	.000	.012	.071	.160	.160	.065	.007	.000
12	.000	.000	.004	.035	.120	.180	.114	.022	.000
13	.000	.000	.001	.015	.074	.166	.164	.055	.002
14	.000	.000	.000	.005	.037	.124	.192	.109	.009
15	.000	.000	.000	.001	.015	.075	.179	.175	.032
16	.000	.000	.000	.000	.005	.035	.130	.218	.090
17	.000	.000	.000	.000	.001	.012	.072	.205	.190
18	.000	.000	.000	.000	.000	.003	.028	.137	.285
19	.000	.000	.000	.000	.000	.000	.007	.058	.270
20	.000	.000	.000	.000	.000	.000	.001	.012	.122

Solve the binomial probability for $n = 20$, $p = .40$, and $X = 10$ by using Table A.2, Appendix A.

SOLUTION

To use Table A.2, first locate the value of n. As $n = 20$ for this problem, the portion of the binomial tables containing values for $n = 20$ presented in Table 5.2 can be used. After locating the value of n, search horizontally across the top of the table for the appropriate value of p. In this problem, $p = .40$. The column under .40 contains the probabilities for the binomial distribution of $n = 20$ and $p = .40$. To get the probability of $X = 10$, find the value of X in the left-most column and locate the probability in the table at the intersection of $p = .40$ and $X = 10$. The answer is .117. Working this problem by the binomial formula yields the same result.

$$_{20}C_{10}(.40)^{10}(.60)^{10} = .1171$$

According to Information Resources, Inc., which publishes data on market share for various products, Oreos control about 10% of the market for cookie brands. Suppose 20 purchasers of cookies are selected randomly from the population. What is the probability that fewer than four purchasers choose Oreos?

SOLUTION

For this problem, $n = 20$, $p = .10$, and $X < 4$. Becaues $n = 20$, the portion of the binomial tables presented in Table 5.2 can be used to work this problem. Search along the row of p values for .10. Determining the probability of getting $X < 4$ involves summing the probabilities for $X = 0$, 1, 2, and 3. The values appear in the X column at the intersection of each X value and $p = .10$.

X VALUE	PROBABILITY
0	.122
1	.270
2	.285
3	.190
	$(X < 4) = .867$

If 10% of all cookie purchasers prefer Oreos and 20 cookie purchasers are randomly selected, about 86.7% of the time fewer than four of the 20 will select Oreos.

Analysis Using Excel

Excel can compute binomial distribution probabilities for both exact and cumulative questions. Begin by selecting the paste function designated as f_x from the Excel standard toolbar. From the left side of the paste function dialog box select **Statistical,** which will change the right side offerings. From the new right side offerings select **BINOMDIST.** The dialog box for **BINOMDIST** is shown in Figure 5.1.

To compute binomial distribution probabilities using **BINOMDIST,** enter the value of X in the first box, the value of n in the second box, and the value of p in the third box. The fourth box, **Cumulative,** allows you to determine whether you want exact probabilities (e.g., $X = 5$) or cumulative probabilities (e.g. $X \le 5$). For exact probabilities, enter **false** in the box; for cumulative probabilities, enter **true** in the box. The cumulative probabilities are always calculated from $X = 0$ through the number entered. For example, if 5 is entered in box one and true is entered in box four, Excel will yield the probability for

Figure 5.1 — Dialog box for **BINOMDIST**

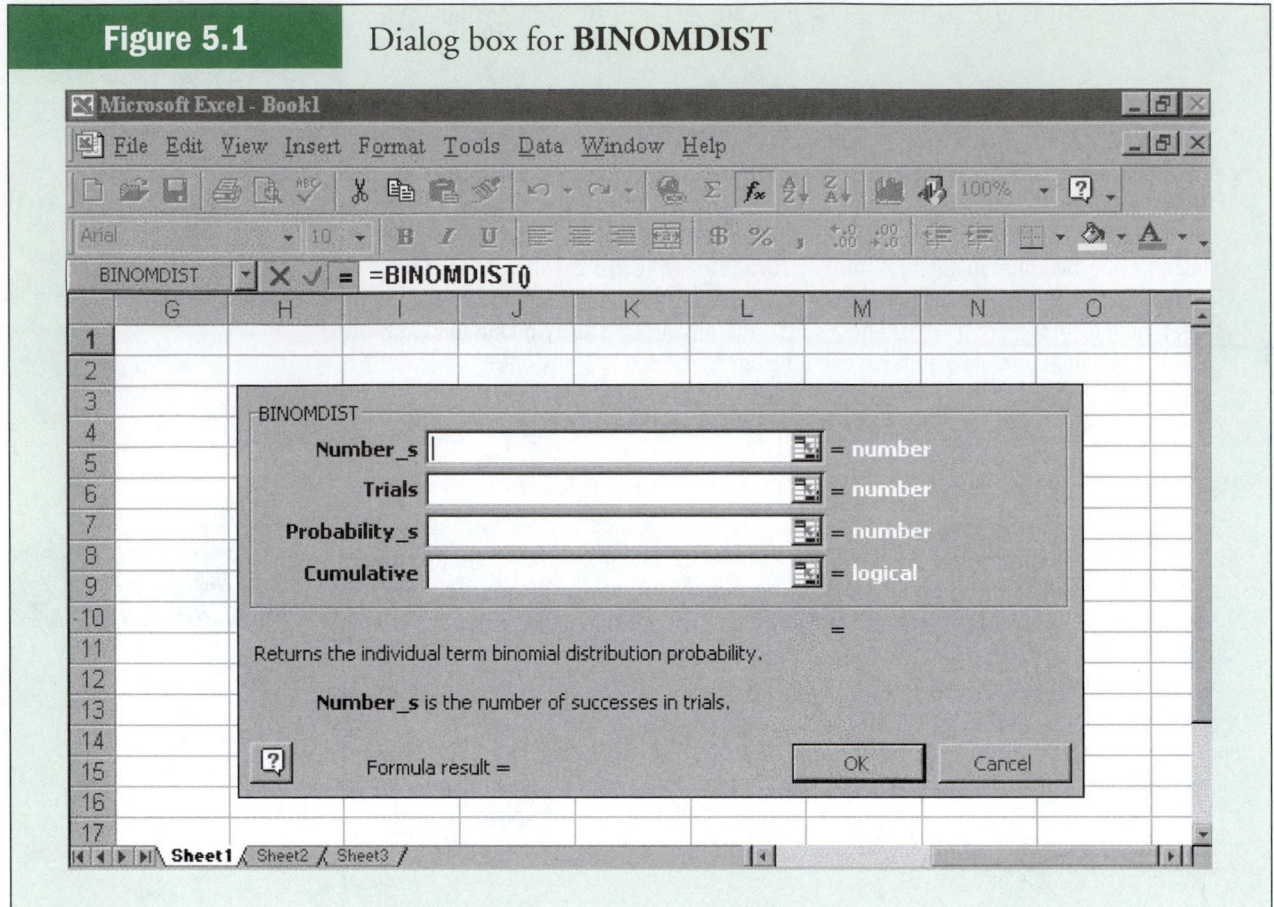

$X \leq 5$ ($X = 0, 1, 2, 3, 4,$ and 5). Suppose you want the probability of $X > 8$. By entering 8 in box one and true in box four, Excel will yield the probability for $X \leq 8$. You would then have to use Excel or manually calculate the probability of $X > 8$ by taking $1 -$ (probability of $X \leq 8$).

Excel output for Demonstration Problems 5.1 and 5.2 is shown in Figure 5.2.

Mean and Standard Deviation of a Binomial Distribution

A binomial distribution has an expected value or a long-run average, which is denoted by μ. The value of μ is determined by $n \cdot p$. For example, if $n = 10$ and $p = .4$, then $\mu = n \cdot p = (10)(.4) = 4$. The long-run average or expected value means that, if n items are sampled over and over for a long time and if p is the probability of getting a success on one trial, the average number of successes per sample is expected to be $n \cdot p$. If 40% of all graduate business students at a large university are women and if random samples of 10 graduate business students are selected many times, the expectation is that, on average, four of the 10 students would be women.

MEAN AND STANDARD DEVIATION OF A BINOMIAL DISTRIBUTION	$\mu = n \cdot p$
	$\sigma = \sqrt{n \cdot p \cdot q}$

Figure 5.2 — Excel output for Demonstration Problems 5.1 and 5.2

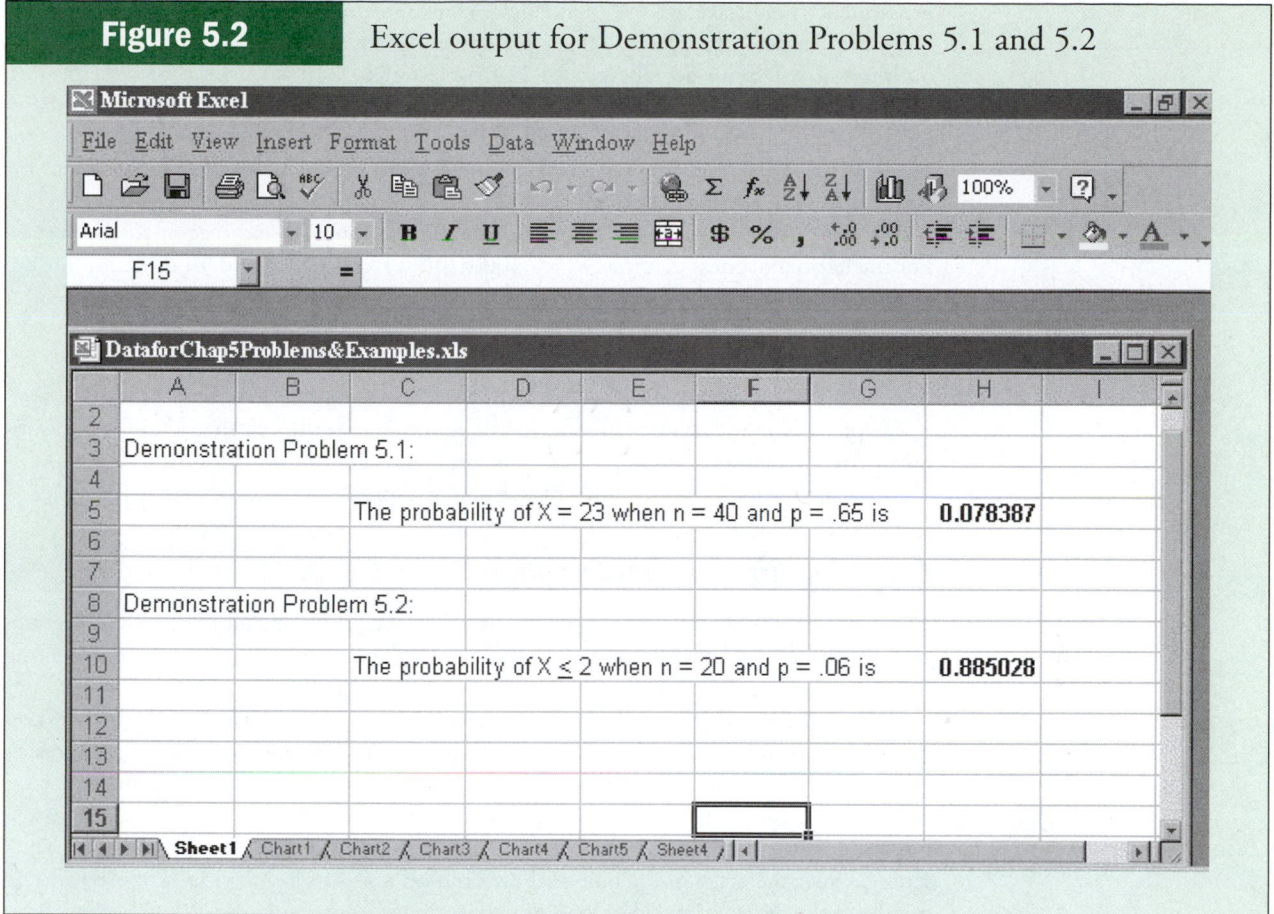

Examining the mean of a binomial distribution gives an intuitive feeling about the likelihood of a given outcome. For example, suppose researchers generally agree that 10% of all people are left-handed. However, suppose a researcher believes, as some have theorized, that this figure is higher for children who are born to women over the age of 35. In an attempt to gather evidence, she randomly selects 100 children who were born to women over the age of 35 and 20 turn out to be left-handed. Is it likely that she would have gotten 20 left-handed people in a sample of 100? How many would she have expected to get in a sample of 100? The mean or expected value for $n = 100$ and $p = .10$ is $(100)(.10) = 10$ left-handed people. Did the 20 left-handed children in a sample of 100 happen by chance or is the researcher drawing from a different population than the general population that produces 10% left-handed people? She can investigate this outcome further by examining the binomial probabilities for this problem. However, the mean of the distribution gives her an expected value from which to work.

According to one study 64% of all financial consumers believe banks are more competitive today than they were five years ago. If 23 financial consumers are selected randomly, what is the expected number who believe banks are more competitive today than they were five years ago? The mean of this binomial distribution yields the expected value for this problem.

$$\mu = n \cdot p = 23(.64) = 14.72$$

In the long run, if 23 financial consumers are selected randomly over and over and if indeed 64% of all financial consumers believe banks are more competitive today, then the experiment should average 14.72 financial consumers out of 23 who believe banks are more competitive today. Realize that because the binomial distribution is a discrete distribution you will *never* actually get 14.72 people out of 23 who believe banks are more competitive today. The mean of the distribution does reveal the relative likelihood of any individual occurrence. Computing probabilities for values of X near the mean results in: $P(X = 15) = .1712$, $P(X = 14) = .1605$, and $P(X = 16) = .1522$. All other probabilities for this distribution are less than these.

The standard deviation of a binomial distribution is denoted σ and is equal to $\sqrt{n \cdot p \cdot q}$. For the left-handedness example, $\sigma = \sqrt{100(.10)(.90)} = 3$. The standard deviation for the financial consumer problem is

$$\sigma = \sqrt{n \cdot p \cdot q} = \sqrt{(23)(.64)(.36)} = 2.30.$$

Chapter 6 shows that some binomial distributions are nearly bell-shaped and can be approximated by using the normal curve. The mean and standard deviation of a binomial distribution are the tools used to convert these binomial problems to normal curve problems.

Graphing Binomial Distributions

The graph of a binomial distribution can be constructed by using all the possible X values of a distribution and their associated probabilities. The X values usually are graphed along the X axis and the probabilities are graphed along the Y axis.

Table 5.3 lists the probabilities for three different binomial distributions; $n = 8$ and $p = .20$, $n = 8$ and $p = .50$, and $n = 8$ and $p = .80$. Figure 5.3 displays Excel graphs for each of these three binomial distributions. Observe how the shape of the distribution changes as the value of p increases. For $p = .50$, the distribution is symmetrical. For $p = .20$ the distribution is skewed right and for $p = .80$ the distribution is skewed left. This pattern makes sense because the mean of the binomial distribution $n = 8$ and $p = .50$ is 4, which is in the middle of the distribution. The mean of the distribution $n = 8$ and $p = .20$ is 1.6, which results in the highest probabilities being near $X = 2$ and $X = 1$. This graph peaks early and stretches toward the higher values of X. The mean of the distribution $n = 8$ and $p = .80$ is 6.4, which results in the highest probabilities being near $X = 6$ and $X = 7$. Thus the peak of the distribution is nearer to 8 than to 0 and the distribution stretches back toward $X = 0$.

In any binomial distribution the largest X value that can occur is n and the smallest value is zero. Thus the graph of any binomial distribution is constrained by zero and n. If the p value of the distribution is not .50, this constraint will result in the graph "piling up" at one end and being skewed at the other end.

TABLE 5.3

Probabilities for Three Binomial Distributions with $n = 8$

	PROBABILITIES FOR		
X	$p = .20$	$p = .50$	$p = .80$
0	.1678	.0039	.0000
1	.3355	.0312	.0001
2	.2936	.1094	.0011
3	.1468	.2187	.0092
4	.0459	.2734	.0459
5	.0092	.2187	.1468
6	.0011	.1094	.2936
7	.0001	.0312	.3355
8	.0000	.0039	.1678

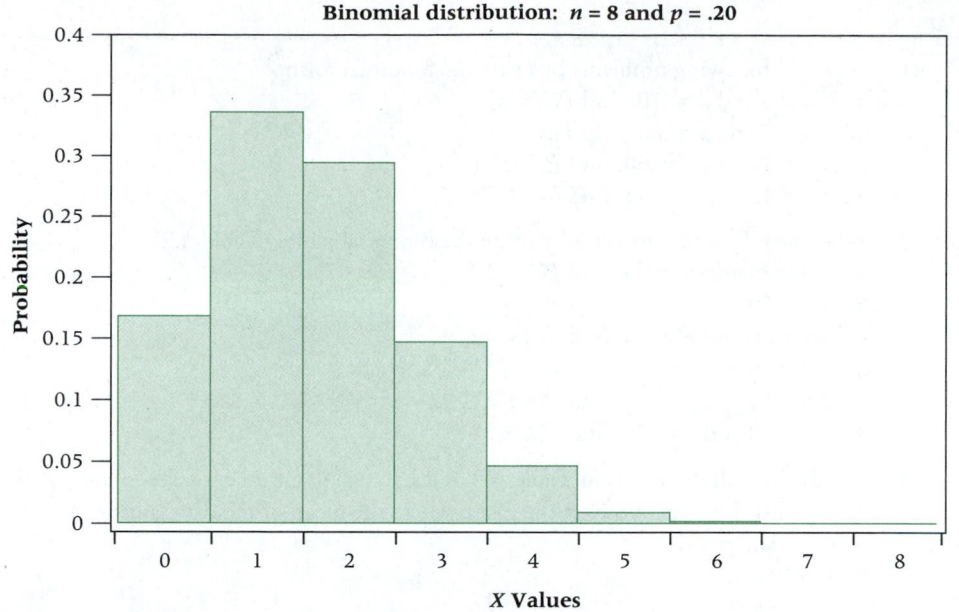

Figure 5.3a

Excel graph of binomial distribution with $n = 8$

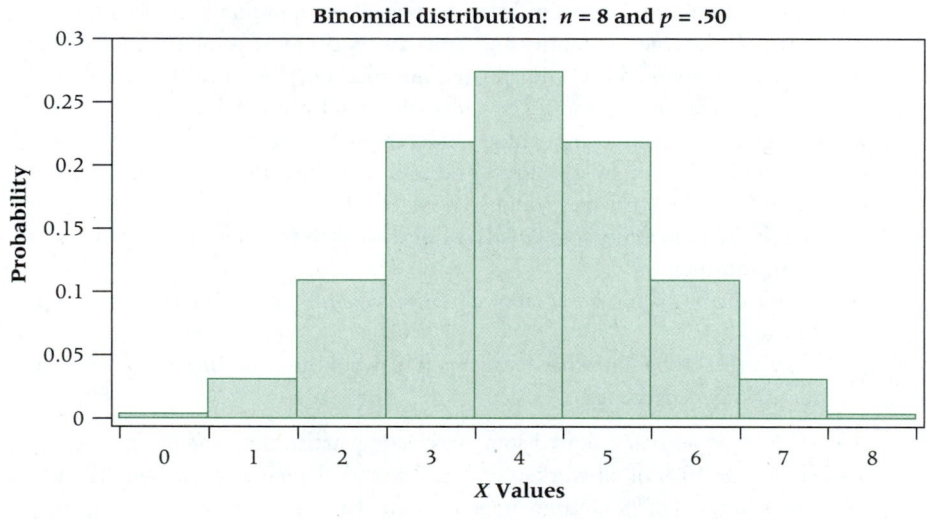

Figure 5.3b

Binomial distribution: $n = 8$ and $p = .50$

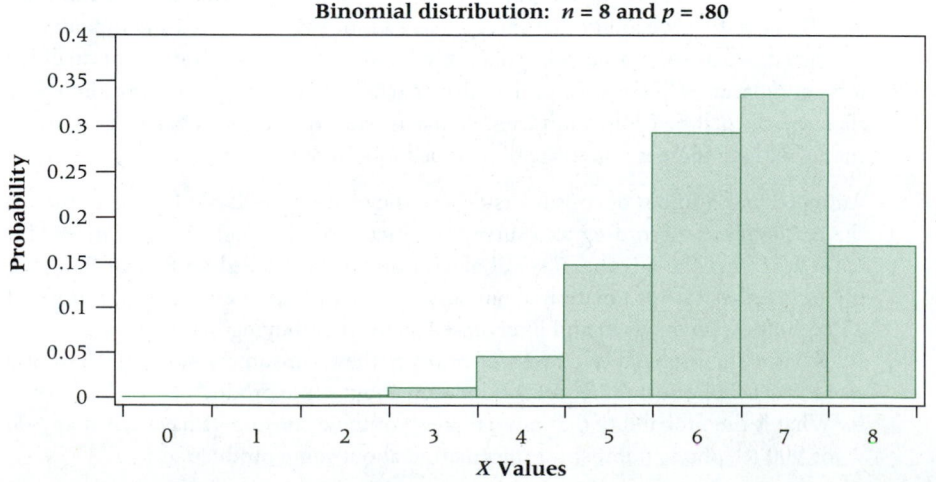

Figure 5.3c

Binomial distribution: $n = 8$ and $p = .80$

5.1　Solve the following problems by using the binomial formula.
　　a. If $n = 4$ and $p = .10$, find $P(X = 3)$.
　　b. If $n = 7$ and $p = .80$, find $P(X = 4)$.
　　c. If $n = 10$ and $p = .60$, find $P(X \geq 7)$.
　　d. If $n = 12$ and $p = .45$, find $P(5 \leq X \leq 7)$.

5.2　Solve the following problems by using the binomial tables (Table A.2).
　　a. If $n = 20$ and $p = .50$, find $P(X = 12)$.
　　b. If $n = 20$ and $p = .30$, find $P(X > 8)$.
　　c. If $n = 20$ and $p = .70$, find $P(X < 12)$.
　　d. If $n = 20$ and $p = .90$, find $P(X \leq 16)$.
　　e. If $n = 15$ and $p = .40$, find $P(4 \leq X \leq 9)$.
　　f. If $n = 10$ and $p = .60$, find $P(X \geq 7)$.

5.3　Use the probability tables in Table A.2 and Excel to graph each of the following binomial distributions. Note on the graph where the mean of the distribution falls.
　　a. $n = 6$ and $p = .70$
　　b. $n = 20$ and $p = .50$
　　c. $n = 8$ and $p = .80$

5.4　*Purchasing* magazine reported the results of a survey in which buyers were asked a series of questions with regard to Internet usage. One question asked was how they would use the Internet if security and other issues could be resolved. Seventy-eight percent said they would use it for pricing information, 75% said they would use it to send purchase orders, and 70% said they would use it for purchase order acknowledgments. Assume that these percentages hold true for all buyers. A researcher randomly samples 20 buyers and asks them how they would use the Internet if security and other issues could be resolved.
　　a. What is the probability that exactly 14 of these buyers would use the Internet for pricing information?
　　b. What is the probability that all of the buyers would use the Internet to send purchase orders?
　　c. What is the probability that fewer than 12 would use the Internet for purchase order acknowledgments?

5.5　*The Wall Street Journal* reported some interesting statistics on the job market. One statistic is that 40% of all workers say they would change jobs for "slightly higher pay." In addition, 88% of companies say that there is a shortage of qualified job candidates. Suppose 16 workers are randomly selected and asked if they would change jobs for "slightly higher pay." What is the probability that nine or more say yes? What is the probability that three, four, five, or six say yes? If 13 companies are contacted, what is the probability that exactly 10 say there is a shortage of qualified job candidates? What is the probability that all of the companies say there is a shortage of qualified job candidates? What is the expected number of companies that would say there is a shortage of qualified job candidates?

5.6　An increasing number of consumers believe they have to look out for themselves in the marketplace. According to a survey conducted by the Yankelovich Partners for *USA WEEKEND* magazine, 60% of all consumers have called an 800 or 900 telephone number for information about some product. Suppose a random sample of 25 consumers is contacted and interviewed about their buying habits.
　　a. What is the probability that 15 or more of these consumers have called an 800 or 900 telephone number for information about some product?
　　b. What is the probability that more than 20 of these consumers have called an 800 or 900 telephone number for information about some product?

c. What is the probability that less than 10 of these consumers have called an 800 or 900 telephone number for information about some product?

5.7 In the past few years outsourcing overseas has become more frequently used than ever before by U.S. companies. However, outsourcing is not without problems. A recent survey by *Purchasing* indicates that 20% of the companies that outsource overseas use a consultant. Suppose 15 companies that outsource overseas are randomly selected.

a. What is the probability that exactly five companies that outsource overseas use a consultant?

b. What is the probability that more than nine companies that outsource overseas use a consultant?

c. What is the probability that none of the companies that outsource overseas use a consultant?

d. What is the probability that between four and seven (inclusive) companies that outsource overseas use a consultant?

e. Construct a graph for this binomial distribution. In light of the graph and the expected value, explain why the probability results from parts (a) through (d) were obtained.

5.8 According to Cerulli Associates of Boston, 30% of all CPA financial advisors have an average client size of between $500,000 and $1 million. Thirty-four percent have an average client size of between $1 million and $5 million. Suppose a complete list of all CPA financial advisors is available, and 18 are randomly selected from that list.

a. What is the expected number of CPA financial advisors that have an average client size of between $500,000 and $1 million? What is the expected number that have an average client size of between $1 million and $5 million?

b. What is the probability that at least eight CPA financial advisors have an average client size of between $500,000 and $1 million?

c. What is the probability that two, three, or four CPA financial advisors have an average client size of between $1 million and $5 million?

d. What is the probability that none of the CPA financial advisors have an average client size of between $500,000 and $1 million? What is the probability that none have an average client size of between $1 million and $5 million? Which probability is higher and why?

The Poisson distribution is another discrete distribution. It is named after Simeon-Denis Poisson (1781–1840), a French mathematician, who published its essentials in a paper in 1837. The Poisson distribution and the binomial distribution have some similarities, but also have several differences. The binomial distribution describes a distribution of two possible outcomes designated as successes and failures from a given number of trials. The **Poisson distribution** *focuses only on the number of discrete occurrences over some interval or continuum*. A Poisson experiment does not have a given number of trials (*n*) as a binomial experiment does. For example, whereas a binomial experiment might be used to determine how many U.S.-made cars there are in a random sample of 20 cars, a Poisson experiment might focus on the number of cars randomly arriving at an automobile repair facility during a 10-minute interval.

The Poisson distribution describes the occurrence of *rare events*. In fact, the Poisson formula has been referred to as the *law of improbable events*. For example, serious accidents at a chemical plant are rare, and the number per month might be described by the Poisson distribution. The Poisson distribution often is used to describe the number of random arrivals per some *time* interval. If the number of arrivals per interval is too frequent, the time interval can be reduced enough so that a rare number of occurrences is

5.3
Poisson Distribution

Poisson distribution
A discrete distribution that is constructed from the probability of occurrence of rare events over an interval; focuses only on the number of discrete occurrences over some interval or continuum.

expected. Another example of a Poisson distribution is the number of random customer arrivals per 5-minute interval at a small boutique on weekday mornings.

The Poisson distribution also has an application in the field of management science. The models used in queuing theory (theory of waiting lines) usually are based on the assumption that the Poisson distribution is the proper distribution to describe random arrival rates over a period of time.

The Poisson distribution has the following characteristics.

- It is a discrete distribution.
- It describes rare events.
- Each occurrence is independent of the other occurrences.
- It describes discrete occurrences over a continuum or interval.
- The occurrences in each interval can range from zero to infinity.
- The expected number of occurrences must hold constant throughout the experiment.

Examples of Poisson-type situations include the following.

1. Number of telephone calls per minute at a small business
2. Number of cases of a rare blood disease per 100,000 people
3. Number of hazardous waste sites per county in the United States
4. Number of major oil spills in the New England region per month
5. Number of arrivals at a turnpike toll booth per minute between 3 A.M. and 4 A.M. in January on the Kansas Turnpike
6. Number of times a 1-year-old personal computer printer breaks down per quarter (3 months)
7. Number of sewing flaws per pair of jeans during production
8. Number of times a tire blows on a commercial airplane per week
9. Number of paint spots per new automobile
10. Number of flaws per bolt of cloth

Each of these examples represents a rare occurrence of events for some interval. Note that, although time is a more common interval for the Poisson distribution, intervals can range from a county in the United States to a pair of jeans. Some of the intervals in these examples might have zero occurrences. Moreover, the average occurrence per interval for many of these examples is probably in the single digits (1–9).

If a Poisson-distributed phenomenon is studied over a long period of time, a *long-run average* can be determined. This average is denoted **lambda** (λ). Each Poisson problem contains a lambda value from which the probabilities of particular occurrences are determined. Although n and p are required to describe a binomial distribution, a Poisson distribution can be described by λ alone. The Poisson formula is used to compute the probability of occurrences over an interval for a given lambda value.

Lambda (λ)
Denotes the long-run average of a Poisson distribution.

POISSON FORMULA

$$P(X) = \frac{\lambda^X e^{-\lambda}}{X!}$$

where:
$X = 0, 1, 2, 3, \ldots$
$\lambda = $ long-run average
$e = 2.718282$

Here, X is the number of occurrences per interval for which the probability is being computed, λ is the long-run average, and $e = 2.718282$ is the base of natural logarithms.

A word of caution about using the Poisson distribution to study various phenomena is necessary. The λ value must hold constant throughout a Poisson experiment. The researcher must be careful not to apply a given lambda to intervals for which lambda changes. For example, the average number of customers arriving at a Sears store during a 1-minute interval will vary from hour to hour, day to day, and month to month. Different times of the day or week might produce different lambdas. The number of flaws per pair of jeans might vary from Monday to Friday. The researcher should be very specific in describing the interval for which λ is being used.

Working Poisson Problems by Formula

Suppose bank customers arrive randomly on weekday afternoons at an average of 3.2 customers every 4 minutes. What is the probability of exactly five customers arriving in a 4-minute interval on a weekday afternoon? The lambda for this problem is 3.2 customers per 4 minutes. The value of X is five customers per 4 minutes. The probability of five customers randomly arriving during a 4-minute interval when the long-run average has been 3.2 customers per 4-minute interval is

$$\frac{(3.2^5)(e^{-3.2})}{5!} = \frac{(335.54)(.0408)}{120} = .1141.$$

If a bank averages 3.2 customers every 4 minutes, the probability of five customers arriving during any one 4-minute interval is .1141.

Bank customers arrive randomly on weekday afternoons at an average of 3.2 customers every 4 minutes. What is the probability of having more than seven customers in a 4-minute interval on a weekday afternoon?

SOLUTION

$$\lambda = 3.2 \text{ customers/4 minutes}$$

$$X > 7 \text{ customers/4 minutes}$$

In theory, the solution requires obtaining the values of X = 8, 9, 10, 11, 12, 13, 14, . . . , ∞. In actuality, each X value is determined until the values ar so far away from $\lambda = 3.2$ that the probabilities approach zero. The exact probabilities are then summed to find $X > 7$.

$$P(X = 8 \mid \lambda = 3.2) = \frac{(3.2^8)(e^{-3.2})}{8!} = .0111$$

$$P(X = 9 \mid \lambda = 3.2) = \frac{(3.2^9)(e^{-3.2})}{9!} = .0040$$

$$P(X = 10 \mid \lambda = 3.2) = \frac{(3.2^{10})(e^{-3.2})}{10!} = .0013$$

$$P(X = 11 \mid \lambda = 3.2) = \frac{(3.2^{11})(e^{-3.2})}{11!} = .0004$$

$$P(X = 12 \mid \lambda = 3.2) = \frac{(3.2^{12})(e^{-3.2})}{12!} = .0001$$

$$P(X = 13 \mid \lambda = 3.2) = \frac{(3.2^{13})(e^{-3.2})}{13!} = .0000$$

$$P(X > 7) = P(X \geq 8) = .0169$$

If the bank has been averaging 3.2 customers every 4 minutes on weekday afternoons, it is unlikely that more than seven people would randomly arrive in any one 4-minute period. This answer indicates that more than seven people would randomly arrive in a 4-minute period only 1.69% of the time. Bank officers could use these results to help them make staffing decisions.

DEMONSTRATION PROBLEM 5.6

A bank has an average random arrival rate of 3.2 customers every 4 minutes. What is the probability of getting exactly 10 customers during an 8-minute interval?

SOLUTION

$$\lambda = 3.2 \text{ customers}/4 \text{ minutes}$$

$$X = 10 \text{ customers}/8 \text{ minutes}$$

This example is different from the first two Poisson examples in that the intervals for lambda and the sample are different. The intervals must be the same in order to use λ and X together in the probability formula. There is a right way and a wrong way to approach this dilemma. The right way is to adjust the interval for lambda so that it and X have the same interval. The interval for X is 8 minutes, so lambda should be adjusted to an 8-minute interval. Logically, if the bank averages 3.2 customers every 4 minutes, it should average twice as many, or 6.4 customers every 8 minutes. If X were for a 2-minute interval, the value of lambda would be halved from 3.2 to 1.6 customers per 2-minute interval. The wrong way is to equalize the intervals by changing the X value. *Never adjust or change X in a problem.* Just because 10 customers arrive in one 8-minute interval does not mean that there would necessarily have been five customers in a 4-minute interval. There is no guarantee how the 10 customers are spread over the 8-minute interval. Always adjust the lambda value. After lambda has been adjusted for an 8-minute interval, the solution is

$$\lambda = 6.4 \text{ customers}/8 \text{ minutes}$$

$$X = 10 \text{ customers}/8 \text{ minutes}$$

$$\frac{(6.4)^{10} e^{-6.4}}{10!} = .0528$$

TABLE 5.4
Poisson Table for $\lambda = 1.6$

X	PROBABILITY
0	.2019
1	.3230
2	.2584
3	.1378
4	.0551
5	.0176
6	.0047
7	.0011
8	.0002
9	.0000

Using the Poisson Tables

Every value of lambda determines a different Poisson distribution. Regardless of the nature of the interval associated with a lambda, the Poisson distribution for a particular lambda is the same. Table A.3, Appendix A, contains the Poisson distributions for selected values of lambda. Probabilities are displayed in the table for each X value associated with a given lambda if the probability has a nonzero value to four decimal places. Table 5.4 is a portion of Table A.3 that contains the probabilities of $X \leq 9$ if lambda is 1.6.

DEMONSTRATION PROBLEM 5.7

If a real estate office sells 1.6 houses on an average weekday and sales of houses on weekdays are Poisson distributed, what is the probability of selling exactly four houses in one day? What is the probability of selling no houses in one day? What is the probability of selling less than five houses in a day? What is the probability of selling 10 or more houses in a day? What is the probability of selling exactly four houses in two days?

SOLUTION

$$\lambda = 1.6 \text{ houses/day}$$
$$P(X = 4 \,|\, \lambda = 1.6) = ?$$

Table 5.4 gives the probabilities for $\lambda = 1.6$. The left column contains the X values. The line $X = 4$ yields the probability .0551. If a real estate firm has been averaging 1.6 houses sold per day, only 5.51% of the days would it sell exactly four houses and still maintain the lambda value. Line 1 of Table 5.4 shows the probability of selling no houses in a day (.2019). That is, on 20.19% of the days, the firm would sell no houses if sales are Poisson distributed with $\lambda = 1.6$ houses per day. Table 5.4 is *not* cumulative. To determine $P(X < 5)$, less than five houses, find the probabilities of $X = 0$, $X = 1$, $X = 2$, $X = 3$, $X = 4$. The answer for $X < 5$ follows.

X	PROBABILITY
0	.2019
1	.3230
2	.2584
3	.1378
4	.0551
$X < 5 =$	.9762

What is the probability of selling 10 or more houses in one day? As the table zeros out at $X = 9$, the probability of $X \geq 10$ is essentially .0000—that is, if the real estate office has been averaging only 1.6 houses sold per day, it is virtually impossible to sell 10 or more houses in a day. What is the probability of selling exactly four houses in two days? In this case, the interval has been changed from one day to two days. Lambda is for one day, so an adjustment must be made: a lambda of 1.6 for one day converts to a lambda of 3.2 for two days. Table 5.4 no longer applies, so Table A.3 must be used to solve this problem. The answer is found by looking up $\lambda = 3.2$ and $X = 4$ in Table A.3: the probability is .1781.

Analysis Using Excel

Both exact and cumulative Poisson probabilities can be computed using Excel. The process begins by selecting the paste function, f_x, from the Excel standard toolbar. From the left side of the paste function dialog box, select **Statistical,** which will change the right side offerings. From the new right side offerings select **POISSON.** The dialog box for **POISSON** is shown in Figure 5.4.

To compute Poisson distribution probabilities using **POISSON,** enter the value of X in the first box and the value of lambda in the second box labeled **Mean.** The third box, **Cumulative,** allows you to determine whether you want exact probabilities or cumulative probabilities. For exact probabilities, enter **false** in the box; and for cumulative probabilities, enter **true** in the box. The cumulative probabilities are always calculated from $X = 0$ through the number entered (from the left).

Excel output for Demonstration Problem 5.7 is shown in Figure 5.5.

Mean and Standard Deviation of a Poisson Distribution

The mean or expected value of a Poisson distribution is λ. It is the long-run average of occurrences for an interval if many random samples are taken. Lambda usually is not a whole number, so most of the time actually observing lambda occurrences in an interval is impossible.

Figure 5.4   Dialog box for **POISSON**

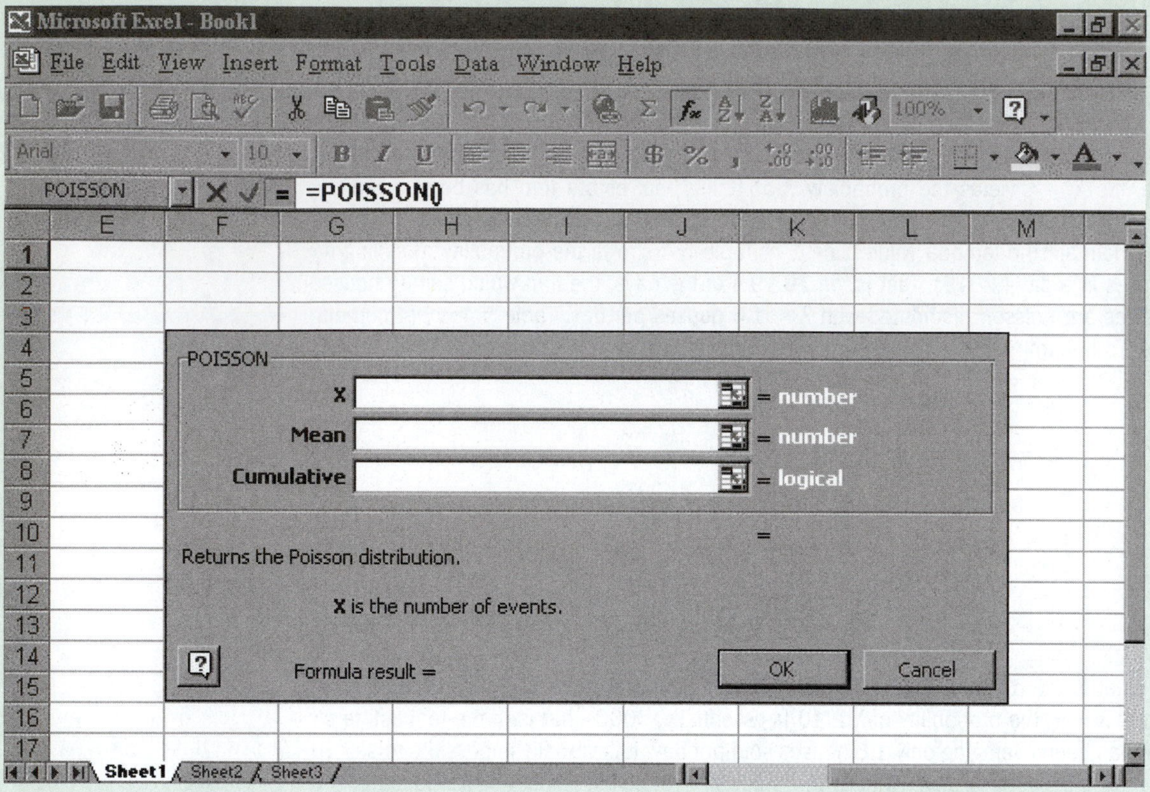

Microsoft Excel - Book1

File Edit View Insert Format Tools Data Window Help

Arial

POISSON   = =POISSON()

POISSON

X [   ] = number

Mean [   ] = number

Cumulative [   ] = logical

=

Returns the Poisson distribution.

X is the number of events.

Formula result =   OK   Cancel

Sheet1 / Sheet2 / Sheet3 /

Figure 5.5   Excel output for Demonstration Problem 5.7

Microsoft Excel - Book1

File Edit View Insert Format Tools Data Window Help

Arial

	A	B	C	D	E	F	G	H	I
3	Demonstration Problem 5.7:								
6		The probability of X = 4 when lambda = 1.6 is					0.055131231		
8		The probability of X < 5 when lambda = 1.6 is					0.976317722		
10		The probability of X ≥ 10 when lambda = 1.6 is					0.000007142		

Sheet1 / Sheet2 / Sheet3 /

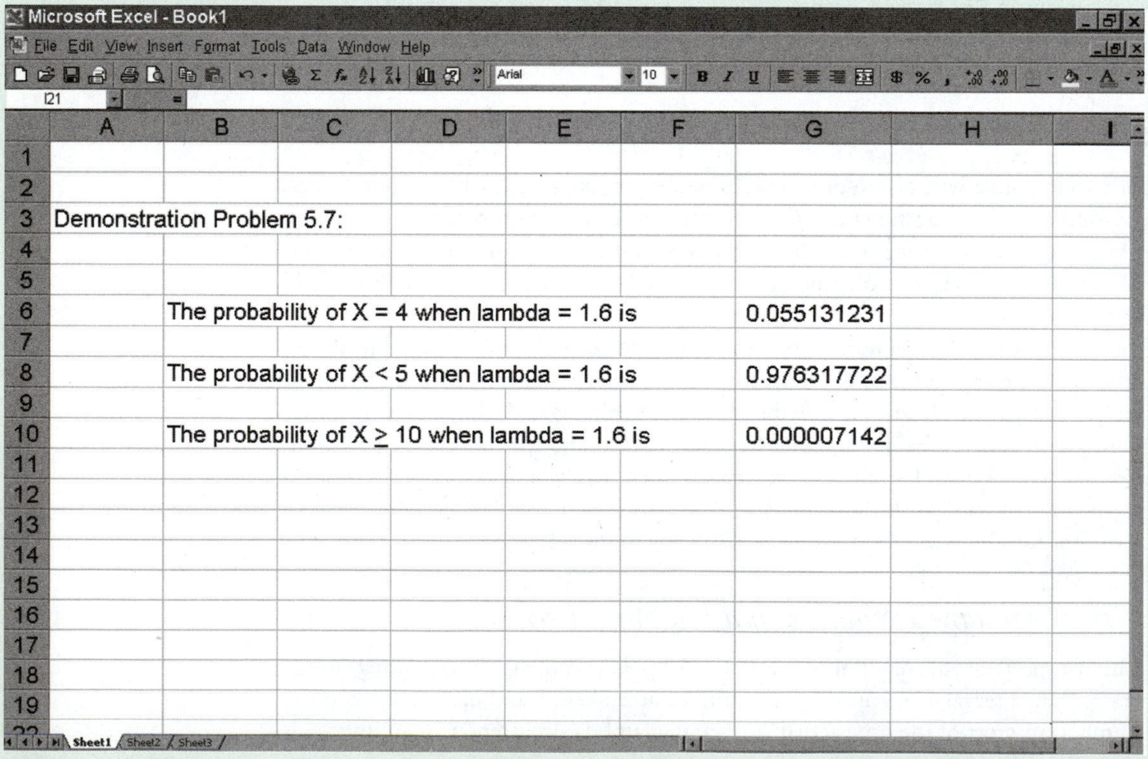

The variance of a Poisson distribution also is λ. The standard deviation is $\sqrt{\lambda}$. For example, if $\lambda = 6.5$, the variance also is 6.5, and the standard deviation is $\sqrt{6.5} = 2.55$.

Graphing Poisson Distributions

Values for graphing the Poisson distribution may be obtained from Excel's **POISSON** function or Table A.3 of Appendix A. The X values are on the X axis and the probabilities are on the Y axis. Figure 5.6 is an Excel graph for the distribution of values for $\lambda = 1.6$.

The graph reveals a Poisson distribution skewed to the right. With a mean of 1.6 and a possible range of X from zero to infinity, the values obviously will "pile up" at 0 and 1. Consider, however, the Excel graph of the Poisson distribution for $\lambda = 6.5$ in Figure 5.7. Note that with $\lambda = 6.5$, the probabilities are greatest for the values of 5, 6, 7, and 8. The graph has less skewness because the probability of occurrence of values near zero is small, as are the probabilities of large values of X.

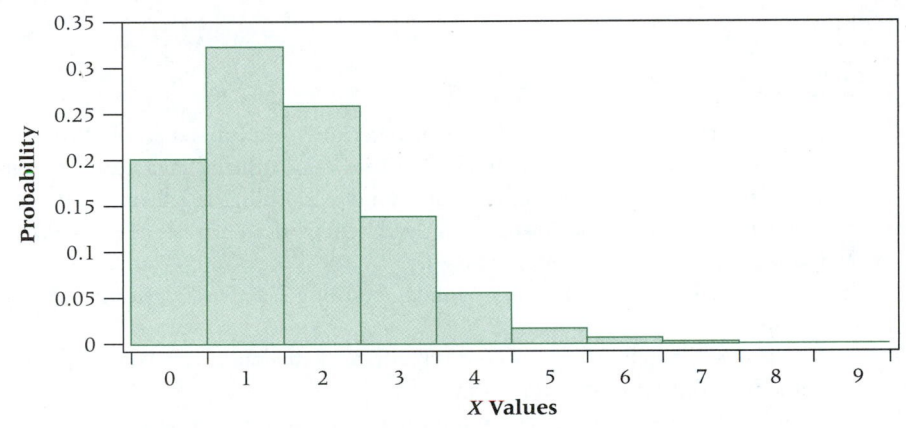

Figure 5.6

Excel graph of Poisson distribution for $\lambda = 1.6$

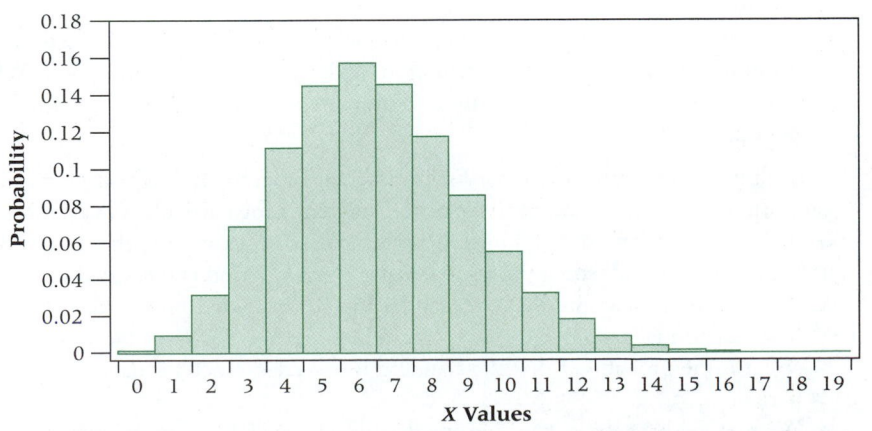

Figure 5.7

Excel graph of Poisson distribution for $\lambda = 6.5$

5.3

Problems

5.9 Find the following values by using the Poisson formula.
a. $P(X = 5 \mid \lambda = 2.3)$
b. $P(X = 2 \mid \lambda = 3.9)$
c. $P(X \leq 3 \mid \lambda = 4.1)$
d. $P(X = 0 \mid \lambda = 2.7)$
e. $P(X = 1 \mid \lambda = 5.4)$
f. $P(4 < X < 8 \mid \lambda = 4.4)$

5.10 Find the following values by using the Poisson tables in Appendix A.
a. $P(X = 6 \mid \lambda = 3.8)$
b. $P(X > 7 \mid \lambda = 2.9)$
c. $P(3 \leq X \leq 9 \mid \lambda = 4.2)$
d. $P(X = 0 \mid \lambda = 1.9)$
e. $P(X \leq 6 \mid \lambda = 2.9)$
f. $P(5 < X \leq 8 \mid \lambda = 5.7)$

5.11 Sketch the graphs of the following Poisson distributions or graph them using Excel. Compute the mean and standard deviation for each distribution. Locate the mean on the graph. Note how the probabilities are graphed around the mean.
a. $\lambda = 6.3$
b. $\lambda = 1.3$
c. $\lambda = 8.9$
d. $\lambda = 0.6$

5.12 On Monday mornings, the First National Bank only has one teller window open for deposits and withdrawals. Experience has shown that the average number of arriving customers in a 4-minute interval on Monday mornings is 2.8, and each teller can serve more than that number efficiently. These random arrivals at this bank on Monday mornings are Poisson distributed.
a. What is the probability that on a Monday morning exactly six customers will arrive in a 4-minute interval?
b. What is the probability that no one will arrive at the bank to make a deposit or withdrawal during a 4-minute interval?
c. Suppose the teller can serve no more than four customers in any 4-minute interval at this window on a Monday morning. What is the probability that, during any given 4-minute interval, the teller will be unable to meet the demand? What is the probability that the teller *will* be able to meet the demand? When demand cannot be met during any given interval, a second window is opened. What percentage of the time will a second window have to be opened?
d. What is the probability that exactly three people will arrive at the bank during a 2-minute period on Monday mornings to make a deposit or a withdrawal? What is the probability that five or more customers will arrive during an 8-minute period?

5.13 According to the United National Environmental Program and World Health Organization, in Bombay, India, air pollution standards for particulate matter are exceeded an average of 5.6 days in every 3-week period. Assume that the distribution of number of days exceeding the standards per 3-week period is Poisson distributed.
a. What is the probability that the standard is not exceeded on any day during a 3-week period?
b. What is the probability that the standard is exceeded exactly 6 days of a 3-week period?
c. What is the probability that the standard is exceeded 15 or more days during a 3-week period? If this outcome actually occurred, what might you conclude?

5.14 A restaurant manager is interested in taking a more statistical approach to predicting customer load. She begins the process by gathering data. One of the restaurant hosts or hostesses is assigned to count customers every 5 minutes from 7 P.M. until 8 P.M. every Saturday night for three weeks. The data are shown here. After the data are gathered, the manager computes lambda using the data from all three weeks as one data set as a basis for probability analysis. What value of lambda did she find? Assume that these customers randomly arrive and that the arrivals are Poisson distributed. Use the value of lambda computed by the manager and help the manager calculate the probabilities in parts (a) through (e) for any given 5-minute interval between 7 P.M. and 8 P.M. on Saturday night.

NUMBER OF ARRIVALS

Week 1	Week 2	Week 3
3	1	5
6	2	3
4	4	5
6	0	3
2	2	5
3	6	4
1	5	7
5	4	3
1	2	4
0	5	8
3	3	1
3	4	3

a. What is the probability that no customers arrive during any given 5-minute interval?
b. What is the probability that six or more customers arrive during any given 5-minute interval?
c. What is the probability that during a 10-minute interval fewer than four customers arrive?
d. What is the probability that between three and six (inclusive) customers arrive in any 10-minute interval?
e. What is the probability that exactly eight customers arrive in any 15-minute interval?

5.15 The average number of annual trips per family to amusement parks in the United States is Poisson distributed, with a mean of 0.6 trips per year. What is the probability of randomly selecting an American family and finding that:
a. The family did not make a trip to an amusement park last year?
b. The family took exactly one trip to an amusement park last year?
c. The family took two or more trips to amusement parks last year?
d. The family took three or fewer trips to amusement parks over a 3-year period?
e. The family took exactly four trips to amusement parks during a 6-year period?

5.16 Ship collisions in the Houston Ship Channel are rare. Suppose the number of collisions are Poisson distributed, with a mean of 1.2 collisions every 4 months.
a. What is the probability of having no collisions occur over a 4-month period?
b. What is the probability of having exactly two collisions in a 2-month period?
c. What is the probability of having one or fewer collisions in a 6-month period? If this outcome occurred, what might you conclude about ship channel conditions during this period? What might you conclude about ship channel safety awareness during this period? What might you conclude about weather conditions during this period? What might you conclude about lambda?

5.4
Hypergeometric Distribution

Hypergeometric distribution
A distribution of probabilities of the occurrence of X items in a sample of n when there are A of that same item in a population of N.

Another discrete statistical distribution is the hypergeometric distribution. Statisticians often use the **hypergeometric distribution** to complement the types of analyses that can be made by using the binomial distribution. Recall that the binomial distribution applies, in theory, only to experiments in which the trials are done with replacement (independent events). The hypergeometric distribution applies only to experiments in which the trials are done *without replacement.*

The hypergeometric distribution, like the binomial distribution, consists of two possible outcomes: success and failure. However, the user must know the size of the population and the proportion of successes and failures in the population to apply the hypergeometric distribution. That is, because the hypergeometric distribution is used when sampling is done without replacement, information about population makeup must be known in order to redetermine the probability of a success in each successive trial as the probability changes.

The hypergeometric distribution has the following characteristics.

- It is discrete distribution.
- Each outcome consists of either a success or a failure.
- Sampling is done *without* replacement.
- The population, N, is finite and known.
- The number of successes in the population, A, is known.

HYPERGEOMETRIC FORMULA	$$P(X) = \frac{{}_A C_X \cdot {}_{N-A} C_{n-X}}{{}_N C_n}$$

where:
N = size of the population
n = sample size
A = number of successes in the population
X = number of successes in the sample; sampling is done *without* replacement

A hypergeometric distribution is characterized or described by three parameters: N, A, and n. Because of the multitude of possible combinations of these three parameters, creating tables for the hypergeometric distribution is practically impossible. Hence, the researcher who selects the hypergeometric distribution for analyzing data must use the hypergeometric formula to calculate each probability. Because this task can be tedious and time-consuming, most researchers use the hypergeometric distribution as a fall-back position when working binomial problems without replacement. Even though the binomial distribution theoretically applies only when sampling is done *with* replacement and p stays constant, recall that, if the population is large enough in comparison with the sample size, the impact of sampling *without* replacement on p is minimal. Thus the binomial distribution can be used in some situations when sampling is done *without* replacement. Because of the tables available, using the binomial distribution instead of the hypergeometric distribution whenever possible is preferable. As a rule of thumb, if the sample size is less than 5% of the population, use of the binomial distribution rather than the hypergeometric distribution is acceptable when sampling is done *without* replacement. The hypergeometric distribution yields the exact probability, and the binomial distribution yields a good approximation of the probability in these situations.

In summary, the hypergeometric distribution should be used instead of the binomial distribution when

1. Sampling is being done *without* replacement *and*
2. $n \geq 5\% \ N$.

Hypergeometric probabilities are calculated under the assumption that there is equally likely sampling of the remaining elements of the sample space.

As an application of the hypergeometric distribution, consider the following problem. Twenty-four people, of whom eight are women, have applied for a job. If five of the applicants are sampled randomly, what is the probability that exactly three of those sampled are women?

This problem contains a small, finite population of 24, or $N = 24$. A sample of five applicants is taken, or $n = 5$. The sampling is being done *without* replacement, because the five applicants selected for the sample are five different people. The sample size is 21% of the population, which is $\geq$ 5% of the population ($n/N = 5/24 = .21$). The hypergeometric distribution is the appropriate distribution to use. The population breakdown is $A = 8$ women (successes) and $N - A = 24 - 8 = 16$ men. The probability of getting $X = 3$ women in the sample of $n = 5$ is

$$\frac{_8C_3 \cdot {}_{16}C_2}{_{24}C_5} = \frac{(56)(120)}{42,504} = .1581$$

Conceptually, the combination in the denominator of the hypergeometric formula yields all the possible ways of getting n samples from a population, N, including the ones with the desired outcome. In this problem, there are 42,504 ways of selecting five people from 24 people. The numerator of the hypergeometric formula computes all the possible ways of getting X successes from the A successes available and $n - X$ failures from the $N - A$ available failures in the population. There are 56 ways of getting three women from a pool of eight and there are 120 ways of getting two men from a pool of 16. The combinations of each are multiplied in the numerator because the joint probability of getting X successes *and* $n - X$ failures is being computed.

Suppose there are 18 major computer companies in the United States and that 12 are located in California's Silicon Valley. If three computer companies are selected randomly from the entire list, what is the probability that one or more of the selected companies are located in the Silicon Valley?

DEMONSTRATION PROBLEM 5.8

SOLUTION

$$N = 18, n = 3, A = 12, \text{ and } X \geq 1$$

This is actually three problems in one: $X = 1$, $X = 2$, and $X = 3$. Sampling is being done without replacement, and the sample size is 16.6% of the population. Hence this problem is a candidate for the hypergeometric distribution. The solution follows.

$$\underset{.2206}{\underbrace{\frac{{}_{12}C_1 \cdot {}_6C_2}{{}_{18}C_3}}^{X=1}} + \underset{.4853}{\underbrace{\frac{{}_{12}C_2 \cdot {}_6C_1}{{}_{18}C_3}}^{X=2}} + \underset{.2696}{\underbrace{\frac{{}_{12}C_3 \cdot {}_6C_0}{{}_{18}C_3}}^{X=3}} = .9755$$

An alternative solution method using the law of complements would be one minus the probability that none of the companies is located in Silicon Valley, or

$$1 - P(X = 0 \mid N = 18, n = 3, A = 12).$$

Thus,

$$1 - \frac{{}_{12}C_0 \cdot {}_6C_3}{{}_{18}C_3} = 1 - .0245 = .9755.$$

Figure 5.8 Dialog box for **HYPGEOMDIST**

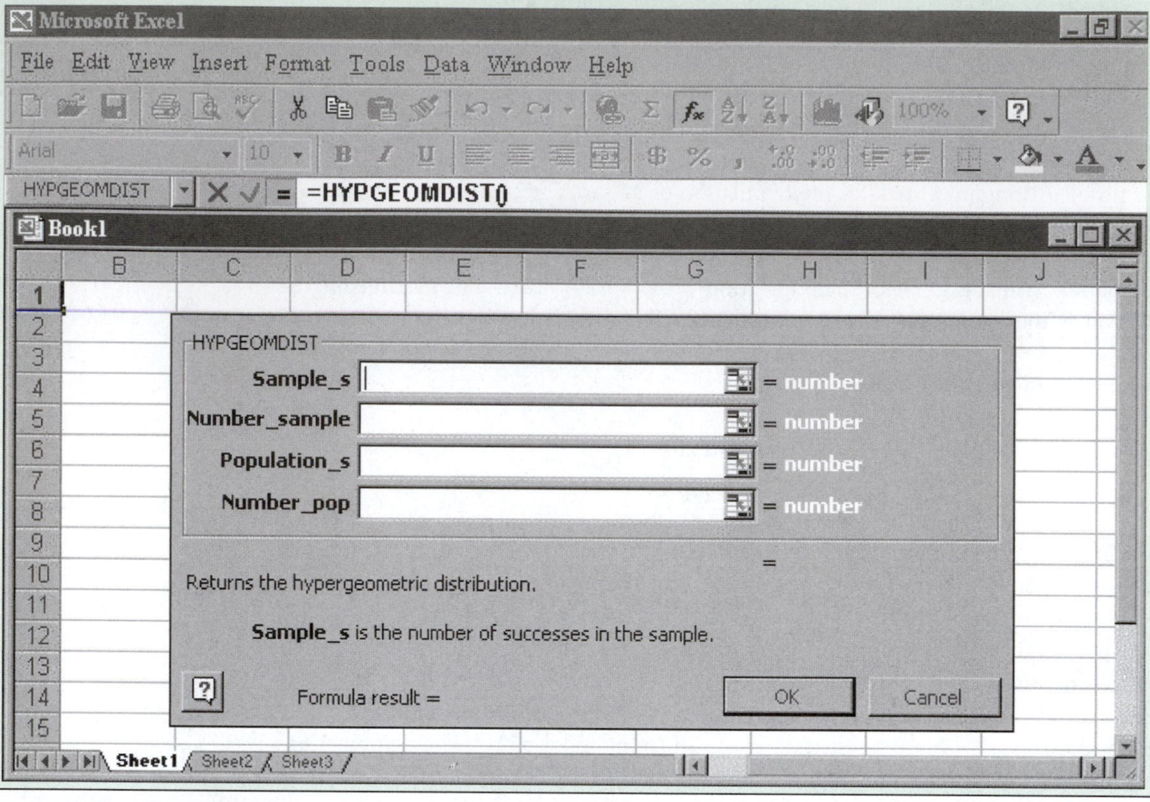

Figure 5.9 Excel output for Demonstration Problem 5.8

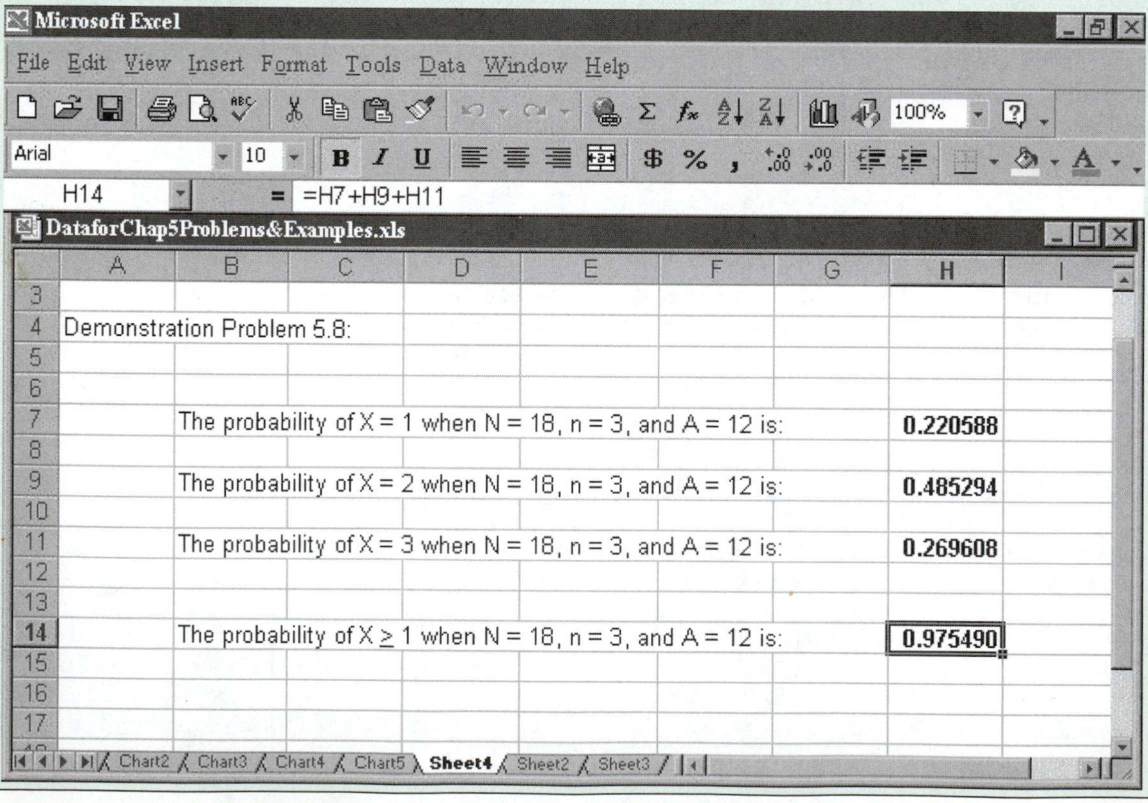

Analysis Using Excel

Excel has the capability of computing exact probabilities for the hypergeometric distribution. Begin by selecting the paste function, f_x, from the Excel standard toolbar. From the left side of the paste function dialog box, select **Statistical,** which will change the right side offerings. From the new right side offerings, select **HYPGEOMDIST.** The dialog box for **HYPGEOMDIST** is shown in Figure 5.8.

There are four boxes that must be correctly completed in the **HYPGEOMDIST** dialog box before Excel can compute the probability. The first box, **Sample_s,** requires the number of successes in the sample, X. The second box, **Number_sample,** requires the sample size, n. The third box, **Population_s,** requires the number of successes in the population, A. The fourth box, **Number_pop,** requires the size of the population, N.

Excel output for Demonstration Problem 5.8. is shown in Figure 5.9.

5.4

Problems

5.17 Compute the following probabilities by using the hypergeometric formula.
   a. The probability of $X = 3$ if $N = 11$, $A = 8$, and $n = 4$
   b. The probability of $X < 2$ if $N = 15$, $A = 5$, and $n = 6$
   c. The probability of $X = 0$ if $N = 9$, $A = 2$, and $n = 3$
   d. The probability of $X > 4$ if $N = 20$, $A = 5$, and $n = 7$

5.18 Shown here are the top 20 companies in the world in terms of oil refining capacity according to the *Petroleum Intelligence Weekly*. Some of the companies are privately owned and others are state owned. Suppose six companies are randomly selected.
   a. What is the probability that exactly one company is privately owned?
   b. What is the probability that exactly four companies are privately owned?
   c. What is the probability that all six companies are privately owned?
   d. What is the probability that none of the companies are privately owned?

COMPANY	OWNERSHIP STATUS
Exxon	Private
Royal Dutch/Shell	Private
China Petrochemical	State
Petroleos de Venezuela	State
Mobil	Private
Saudi Arabian Oil Co.	State
British Petroleum	Private
Chevron	Private
Petrobras	State
Texaco	Private
Pemex	State
National Iranian Oil Co.	State
Amoco	Private
Pertamina	State
Kuwait Petroleum Corp.	State
Elf Aquitaine	Private
Idemitsu	Private
Ente Nazionale Idrocarburi	State
Total	Private
Nippon Oil	Private

5.19 *Catalog Age* lists the top 17 U.S. firms in annual catalog sales. The Dell Computer Corp. is number one with almost $12 billion in annual sales followed by Gateway 2000 with $6.3 billion. JCPenney Co. is third with $3.88 billion in annual sales.

Of these 17 firms on the list, eight are in some type of computer-related business. Suppose four of these firms are randomly selected.

a. What is the probability that none of the firms are in some type of computer-related business?

b. What is the probability that all four firms are in some type of computer-related business?

c. What is the probability that exactly two are in non-computer-related business?

5.20 W. Edwards Deming in his red bead experiment had a box of 4000 beads, of which 800 were red and 3200 were white.* Suppose a researcher were to conduct a modified version of the red bead experiment. In her experiment, she has a bag of 20 beads, of which four are red and 16 are white. This experiment requires a participant to reach into the bag and randomly select five beads without replacement.

a. What is the probability that the participant will select exactly four white beads?

b. What is the probability that the participant will select exactly four red beads?

c. What is the probability that the participant will select all red beads?

5.21 Shown here are the top 10 U.S. cities ranked by number of hotel rooms as compiled by Smith Travel Research.

RANK	CITY	NUMBER OF ROOMS
1	Las Vegas, NV	106,100
2	Orlando, FL	92,200
3	Los Angeles-Long Beach, CA	80,000
4	Atlanta, GA	73,100
5	Chicago, IL	71,000
6	Washington, DC	68,700
7	New York, NY	66,600
8	Dallas, TX	48,500
9	San Diego, CA	47,200
10	Anaheim-Santa Ana, CA	44,600

Suppose four of these cities are selected randomly.

a. What is the probability that exactly two cities are in California?

b. What is the probability that none of the cities are east of the Mississippi River?

c. What is the probability that exactly three of the cities are ones with more than 70,000 rooms?

5.22 A company produces and ships 16 personal computers knowing that four of them have defective wiring. The company that has purchased the computers is going to thoroughly test three of the computers. The purchasing company can detect the defective wiring. What is the probability that the purchasing company will find:

a. No defective computers?

b. Exactly three defective computers?

c. Two or more defective computers?

d. One or fewer defective computers?

5.23 A western city has 18 police officers who are eligible for promotion. Eleven of the 18 are Hispanic. Suppose only five of the police officers are chosen for promotion and that one is Hispanic. If the officers chosen for promotion had been selected by chance alone, what is the probability that one or fewer of the five promoted officers would have been Hispanic? What might this result indicate?

---

*Mary Walton, "Deming's Parable of Red Beads," *Across the Board* (February 1987):43–48.

Summary

Probability experiments produce random outcomes. A variable that contains the outcomes of a random experiment is called a random variable. Random variables such that the set of all possible values is at most a finite or countably infinite number of possible values are called discrete random variables. Random variables that take on values at all points over a given interval are called continuous random variables. Discrete distributions are constructed from discrete random variables. Continuous distributions are constructed from continuous random variables. Three discrete distributions are the binomial distribution, Poisson distribution, and hypergeometric distribution.

The binomial distribution fits experiments when only two mutually exclusive outcomes are possible. In theory, each trial in a binomial experiment must be independent of the other trials. However, if the population size is large enough in relation to the sample size ($n < 5\%N$), the binomial distribution can be used where applicable in cases where the trials are not independent. The probability of getting a desired outcome on any one trial is denoted as p, which is the probability of getting a success. The binomial distribution can be used to analyze discrete studies involving such things as heads/tails, defective/good, and male/female. The binomial formula is used to determine the probability of obtaining X outcomes in n trials. Binomial distribution problems can be solved more rapidly with the use of binomial tables than by formula. A binomial table can be constructed for every different pair of n and p values. Table A.2 of Appendix A contains binomial tables for selected values of n and p. The mean, or long-run average, of a binomial distribution is $\mu = n \cdot p$. The standard deviation of a binomial distribution is $\sqrt{n \cdot p \cdot q}$.

The Poisson distribution usually is used to analyze phenomena that produce rare occurrences. The only information required to generate a Poisson distribution is the long-run average, which is denoted by lambda (λ). The Poisson distribution pertains to occurrences over some interval. The assumptions are that each occurrence is independent of other occurrences and that the value of lambda remains constant throughout the experiment. Some examples of Poisson-type experiments are number of flaws per page of paper, number of crashes per 1000 commercial airline flights, and number of calls per minute to a switchboard. Poisson probabilities can be determined by either the Poisson formula or the Poisson tables in Table A.3 of Appendix A. Lambda is both the mean and the variance of a Poisson distribution.

The hypergeometric distribution is a discrete distribution that is usually used for binomial-type experiments when the population is small and finite and sampling is done without replacement. Because using the hypergeometric distribution is a tedious process, using the binomial distribution whenever possible is generally more advantageous.

Key Terms

binomial distribution	hypergeometric distribution
continuous distributions	lambda (λ)
continuous random variables	Poisson distribution
discrete distributions	random variable
discrete random variables	

SUPPLEMENTARY PROBLEMS

5.24 In a study by Peter D. Hart Research Associates for the Nasdaq Stock Market, it was determined that 20% of all stock investors are retired people. In addition, 40% of all U.S. adults have invested in mutual funds. Suppose a random sample of 25 stock investors is taken. What is the probability that exactly seven are retired people? What is the probability that 10 or more are retired people? How many retired people would you expect to find in a random sample of 25 stock investors? Suppose a random sample of 20 U.S. adults is taken. What is the probability that exactly eight adults have invested in mutual funds? What is the probability that fewer than six adults have invested in mutual funds? What is the probability that none of the adults have invested in mutual funds? What is the probability that 12 or more adults have invested in mutual funds? For which exact number of adults is the probability the highest? How does this compare to the expected number?

5.25 A service station has a pump that distributes diesel fuel to automobiles. The station owner estimates that only about 3.2 cars use the diesel pump every 2 hours. Assume the arrivals of diesel pump users are Poisson distributed.
   a. What is the probability that exactly three cars will arrive to use the diesel pump during a 1-hour period?
   b. Suppose the owner needs to shut down the diesel pump for half an hour to make repairs. However, the owner hates to lose any business. What is the probability that no cars will arrive to use the diesel pump during a half-hour period?
   c. Suppose five cars arrive during a 1-hour period to use the diesel pump. What is the probability of five or more cars arriving during a 1-hour period to use the diesel pump? If this outcome actually occurred, what might you conclude?

5.26 In a particular manufacturing plant, two machines (A and B) produce a particular part. One machine (B) is newer and faster. In one 5-minute period, a lot consisting of 32 parts is produced. Twenty-two are produced by machine B and the rest by machine A. Suppose an inspector randomly samples a dozen of the parts from this lot.
   a. What is the probability that exactly three parts were produced by machine A?
   b. What is the probability that half of the parts were produced by each machine?

   c. What is the probability that all of the parts were produced by machine B?
   d. What is the probability that seven, eight, or nine parts were produced by machine B?

5.27 Suppose that, for every lot of 100 computer chips a company produces, an average of 1.4 are defective. Another company buys many lots of these chips at a time, from which one lot is selected randomly and tested for defects. If the tested lot contains more than three defects, the buyer will reject all the lots sent in that batch. What is the probability that the buyer will accept the lots? Assume that the defects per lot are Poisson distributed.

5.28 The National Center for Health Statistics reports that 25% of all Americans between the ages of 65 and 74 have a chronic heart condition. Suppose you live in a state where the environment is conducive to good health and low stress and you believe the conditions in your state promote healthy hearts. To investigate this theory, you conduct a random telephone survey of 20 persons 65 to 74 years of age in your state.
   a. On the basis of the figure from the National Center for Health Statistics, what is the expected number of persons 65 to 74 years of age in your survey who have a chronic heart condition?
   b. Suppose only one person in your survey has a chronic heart condition. What is the probability of getting one or fewer people with a chronic heart condition in a sample of 20 if 25% of the population in this age bracket has this health problem? What do you conclude about your state from the sample data?

5.29 A survey conducted for the Northwestern National Life Insurance Company revealed that 70% of American workers say job stress caused frequent health problems. One in three said they expected to burn out in the job in the near future. Thirty-four percent said they thought seriously about quitting their job last year because of workplace stress. Fifty-three percent said they were required to work more than 40 hours a week very often or somewhat often.
   a. Suppose a random sample of 10 American workers is selected. What is the probability that more than seven of them say job stress caused frequent health problems? What is the expected number of workers who say job stress caused frequent health problems?

b. Suppose a random sample of 15 American workers is selected. What is the expected number of these sampled workers who say they will burn out in the near future? What is the probability that none of the workers say they will burn out in the near future?

c. Suppose a sample of seven workers is selected randomly. What is the probability that all seven say they are asked very often or somewhat often to work more than 40 hours a week? If this outcome actually happened, what might you conclude?

5.30 According to Padgett Business Services, 20% of all small-business owners say the most important advice for starting a business is to prepare for long hours and hard work. Twenty-five percent say the most important advice is to have good financing ready. Nineteen percent say having a good plan is the most important advice; 18% say studying the industry is the most important advice; and 18% list other advice. Suppose 12 small-business owners are contacted, and assume that the percentages hold for all small-business owners.

a. What is the probability that none of the owners would say preparing for long hours and hard work is the most important advice?

b. What is the probability that six or more owners would say preparing for long hours and hard work is the most important advice?

c. What is the probability that exactly five owners would say having good financing ready is the most important advice?

d. What is the expected number of owners who would say having a good plan is the most important advice?

5.31 A hair stylist has been in business one year. Sixty percent of his customers are walk-in business. If he randomly samples eight of the people from last week's list of customers, what is the probability that three or fewer were walk-ins? If this outcome actually occurred, what would be some of the explanations for it?

5.32 According to the U.S. Bureau of the Census, about 20% of Idaho residents live in metropolitan areas. This percentage is the lowest of all 50 states. A catalog sales company in Georgia has just purchased a list of Idaho consumers. Its market analyst randomly selects 25 people from this list.

a. What is the probability that exactly eight people live in metropolitan areas?

b. What is the probability that the analyst would get more than 10 people in this sample who live in metropolitan areas?

c. Suppose the analyst got more than 10 people who live in metropolitan areas from the group of 25. What might she conclude about the company's list of Idaho consumers? What might she conclude about the census figure?

5.33 Suppose that, for every family vacation trip by car of more than 2000 miles, an average of .60 flat tires occurs. Suppose also that the distribution of the number of flat tires per trip of more than 2000 miles is Poisson. What is the probability that a family will take a trip of more than 2000 miles and have no flat tires? What is the probability that the family will have three or more flat tires on such a trip? Suppose trips are independent and the value of lambda holds for all trips of more than 2000 miles. If a family takes two trips of more than 2000 miles during a summer, what is the probability that the family will have no flat tires on either trip?

5.34 *Editor and Publisher Yearbook* releases figures on the top newspapers in the United States. Shown here are the top 25 Sunday newspapers in the United States ranked according to circulation.

RANK	NEWSPAPER
1	*New York Times* (NY)
2	*Los Angeles Times* (CA)
3	*Washington Post* (DC)
4	*Chicago Tribune* (IL)
5	*Philadelphia Inquirer* (PA)
6	*Detroit News and Free Press* (MI)
7	*New York Daily News* (NY)
8	*Dallas Morning News* (TX)
9	*Boston Sunday Globe* (MA)
10	*Houston Chronicle* (TX)
11	*Atlanta Journal and Constitution* (GA)
12	*Minneapolis Star Tribune* (MN)
13	*Long Island Newsday* (NY)
14	*San Francisco Examiner and Chronicle* (CA)
15	*Newark Star-Ledger* (NJ)
16	*Phoenix Arizona Republic* (AZ)
17	*St. Louis Post-Dispatch* (MO)
18	*Cleveland Plain Dealer* (OH)
19	*Seattle Times/Post-Intelligencer* (WA)
20	*Baltimore Sunday Sun* (MD)
21	*Miami Herald* (FL)
22	*Denver Post* (CO)
23	*Milwaukee Journal Sentinel* (WI)
24	*San Diego Union-Tribune* (CA)
25	*Portland Oregonian* (OR)

Suppose a researcher wants to sample a portion of these newspapers and compare the sizes of the business sections of the Sunday papers. She randomly samples eight of these newspapers.

a. What is the probability that the sample contains exactly one newspaper located in New York state?

b. What is the probability that half of the newspapers are ranked in the top ten by circulation?

c. What is the probability that none of the newspapers are located in California?

d. What is the probability that exactly three of the newspapers are located in states that begin with the letter *M*?

5.35 An office in Albuquerque has 24 workers including management. Eight of the workers commute to work from the west side of the Rio Grande River. Suppose six of the office workers are randomly selected.

a. What is the probability that all six workers commute from the west side of the Rio Grande?

b. What is the probability that none of the workers commute from the west side of the Rio Grande?

c. Which probability from parts (a) and (b) was greatest? Why do think this is?

d. What is the probability that half of the workers do not commute from the west side of the Rio Grande?

5.36 According to the U.S. Bureau of the Census, 20% of the workers in Atlanta use public transportation. If 25 Atlanta workers are randomly selected, what is the expected number who use public transportation? Graph the binomial distribution for this sample. What is the mean and the standard deviation for this distribution? What is the probability that more than 12 of the selected workers use public transportation? Explain conceptually and from the graph why you would get this probability. Suppose you randomly sample 25 Atlanta workers and actually get 14 who use public transportation. Is this likely? How might you explain this result?

5.37 One of the earliest applications of the Poisson distribution was in analyzing incoming calls to a telephone switchboard. Analysts generally believe that random phone calls are Poisson distributed. Suppose phone calls to a switchboard arrive at an average rate of 2.4 calls per minute.

a. If an operator wants to take a 1-minute break, what is the probability that there will be no calls during a 1-minute interval?

b. If an operator can handle at most five calls per minute, what is the probability that the operator will be unable to handle the calls in any 1-minute period?

c. What is the probability that exactly three calls will arrive in a 2-minute interval?

d. What is the probability that one or fewer calls will arrive in a 15-second interval?

5.38 According to the American Medical Association, about 36% of all U.S. physicians under the age of 35 are women. Your company has just hired eight physicians under the age of 35 and none is a woman. If a group of women physicians under the age of 35 want to sue your company for discriminatory hiring practices, would they have a strong case based on these numbers? Use the binomial distribution to determine the probability of the company's hiring result occurring randomly and comment on the potential justification for a lawsuit.

5.39 The following table lists the 26 largest U.S. universities according to enrollment figures from the *World Almanac*.

UNIVERSITY	ENROLLMENT
University of Texas at Austin	48,857
Ohio State University-Columbus	48,278
University of Minnesota	45,410
Arizona State University	44,255
Michigan State University	42,603
University of Florida	41,713
University of Phoenix	41,467
Texas A&M University-College Station	41,461
Pennsylvania State-University Park	40,538
University of Wisconsin-Madison	40,196
University of Michigan	36,995
New York University	36,684
University of Illinois-Champaign	36,019
Purdue University-West Lafayette, Indiana	35,715
University of California at Los Angeles	35,558
University of Washington	35,367
Indiana University-Bloomington	34,937
University of South Florida	34,036
University of Arizona	33,737
University of Maryland-College Park	32,711
Brigham Young University	32,161
University of Houston	31,602
Wayne State University (Detroit)	30,729
University of California at Berkeley	30,584
Florida State University	30,401
Florida International University	30,012

a. If five different universities are selected randomly from the list, what is the probability that exactly three of them have enrollments of 40,000 or more?

b. If eight different universities are selected randomly from the list, what is the probability that two or fewer are universities in Michigan or Arizona?

c. Suppose universities are being selected randomly from this list *with* replacement. If five universities are sampled, what is the probability that the sample will contain exactly two universities in Texas?

5.40 In one midwestern city, the government has 14 repossessed houses, all of which are evaluated to be worth about the same. Ten of the houses are on the north side of town and the rest are on the west side. A local contractor has submitted a bid in which he offered to purchase four of the houses. Which houses the contractor will get is subject to a random draw.
   a. What is the probability that all four houses selected for the contractor will be on the north side of town?
   b. What is the probability that all four houses selected for the contractor will be on the west side of town?
   c. What is the probability that half of the houses selected for the contractor will be on the west side and half on the north side of town?

5.41 The Public Citizen's Health Research Group studied the serious disciplinary actions that were taken during a recent year on nonfederal medical doctors in the United States. The national average was 3.84 serious actions per 1000 doctors. The state with the lowest number was Minnesota, with 1.6 serious actions per 1000 doctors. Assume that the number of serious actions per 1000 doctors in both the United States and in Minnesota are Poisson distributed.
   a. What is the probability of randomly selecting 1000 U.S. doctors and finding out that there were no serious actions taken?
   b. What is the probability of randomly selecting 2000 U.S. doctors and finding out that there were six serious actions taken?
   c. What is the probability of randomly selecting 3000 Minnesota doctors and finding out that there were fewer than seven serious actions taken?

ANALYZING THE DATABASES

1. Use the manufacturing database. What is the probability that a randomly selected SIC code industry has a value of industry shipments equal to 2? Use this as the p value for a binomial experiment. If you were to randomly select 12 SIC code industries, what is the probability that fewer than three would have a value of industry shipments equal to 2? If you were to randomly select 25 SIC code industries, what is the probability that exactly eight would have a value of industry shipments equal to 2?

2. Use the hospital database. In this population of 200, what is the breakdown between hospitals that are general medical hospitals and psychiatric hospitals? Using those figures as a breakdown of the population and the hypergeometric distribution, what is the probability of randomly selecting 16 hospitals from this database and getting exactly nine that are psychiatric hospitals? Using the number of hospitals in this database that are for-profit, compute p = probability that a hospital is for-profit. Now use the binomial formula to determine the probability of randomly selecting 30 hospitals and getting exactly 10 that are for-profit.

3. Use the financial database of chemical companies. If five of these companies are selected randomly, what is the probability that exactly three have a return on equity of 15% or more? *Hint:* Use the hypergeometric distribution and a breakdown of this population of 19 companies to compute this probability. What is the probability of randomly selecting eight insurance companies and getting exactly four of them with average yields of less than 1%?

CASE

FUJI FILM INTRODUCES APS
In the early 1990s, Fuji Photo Film, USA, joined forces with four of its rivals to create the Advanced Photo System (APS), which is hailed as the first major development in the film industry since 35-millimeter technology was introduced. In February 1996, the new 24-millimeter system, promising clearer and sharper pictures, was launched. By the end of the year, the lack of communications and a limited supply of products had made retailers angry and consumers baffled. Advertising was almost nonexistent.

Because the product was developed by five industry rivals, the companies had enacted a secrecy agreement in which no one outside of company management, including the company's sales force, would know details about the product until each company introduced its APS products on the same day. When the product was actually introduced, there was little communication with retailers about the product, virtually no training of sales representatives on the product (so that they could demonstrate and explain the features), and a great underestimation of demand for the product. Fortunately, Fuji pressed on by taking an "honesty is the best policy" stance and explaining to retailers and other customers what had happened and asking for patience. In addition, Fuji increased its research to better ascertain market positioning and size. By 1997, Fuji had geared up production to meet the demand and was increasing customer promotion. APS products were on the road to success. By 1998, APS cameras owned 20% of the point-and-shoot camera market with the number expected to more than double by the year 2000.

DISCUSSION

Suppose you are a part of a Fuji team whose task it is to examine issues about market share, customer acceptance, complaints, and the reasons why new products are successful.

1. As stated, by 1998 APS cameras owned 20% of the point-and-shoot camera market with the number expected to more than double by the year 2002. Now it is the year 2002 and the market share might be nearer to 40%. Suppose 30 customers from the point-and-shoot camera market are randomly selected. If the market share is really .40, what is the expected number of point-and-shoot camera customers who purchase an APS camera? What is the probability that six or fewer (≤20%) purchase an APS camera? Suppose you actually got six or fewer APS customers in the sample of 30. Based on the probability just calculated, is this enough evidence to convince you that the market share is not 40%? Why or why not?

2. Suppose customer complaints on the 24-millimeter film are Poisson distributed at an average rate of 2.4 complaints/100,000 rolls sold. Suppose further that Fuji is having trouble with shipments being late and one batch of 100,000 rolls yields seven complaints from customers. Assuming that it is unacceptable to management for the average rate of complaints to increase, is this enough evidence to convince management that the average rate of complaints has increased, or can this be written off as a random occurrence that happens quite frequently? Produce the Poisson distribution for this question and discuss its implication for this problem.

3. One study of 52 product launches found that those undertaken with revenue growth as the main objective are more likely to fail than those undertaken to increase customer satisfaction or to create a new market such as the APS system. Suppose of the 52 products launched, 34 were launched with revenue growth as the main objective and the rest were launched to increase customer satisfaction or to create a new market. Now suppose only 10 of these products were successful (the rest failed) and seven were products that were launched to increase customer satisfaction or to create a new market. What is the probability of this occurring by chance? What does this probability tell you about the basic premise regarding the importance of the main objective?

6
Continuous Probability Distributions

Learning Objectives

The primary learning objective of Chapter 6 is to help you understand continuous distributions, thereby enabling you to:

1. Understand concepts of the uniform distribution.

2. Appreciate the importance of the normal distribution.

3. Recognize normal distribution problems and know how to solve such problems.

4. Decide when to use the normal distribution to approximate binomial distribution problems and know how to work such problems.

5. Decide when to use the exponential distribution to solve problems in business and know how to work such problems.

Whereas Chapter 5 focused on the characteristics and applications of discrete distributions, Chapter 6 concentrates on information about continuous distributions. Continuous distributions are constructed from continuous random variables in which values are taken on for every point over a given interval and are usually generated from experiments in which things are "measured" as opposed to "counted." With continuous distributions, probabilities of outcomes occurring between particular points are determined by calculating the area under the curve between those points. In addition, the entire area under the whole curve is 1. The many continuous distributions in statistics include the uniform distribution, the normal distribution, the exponential distribution, the t distribution, the chi-square distribution, and the F distribution. This chapter presents the uniform distribution, the normal distribution, and the exponential distribution.

6.1
The Uniform Distribution

The **uniform distribution,** sometimes referred to as the **rectangular distribution,** is *a relatively simple continuous distribution in which the same height, or $f(X)$, is obtained over a range of values.* The following probability density function defines a uniform distribution.

PROBABILITY DENSITY FUNCTION OF A UNIFORM DISTRIBUTION	$f(X) = \begin{cases} \dfrac{1}{b-a} & \text{for } a \le X \le b \\ 0 & \text{for all other values} \end{cases}$

Uniform distribution; Rectangular distribution
A relatively simple continuous distribution in which the same height is obtained over a range of values.

Figure 6.1 is an example of a uniform distribution. In a uniform, or rectangular, distribution, the area under the curve is equal to the product of the length and the width of the rectangle and equals 1. Because the distribution lies, by definition, between the X values of a and b, the length of the rectangle is $(b - a)$. Combining this with the fact that the area equals 1, the height of the rectangle can be solved as follows.

$$\text{Area of Rectangle} = (\text{Length})(\text{Height}) = 1$$

But

$$\text{Length} = (b - a)$$
$$(b - a)(\text{Height}) = 1$$
$$\text{Height} = \frac{1}{(b - a)}.$$

Figure 6.1

Uniform distribution

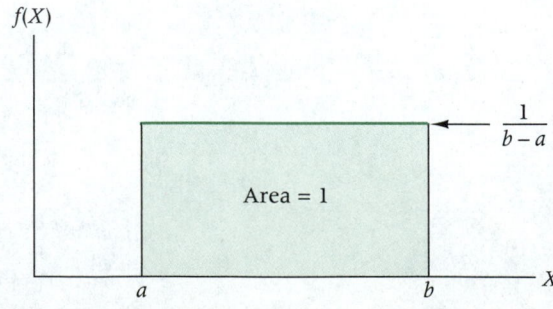

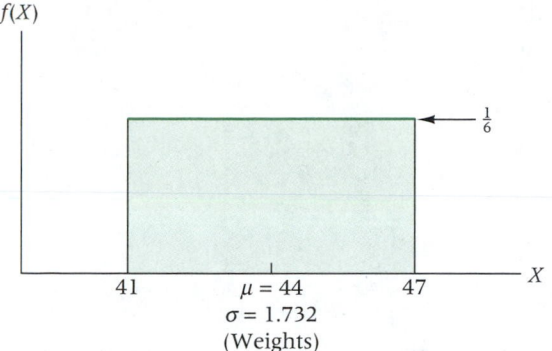

Figure 6.2

Distribution
of lot weights

From this, it can be seen that between the X values of a and b, the distribution has a constant height of $1/(b - a)$.

The mean and standard deviation of a uniform distribution are given as follows.

$$\mu = \frac{a + b}{2}$$

$$\sigma = \frac{b - a}{\sqrt{12}}$$

MEAN AND
STANDARD
DEVIATION
OF A UNIFORM
DISTRIBUTION

There are many possible situations in which data might be uniformly distributed. As an example, suppose a production line is set up to manufacture machine braces in lots of five per minute during a shift. When the lots are weighed, there is variation among the weights, with lot weights ranging from 41 to 47 grams in a uniform distribution. The height of this distribution is

$$f(X) = \text{Height} = \frac{1}{(b - a)} = \frac{1}{(47 - 41)} = \frac{1}{6}.$$

The mean and standard deviation of this distribution are

$$\text{Mean} = \frac{a + b}{2} = \frac{41 + 47}{2} = \frac{88}{2} = 44$$

$$\text{Standard Deviation} = \frac{b - a}{\sqrt{12}} = \frac{47 - 41}{\sqrt{12}} = \frac{6}{3.464} = 1.732.$$

Figure 6.2 is the uniform distribution for this example, with its mean, standard deviation, and the height of the distribution.

Determining Probabilities in a Uniform Distribution

With discrete distributions, the probability function yields the value of the probability. For continuous distributions, probabilities are calculated by determining the area over an interval of the function. With continuous distributions, any single value is possible but it

Figure 6.3

Solved probability
in a uniform
distribution

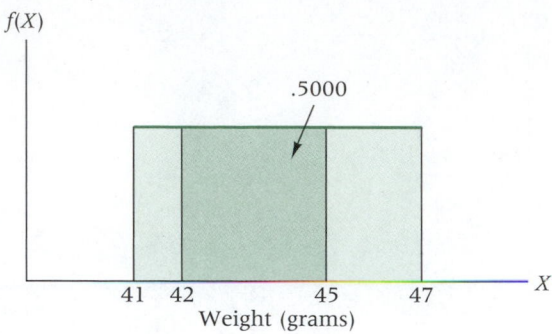

has a probability of zero. There is no area under the curve for a single point. The following equation is used to determine the probabilities of X for a uniform distribution between a and b.

PROBABILITIES IN A UNIFORM DISTRIBUTION	$$P(X) = \frac{X_2 - X_1}{b - a}$$

where:
$$a \leq X_1 \leq X_2 \leq b$$

Remember that the area between a and b is equal to 1. The probability for any interval that includes a and b is 1. The probability of $X \geq b$ or of $X \leq a$ is zero because there is no area above b or below a.

Suppose that on the machine braces problem we want to determine the probability that a lot weighs between 42 and 45 grams. This probability is computed as

$$P(X) = \frac{X_2 - X_1}{b - a} = \frac{45 - 42}{47 - 41} = \frac{3}{6} = .5000.$$

Figure 6.3 displays this solution.

The probability that a lot weighs more than 48 grams is zero, because $X = 48$ is greater than the upper value, $X = 47$, of the uniform distribution. A similar argument gives the probability of a lot weighing less than 40 grams. Because 40 is less than the lowest value of the uniform distribution range, 41, the probability is zero.

DEMONSTRATION PROBLEM 6.1	Suppose the amount of time it takes to assemble a plastic module ranges from 27 to 39 seconds and that assembly times are uniformly distributed. Describe the distribution. What is the probability that a given assembly will take between 30 and 35 seconds? Less than 30 seconds?

SOLUTION

$$f(X) = \frac{1}{39 - 27} = \frac{1}{12}$$

$$\mu = \frac{a + b}{2} = \frac{39 + 27}{2} = 33$$

$$\sigma = \frac{b - a}{\sqrt{12}} = \frac{39 - 27}{\sqrt{12}} = \frac{12}{\sqrt{12}} = 3.464$$

The height of the distribution is 1/12. The mean time is 33 seconds with a standard deviation of 3.464 seconds.

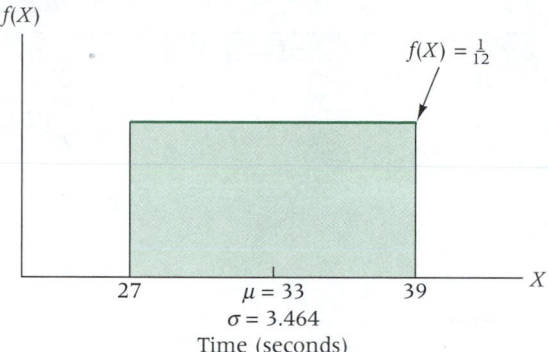

$$P(30 \leq X \leq 35) = \frac{35 - 30}{39 - 27} = \frac{5}{12} = .4167$$

There is a .4167 probability that it will take between 30 and 35 seconds to assemble the module.

$$P(X < 30) = \frac{30 - 27}{39 - 27} = \frac{3}{12} = .2500$$

There is a .2500 probability that it will take less than 30 seconds to assemble the module. Because there is no area less than 27 seconds, $P(X < 30)$ is determined by using only the interval $27 \leq X < 30$. In a continuous distribution, there is no area at any one point (only over an interval). Thus the probability $X < 30$ is the same as the probability of $X \leq 30$.

Analysis Using Excel

Even though Excel does not have a tool for directly computing uniform distribution probabilities, such probabilities can be obtained using **FAST ＼ STAT**. The **FAST ＼ STAT** dialog box for computing uniform distribution probabilities is shown in Figure 6.4. This

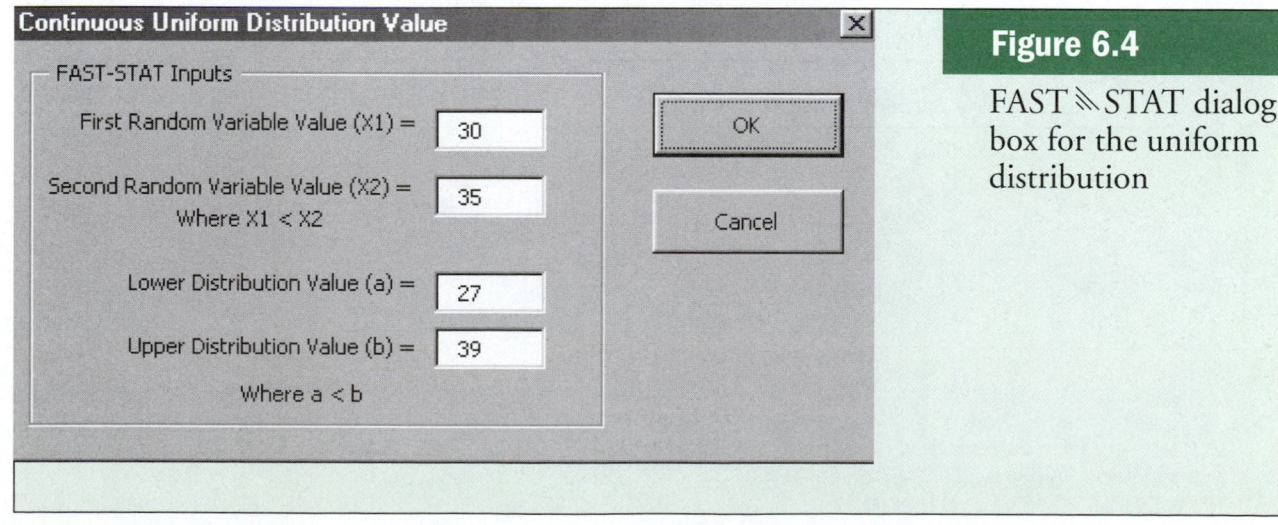

Figure 6.4

FAST ＼ STAT dialog box for the uniform distribution

Figure 6.5a

FAST \ STAT output for Demonstration Problem 6.1

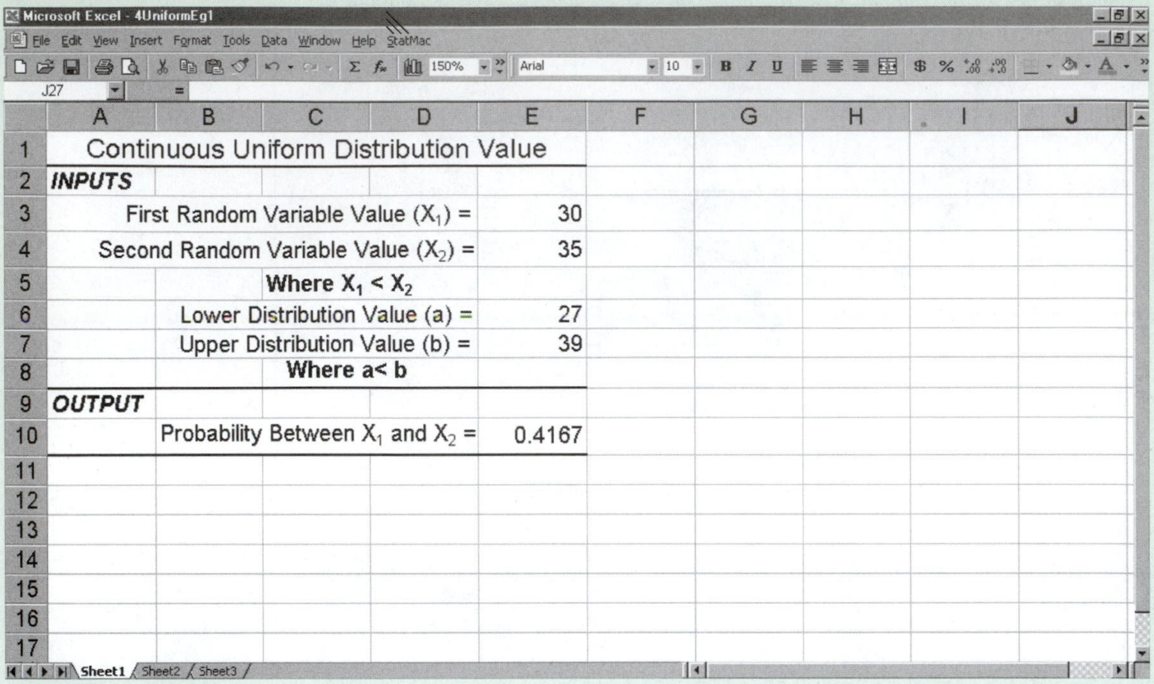

	A	B	C	D	E
1	Continuous Uniform Distribution Value				
2	*INPUTS*				
3	First Random Variable Value (X_1) =				30
4	Second Random Variable Value (X_2) =				35
5	**Where $X_1 < X_2$**				
6	Lower Distribution Value (a) =				27
7	Upper Distribution Value (b) =				39
8	**Where a< b**				
9	*OUTPUT*				
10	Probability Between X_1 and X_2 =				0.4167

Microsoft Excel - 4UniformEg1
File Edit View Insert Format Tools Data Window Help StatMac

Figure 6.5b

	A	B	C	D	E
1	Continuous Uniform Distribution Value				
2	*INPUTS*				
3	First Random Variable Value (X_1) =				27
4	Second Random Variable Value (X_2) =				30
5	**Where $X_1 < X_2$**				
6	Lower Distribution Value (a) =				27
7	Upper Distribution Value (b) =				39
8	**Where a< b**				
9	*OUTPUT*				
10	Probability Between X_1 and X_2 =				0.2500

Microsoft Excel - 4UniformEg2
File Edit View Insert Format Tools Data Window Help StatMac

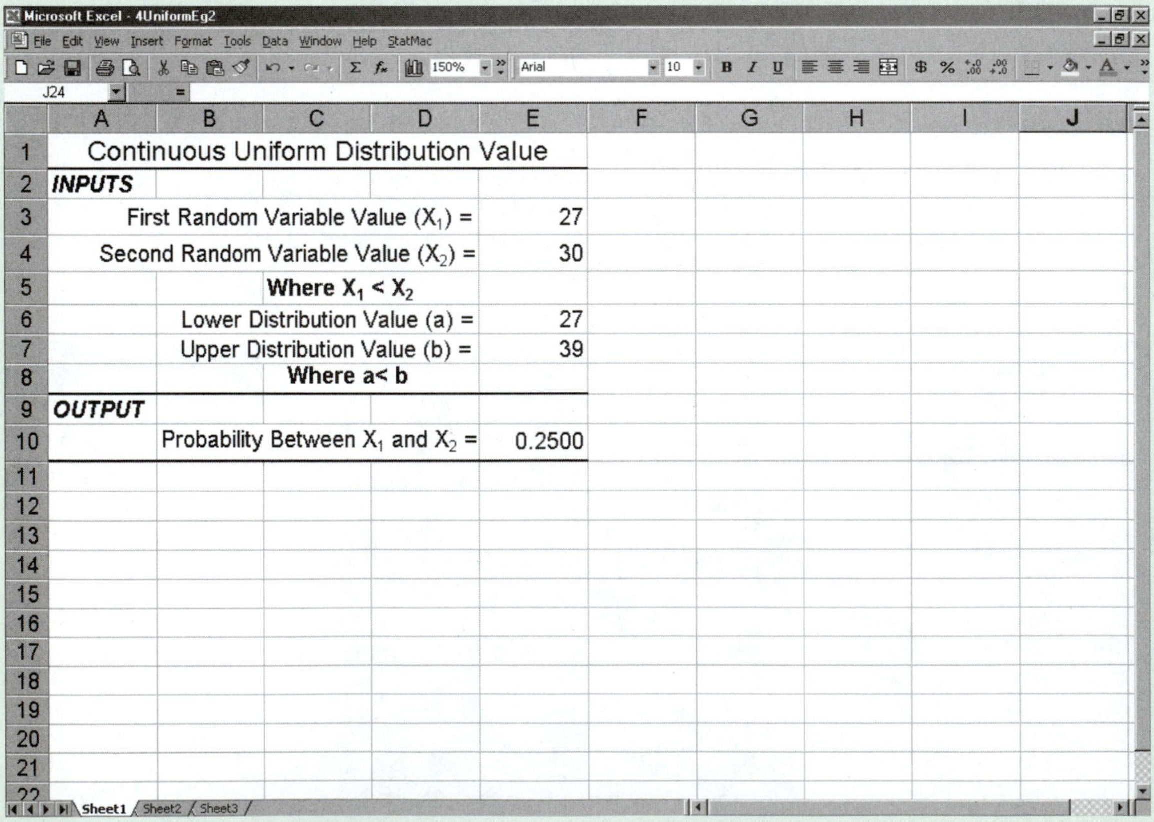

dialog box requires the values of *a, b, X₁* and *X₂.* The resulting worksheet repeats the inputs and presents the probability between the two values for X, $P(X)$. Figures 6.5a and b display the **FAST** **STAT** output for the two probability computations of Demonstration Problem 6.1.

According to the National Association of Insurance Commissioners, the average annual cost for automobile insurance in the United States is $691. Suppose automobile insurance costs are uniformly distributed in the United States with a range of from $200 to $1182. What is the standard deviation of this uniform distribution? What is the height of the distribution? What is the probability that a person's annual cost for automobile insurance in the United States is between $410 and $825?

SOLUTION

The mean is given as $691. The value of *a* is $200 and *b* is $1182.

$$\sigma = \frac{b-a}{\sqrt{12}} = \frac{1182-200}{\sqrt{12}} = 283.479$$

The height of the distribution is $1/(1182 - 200) = 1/982 = .001$. $X_1 = 410$ and $X_2 = 825$.

$$P(410 \leq X \leq 825) = \frac{825 - 410}{1182 - 200} = \frac{415}{982} = .4226$$

The probability that a randomly selected person pays between $410 and $825 annually for automobile insurance in the United States is .4226. That is, about 42.26% of all people in the United States pay in that range. Figure 6.6 displays the **FAST** **STAT** output for this problem.

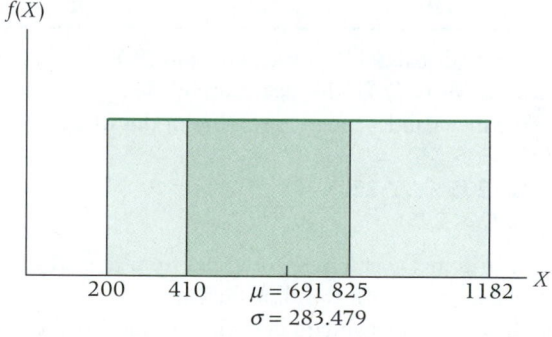

Figure 6.6	FAST ＼ STAT output for Demonstration Problem 6.2

Microsoft Excel - 4UniformEg3

File Edit View Insert Format Tools Data Window Help StatMac

J24

	A	B	C	D	E
1	Continuous Uniform Distribution Value				
2	INPUTS				
3	First Random Variable Value (X₁) =				410
4	Second Random Variable Value (X₂) =				825
5	Where X₁ < X₂				
6	Lower Distribution Value (a) =				200
7	Upper Distribution Value (b) =				1182
8	Where a< b				
9	OUTPUT				
10	Probability Between X₁ and X₂ =				0.4226

Sheet1 / Sheet2 / Sheet3 /

6.1   Values are uniformly distributed between 200 and 240.
  a. What is the value of $f(X)$ for this distribution?
  b. Determine the mean and standard deviation of this distribution.
  c. Probability of $(X > 230) = ?$
  d. Probability of $(205 \leq X \leq 220) = ?$
  e. Probability of $(X \leq 225) = ?$

6.2 X is uniformly distributed over a range of values from 8 to 21.
  a. What is the value of $f(X)$ for this distribution?
  b. Determine the mean and standard deviation of this distribution.
  c. Probability of $(10 \leq X < 17) = ?$
  d. Probability of $(X < 22) = ?$
  e. Probability of $(X \geq 7) = ?$

6.3   The retail price of a medium-size box of a well-known brand of cornflakes ranges from \$2.80 to \$3.14. Assume these prices are uniformly distributed. What is the average price and the standard deviation of prices in this distribution? If a price is randomly selected from this list, what is the probability that it will be between \$3.00 and \$3.10?

6.4   The average fill volume of a regular can of soft drink is 12 oz. Suppose the fill volume of these cans ranges from 11.97 to 12.03 oz and is uniformly distributed.

What is the height of this distribution? What is the probability that a randomly selected can contains more than 12.01 oz of fluid? What is the probability that the fill volume is between 11.98 and 12.01 oz?

6.5   The average U.S. household spends $2100 a year on all types of insurance. Suppose the figures are uniformly distributed between the values of $400 and $3800. What is the standard deviation and the height of this distribution? What proportion of households spend more than $3000 a year on insurance? More than $4000? Between $700 and $1500?

Probably the most widely known and used of all distributions is the **normal distribution.** It fits many human characteristics, such as height, weight, length, speed, IQ, scholastic achievement, and years of life expectancy, among others. Like their human counterparts, living things in nature, such as trees, animals, insects, and others, have many characteristics that are normally distributed.

Many variables in business and industry also are normally distributed. Some examples of variables that could produce normally distributed measurements include the annual cost of household insurance, the cost per square foot of renting warehouse space, and the satisfaction of managers with their support from ownership on a five-point scale. In addition, most items produced or filled by machines are normally distributed.

Because of its many applications, the normal distribution is an extremely important distribution. Besides the many variables mentioned that are normally distributed, the normal distribution and its associated probabilities are an integral part of statistical process control (see Chapter 15). Chapter 7 addresses the fact that, when large enough sample sizes are taken, many statistics are normally distributed regardless of the shape of the underlying distribution. Figure 6.7 is the graphic representation of the normal distribution: the normal curve.

History of the Normal Distribution

Discovery of the normal curve of errors is generally credited to mathematician and astronomer Karl Gauss (1777–1855), who recognized that the errors of repeated measurement of objects are often normally distributed.* Thus the normal distribution is sometimes referred to as the Gaussian distribution or the normal curve of error. A modern-day analogy of Gauss's work might be the distribution of measurements of machine-produced parts, which often yield a normal curve of error around a mean specification.

6.2
Normal Distribution

Normal distribution
A widely known and much-used continuous distribution that fits the measurements of many human characteristics and many machine-produced items.

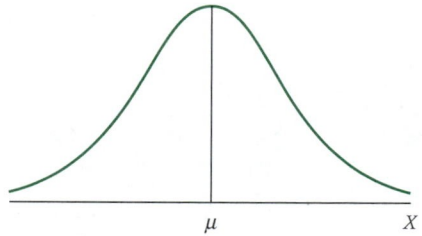

Figure 6.7

The normal curve

---

*John A. Ingram and Joseph G. Monks, *Statistics for Business and Economics* (San Diego: Harcourt Brace Jovanovich, Publishers, 1989).

To a lesser extent, some credit has been given to Pierre-Simon de Laplace (1749–1827) for discovering the normal distribution. However, many people now believe that Abraham de Moivre (1667–1754), a French mathematician, first understood the normal distribution. De Moivre determined that the binomial distribution approached the normal distribution as a limit. De Moivre worked with remarkable accuracy. His published table values for the normal curve are only a few ten-thousandths off the values of currently published tables.*

The normal distribution has the following characteristics.

- It is a continuous distribution.
- It is a symmetrical distribution.
- It is asymptotic to the axis.
- It is unimodal.
- It is a family of curves.
- Area under the curve is 1.

The normal distribution is symmetrical. Each half of the distribution is a mirror image of the other half. Many normal distribution tables contain probability values for only one side of the distribution because probability values for the other side of the distribution are identical owing to symmetry.

In theory, the normal distribution is asymptotic to the axis. That is, it does not touch the X axis and it goes forever in each direction. The reality is that most applications of the normal curve are experiments that have finite limits of potential outcomes. For example, even though SAT scores are analyzed by the normal distribution, the range of SAT scores is only from 200 to 800.

The normal curve sometimes is referred to as the *bell-shaped curve*. It is unimodal in that values *mound up* in only one portion of the graph—the center of the curve. The normal distribution actually is a family of curves. Every unique value of the mean and every unique value of the standard deviation has a different normal curve. In addition, *the total area under any normal distribution is 1*. The area under the curve yields the probabilities, so the total of all probabilities for a normal distribution is 1. Because the distribution is symmetric, the area of the distribution on each side of the mean is 0.5.

Probability Density Function of the Normal Distribution

The normal distribution is described or characterized by two parameters: μ and σ. The values of μ and σ produce a normal distribution. The function of the normal distribution is

$$f(X) = \left(\frac{1}{\sigma\sqrt{2\pi}} \right) e^{-(1/2)[(X-\mu)/\sigma]^2}$$

where:

μ = mean of X,
σ = standard deviation of X,
π = 3.14159 . . . , and
e = 2.71828. . . .

Because the formula is so complex, using it to determine areas under the curve is cumbersome and time-consuming. Virtually all researchers use table values to analyze normal distribution problems rather than this formula.

---

*Roger E. Kirk, *Statistical Issues: A Reader for the Behavioral Sciences* (Monterey, CA: Brooks/Cole Publishing Co., 1972).

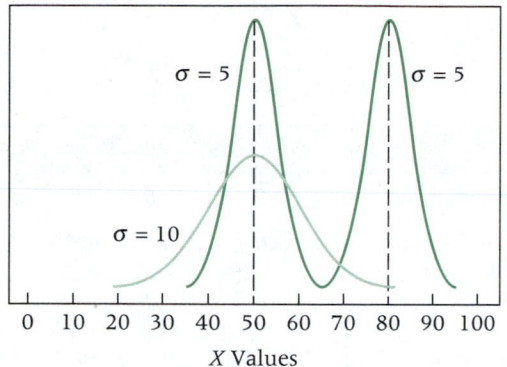

Figure 6.8

Normal curves
for three different
combinations of
means and standard
deviations

Standardized Normal Distribution

Every unique pair of μ and σ values defines a different normal distribution. Figure 6.8 shows graphs of normal distributions for the following three pairs of parameters.

1. $\mu = 50$ and $\sigma = 5$
2. $\mu = 80$ and $\sigma = 5$
3. $\mu = 50$ and $\sigma = 10$

Note that every change in a parameter (μ or σ) determines a different normal distribution. This characteristic of the normal curve (a *family* of curves) could make analysis by the normal distribution tedious because volumes of normal curve tables—one for each different combination of μ and σ—would be required. Fortunately, a mechanism was developed by which all normal distributions can be converted into a single distribution: the Z distribution. This process yields the **standardized normal distribution** (or curve). The conversion formula for any X value of a given normal distribution follows.

Standardized normal distribution
Z distribution; a distribution of Z scores produced for values from a normal distribution with a mean of 0 and a standard deviation of 1.

$$Z = \frac{X - \mu}{\sigma}$$

Z FORMULA

A **Z score** is *the number of standard deviations that a value, X, is above or below the mean.* If the value of X is less than the mean, the Z score is negative; if the value of X is more than the mean, the Z score is positive. This formula allows conversion of the distance of any X value from its mean into standard deviation units. A standard Z table can be used to find probabilities for any normal curve problem that has been converted to Z scores. The **Z distribution** is *a normal distribution with a mean of 0 and a standard deviation of 1.* That is, any value of X at the mean of a normal curve is zero standard deviations from the mean. Any value of X that is one standard deviation above the mean has a Z value of 1. The empirical rule, introduced in Chapter 3, is based on the normal distribution in which about 68% of all values are within one standard deviation of the mean. In a Z distribution, about 68% of the Z values are between $Z = -1$ and $Z = +1$.

The Z distribution probability values are given in Table A.5. Because it is so frequently used, the Z distribution is also printed inside the back cover of this text. For discussion purposes, a list of Z distribution values is presented in Table 6.1.

Z score
The number of standard deviations a value (X) is above or below the mean of a set of numbers when the data are normally distributed.

Z distribution
A distribution of Z scores; a normal distribution with a mean of 0 and a standard deviation of 1.

TABLE 6.1
Z Distribution

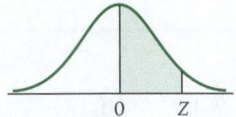

	SECOND DECIMAL PLACE IN *Z*								
Z	*0.00*	*0.01*	*0.02*	*0.03*	*0.04*	*0.05*	*0.06*	*0.07*	*0.08*	*0.09*
0.0	.0000	.0040	.0080	.0120	.0160	.0199	.0239	.0279	.0319	.0359
0.1	.0398	.0438	.0478	.0517	.0557	.0596	.0636	.0675	.0714	.0753
0.2	.0793	.0832	.0871	.0910	.0948	.0987	.1026	.1064	.1103	.1141
0.3	.1179	.1217	.1255	.1293	.1331	.1368	.1406	.1443	.1480	.1517
0.4	.1554	.1591	.1628	.1664	.1700	.1736	.1772	.1808	.1844	.1879
0.5	.1915	.1950	.1985	.2019	.2054	.2088	.2123	.2157	.2190	.2224
0.6	.2257	.2291	.2324	.2357	.2389	.2422	.2454	.2486	.2517	.2549
0.7	.2580	.2611	.2642	.2673	.2704	.2734	.2764	.2794	.2823	.2852
0.8	.2881	.2910	.2939	.2967	.2995	.3023	.3051	.3078	.3106	.3133
0.9	.3159	.3186	.3212	.3238	.3264	.3289	.3315	.3340	.3365	.3389
1.0	.3413	.3438	.3461	.3485	.3508	.3531	.3554	.3577	.3599	.3621
1.1	.3643	.3665	.3686	.3708	.3729	.3749	.3770	.3790	.3810	.3830
1.2	.3849	.3869	.3888	.3907	.3925	.3944	.3962	.3980	.3997	.4015
1.3	.4032	.4049	.4066	.4082	.4099	.4115	.4131	.4147	.4162	.4177
1.4	.4192	.4207	.4222	.4236	.4251	.4265	.4279	.4292	.4306	.4319
1.5	.4332	.4345	.4357	.4370	.4382	.4394	.4406	.4418	.4429	.4441
1.6	.4452	.4463	.4474	.4484	.4495	.4505	.4515	.4525	.4535	.4545
1.7	.4554	.4564	.4573	.4582	.4591	.4599	.4608	.4616	.4625	.4633
1.8	.4641	.4649	.4656	.4664	.4671	.4678	.4686	.4693	.4699	.4706
1.9	.4713	.4719	.4726	.4732	.4738	.4744	.4750	.4756	.4761	.4767
2.0	.4772	.4778	.4783	.4788	.4793	.4798	.4803	.4808	.4812	.4817
2.1	.4821	.4826	.4830	.4834	.4838	.4842	.4846	.4850	.4854	.4857
2.2	.4861	.4864	.4868	.4871	.4875	.4878	.4881	.4884	.4887	.4890
2.3	.4893	.4896	.4898	.4901	.4904	.4906	.4909	.4911	.4913	.4916
2.4	.4918	.4920	.4922	.4925	.4927	.4929	.4931	.4932	.4934	.4936
2.5	.4938	.4940	.4941	.4943	.4945	.4946	.4948	.4949	.4951	.4952
2.6	.4953	.4955	.4956	.4957	.4959	.4960	.4961	.4962	.4963	.4964
2.7	.4965	.4966	.4967	.4968	.4969	.4970	.4971	.4972	.4973	.4974
2.8	.4974	.4975	.4976	.4977	.4977	.4978	.4979	.4979	.4980	.4981
2.9	.4981	.4982	.4982	.4983	.4984	.4984	.4985	.4985	.4986	.4986
3.0	.4987	.4987	.4987	.4988	.4988	.4989	.4989	.4989	.4990	.4990
3.1	.4990	.4991	.4991	.4991	.4992	.4992	.4992	.4992	.4993	.4993
3.2	.4993	.4993	.4994	.4994	.4994	.4994	.4994	.4995	.4995	.4995
3.3	.4995	.4995	.4995	.4996	.4996	.4996	.4996	.4996	.4996	.4997
3.4	.4997	.4997	.4997	.4997	.4997	.4997	.4997	.4997	.4997	.4998
3.5	.4998									
4.0	.49997									
4.5	.499997									
5.0	.4999997									

Working Normal Curve Problems

The mean and standard deviation of a normal distribution and the *Z* formula and table enable a researcher to determine the probabilities for intervals of any particular values of a normal curve. One example is the many possible probability values of Graduate Management Aptitude Test (GMAT) scores examined next.

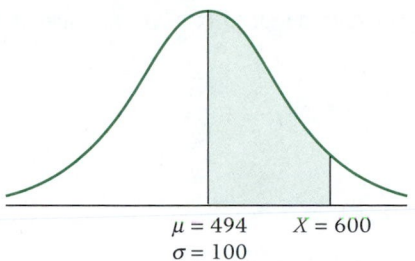

Figure 6.9

Graphical depiction of the area between a score of 600 and a mean on a GMAT

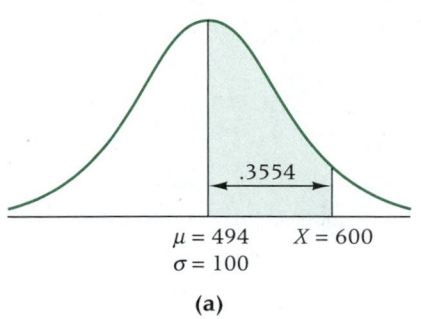

(a)

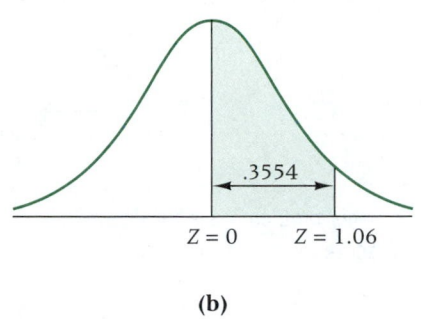

(b)

Figure 6.10

Graphical solutions to the GMAT problem

The Graduate Management Aptitude Test (GMAT), produced by the Educational Testing Service in Princeton, New Jersey, is widely used by graduate schools of business in the United States as an entrance requirement. Assuming that the scores are normally distributed, probabilities of achieving scores over various ranges of the GMAT can be determined. In a recent year, the mean GMAT score was 494 and the standard deviation was about 100. What is the probability that a randomly selected score from this administration of the GMAT is between 600 and the mean? That is,

$$P(494 \leq X \leq 600 \mid \mu = 494 \text{ and } \sigma = 100) = ?$$

Figure 6.9 is a graphical representation of this problem.

The Z formula yields the number of standard deviations that the X value, 600, is away from the mean.

$$Z = \frac{X - \mu}{\sigma} = \frac{600 - 494}{100} = \frac{106}{100} = 1.06$$

The Z value of 1.06 reveals that the GMAT score of 600 is 1.06 standard deviations more than the mean. The Z distribution values in Table 6.1 give the probability of a value being between this value of X and the mean. The whole-number and tenths-place portion of the Z score appear in the first column of Table 6.1 (the 1.0 portion of this Z score). Across the top of the table are the values of the hundredths-place portion of the Z score. For this Z score, the hundredths-place value is 6. The probability value in Table 6.1 for $Z = 1.06$ is .3554. The shaded portion of the curve at the top of the table indicates that the probability value given *always* is the probability or area between an X value and the mean. In this particular example, that is the desired area. Thus the answer is that .3554 of the scores on the GMAT are between a score of 600 and the mean of 494. Figure 6.10 (a) depicts graphically the solution in terms of X values. Figure 6.10 (b) shows the solution in terms of Z values.

What is the probability of obtaining a score greater than 700 on a GMAT test that has a mean of 494 and a standard deviation of 100? Assume GMAT scores are normally distributed.

$$P(X > 700 \mid \mu = 494 \text{ and } \sigma = 100) = \,?$$

SOLUTION
Examine the following diagram.

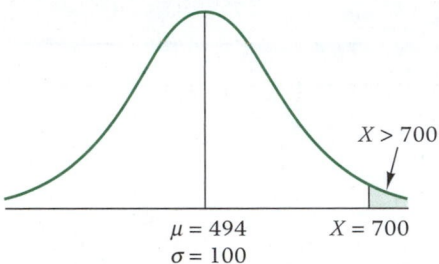

This problem calls for determining the area of the upper tail of the distribution. The *Z* score for this problem is

$$Z = \frac{X - \mu}{\sigma} = \frac{700 - 494}{100} = \frac{206}{100} = 2.06.$$

Table 6.1 gives a probability of .4803 for this *Z* score. However, the table gives the area between an *X* value and the mean, which in this case is .4803. This value is the probability of randomly drawing a GMAT with a score between the mean and 700. Finding the probability of getting a score greater than 700, which is the tail of the distribution, requires subtracting the probability value of .4803 from .5000, because each half of the distribution contains .5000 of the area. The result is .0197. Note that an attempt to determine the area of $X \geq 700$ instead of $X > 700$ would have made no difference because, in continuous distributions, the area under an exact number such as $X = 700$ is zero. A line segment has no width and hence no area.

.5000 (probability of *X* greater than the mean)

$-$.4803 (probability of *X* between 700 and the mean)

.0197 (probability of *X* greater than 700)

The solution is depicted graphically in (a) for *X* values and in (b) for *Z* values.

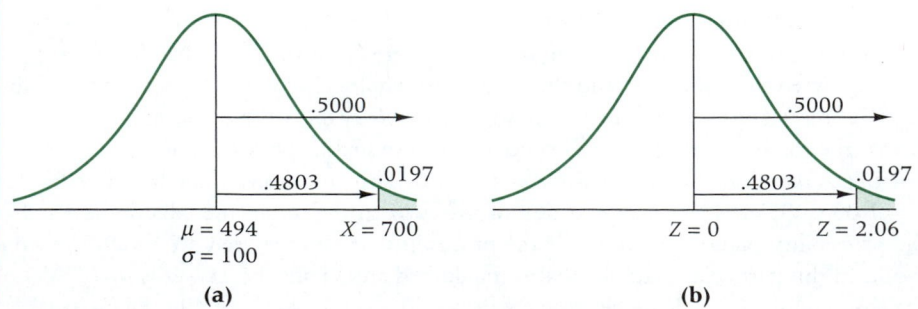

(a)   (b)

For the same GMAT examination, what is the probability of randomly drawing a score that is 550 or less?

$$P(X \leq 550 \mid \mu = 494 \text{ and } \sigma = 100) = ?$$

SOLUTION
A sketch of this problem is shown here. Determine the area under the curve for all values less than or equal to 550.

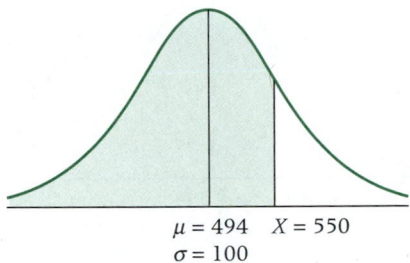

$$\mu = 494 \quad X = 550$$
$$\sigma = 100$$

The Z formula yields the area between 550 and the mean.

$$Z = \frac{X - \mu}{\sigma} = \frac{550 - 494}{100} = \frac{56}{100} = 0.56$$

The area under the curve for $Z = 0.56$ is .2123, which is the probability of getting a score between 550 and the mean. However, obtaining the probability for all values less than or equal to 550 also requires including the values less than the mean. Because one-half or .5000 of the values are less than the mean, the probability of $X \leq 550$ is found as follows.

.5000 (probability of values less than the mean)

+.2123 (probability of values between 550 and the mean)

.7123 (probability of values $\leq$ 550)

This solution is depicted graphically in (a) for X values and in (b) for Z values.

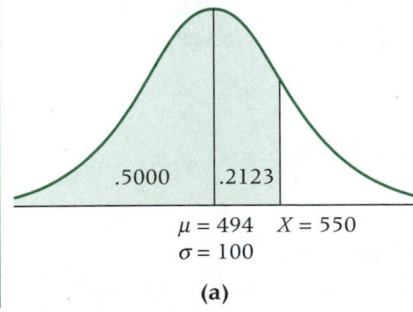

$$\mu = 494 \quad X = 550$$
$$\sigma = 100$$

(a)

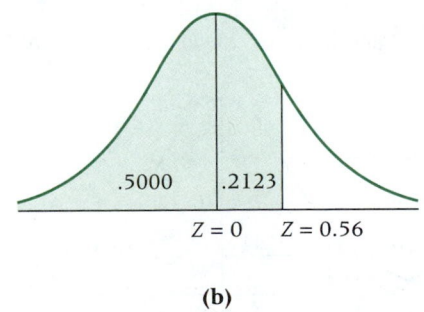

$$Z = 0 \quad Z = 0.56$$

(b)

What is the probability of getting a score of less than 400 on the same GMAT test?

$$P(X < 400 \mid \mu = 494 \text{ and } \sigma = 100) = ?$$

SOLUTION

The following sketch reveals that the problem is to determine the area of the lower tail of the distribution.

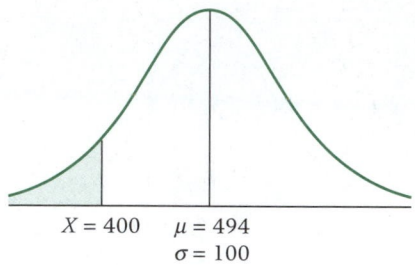

$$X = 400 \qquad \mu = 494$$
$$\sigma = 100$$

The Z score for this problem is

$$Z = \frac{X - \mu}{\sigma} = \frac{400 - 494}{100} = \frac{-94}{100} = -0.94.$$

Note that this Z value is negative. A negative Z value indicates that the X value is below the mean and the Z value is on the left side of the distribution. None of the Z values in Table 6.1 is negative. However, because the normal distribution is symmetric, probabilities for Z values on the left side of the distribution are the same as the values on the right side of the distribution. The negative sign in the Z value merely indicates that the area is on the left side of the distribution. The probability is always positive. Table 6.1 yields a probability of .3264 for a Z value of 0.94. The problem is to find the area in the lower tail of the distribution, so the probability, .3264, must be subtracted from .5000 to obtain the answer.

.5000 (probability of value less than the mean)

−.3264 (probability of value between 400 and the mean)

.1736 (probability of value less than 400)

Graphically, the solution is shown in (a) for X values and in (b) for Z values.

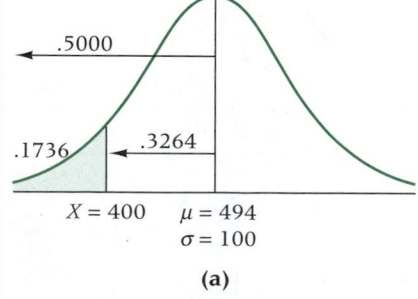

$$X = 400 \qquad \mu = 494$$
$$\sigma = 100$$

(a)

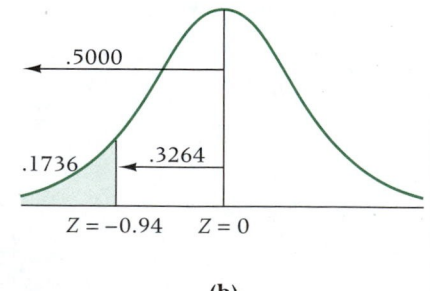

$$Z = -0.94 \qquad Z = 0$$

(b)

What is the probability of randomly obtaining a score between 300 and 600 on the same GMAT exam?

$$P(300 < X < 600 \mid \mu = 494 \text{ and } \sigma = 100) = ?$$

SOLUTION

The following sketch depicts the problem graphically: determine the area between $X = 300$ and $X = 600$, which spans the mean value. As areas in the Z distribution are given in relation to the mean, this problem must be worked as two separate problems and the results combined.

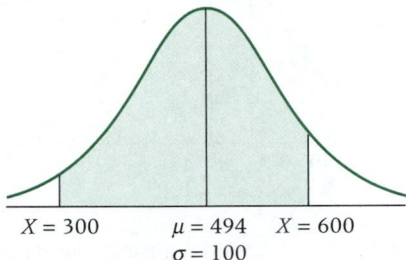

$X = 300 \qquad \mu = 494 \quad X = 600$
$\sigma = 100$

A Z score is determined for each X value.

$$Z = \frac{X - \mu}{\sigma} = \frac{600 - 494}{100} = \frac{106}{100} = 1.06$$

and

$$Z = \frac{X - \mu}{\sigma} = \frac{300 - 494}{100} = \frac{-194}{100} = -1.94$$

The probability for $Z = 1.06$ is .3554; the probability for $Z = -1.94$ is .4738. The solution of $P(300 < X < 600)$ is obtained by summing the probabilities.

.3554 (probability of a value between the mean and 600)

+.4738 (probability of a value between the mean and 300)

.8292 (probability of a value between 300 and 600)

Graphically, the solution is shown in (a) for X values and in (b) for Z values.

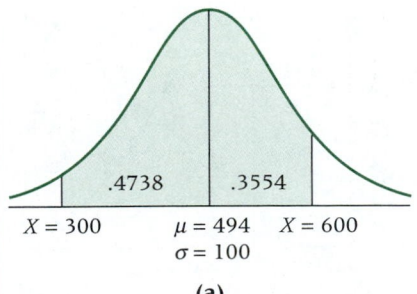

.4738   .3554

$X = 300 \qquad \mu = 494 \quad X = 600$
$\sigma = 100$

(a)

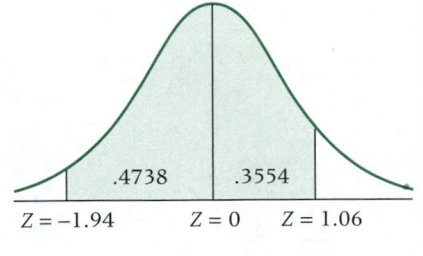

.4738   .3554

$Z = -1.94 \qquad Z = 0 \qquad Z = 1.06$

(b)

DEMONSTRATION PROBLEM 6.7

What is the probability of getting a score between 350 and 450 on the same GMAT exam?

$$P(350 < X < 450 \mid \mu = 494 \text{ and } \sigma = 100) = ?$$

SOLUTION

The following sketch reveals that the solution to the problem involves determining the area of the shaded slice in the lower half of the curve.

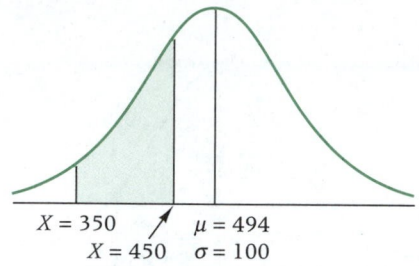

In this problem, the two X values are on the same side of the mean. The areas or probabilities of each X value must be determined and the final probability found by determining the difference between the two areas.

$$Z = \frac{X - \mu}{\sigma} = \frac{350 - 494}{100} = \frac{-144}{100} = -1.44$$

and

$$Z = \frac{X - \mu}{\sigma} = \frac{450 - 494}{100} = \frac{-44}{100} = -0.44$$

The probability associated with $Z = -1.44$ is .4251.
The probability associated with $Z = -0.44$ is .1700.
Subtracting gives the solution.

    .4251 (probability of a value between 350 and the mean)

   −.1700 (probability of a value between 450 and the mean)

    .2551 (probability of a value between 350 and 450)

Graphically, the solution is shown in (a) for X values and in (b) for Z values.

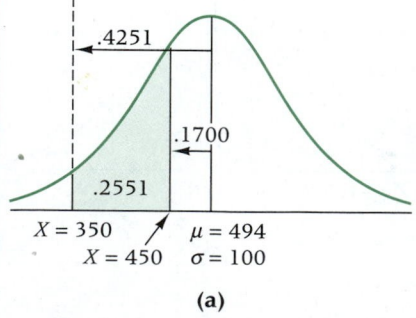

(a)

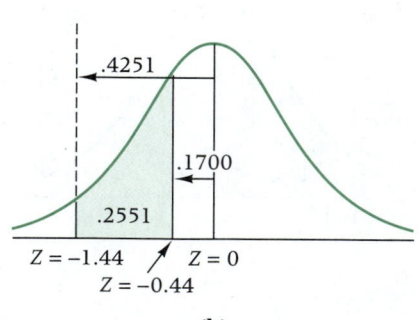

(b)

Analysis Using Excel

Excel can compute probabilities for normal curve problems using the function name, **NORMDIST,** in the **Statistical** function category of the Paste Function (f_x). Shown in Figure 6.11 is the dialog box for **NORMDIST.**

The **NORMDIST** dialog box requires four entries: the value of X, the mean (μ), the standard deviation (σ), and a logical value. The logical value requires either the word, TRUE, or the word, FALSE. If TRUE is inserted into this line, Excel will compute a cumulative probability from the left of the normal distribution. If FALSE is inserted into this line, Excel will compute the probability mass function. In virtually all cases for this book, we will use TRUE in this line.

Since all cumulative normal curve probabilities using **NORMDIST** cumulate from the left, we will have to sometimes manipulate the Excel answers to solve the problem at hand. For Demonstration Problem 6.3, we will take 1 – the answer provided by Excel since we are solving for the upper tail of the distribution. On both Demonstration Problems 6.4 and 6.5, the cumulative Excel answer is correct as it is. To solve for the probability in Demonstration Problem 6.6, subtract the Excel computed probabilities for $X < 600$ and $X < 300$. To solve for the probability in Demonstration Problem 6.7, subtract the Excel computed probabilities for $X < 450$ and $X < 350$. See Figure 6.12.

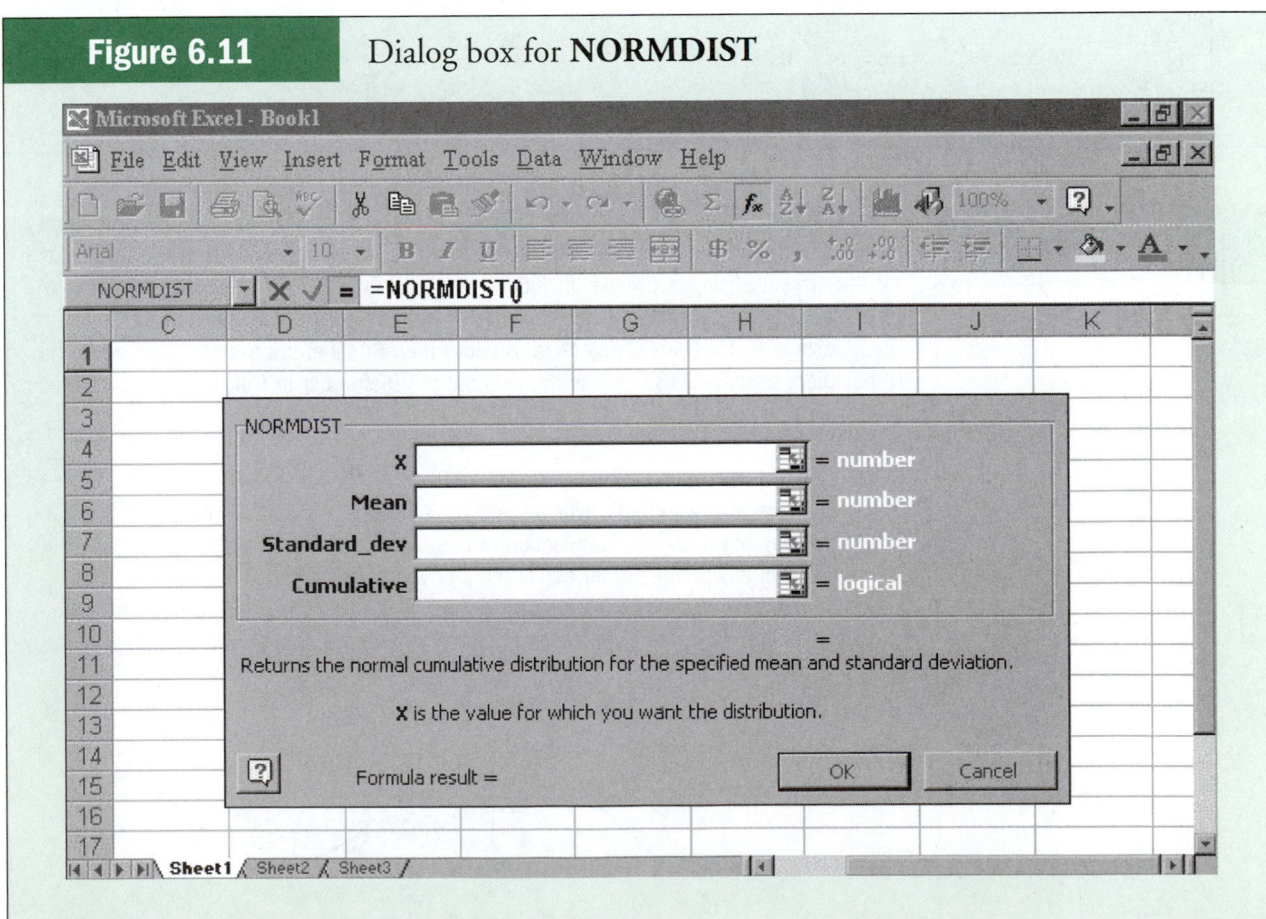

Figure 6.11   Dialog box for **NORMDIST**

Figure 6.12	Excel output for Demonstration Problems 6.3–6.7

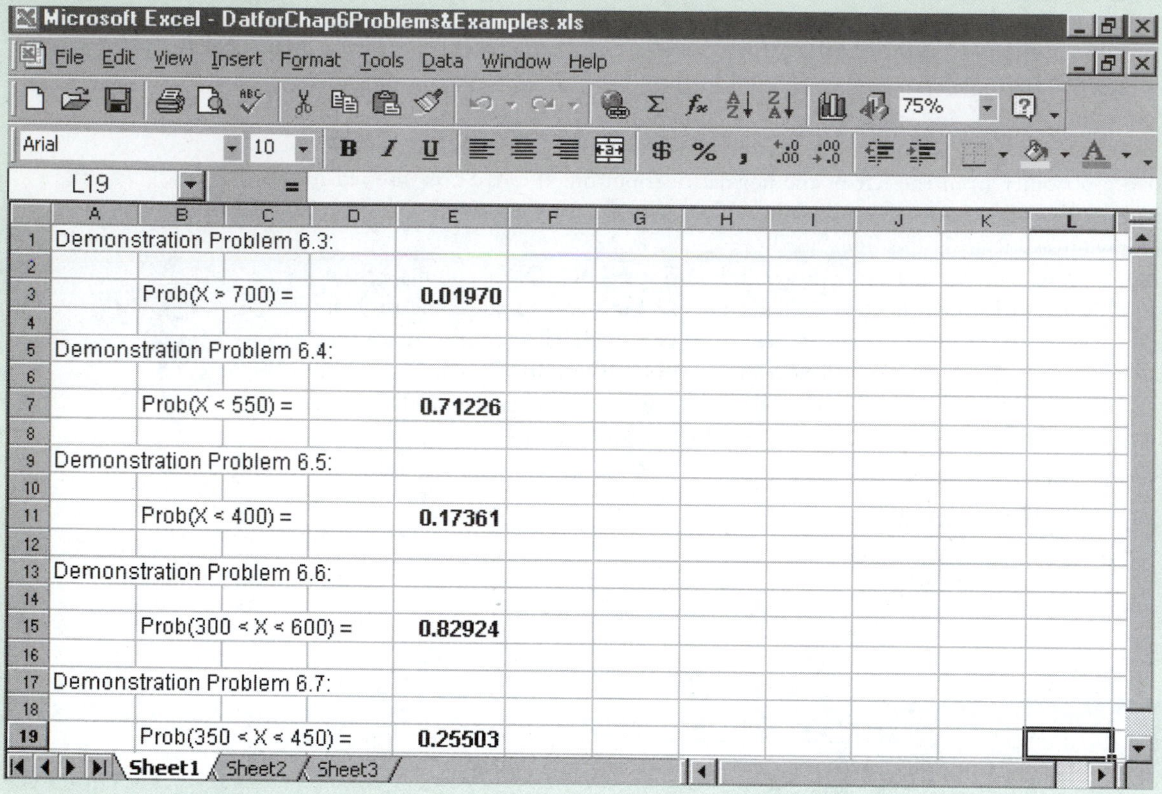

Microsoft Excel - DatforChap6Problems&Examples.xls

File Edit View Insert Format Tools Data Window Help

Arial   10   **B** *I* U ≡ ≡ ≡ $ % ,

L19

	A	B	C	D	E	F	G	H	I	J	K	L
1	Demonstration Problem 6.3:											
2												
3		Prob(X > 700) =			**0.01970**							
4												
5	Demonstration Problem 6.4:											
6												
7		Prob(X < 550) =			**0.71226**							
8												
9	Demonstration Problem 6.5:											
10												
11		Prob(X < 400) =			**0.17361**							
12												
13	Demonstration Problem 6.6:											
14												
15		Prob(300 < X < 600) =			**0.82924**							
16												
17	Demonstration Problem 6.7:											
18												
19		Prob(350 < X < 450) =			**0.25503**							

Sheet1 / Sheet2 / Sheet3 /

DEMONSTRATION PROBLEM 6.8

Runzheimer International publishes business travel costs for various cities throughout the world. In particular, they publish per diem totals, which represent the average costs for the typical business traveler including three meals a day in business-class restaurants and single-rate lodging in business-class hotels and motels. If 86.65% of the per diem costs in Buenos Aires, Argentina, are less than $449 and if the standard deviation of per diem costs is $36, what is the average per diem cost in Buenos Aires? Assume that per diem costs are normally distributed.

SOLUTION

In this problem, the standard deviation and an X value are given; the object is to determine the value of the mean. Examination of the Z score formula reveals four variables: X, μ, σ, and Z. In this problem, only two of the four variables are given. Because solving one equation with two unknowns is impossible, one of the other unknowns must be determined. The value of Z can be determined from the normal distribution table (Table 6.1).

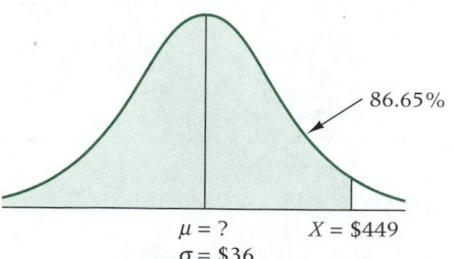

86.65%

$\mu = ?$ $X = \$449$
$\sigma = \$36$

As 86.65% of the values are less than $X = \$449$, 36.65% of the per diem costs are between $449 and the mean. The other 50% of the per diem costs are in the lower half of the distribution. Converting the percentage to a proportion yields .3665 of the values between the X value and the mean. What Z value is associated with this area? This area, or probability, of .3665 in Table 6.1 is associated with the Z value of 1.11. This Z value is positive, because it is in the upper half of the distribution. Using the Z value of 1.11, the X value of $449, and the σ value of $36 allows solving for the mean algebraically.

$$Z = \frac{X - \mu}{\sigma}$$

$$1.11 = \frac{\$449 - \mu}{\$36}$$

and

$$\mu = \$449 - (\$36)(1.11) = \$449 - \$39.96 = \$409.04$$

The mean per diem cost for business travel in Buenos Aires is $409.04.

The U.S. Environmental Protection Agency publishes figures on solid waste generation in the United States. One year, the average number of waste generated per person per day was 3.58 pounds. Suppose the daily amount of waste generated per person is normally distributed, with a standard deviation of 1.04 pounds. Of the daily amounts of waste generated per person, 67.72% would be greater than what amount?

SOLUTION

The mean and standard deviation are given, but X and Z are unknown. The problem is to solve for a specific X value when .6772 of the X values are greater than that value.

If .6772 of the values are greater than X, then .1772 are between X and the mean (.6772 − .5000). Table 6.1 shows that the probability of .1772 is associated with a Z value of 0.46. As X is less than the mean, the Z value actually is −0.46. Whenever an X value is less than the mean, its associated Z value is negative and should be reported that way.

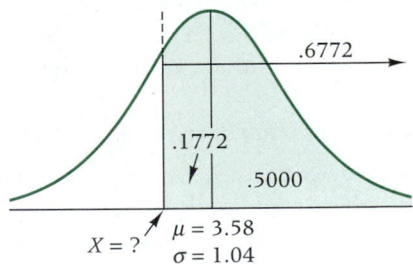

Solving the Z equation yields

$$Z = \frac{X - \mu}{\sigma}$$

$$-0.46 = \frac{X - 3.58}{1.04}$$

and

$$X = 3.58 + (-0.46)(1.04) = 3.10.$$

Thus 67.72% of the daily average amount of solid waste per person weighs more than 3.10 pounds.

6.2
Problems

6.6 Determine the probabilities for the following normal distribution problems.
  a. $\mu = 604$, $\sigma = 56.8$, $X \le 635$
  b. $\mu = 48$, $\sigma = 12$, $X < 20$
  c. $\mu = 111$, $\sigma = 33.8$, $100 \le X < 150$
  d. $\mu = 264$, $\sigma = 10.9$, $250 < X < 255$
  e. $\mu = 37$, $\sigma = 4.35$, $X > 35$
  f. $\mu = 156$, $\sigma = 11.4$, $X \ge 170$

6.7 Tompkins Associates reports that the mean clear height for a Class A warehouse in the United States is 22 feet. Suppose clear heights are normally distributed and that the standard deviation is 4 feet. A Class A warehouse in the United States is randomly selected.
  a. What is the probability that the clear height is greater than 17 feet?
  b. What is the probability that the clear height is less than 13 feet?
  c. What is the probability that the clear height is between 25 and 31 feet?

6.8 According to the Cellular Telecommunications Industry Association, the average local monthly cell phone bill is $42.78. Suppose local monthly cell phone bills are normally distributed, with a standard deviation of $11.35.
  a. What is the probability that a randomly selected cell phone bill is more than $67.75?
  b. What is the probability that a randomly selected cell phone bill is between $30 and $50?
  c. What is the probability that a randomly selected cell phone bill is no more than $25?
  d. What is the probability that a randomly selected cell phone bill is between $45 and $55?

6.9 According to the Internal Revenue Service, income tax returns one year averaged $1332.00 in refunds for taxpayers. One explanation of this figure is that taxpayers would rather have the government keep back too much money during the year than to owe it money at the end of the year. Suppose the average amount of tax at the end of a year is a refund of $1332.00, with a standard deviation of $725. Assume that amounts owed or due on tax returns are normally distributed.
  a. What proportion of tax returns show a refund greater than $2,000?
  b. What proportion of the tax returns show that the taxpayer owes money to the government?
  c. What proportion of the tax returns show a refund between $100 and $700?

6.10 Toolworkers are subject to work-related injuries. One disorder, caused by strains to the hands and wrists, is called carpal tunnel syndrome. It strikes as many as 23,000 workers per year. The U.S. Labor Department estimates that the average cost of this disorder to employers and insurers is approximately $30,000 per injured worker. Suppose these costs are normally distributed, with a standard deviation of $9000.
  a. What proportion of the costs are between $15,000 and $45,000?
  b. What proportion of the costs are greater than $50,000?
  c. What proportion of the costs are between $5000 and $20,000?
  d. Suppose the standard deviation is unknown, but 90.82% of the costs are more than $7000. What would be the value of the standard deviation?
  e. Suppose the mean value is unknown, but the standard deviation is still $9000. How much would the average cost be if 79.95% of the costs were less than $33,000?

6.11 Suppose you are working with a data set that is normally distributed, with a mean of 200 and a standard deviation of 47. Determine the value of X from the following information.

  a. 60% of the values are greater than X.
  b. X is less than 17% of the values.
  c. 22% of the values are less than X.
  d. X is greater than 55% of the values.

6.12 Suppose the standard deviation for Problem 6.7 is unknown but the mean is still 22 feet. If 72.4% of all U.S. Class A warehouses have a clear height greater than 18.5 feet, what is the standard deviation?

6.13 Suppose the mean clear height of all U.S. Class A warehouses is unknown but the standard deviation is known to be 4 feet. What is the value of the mean clear height if 29% of U.S. Class A warehouses have a clear height less than 20 feet?

6.14 Data accumulated by the National Climatic Data Center shows that the average wind speed in miles per hour for St. Louis, Missouri, is 9.7. Suppose wind speed measurements are normally distributed for a given geographic location. If 22.45% of the time the wind speed measurements are more than 11.6 miles per hour, what is the standard deviation of wind speed in St. Louis?

6.3 Using the Normal Curve to Work Binomial Distribution Problems

For certain types of binomial distribution problems, the normal distribution can be used to approximate the probabilities. As sample sizes become large, binomial distributions approach the normal distribution in shape regardless of the value of p. This phenomenon occurs faster (for smaller values of n) when p is near .50. Figures 6.13 through 6.15 show three binomial distributions. Note in Figure 6.13 that even though the sample size, n, is only 10, the binomial graph bears a strong resemblance to a normal curve.

The graph in Figure 6.14 ($n = 10$ and $p = .20$) is skewed to the right because of the low p value and the small size. For this distribution, the expected value is only 2 and the probabilities pile up at $X = 0$ and 1. However, when n becomes large enough, as in the binomial distribution ($n = 100$ and $p = .20$) presented in Figure 6.15, the graph is relatively symmetric around the mean ($\mu = n \cdot p = 20$) because there are enough possible outcome values to the left of $X = 20$ to allow the curve to fall back to the X axis.

For large n values, the binomial distribution is cumbersome to analyze without a computer. Table A.2 goes only to $n = 25$. Because of the size of the factorials involved, using calculators to work binomial problems when n is large is difficult or impossible. Fortunately, the normal distribution is a good approximation for binomial distribution problems for large values of n.

To work a binomial problem by the normal curve requires a translation process. The first part of this process is to convert the two parameters of a binomial distribution, n and p, to the two parameters of the normal distribution, μ and σ. This conversion utilizes formulas from Chapter 5.

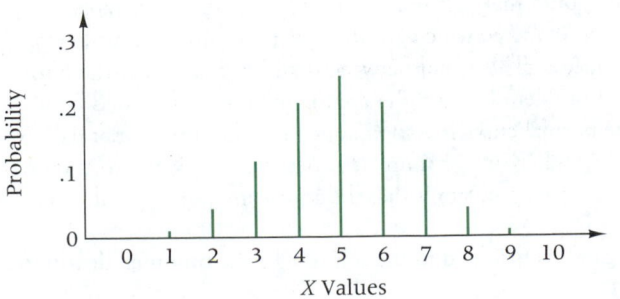

Figure 6.13

The binomial distribution for $n = 10$ and $p = .50$

Figure 6.14

The binomial
distribution for
$n = 10$ and $p = .20$

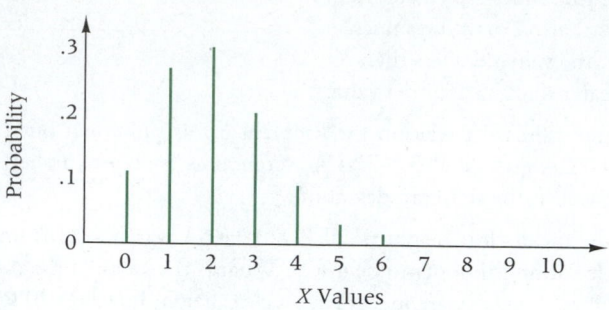

Figure 6.15

The binomial
distribution for
$n = 100$ and $p = .20$

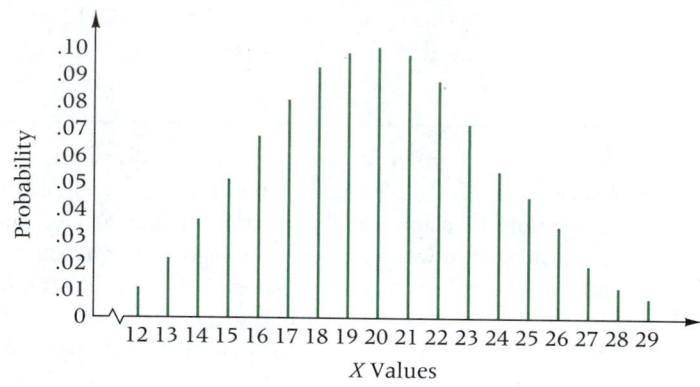

CONVERSION OF A BINOMIAL PROBLEM TO THE NORMAL CURVE	$\mu = n \cdot p$ and $\sigma = \sqrt{n \cdot p \cdot q}$

After completion of this conversion, a test must be made to determine whether the normal distribution is a good enough approximation of the binomial distribution:

Does the interval $\mu \pm 3\sigma$ lie between 0 and n?

Recall that the empirical rule states that approximately 99.7%, or almost all, of the values of a normal curve are within three standard deviations of the mean. For a normal curve approximation of a binomial distribution problem to be acceptable, all possible X values should be between 0 and n, which are the lower and upper limits, respectively, of a binomial distribution. If $\mu \pm 3\sigma$ is not between 0 and n, do *not* use the normal distribution to work a binomial problem because the approximation is not good enough. Upon demonstration that the normal curve is a good approximation for a binomial problem, the procedure continues. Another rule of thumb for determining when to use the normal curve to approximate a binomial problem is that the approximation is good enough if both $n \cdot p > 5$ and $n \cdot q > 5$.

The process can be illustrated in the solution of the binomial distribution problem.

$$P(X \geq 25 \mid n = 60 \text{ and } p = .30) = ?$$

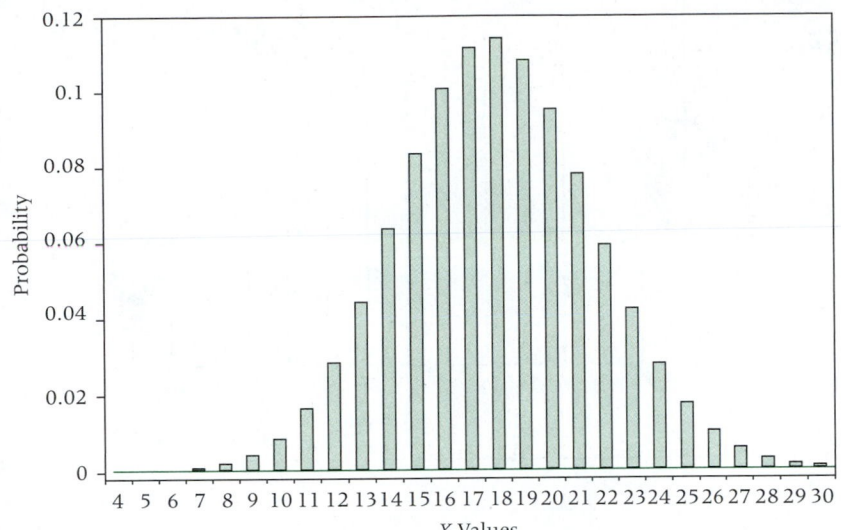

Figure 6.16

Excel graph of the binomial problem: $n = 60$ and $p = .30$

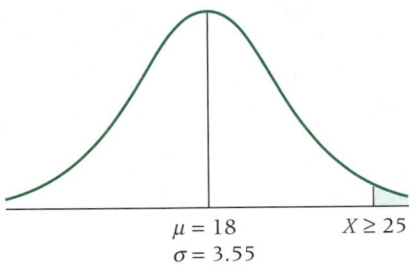

$\mu = 18$
$\sigma = 3.55$
$X \geq 25$

Figure 6.17

Graph of apparent solution of binomial problem worked by the normal curve

Note that this binomial problem contains a relatively large sample size and that none of the binomial tables in Appendix A.2 can be used to solve the problem. This problem is a good candidate for use of the normal distribution.

Translating from a binomial problem to a normal curve problem gives

$$\mu = n \cdot p = (60)(.30) = 18 \quad \text{and} \quad \sigma = \sqrt{n \cdot p \cdot q} = 3.55.$$

The binomial problem becomes a normal curve problem.

$$P(X \geq 25 \mid \mu = 18 \text{ and } \sigma = 3.55) = ?$$

Next, the test is made to determine whether the normal curve sufficiently fits this binomial distribution to justify the use of the normal curve.

$$\mu \pm 3\sigma = 18 \pm 3(3.55) = 18 \pm 10.65$$
$$7.35 \leq \mu \pm 3\sigma \leq 28.65$$

This interval is between 0 and 60, so the approximation is sufficient to allow use of the normal curve. Figure 6.16 is an Excel graph of this binomial distribution. Notice how closely it resembles the normal curve. Figure 6.17 is the *apparent* graph of the normal curve version of this problem.

Figure 6.18

Graph of a portion
of the binomial
problem: $n = 60$
and $p = .30$

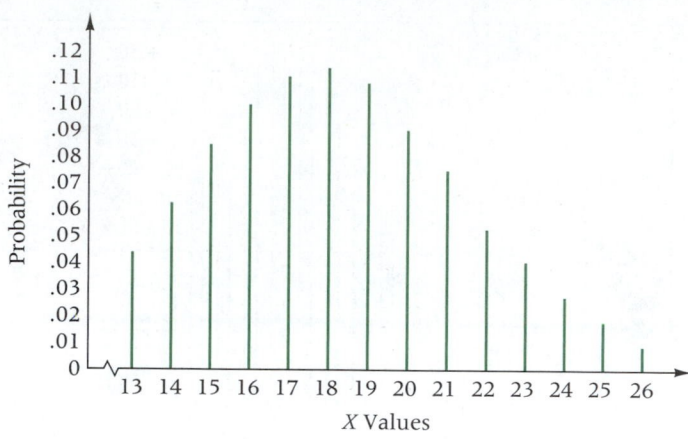

Correcting for Continuity

The translation of a discrete distribution to a continuous distribution is not completely straightforward. A correction of +.50 or −.50 or ±.50, depending on the problem, is required. This correction ensures that most of the binomial problem's information is correctly transferred to the normal curve analysis. This correction is called the **correction for continuity,** which is *made during conversion of a discrete distribution into a continuous distribution.*

Correction for continuity

A correction made when a binomial distribution problem is approximated by the normal distribution because a discrete distribution problem is being approximated by a continuous distribution.

Figure 6.18 is a portion of the graph of the binomial distribution, $n = 60$ and $p = .30$. Note that with a binomial distribution, all the probabilities are concentrated on the whole numbers. Thus, the answers for $X \geq 25$ are found by summing the probabilities for $X = 25, 26, 27, \ldots, 60$. There are *no* values between 24 and 25, 25 and 26, . . . , 59 and 60. Yet, the normal distribution is continuous, and values are present all along the X axis. A correction must be made for this discrepancy for the approximation to be as accurate as possible.

As an analogy, visualize the process of melting iron rods in a furnace. The iron rods are like the probability values on each whole number of a binomial distribution. Note that the binomial graph in Figure 6.18 looks like a series of iron rods in a line. When the rods are placed in a furnace, they melt down and spread out. Each rod melts and moves to fill the area between it and the adjacent rods. The result is a continuous sheet of solid iron (continuous iron) that looks like the normal curve. The melting of the rods is analogous to spreading the binomial distribution to approximate the normal distribution.

How far does each rod spread toward the others? A good estimate is that each rod goes about halfway toward the adjacent rods. In other words, a rod that was concentrated at $X = 25$ spreads to cover the area from 24.5 to 25.5; $X = 26$ becomes continuous from 25.5 to 26.5; and so on. For the problem $P(X \geq 25 \mid n = 60$ and $p = .30)$, conversion to a continuous normal curve problem yields $P(X \geq 24.5 \mid \mu = 18$ and $\sigma = 3.55)$. The correction for continuity was −.50 because the problem called for the inclusion of the value of 25 along with all greater values; the binomial value of $X = 25$ translates to the normal curve value of from 24.5 to 25.5. Had the binomial problem been to analyze $P(X > 25)$, the correction would have been +.50, resulting in a normal curve problem of $P(X \geq 25.5)$. The latter case would begin at more than 25 because the value of 25 would not be included.

The decision as to how to correct for continuity depends on the equality sign and the direction of the desired outcomes of the binomial distribution. Table 6.2 lists some rules of thumb that can help in the application of the correction for continuity.

TABLE 6.2
Some Rules of Thumb for the Correction for Continuity

VALUES BEING DETERMINED	CORRECTION
$X >$	$+.50$
$X \geq$	$-.50$
$X <$	$-.50$
$X \leq$	$+.50$
$\leq X \leq$	$-.50$ and $+.50$
$< X <$	$+.50$ and $-.50$
$X =$	$-.50$ and $+.50$

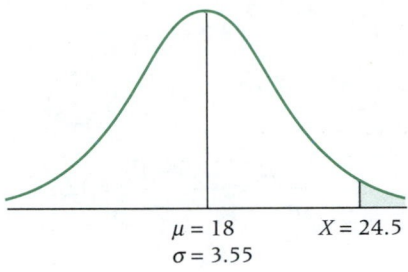

$\mu = 18$ $X = 24.5$
$\sigma = 3.55$

Figure 6.19

Graph of the solution to the binomial problem worked by the normal curve

For the binomial problem $P(X \geq 25 \mid n = 60$ and $p = .30)$, the normal curve becomes $P(X \geq 24.5 \mid \mu = 18$ and $\sigma = 3.55)$, as shown in Figure 6.19, and

$$Z = \frac{X - \mu}{\sigma} = \frac{24.5 - 18}{3.55} = 1.83.$$

The probability (Table 6.1) of this Z value is .4664. The answer to this problem lies in the tail of the distribution, so the final answer is obtained by subtracting.

$$
\begin{array}{r}
.5000 \\
-.4664 \\
\hline
.0336
\end{array}
$$

Had this problem been worked by using the binomial formula, the solution would have been as shown in Table 6.3. The difference between the normal distribution approximation and the actual binomial values is only .0025 (.0361 − .0336).

TABLE 6.3
Probability Values for the Binomial Problem: $n = 60$, $p = .30$, and $X \geq 25$

X VALUE	PROBABILITY
25	.0167
26	.0096
27	.0052
28	.0026
29	.0012
30	.0005
31	.0002
32	.0001
33	.0000
$X \geq 25$	.0361

Work the following binomial distribution problem by using the normal distribution.

$$P(X = 12 \mid n = 25 \text{ and } p = .40) = ?$$

DEMONSTRATION PROBLEM 6.10

SOLUTION
Find μ and σ.

$$\mu = n \cdot p = (25)(.40) = 10.0$$

$$\sigma = \sqrt{n \cdot p \cdot q} = \sqrt{(25)(.40)(.60)} = 2.45$$

Test: $\mu \pm 3\sigma = 10.0 \pm 3(2.45) = 2.65$ to 17.35

This range is between 0 and 25, so the approximation is close enough. Correct for continuity next. Because the problem is to determine the probability of X being exactly 12, the correction entails both −.50 and +.50. That is, a binomial probability at $X = 12$ translates to a continuous normal curve area that lies between 11.5 and 12.5. The graph of the problem follows:

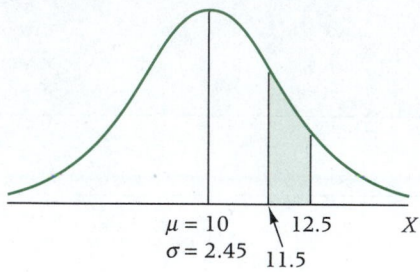

$$\mu = 10$$
$$\sigma = 2.45 \quad 11.5 \qquad 12.5 \qquad X$$

Then,

$$Z = \frac{X - \mu}{\sigma} = \frac{12.5 - 10}{2.45} = 1.02$$

and

$$Z = \frac{X - \mu}{\sigma} = \frac{11.5 - 10}{2.45} = 0.61.$$

$Z = 1.02$ produces a probability of .3461.

$Z = 0.61$ produces a probability of .2291.

The difference in areas yields the answer:

$$.3461 - .2291 = .1170.$$

Had this problem been worked by using the binomial tables, the resulting answer would have been .114. The difference between the normal curve approximation and the value obtained by using binomial tables is only .003.

DEMONSTRATION PROBLEM 6.11

Solve the following binomial distribution problem by using the normal distribution.

$$P(X < 27 \,|\, n = 100 \text{ and } p = .37) = ?$$

SOLUTION
As neither the sample size nor the p value is contained in Table A.2, working this problem by using binomial distribution techniques is impractical. It is a good candidate for the normal curve. Calculating μ and σ yields

$$\mu = n \cdot p = (100)(.37) = 37.0$$
$$\sigma = \sqrt{n \cdot p \cdot q} = \sqrt{(100)(.37)(.63)} = 4.83.$$

Testing to determine the closeness of the approximation gives

$$\mu \pm 3\sigma = 37 \pm 3(4.83) = 37 \pm 14.49.$$

The range 22.51 to 51.49 is between 0 and 100. This problem satisfies the conditions of the test. Next, correct for continuity: $X < 27$ as a binomial problem translates to $X \leq 26.5$ as a normal distribution problem. The graph of the problem follows.

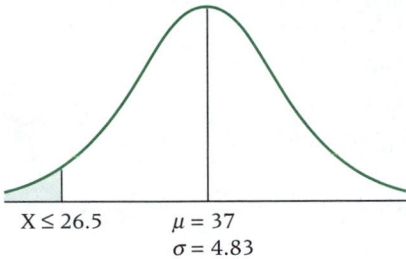

$X \leq 26.5$ $\mu = 37$
$\sigma = 4.83$

Then,

$$Z = \frac{X - \mu}{\sigma} = \frac{26.5 - 37}{24.83} = -2.17$$

Table 6.1 shows a probability of .4850. Solving for the tail of the distribution gives

$$.5000 - .4850 = .0150,$$

which is the answer.

Had this problem been solved by using the binomial formula, the probabilities would have been the following.

X VALUE	PROBABILITY
26	.0059
25	.0035
24	.0019
23	.0010
22	.0005
21	.0002
20	.0001
$X < 27$	.0131

The answer obtained by using the normal curve approximation (.0150) compares favorably to this exact binomial answer. The difference is only .0019.

6.15 Convert the following binomial distribution problems to normal distribution problems. Use the correction for continuity.

a. $P(X \leq 16 \mid n = 30 \text{ and } p = .70)$
b. $P(10 < X \leq 20) \mid n = 25 \text{ and } p = .50)$
c. $P(X = 22 \mid n = 40 \text{ and } p = .60)$
d. $P(X > 14 \mid n = 16 \text{ and } p = .45)$

6.16. Use the test $\mu \pm 3\sigma$ to determine whether the following binomial distributions can be approximated by using the normal distribution.

a. $n = 8$ and $p = .50$
b. $n = 18$ and $p = .80$
c. $n = 12$ and $p = .30$
d. $n = 30$ and $p = .75$
e. $n = 14$ and $p = .50$

6.3

Problems

6.17 Where appropriate, work the following binomial distribution problems by using the normal curve. Also, use Table A.2 to find the answers by using the binomial distribution and compare the answers obtained by the two methods.
  a. $P(X = 8 \mid n = 25$ and $p = .40) = ?$
  b. $P(X \geq 13 \mid n = 20$ and $p = .60) = ?$
  c. $P(X = 7 \mid n = 15$ and $p = .50) = ?$
  d. $P(X < 3 \mid n = 10$ and $p = .70) = ?$

6.18 The Zimmerman Agency conducted a study, for Residence Inn by Marriott, of business travelers who take trips of five nights or more. According to this study, 37% of these travelers enjoy sightseeing more than any other activity that they do not get to do as much at home. Suppose 120 randomly selected business travelers who take trips of five nights or more are contacted. What is the probability that fewer than 40 enjoy sightseeing more than any other activity that they do not get to do as much at home?

6.19 One study on managers' satisfaction with management tools reveals that 59% of all managers use self-directed work teams as a management tool. Suppose 70 managers selected randomly in the United States are interviewed. What is the probability that fewer than 35 use self-directed work teams as a management tool?

6.20 According to The Yankee Group, 53% of all cable households rate cable companies as good or excellent in quality transmission. Sixty percent of all cable households rate cable companies as good or excellent in having professional personnel. Suppose 300 cable households are randomly contacted.
  a. What is the probability that more than 175 cable households rate cable companies as good or excellent in quality transmission?
  b. What is the probability that between 165 and 170 inclusive cable households rate cable companies as good or excellent in quality transmission?
  c. What is the probability that between 155 and 170 inclusive cable households rate cable companies as good or excellent in having professional personnel?
  d. What is the probability that less than 200 cable households rate cable companies as good or excellent in having professional personnel?

6.21 The International Data Corporation reports that Compaq is number one in PC market share in the United States with 16.0% of the market. Suppose a researcher randomly selects 130 recent purchasers of PCs.
  a. What is the probability that more than 25 PC purchasers bought a Compaq?
  b. What is the probability that between 15 and 23 (inclusive) PC purchasers bought a Compaq?
  c. What is the probability that fewer than 12 PC purchasers bought a Compaq?
  d. What is the probability that exactly 22 PC purchasers bought a Compaq?

6.22 A study about strategies for competing in the global marketplace states that 52% of the respondents agreed that companies need to make direct investments in foreign countries. It also states that about 70% of those responding agree that it is attractive to have a joint venture to increase global competitiveness. Suppose CEOs of 95 manufacturing companies are randomly contacted about global strategies.
  a. What is the probability that between 44 and 52 (inclusive) CEOs agree that companies should make direct investments in foreign countries?
  b. What is the probability that more than 56 CEOs agree with that assertion?
  c. What is the probability that fewer than 60 CEOs agree that it is attractive to have a joint venture to increase global competitiveness?
  d. What is the probability that between 55 and 62 (inclusive) CEOs agree with that assertion?

Another useful continuous distribution is the exponential distribution. It is closely related to the Poisson distribution. Whereas the Poisson distribution is discrete and describes random occurrences over some interval, the **exponential distribution** *is continuous and describes the times between random occurrences.* The following are the characteristics of the exponential distribution.

- It is a continuous distribution.
- It is a family of distributions.
- It is skewed to the right.
- The X values range from zero to infinity.
- Its apex is always at $X = 0$.
- The curve steadily decreases as X gets larger.

The exponential probability distribution is determined by the following.

$$f(X) = \lambda e^{-\lambda X}$$

where:
$X \geq 0$
$\lambda > 0$

An exponential distribution can be characterized by the one parameter, λ. Each unique value of λ determines a different exponential distribution. Thus there is actually a family of exponential distributions. Figure 6.20 shows graphs of exponential distributions for four values of λ. The points on the graph are determined by using λ and various values of X in the probability density formula. The mean of an exponential distribution is $\mu = 1/\lambda$, and the standard deviation of an exponential distribution is $\sigma = 1/\lambda$.

6.4
Exponential Distribution

Exponential distribution
A continuous distribution closely related to the Poisson distribution that describes the times between random occurrences.

EXPONENTIAL
PROBABILITY DENSITY
FUNCTION

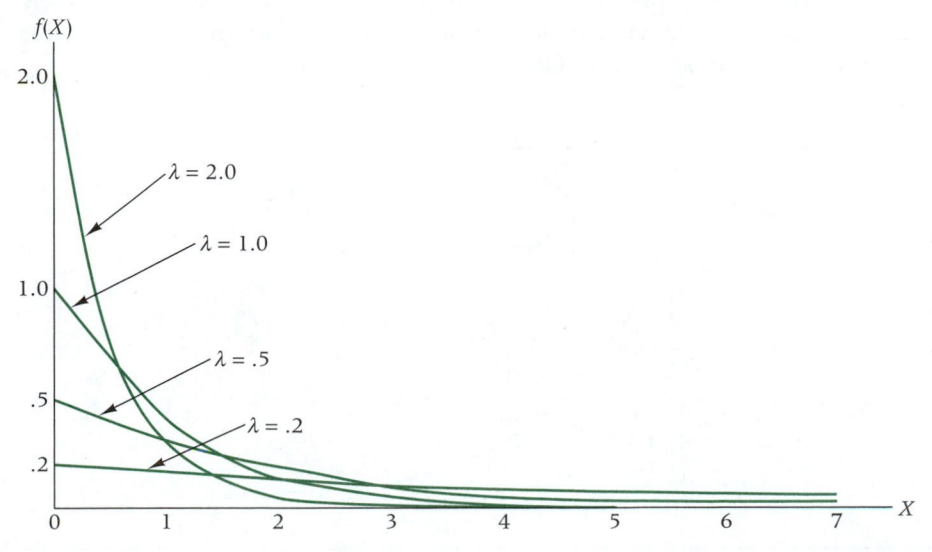

Figure 6.20

Graphs of some exponential distributions

Probabilities of the Exponential Distribution

Probabilities are computed for the exponential distribution by determining the area under the curve between two points. Applying calculus to the exponential probability density function produces a formula that can be used to calculate the probabilities of an exponential distribution.

PROBABILITIES OF THE RIGHT TAIL OF THE EXPONENTIAL DISTRIBUTION	$$P(X \geq X_0) = e^{-\lambda \cdot X_0}$$ where: $X_0 \geq 0$

To use this formula requires finding values of e^{-X}. These values can be computed on most calculators or obtained from Table A.4, which contains the values of e^{-X} for selected values of X. X_0 is the fraction of the interval or the number of intervals between arrivals in the probability question.

For example, arrivals at a bank are Poisson distributed with a λ of 1.2 customers every minute. What is the average time between arrivals and what is the probability that at least 2 minutes will elapse between one arrival and the next arrival? X_0 is 2 because we want to know the probability that at least two intervals transpire between arrivals.

Interarrival times of random arrivals are exponentially distributed. The mean of this exponential distribution is $\mu = 1/\lambda = 1/1.2 = .833$ minute (50 seconds). On average, .833 minute, or 50 seconds, will elapse between arrivals at the bank. The probability of an interval of 2 minutes or more between arrivals can be calculated by

$$P(X \geq 2 \mid \lambda = 1.2) = e^{-1.2(2)} = .0907.$$

About 9.07% of the time when the rate of random arrivals is 1.2 per minute, 2 minutes or more will elapse between arrivals, as shown in Figure 6.21.

This problem underscores the potential of using the exponential distribution in conjunction with the Poisson distribution to solve problems. In operations research and management science, these two distributions are used together to solve queuing problems (theory of lines). The Poisson distribution is used to analyze the arrivals to the queue, and the exponential distribution is used to analyze interarrival time.

Figure 6.21

Exponential distribution for $\lambda = 1.2$ and solution for $X \geq 2$

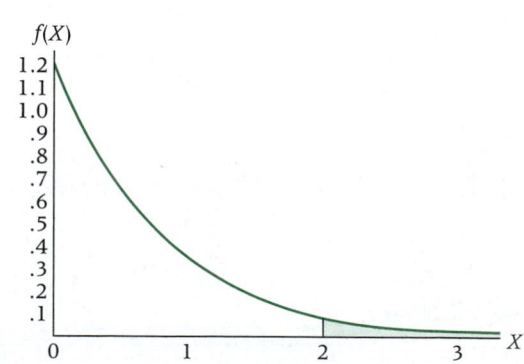

DEMONSTRATION
PROBLEM 6.12

A manufacturing firm has been involved in statistical quality control for several years. As part of the production process, parts are randomly selected and tested. From the records of these tests, it has been established that a defective part occurs in a pattern that is Poisson distributed on the average of 1.38 defects every 20 minutes during production runs. Use this information to determine the probability that less than 15 minutes will elapse between any two defects.

SOLUTION

The value of λ is 1.38 defects per 20-minute interval. The value of μ can be determined by

$$\mu = \frac{1}{\lambda} = \frac{1}{1.38} = .7246.$$

On the average, it is .7246 of the interval, or $(.7246)(20 \text{ minutes}) = 14.49$ minutes, between defects. The value of X_0 represents the desired number of intervals between arrivals or occurrences for the probability question. In this problem, the probability question involves 15 minutes and the interval is 20 minutes. Thus X_0 is 15/20, or .75 of an interval. The question here is to determine the probability of there being less than 15 minutes between defects. The probability formula always yields the right tail of the distribution—in this case, the probability of there being 15 minutes or more between arrivals. By using the value of X_0 and the value of λ, the probability of there being 15 minutes or more between defects can be determined.

$$P(X \geq X_0) = P(X \geq .75) = e^{-\lambda X_0} = e^{(-1.38)(.75)} = e^{-1.035} = .3552$$

The probability of .3552 is the probability that at least 15 minutes will elapse between defects. To determine the probability of there being less than 15 minutes between defects, compute $1 - P(X)$. In this case, $1 - .3552 = .6448$. There is a probability of .6448 that less than 15 minutes will elapse between two defects when there is an average of 1.38 defects per 20-minute interval or an average of 14.49 minutes between defects.

DEMONSTRATION
PROBLEM 6.13

A company that manufactures washing machines makes them to last an average of 6 years before they have a major breakdown. The manufacturer offers a free warranty against major breakdowns. However, the company wants to be sure that no more than 20% of the machines have a major breakdown during the warranty period. For how many years should the warranty be promised? Assume that breakdown occurrence is Poisson distributed.

SOLUTION

The exponential distribution is used to analyze the times between major breakdowns. The mean time is 6 years, so $\mu = 6$ for this exponential distribution. Lambda is then

$$\lambda = \frac{1}{\mu} = \frac{1}{6} = .167/\text{year}.$$

The problem is to find a number of years, X_0, so that no more than 20% of the machines have a major breakdown before that time. Thus X_0 in the exponential probability formula represents the time after which 80% or more of the machines can be expected to break down. That is,

$$P(X \geq X_0) = e^{-\lambda X_0} = e^{-(0.167)X_0} \geq .80.$$

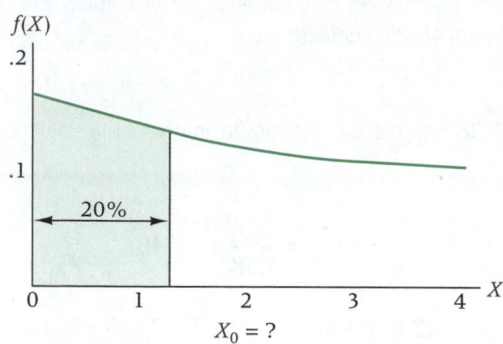

The table shows several values of X_0 and their associated probabilities. If a 1-year warranty is given with the product, 84.62% of the breakdowns will occur after that time and less than 20% of the machines will have major breakdowns before the warranty has expired. For other warranty periods, less than 80% of breakdowns will occur after warranty expiration (X_0 value) and more than 20% will occur before expiration, which is unacceptable to the company.

X_0	$P(X \geq X_0)$	$P(X < X_0) = 1 - P(X \geq X_0)$
1	.8462	.1538 (less than 20%)
2	.7161	.2839
3	.6059	.3941
4	.5127	.4873

Using natural logarithms, the equation $e^{-(0.167)X_0} = .80$ can be solved yielding $X_0 = 1.336$ years. For values of X_0 greater than 1.336, more than 20% of the machines will have a breakdown during the warranty period. Therefore, the company should offer a 1-year warranty.

Analysis Using Excel

Excel has the capability of computing exponential probabilities using a feature called **EXPONDIST.** This feature is found in the right column of selections under **Statistical** in the Paste Function. The dialog box for **EXPONDIST** is shown in Figure 6.22.

EXPONDIST requires three inputs, the value of X, the value of λ, and a logical value. As with **NORMDIST,** the logical value is TRUE if you want to solve for cumulative probabilities from the left ($X \leq X_0$) and FALSE for the probability density function. In virtually all cases, we will be interested only in the TRUE case in this text. Since cumulative probabilities are only generated by **EXPONDIST** from the left, if you want to solve for the right portion of the distribution you must have Excel compute 1 − (answer obtained using TRUE). Solving for probabilities over an interval between two values of X ($X_a \leq X \leq X_b$) will require computing the difference between the Excel-produced probability answers for each of the X values. Figure 6.23 contains Excel output for the bank problem (probability that $X \geq 2$ when $\lambda = 1.2$) and for Demonstration Problem 6.12.

Figure 6.22 — Dialog box for **EXPONDIST**

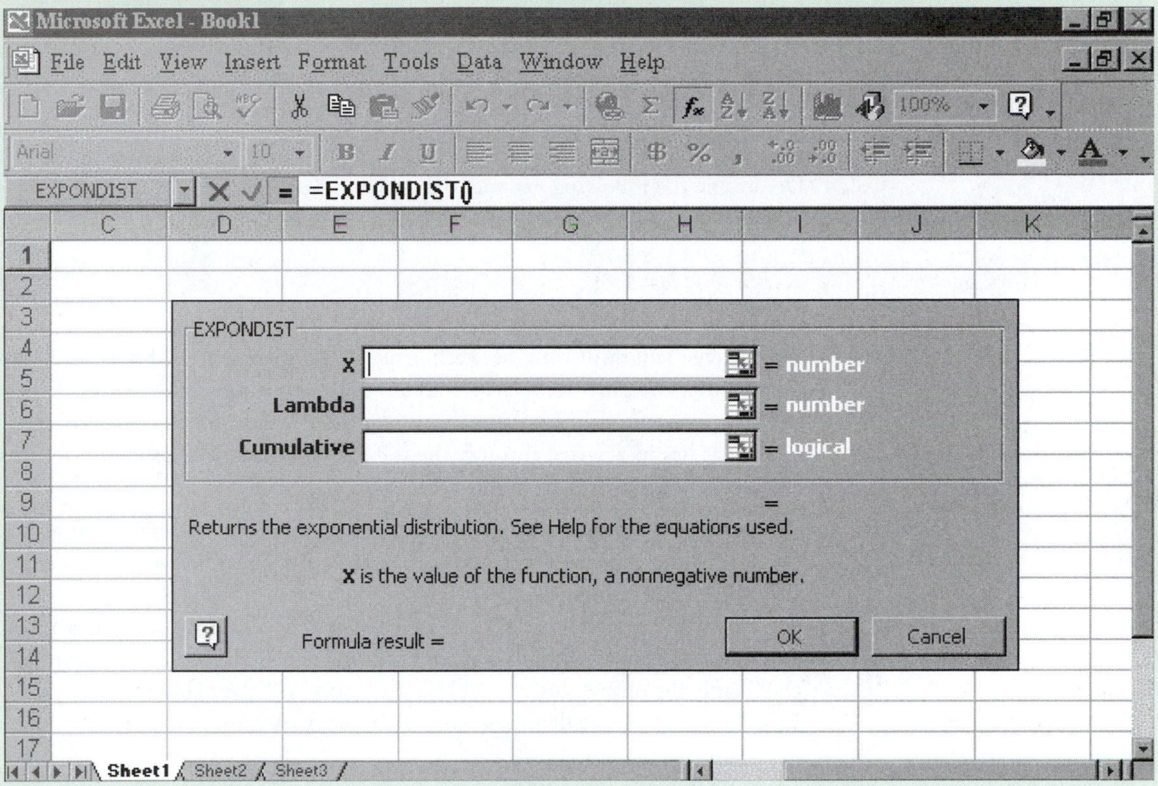

Microsoft Excel - Book1

File Edit View Insert Format Tools Data Window Help

EXPONDIST = =EXPONDIST()

EXPONDIST

X [ ] = number

Lambda [ ] = number

Cumulative [ ] = logical

=

Returns the exponential distribution. See Help for the equations used.

X is the value of the function, a nonnegative number.

Formula result =   OK   Cancel

Sheet1 / Sheet2 / Sheet3 /

Figure 6.23 — Excel output for bank problem and Demonstration Problem 6.12

Microsoft Excel - DatforChap6Problems&Examples.xls

File Edit View Insert Format Tools Data Window Help

J16 =

	A	B	C	D	E	F	G	H	I
1									
2									
3	Bank Problem:								
4									
5		Probability X ≥ 2 minutes given lambda = 1.2 per minute is:						**0.090718**	
6									
7									
8	Demonstration Problem 6.12:								
9									
10		Probability X < 15 minutes given lambda = 1.38 per 20 minutes is:						**0.644774**	
11									
12									
13									
14									
15									
16									
17									

Sheet1 / Chart1 / Sheet2 \ **Sheet3** /

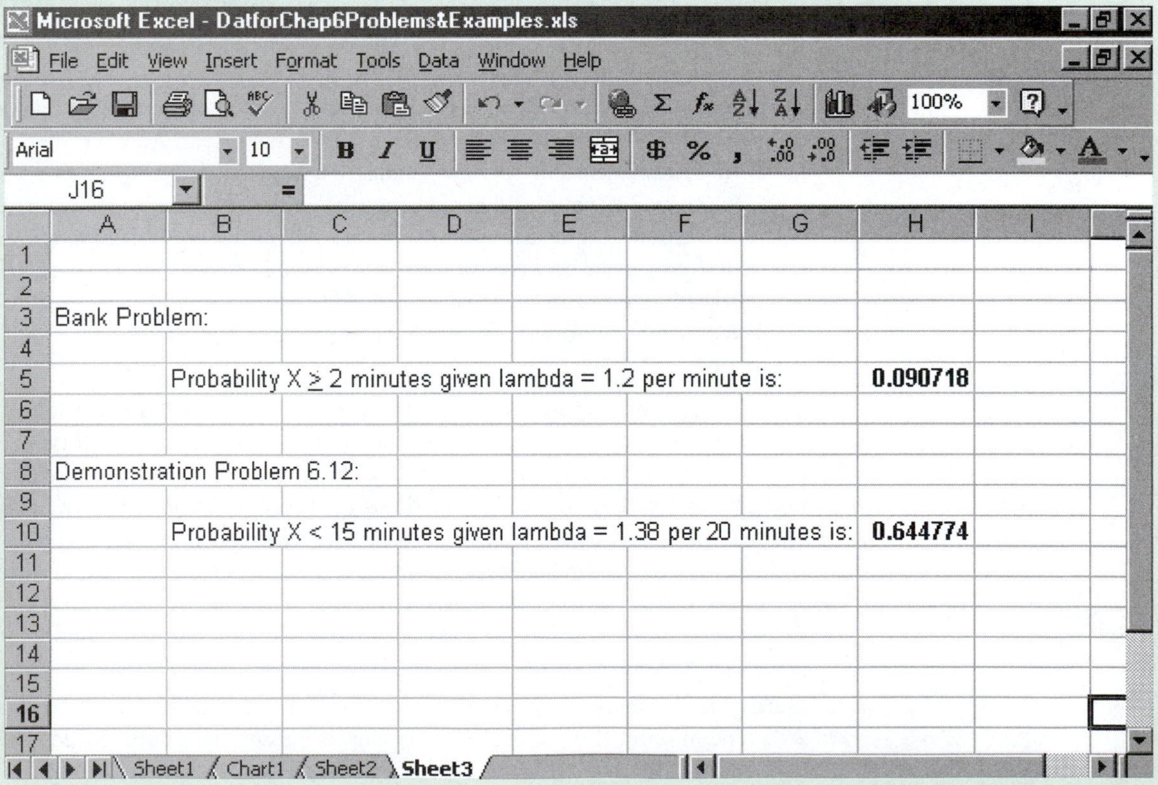

6.4
Problems

6.23 Determine the mean and standard deviation of the following exponential distributions.
   a. $\lambda = 3.25$
   b. $\lambda = 0.7$
   c. $\lambda = 1.1$
   d. $\lambda = 6.0$

6.24 Determine the following exponential probabilities.
   a. $P(X \geq 5 \mid \lambda = 1.35)$
   b. $P(X < 3 \mid \lambda = 0.68)$
   c. $P(X > 4 \mid \lambda = 1.7)$
   d. $P(X < 6 \mid \lambda = 0.80)$

6.25 The average length of time between arrivals at a turnpike toll booth is 23 seconds. Assume that the time between arrivals at the toll booth is exponentially distributed.
   a. What is the probability that a minute or more will elapse between arrivals?
   b. If a car has just passed through the toll booth, what is the probability that no car will show up for at least 3 minutes?

6.26 A busy city restaurant has determined that between 6:30 P.M. and 9:00 P.M. on Friday nights, the arrivals of customers are Poisson distributed with an average arrival rate of 2.44 per minute.
   a. What is the probability that at least 10 minutes will elapse between arrivals?
   b. What is the probability that at least 5 minutes will elapse between arrivals?
   c. What is the probability that at least 1 minute will elapse between arrivals?
   d. What is the expected amount of time between arrivals?

6.27 During the summer at a small private airport in western Nebraska, the unscheduled arrival of airplanes is Poisson distributed with an average arrival rate of 1.12 planes per hour.
   a. What is the average interarrival time between planes?
   b. What is the probability that at least 2 hours will elapse between plane arrivals?
   c. What is the probability of two planes arriving less than 10 minutes apart?

6.28 The exponential distribution can be used to solve Poisson-type problems in which the intervals are not time. The Air Travel Consumer Report published by the U.S. Department of Transportation reported that in a recent year, America West led the nation in fewest occurrences of mishandled baggage, with a mean rate of 3.39 per 1000 passengers. Assume mishandled baggage occurrences are Poisson distributed. Using the exponential distribution to analyze this problem, determine the average number of passengers between occurrences. Suppose baggage has just been mishandled. What is the probability that at least 500 passengers will have their baggage handled properly before the next mishandling occurs? What is the probability that the number will be fewer than 200 passengers?

6.29 The Foundation Corporation specializes in constructing the concrete foundations for new houses in the South. The company knows that because of soil types, moisture conditions, variable construction, and other factors, eventually most foundations will need major repair. On the basis of its records, the company's president believes that a new house foundation on average will not need major repair for 20 years. If she wants to guarantee the company's work against major repair but wants to have to honor no more than 10% of its guarantees, for how many years should the company guarantee its work? Assume that occurrences of major foundation repairs are Poisson distributed.

6.30 During the dry month of August, one U.S. city has measurable rain on average only 2 days per month. If the arrival of rainy days is Poisson distributed in this city during the month of August, what is the average number of days that will pass between measurable rain? What is the standard deviation? What is the probability during this month that there will be a period of less than 2 days between rain?

Summary

This chapter discussed three different continuous distributions: the uniform distribution, the normal distribution, and the exponential distribution. With continuous distributions, the value of the probability density function does not yield the probability, but instead gives the height of the curve at any given point. In fact, with continuous distributions, the probability at any discrete point is .0000. Probabilities are determined over an interval. In each case, the probability is the area under the curve for the interval being considered. In each distribution, the probability or total area under the curve is 1.

Probably the simplest of these distributions is the uniform distribution, sometimes referred to as the rectangular distribution. The uniform distribution is determined from a probability density function that contains equal values along some interval between the points a and b. Basically, the height of the curve is the same everywhere between these two points. Probabilities are determined by calculating the portion of the rectangle between the two points a and b that is being considered.

The most widely used of all distributions is the normal distribution. Many phenomena are normally distributed, including characteristics of most machine-produced parts, many measurements of the biological and natural environment, and many human characteristics such as height, weight, IQ, and achievement test scores. The normal curve is continuous, symmetrical, unimodal, and asymptotic to the axis; actually, it is a family of curves.

The parameters necessary to describe a normal distribution are the mean and the standard deviation. For convenience, data that are being analyzed by the normal curve should be standardized by using the mean and the standard deviation to compute Z scores. A Z score is the distance that a value is from the mean in units of standard deviations. With the Z score of a value, the probability of that value occurring by chance from a given normal distribution can be determined by using a table of Z scores and their associated probabilities.

The normal distribution can be used to work certain types of binomial distribution problems. Doing so requires converting the n and p values of the binomial distribution to μ and σ of the normal distribution. When worked by using the normal distribution, the binomial distribution solution is only an approximation. If the values of $\mu \pm 3\sigma$ are within a range from 0 to n, the approximation is reasonably accurate. Adjusting for the fact that a discrete distribution problem is being worked by using a continuous distribution requires a correction for continuity. The correction for continuity involves adding and/or subtracting .50 to the X value being analyzed. This correction usually improves the normal curve approximation.

Another continuous distribution is the exponential distribution. It complements the discrete Poisson distribution. The exponential distribution is used to compute the probabilities of times between random occurrences. The exponential distribution is a family of distributions described by one parameter, λ. The distribution is skewed to the right and always has its highest value at $X = 0$.

Key Terms

correction for continuity	standardized normal distribution
exponential distribution	uniform distribution
normal distribution	Z distribution
rectangular distribution	Z score

SUPPLEMENTARY PROBLEMS

6.31 The U.S. Bureau of Labor Statistics reports that of persons who usually work full time, the average number of hours worked per week is 43.4. Assume that the number of hours worked per week for those who usually work full time is normally distributed. Suppose 12% of these workers work more than 48 hours. Based on this, what is the standard deviation of number of hours worked per week for these workers?

6.32 A U.S. Bureau of Labor Statistics survey showed that one in five people 16 years of age or older volunteers some of his or her time. If this figure holds for the entire population and if a random sample of 150 people 16 years of age or older is taken, what is the probability that more than 50 of those sampled do volunteer work?

6.33 An entrepreneur opened a small hardware store in a strip mall. During the first few weeks, business was slow, with the store averaging only one customer every 20 minutes in the morning. Assume that the random arrival of customers is Poisson distributed.
   a. What is the probability that at least 1 hour would elapse between customers?
   b. What is the probability that 10 to 30 minutes would elapse between customers?
   c. What is the probability that less than 5 minutes would elapse between customers?

6.34 In a recent year, the average price of a Microsoft Windows Upgrade was $90.28 according to *PC Data*. Assume that prices of the Microsoft Windows Upgrade that year were normally distributed, with a standard deviation of $8.53. If a retailer of computer software was randomly selected that year, what is the probability that the price of a Microsoft Windows Upgrade was below $80? What is the probability that the price was above $95? What is the probability that the price was between $83 and $87?

6.35 According to the U.S. Department of Agriculture, Alabama egg farmers produce millions of eggs every year. Suppose egg production per year in Alabama is normally distributed, with a standard deviation of 83 million eggs. If during only 3% of the years Alabama egg farmers produce more than 2655 million eggs, what is the mean egg production by Alabama farmers?

6.36 The U.S. Bureau of Labor Statistics releases figures on the number of full-time wage and salary workers with flexible schedules. The numbers of full-time wage and salary workers in each age category are al-most uniformly distributed by age, with ages ranging from 18 to 65 years. If a worker with a flexible schedule is randomly drawn from the U.S. work force, what is the probability that he or she will be between 25 and 50 years of age? What is the mean value for this distribution? What is the height of the distribution?

6.37 A business convention holds its registration on Wednesday morning from 9:00 A.M. until 12:00 noon. Past history has shown that registrant arrivals follow a Poisson distribution at an average rate of 1.8 every 15 seconds. Fortunately, several facilities are available to register convention members.
   a. What is the average number of seconds between arrivals to the registration area for this conference based on past results?
   b. What is the probability that 25 seconds or more would pass between registration arrivals?
   c. What is the probability that less than 5 seconds will elapse between arrivals?
   d. Suppose the registration computers went down for a 1-minute period. Would this condition pose a problem? What is the probability that at least 1 minute will elapse between arrivals?

6.38 *M/PF Research, Inc.* lists the average monthly apartment rent in some of the most expensive apartment rental locations in the United States. According to their report, the average cost of renting an apartment in Minneapolis is $951. Suppose that the standard deviation of the cost of renting an apartment in Minneapolis is $96 and that apartment rents in Minneapolis are normally distributed. If a Minneapolis apartment is randomly selected, what is the probability that the price is:
   a. $1000 or more?
   b. Between $900 and $1100?
   c. Between $825 and $925?
   d. Less than $700?

6.39 According to *The Wirthlin Report*, 24% of all workers say that their job is very stressful. If 60 workers are randomly selected, what is probability that 17 or more say that their job is very stressful? What is the probability that more than 22 say that their job is very stressful? What is the probability that between 8 and 12 (inclusive) say that their job is very stressful?

6.40 The U.S. Bureau of Labor Statistics reports that the average annual salary in the metropolitan Boston area is $34,383. Suppose annual salaries in the metropolitan Boston area are normally distributed, with a standard deviation of $4097. A Boston area worker is randomly selected.

a. What is the probability that the worker's annual salary is more than $40,000?

b. What is the probability that the worker's annual salary is less than $30,000?

c. What is the probability that the worker's annual salary is more than $20,000?

d. What is the probability that the worker's annual salary is between $27,000 and $36,000?

6.41 Suppose interarrival times at a hospital emergency room during a weekday are exponentially distributed, with an average interarrival time of 9 minutes. If the arrivals are Poisson distributed, what would the average number of arrivals per hour be? What is the probability that less than 5 minutes will elapse between any two arrivals?

6.42 Suppose the average speeds of passenger trains traveling from Newark, New Jersey, to Philadelphia, Pennsylvania, are normally distributed, with a mean average speed of 88 mph and a standard deviation of 6.4 mph.

a. What is the probability that a train will average less than 70 mph?

b. What is the probability that a train will average more than 80 mph?

c. What is the probability that a train will average between 90 and 100 mph?

6.43 The Conference Board published information on why companies expect to increase the number of part-time jobs and reduce full-time positions. Eighty-one percent of the companies said the reason was to get a flexible workforce. Suppose 200 companies that expect to increase the number of part-time jobs and reduce full-time positions are identified and contacted. What is the expected number of these companies that would agree that the reason is to get a flexible workforce? What is the probability that between 150 and 155 (not including the 150 or the 155) would give that reason? What is the probability that more than 158 would give that reason? What is the probability that fewer than 144 would give that reason?

6.44 According to the U.S. Bureau of the Census, about 75% of a commuters in the United States drive to work alone. Suppose 150 U.S. commuters are randomly sampled.

a. What is the probability that fewer than 105 commuters drive to work alone?

b. What is the probability that between 110 and 120 (inclusive) commuters drive to work alone?

c. What is the probability that more than 95 commuters drive to work alone?

6.45 According to figures released by the National Agricultural Statistics Service of the U.S. Department of Agriculture, the U.S. production of wheat over the past 20 years has been approximately uniformly distributed. Suppose the mean production over this period was 2.165 billion bushels. If the height of this distribution is .862 billion bushels, what are the values of a and b for this distribution?

6.46 The Federal Reserve System publishes data on family income based on its Survey of Consumer Finances. When the head of the household has a college degree, the mean before-tax family income is $70,400. Suppose that 60% of the before-tax family incomes when the head of the household has a college degree are between $61,200 and $79,600 and that these incomes are normally distributed. What is the standard deviation of before-tax family incomes when the head of the household has a college degree?

6.47 According to The Polk Company, a survey of households using the Internet in buying or leasing cars reported that 81% were seeking information about prices. In addition, 44% were seeking information about products offered. Suppose 75 randomly selected households are contacted who are using the Internet in buying or leasing cars.

a. What is the expected number of households who are seeking price information?

b. What is the expected number of households who are seeking information about products offered?

c. What is the probability that 67 or more households are seeking information about prices?

d. What is the probability that less than 23 households are seeking information about products offered?

6.48 Coastal businesses along the Gulf of Mexico from Texas to Florida worry about the threat of hurricanes during the season from June through October. Businesses become especially nervous when hurricanes enter the Gulf of Mexico. Suppose the arrival of hurricanes during this season is Poisson distributed, with an average of three hurricanes entering the Gulf of Mexico during the 5-month season. If a hurricane has just entered the Gulf of Mexico, what is the probability that at least 1 month will pass before the next hurricane enters the Gulf? What is the probability that another hurricane will enter the Gulf of Mexico in 2 weeks or less? What is the average amount of time between hurricanes entering the Gulf of Mexico?

6.49 With the growing emphasis on technology and the changing business environment, many workers are discovering that training such as reeducation, skill development, and personal growth are of great assistance in the job marketplace. A recent Gallup survey found that 80% of Generation Xers considered the availability of company-sponsored training as a factor to weigh in taking a job. If 50 Generation Xers are randomly sampled, what is the probability that fewer than 35 consider the availability of company-sponsored training as a factor to weigh in taking a job? What is the expected number? What is the probability that between 42 and 47 (inclusive) consider the availability of company-sponsored training as a factor to weigh in taking a job?

6.50 According to the Air Transport Association of America, the average operating cost of an MD-80 jet airliner is $2087 per hour. Suppose the operating costs of an MD-80 jet airliner are normally distributed with a standard deviation of $175 per hour. For what operating cost would only 20% of the operating costs be less? For what operating cost would 65% of the operating costs be more? What operating cost would be more than 85% of operating costs?

6.51 Supermarkets usually get very busy at about 5 P.M. on weekdays, because many workers stop by on the way home to shop. Suppose at that time arrivals at a supermarket's express checkout station are Poisson distributed, with an average of .8 person/minute. If the clerk has just checked out the last person in line, what is the probability that at least 1 minute will elapse before the next customer arrives? Suppose the clerk wants to go to the manager's office to ask a quick question and needs 2.5 minutes to do so. What is the probability that the clerk will get back before the next customer arrives?

6.52 According to *Editor and Publisher Yearbook,* the average daily circulation for a recent year of *The Wall Street Journal* is 1,774,880. The standard deviation is 50,940. Assume the paper's daily circulation is normally distributed. On what percentage of days would it surpass a circulation of 1,850,000? Suppose the paper cannot support the fixed expenses of a full-production setup if the circulation drops below 1,620,000. If the probability of this event occurring is low, the production manager might try to keep the full crew in place and not disrupt operations. How often will this event happen, based on the historical information?

6.53 Incoming phone calls generally are thought to be Poisson distributed. If an operator averages 2.2 phone calls every 30 seconds, what is the expected (average) amount of time between calls? What is the probability that a minute or more would elapse between incoming calls? Two minutes?

ANALYZING THE DATABASES

1. Select the agri-business time-series database. Create a histogram graph for onions and for broccoli. Each of these variables is approximately normally distributed. Compute the mean and the standard deviation for each distribution. The data in this database represent the monthly weight (in 1000 lb) of each vegetable. In terms of monthly weight, describe each vegetable (onions and broccoli). If a month were randomly selected from the onion distribution, what is the probability that the weight would be more than 50,000 (1000 lb)? What is the probability that the weight would be between 25,000 (1000 lb) and 35,000 (1000 lb)? If a month were randomly selected from the broccoli distribution, what is the probability that the weight would be more than 100,000 (1000 lb)? What is the probability that the weight would be between 135,000 (1000 lb) and 170,000 (1000 lb)?

2. Use the manufacturing database. The industry group variable is nearly uniformly distributed in this database, with values from $a = 1$ to $b = 20$. What is the height of this distribution? What is the probability of randomly selecting an industry group from 7 to 13 (inclusive) from this population if the distribution is uniform? (Use the uniform distribution theory to work this problem, not the actual numbers from the database).

3. Construct histogram graphs of all variables in the manufacturing database. Find at least one graph that appears to take on the shape of an exponential distribution. Compute descriptive statistics for that variable. Study the statistics and discuss what information relayed by the statistics would indicate that the shape of the distribution might be exponential.

MERCEDES GOES AFTER YOUNGER BUYERS

Mercedes and BMW have been competing head-to-head for market share in the luxury-car market for over three decades. Back in 1959, BMW (Bayerische Motoren Werke) almost went bankrupt and nearly sold out to Daimler-Benz, the maker of Mercedes-Benz cars. BMW was able to recover to the point that in 1992 it passed Mercedes in worldwide sales. Among the reasons for BMW's success was its ability to sell models that were more luxurious than previous models but still held an eye toward consumer quality and environmental responsibility. In particular, BMW targeted its sales pitch to the younger market, whereas Mercedes-Benz retained a more mature customer base.

In response to BMW's success, Mercedes has been trying to change their image by launching several products in an effort to attract younger buyers who are interested in sporty, performance-oriented cars. Several Mercedes models are designed specifically to lure buyers away from BMW. In particular, Mercedes hopes their CLK will attract buyers who have been purchasing the BMW 328is. However, according to one recent automotive expert, the focus is still on luxury and comfort for Mercedes while BMW focuses on performance and driving dynamics. As of mid-1998, the average price for a CLK was $39,850 as compared to $34,745 for a 328is. Gas mileage for the CLK is 20 mpg in town and 27 mpg on the road as compared to 18 mpg in town and 26 mpg on the road for the 328is.

DISCUSSION

1. Suppose Mercedes-Benz is concerned that dealer prices of the CLK are not consistent and that while the average price is $39,850, the prices are actually normally distributed with a standard deviation of $2005. Suppose also that Mercedes-Benz believes that at $42,000, the CLK is priced out of the BMW 328is market. What percentage of the dealer prices are more than $42,000? The average price for a BMW 328is is $34,745. Suppose these prices are also normally distributed with a standard deviation of $1780. What percentage of BMW dealers are pricing the 328is at more than the average price for a CLK? What percentage of Mercedes-Benz dealers are pricing the CLK at less than the average price for a 328is? Suppose a dealer is selling a 328is for $37,059. What percentage of Mercedes-Benz dealers price the CLK less than this? In terms of the CLK competing with the 328is price-wise, what do these data tell you?

2. Suppose that gas mileage rates for various CLK cars (including the fact that some drivers are less efficient than others) are uniformly distributed over a range from 20 mpg to 34 mpg on the road. What proportion of cars fall into the 25 mpg to 30 mpg range? Suppose that gas mileage rates for various 328is cars are uniformly distributed over a range from 21 mpg to 31 mpg on the road. What proportion of 328is cars fall into the 25 mpg to 30 mpg range? How does this compare to the figure for the CLK? What does that mean? Suppose these figures were true and Mercedes-Benz wanted to appeal to environmentally conscious shoppers on the basis of fuel economy. Compute the proportion of each car that gets 30 or more mpg according to these figures, and compare the results.

3. Suppose that in one dealership an average of 1.37 CLKs is sold every three hours (during a 12-hour showroom day) and that sales are Poisson distributed. Below are Excel-produced probabilities of the occurrence of different intersales times based on this information. Study the output and interpret it for the salespeople. For example, what is the probability that less than an hour will elapse between sales? What is the

probability that more than a day (12-hour day) will pass before the next sale after a car has been sold? What can the dealership managers do with such information? How can it help in staffing? How can such information be used as a tracking device for the impact of advertising? Is there a chance that these probabilities would change during the year? If so, why?

PORTION OF 3 HR TIME FRAME	CUM. EXPONENTIAL PROB. FROM LEFT
0.167	0.2045
0.333	0.3663
0.667	0.5990
1	0.7459
2	0.9354
3	0.9836
4	0.9958
5	0.9989

7

Sampling and Sampling Distributions

Learning Objectives

The two main objectives for Chapter 7 are to give you an appreciation for the proper application of sampling techniques and an understanding of the sampling distributions of two statistics, thereby enabling you to:

1. Determine when to use sampling instead of a census.

2. Distinguish between random and nonrandom sampling.

3. Decide when and how to use various sampling techniques.

4. Be aware of the different types of error that can occur in a study.

5. Understand the impact of the central limit theorem on statistical analysis.

6. Use the sampling distributions of $\bar{X}$ and $\hat{p}$.

This chapter explores the process of sampling and the sampling distributions of some statistics. How do we obtain the data used in statistical analysis? Why do researchers often take a sample rather than conduct a census? What are the differences between random and nonrandom sampling? This chapter addresses these and other questions about sampling.

Also presented are the distributions of two statistics: the sample mean and the sample proportion. It has been determined that statistics such as these are approximately normally distributed under certain conditions. Knowledge of this is important in the study of statistics and is basic to much of statistical analysis.

7.1
Sampling

Sampling is widely used in business as a means of gathering useful information about a population. Data are gathered from samples and conclusions are drawn about the population as a part of the inferential statistics process. For example, suppose a researcher wants to ascertain the viewpoints of maquiladora workers along the U.S.-Mexico border. To do this, a random sample of workers could be taken from a wide selection of companies in several industries in many of the key border cities. A carefully constructed questionnaire that is culturally sensitive to Mexicans could be administered to the selected workers to determine work attitudes, expectations, and cultural differences between workers and companies. The researchers could compile and analyze the data gleaned from the responses. Summaries and observations could be made about worker outlook and culture in the maquiladora program. Management and decision makers could then attempt to use the results of the study to improve worker performance and motivation. Often, a sample provides a reasonable means for gathering such useful decision-making information that might be otherwise unattainable and unaffordable.

Reasons for Sampling

There are several good reasons for taking a sample instead of conducting a census.

1. The sample can save money.
2. The sample can save time.
3. For given resources, the sample can broaden the scope of the study.
4. Because the research process is sometimes destructive, the sample can save product.
5. If accessing the population is impossible, the sample is the only option.

A sample can be cheaper to obtain than a census for a given magnitude of questions. For example, if an 8-minute telephone interview is being undertaken, conducting the interviews with a sample of 100 customers rather than with a population of 100,000 customers obviously is less expensive. In addition to the cost savings, the significantly smaller number of interviews usually requires less total time. Thus, if there is an urgency about obtaining the results, sampling can provide them more quickly. With the volatility of some markets and the constant barrage of new competition and new ideas, sampling has a strong advantage over a census in terms of research turnaround time.

If the resources allocated to a research project are fixed, more detailed information can be gathered by taking a sample than by conducting a census. With resources concentrated on fewer individuals or items, the study can be broadened in scope to allow for more specialized questions. One organization budgeted $100,000 for a study and opted to take a census instead of a sample by using a mail survey. The researchers mass-mailed thousands of copies of a computer card that looked like a major league all-star ballot. The card contained 20 questions to which the respondent could answer yes or no by punching out a perforated hole. The information retrieved amounted to the percentages of respondents who answered yes and no on the 20 questions. For the same amount of money, the company could have taken a random sample from the population, held interactive one-on-one sessions with highly trained interviewers, and gathered detailed information about the process being studied. By using the money on a sample, the researchers could have spent significantly more time with each respondent and thus increased the potential for gathering useful information.

Some research processes are destructive to the product or item being studied. For example, if light bulbs are being tested to determine how long they burn or if candy bars are being taste tested to determine whether the taste is acceptable, the product is destroyed. If a census were conducted for this type of research, there would be no product to sell. Hence, taking a sample is the only realistic option for testing such products.

Sometimes a population is virtually impossible to access for research. For example, some people refuse to answer sensitive questions, and some telephone numbers are unlisted. Some items of interest (like a 1957 Chevrolet) are so scattered that locating all of them would be extremely difficult. When the population is inaccessible for these or other reasons, sampling is the only option.

Reasons for Taking a Census

Sometimes taking a census makes more sense than using a sample. One reason to take a census is to eliminate the possibility that by chance a randomly selected sample might not be representative of the population. Even when all the proper sampling techniques are implemented, a sample that is nonrepresentative of the population can be selected by chance. For example, if the population of interest is all truck owners in the state of Colorado, a random sample of owners could yield mostly ranchers, when in fact many of the truck owners in Colorado are urban dwellers.

A second reason to take a census is that the client (person authorizing and/or underwriting the study) does not have an appreciation for random sampling and feels more comfortable with conducting a census. Both of these reasons for taking a census are based on the assumption that enough time and money are available to conduct such a census.

Frame

Every research study has a target population that consists of the individuals, institutions, or entities that are the object of investigation. The sample is taken from a population *list, map, directory, or other source that is being used to represent the population.* This list, map, or directory is called the **frame,** which can be school lists, trade association lists, or even lists sold by list brokers. Ideally, there is a one-to-one correspondence between the frame units and the population units. In reality, the frame and the target population are often different. For example, suppose the target population is all families living in Detroit. A feasible frame would be the residential pages of the Detroit telephone books. How would the frame differ from the target population? Some families have no telephone. Other families have unlisted numbers. Still other families might have moved and/or changed numbers since the directory was printed. Some families even have multiple listings under different names.

Frames that have *overregistration* contain all the target population units plus some additional units. Frames that have *underregistration* contain fewer units than does the target population. Sampling is done from the frame, not the target population. In theory, the target population and the frame are the same. In reality, a researcher's goal is to minimize the differences between the frame and the target population.

Random versus Nonrandom Sampling

The two main types of sampling are random and nonrandom. In **random sampling** *every unit of the population has the same probability of being selected into the sample.* Random sampling implies that chance enters into the process of selection. For example, most

Frame
A list, map, directory, or some other source that is being used to represent the population in the process of sampling.

Random sampling
Sampling in which every unit of the population has the same probability of being selected for the sample.

Americans would like to believe that winners of nationwide magazine sweepstakes are selected by some random draw of numbers. Late in the 1960s when the military draft lottery was being used, most people eligible for the draft trusted that a given birthdate was selected by chance as the first date to use to draft people. In both of these situations, members of the population believed that selections were made by chance.

In **nonrandom sampling** *not every unit of the population has the same probability of being selected into the sample.* Members of nonrandom samples are not selected by chance. For example, they might be selected because they are at the right place at the right time or because they know the people conducting the research.

Sometimes random sampling is called *probability sampling,* and nonrandom sampling is called *nonprobability sampling.* Because every unit of the population is not equally likely to be selected, assigning a probability of occurrence in nonrandom sampling is impossible. The statistical methods presented and discussed in this text are based on the assumption that the data come from random samples. *Nonrandom sampling methods are not appropriate techniques for gathering data to be analyzed by most of the statistical methods presented in this text.* However, several nonrandom sampling techniques are described in this section, primarily to alert you to their characteristics and limitations.

> **Nonrandom sampling**
> Sampling in which not every unit of the population has the same probability of being selected into the sample.

Random Sampling Techniques

The four basic random sampling techniques are simple random sampling, stratified random sampling, systematic random sampling, and cluster (or area) random sampling. Each technique has advantages and disadvantages. Some techniques are simpler to use, some are less costly, and others have the potential for reducing sampling error.

SIMPLE RANDOM SAMPLING The most elementary random sampling technique is **simple random sampling.** Simple random sampling can be viewed as the basis for the other three random sampling techniques. With simple random sampling, each unit of the frame is numbered from 1 to N (where N is the size of the population). Next, a table of random numbers or a random number generator is used to select n items into the sample. A *random number generator* is usually a computer program that allows computer to calculator output to yield random numbers. Table 7.1 contains a brief table of random numbers. Table A.1 in Appendix A contains a full table of random numbers. These numbers are random in all directions. The spaces in the table are there only for ease of reading the values. For each number, any of the 10 digits (0–9) is equally likely, so getting the same digit twice or more in a row is possible.

> **Simple random sampling**
> The most elementary of the random sampling techniques; involves numbering each item in the population and using a list or roster of random numbers to select items for the sample.

As an example, from the population frame of companies listed in Table 7.2, we will use simple random sampling to select a sample of six companies. First, we number every member of the population. We select as many digits for each unit sampled as there are in the largest number in the population. For example, if a population has 2000 members, we select four-digit numbers. Because the population in Table 7.2 contains 30 members, only two digits need be selected for each number. The population is numbered from 01 to 30, as shown in Table 7.3.

TABLE 7.1
A Brief Table of Random Numbers

91567	42595	27958	30134	04024	86385	29880	99730
46503	18584	18845	49618	02304	51038	20655	58727
34914	63976	88720	82765	34476	17032	87589	40836
57491	16703	23167	49323	45021	33132	12544	41035
30405	83946	23792	14422	15059	45799	22716	19792
09983	74353	68668	30429	70735	25499	16631	35006
85900	07119	97336	71048	08178	77233	13916	47564

The object is to sample six companies, so six different two-digit numbers must be selected from the table of random numbers. Because this population contains only 30 companies, all numbers greater than 30 (31–99) must be ignored. If, for example, the number 67 is selected, the process is continued until a value between 1 and 30 is obtained. If the same number occurs more than once, we proceed to another number. For ease of understanding, we start with the first pair of digits in Table 7.1 and proceed across the first row until $n = 6$ different values between 01 and 30 are selected. If additional numbers are needed, we proceed across the second row, and so on. Often a researcher will start at some randomly selected location in the table and proceed in a predetermined direction to select numbers.

In the first row of digits in Table 7.1, the first number is 91. This number is out of range so it is cast out. The next two digits are 56. Next is 74, followed by 25, which is the first usable number. From Table 7.3, we see that 25 is the number associated with Occidental Petroleum, so Occidental Petroleum is the first company selected into the sample. The next number is 95, unusable, followed by 27, which is usable. Twenty-seven is the number for Philadelphia Electric, so this company is selected. Continuing the process, we pass over the numbers 95 and 83. The next usable number is 01, which is the value for Alaska Airlines. Thirty-four is next, followed by 04 and 02, both of which are usable. These numbers are associated with Atlantic Richfield and Alcoa, respectively. Continuing along the first row, the next usable number is 29, which is associated with Sears. As this is the sixth selection, the sample is complete. The following companies constitute the final sample.

Alaska Airlines

Alcoa

Atlantic Richfield

Occidental Petroleum

Philadelphia Electric

Sears

Alaska Airlines	DuPont	LTV
Alcoa	Exxon	Litton
Amoco	Farah	Mead
Atlantic Richfield	GTE	Mobil
Bank of America	General Electric	Occidental Petroleum
Bell of Pennsylvania	General Mills	JCPenney
Chevron	General Dynamics	Philadelphia Electric
Chrysler	Grumman	Ryder
Citicorp	IBM	Sears
Disney	Kmart	Time

TABLE 7.2
A Population Frame of 30 Companies

01	Alaska Airlines	11	DuPont	21	LTV
02	Alcoa	12	Exxon	22	Litton
03	Amoco	13	Farah	23	Mead
04	Atlantic Richfield	14	GTE	24	Mobil
05	Bank of America	15	General Electric	25	Occidental Petroleum
06	Bell of Pennsylvania	16	General Mills	26	JCPenney
07	Chevron	17	General Dynamics	27	Philadelphia Electric
08	Chrysler	18	Grumman	28	Ryder
09	Citicorp	19	IBM	29	Sears
10	Disney	20	Kmart	30	Time

TABLE 7.3
Numbered Population of 30 Companies

Simple random sampling is easier to perform on small than on large populations. The process of numbering all the members of the population and selecting items is cumbersome for large populations.

Stratified random sampling
A type of random sampling in which the population is divided into various nonoverlapping strata and then items are randomly selected into the sample from each stratum.

STRATIFIED RANDOM SAMPLING   A second type of random sampling is **stratified random sampling,** in which the population is divided into nonoverlapping subpopulations called *strata*. The researcher then extracts a simple random sample from each of the subpopulations. The main reason for using stratified random sampling is that it has the potential for reducing sampling error. Sampling error occurs when, by chance, the sample does not represent the population. With stratified random sampling, the potential to match the sample closely to the population is greater than it is with simple random sampling because portions of the total sample are taken from different population subgroups. However, stratified random sampling is generally more costly than simple random sampling because each unit of the population must be assigned to a stratum before the random selection process begins.

Strata selection is usually based on available information. Such information may have been gleaned from previous censuses or surveys. Stratification benefits increase as the strata differ more. Internally, a stratum should be relatively homogeneous; externally, strata should contrast with each other. Stratification is often done by using demographic variables, such as gender, socioeconomic class, geographic region, religion, and ethnicity. For example, if a U.S. presidential election poll is to be conducted by a market research firm, what important variables should be stratified? The gender of the respondent might make a difference because a gender gap in voter preference has been noted in past elections. That is, men and women have tended to vote differently in national elections. Geographic region also has been an important variable in national elections as voters are influenced by local cultural values that differ from region to region. Voters in the South voted almost exclusively for Democrats in the past, but recently they have tended to vote for Republican candidates in national elections. Voters in the Rocky Mountain states have supported Republican presidential candidates; in the industrial Northeast, voters have been more inclined toward Democratic candidates.

In FM radio markets, age of listener is an important determinant of the type of programing used by a station. Figure 7.1 contains a stratification by age with three strata, based on the assumption that age makes a difference in preference of programing. This stratification implies that listeners 20 to 30 years of age tend to prefer the same type of programing, which is different from that preferred by listeners 30 to 40 and 40 to 50 years of

Figure 7.1

Stratified random sampling of FM radio listeners

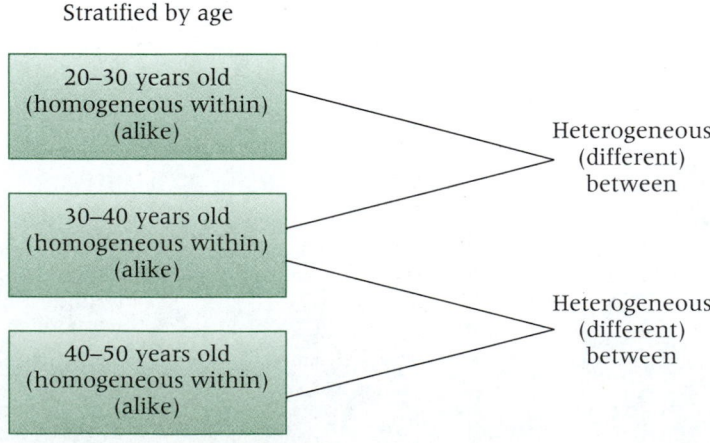

age. Within each age subgroup (stratum), *homogeneity* or alikeness is present; between each pair of subgroups a difference, or *heterogeneity,* is present.

Stratified random sampling can be either proportionate or disproportionate. **Proportionate stratified random sampling** occurs *when the percentage of the sample taken from each stratum is proportionate to the percentage that each stratum is within the whole population.* For example, suppose voters are being surveyed in Boston and the sample is being stratified by religion as Catholic, Protestant, Jewish, and others. If Boston's population is 90% Catholic and if a sample of 1000 voters is being taken, the sample would require inclusion of 900 Catholics to achieve proportionate stratification. Any other number of Catholics would be disproportionate stratification. The sample proportion of other religions would also have to follow population percentages. Or consider the city of El Paso, Texas, where the population is approximately 69% Hispanic. If a researcher is conducting a citywide poll in El Paso and if stratification is by ethnicity, a proportionate stratified random sample should contain 69% Hispanics. Hence, an ethnically proportionate stratified sample of 160 residents from El Paso's 600,000 residents should contain approximately 110 Hispanics. Whenever *the proportions of the strata in the sample are different than the proportions of the strata in the population,* **disproportionate stratified random sampling** occurs.

SYSTEMATIC SAMPLING   Systematic sampling is a third random sampling technique. Unlike stratified random sampling, systematic sampling is not done in an attempt to reduce sampling error. Rather, **systematic sampling** is used because of its convenience and relative ease of administration. With systematic sampling, *every kth item is selected to produce a sample of size* n *from a population of size* N. The value of k can be determined by the following formula. If k is not an integer value, the whole-number value should be used.

$$k = \frac{N}{n}$$

DETERMINING
THE VALUE OF k

where:

n = sample size
N = population size
k = size of interval for selection

As an example of systematic sampling, a management information systems researcher wanted to sample the manufacturers in Texas. He had enough financial support to sample 1000 companies (*n*). The *Directory of Texas Manufacturers* listed approximately 17,000 total manufacturers in Texas (*N*) in alphabetical order. The value of k was 17 (17,000/1000) and the researcher selected every 17th company in the directory for his sample.

Did the researcher begin with the first company listed or the 17th or one somewhere between? In selecting every kth value, a simple random number table should be used to select a value between 1 and k inclusive as a starting point. The second element for the sample is the starting point plus k. In the example, $k = 17$, so the researcher would have gone to a table of random numbers to determine a starting point between 1 and 17. Suppose he selected the number 5. He would have started with the 5th company, then selected the 22nd (5 + 17), and then the 39th, and so on.

Besides convenience, systematic sampling has other advantages. Because systematic sampling is evenly distributed across the frame, a knowledgeable person can easily determine whether a sampling plan has been followed in a study. However, a problem with systematic sampling can occur if there is periodicity in the data, and the sampling interval is in syncopation with it. For example, if a list of 150 college students is actually a merged

Sidebar (glossary)

Proportionate stratified random sampling
A type of stratified random sampling in which the proportions of the items selected for the sample from the strata reflect the proportions of the strata in the population.

Disproportionate stratified random sampling
A type of stratified random sampling in which the proportions of items selected from the strata for the final sample do not reflect the proportions of the strata in the population.

Systematic sampling
A random sampling technique in which every kth item or person is selected from the population.

Figure 7.2

Some test market cities

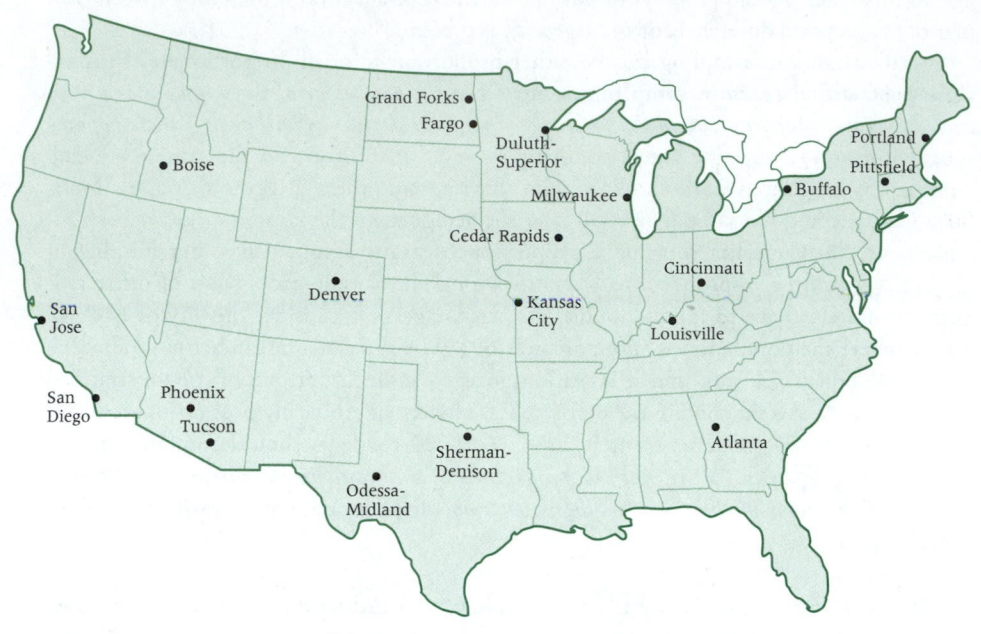

list of five classes with 30 students in each class and if each of the lists of the five classes has been ordered with the names of top students first and bottom students last, systematic sampling of every 30th student could cause selection of all top students, all bottom students, or all mediocre students. That is, there is a cyclical or periodic organization of the original list. Systematic sampling methodology is based on the assumption that the source of population elements is random.

Cluster (or area) sampling
A type of random sampling in which the population is divided into nonoverlapping areas or clusters and elements are randomly sampled from the areas or clusters.

CLUSTER (OR AREA) SAMPLING   Cluster (or area) sampling is a fourth type of random sampling. **Cluster (or area) sampling** involves dividing the population into nonoverlapping areas or clusters. However, in contrast to stratified random sampling where strata are homogeneous, cluster sampling identifies clusters that tend to be internally heterogeneous. In theory, each cluster contains a wide variety of elements, and the cluster is a miniature, or microcosm, of the population. Examples of clusters are towns, companies, homes, colleges, areas of a city, and geographic regions. Often clusters are naturally occurring groups of the population and are already identified, such as states or Standard Metropolitan Statistical Areas. Although area sampling usually refers to clusters that are areas of the population, such as geographic regions and cities, the terms *cluster sampling* and *area sampling* are used interchangeably in this text.

After choosing the clusters, the researcher randomly selects individual elements into the sample from the clusters. One example of business research that makes use of clustering is test marketing of new products. Often in test marketing, the United States is divided into clusters of test market cities, and individual consumers within the test market cities are surveyed.* Figure 7.2 shows some U.S. test market cities that are used as clusters to test products.

Two-stage sampling
Cluster sampling done in two stages: A first round of samples is taken and then a second round is taken from within the first samples.

Sometimes the clusters are too large, and a second set of clusters is taken from each original cluster. This technique is called **two-stage sampling.** For example, a researcher could divide the United States into clusters of cities. She could then divide the cities into clusters of blocks and randomly select individual houses from the block clusters. The first stage is selecting the test cities and the second stage is selecting the blocks.

*Bristol Voss, "The Nation's Most Popular Test Markets," *Sales and Marketing Management,* March 1989, 141.

Cluster or area sampling has several advantages. Two of the foremost advantages are convenience and cost. Clusters are usually convenient to obtain, and the cost of sampling from the entire population is reduced because the scope of the study is reduced to the clusters. The cost per element is usually lower in cluster or area sampling than in stratified sampling because of lower element listing or locating costs. The time and cost of contacting elements of the population can be reduced, especially if travel is involved, because clustering reduces the distance to the sampled elements. In addition, administration of the sample survey can be simplified. Sometimes cluster or area sampling is the only feasible approach because the sampling frames of the individual elements of the population are unavailable and therefore other random sampling techniques cannot be used.

Cluster or area sampling also has several disadvantages. If the elements of a cluster are similar, cluster sampling may be statistically less efficient than simple random sampling. In an extreme case—when the elements of a cluster are the same—sampling from the cluster may be no better than sampling a single unit from the cluster. Moreover, the costs and problems of statistical analysis are greater with cluster or area sampling than with simple random sampling.

Nonrandom Sampling

Sampling techniques used to select elements from the population by any mechanism that does not involve a random selection process are called **nonrandom sampling techniques.** Because chance is not used to select items from the samples, these techniques are nonprobability techniques and are *not desirable for use in gathering data to be analyzed by the methods of inferential statistics presented in this text.* Sampling error cannot be determined objectively for these sampling techniques. Four nonrandom sampling techniques are presented here: convenience sampling, judgmental sampling, quota sampling, and snowball sampling.

CONVENIENCE SAMPLING   In **convenience sampling,** *elements for the sample are selected for the convenience of the researcher.* The researcher typically chooses items that are readily available, nearby, and/or willing to participate. The sample tends to be less variable than the population because in many environments the extreme elements of the population are not readily available. The researcher will select more elements from the middle of the population. For example, a convenience sample of homes for door-to-door interviews might include houses where people are at home, houses with no dogs, houses near the street, first-floor apartments, and houses with friendly people. In contrast, a random sample would require the researcher to gather data only from houses and apartments that have been selected randomly, no matter how inconvenient or unfriendly the location. If a research firm is located in a mall, a convenience sample might be selected by interviewing only shoppers who pass the shop and look friendly.

JUDGMENT SAMPLING   **Judgment sampling** occurs when *elements selected for the sample are chosen by the judgment of the researcher.* Researchers often believe they can obtain a representative sample by using sound judgment, which will result in saving time and money. Sometimes ethical, professional researchers might believe they can select a more representative sample than the random process will provide. They might be right! However, some studies have shown that random sampling methods outperform judgment sampling in estimating the population mean even when the researcher who is administering the judgment sampling is trying to put together a very representative sample. When sampling is done by judgment, calculating the probability that an element is going to be selected into the sample is not possible. The sampling error cannot be determined objectively because probabilities are based on *nonrandom* selection.

Other problems are associated with judgment sampling. The researcher tends to make errors of judgment in one direction. These systematic errors lead to what are called *biases.* The researcher also is unlikely to include extreme elements. With judgment sampling, there is no objective method for determining whether one person's judgment is better than another's.

Quota sampling

A nonrandom sampling technique in which the population is stratified on some characteristic and then elements selected for the sample are chosen by nonrandom processes.

QUOTA SAMPLING A third nonrandom sampling technique is **quota sampling,** which appears to be similar to stratified random sampling. Certain population subclasses, such as age group, gender, or geographic region, are used as strata. However, instead of randomly sampling from each stratum, the researcher *uses a nonrandom sampling method to gather data from one stratum until the desired quota of samples is filled.* Quotas are described by quota controls, which set the sizes of the samples to be obtained from the subgroups. Generally, a quota is based on the proportions of the subclasses in the population. In this case, the quota concept is similar to that of proportional stratified sampling.

Quotas often are filled by using available, recent, or applicable elements. For example, instead of randomly interviewing people to obtain a quota of Italian Americans, the researcher would go to the Italian area of the city and interview there until enough responses are obtained to fill the quota. In quota sampling, an interviewer would begin by asking a few filter questions; if the respondent represents a subclass whose quota has been filled, the interviewer would terminate the interview.

Quota sampling can be useful if no frame is available for the population. For example, suppose a researcher wants to stratify the population into owners of different types of cars but fails to find any lists of Toyota van owners. Through quota sampling, the researcher would proceed by interviewing all car owners and casting out non-Toyota van owners until the quota of Toyota van owners is filled.

Quota sampling is less expensive than most random sampling techniques because it essentially is a technique of convenience. However, cost may have no meaning because the quality of nonrandom and random sampling techniques cannot be compared. Another advantage of quota sampling is the speed of data gathering. The researcher does not have to call back or send out a second questionnaire if there is no response; he just moves on to the next element. Also, preparatory work for quota sampling is minimal.

The main problem with quota sampling is that, when all is said and done, it still is only a *nonrandom* sampling technique. Some researchers have said that if the quota is filled by *randomly* selecting elements and discarding those that are not from a stratum, quota sampling is essentially a version of stratified random sampling. However, most quota sampling is carried out by the researcher going where the quota can be filled quickly. The object is to gain the benefits of stratification without the high field costs of stratification. Ultimately, it remains a nonprobability sampling method.

Snowball sampling

A nonrandom sampling technique in which survey subjects who fit a desired profile are selected based on referral from other survey respondents who also fit the desired profile.

SNOWBALL SAMPLING Another nonrandom sampling technique is **snowball sampling,** in which *survey subjects are selected based on referral from other survey respondents.* The researcher identifies a person who fits the profile of subjects wanted for the study. The researcher then asks this person for the names and locations of others who would also fit the profile of subjects wanted for the study. Through these referrals, survey subjects can be identified cheaply and efficiently, which is particularly useful when survey subjects are difficult to locate. This is the main advantage of snowball sampling; its main disadvantage is that it is nonrandom.

Sampling Error

Sampling error

Error that occurs when the sample is not representative of the population.

Sampling error occurs *when the sample is not representative of the population.* When random sampling techniques are used to select elements for the sample, sampling error occurs by chance. Many times the statistic computed on the sample is not an accurate estimate of the population parameter because the sample was not representative of the population. This result is caused by sampling error. With random samples, sampling error can be computed and analyzed.

Nonsampling Errors

All errors other than sampling errors are **nonsampling errors.** The many possible nonsampling errors include missing data, recording errors, input processing errors, and analysis errors. Other nonsampling errors have to do with the measurement instrument, such as errors of unclear definitions, defective questionnaires, and poorly conceived concepts. Improper definition of the frame is a nonsampling error. In many cases, finding a frame that perfectly fits the population is impossible. Insofar as it does not fit, a nonsampling error has been committed.

Response errors are also nonsampling errors. They occur when people do not know, will not say, or overstate. There is virtually no statistical way to measure or control for nonsampling errors. The statistical techniques presented in this text are based on the assumption that none of these nonsampling errors has been committed. The researcher must eliminate these errors through carefully planning and executing the research study.

Nonsampling errors
All errors other than sampling errors.

Analysis Using Excel

It is possible to generate random numbers from six different types of distributions using Excel, including the binomial, the Poisson, the uniform, and the normal distributions. The process begins by selecting the **Random Number Generation** feature from the **Data Analysis** dialog box under **Tools.** The **Random Number Generation** dialog box is shown in Figure 7.3. Place the number of variables for which random numbers are to be

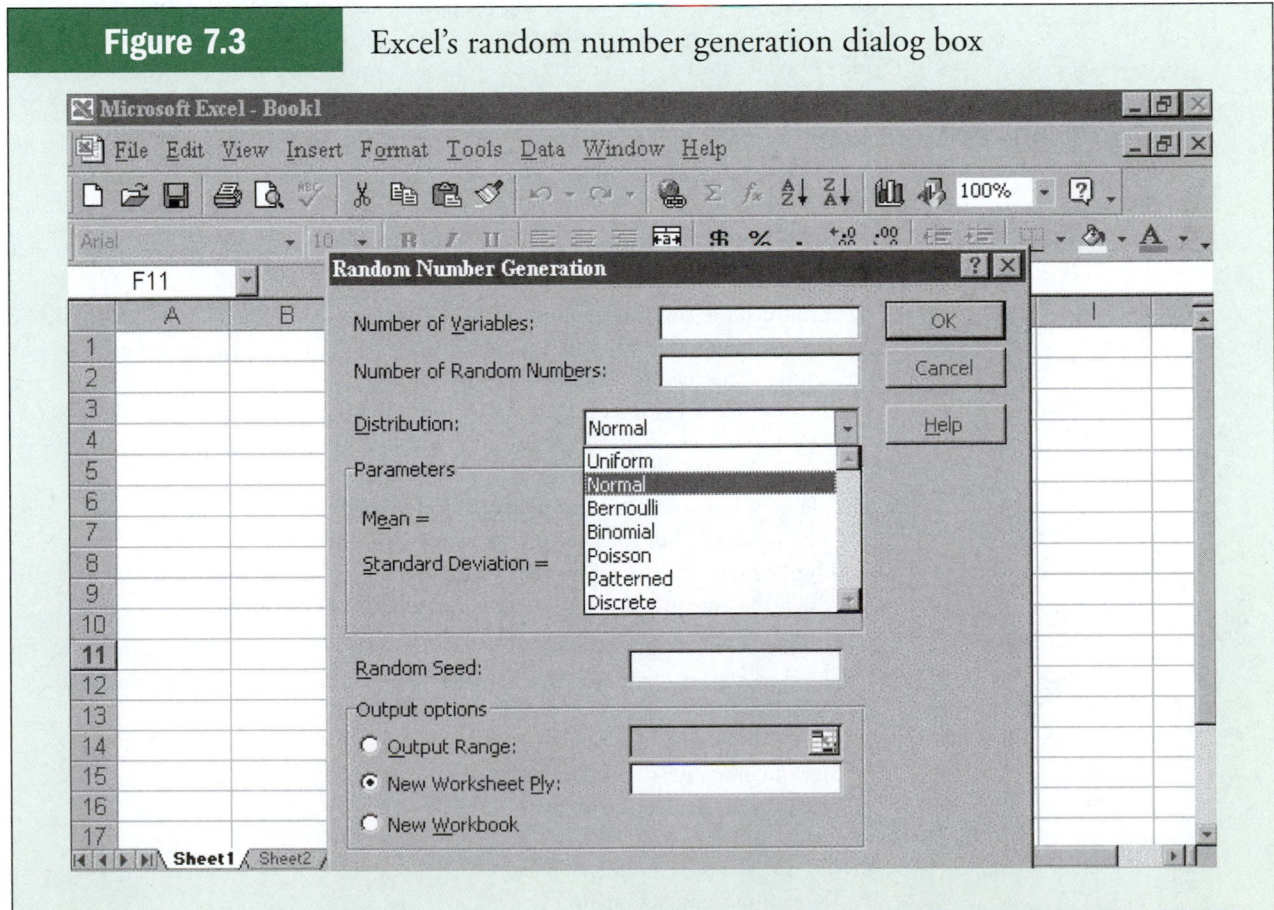

Figure 7.3 Excel's random number generation dialog box

generated in the first line of the dialog box. In the second line, insert the number of random numbers to be generated for each variable. The third line of the dialog box contains a pull-down menu with a list of the six types of distributions from which a distribution can be selected for generating the random numbers (this list also includes a non-random selection called patterned). You will be given different parameters to complete dependent upon which of the different distributions you select (e.g. for normal distribution, mean and standard deviation). You have the option of entering a seed number in the line denoted as **Random Seed** or leaving it blank (default option). Computerized random numbers are generated with a mathematical function. The starting value for the mathematical function is specified by the Random Seed. If the same seed number is used again, Excel will produce the same random numbers.

7.1 Problems

7.1 Develop a frame for the population of each of the following research projects.
   a. Measuring the job satisfaction of all union employees in a company
   b. Conducting a telephone survey in Utica, New York, to determine whether there is any interest in having a new hunting and fishing specialty store in the mall
   c. Interviewing passengers of a major airline about its food service
   d. Studying the quality control programs of boat manufacturers
   e. Attempting to measure the corporate culture of cable television companies

7.2 Make a list of 20 people you know. Include men and women, various ages, various educational levels, and so on. Number the list and then use the random number list in Table 7.1 to select six people randomly from your list. How representative of the population is the sample? Find the proportion of men in your population and in your sample. How do the proportions compare? Find the proportion of 20-year-olds in your sample and the proportion in the population. How do they compare?

7.3 Use the random numbers in Table A.1 of Appendix A to select 10 of the companies from the 30 companies listed in Table 7.2. Compare the types of companies in your sample with the types in the population. How representative of the population is your sample?

7.4 For each of the following research projects, list three variables for stratification of the sample.
   a. A nationwide study of motels and hotels is being conducted. An attempt will be made to determine the extent of the availability of online links for customers. A sample of motels and hotels will be taken.
   b. A consumer panel is to be formed by sampling people in Michigan. Members of the panel will be interviewed periodically in an effort to understand current consumer attitudes and behaviors.
   c. A large soft-drink company wants to study the characteristics of the U.S. bottlers of its products, but the company does not want to conduct a census.
   d. The business research bureau of a large university is conducting a project in which the bureau will sample paper-manufacturing companies.

7.5 In each of the following cases, the variable represents one way that a sample can be stratified in a study. For each variable, list some strata into which the variable can be divided.
   a. Age of respondent (person)
   b. Size of company (sales volume)
   c. Size of retail outlet (square feet)
   d. Geographic location
   e. Occupation of respondent (person)
   f. Type of business (company)

7.6   A city's telephone book lists 100,000 people. If the telephone book is the frame for a study, how large would the sample size be if systematic sampling were done on every 200th person?

7.7   If every 11th item is systematically sampled to produce a sample size of 75 items, approximately how large is the population?

7.8   If a company employs 3500 people and if a random sample of 175 of these employees has been taken by systematic sampling, what is the value of k? The researcher would start the sample selection between what two values? Where could the researcher obtain a frame for this study?

7.9   For each of the following research projects, list at least one area or cluster that could be used in obtaining the sample.
     a. A study of road conditions in the state of Missouri
     b. A study of U.S. offshore oil wells
     c. A study of the environmental effects of petrochemical plants west of the Mississippi River

7.10   Give an example of how judgment sampling could be used in a study to determine how district attorneys feel about attorneys advertising on television.

7.11   Give an example of how convenience sampling could be used in a study of *Fortune 500* executives to measure corporate attitude toward paternity leave for employees.

7.12   Give an example of how quota sampling could be used to conduct sampling by a company test marketing a new personal computer.

In the inferential statistics process, a random sample is selected from the population, a statistic is computed on the sample, and conclusions are reached about the population parameter from the statistic. In attempting to analyze the sample statistic, it is essential to know the distribution of the statistic. So far we have studied several distributions, including the binomial distribution, the Poisson distribution, the hypergeometric distribution, the uniform distribution, the normal distribution, and the exponential distribution.

    In this section we explore the sample mean, $\overline{X}$, as the statistic. The sample mean is one of the more common statistics used in the inferential process. To compute and assign the probability of occurrence of a particular value of a sample mean, the researcher must know the distribution of the sample means. One way to examine the distribution possibilities is to take a population with a particular distribution, randomly select samples of a given size, compute the sample means, and attempt to determine how the means are distributed. Suppose a small finite population consists of only $N = 8$ numbers:

$$54, 55, 59, 63, 64, 68, 69, \text{ and } 70.$$

Using an Excel-produced histogram, we can see the shape of the distribution of this population of data shown in Figure 7.4

    Suppose we take all possible samples of size $n = 2$ from this population with replacement. The result is the following pairs of data.

(54,54)	(55,54)	(59,54)	(63,54)
(54,55)	(55,55)	(59,55)	(63,55)
(54,59)	(55,59)	(59,59)	(63,59)
(54,63)	(55,63)	(59,63)	(63,63)
(54,64)	(55,64)	(59,64)	(63,64)
(54,68)	(55,68)	(59,68)	(63,68)

7.2

Sampling Distribution of $\overline{X}$

continued

Figure 7.4

Histogram of eight numbers

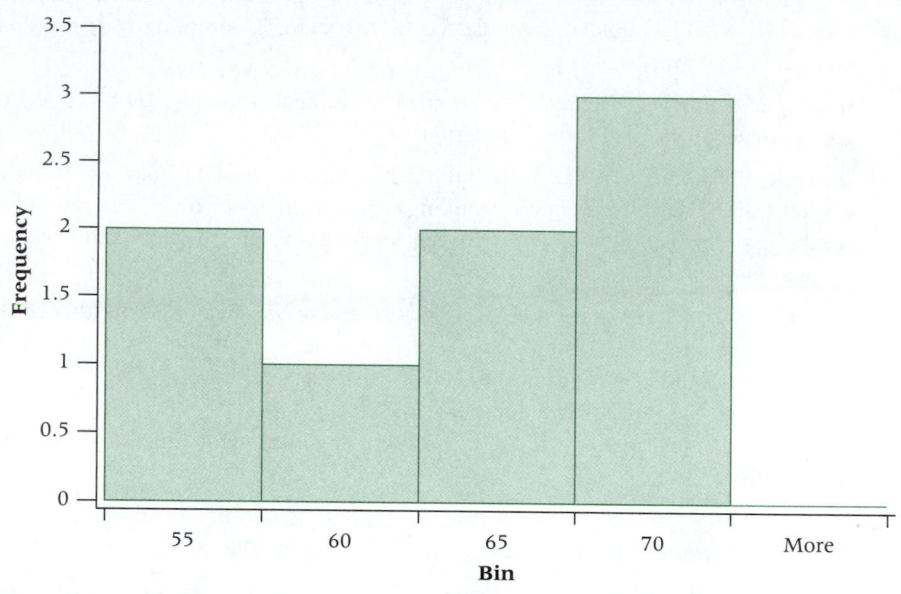

(54,69)	(55,69)	(59,69)	(63,69)
(54,70)	(55,70)	(59,70)	(63,70)
(64,54)	(68,54)	(69,54)	(70,54)
(64,55)	(68,55)	(69,55)	(70,55)
(64,59)	(68,59)	(69,59)	(70,59)
(64,63)	(68,63)	(69,63)	(70,63)
(64,64)	(68,64)	(69,64)	(70,64)
(64,68)	(68,68)	(69,68)	(70,68)
(64,69)	(68,69)	(69,69)	(70,69)
(64,70)	(68,70)	(69,70)	(70,70)

The means of each of these samples follow.

54	54.5	56.5	58.5	59	61	61.5	62
54.5	55	57	59	59.5	61.5	62	62.5
56.5	57	59	61	61.5	63.5	64	64.5
58.5	59	61	63	63.5	65.5	66	66.5
59	59.5	61.5	63.5	64	66	66.5	67
60	61.5	63.5	65.5	66	68	68.5	69
61.5	62	64	66	66.5	68.5	69	69.5
62	62.5	64.5	66.5	67	69	69.5	70

Again using an Excel histogram, we can see the shape of the distribution of these sample means in Figure 7.5. Notice that the shape of the histogram for sample means is quite unlike the shape of the histogram for the population. The sample means appear to "pile up" toward the middle of the distribution and "tail off" toward the extremes.

Figure 7.6 is an Excel histogram of the data from a Poisson distribution of values with a population mean of 1.25. Note that the histogram is skewed to the right. Suppose 90 samples of size $n = 30$ are taken randomly from a Poisson distribution with $\lambda = 1.25$ and the means are computed on each sample. The resulting distribution of sample means is

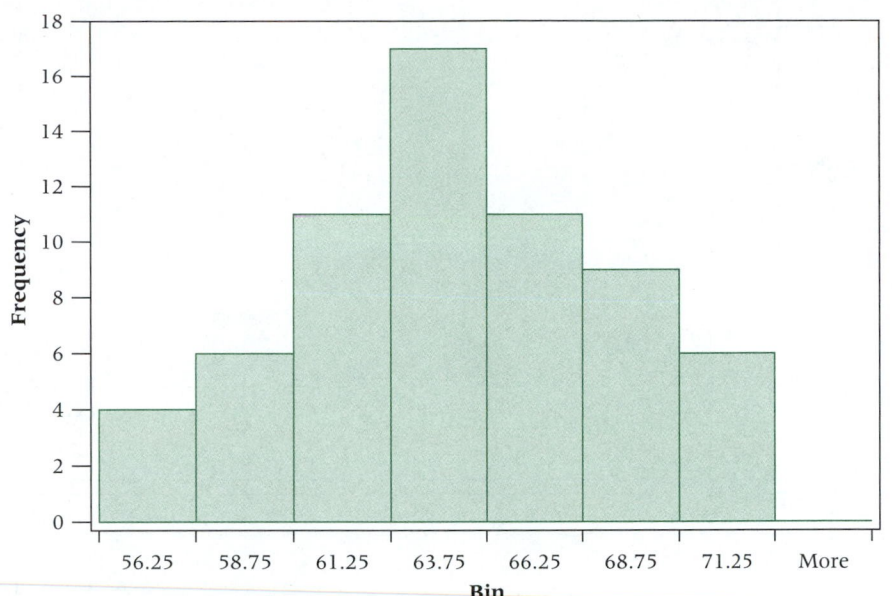

Figure 7.5

Histogram of sample means

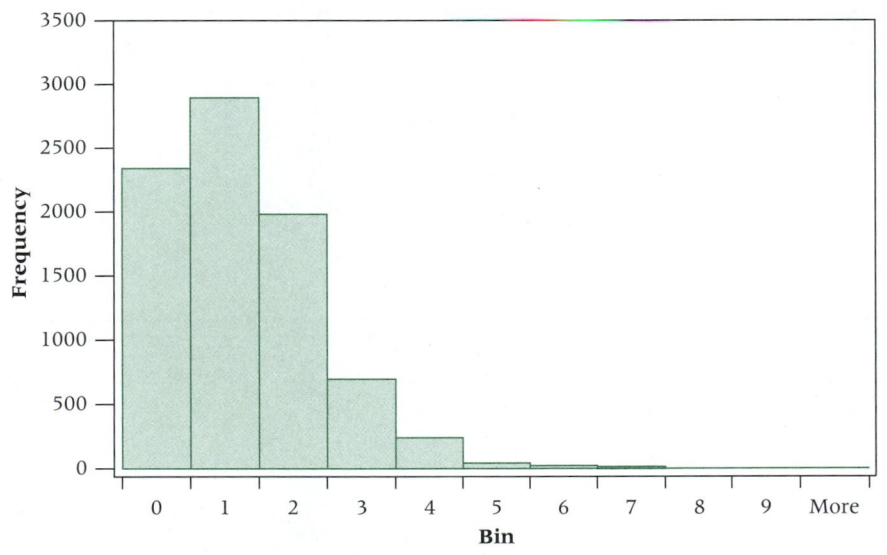

Figure 7.6

An Excel-produced histogram of a Poisson distributed population, $\lambda = 1.25$

displayed in Figure 7.7. Notice that although the samples were drawn from an exponential distribution, which is skewed to the right, the sample means form a distribution that approaches a symmetrical, nearly normal-curve-type distribution.

Suppose a population is uniformly distributed. If samples are selected randomly from a population with a uniform distribution, how are the sample means distributed? Figures 7.8a through e display the Excel histogram distributions of sample means from five different sample sizes. Each of these histograms represents the distribution of sample means from 90 samples generated randomly from a uniform distribution in which a was 10 and b was

Figure 7.7

An Excel histogram of 90 sample means from samples of size $n = 30$ from a Poisson distributed population

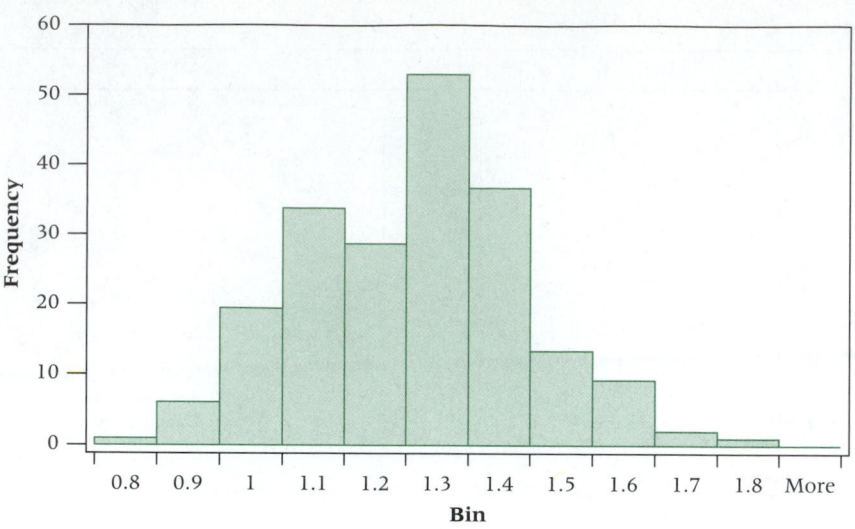

Figure 7.8a

Excel histogram outputs for sample means from 90 samples ranging in size from $n = 2$ to $n = 30$ from a uniformly distributed population with $a = 10$ and $b = 30$

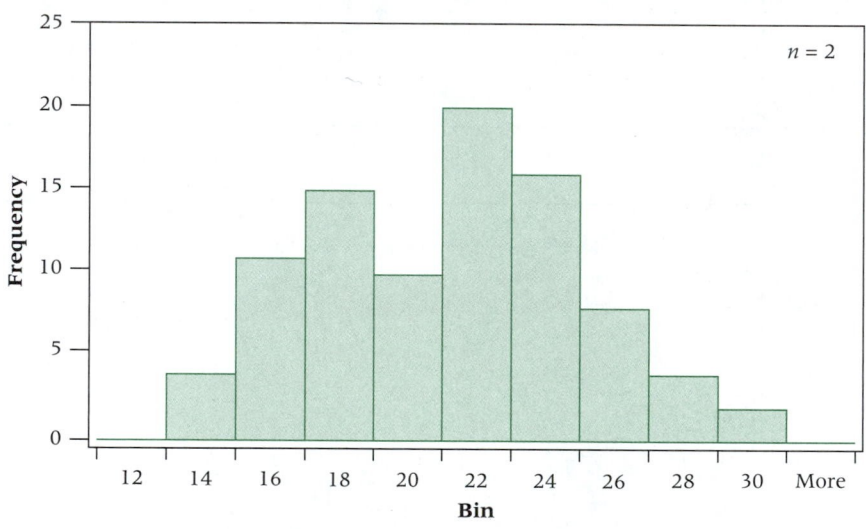

Figure 7.8b

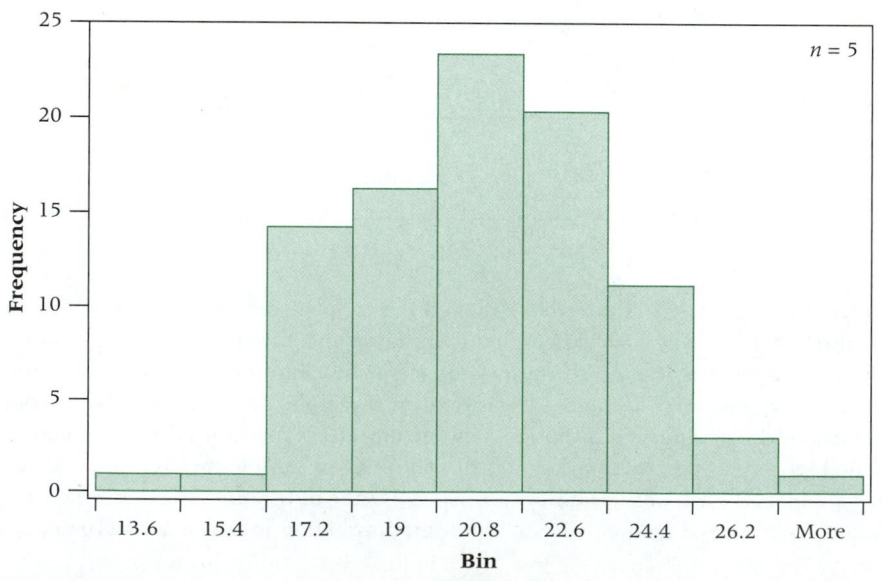

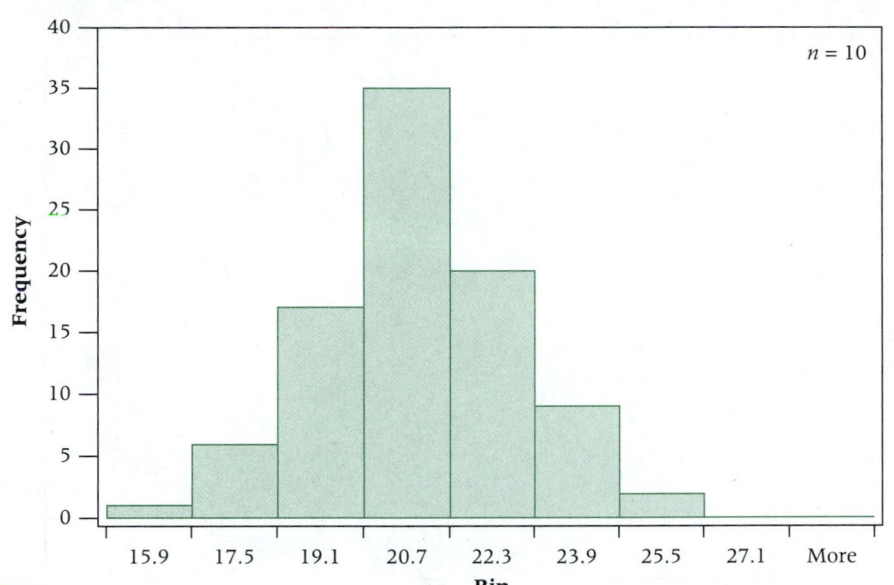

Figure 7.8c

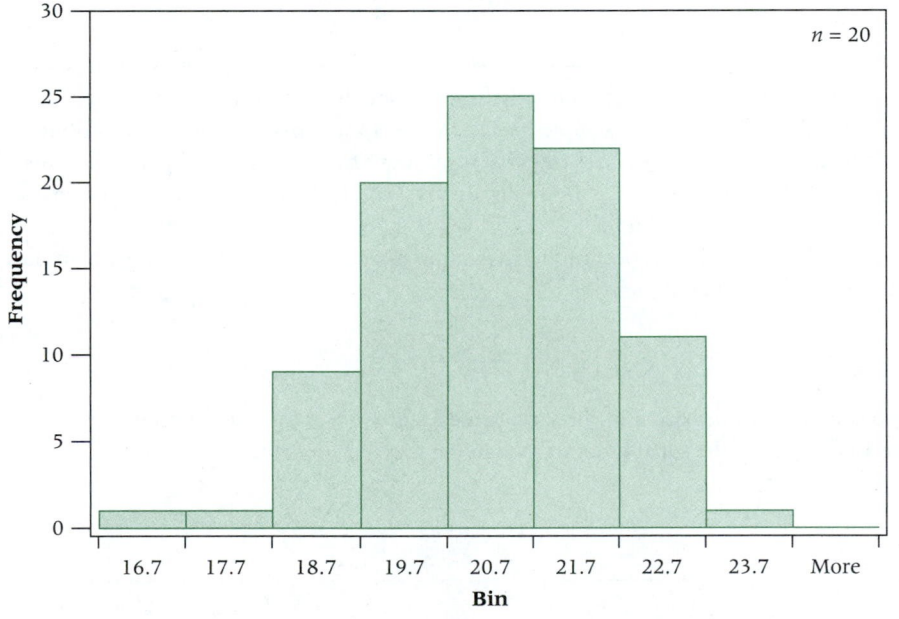

Figure 7.8d

30. Observe the shape of the distributions. Notice that even for small sample sizes, the distributions of sample means for samples taken from the uniformly distributed population begin to "pile up" in the middle. As sample sizes become much larger, the sample mean distributions begin to approach a normal distribution and the variation among the means decreases.

So far, we have examined three populations with different distributions. However, the sample means for samples taken from these populations appear to be approximately normally distributed, especially as the sample sizes become larger. What would happen to the distribution of sample means if we studied populations that have differently shaped distributions? The answer to that question is given in the **central limit theorem.**

Central limit theorem
A theorem that states that regardless of the shape of a population, the distributions of sample means and proportions are normal if sample sizes are large.

Figure 7.8e

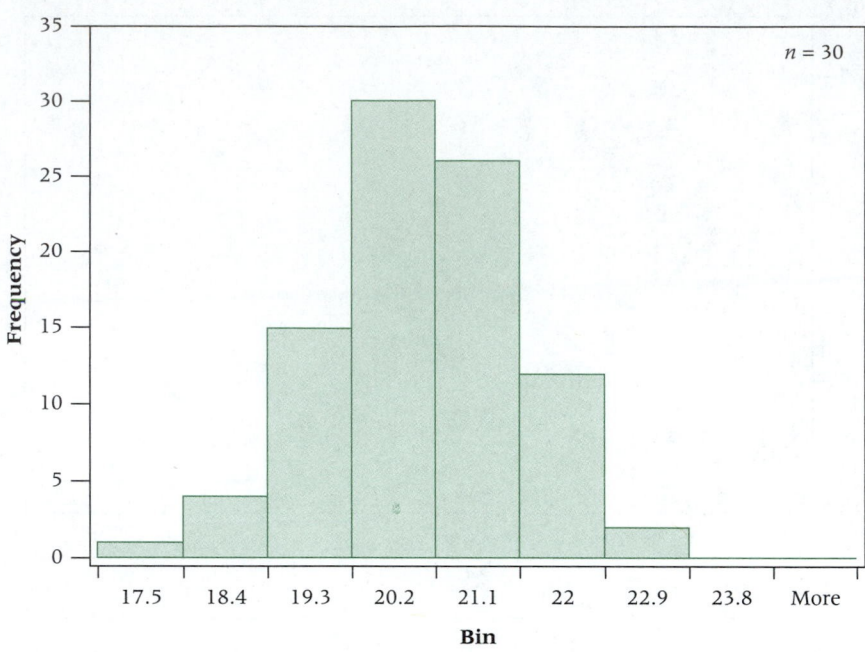

CENTRAL LIMIT THEOREM	If samples of size n are drawn randomly from a population that has a mean of μ and a standard deviation of σ, the sample means, $\overline{X}$, are approximately normally distributed for sufficiently large sample sizes ($n \geq 30$) regardless of the shape of the population distribution. If the population is normally distributed, the sample means are normally distributed for any size sample.

From mathematical expectation,* it can be shown that the mean of the sample means is the population mean,

$$\mu_{\overline{X}} = \mu,$$

and the standard deviation of the sample means is the standard deviation of the population divided by the square root of the sample size,

$$\sigma_{\overline{X}} = \frac{\sigma}{\sqrt{n}}.$$

The central limit theorem creates the potential for applying the normal distribution to many problems when sample size is sufficiently large. Sample means that have been computed for random samples drawn from normally distributed populations are normally distributed. However, the real advantage of the central limit theorem is that sample data drawn from populations not normally distributed or from populations of unknown shape also can be analyzed by using the normal distribution because the sample means are nor-

*The derivations are beyond the scope of this text and are not shown.

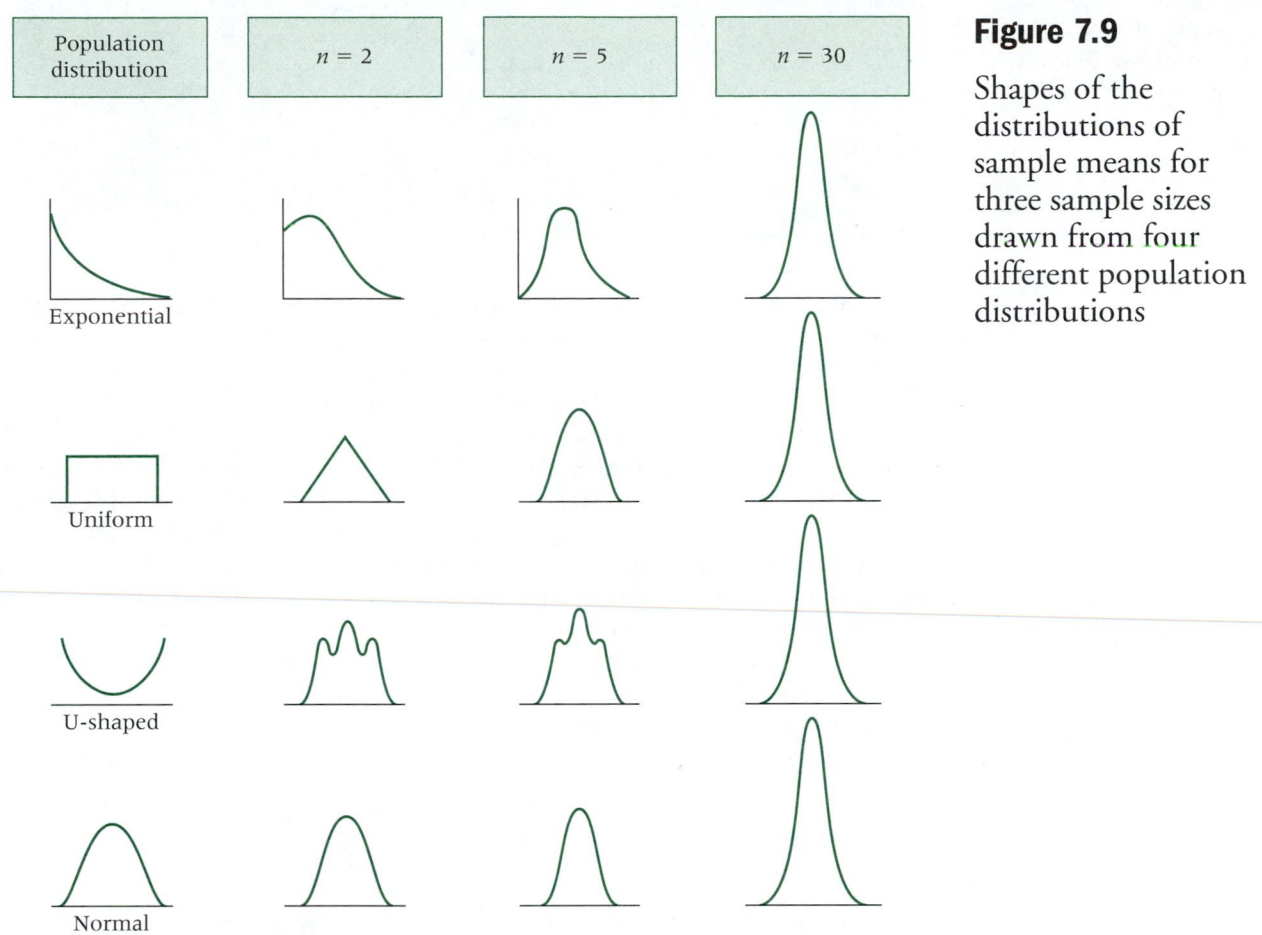

Figure 7.9

Shapes of the distributions of sample means for three sample sizes drawn from four different population distributions

mally distributed for sufficiently large sample sizes.* Column 1 of Figure 7.9 shows four different population distributions. Each succeeding column displays the shape of the distribution of the sample means for a particular sample size. Note in the bottom row for the normally distributed population that the sample means are normally distributed even for $n = 2$. Note also that with the other population distributions, the distribution of the sample means begins to approximate the normal curve as n becomes larger. For all four distributions, the distribution of sample means is approximately normal for $n = 30$.

How large must a sample be for the central limit theorem to apply? The sample size necessary varies according to the shape of the population. However, in this text (as in many others), a sample of *size 30 or larger* will suffice. Recall that if the population is normally distributed, the sample means are normally distributed for sample sizes as small as $n = 1$.

The shapes displayed in Figure 7.9 coincide with the results obtained empirically from the random sampling shown in Figures 7.6 and 7.7. As shown in Figure 7.9, and as indicated in Figure 7.7, as sample size increases, the distribution narrows, or becomes more leptokurtic. This makes sense because the standard deviation of the mean is $\sigma/\sqrt{n}$. This value will become smaller as the size of n increases.

---

*The actual form of the central limit theorem is a limit function of calculus. As the sample size increases to infinity, the distribution of sample means literally becomes normal in shape. The central limit theorem ensures that the sample mean is both unbiased and consistent, two important characteristics of estimators that are not discussed in this text.

TABLE 7.4

$\mu_{\overline{X}}$ and $\sigma_{\overline{X}}$ of 90 Random Samples for Five Different Sample Sizes

SAMPLE SIZE	MEAN OF SAMPLE MEANS	STANDARD DEVIATION OF SAMPLE MEANS	μ	$\sigma/\sqrt{n}$
$n = 2$	20.20	3.87	20	4.08
$n = 5$	20.00	2.63	20	2.58
$n = 10$	20.17	1.76	20	1.83
$n = 20$	20.20	1.29	20	1.29
$n = 30$	20.05	1.07	20	1.05

In Table 7.4, the means and standard deviations of the means are displayed for random samples of various sizes ($n = 2$ through $n = 30$) drawn from the uniform distribution of $a = 10$ and $b = 30$ shown in Figure 7.8. The population mean is 20, and the standard deviation of the population is 5.774. Note that the mean of the sample means for each sample size is approximately 20 and that the standard deviation of the sample means for each set of 90 samples is approximately equal to $\sigma/\sqrt{n}$. There is a small discrepancy between the standard deviation of the sample means and $\sigma/\sqrt{n}$, because not all possible samples of a given size were taken from the population (only 90). In theory, if all possible samples for a given sample size are taken, the mean of the sample means will equal the population mean and the standard deviation of the sample means will equal the population standard deviation divided by the square root of n.

The central limit theorem states that sample means are normally distributed regardless of the shape of the population for large samples and for any sample size with normally distributed populations. Thus sample means can be analyzed by using Z scores. Recall from Chapter 6 that

$$Z = \frac{X - \mu}{\sigma}.$$

If sample means are normally distributed, the Z score formula applied to sample means would be

$$Z = \frac{\overline{X} - \mu_{\overline{X}}}{\sigma_{\overline{X}}}.$$

This result follows the general pattern of Z scores: the difference between the statistic and its mean divided by the statistic's standard deviation. In this formula, the mean of the statistic of interest is $\mu_{\overline{X}}$ and *the standard deviation of the statistic of interest is* $\sigma_{\overline{X}}$, sometimes referred to as the **standard error of the mean.** To determine $\mu_{\overline{X}}$, the researcher would have to randomly draw out all possible samples of the given size from the population, compute the sample means, and average them. This task is virtually impossible to accomplish in any realistic period of time. Fortunately, $\mu_{\overline{X}}$ equals the population mean, μ, which is easier to access. Likewise, to determine directly the value of $\sigma_{\overline{X}}$, the researcher would have to take all possible samples of a given size from a population, compute the sample means, and determine the standard deviation of sample means. This task also is practically impossible. Fortunately, $\sigma_{\overline{X}}$ can be computed by using the population standard deviation divided by the square root of the sample size.

Standard error of the mean

The standard deviation of the distribution of sample means.

As sample size increases, the standard deviation of the sample means becomes smaller and smaller because the population standard deviation is being divided by larger and larger values of the square root of n. The ultimate benefit of the central limit theorem is a practical, useful version of the Z formula for sample means.

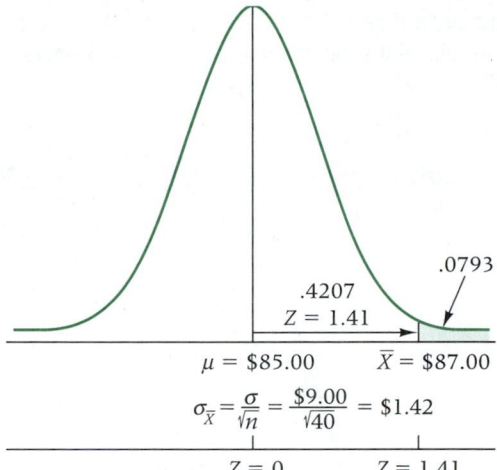

Figure 7.10

Graphical solution to
the tire store example

$$Z = \frac{\overline{X} - \mu}{\frac{\sigma}{\sqrt{n}}}$$

Z FORMULA FOR
SAMPLE MEANS

When the population is normally distributed and the sample size is 1, this formula for
sample means becomes the original *Z* formula for individual values. The reason is that the
mean of one value is that value, and when $n = 1$ the value of $\sigma/\sqrt{n} = \sigma$.

Suppose, for example, that the mean expenditure per customer at a tire store is $85.00,
with a standard deviation of $9.00. If a random sample of 40 customers is taken, what is
the probability that the sample average expenditure per customer for this sample will be
$87.00 or more? Because the sample size is greater than 30, the central limit theorem can
be used, and the sample means are normally distributed. With $\mu = \$85.00$, $\sigma = \$9.00$,
and the *Z* formula for sample means, *Z* is computed as

$$Z = \frac{\overline{X} - \mu}{\frac{\sigma}{\sqrt{n}}} = \frac{\$87.00 - \$85.00}{\frac{\$9.00}{\sqrt{40}}} = \frac{\$2.00}{\$1.42} = 1.41.$$

For $Z = 1.41$ in the *Z* distribution (Table A.5), the probability is .4207. It is the probability of getting a mean between $87.00 and the population mean, $85.00. Solving for the
tail of the distribution yields

$$.5000 - .4207 = .0793,$$

which is the probability of $\overline{X} \geq \$87.00$. That is, 7.93% of the time, a random sample of
40 customers from this population will yield a mean expenditure of $87.00 or more. Fig-
ure 7.10 shows the problem and its solution.

DEMONSTRATION PROBLEM 7.1

Suppose that during any hour in a large department store, the average number of shoppers is 448, with a standard deviation of 21 shoppers. What is the probability that a random sample of 49 different shopping hours will yield a sample mean between 441 and 446 shoppers?

SOLUTION

For this problem, $\mu = 448$, $\sigma = 21$, and $n = 49$. The problem is to determine $P(441 \leq \bar{X} \leq 446)$. The following diagram depicts the problem.

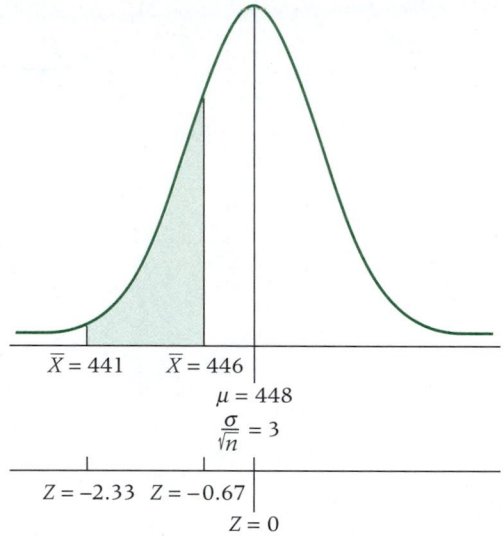

$$\bar{X} = 441 \qquad \bar{X} = 446$$
$$\mu = 448$$
$$\frac{\sigma}{\sqrt{n}} = 3$$

$$Z = -2.33 \quad Z = -0.67$$
$$Z = 0$$

Solve this problem by calculating the Z scores and using Table A.5 to determine the probabilities.

$$Z = \frac{441 - 448}{\dfrac{21}{\sqrt{49}}} = \frac{-7}{3} = -2.33$$

and

$$Z = \frac{446 - 448}{\dfrac{21}{\sqrt{49}}} = \frac{-2}{3} = -0.67$$

Z VALUE	PROBABILITY
−2.33	.4901
−0.67	−.2486
	.2415

The probability of a value being between $Z = -2.33$ and -0.67 is .2415. That is, there is a 24.15% chance of randomly selecting 49 hourly periods for which the sample mean is between 441 and 446 shoppers.

POPULATION SIZE	SAMPLE SIZE	VALUE OF CORRECTION FACTOR
2000	30 (<5%N)	.993
2000	500	.866
500	30	.971
500	200	.775
200	30	.924
200	75	.793

TABLE 7.5
Finite Correction Factor for Some Sample Sizes

Sampling from a Finite Population

The example shown in this section and Demonstration Problem 7.1 was based on the assumption that the population was infinitely or extremely large. In cases of a finite population, *a statistical adjustment can be made to the Z formula for sample means*. The adjustment is called the **finite correction factor:** $\sqrt{(N - n)/(N - 1)}$. It operates on the standard deviation of sample means, $\sigma_{\bar{X}}$. Following is the Z formula for sample means when samples are drawn from finite populations.

Finite correction factor
A statistical adjustment made to the Z formula for sample means; adjusts for the fact that a population is finite and the size is known.

$$Z = \frac{\bar{X} - \mu}{\frac{\sigma}{\sqrt{n}}\sqrt{\frac{N - n}{N - 1}}}$$

Z FORMULA
FOR SAMPLE MEANS
WHEN THERE IS A
FINITE POPULATION

If a random sample of size 35 were taken from a finite population of only 500, the sample mean would be less likely to deviate from the population mean than would be the case if a sample of size 35 were taken from an infinite population. For a sample of size 35 taken from a finite population of size 500, the finite correction factor is

$$\sqrt{\frac{500 - 35}{500 - 1}} = \sqrt{\frac{465}{499}} = .965.$$

Thus the standard deviation of the mean—sometimes referred to as the standard error of the mean—is adjusted downward by using .965. As the size of the finite population becomes larger in relation to sample size, the finite correction factor approaches 1. In theory, whenever researchers are working with a finite population, they can use the finite correction factor. A rough rule of thumb for many researchers is that, if the sample size is less than 5% of the finite population size, the finite correction factor does not significantly modify the solution. Table 7.5 contains some illustrative finite correction factors.

A production company's 350 hourly employees average 37.6 years of age, with a standard deviation of 8.3 years. If a random sample of 45 hourly employees is taken, what is the probability that the sample will have an average age of less than 40 years?

DEMONSTRATION PROBLEM 7.2

SOLUTION
The population mean is 37.6, with a population standard deviation of 8.3; that is, $\mu = 37.6$ and $\sigma = 8.3$. The sample size is 45, but it is being drawn from a finite population of 350;

that is, $n = 45$ and $N = 350$. The sample mean under consideration is 40, or $\bar{X} = 40$. The following diagram depicts the problem on a normal curve.

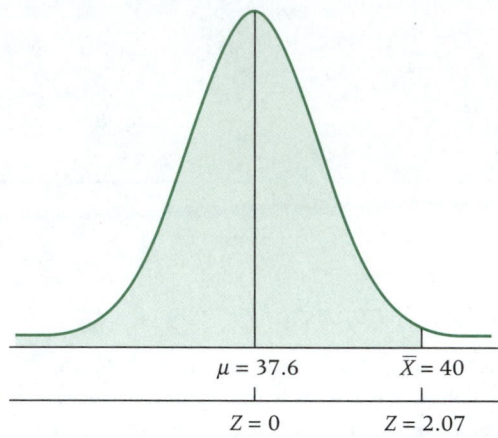

$$\mu = 37.6 \qquad \bar{X} = 40$$
$$Z = 0 \qquad Z = 2.07$$

Using the Z formula with the finite correction factor gives

$$Z = \frac{40 - 37.6}{\dfrac{8.3}{\sqrt{45}}\sqrt{\dfrac{350 - 45}{350 - 1}}} = \frac{2.4}{1.157} = 2.07.$$

This Z value yields a probability (Table A.5) of .4808. Therefore, the probability of getting a sample average age of less than 40 years is .4808 + .5000 = .9808. Had the finite correction factor not been used, the Z value would have been 1.94, and the final answer would have been .9738.

7.2

Problems

7.13 A population has a mean of 50 and a standard deviation of 10. If a random sample of 64 is taken, what is the probability that the sample mean is each of the following?
  a. Greater than 52
  b. Less than 51
  c. Less than 47
  d. Between 48.5 and 52.4
  e. Between 50.6 and 51.3

7.14 A population is normally distributed, with a mean of 23.45 and a standard deviation of 3.8. What is the probability of each of the following?
  a. Taking a sample of size 10 and obtaining a sample mean of 22 or more
  b. Taking a sample of size 4 and getting a sample mean of more than 26

7.15 Suppose a random sample of size 36 is being drawn from a population with a mean of 278. If 86% of the time the sample mean is less than 280, what is the population standard deviation?

7.16 A random sample of size 81 is being drawn from a population with a standard deviation of 12. If only 18% of the time a sample mean greater than 300 is obtained, what is the mean of the population?

7.17 Find the probability in each case.
  a. $N = 1000$, $n = 60$, $\mu = 75$, and $\sigma = 6$; $P(\bar{X} < 76.5) = ?$
  b. $N = 90$, $n = 36$, $\mu = 108$, and $\sigma = 3.46$; $P(107 < \bar{X} < 107.7) = ?$

   c. $N = 250$, $n = 100$, $\mu = 35.6$, and $\sigma = 4.89$; $P(\overline{X} \geq 36) = ?$

   d. $N = 5000$, $n = 60$, $\mu = 125$, and $\sigma = 13.4$; $P(\overline{X} \leq 125) = ?$

7.18 The *Statistical Abstract of the United States* published by the U.S. Bureau of the Census reports that the average annual consumption of fresh fruit per person is 99.9 pounds. The standard deviation of fresh fruit consumption is about 30 pounds. Suppose a researcher took a random sample of 38 people and had them keep a record of the fresh fruit they ate for 1 year.

   a. What is the probability that the sample average would be less than 90 pounds?

   b. What is the probability that the sample average would be between 98 and 105 pounds?

   c. What is the probability that the sample average would be less than 112 pounds?

   d. What is the probability that the sample average would be between 93 and 96 pounds?

7.19 Suppose a subdivision on the southwest side of Denver, Colorado, contains 1500 houses. The subdivision was built in 1983. A sample of 100 houses is selected randomly and evaluated by an appraiser. If the mean appraised value of a house in this subdivision for all houses is $147,000, with a standard deviation of $8500, what is the probability that the sample average is greater than $155,000?

7.20 Suppose the average checkout tab at a large supermarket is $65.12, with a standard deviation of $21.45. Twenty-three percent of the time when a random sample of 45 customer tabs is examined, the sample average should exceed what value?

7.21 According to Nielsen Media Research, the average number of hours of TV viewing per household per week in the United States is 50.4 hours. Suppose the standard deviation is 11.8 hours and a random sample of 42 U.S. households is taken.

   a. What is the probability that the sample average is more than 52 hours?

   b. What is the probability that the sample average is less than 47.5 hours?

   c. What is the probability that the sample average is less than 40 hours? If the sample average actually is less than 40 hours, what could this mean in terms of the Nielsen Media Research figures?

   d. Suppose the population standard deviation is unknown. If 71% of all sample means are greater than 49 hours and the population mean is still 50.4 hours, what is the value of the population standard deviation?

7.3 Sampling Distribution of $\hat{p}$

Sometimes in analyzing a sample, a researcher will choose to use the sample proportion, denoted $\hat{p}$, instead of the sample mean. If research produces *measurable* data such as weight, distance, time, and income, the sample mean is often the statistic of choice. However, if research results in *countable* items such as how many people in a sample choose Dr. Pepper as their soft drink or how many people in a sample have a flexible work schedule, the sample proportion is often the statistic of choice. Whereas the mean is computed by averaging a set of values, the **sample proportion** is *computed by dividing the frequency with which a given characteristic occurs in a sample by the number of items in the sample.*

$$\hat{p} = \frac{X}{n}$$

SAMPLE PROPORTION

where:

   $X =$ number of items in a sample that have the characteristic

   $n =$ number of items in the sample

Sample proportion
The quotient of the frequency at which a given characteristic occurs in a sample and the number of items in the sample.

For example, in a sample of 100 factory workers, 30 workers might belong to a union. The value of $\hat{p}$ for this characteristic, union membership, is $30/100 = .30$. Or, in a sample of 500 businesses in suburban malls, 10 might be shoe stores. The sample proportion of shoe stores is $10/500 = .02$. The sample proportion is a widely used statistic and is usually computed on questions involving *yes* or *no* answers. For example, do you have at least a high school education? Are you predominantly right-handed? Are you female? Do you belong to the student accounting association?

How does a researcher use the sample proportion in analysis? The central limit theorem applies to sample proportions in that the normal distribution approximates the shape of the distribution of sample proportions if $n \cdot P > 5$ and $n \cdot Q > 5$ (P is the population proportion and $Q = 1 - P$). The mean of sample proportions for all samples of size n randomly drawn from a population is P (the population proportion) and *the standard deviation of sample proportions is* $\sqrt{(P \cdot Q)/n}$, sometimes referred to as the **standard error of the proportion.** Sample proportions also have a Z formula.

Standard error of the proportion
The standard deviation of the distribution of sample proportions.

Z FORMULA FOR SAMPLE PROPORTIONS FOR $n \cdot P > 5$ AND $n \cdot Q > 5$	$$Z = \dfrac{\hat{p} - P}{\sqrt{\dfrac{P \cdot Q}{n}}}$$

where
$\hat{p}$ = sample proportion
n = sample size
P = population proportion
$Q = 1 - P$

Suppose 60% of the electrical contractors in a region use a particular brand of wire. What is the probability of taking a random sample of size 120 from these electrical contractors and finding that .50 or less use that brand of wire? For this problem,

$$P = .60, \qquad \hat{p} = .50, \quad \text{and} \quad n = 120.$$

The Z formula yields

$$Z = \frac{.50 - .60}{\sqrt{\dfrac{(.60)(.40)}{120}}} = \frac{-.10}{.0447} = -2.24.$$

From Table A.5, the probability corresponding to $Z = -2.24$ is .4875. For $Z < -2.24$ (the tail of the distribution), the answer is $.5000 - .4875 = .0125$. Figure 7.11 shows the problem and solution graphically.

This answer indicates that a researcher would have difficulty (probability of .0125) finding that 50% or less of a sample of 120 contractors use a given brand of wire if indeed the population market share for that wire is .60. If this sample result actually occurs, either it is a rare chance result or perhaps the .60 proportion does not hold for this population.

Figure 7.11

Graphical solution
to the electrical
contractor example

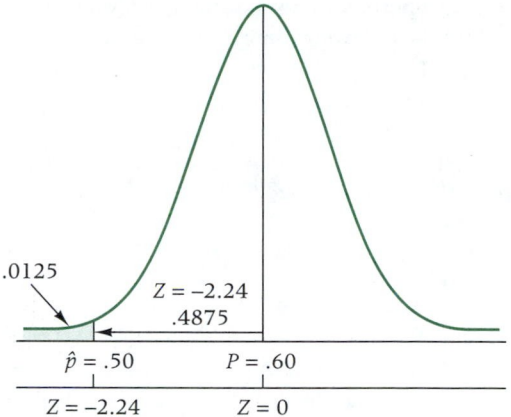

**DEMONSTRATION
PROBLEM 7.3**

If 10% of a population of parts is defective, what is the probability of randomly selecting
80 parts and finding that 12 or more parts are defective?

SOLUTION
Here, $P = .10$, $\hat{p} = 12/80 = .15$, and $n = 80$. Using the Z formula gives

$$Z = \frac{.15 - .10}{\sqrt{\dfrac{(.10)(.90)}{80}}} = \frac{.05}{.0335} = 1.49.$$

Table A.5 gives a probability of .4319 for a Z value of 1.49, which is the area between
the sample proportion, .15, and the population proportion, .10. The answer to the question is

$$P(\hat{p} \geq .15) = .5000 - .4319 = .0681.$$

Thus, about 6.81% of the time, 12 or more defective parts would appear in a random
sample of 80 parts when the population proportion is .10. If this result actually occurred, the 10% proportion for population defects would be open to question. The diagram shows the problem graphically.

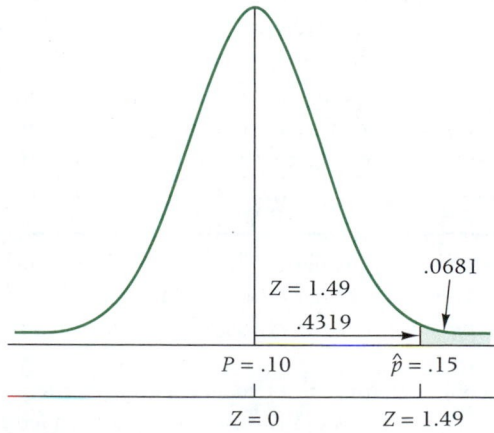

7.3
Problems

7.22 A given population proportion is .25. For the given value of n, what is the probability of getting each of the following sample proportions?
   a. $n = 110$ and $\hat{p} \le .21$
   b. $n = 33$ and $\hat{p} > .24$
   c. $n = 59$ and $.24 \le \hat{p} < .27$
   d. $n = 80$ and $\hat{p} > .30$
   e. $n = 800$ and $\hat{p} > .30$

7.23 A population proportion is .58. Suppose a random sample of 660 items is sampled randomly from this population.
   a. What is the probability that the sample proportion is greater than .60?
   b. What is the probability that the sample proportion is between .55 and .65?
   c. What is the probability that the sample proportion is greater than .57?
   d. What is the probability that the sample proportion is between .53 and .56?
   e. What is the probability that the sample proportion is less than .48?

7.24 Suppose a population proportion is .40 and 80% of the time when you draw a random sample from this population you get a sample proportion of .35 or more. How large a sample were you taking?

7.25 If a population proportion is .28 and if the sample size is 140, 30% of the time the sample proportion will be less than what value if you are taking random samples?

7.26 A *USA Today*/IntelliQuest survey of computer users revealed that 23% log onto the Internet and/or an online service user more than 20 times per month. Suppose a random sample of 600 computer users is taken. What is the probability that more than 150 computer users log onto the Internet and/or an online service user more than 20 times per month?

7.27 According to the Consumer Electronics Manufacturers Association, 39% of all U.S. households have a cellular phone. Suppose 200 U.S. households are randomly surveyed.
   a. What is the probability that fewer than 70 households have a cellular phone?
   b. What is the probability that more than 90 households have a cellular phone?
   c. What is the probability that more than 65 households do not have a cellular phone?

7.28 The *Travel Weekly International Air Transport Association* survey asked business travelers about the purpose for their most recent business trip. Nineteen percent responded that it was for an internal company visit. Suppose 950 business travelers are randomly selected.
   a. What is the probability that more than 25% of the business travelers say that the reason for their most recent business trip was an internal company visit?
   b. What is the probability that between 15% and 20% of the business travelers say that the reason for their most recent business trip was an internal company visit?
   c. What is the probability that between 133 and 171 of the business travelers say that the reason for their most recent business trip was an internal company visit?

Summary

For much business research, successfully conducting a census is virtually impossible and the sample is a feasible alternative. Other reasons for sampling include cost reduction, potential for broadening the scope of the study, and loss reduction when the testing process destroys the product.

To take a sample, a population has to be identified. Often the researcher has no exact roster or list of the population and so must find some way to identify the population as closely as possible. The final list or directory used to represent the population and from which the sample is drawn is called the frame.

The two main types of sampling are random and nonrandom. Random sampling occurs when each unit of the population has the same probability of being selected for the sample. Nonrandom sampling is any sampling that is not random. The four main types of random sampling discussed are simple random sampling, stratified sampling, systematic sampling, and cluster or area sampling.

In simple random sampling, every unit of the population is numbered. A table of random numbers or a random number generator is used to select n units from the population for the sample.

Stratified random sampling uses the researcher's prior knowledge of the population to stratify the population into subgroups. Each subgroup is internally homogeneous but different from the others. Stratified random sampling is an attempt to reduce sampling error and ensure that at least some of each of the subgroups appear in the sample. After the strata have been identified, units are sampled randomly from each stratum. If the proportions of units selected from each subgroup for the sample are the same as the proportions of the subgroups in the population, the process is called proportionate stratified sampling. If not, it is called disproportionate stratified sampling.

With systematic sampling, every kth item of the population is sampled until n units have been selected. Systematic sampling is used because of its convenience and ease of administration.

Cluster or area sampling involves subdividing the population into nonoverlapping clusters or areas. Each cluster or area is a microcosm of the population and is usually heterogeneous within. Individual units are then selected randomly from the clusters or areas to get the final sample. Cluster or area sampling is usually done to reduce costs. If a set of second clusters or areas is selected from the first set, the method is called two-stage sampling.

Four types of nonrandom sampling were discussed: convenience, judgment, quota, and snowball. In convenience sampling, the researcher selects units from the population to be in the sample for convenience. In judgment sampling, units are selected according to the judgment of the researcher. Quota sampling is similar to stratified sampling, with the researcher identifying subclasses or strata. However, the researcher selects units from each stratum by some nonrandom technique until a specified quota from each stratum is filled. With snowball sampling, the researcher obtains additional sample members by asking current sample members for referral information.

Sampling error occurs when the sample does not represent the population. With random sampling, sampling error occurs by chance. Nonsampling errors are all other research and analysis errors that occur in a study. They include recording errors, input errors, missing data, and incorrect definition of the frame.

According to the central limit theorem, if a population is normally distributed, the sample means for samples taken from that population also are normally distributed regardless of sample size. The central limit theorem also says that if the sample sizes are large ($n \geq 30$), the sample mean is approximately normally distributed regardless of the distribution shape of the population. This theorem is extremely useful because it enables researchers to analyze sample data by using the normal distribution for virtually any type of study in which means are an appropriate statistic, so long as the sample size is large enough. The central limit theorem states that sample proportions are normally distributed for large sample sizes.

Key Terms

central limit theorem
cluster (or area) sampling
convenience sampling
disproportionate stratified random sampling
finite correction factor

frame
judgment sampling
nonrandom sampling
nonrandom sampling techniques
nonsampling errors

proportionate stratified random sampling

quota sampling

random sampling

sample proportion

sampling error

simple random sampling

snowball sampling

standard error of the mean

standard error of the proportion

stratified random sampling

systematic sampling

two-stage sampling

SUPPLEMENTARY PROBLEMS

7.29 The mean of a population is 76 and the standard deviation is 14. The shape of the population is unknown. Determine the probability of each of the following occurring from this population.
  a. A random sample of size 35 yielding a sample mean of 79 or more
  b. A random sample of size 140 yielding a sample mean of between 74 and 77
  c. A random sample of size 219 yielding a sample mean of less than 76.5

7.30 Forty-six percent of a population possess a particular characteristic. Random samples are taken from this population. Determine the probability of each of the following occurrences.
  a. The sample size is 60 and the sample proportion is between .41 and .53
  b. The sample size is 458 and the sample proportion is less than .40
  c. The sample size is 1350 and the sample proportion is greater than .49

7.31 Suppose the age distribution in a city is as follows.

Under 18	22%
18–25	18%
26–50	36%
51–65	10%
Over 65	14%

A researcher is conducting proportionate stratified random sampling with a sample size of 250. Approximately how many people should he sample from each stratum?

7.32 Candidate Jones believes she will receive .55 of the total votes cast in her county. However, in an attempt to validate this figure, she has her pollster contact a random sample of 600 registered voters in the county. The poll results show that 298 of the voters say they are committed to voting for her. If she actually has .55 of the total vote, what is the probability of getting a sample proportion this small or smaller? Do you think she actually has 55% of the vote? Why or why not?

7.33 Determine a possible frame for conducting random sampling in each of the following studies.
  a. The average amount of overtime per week for production workers in a plastics company in Pennsylvania
  b. The average number of employees in all Alpha/Beta supermarkets in California
  c. A survey of commercial lobster catchers in Maine

7.34 A particular automobile costs an average of $17,755 in the Pacific Northwest. The standard deviation of prices is $650. Suppose a random sample of 30 dealerships in Washington and Oregon is taken and their managers are asked what they charge for this automobile. What is the probability of getting a sample average cost of less than $17,500? Assume that only 120 dealerships in the entire Pacific Northwest sell this automobile.

7.35 A company has 1250 employees, and you want to take a simple random sample of $n = 60$ employees. Explain how you would go about selecting this sample by using the table of random numbers. Are there numbers that you cannot use? Explain.

7.36 Suppose the average client charge per hour for out-of-court work by lawyers in the state of Iowa is $125. Suppose further that a random telephone sample of 32 lawyers in Iowa is taken and that the sample average charge per hour for out-of-court work is $110. If the population variance is $525, what is the probability of getting a sample mean this large or larger? What is the probability of getting a sample mean larger than $135 per hour? What is the probability of getting a sample mean of between $120 and $130 per hour?

7.37 A survey of 2645 consumers by DDB Needham Worldwide of Chicago for public relations agency Porter/Novelli showed that how a company handles a crisis when at fault is one of the top influences in consumer buying decisions, with 73% claiming it is an influence. Quality of product was the number-one influence, with 96% of consumers stating that quality has an influence on their buying decisions. How a company handles complaints was number two, with

85% of consumers reporting it as an influence in their buying decisions. Suppose a random sample of 1100 consumers is taken and each is asked which of these three factors influence their buying decisions.

a. What is the probability that more than 810 consumers claim that how a company handles a crisis when at fault is an influence in their buying decisions?

b. What is the probability that fewer than 1030 consumers claim that quality of product is an influence in their buying decisions?

c. What is the probability that between 82% and 84% of consumers claim that how a company handles complaints is an influence in their buying decisions?

7.38 Suppose you are sending out questionnaires to a randomly selected sample of 100 managers. The frame for this study is the membership list of the American Managers Association. The questionnaire contains demographic questions about the company and its top manager. In addition, it asks questions about the manager's leadership style. Research assistants are to score and enter the responses into the computer as soon as they are received. You are to conduct a statistical analysis of the data. Name and describe four nonsampling errors that could occur in this study.

7.39 A researcher is conducting a study of a *Fortune 500* company that has factories, distribution centers, and retail outlets across the country. How can she use cluster or area sampling to take a random sample of employees of this firm?

7.40 A directory of personal computer retail outlets in the United States contains 12,080 alphabetized entries. Explain how systematic sampling could be used to select a sample of 300 outlets.

7.41 In an effort to cut costs and improve profits, many U.S. companies have been turning to outsourcing. In fact, according to *Purchasing* magazine, 54% of companies surveyed outsourced some part of their manufacturing process in the past two to three years. Suppose 565 of these companies are contacted.

a. What is the probability that 339 or more companies have outsourced some part of their manufacturing process in the past two to three years?

b. What is the probability that 288 or more companies have outsourced some part of their manufacturing process in the past two to three years?

c. What is the probability that 50% or less of these companies have outsourced some part of their manufacturing process in the past two to three years?

7.42 The average cost of a one-bedroom apartment in a town is $550 per month. What is the probability of randomly selecting a sample of 50 one-bedroom apartments in this town and getting a sample mean of less than $530 if the population standard deviation is $100?

7.43 The Aluminum Association reports that the average American uses 56.8 pounds of aluminum in a year. A random sample of 51 households is monitored for 1 year to determine aluminum usage. If the population standard deviation of annual usage is 12.3 pounds, what is the probability that the sample mean will be each of the following?

a. More than 60 pounds
b. More than 58 pounds
c. Between 56 and 57 pounds
d. Less than 55 pounds
e. Less than 50 pounds

7.44 Use Table A.1 to select 20 three-digit random numbers. Did any of the numbers occur more than once? How can this happen? Make a histogram of the numbers. Do the numbers seem to be equally distributed, or are they bunched together?

7.45 Direct marketing companies are turning to the Internet for new opportunities. A recent study by Gruppo, Levey, & Co. showed that 73% of all direct marketers conduct transactions on the Internet. Suppose a random sample of 300 direct marketing companies is taken.

a. What is the probability that between 210 and 234 (inclusive) direct marketing companies are turning to the Internet for new opportunities?

b. What is the probability that 78% or more of direct marketing companies are turning to the Internet for new opportunities?

c. Suppose a random sample of 800 direct marketing companies is taken. Now what is the probability that 78% or more are turning to the Internet for new opportunities? How does this answer differ from the answer in part (b)? Why do the answers differ?

7.46 According to the U.S. Bureau of Labor Statistics, 20% of all people 16 years of age or older do volunteer work. Women volunteer slightly more than men, with 22% of women volunteering and 19% of men volunteering. What is the probability of randomly sampling 140 women 16 years of age or older and getting 35 or more who do volunteer work? What is the probability of getting 21 or fewer from this group? Suppose a sample of 300 men and women 16 years of age or older is selected randomly from the U.S. population. What is the probability that the sample proportion who do volunteer work is between 18% and 25%?

7.47 Suppose you work for a large firm that has 20,000 employees. The CEO calls you in and asks you to determine employee attitudes toward the company. She is willing to commit $100,000 to this project. What are the advantages of taking a sample versus conducting a census? What are the trade-offs?

7.48 In a particular area of the Northeast, an estimated 75% of the homes use heating oil as the principal heating fuel during the winter. A random telephone survey of 150 homes is taken in an attempt to determine whether this figure is correct. Suppose 120 of the 150 homes surveyed use heating oil as the principal heating fuel. What is the probability of getting a sample proportion this large or larger if the population estimate is true?

7.49 The U.S. Bureau of Labor Statistics released hourly wage figures for western countries in 1996 for workers in the manufacturing sector. The hourly wage was $28.34 in Switzerland, $20.84 in Japan, and $17.70 in the United States. Suppose 40 manufacturing workers are selected randomly from across Switzerland and asked what their hourly wage is. What is the probability that the sample average will be between $28 and $29? Suppose 35 manufacturing workers are selected randomly from across Japan. What is the probability that the sample aver-

age will exceed $22? Suppose 50 manufacturing workers are selected randomly from across the United States. What is the probability that the sample average will be less than $16.50? Assume that in all three countries, the standard deviation of hourly labor rates is $3.

7.50 Give a variable that could be used to stratify the population for each of the following studies. List at least four subcategories for each variable.
   a. A political party wants to conduct a poll prior to an election for the office of U.S. senator in Minnesota.
   b. A soft-drink company wants to take a sample of soft-drink purchases in an effort to estimate market share.
   c. A retail outlet wants to interview customers over a 1-week period.
   d. An eyeglasses manufacturer and retailer wants to determine the demand for prescription eyeglasses in its marketing region.

7.51 According to Runzheimer International, a typical business traveler spends an average of $281 per day in Chicago. This cost includes hotel, meals, car rental, and incidentals. A survey of 65 randomly selected business travelers who have been to Chicago on business recently is taken. For the population mean of $281 per day, what is the probability of getting a sample average of more than $273 per day if the population standard deviation is $47?

ANALYZING THE DATABASES

1. Let the manufacturing database be the frame for a population of manufacturers that are to be studied. This database has 140 different SIC codes. How would you proceed to take a simple random sample of size 6 from these industries? Explain how you would take a systematic sample of size 10 from this frame. Examine the variables in the database. Name two variables that could be used to stratify the population. Explain how these variables could be used in stratification and why they might be important strata.

2. Assume the manufacturing database is the population of interest. Compute the mean and standard deviation for cost of materials on this population. Take a random sample of 32 of the SIC code categories and compute the sample mean cost of materials on this sample. Using techniques presented in this chapter, determine

the probability of getting a mean this large or larger from the population. Note that the population contains only 140 items. Work this problem with and without the finite correction factor. Compare the results and discuss the differences in answers.

3. Use the hospital database to calculate the mean and standard deviation of personnel. Assume that these figures are true for the population of hospitals in the United States. Suppose a random sample of 36 hospitals is taken from hospitals in the United States. What is the probability that the sample mean of personnel is less than 650? What is the probability that the sample mean of personnel is between 700 and 1100? What is the probability that the sample mean is between 900 and 950?

   Determine the proportion of the hospital database that is under the control of nongovernment-not-for-profit organizations (category 2). Assume that this proportion represents the entire population of hospitals. If you randomly selected 500 hospitals from across the United States, what is the probability that 45% or more

are under the control of nongovernment-not-for-profit organizations? If you randomly selected 100 hospitals, what is the probability that less than 40% are under the control of nongovernment-not-for-profit organizations?

SHELL ATTEMPTS TO RETURN TO PREMIER STATUS

The Shell Oil Company, which began around 1912, had been for decades a household name as a top-flight oil company in the United States. However, by the late 1970s much of its prestige as a premier company had disappeared. How could Shell regain its high status?

In the 1990s, Shell undertook an extensive research effort to find out what it needed to do to improve its image. As a first step, Shell hired Responsive Research, Inc. and the Opinion Research Corp. to conduct a series of focus-group and personal interviews among various segments of the population. Included in these were youths, minorities, residents in neighborhoods near Shell plants, legislators, academics, and present and past employees of Shell. The researchers learned that people believe that top companies are integral parts of the communities in which the companies are located rather than separate entities. These studies and others led to the development of materials that Shell used to explain their core values to the general public.

Next, PERT Survey Research ran a large quantitative study to determine which values were best received by the target audience. Social issues emerged as the theme with the most support. During the next few months, the advertising agency Ogilvy & Mather was hired by Shell to develop several campaigns with social themes. Two market research companies were hired to evaluate the receptiveness of the various campaigns. The result was the "Count on Shell" campaign, which featured safety messages with useful information about what to do in various dangerous situations.

A public "Count on Shell" campaign was launched in February 1998 and has met with considerable success: the ability to recall Shell advertising has jumped from 20% to 32% among opinion influencers; over a million copies of Shell's free safety brochures have been distributed; and activity on Shell's Internet "Count on Shell" site has been extremely strong. By promoting itself as a reliable company that cares, Shell seems to be regaining its premier status.

DISCUSSION

1. Suppose you were asked to develop a sampling plan to determine what a "premier company" is to the general public. What sampling plan would you use? What is the target population? What would you use for a frame? Which of the four types of random sampling discussed in this chapter would you use? Could you use a combination of two or more of the types (two-stage sampling)? If so, how?

2. It appears that at least one of the research companies hired by Shell used some stratification in their sampling. What are some of the variables on which they stratified? If you were truly interested in ascertaining opinions from a variety of segments of the population with regard to opinions on "premier" companies or about Shell, what strata might make sense? Name at least five and justify why you would include them.

3. Suppose that in 1979 only 12% of the general adult U.S. public believed that Shell was a "premier" company. Suppose further that you randomly selected 350 people from the general adult U.S. public this year and 25% said that Shell was a "premier"

company. If only 12% of the general adult U.S. public still believes that Shell is a "premier" company, how likely is it that the 25% figure is a chance result in sampling 350 people? *Hint:* Use the techniques in this chapter to determine the probability of the 25% figure occurring by chance.

4. PERT Survey Research conducted quantitative surveys in an effort to measure the effectiveness of various campaigns. Suppose they used a 1-to-5 scale where 1 denotes that the campaign is not effective at all, 5 denotes that the campaign is extremely effective, and 2, 3, and 4 fall in between on an interval scale. Suppose also that a particular campaign received an average of 1.8 on the scale with a standard deviation of .7 early in the tests. Later, after the campaign had been critiqued and improved, a survey of 35 people was taken and a sample mean of 2.0 was recorded. What is the probability of a mean this large or larger occurring if the actual population mean is still just 1.8? Based on this, do you think that a sample mean of 2.0 is probably just a chance fluctuation on the 1.8 population mean, or do you think that perhaps it is evidence that the population mean is now greater than 1.8? Support your conclusion. Suppose a sample mean of 2.5 is attained. What is the likelihood of a mean this large or larger occurring by chance when the population mean is 1.8? Suppose this actually happens after the campaign has been improved. What does this mean?

ADAPTED FROM: "Count on It," *American Demographics,* March 1999, 60.

8
Estimation with Single Samples

Learning Objectives

The overall learning objective of Chapter 8 is to help you understand estimating parameters of single populations, thereby enabling you to:

1. Know the difference between point and interval estimation.

2. Estimate a population mean from a sample mean for large sample sizes.

3. Estimate a population mean from a sample mean for small sample sizes.

4. Estimate a population proportion from a sample proportion.

5. Estimate the minimum sample size necessary to achieve given statistical goals.

The central limit theorem presented in Chapter 7 states that certain statistics of interest, such as the sample mean and the sample proportion, are approximately normally distributed for large sample sizes regardless of the shape of the population distribution. The Z formulas for each statistic that were developed and discussed have the potential for use in parametric estimation, hypothesis testing, and determination of sample size. This chapter describes how these Z formulas for large statistics can be manipulated algebraically into a format that can be used to estimate population parameters and to determine the size of samples necessary to conduct research. In addition, mechanisms are introduced for estimation with small sample sizes.

8.1
Estimating the Population Mean with Large Sample Sizes

On many occasions estimating the population mean is useful in business research. For example, the manager of human resources in a company might want to estimate the average number of days of work an employee misses per year because of illness. If the firm has thousands of employees, direct calculation of a population mean such as this may be practically impossible. Instead, a random sample of employees can be taken, and the sample mean number of sick days can be used to estimate the population mean. Suppose another company has developed a new process for prolonging the shelf life of a loaf of bread. The company wants to be able to date each loaf for freshness, but company officials do not know exactly how long the bread will stay fresh. By taking a random sample and determining the sample mean shelf life, they can estimate the average shelf life for the population of bread.

As the cellular telephone industry matures, a cellular telephone company is rethinking its pricing structure. Users appear to be spending more time on the phone and are shopping around for the best deals. To do better planning, the cellular company wants to ascertain the average number of minutes of time used per month by each of its residential users but does not have the resources available to examine all monthly bills and extract the information. The company decides to take a sample of customer bills and estimate the population mean from sample data. A business analyst for the company takes a random sample of 85 bills for a recent month and from these bills computes a sample mean of 153 minutes. This sample mean, which is a statistic, is being used to estimate the population mean, which is a parameter. If the company uses the sample mean of 153 minutes as an estimate for the population mean, then the sample mean is being used as a *point estimate*.

Point estimate

An estimate of a population parameter constructed from a statistic taken from a sample.

Interval estimate

A range of values within which it is estimated with some confidence the population parameter lies.

A **point estimate** is *a statistic taken from a sample and is used to estimate a population parameter.* However, a point estimate is only as good as the representativeness of its sample. If other random samples are taken from the population, the point estimates derived from those samples are likely to vary. Because of variation in sample statistics, estimating a population parameter with an interval estimate is often preferable to using a point estimate. An **interval estimate** (confidence interval) is *a range of values within which the analyst can declare with some confidence the population parameter lies.* Confidence intervals can be two sided or one sided. This text presents only two-sided confidence intervals. How are confidence intervals constructed?

As a result of the central limit theorem, the following Z formula for sample means can be used when sample sizes are large, regardless of the shape of the population distribution, or for smaller sizes if the population is normally distributed.

$$Z = \frac{\overline{X} - \mu}{\frac{\sigma}{\sqrt{n}}}$$

Rearranging this formula algebraically to solve for μ gives

$$\mu = \bar{X} - Z\frac{\sigma}{\sqrt{n}}.$$

Because a sample mean can be greater than or less than the population mean, Z can be positive or negative. Thus the preceding expression takes the following form.

$$\bar{X} \pm Z\frac{\sigma}{\sqrt{n}}$$

Rewriting this expression yields the confidence interval formula for estimating μ with large sample sizes.

$$\bar{X} \pm Z_{\alpha/2}\frac{\sigma}{\sqrt{n}}$$

$100(1-\alpha)\%$ CONFIDENCE INTERVAL TO ESTIMATE μ

or

(8.1) $$\bar{X} - Z_{\alpha/2}\frac{\sigma}{\sqrt{n}} \leq \mu \leq \bar{X} + Z_{\alpha/2}\frac{\sigma}{\sqrt{n}}$$

where:

α = the area under the normal curve outside the confidence interval area

$\alpha/2$ = the area in one end (tail) of the distribution outside the confidence interval

Alpha (α) is the area under the normal curve in the tails of the distribution outside the area defined by the confidence interval. We will focus more on α in Chapter 9. Here we use α to locate the Z value in constructing the confidence interval as shown in Figure 8.1. Because the standard normal table is based on areas between a Z of 0 and $Z_{\alpha/2}$, the table Z value is found by locating the area of $.5000 - \alpha/2$, which is the part of the normal curve between the middle of the curve and one of the tails. Another way to locate this Z value is to change the confidence level from percentage to proportion, divide it in half, and go to the table with this value. The results are the same.

The confidence interval formula (8.1) yields a range (interval) within which we feel with some confidence the population mean is located. It is not certain that the population mean is in the interval unless we have a 100% confidence interval that is infinitely wide.

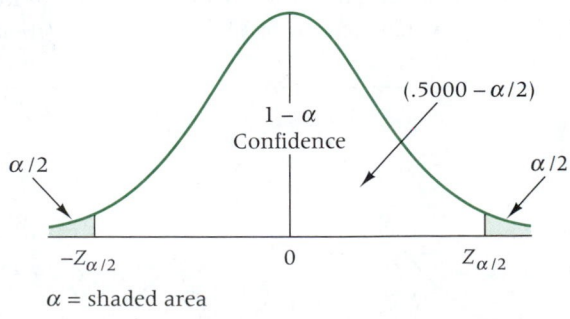

Figure 8.1

Z scores for confidence intervals in relation to α

However, we can assign a probability that the parameter (in this case, μ) is located within the interval. Formula 8.1 can be presented as a probability statement.

$$\text{Prob}\left[\bar{X} - Z_{\alpha/2}\frac{\sigma}{\sqrt{n}} \leq \mu \leq \bar{X} + Z_{\alpha/2}\frac{\sigma}{\sqrt{n}}\right] = 1 - \alpha$$

If we want to construct a 95% confidence interval, the level of confidence is 95% or .95. The probability statement shown tells us there is a .95 probability that the population mean is in this interval. If 100 such intervals are constructed by taking random samples from the population, it is likely that 95 of the intervals would include the population mean and five would not. The probability tells us the likelihood that a particular interval is one that does include the population mean.

As an example, in the cellular telephone company problem of estimating the population mean number of minutes called per residential user per month, from the sample of 85 bills it was determined that the sample mean is 153 minutes. Using this sample mean, a confidence interval can be calculated within which the researcher is relatively confident the actual population mean is located. To do this using Formula 8.1, the value of the population standard deviation and the value of Z (in addition to the sample mean, 153, and the sample size, 85) must be known. Suppose past history and similar studies indicate that the population standard deviation is 46 minutes.

The value of Z is determined by the level of confidence desired. An interval with 100% confidence is so wide that it is meaningless. Some of the more common levels of confidence used by business analysts are 90%, 95%, 98%, and 99%. Why would a business analyst not just select the highest confidence and always use that level? The reason is that trade-offs between sample size, interval width, and level of confidence must be considered. For example, as the level of confidence is increased, the interval gets wider, provided the sample size and standard deviation remain constant.

For the cellular telephone problem, suppose the business analyst decided on a 95% confidence interval for the results. Figure 8.2 shows a normal distribution of sample means about the population mean. When using a 95% level of confidence, he is selecting an interval centered on μ within which 95% of all sample mean values will fall and then using the width of that interval to create an interval around the *sample mean* within which he has some confidence the population mean will fall.

For 95% confidence, $\alpha = .05$ and $\alpha/2 = .025$. The value of $Z_{\alpha/2}$ or $Z_{.025}$ is found by looking in the standard normal table under $.5000 - .0250 = .4750$. This area in the table is associated with a Z value of 1.96. There is another way to locate the table Z value. Because the distribution is symmetric and the intervals are equal on each side of the population mean, $(1/2)(95\%)$, or .4750, of the area is on each side of the mean. Table A.5 yields a Z value of 1.96 for this portion of the normal curve. Thus the Z value for a 95% confidence interval is always 1.96. In other words, of all the possible $\bar{X}$ values along the horizontal axis of the diagram, 95% of them should be within a Z score of 1.96 from the population mean.

Figure 8.2

Distribution of
sample means for
95% confidence

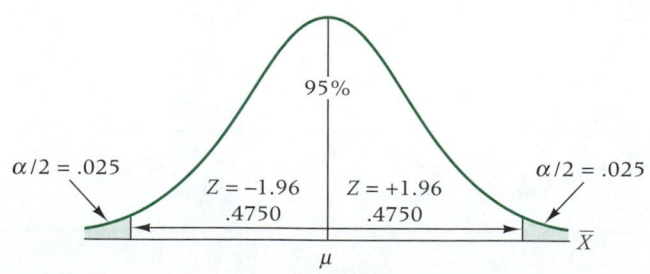

The business analyst can now complete the cellular telephone problem. To determine a 95% confidence interval for $\bar{X} = 153$, $\sigma = 46$, $n = 85$, and $Z = 1.96$, he estimates the average call length by including the value of Z in Formula 8.1.

$$153 - 1.96\frac{46}{\sqrt{85}} \le \mu \le 153 + 1.96\frac{46}{\sqrt{85}}$$

$$153 - 9.78 \le \mu \le 153 + 9.78$$

$$143.22 \le \mu \le 162.78$$

The confidence interval is constructed from the point estimate, which in this problem is 153 minutes, and the error of this estimate, which is ±9.78 minutes. The resulting confidence interval is $143.22 \le \mu \le 162.78$. The cellular telephone company business analyst is 95% confident that the average length of a call for the population is between 143.22 and 162.78 minutes.

What does being 95% confident that the population mean is in an interval actually indicate? It indicates that, if the company researcher were to randomly select 100 samples of 85 calls and use the results of each sample to construct a 95% confidence interval, approximately 95 of the 100 intervals would contain the population mean. It also indicates that 5% of the intervals would not contain the population mean. The company business analyst is likely to take only a single sample and compute the confidence interval from that sample information. That interval either contains the population mean or it does not. The odds are in his favor.

Figure 8.3 depicts the meaning of a 95% confidence interval for the mean. Note that if 20 random samples are taken from the population, 19 of the 20 are likely to contain the

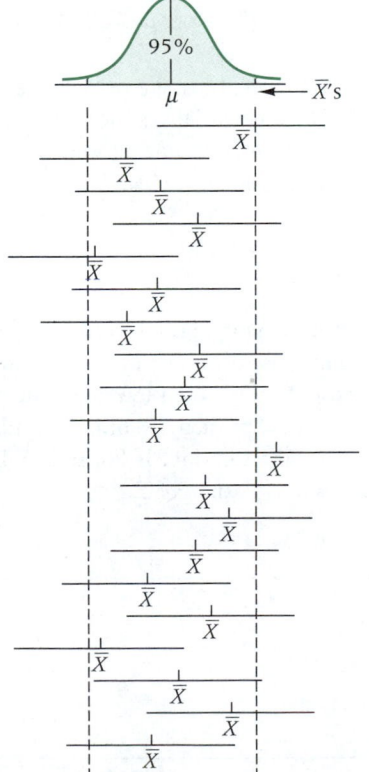

Figure 8.3

Twenty 95% confidence intervals of μ

population mean if a 95% confidence interval is used (19/20 = 95%). If a 90% confidence interval is constructed, only 18 of the 20 intervals are likely to contain the population mean.

<table>
<tr><td>

**DEMONSTRATION
PROBLEM 8.1**

</td><td>

A survey was taken of U.S. companies that do business with firms in India. One of the questions on the survey was: Approximately how many years has your company been trading with firms in India? A random sample of 44 responses to this question yielded a mean of 10.455 years. Suppose the population standard deviation for this question is 7.7 years. Using this information, construct a 90% confidence interval for the mean number of years that a company has been trading in India for the population of U.S. companies trading with firms in India.

</td></tr>
</table>

SOLUTION

Here, $n = 44$, $\bar{X} = 10.455$, and $\sigma = 7.7$. To determine the value of $Z_{\alpha/2}$, divide the 90% confidence in half or take $.5000 - \alpha/2 = .5000 - .0500$. The Z distribution of $\bar{X}$ around μ contains .4500 of the area on each side of μ, or (1/2)(90%). Table A.5 yields a Z value of 1.645 for the area of .4500 (interpolating between .4495 and .4505). The confidence interval is

$$\bar{X} - Z\frac{\sigma}{\sqrt{n}} \leq \mu \leq \bar{X} + Z\frac{\sigma}{\sqrt{n}}$$

$$10.455 - 1.645\frac{7.7}{\sqrt{44}} \leq \mu \leq 10.455 + 1.645\frac{7.7}{\sqrt{44}}$$

$$10.455 - 1.91 \leq \mu \leq 10.455 + 1.91$$

$$8.545 \leq \mu \leq 12.365$$

$$\text{Prob}\,[8.545 \leq \mu \leq 12.365] = .90.$$

That is, the analyst is 90% confident that if a census of all U.S. companies trading with firms in India were taken at the time of this survey, the actual population mean number of years a company would have been trading with firms in India would be between 8.545 and 12.365. The point estimate is 10.455 years.

Finite Correction Factor

Recall from Chapter 7 that if the sample is taken from a finite population, a finite correction factor may be used to increase the accuracy of the solution. In the case of interval estimation, the finite correction factor is used to reduce the width of the interval. As stated in Chapter 7, if the sample size is less than 5% of the population, the finite correction factor does not significantly alter the solution. If Formula 8.1 is modified to include the finite correction factor, the result is Formula 8.2.

<table>
<tr><td>

CONFIDENCE
INTERVAL TO
ESTIMATE μ
USING THE FINITE
CORRECTION FACTOR

</td><td>

$$\bar{X} - Z_{\alpha/2}\frac{\sigma}{\sqrt{n}}\sqrt{\frac{N-n}{N-1}} \leq \mu \leq \bar{X} + Z_{\alpha/2}\frac{\sigma}{\sqrt{n}}\sqrt{\frac{N-n}{N-1}}$$

</td><td>

(8.2)

</td></tr>
</table>

Demonstration Problem 8.2 shows how the finite correction factor can be used.

A study is being conducted in a company that has 800 engineers. A random sample of 50 of these engineers reveals that the average sample age is 34.3 years. Historically, the population standard deviation of the age of the company's engineers is approximately 8 years. Construct a 98% confidence interval to estimate the average age of all the engineers in this company.

SOLUTION

This problem has a finite population. The sample size, 50, is greater than 5% of the population, so the finite correction factor may be helpful. In this case $N = 800$, $n = 50$, $\bar{X} = 34.3$, and $\sigma = 8$. The Z value for a 98% confidence interval is 2.33 (.98 divided into two equal parts yields .4900; the Z value is obtained from Table A.5 by using .4900). Substituting into Formula 8.2 and solving for the confidence interval gives

$$34.3 - 2.33 \frac{8}{\sqrt{50}} \sqrt{\frac{750}{799}} \leq \mu \leq 34.3 + 2.33 \frac{8}{\sqrt{50}} \sqrt{\frac{750}{799}}$$

$$34.3 - 2.55 \leq \mu \leq 34.3 + 2.55$$

$$31.75 \leq \mu \leq 36.85.$$

Without the finite correction factor, the result would have been

$$34.3 - 2.64 \leq \mu \leq 34.3 + 2.64$$

$$31.66 \leq \mu \leq 36.94$$

The finite correction factor takes into account the fact that the population is only 800 instead of being infinitely large. The sample, $n = 50$, is a greater proportion of the 800 than it would be of a larger population, and thus the width of the confidence interval is reduced.

Confidence Interval to Estimate μ When σ Is Unknown

In the formulas and problems presented so far in this section, the population standard deviation was known. Estimating the population mean when the population standard deviation is known may seem strange. Sometimes the population standard deviation is estimated from past records or from industry standards. However, the reality is that in most instances, the population standard deviation is unknown. For example, in the cellular telephone example, the average length of a call is unknown. The likelihood is high that the population standard deviation also is unknown. So how does the business analyst get around this dilemma?

When samples sizes are large ($n \geq 30$), the sample standard deviation is a good estimate of the population standard deviation and can be used as an acceptable approximation of the population standard deviation in the Z formula for a mean. Because formulas based on the central limit theorem require large samples for nonnormal populations, it makes sense to modify Formula 8.1 to use the sample standard deviation, S. Beware, however, not to use this modified formula for small samples when the population standard deviation is unknown, even when the population is normally distributed. Section 8.2 presents techniques for handling the case of the small samples when the population standard deviation is unknown and X is normally distributed.

Shown next is the confidence interval formula to estimate μ with large samples when using the sample standard deviation.

CONFIDENCE INTERVAL TO ESTIMATE μ WHEN POPULATION STANDARD DEVIATION IS UNKNOWN AND n IS LARGE	$$\bar{X} \pm Z_{\alpha/2} \frac{S}{\sqrt{n}}$$ or $$\bar{X} - Z_{\alpha/2} \frac{S}{\sqrt{n}} \leq \mu \leq \bar{X} + Z_{\alpha/2} \frac{S}{\sqrt{n}} \qquad (8.3)$$

As an example, suppose a U.S. car rental firm wants to estimate the average number of miles traveled per day by each of its cars rented in California. A random sample of 110 cars rented in California reveals that the sample mean travel distance per day is 85.5 miles, with a sample standard deviation of 19.3 miles. Compute a 99% confidence interval to estimate μ.

Here, $n = 110$, $\bar{X} = 85.5$, and $S = 19.3$. For a 99% level of confidence, a Z value of 2.575 is obtained. The confidence interval is

$$\bar{X} - Z_{\alpha/2} \frac{S}{\sqrt{n}} \leq \mu \leq \bar{X} + Z_{\alpha/2} \frac{S}{\sqrt{n}}$$

$$85.5 - 2.575 \frac{19.3}{\sqrt{110}} \leq \mu \leq 85.5 + 2.575 \frac{19.3}{\sqrt{110}}$$

$$85.5 - 4.7 \leq \mu \leq 85.5 + 4.7$$

$$80.8 \leq \mu \leq 90.2.$$

The point estimate indicates that the average number of miles traveled per day by a rental car in California is 85.5. With 99% confidence, we estimate that the population mean is somewhere between 80.8 and 90.2 miles per day.

For convenience, Table 8.1 contains some of the more common levels of confidence and their associated Z values.

TABLE 8.1

Values of Z for Some of the More Common Levels of Confidence

CONFIDENCE LEVEL	Z VALUE
90%	1.645
95%	1.96
98%	2.33
99%	2.575

Analysis Using Excel

Confidence intervals for a population mean with the normal distribution can be constructed using three statistical functions and one mathematical function from Excel. However to simplify the process, **FAST ⦚ STAT** has a macro procedure for constructing such a confidence interval. In addition, the procedure performs the computations necessary for conducting a hypothesis test as discussed in Chapter 9. This feature is available from FAST ⦚ STAT's pull-down menu with the name **1 Mean Using Z Dist.** The dialog box for it is shown in Figure 8.4.

As you may note from the figure, the inputs are grouped into three segments. The first group is for the sample itself. This input is the worksheet cell range for the sample values. The second grouping also has just one input. It is the level of confidence for the confidence interval. It is to be given as a percentage such as 99%, 95%, 90%, and so on.

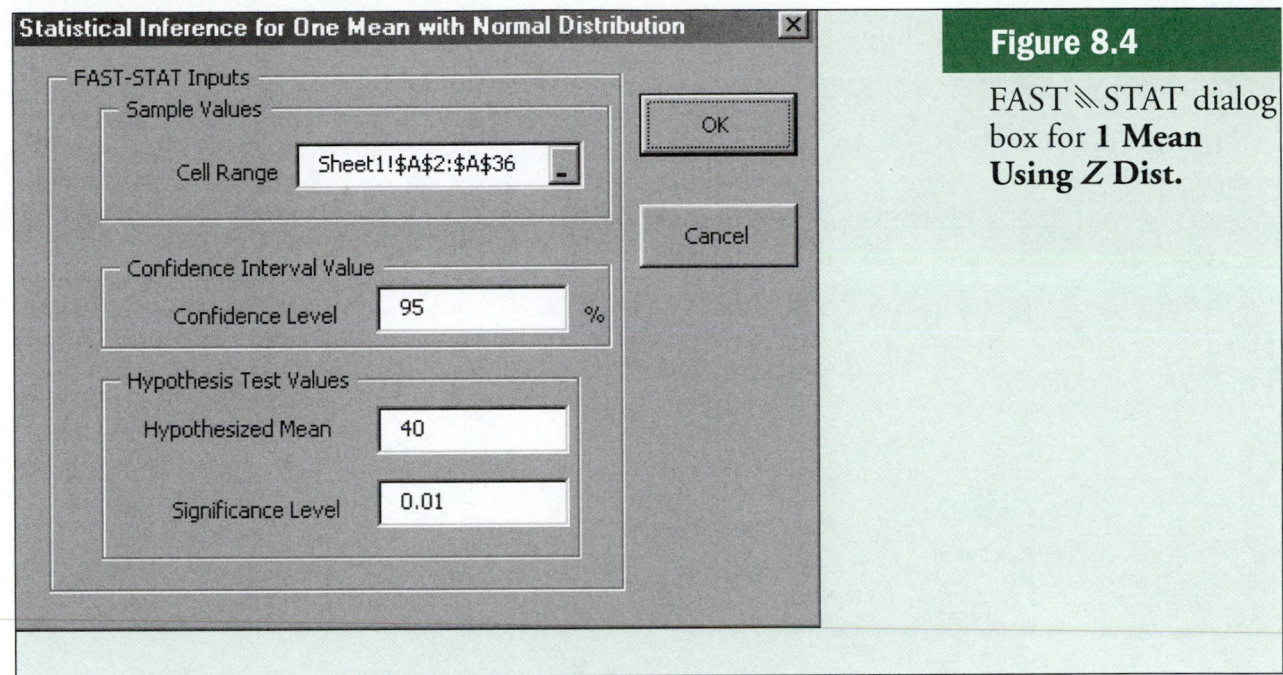

Figure 8.4

FAST ＼ STAT dialog box for **1 Mean Using *Z* Dist.**

The third grouping includes two inputs for conducting a hypothesis test. These will be discussed later in Section 9.2 of Chapter 9. If you only wish to compute a confidence interval, enter *any number* for the *hypothesized mean* and *a number between 0 and 100* for the *significance level*. We should note, however, if these two values are left blank or are zero, Excel will indicate the **#NUM!** error condition. This doesn't invalidate the confidence interval results, but some users may find this error message distracting. To avoid the message, enter a number that is greater than zero and less than one for the significance level.

To demonstrate this FAST ＼ STAT feature, consider this example. A newly organized dot-com company wishes to construct a 95% confidence interval for the age of their employees. A sample of 35 employees produces the following ages.

44	37	49	30	56	48	53
42	51	38	39	45	47	52
59	50	46	34	39	46	27
35	52	51	46	45	58	51
37	45	52	51	54	39	48

These values are entered into an Excel worksheet in cells A2 through A36. Figure 8.5 shows the results for this example.

In column A you will find the sample values. In columns C through G, the values for the four dialog box inputs are repeated. The outputs that follow are in four parts: sample statistic values, the point estimate, the confidence interval, and the hypothesis test for the sample. The sample statistic values include the sample mean, standard deviation, and sample size. The confidence interval results include the *Z* value for the specified confidence level and the standard error of the mean. In addition, it includes the error of estimation (sometimes termed the interval half width), the endpoints of the interval, and the complete interval.

| | Figure 8.5 | | FAST ﹨ STAT output for Dot-Com example problem |

Microsoft Excel - Book1

File Edit View Insert Format Tools Data Window FAST-STAT Help

R34 =

	A	B	C	D	E	F	G	H
1	**Sample Values**		**Statistical Inference for One Mean-Normal Distribution**					
2	44		INPUTS					
3	42			Cell Range for Sample Values			Sheet1!$A$2:$A$36	
4	59							
5	35			Level of Confidence			95	%
6	37							
7	37			Hypothesized Mean			40	
8	51			Significance Level of Test			0.010	
9	50		OUTPUTS					
10	52		**Sample Statistics**					
11	45			Sample Mean			45.6	
12	49			Sample Standard Deviation			7.746726	
13	38			Sample Size			35	
14	46							
15	51		**Point Estimate**				45.6	
16	52							
17	30		**Confidence Interval**					
18	39							
19	34			Z Value for Level of Confidence			1.960	
20	46			Standard Error of the Mean			1.309436	
21	51			Error of Estimation			2.566443	
22	56			Interval Lower Limit			43.03356	
23	45			Interval Upper Limit			48.16644	
24	39			Confidence Interval		43.03355699	to	48.16644
25	45		**Hypothesis Test**					
26	54			Test Statistic			4.277	
27	48			Critical Z Value for Two-Tailed Test			2.576	
28	47			Critical Z Value for Upper-Tailed Test			2.326	
29	46			Critical Z Value for Lower-Tailed Test			-2.326	
30	58			p-Value for Two-Tailed Test			0.000	
31	39			p-Value for Upper-Tailed Test			0.000	
32	53			p-Value for Lower-Tailed Test			1.000	
33	52							
34	27							
35	51							

CI-HT / Sheet1 / Sheet2 / Sheet3 /

8.1 Problems

8.1 Use the following information to construct the confidence intervals specified to estimate μ.
  a. 95% confidence; $\overline{X} = 25$, $\sigma = 3.5$, and $n = 60$
  b. 98% confidence; $\overline{X} = 119.6$, $S = 23.89$, and $n = 75$
  c. 90% confidence; $\overline{X} = 3.419$, $S = 0.974$, and $n = 32$
  d. 80% confidence; $\overline{X} = 56.7$, $\sigma = 12.1$, $N = 500$, and $n = 47$

8.2 For a random sample of 36 items and a sample mean of 211, compute a 95% confidence interval for μ if the population standard deviation is 23.

8.3 A random sample of 81 items is taken, producing a sample mean of 47 and a sample standard deviation of 5.89. Construct a 90% confidence interval to estimate the population mean.

8.4 A candy company fills a 20-ounce package of Halloween candy with individually wrapped pieces of candy. The number of pieces of candy per package varies because the package is sold by weight. The company wants to estimate the number of pieces per package. Inspectors randomly sample 120 packages of this candy and count the number of pieces in each package. They find that the sample mean number of pieces is 18.72, with a sample standard deviation of .8735. What is the point estimate of the number of pieces per package? Construct a 99% confidence interval to estimate the mean number of pieces per package for the population.

8.5   A small lawnmower company produced 1500 lawnmowers in 1990. In an effort to determine how maintenance-free these units were, the company decided to conduct a multi-year study of the 1990 lawnmowers. A sample of 200 owners of these lawnmowers was drawn randomly from company records and contacted. The owners were given an 800 number and asked to call the company when the first major repair was required for the lawnmowers. Owners who no longer used the lawnmower to cut their grass were disqualified. After many years, 187 of the owners had reported. The other 13 disqualified themselves. The average number of years until the first major repair was 5.3 for the 187 owners reporting, and the sample standard deviation was 1.28 years. If the company wants to advertise an average number of years of repair-free lawnmowing for this lawnmower, what is the point estimate? Construct a 95% confidence interval for the average number of years until the first major repair.

8.6   The average total dollar purchase at a convenience store is less than that at a supermarket. Despite smaller-ticket purchases, convenience stores can still be profitable because of the size of operation, volume of business, and the markup. A researcher is interested in estimating the average purchase amount for convenience stores in suburban Long Island. To do so, she randomly sampled 32 purchases from several convenience stores in suburban Long Island and tabulated the amounts to the nearest dollar. Use the following data to construct a 90% confidence interval for the population average amount of purchases.

$2	$11	$8	$7	$9	$3	$3	$6
5	4	2	1	10	8	5	4
14	7	6	3	7	2	3	6
4	1	3	6	8	4	7	12

8.7   A community health association is interested in estimating the average number of maternity days women stay in the local hospital. A random sample is taken of 36 women who had babies in the hospital during the past year. The following numbers of maternity days each woman was in the hospital are rounded to the nearest day.

3	3	4	3	2	5	3	1	4	3	4	2
3	5	3	2	4	3	2	4	1	6	3	4
3	3	5	2	3	2	3	5	4	3	5	4

Use these data to construct a 98% confidence interval to estimate the average maternity stay in the hospital for all women who have babies in this hospital.

8.8   A meat-processing company in the Midwest produces and markets a package of eight small sausage sandwiches. The product is nationally distributed, and the company is interested in knowing the average retail price charged for this item in stores across the country. The company cannot justify a national census to generate this information. The company information system produces a list of all retailers who carry the product. A researcher for the company contacts 36 of these retailers and ascertains the selling prices for the product. Use the following price data to determine a point estimate for the national retail price of the product. Construct a 90% confidence interval to estimate this price.

$2.23	$2.11	$2.12	$2.20	$2.17	$2.10
2.16	2.31	1.98	2.17	2.14	1.82
2.12	2.07	2.17	2.30	2.29	2.19
2.01	2.24	2.18	2.18	2.32	2.02
1.99	1.87	2.09	2.22	2.15	2.19
2.23	2.10	2.08	2.05	2.16	2.26

8.9 According to the U.S. Bureau of the Census, the average travel time to work in Philadelphia is 27.4 minutes. Suppose a researcher wants to estimate the average travel time to work in Cleveland using a 95% level of confidence. A random sample of 45 Cleveland commuters is taken and the travel time to work is obtained from each. The data are entered into an Excel spreadsheet with a population standard deviation of 5.1 minutes. Excel calculates the mean as 25 and then computes the error of the interval around the mean. Excel yielded 1.49 as the error of the confidence estimation around the mean. The data used in the calculation follow. Compute a 95% confidence interval on the data. How do your calculations compare with Excel's figures?

27	25	19	21	24	27	29	34	18	29	16	28
20	32	27	28	22	20	14	15	29	28	29	33
16	29	28	28	27	23	27	20	27	25	21	18
26	14	23	27	27	21	25	28	30			

8.2
Estimating the Population Mean: Small Sample Sizes, σ Unknown

In Section 8.1 we learned how to estimate the population mean by using the sample mean. The central limit theorem, presented in Chapter 7, guarantees that the sample means are normally distributed when sample size is large. These procedures can still be used when the sample size is small if the population is normally distributed and if σ is known.

When the population standard deviation is unknown, the sample standard deviation is an acceptable estimate of and substitute for the population standard deviation and can be used in the confidence interval, along with the sample mean, to estimate the population mean if sample size is *large*. A value of $n \geq 30$ is generally considered a lower limit for large sample size.

On the other hand, in many real-life situations, sample sizes of less than 30 are the norm. For example, a researcher is interested in studying the average flying time of a DC-10 from New York to Los Angeles, but a sample of only 21 flights is available. Another researcher is studying the impact of movie video advertisements on consumers, but the group used in the study contains only 11 people.

If the population is known to be normally distributed and the population standard deviation is known, the Z values computed from the sample means are normally distributed regardless of sample size. Thus, by assuming that the *population* is normally distributed (many phenomena are normally distributed) and that the population standard deviation is known, a researcher could theoretically continue to use the techniques presented in Section 8.1 for confidence interval estimation, even with small samples. This formula for sample means is

$$\bar{X} - Z\frac{\sigma}{\sqrt{n}} \leq \mu \leq \bar{X} + Z\frac{\sigma}{\sqrt{n}}.$$

In many research situations, the population standard deviation is not known and must be estimated by using the sample standard deviation. In these situations, S is substituted into the preceding formula.

$$\bar{X} - Z\frac{S}{\sqrt{n}} \leq \mu \leq \bar{X} + Z\frac{S}{\sqrt{n}}$$

However, the sample standard deviation, S, is only a good approximation for the population standard deviation, σ, for large samples. The Z formulas that use S therefore are not applicable to small-sample analysis. This problem was considered and solved by a British statistician, William S. Gosset.

Gosset was born in 1876 in Canterbury, England. He studied chemistry and mathematics and in 1899 went to work for the Guinness Brewery in Dublin, Ireland. Gosset was involved in quality control at the brewery, studying variables such as raw materials and temperature. Because of the circumstances of his experiments, Gosset conducted many studies with small samples. He discovered that using the standard Z test with a *sample* standard deviation produced inexact and incorrect distributions for small sample sizes. This finding led to his development of the distribution of the sample standard deviation and the t test.

Gosset was a student and close personal friend of Karl Pearson. When Gosset's first work on the t test was published, he used the pen name "Student." As a result, the t test is sometimes referred to as the Student's t test. Gosset's contribution was significant because it led to more exact statistical tests, which some scholars say marked the beginning of the modern era in mathematical statistics.*

The t Distribution

Gosset developed the **t distribution,** which *describes the sample data in small samples when the population standard deviation is unknown and the population is normally distributed.* The formula for the **t value** is

$$t = \frac{\bar{X} - \mu}{\frac{S}{\sqrt{n}}}$$

t distribution

A distribution that describes the sample data in small samples when the standard deviation is unknown and the population is normally distributed.

This formula is essentially the same as the Z formula, but the distribution table values are different. The t distribution values are contained in Table A.6 and, for convenience, inside the front cover of the text.

The t distribution actually is a series of distributions because every sample size has a different distribution, thereby creating the potential for many t tables. To make these t values more manageable, only select key values are presented; each line in the table contains values from a different t distribution. *The assumption underlying the use of the techniques discussed in this chapter for small sample sizes is that the population is normally distributed.* If the population distribution is not normal or is unknown, nonparametric techniques should be used.

t value

The computed value of t used to reach statistical conclusions regarding the null hypothesis in small-sample analysis.

Robustness

Most statistical techniques have one or more underlying assumptions. If a statistical technique *is relatively insensitive to minor violations in one or more of its underlying assumptions,* the technique is said to be **robust** to that assumption. The t statistic for estimating a population mean is relatively robust to the assumption that the population is normally distributed.

Some statistical techniques are not robust, and a business analyst should exercise extreme caution to be certain that the assumptions underlying a technique are being met

Robust

Describes a statistical technique that is relatively insensitive to minor violations in one or more of its underlying assumptions.

---

*ADAPTED FROM: Arthur L. Dudycha and Linda W. Dudycha, "Behavioral Statistics: An Historical Perspective," in *Statistical Issues: A Reader for the Behavioral Sciences,* edited by Roger Kirk (Monterey, CA: Brooks/Cole, 1972).

Figure 8.6

Comparison of two
t distributions to
the standard
normal curve

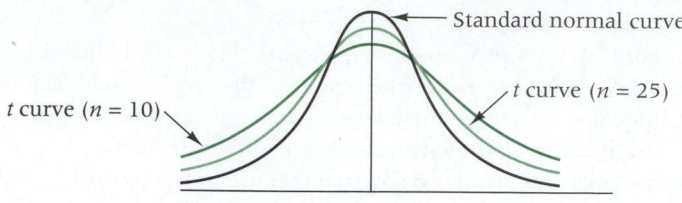

before using it or interpreting statistical output resulting from its use. A business analyst should always beware of statistical assumptions and the robustness of techniques being used in an analysis.

Characteristics of the t Distribution

Figure 8.6 displays two *t* distributions superimposed on the standard normal distribution. Like the normal curve, *t* distributions are symmetric, unimodal, and a family of curves. The *t* distributions are flatter in the middle and have more area in their tails than the standard normal distribution.

An examination of *t* distribution values reveals that the *t* distribution approaches the standard normal curve as *n* becomes large. The *t* distribution is the appropriate distribution to use any time the population variance or standard deviation is unknown, regardless of sample size. However, because the difference between the table values for Z and *t* becomes negligible for large samples, many researchers use the Z distribution for large-sample analysis even when the standard deviation or variance is unknown. In this text, the *t* distribution is reserved for use with small sample size problems ($n < 30$) because, as *n* nears size 30, the *t* table values approach the Z table values.

Reading the t Distribution Table

Degrees of freedom
A mathematical adjustment made to the size of the sample; used along with α to locate values in statistical tables.

To find a value in the *t* distribution table requires knowing the sample size. The *t* distribution table is a compilation of many *t* distributions, with each line of the table representing a different sample size. However, the sample size must be converted to **degrees of freedom (df)** before a table value can be determined. The degrees of freedom vary according to which *t* formula is being used, so a df formula is given along with each *t* formula. The concept of degrees of freedom is difficult and beyond the scope of this text. A brief explanation is that *t* formulas are used because the population variance or standard deviation, which is part of the Z formula, is unknown and must be estimated by a sample standard deviation or variance. For every parameter (such as variance or standard deviation) of a statistical formula that is unknown and must be estimated by a statistic (e.g., sample variance or standard deviation) in the formula, one degree of freedom is lost.

In Table A.6 the degrees of freedom are located in the left column. The *t* distribution table in this text does not use the area between the statistic and the mean as does the Z distribution (standard normal distribution). Instead, the *t* table uses the area in the tail of the distribution. The emphasis in the *t* table is on α, and each tail of the distribution contains $\alpha/2$ of the area under the curve when confidence intervals are constructed. (In Chapter 9, sometimes a single tail of the distribution will contain α proportion of the area.) For confidence intervals, the table *t* value is in the column under the value of $\alpha/2$ at the intersection of the df value.

For example, if a 90% confidence interval is being computed, the total area in the two tails is 10%. Thus, α is .10 and $\alpha/2$ is .05, as indicated in Figure 8.7. The *t* distribution

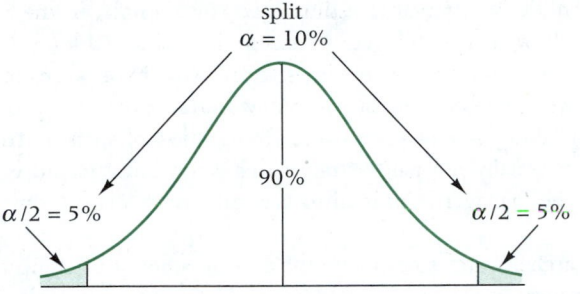

Figure 8.7

Distribution with alpha for 90% confidence

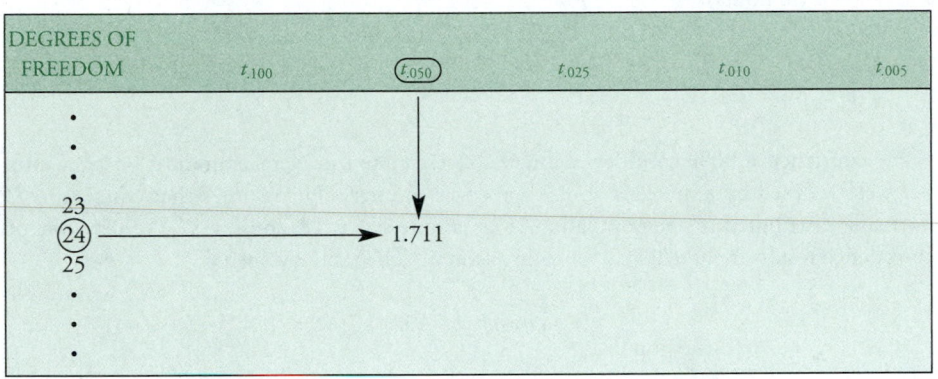

TABLE 8.2
t Distribution

table shown in Table 8.2 contains only five values of $\alpha/2$ (.10, .05, .025, .01, .005). The *t* value is located at the intersection of the df value and the selected $\alpha/2$ value. So if the degrees of freedom for a given *t* statistic are 24 and the desired $\alpha/2$ value is .05, the *t* value is 1.711.

Confidence Intervals to Estimate μ When σ Is Unknown and Sample Size Is Small

The *t* formula,

$$t = \frac{\overline{X} - \mu}{\frac{S}{\sqrt{n}}},$$

can be manipulated algebraically to produce a formula for estimating the population mean using small samples when σ is unknown and the population is normally distributed. The result is the formulas given next.

$$\overline{X} \pm t_{\alpha/2,n-1}\frac{S}{\sqrt{n}}$$

(8.4)

$$\overline{X} - t_{\alpha/2,n-1}\frac{S}{\sqrt{n}} \leq \mu \leq \overline{X} + t_{\alpha/2,n-1}\frac{S}{\sqrt{n}}$$

$$df = n - 1$$

CONFIDENCE INTERVAL TO ESTIMATE μ: SMALL SAMPLES AND POPULATION STANDARD DEVIATION UNKNOWN

Formula 8.4 can be used in a manner similar to methods presented in Section 8.1 for constructing a confidence interval to estimate μ. For example, in the aerospace industry some companies allow their employees to accumulate extra working hours beyond their 40-hour week. These extra hours sometimes are referred to as *green* time, or *comp* time. Many managers work longer than the 8-hour workday, preparing proposals, overseeing crucial tasks, and taking care of paperwork. Recognition of such overtime is important. Most managers are usually not paid extra for this work, but a record is kept of this time and occasionally the manager is allowed to use some of this comp time as extra leave or vacation time.

Suppose a researcher wants to estimate the average amount of comp time accumulated per week for managers in the aerospace industry. He randomly samples 18 managers and measures the amount of extra time they work during a specific week and obtains the results shown (in hours).

6	21	17	20	7	0	8	16	29
3	8	12	11	9	21	25	15	16

He constructs a 90% confidence interval to estimate the average amount of extra time per week worked by a manager in the aerospace industry. He assumes that comp time is normally distributed in the population. The sample size is 18, so df = 17. A 90% level of confidence results in an $\alpha/2 = .05$ area in each tail. The table t value is

$$t_{.05,17} = 1.740.$$

The subscripts in the t value denote to other researchers the area in the right tail of the t distribution (for confidence intervals $\alpha/2$) and the number of degrees of freedom. The sample mean is 13.56 hours, and the sample standard deviation is 7.8 hours. The confidence interval is computed from this information as

$$\bar{X} \pm t_{\alpha/2,n-1} \frac{S}{\sqrt{n}}$$

$$13.56 \pm 1.740 \frac{7.8}{\sqrt{18}} = 13.56 \pm 3.20$$

$$10.36 \leq \mu \leq 16.76$$

$$\text{Prob}[10.36 \leq \mu \leq 16.76] = .90.$$

The point estimate for this problem is 13.56 hours, with an error of ±3.20 hours. The researcher is 90% confident that the average amount of comp time accumulated by a manager per week in this industry is between 10.36 and 16.76 hours.

From these figures, aerospace managers could attempt to build a reward system for such extra work or evaluate the regular 40-hour week to determine how to use the normal work hours more effectively and thus reduce comp time.

DEMONSTRATION PROBLEM 8.3

The owner of a large equipment rental company wants to make a rather quick estimate of the average number of days a piece of ditchdigging equipment is rented out per person per time. The company has records of all rentals, but the amount of time required to conduct an audit of *all* accounts would be prohibitive. The owner decides to take a random sample of rental invoices. Fourteen different rentals of ditchdiggers are selected randomly from the files, yielding the following data. She uses these data to construct a 99% confidence interval to estimate the average number of days that a ditchdigger is rented and assumes that the number of days per rental is normally distributed in the population.

| 3 | 1 | 3 | 2 | 5 | 1 | 2 | 1 | 4 | 2 | 1 | 3 | 1 | 1 |

SOLUTION

As $n = 14$, the df $= 13$. The 99% level of confidence results in $\alpha/2 = .005$ area in each tail of the distribution. The table t value is

$$t_{.005,13} = 3.012.$$

The sample mean is 2.14 and the sample standard deviation is 1.29. The confidence interval is

$$\bar{X} \pm t \frac{S}{\sqrt{n}}$$

$$2.14 \pm 3.012 \frac{1.29}{\sqrt{14}} = 2.14 \pm 1.04$$

$$1.10 \le \mu \le 3.18$$

$$\text{Prob}[1.10 \le \mu \le 3.18] = .99.$$

The point estimate of the average length of time per rental is 2.14 days, with an error of ± 1.04. With a 99% level of confidence, the company's owner can estimate that the average length of time per rental is between 1.10 and 3.18 days. Combining this figure with variables such as frequency of rentals per year can help the owner estimate potential profit or loss per year for such a piece of equipment.

Analysis Using Excel

Confidence intervals for a population mean with the t distribution can be constructed using a combination of statistical and mathematical functions from Excel. However to simplify the process, **FAST ⟍ STAT** has a macro procedure for constructing such a confidence interval. In addition, the procedure performs the computations necessary for conducting a hypothesis test as discussed in Chapter 9. This feature is available from FAST ⟍ STAT's pull-down menu with the name **1 Mean Using t Dist.** The dialog box for it is shown in Figure 8.8

As you may note from the figure, the inputs are grouped into three segments. The first group is for the sample itself. This input is the worksheet cell range for the sample values. The second grouping also has just one input. It is the level of confidence for the confidence interval. It is to be given as a percentage such as 99%, 95%, 90%, and so on.

The third grouping includes two inputs for conducting a hypothesis test. These will be discussed later in Section 9.3 of Chapter 9. If you only wish to compute a confidence interval, enter *any number* for the *hypothesized mean* and *a number between 0 and 100* for the *significance level*.

Figure 8.9 presents the **1 Mean Using t Dist.** results for Demonstration Problem 8.3.

In column A you will find the sample values. In columns C through G, the values for the four dialog box inputs are repeated. The outputs that follow are in four parts: sample statistic values, the point estimate, the confidence interval, and the hypothesis test for the sample. The sample statistic values include the sample mean, standard deviation, and sample size. The confidence interval results include the t value for the specified confidence

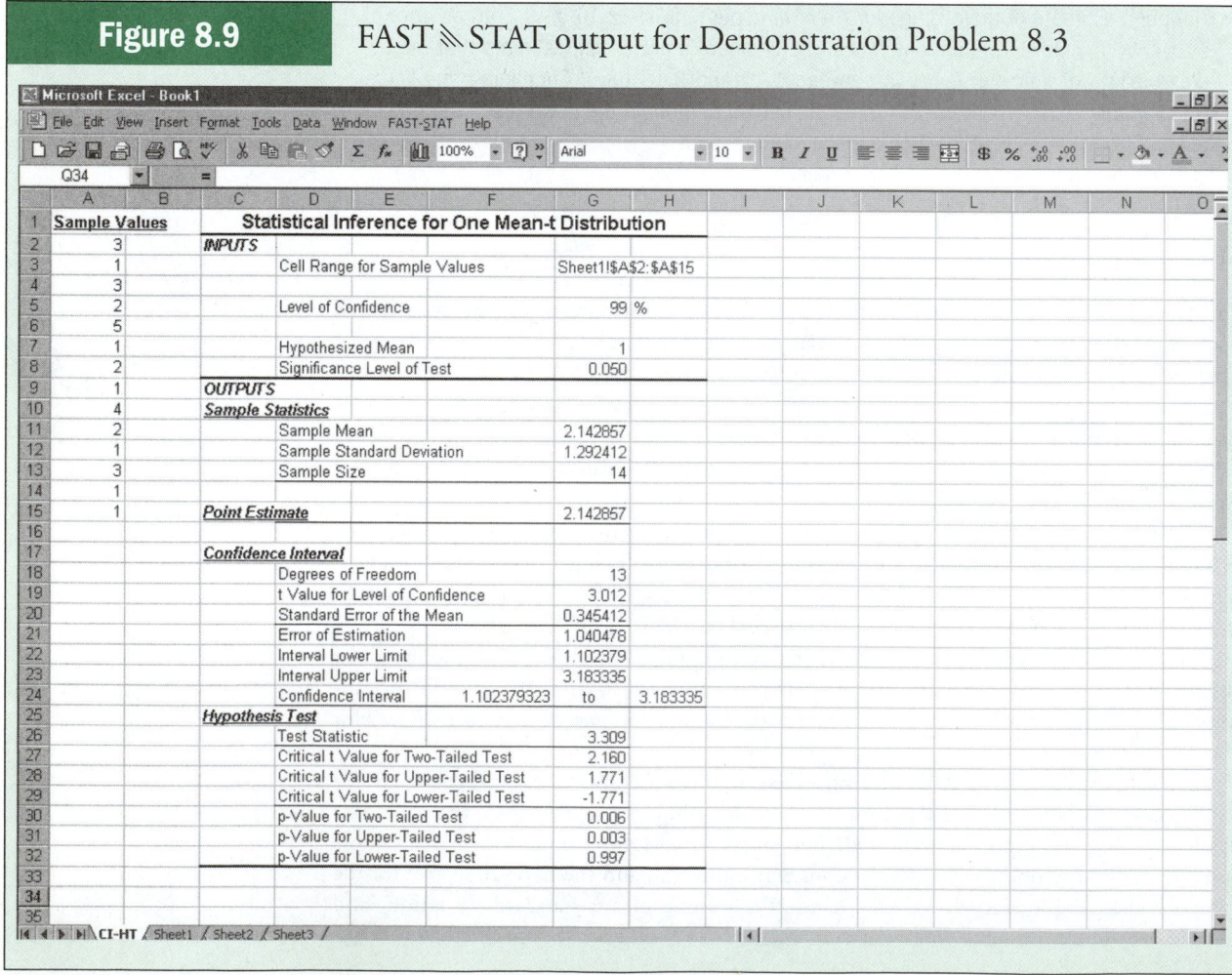

Figure 8.8

FAST\STAT dialog box for **1 Mean Using *t* Dist.**

Statistical Inference for One Mean with t Distribution

FAST-STAT Inputs

Sample Values

Cell Range: Sheet1!$A$2:$A$15

Confidence Interval Value

Confidence Level: 99 %

Hypothesis Test Values

Hypothesized Mean: 1

Significance Level: 0.05

OK

Cancel

Figure 8.9   FAST\STAT output for Demonstration Problem 8.3

Microsoft Excel - Book1

File Edit View Insert Format Tools Data Window FAST-STAT Help

	A	C D E F	G	H
1	**Sample Values**	**Statistical Inference for One Mean-t Distribution**		
2	3	**INPUTS**		
3	1	Cell Range for Sample Values	Sheet1!$A$2:$A$15	
4	3			
5	2	Level of Confidence	99 %	
6	5			
7	1	Hypothesized Mean	1	
8	2	Significance Level of Test	0.050	
9	1	**OUTPUTS**		
10	4	**Sample Statistics**		
11	2	Sample Mean	2.142857	
12	1	Sample Standard Deviation	1.292412	
13	3	Sample Size	14	
14	1			
15	1	**Point Estimate**	2.142857	
16				
17		**Confidence Interval**		
18		Degrees of Freedom	13	
19		t Value for Level of Confidence	3.012	
20		Standard Error of the Mean	0.345412	
21		Error of Estimation	1.040478	
22		Interval Lower Limit	1.102379	
23		Interval Upper Limit	3.183335	
24		Confidence Interval 1.102379323 to		3.183335
25		**Hypothesis Test**		
26		Test Statistic	3.309	
27		Critical t Value for Two-Tailed Test	2.160	
28		Critical t Value for Upper-Tailed Test	1.771	
29		Critical t Value for Lower-Tailed Test	-1.771	
30		p-Value for Two-Tailed Test	0.006	
31		p-Value for Upper-Tailed Test	0.003	
32		p-Value for Lower-Tailed Test	0.997	

CI-HT / Sheet1 / Sheet2 / Sheet3 /

level and the standard error of the mean. In addition, it includes the error of estimation (sometimes termed the interval half width), the endpoints of the interval, and the complete interval.

8.10 Suppose the following data are selected randomly from a population of normally distributed values.

40	51	43	48	44	57	54
39	42	48	45	39	43	

Construct a 95% confidence interval to estimate the population mean value.

8.11 Assuming X is normally distributed, use the following information to compute a 90% confidence interval to estimate μ.

313	320	319	340
325	310	321	329
317	311	307	318

8.12 If a random sample of 27 items produces $\overline{X} = 128.4$ and $S = 20.6$, what is the 98% confidence interval for μ? Assume X is normally distributed for the population. What is the point estimate?

8.13 Use the following data to construct a 99% confidence interval for μ.

16.4	17.1	17.0	15.6	16.2
14.8	16.0	15.6	17.3	17.4
15.6	15.7	17.2	16.6	16.0
15.3	15.4	16.0	15.8	17.2
14.6	15.5	14.9	16.7	16.3

Assume X is normally distributed. What is the point estimate for μ?

8.14 According to Runzheimer International, the average cost of a domestic trip for business travelers in the financial industry is $1250. Suppose another travel industry research company takes a random sample of 22 business travelers in the financial industry and determines that the sample average cost of a domestic trip is $1192, with a sample standard deviation of $279. Construct a 98% confidence interval for the population mean from these sample data. Assume that the data are normally distributed in the population. Now go back and examine the $1250 figure published by Runzheimer International. Does it fall into the confidence interval computed from the sample data? What does that tell you?

8.15 A valve manufacturer produces a butterfly valve composed of two semicircular plates on a common spindle and is used to permit flow in one direction only. The semicircular plates are supplied by a vendor with specifications that the plates be 2.37 mm thick and have a tensile strength of 5 lb/mm. A random sample of 20 such plates is taken. Electronic calipers are used to measure the thickness of each plate; the measurements are given here. Assuming that the thicknesses of such plates are normally distributed, use the data to construct a 95% level of confidence for the population mean thickness of these plates. What is the point estimate? How much is the error of the interval?

2.4066	2.4579	2.6724	2.1228	2.3238
2.1328	2.0665	2.2738	2.2055	2.5267
2.5937	2.1994	2.5392	2.4359	2.2146
2.1933	2.4575	2.7956	2.3353	2.2699

8.16 Some fast-food chains have been offering a lower-priced combination meal in an effort to attract budget-conscious customers. One chain test marketed an offer of a burger, fries, and a drink for $1.71. The weekly sales volume for these meals was impressive. Suppose the chain wants to estimate the average amount its customers spent on a meal at their restaurant while this offer was in effect. An analyst gathers data from 28 randomly selected customers. The following data represent the sample meal totals.

$3.21	$5.40	$3.50	$4.39	$5.60	$8.65	$5.02
4.20	1.25	7.64	3.28	5.57	3.26	3.80
5.46	9.87	4.67	5.86	3.73	4.08	5.47
4.49	5.19	5.82	7.62	4.83	8.42	9.10

Use these data to construct a 90% confidence interval to estimate the population mean value. Assume the amounts spent are normally distributed.

8.17 The marketing director of a large department store wants to estimate the average number of customers who enter the store every 5 minutes. She has a research assistant randomly select 5-minute intervals and count the number of arrivals at the store. The assistant obtains the figures 58, 32, 41, 47, 56, 80, 45, 29, 32, and 78. The analyst assumes the number of arrivals is normally distributed. Using these data, the analyst computes a 95% confidence interval to estimate the mean value for all 5-minute intervals. What interval values does she get?

8.18 Runzheimer International publishes results of studies on overseas business travel costs. Suppose as a part of one of these studies the following per diem travel accounts (in dollars) are obtained for 14 business travelers staying in Johannesburg, South Africa. Use these data to construct a 98% confidence interval to estimate the average per diem expense for business people traveling to Johannesburg. What is the point estimate? Assume per diem rates for any locale are approximately normally distributed.

142.59	148.48	159.63	171.93	146.90	168.87	141.94
159.09	156.32	142.49	129.28	151.56	132.87	178.34

8.3 Estimating the Population Proportion

Business decision makers and researchers often need to be able to estimate the population proportion. For most businesses, estimating market share (their proportion of the market) is important because many company decisions evolve from market share information. Political pollsters are interested in estimating the proportion of the vote that their candidates will receive. Companies spend thousands of dollars estimating the proportion of produced goods that are defective. Market segmentation opportunities come from a knowledge of the proportion of various demographic characteristics among potential customers or clients.

Methods similar to those in Section 8.1 can be used to estimate the population proportion. The central limit theorem for sample proportions led to the following formula in Chapter 7.

$$Z = \frac{\hat{p} - P}{\sqrt{\dfrac{P \cdot Q}{n}}}$$

where $Q = 1 - P$. Recall that this formula can be applied only when $n \cdot P$ and $n \cdot Q$ are greater than 5.

Algebraically manipulating this formula to estimate P involves solving for P. However, P is in both the numerator and the denominator, which complicates the resulting formula. For this reason—for confidence interval purposes only and for large sample sizes—$\hat{p}$ is substituted for P in the denominator, yielding

$$Z = \frac{\hat{p} - P}{\sqrt{\dfrac{\hat{p}\hat{q}}{n}}}$$

where $\hat{q} = 1 - \hat{p}$. Solving for P results in the confidence interval in Formula 8.5*

(8.5)	$\hat{p} - Z_{\alpha/2}\sqrt{\dfrac{\hat{p}\hat{q}}{n}} \le P \le \hat{p} + Z_{\alpha/2}\sqrt{\dfrac{\hat{p}\hat{q}}{n}}$	CONFIDENCE INTERVAL TO ESTIMATE P

where:
$\hat{p}$ = sample proportion
$\hat{q} = 1 - \hat{p}$
P = population proportion
n = sample size

In this formula, $\hat{p}$ is the point estimate and $\pm Z_{\alpha/2}\sqrt{\dfrac{\hat{p}\hat{q}}{n}}$ is the error of the estimation.

As an example, a study of 87 randomly selected companies with a telemarketing operation revealed that 39% of the sampled companies had used telemarketing to assist them in order processing. Using this information, how could a researcher estimate the *population* proportion of telemarketing companies that use their telemarketing operation to assist them in order processing?

The sample proportion, $\hat{p} = .39$, is the *point estimate* of the population proportion, P. For $n = 87$ and $\hat{p} = .39$, a 95% confidence interval can be computed to determine the interval estimation of P. The Z value for 95% confidence is 1.96. The value of $\hat{q} = 1 - \hat{p} = 1 - .39 = .61$. The confidence interval estimate is

$$.39 - 1.96\sqrt{\frac{(.39)(.61)}{87}} \le P \le .39 + 1.96\sqrt{\frac{(.39)(.61)}{87}}$$

$$.39 - .10 \le P \le .39 + .10$$

$$.29 \le P \le .49$$

$$\text{Prob}\,[.29 \le P \le .49] = .95.$$

This interval suggests a .95 probability that the population proportion of telemarketing firms that use their operation to assist order processing is somewhere between .29 and .49. There is a point estimate of .39 with an error of $\pm.10$. This result has a 95% level of confidence.

---

*Because we are not using the true standard deviation of $\hat{p}$, the correct divisor of the standard error of $\hat{p}$ is $n - 1$. However, for large sample sizes, the effect is negligible. Although technically the minimal sample size for the techniques presented in this section is $n \cdot P$ and $n \cdot Q$ greater than 5, in actual practice sample sizes of several hundred are more commonly used. As an example, for $\hat{p}$ and $\hat{q}$ of .50 and $n = 300$, the standard error of $\hat{p}$ is .02887 using n and .02892 using $n - 1$, a difference of only .00005.

DEMONSTRATION PROBLEM 8.4

Coopers & Lybrand surveyed 210 chief executives of fast-growing small companies. Only 51% of these executives had a management-succession plan in place. A spokesman for Cooper & Lybrand said that many companies do not worry about management succession unless it is an immediate problem. However, the unexpected exit of a corporate leader can disrupt and unfocus a company for long enough to cause it to lose its momentum.

Use the data given to compute a 92% confidence interval to estimate the proportion of *all* fast-growing small companies that have a management-succession plan.

SOLUTION

The point estimate is the sample proportion given to be .51. It is estimated that .51, or 51% of all fast-growing small companies have a management-succession plan. Realizing that the point estimate might change with another sample selection, we calculate a confidence interval.

The value of n is 210; $\hat{p}$ is .51 and $\hat{q} = 1 - \hat{p} = .49$. Because the level of confidence is 92%, the value of $Z_{.04} = 1.75$. The confidence interval is computed as

$$.51 - 1.75\sqrt{\frac{(.51)(.49)}{210}} \le P \le .51 + 1.75\sqrt{\frac{(.51)(.49)}{210}}$$

$$.51 - .06 \le P \le .51 + .06$$

$$.45 \le P \le .57$$

$$\text{Prob}[.45 \le P \le .57] = .92.$$

It is estimated with 92% confidence that the proportion of the population of fast-growing small companies that have a management-succession plan is between .45 and .57.

DEMONSTRATION PROBLEM 8.5

A clothing company produces men's jeans. The jeans are made and sold with either a regular cut or a boot cut. In an effort to estimate the proportion of their men's jeans market in Oklahoma City that is for boot-cut jeans, the analyst takes a random sample of 212 jeans sales from the company's two Oklahoma City retail outlets. Only 34 of the sales were for boot-cut jeans. Construct a 90% confidence interval to estimate the proportion of the population in Oklahoma City who prefer boot-cut jeans.

SOLUTION

The sample size is 212, and the number preferring boot-cut jeans is 34. The sample proportion is $\hat{p} = 34/212 = .16$. A point estimate for boot-cut jeans in the population is .16, or 16%. The Z value for a 90% level of confidence is 1.645, and the value of $\hat{q} = 1 - \hat{p} = 1 - .16 = .84$. The confidence interval estimate is

$$.16 - 1.645\sqrt{\frac{(.16)(.84)}{212}} \le P \le .16 + 1.645\sqrt{\frac{(.16)(.84)}{212}}$$

$$.16 - .04 \le P \le .16 + .04$$

$$.12 \le P \le .20$$

$$\text{Prob}[.12 \le P \le .20] = .90.$$

The analyst estimates with a probability of .90 that the population proportion of boot-cut jeans purchases is between .12 and .20. The level of confidence in this result is 90%.

Analysis Using Excel

Excel itself does not have a feature for computing a confidence interval for a population proportion. However, a confidence interval to estimate a population proportion can be constructed using the **FAST ⟍ STAT** feature, **1 Proportion C.I. and H.T.,** which is available from FAST ⟍ STAT's pull-down menu. The dialog box for **1 Proportion C.I. and H.T.** is shown in Figure 8.10.

As you may note from the figure, the inputs are grouped into three segments. The first group is for the sample itself. It includes the sample size and the number of items in the sample that have the characteristic of interest. For some situations, the business analyst may know the proportion of the items that have the characteristic instead of the number. For such instances, the number with the characteristic can be found by multiplying the known proportion times the sample size.

The second grouping has just one input. It is the level of confidence for the confidence interval. It is to be given as a percentage such as 99%, 95%, 90%, and so on.

The third grouping includes two inputs for conducting a hypothesis test. These will be discussed in Section 9.4 of Chapter 9. If you wish to compute a confidence interval, enter a number greater than zero and less than one for these two entries.

Figure 8.11 gives the **1 Proportion** results for Demonstration Problem 8.5. At the top of the figure, the values for the five inputs are repeated. The outputs are in three parts: the point estimate, the confidence interval, and the hypothesis test for the sample. The confidence interval results include the Z value for the specified confidence level and the standard error of the proportion. In addition, it includes the error of estimation (sometimes termed the interval half width), the endpoints of the interval, and the complete interval.

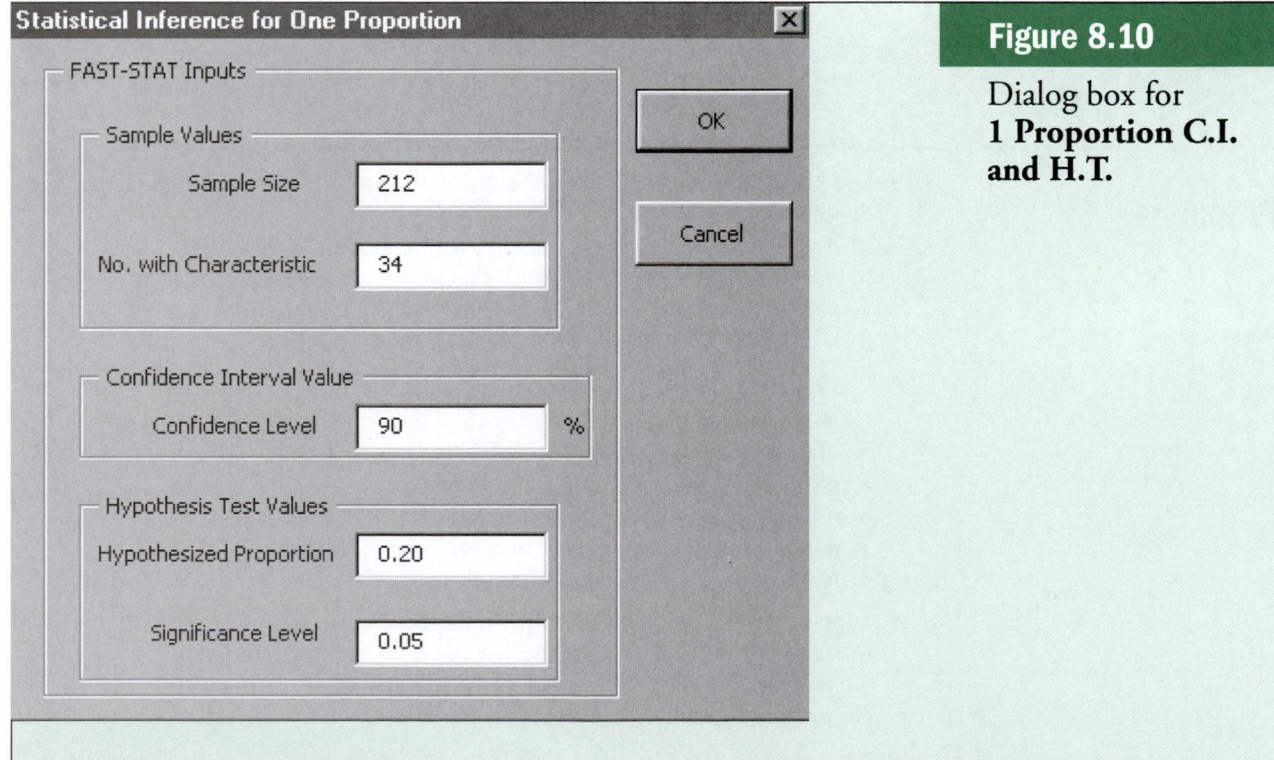

Figure 8.10

Dialog box for **1 Proportion C.I. and H.T.**

Figure 8.11	FAST \ STAT output for Demonstration Problem 8.5

Microsoft Excel - Book1

File Edit View Insert Format Tools Data Window FAST-STAT Help

S34

	A	B	C	D	E	F	G
1	**Statistical Inference for One Proportion**						
2	*INPUTS*						
3				Sample Size		212	
4		Number of Items with Characteristic				34	
5							
6			Level of Confidence			90 %	
7							
8			Hypothesized Proportion			0.2	
9			Level of Significance			0.05	
10	*OUTPUTS*						
11	*Point Estimate*					0.160377	
12							
13	*Confidence Interval*						
14		Z Value for Level of Confidence				1.645	
15		Standard Error of the Proportion				0.0252	
16		Error of Estimation				0.0415	
17		Interval Lower Limit				0.1189	
18		Interval Upper Limit				0.2018	
19		Confidence Interval			0.1189	to	0.2018
20	*Hypothesis Test*						
21		Test Statistic				-1.4423	
22		Critical Z Value for Two-Tailed Test				1.960	
23		Critical Z Value for Upper-Tailed Test				1.645	
24		Critical Z Value for Lower-Tailed Test				-1.645	
25		p-Value for Two-Tailed Test				0.149	
26		p-Value for Upper-Tailed Test				0.925	
27		p-Value for Lower-Tailed Test				0.075	

1 Proportion / Sheet1 / Sheet2 / Sheet3 /

8.3
Problems

8.19 Use the information about each of the following samples to compute the confidence interval to estimate P.

a. $n = 44$ and $\hat{p} = .51$; compute a 99% confidence interval.
b. $n = 300$ and $\hat{p} = .82$; compute a 95% confidence interval.
c. $n = 1,150$ and $\hat{p} = .48$; compute a 90% confidence interval.
d. $n = 95$ and $\hat{p} = .32$; compute an 88% confidence interval.

8.20 Use the following sample information to calculate the confidence interval to estimate the population proportion. Let X be the number of items in the sample having the characteristic of interest.

a. $n = 116$ and $X = 57$, with 99% confidence
b. $n = 800$ and $X = 479$, with 97% confidence
c. $n = 240$ and $X = 106$, with 85% confidence
d. $n = 60$ and $X = 21$, with 90% confidence

8.21 Suppose a random sample of 85 items has been taken from a population and 40 of the items contain the characteristic of interest. Use this information to calculate a 90% confidence interval to estimate the proportion of the population that has the characteristic of interest. Calculate a 95% confidence interval. Calculate a 99%

confidence interval. As the level of confidence changes and the other sample information stays constant, what happens to the confidence interval?

8.22 A study released by Scoop Marketing early in 1999 showed that Universal/PolyGram held a 24.5% share of the music CD market. Suppose this figure is actually a point estimate obtained by interviewing 1003 people who purchased a music CD in January of 1999. Use this information to compute a 99% confidence interval for the proportion of the market that is held by Universal/PolyGram. Suppose the figure was obtained from a survey of 10,000 people. Recompute the confidence interval and compare your results with the first confidence interval. How did they differ? What might you conclude about sample size and confidence intervals?

8.23 According to the Stern Marketing Group, 9 out of 10 professional women say that financial planning is more important today than it was 5 years ago. Where do these women go for help in financial planning? Forty-seven percent use a financial advisor (broker, tax consultant, financial planner). Twenty-eight percent use written sources such as magazines, books, and newspapers. Suppose these figures were obtained by taking a sample of 560 professional women who said that financial planning is more important today than it was 5 years ago. Construct a 95% confidence interval for the proportion who use a financial advisor. Use the percentage given in this problem as the point estimate. Construct a 90% confidence interval for the proportion who use written sources. Use the percentage given in this problem as the point estimate.

8.24 What proportion of pizza restaurants that are primarily for walk-in business have a salad bar? Suppose that, in an effort to determine this figure, a random sample of 1250 of these restaurants across the U.S. based on the Yellow Pages is called. If 997 of the restaurants sampled have a salad bar, what is the 98% confidence interval for the population proportion?

8.25 The highway department wants to estimate the proportion of vehicles on Interstate 25 between the hours of midnight and 5:00 A.M. that are 18-wheel tractor trailers. The estimate will be used to determine highway repair and construction considerations and in highway patrol planning. Suppose researchers for the highway department counted vehicles at different locations on the interstate for several nights during this time period. Of the 3481 vehicles counted, 927 were 18-wheelers.
   a. Determine the point estimate for the proportion of vehicles traveling Interstate 25 during this time period that are 18-wheelers.
   b. Construct a 99% confidence interval for the proportion of vehicles on Interstate 25 during this time period that are 18-wheelers.

8.26 What proportion of commercial airline pilots are more than 40 years of age? Suppose a researcher has access to a list of all pilots who are members of the Commercial Airline Pilots Association. If this list is used as a frame for the study, she can randomly select a sample of pilots, contact them, and ascertain their ages. From 89 of these pilots so selected, she learns that 48 are more than 40 years of age. Construct an 85% confidence interval to estimate the population proportion of commercial airline pilots who are more than 40 years of age.

8.27 According to Runzheimer International, in a survey of relocation administrators 63% of all workers who rejected relocation offers did so for family considerations. Suppose this figure was obtained by using a random sample of the files of 672 workers who had rejected relocation offers. Use this information to construct a 95% confidence interval to estimate the population proportion of workers who reject relocation offers for family considerations.

8.4
Estimating Sample Size

Sample-size estimation
An estimate of the size of sample necessary to fulfill the requirements of a particular level of confidence and to be within a specified amount of error.

In most business research that uses sample statistics to infer about the population, being able to *estimate the size of sample necessary to accomplish the purposes of the study* is important. The need for this **sample-size estimation** is the same for the large corporation investing tens of thousands of dollars in a massive study of consumer preference and for students undertaking a small case study and wanting to send questionnaires to local business people. In either case, such things as level of confidence, sampling error, and width of estimation interval are closely tied to sample size. If the large corporation is undertaking a market study, should it sample 40 people or 4000 people? The question is an important one. In most cases, because of cost considerations, researchers do not want to sample any more units or individuals than necessary.

Sample Size When Estimating μ

In research studies when μ is being estimated, the size of sample can be determined by using the Z formula for sample means to solve for n.

$$Z = \frac{\overline{X} - \mu}{\frac{\sigma}{\sqrt{n}}}$$

Error of estimation
The difference between the statistic computed to estimate a parameter and the parameter.

The difference between $\overline{X}$ and μ is the **error of estimation** resulting from the sampling process. Let $E = (\overline{X} - \mu) =$ the error of estimation. Substituting E into the preceding formula yields

$$Z = \frac{E}{\frac{\sigma}{\sqrt{n}}}.$$

Solving for n produces the sample size.

SAMPLE SIZE WHEN
ESTIMATING μ

$$n = \frac{Z_{\alpha/2}^2 \sigma^2}{E^2} = \left(\frac{Z_{\alpha/2} \sigma}{E} \right)^2 \qquad (8.6)$$

Sometimes in estimating sample size the population variance is known or can be determined from past studies. Other times, the population variance is unknown and must be estimated to determine the sample size. In such cases, it is acceptable to use the following estimate to represent σ.

$$\sigma = \frac{1}{4}(\text{range})$$

This estimate is derived from the empirical rule (Chapter 3) stating that approximately 95% of the values in a normal distribution are within $\pm 2\sigma$ of the mean, giving a range within which most of the values are located.

Using Formula 8.6, the researcher can estimate the sample size needed to achieve the goals of the study before gathering data. For example, suppose a researcher wants to estimate the average monthly expenditure on bread by a family in Chicago. She wants to be 90% confident of her results. How much error is she willing to tolerate in the results? Suppose she wants the estimate to be within $1.00 of the actual figure and the standard deviation of average monthly bread purchases is $4.00. What is the sample size estimation for this problem? The value of Z for a 90% level of confidence is 1.645. Using Formula 8.6 with $E = \$1.00$, $\sigma = \$4.00$, and $Z = 1.645$ gives

$$n = \frac{Z_{\alpha/2}^2 \sigma^2}{E^2} = \frac{(1.645)^2(4)^2}{1^2} = 43.30.$$

That is, at least $n = 43.3$ must be sampled randomly to attain a 90% level of confidence and produce an error within $1.00 for a standard deviation of $4.00. Sampling 43.3 units is impossible, so this result should be rounded up to $n = 44$ units.

In this approach to estimating sample size, we view the error of the estimation as the amount of difference between the statistic (in this case, $\bar{X}$) and the parameter (in this case, μ). The error could be in either direction, that is, the statistic could be over or under the parameter. Thus, the error, E, is actually $\pm E$ as we view it. So when a problem states that the researcher wants to be within $1.00 of the actual monthly family expenditure for bread, it means that the researcher is willing to allow a tolerance within $\pm\$1.00$ of the actual figure. Another name for this error is the **bounds** of the interval. Some business analysts prefer to view the error of the confidence interval in terms of the total distance across the interval (width of the interval). In such a case, the value of E is equal to one-half this distance.

Bounds
The error portion of the confidence interval that is added and/or subtracted from the point estimate to form the confidence interval.

DEMONSTRATION PROBLEM 8.6

Suppose you want to estimate the average age of all Boeing 727 airplanes now in active domestic U.S. service. You want to be 95% confident, and you want your estimate to be within 2 years of the actual figure. The 727 was first placed in service about 30 years ago, but you believe that no active 727s in the U.S. domestic fleet are more than 25 years old. How large a sample should you take?

SOLUTION
Here, $E = 2$ years, the Z value for 95% is 1.96, and σ is unknown, so it must be estimated by using $\sigma \approx (1/4) \cdot (\text{range})$. As the range of ages is 0 to 25 years, $\sigma = (1/4)(25) = 6.25$. Use Formula 8.6.

$$n = \frac{Z^2 \sigma^2}{E^2} = \frac{(1.96)^2(6.25)^2}{2^2} = 37.52$$

Because you cannot sample 37.52 units, the required sample size is 38. If you randomly sample 38 units, you have an opportunity to estimate the average age of active 727s within 2 years and be 95% confident of the results. If you want to be within 1 year for the estimate ($E = 1$), the sample-size estimate changes to

$$n = \frac{Z^2 \sigma^2}{E^2} = \frac{(1.96)^2(6.25)^2}{1^2} = 150.1.$$

Note that cutting the error by a factor of ½ increases the required sample size by a factor of 4. The reason is the squaring factor in Formula 8.6. If you want to reduce the error to one-half of what you used before, you must be willing to incur the cost of a sample that is four times larger, for the same level of confidence.

Note: Sample-size estimates for the population mean with small samples where σ is unknown using the t distribution are not shown here. Because a sample size must be known to determine the table value of t, which in turn is used to estimate the sample size, this procedure usually involves an iterative process.

Determining Sample Size When Estimating P

Determining the sample size required to estimate the population proportion, P, also is possible. The process begins with the Z formula for sample proportions.

$$Z = \frac{\hat{p} - P}{\sqrt{\dfrac{P \cdot Q}{n}}}$$

where $Q = 1 - P$.

As various samples are taken from the population, $\hat{p}$ will rarely equal the population proportion, P, resulting in an error of estimation. That is, the difference between $\hat{p}$ and P is the error of estimation, so $E = \hat{p} - P$.

$$Z = \frac{E}{\sqrt{\dfrac{P \cdot Q}{n}}}$$

Solving for n yields the sample size.

SAMPLE SIZE WHEN ESTIMATING P	$n = \dfrac{Z^2 PQ}{E^2}$	(8.7)

where:
  P = population proportion
  $Q = 1 - P$
  E = error of estimation
  n = sample size

TABLE 8.3
PQ for Various Selected Values of *P*

P	PQ
.5	.25
.4	.24
.3	.21
.2	.16
.1	.09

How can the value of n be determined prior to a study if the formula requires the value of P and the study is being done to estimate P? Although the actual value of P is not known prior to the study, similar studies might have generated a good approximation for P. If no previous value is available for use in estimating P, some possible P values, as shown in Table 8.3, might be considered.

Note that, as PQ is in the numerator of the sample-size formula, $P = .5$ will result in the largest sample sizes. Often *if P is unknown, business analysts use .5 as an estimate of P in* Formula 8.7. This selection results in the largest sample size that could be determined from Formula 8.7 for a given Z value and a given error value.

Hewitt Associates conducted a national survey to determine the extent to which employers are promoting health and fitness among their employees. One of the questions asked was, Does your company offer on-site exercise classes? Suppose it was estimated before the study that no more than 40% of the companies would answer yes. How large a sample would Hewitt Associates have to take in estimating the population proportion to ensure a 98% confidence in the results and to be within .03 of the true population proportion?

SOLUTION

The value of E for this problem is .03. Because it is estimated that no more than 40% of the companies would say yes, $P = .40$ can be used. A 98% confidence interval results in a Z value of 2.33. Inserting these values into Formula 8.9 yields

$$n = \frac{(2.33)^2(.40)(.60)}{(.03)^2} = 1447.7.$$

Hewitt Associates would have to sample 1448 companies to be 98% confident in the results and maintain an error of .03.

The Packer, a produce industry trade publication, reports on American produce-eating habits. One result of a survey published in the journal showed that about two-thirds of all Americans tried a new type of produce in the past 12 months. Suppose a produce industry organization wants to survey Americans and ask whether they are eating more fresh fruit and vegetables than they did 1 year ago. The organization wants to be 90% confident in its results and maintain an error within .05. How large a sample should it take?

SOLUTION

The value of E is .05. Because no approximate figure has been given as to what proportion of people might answer yes to the question of eating more fresh fruit and vegetables, we shall use $P = .50$. The Z value for a 90% confidence interval is ± 1.645. Solving for n gives

$$n = \frac{Z^2 PQ}{E^2} = \frac{(1.645)^2(.50)(.50)}{(.05)^2} = 270.6.$$

The organization should sample at least 271 consumers to achieve 90% confidence and have an error within .05.

Analysis Using Excel

Excel can determine the sample size needed to estimate a population mean through the use of a statistical and a mathematical function. This process has been implemented in **FAST ⫽ STAT** in a macro procedure entitled **Sample Size for 1 Mean.** The procedure is selected from FAST ⫽ STAT's pull-down menu. The resulting dialog box is as shown in Figure 8.12.

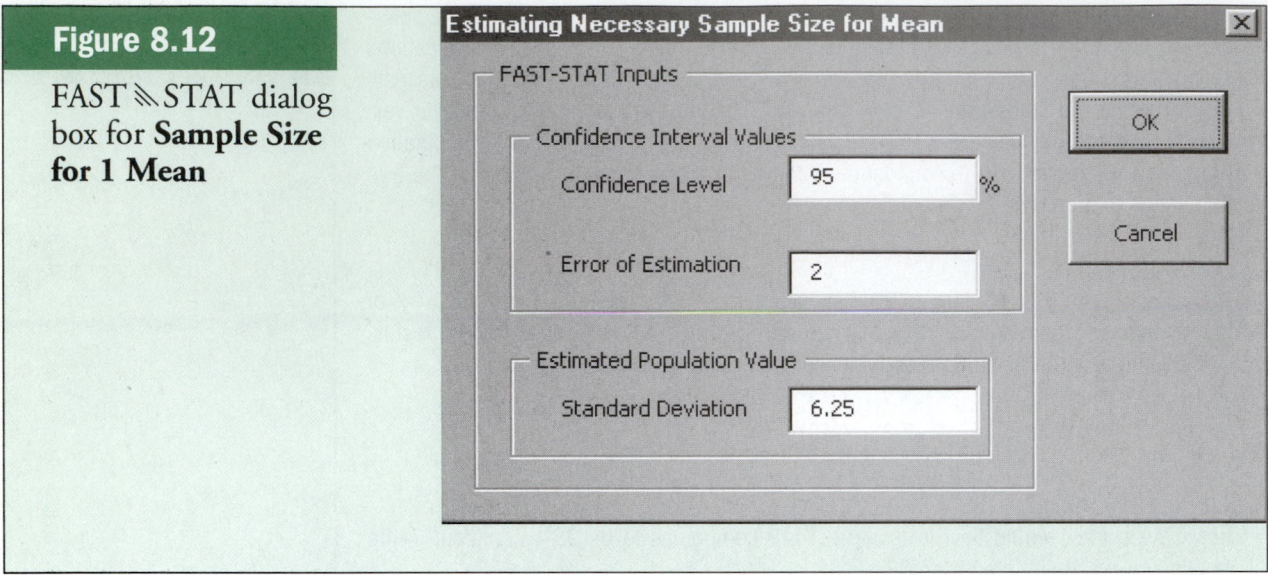

Figure 8.12

FAST ⟍ STAT dialog box for **Sample Size for 1 Mean**

The **Sample Size for 1 Mean** procedure requires three inputs. The first two specify the desired confidence interval through the value for the confidence level and the value for the error of estimation or bounds for the interval (E). The remaining input is an estimation of the standard deviation for the population. This estimate could come from a prior study or sample or could be set equal to ¼ the range as previously discussed.

Figure 8.13 presents the results of this procedure for Demonstration Problem 8.6. The results include a summary of the input values together with the minimal recommended sample size. The procedure always rounds the sample size up to the next higher integer.

To use **FAST ⟍ STAT** to determine sample size when estimating a population proportion, select **Sample Size—1 Proportion** from the FAST ⟍ STAT pull-down menu. The dialog box for this procedure, which is quite similar to that for the mean, is displayed in Figure 8.14.

The **Sample Size—1 Proportion** procedure requires three inputs. The first two specify the desired confidence interval with the values for the confidence level and the error of estimation or bounds for the interval (E). The remaining input is an estimation of the population proportion. This estimate could come from a prior study or sample. If no such estimate is available, a value of 0.50 is oftentimes used for the estimate as previously mentioned.

Figure 8.15 presents the results of this procedure for Demonstration Problem 8.7. As you will note, the results include a summary of the input values together with the minimal recommended sample size. The procedure always rounds the sample size up to the next higher integer.

You may have noted that the sample size computed by Excel in Figure 8.15 differs slightly from the hand-calculated value given for Demonstration Problem 8.7. The difference results from the value used for Z. When computing by hand we use the value 2.33 from the normal table. When computing with Excel we use the statistical function **NORMSINV**. The resulting computer-generated Z value is not limited to two decimal places, but has 14 places to the right of the decimal point. Specifically, the value is 2.32634192798286. Obviously not all these digits are meaningful. However, it does differ slightly from 2.33. As a result, when computing necessary sample sizes for proportions that usually are in the neighborhood of 1000 or more, a small difference in the Z value results in a small, but noticeable, difference in values for n. If you had to choose between the answers, most persons would elect the computer-generated value.

Figure 8.13   FAST⧫STAT output for Demonstration Problem 8.6

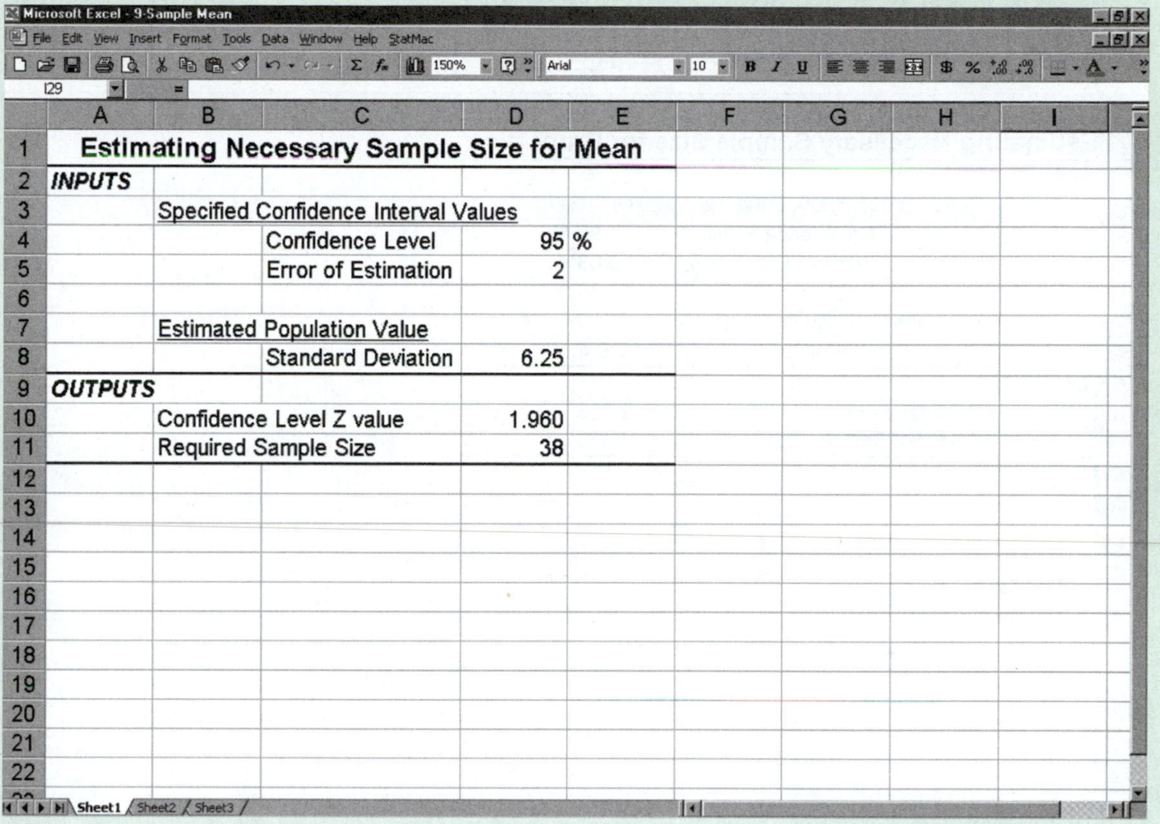

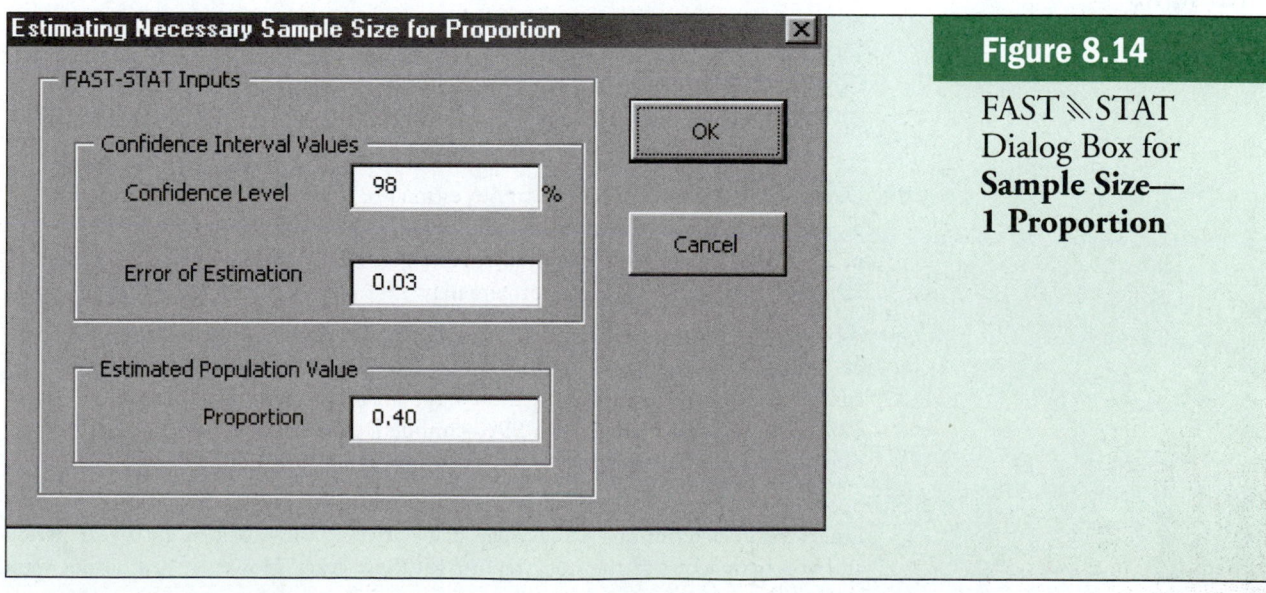

Figure 8.14

FAST⧫STAT Dialog Box for **Sample Size— 1 Proportion**

Figure 8.15	FAST ⟍ STAT output for Demonstration Problem 8.7

	A	B	C	D	E	F	G	H	I
1	**Estimating Necessary Sample Size for Proportion**								
2	*INPUTS*								
3			Specified Confidence Interval Values						
4			Confidence Level	98 %					
5			Error of Estimation	0.03					
6									
7			Estimated Population Value						
8			Proportion	0.4					
9	*OUTPUTS*								
10			Confidence Level Z value	2.326					
11			Required Sample Size	1444					

8.4

Problems

8.28 Determine the sample size necessary to estimate μ for the following information.
  a. $\sigma = 36$ and $E = 5$ at 95% confidence
  b. $\sigma = 4.13$ and $E = 1$ at 99% confidence
  c. Values range from 80 to 500, error is to be within 10, and the confidence level is 90%
  d. Values range from 50 to 108, error is to be within 3, and the confidence level is 88%

8.29 Determine the sample size necessary to estimate P for the following information.
  a. $E = .02$, P is approximately .40, and confidence level is 96%
  b. E is to be within .04, P is unknown, and confidence level is 95%
  c. E is to be within 5%, P is approximately 55%, and confidence level is 90%

8.30 A bank officer wants to determine the amount of the average total monthly deposits per customer at the bank. He believes an estimate of this average amount using a confidence interval is sufficient. How large a sample should he take to be within $200 of the actual average with 99% confidence? He assumes the standard deviation of total monthly deposits for all customers is about $1000.

8.31 Suppose you have been following a particular airline stock for many years. You are interested in determining the average daily price of this stock in a 10-year period and you have access to the stock reports for these years. However, you do not want to average all the daily prices over 10 years because of the more than 2500 data points, so you decide to take a random sample of the daily prices and estimate the average. You want to be 90% confident of your results, you want the estimate to be within $2.00 of the true average, and you believe the standard deviation of the price of this stock is about $12.50 over this period of time. How large a sample should you take?

8.32 A group of investors wants to develop a chain of fast-food restaurants. In determining potential costs for each facility, they must consider, among other expenses, the average monthly electric bill. They decide to sample some fast-food restaurants currently operating to estimate the monthly cost of electricity. They want to be 90% confident of their results and want the error of the interval estimate to be no more than $100. They estimate that such bills range from $600 to $2500. How large a sample should they take?

8.33 Suppose a production facility purchases a particular component part in large lots from a supplier. The production manager wants to estimate the proportion of defective parts received from this supplier. She believes the proportion defective is no more than .20 and wants to be within .02 of the true proportion of defective parts with a 90% level of confidence. How large a sample should she take?

8.34 What proportion of secretaries of *Fortune 500* companies has a personal computer at his or her work station? You want to answer this question by conducting a random survey. How large a sample should you take if you want to be 95% confident of the results and you want the error of the confidence interval to be no more than .05? Assume no one has any idea of what the proportion actually is.

8.35 What proportion of shoppers at a large appliance store actually make a large-ticket purchase? To estimate this proportion within 10% and be 95% confident of the results, how large a sample should you take? Assume that no more than 50% of all shoppers actually make a large-ticket purchase.

Summary

Techniques for estimating population parameters from sample statistics are important tools for business research. These tools include techniques for estimating population means by using both large- and small-sample statistics, techniques for estimating the population proportion, and methodology for determining how large a sample to take is presented.

At times in business research a product is new and/or untested or information about the population is unknown. In such cases, gathering data from a sample and making estimates about the population is useful and can be done with a point estimate or an interval estimate. A point estimate is the use of a statistic from the sample as an estimate for a parameter of the population. The problem with point estimates is that they are likely to change with every sample taken. An interval estimate is a range of values computed from the sample within which the researcher believes with some confidence that the population parameter lies. Certain levels of confidence seem to be used more than others: 90%, 95%, 98%, and 99%.

In estimating the population mean when sample size is large, it is permissible to use the sample standard deviation as an approximation for the population standard deviation when the population standard deviation is unknown.

However, if sample size is small and the population standard deviation is unknown, the t distribution should be used instead of the Z distribution. It is assumed when using the t distribution that the population from which the samples are drawn is normally distributed. However, the technique for estimating a population mean by using the t test is robust, which means it is relatively insensitive to minor violations to the assumption.

The formulas in Chapter 7 resulting from the central limit theorem can be manipulated to produce formulas for estimating sample size for large samples. Determining the sample size necessary to estimate a population mean, if the population standard deviation is unavailable, can be based on one-fourth the range as an estimate of the population standard deviation. Estimating the population proportion calls for the value of the population proportion. The population proportion is the parameter to be estimated, so it is unknown. A value of the population proportion from a similar study can be used in sample-size estimation. If none is available, using a value of .50 will result in the largest sample-size estimation for the problem if other variables are held constant. Sample-size estimation is used mostly to provide a ballpark figure to give researchers some guidance. Costs are associated with research, and larger sample sizes usually result in greater costs.

Key Terms

bounds
degrees of freedom (df)
error of estimation
interval estimate
point estimate

robust
sample-size estimation
t distribution
t value

SUPPLEMENTARY PROBLEMS

8.36 In planning both market opportunity and production levels, being able to estimate the size of a market can be important. Suppose a diaper manufacturer wants to know how many diapers a 1-month-old baby uses during a 24-hour period. To determine this, the manufacturer's analyst randomly selects 17 parents of 1-month-olds and asks them to keep track of diaper usage for 24 hours. The results are shown. Construct a 99% confidence interval to estimate the average daily diaper usage of a 1-month-old baby. Assume diaper usage is normally distributed.

12	8	11	9	13	14	10
10	9	13	11	8	11	15
10	7	12				

8.37 Suppose you want to estimate the proportion of cars that are sport utility vehicles (SUVs) being driven in Kansas City, Missouri, at rush hour by standing on the corner of I-70 and I-470 and counting SUVs. You believe the figure is no higher than .40. If you want the error of the confidence interval to be no greater than .03, how many cars should you randomly sample? Use a 90% level of confidence.

8.38 What is the average length of a company's policy book? Suppose policy books are sampled from 45 medium-size companies. The average number of pages in the sample books is 213, with a sample standard deviation of 48. Use this information to construct a 98% confidence interval to estimate the mean number of pages for the population of medium-size company policy books.

8.39 A random sample of small-business managers was given a leadership style questionnaire. The results were scaled so that each manager received a score for initiative. Suppose the following data are a random sample of these scores.

37	42	40	39	38	31	40
37	35	45	30	33	35	44
36	37	39	33	39	40	41
33	35	36	41	33	37	38
40	42	44	35	36	33	38
32	30	37	42			

Use these data to construct a 90% confidence interval to estimate the average score on initiative for all small-business managers.

8.40 A national beauty salon chain wants to estimate the number of times per year a woman has her hair done at a beauty salon if she uses one at least once a year. The chain's researcher estimates that, of those women who use a beauty salon at least once a year, the standard deviation of number of times of usage is approximately 6. The national chain wants the estimate to be within one time of the actual mean value. How large a sample should the researcher take to obtain a 98% confidence level?

8.41 Is the environment a major issue with Americans? To answer that question, a researcher conducts a survey of 1255 randomly selected Americans. Suppose 714 of the sampled people replied that the environment is a major issue with them. Construct a 95% confidence interval to estimate the proportion of Americans who feel that the environment is a major issue with them. What is the point estimate of this proportion?

8.42 According to a survey by Topaz Enterprises, a travel auditing company, the average error by travel agents is $128. Suppose this figure was obtained from a random sample of 25 travel agents and the sample standard deviation is $21. What is the point estimate of the national average error for all travel agents? Compute a 98% confidence interval for the national average error based on these sample results. Assume the travel agent errors are normally distributed in the population. How wide is the interval? Interpret the interval.

8.43 A national survey on telemarketing was undertaken. One of the questions asked was: How long has your organization had a telemarketing operation? Suppose the following data represent some of the answers received to this question. Suppose further that only 300 telemarketing firms comprised the population when this survey was taken. Use the following data to compute a 98% confidence interval to estimate the average number of years a telemarketing organization has had a telemarketing operation.

5	5	6	3	6	7	5
5	6	8	4	9	6	4
10	5	10	11	5	14	7
5	9	6	7	3	4	3
7	5	9	3	6	8	16
12	11	5	4	3	6	5
8	3	5	9	7	13	4
6	5	8	3	5	8	7
11	5	14	4			

8.44 An entrepreneur wants to open an appliance service repair shop. She would like to know about what the average home repair bill is, including the charge for the service call for appliance repair in the area. She wants the estimate to be within $20 of the actual figure. She believes the range of such bills is between $30 and $600. How large a sample should the entrepreneur take if she wants to be 95% confident of the results?

8.45 A national survey of insurance offices was taken, resulting in a random sample of 245 companies. Of these 245 companies, 189 responded that they were going to purchase new software for their offices in the next year. Construct a 90% confidence interval to estimate the population proportion of insurance offices that intend to purchase new software during the next year.

8.46 A national survey of companies included a question that asked whether the company had at least one bilingual telephone operator. The sample results of 90 companies follow (y denotes that the company does have at least one bilingual operator; n denotes that it does not).

n	n	n	n	y	n	y	n	n
y	n	n	n	y	y	n	n	n
n	n	y	n	y	n	y	n	y
y	y	n	y	n	n	n	y	n
n	y	n	n	n	n	n	n	n
y	n	y	y	n	n	y	n	y
n	n	y	y	n	n	n	n	n
y	n	n	n	n	y	n	n	n
y	y	y	n	n	y	n	n	n
n	n	n	y	y	n	n	y	n

Use this information to estimate with 95% confidence the proportion of the population that does have at least one bilingual operator.

8.47 A movie theater has had a poor accounting system. The manager has no idea how many large containers of popcorn are sold per movie showing. She knows that the amounts vary by day of the week and hour of the day. However, she wants to estimate the overall average per movie showing. To do so, she randomly selects 12 movie performances and counts the number of large containers of popcorn sold between 1/2 hour before the movie showing and 15 minutes after the movie showing. The sample average was 43.7 containers, with a variance of 228. Construct a 95% confidence interval to estimate the mean number of large containers of popcorn sold during a movie showing. Assume the number of large containers of popcorn sold per movie is normally distributed in the population.

8.48 According to a survey by Runzheimer International, the average cost of a fast-food meal (quarter-pound cheeseburger, large fries, medium soft drink, excluding taxes) in Seattle is $4.82. Suppose this was based on a sample of 27 different establishments and the standard deviation was $0.37. Construct a 95% confidence interval for the population mean cost for all fast-food meals in Seattle. Assume the costs of a fast-food meal in Seattle are normally distributed. Using the interval as a guide, is it likely that the population mean is really $4.50? Why or why not?

8.49 A survey of 77 commercial airline flights of under 2 hours resulted in a sample average late time for a flight of 2.48 minutes. The sample standard deviation was 12 minutes. Construct a 95% confidence interval for the average time that a commercial flight of under 2 hours is late. What is the point estimate? What does the interval tell about whether the average flight is late?

8.50 A regional survey of 560 companies asked the vice president of operations how satisfied he or she was with the software support being received from the computer staff of the company. Suppose 33% of the 560 vice presidents said they were satisfied. Construct a 99% confidence interval for the proportion of the population of vice presidents who would have said they were satisfied with the software support if a census had been taken.

8.51 A research firm has been asked to determine the proportion of all restaurants in the state of Ohio that serve alcoholic beverages. The firm wants to be 98% confident of its results but has no idea of what the actual proportion is. The firm would like to report an error of no more than .05. How large a sample should it take?

8.52 A national magazine marketing firm attempts to win subscribers with a mail campaign that involves a contest using magazine stickers. Often when people subscribe to magazines in this manner they sign up for multiple magazine subscriptions. Suppose the marketing firm wants to estimate the average number of subscriptions per customer of those who purchase at least one subscription. To do so, the marketing firm's researcher randomly selects 65 returned contest entries. Twenty-seven contain subscription requests. Of the 27, the average number of subscriptions is 2.10, with a standard deviation of

.86. The researcher uses this information to compute a 98% confidence interval to estimate μ and assumes that X is normally distributed. What does he find?

8.53 Suppose a national survey of 23 retail outlets of Hillshire Farm Deli Select cold cuts was taken to estimate the price per pound. If the data below represent these prices, what is a 90% confidence interval for the population mean of these prices? Assume prices are normally distributed in the population.

5.18	5.22	5.25	5.19	5.30
5.17	5.15	5.28	5.20	5.14
5.05	5.19	5.26	5.23	5.19
5.22	5.08	5.21	5.24	5.33
5.22	5.19	5.19		

8.54 The price of a head of iceberg lettuce varies greatly with the season and the geographic location of a store. During February a researcher contacts a random sample of 39 grocery stores across the United States and asks the produce manager of each to state the current price charged for a head of iceberg lettuce. Using the researcher's results that follow, construct a 99% confidence interval to estimate the mean price of a head of iceberg lettuce in February in the United States.

$1.59	$1.25	$1.65	$1.40	$0.89
1.19	1.50	1.49	1.30	1.39
1.29	1.60	0.99	1.29	1.19
1.20	1.50	1.49	1.29	1.35
1.10	0.89	1.10	1.39	1.39
1.50	1.50	1.55	1.20	1.15
0.99	1.00	1.30	1.25	1.10
1.00	1.55	1.29	1.39	

ANALYZING THE DATABASES

1. Construct a 95% confidence interval for the population mean number of production workers using the manufacturing database as a sample. What is the point estimate? How much is the error of the estimation? Comment on the results.

2. Construct a 90% confidence interval to estimate the average census for hospitals using the hospital database. State the point estimate and the error of the estimation. Change the level of confidence to 99%. What happened to the interval? Did the point estimate change?

3. The financial database contains financial data on 100 companies. Use this database as a sample and estimate

the earnings per share for all corporations from these data. Select several levels of confidence and compare the results.

4. Using the tally or frequency feature of the computer software, determine the sample proportion of the hospital database under the variable "service" that are "general medical" (category 1). From this statistic, construct a 95% confidence interval to estimate the population proportion of hospitals that are "general medical." What is the point estimate? How much error is there in the interval?

THERMATRIX

In 1985, a company called In-Process Technology was set up to produce and sell a thermal oxidation process that could be used to reduce industrial pollution. The initial investors had acquired the rights to technology that had been developed at a federal government laboratory. However, for years the company performed dismally and by 1991 was still only earning $264,000 annually.

In 1992, current CEO John Schofield was hired to turn things around. Under his tutelage, the company was reorganized and renamed Thermatrix. Schofield realized the potential of the technology in the environmental marketplace. He was able to raise more than $20 million in private equity offerings over several years to produce, market, and distribute the product. In June of 1996, there was a successful public offering of Thermatrix in the financial markets.

Thermatrix's philosophy was to give customers more than competitors gave without charging more. The company targeted large corporations as customers, hoping to use its client list as a selling tool. In addition, realizing that they were a small, thinly capitalized company, Thermatrix partnered with many of its clients in developing solutions to the clients' specific environment problems.

Eventually, Schofield located the Thermatrix operations group in Knoxville, Tennessee because of the low cost of living and the large pool of highly trained professional workers. Thermatrix was able to attract good employees using stock options and other competitive compensation. There are presently 60 employees in the company, and annual sales have risen to around $15 million.

Thermatrix also has become a player in the international marketplace. In 1997, 35% of company revenue came from overseas business. By 1998, it was expected that over 60% of its revenues would be derived from overseas customers. One of the main keys to the company's success has been its customer satisfaction.

DISCUSSION

1. Thermatrix has grown and flourished because of its good customer relationships, which include partnering, delivering a quality product on time, and listening to the customer's needs. Suppose company management wants to formally measure customer satisfaction at least once a year and develops a brief survey that includes the following four questions. Suppose 115 customers participated in this survey with the results shown. Use techniques presented in this chapter to analyze the data to estimate population responses to these questions.

Question	Yes	No
1. In general, were deliveries on time?	63	52
2. Were the contact people at Thermatrix helpful and courteous?	86	29
3. Was the pricing structure fair to your company?	101	14
4. Would you recommend Thermatrix to other companies?	105	10

2. Now suppose Thermatrix officers want to ascertain employee satisfaction with the company. To do this, they randomly sample nine employees and ask them to complete a satisfaction survey under the supervision of an independent testing organization. As part of this survey, employees are asked to respond to questions on a 5-point scale where 1 is low satisfaction and 5 is high satisfaction. Assume the data are at least interval and that the overall responses on questions are normally distributed. The questions and the results of the survey are shown here. Analyze the results by using techniques from this chapter.

QUESTION	MEAN	STANDARD DEVIATION
1. Are you treated fairly as an employee?	3.79	.86
2. Has the company given you the training you need to do the job adequately?	2.74	1.27
3. Does management seriously consider your input in making decisions about production?	4.18	.63
4. Is your physical work environment acceptable?	3.34	.81
5. Is the compensation for your work adequate and fair?	3.95	.21

ADAPTED FROM: "Thermatrix: Selling Products, Not Technology," *Insights and Inspiration: How Businesses Succeed*. Published by *Nation's Business* on behalf of MassMutual—The Blue Chip Company and the U.S. Chamber of Commerce in association with The Blue Chip Enterprise Initiative, 1997; and Thermatrix, Inc. website accessed by http://www.thermatrix.com under the title "Company Background."

9

Hypothesis Testing with Single Samples

Learning Objectives

The main objective of Chapter 9 is to help you to learn how to test hypotheses on single populations, thereby enabling you to:

1. Understand the logic of hypothesis testing and know how to establish null and alternative hypotheses.

2. Understand Type I and Type II errors.

3. Use large samples to test hypotheses about a single population mean and about a single population proportion.

4. Test hypotheses about a single population mean using small samples when σ is unknown and the population is normally distributed.

The concept of hypothesis testing lies at the very heart of inferential statistics, and the use of statistics to "prove" or "disprove" claims hinges on it. Applications of statistical hypothesis testing run the gamut from determining whether a production line process is out of control to providing conclusive evidence that a new medicine is significantly more effective than the old. The process of hypothesis testing is used in the legal system to provide evidence in civil suits. Hypotheses are tested in virtually all areas of life, including education, psychology, marketing, science, law, and medicine. How does the hypothesis testing process begin?

9.1
Introduction to Hypothesis Testing

Hypothesis testing
A process of testing hypotheses about parameters by setting up null and alternative hypotheses, gathering sample data, computing statistics from the samples, and using statistical techniques to reach conclusions about the hypotheses.

One of the foremost statistical mechanisms for decision making is the *hypothesis test*. With **hypothesis testing,** business analysts are able to structure problems in such a way that they can use statistical evidence to test various theories about business phenomena. For example, a U.S. Bureau of Labor Statistics report published in March 1994 stated that the average number of vacation days awarded to manufacturing workers in Germany was 30. Suppose international business analysts hypothesize that the figure is not the same this year. How do they go about testing this hypothesis? If they have the necessary resources, they might choose to interview every manufacturing worker in Germany at the end of the year and compute the national mean number of vacation days from the census. However, the business analysts are likely to take a random sample of German workers, gather sample data, and attempt to reach some conclusion about the population from the sample data. They would probably use a hypothesis-testing approach. Hypothesis testing is a process that consists of several steps.

Steps in Hypothesis Testing

Most business analysts take the following steps when testing hypotheses.

1. Establish the hypotheses; state the null and alternative hypotheses.
2. Determine the appropriate statistical test and sampling distribution.
3. Specify the Type I error rate.
4. State the decision rule.
5. Gather sample data.
6. Calculate the value of the test statistic.
7. State the statistical conclusion.
8. Make a managerial decision.

After stating the null and alternative hypotheses, the business analyst selects the appropriate statistical test and sampling distribution. This selection involves matching the level of data collected and the type of statistic being analyzed with the statistical tests available. In this chapter and Chapter 10, various statistical tests and their associated sampling distribution are presented, along with their applicability to specific situations. Particular statistical tests have certain assumptions that must be met for the tests to be valid.

The next step is for the business analyst to specify the Type I error rate, α. Alpha, sometimes referred to as the amount of risk, is the probability for committing a Type I error and is discussed later. The fourth step is to state a decision rule. In conjunction with alpha and the type of statistical test and sampling distribution being used, a critical value is established. The critical value is obtained from tables and is used as a standard against which gathered data are compared to reach a statistical decision about whether to reject the null hypothesis or not. *The business analyst should not gather data before taking the first four hypothesis-testing steps.* Too often, the business analyst gathers the data first and then tries to determine what to do with them.

In gathering data, the business analyst is cautioned to recall the proper techniques of random sampling (presented in Chapter 7). Care should be taken in establishing a frame, determining the sampling technique, and constructing the measurement device. A strong effort should be made to avoid all nonsampling errors. After the data are gathered, the value of the test statistic can be calculated by using both the data from the study and the hypothesized value of the parameter(s) being studied (mean, proportion, etc.). Using the previously established decision rule and the value of the test statistic, the business analyst can draw a statistical conclusion. In *all* hypothesis tests, the business analyst needs to conclude whether the null hypothesis is rejected or cannot be rejected. From this information, a managerial decision about the phenomenon being studied can be reached. For example, if the hypothesis-testing procedure results in a conclusion that train passengers are significantly older today than they were in the past, the manager may decide to cater to these older customers or to draw up a strategy to make ridership more appealing to younger people.

Null and Alternative Hypotheses

The first step in testing a hypothesis is to establish a **null hypothesis** and an **alternative hypothesis.** The null hypothesis is represented by H_0 and the alternative hypothesis by H_a. Establishing null and alternative hypotheses can be a frustrating and confusing process.

Null hypothesis
The hypothesis that assumes the status quo—that the old theory, method, or standard is still true, the complement of the alternative hypothesis.

Alternative hypothesis
The hypothesis that complements the null hypothesis; usually it is the hypothesis that the business analyst is interested in proving.

The null and alternative hypotheses are set up in opposition to each other. The alternative hypothesis often contains the research question and the null hypothesis can be seen as the negation of the alternative hypothesis. The hypothesis-testing process is structured so that either the null hypothesis is true or the alternative hypothesis is true, but not both. The null hypothesis initially is assumed to be true. The data are gathered and examined to determine whether the evidence is strong enough away from the null hypothesis to reject it. When the business analyst is testing an industry standard or a widely accepted value, the standard or accepted value is assumed true in the null hypothesis. *Null* in this sense means that nothing is new, or there is no new value or standard. The burden is then placed on the business analyst to demonstrate through gathered data that the null hypothesis is false. This task is analogous to the courtroom ideal of innocent until proven guilty. In the courtroom the accused is assumed to be innocent before the trial (null is assumed true). Evidence is presented during the trial (data are gathered). If there is enough evidence against innocence, the accused is found guilty (null is rejected). If there is not enough evidence to prove guilt, the prosecutors have failed to prove the accused guilty. However, they have not "proven" the accused innocent. Typically, what the business analyst is interested in "proving" is formulated into the alternative hypothesis, although in some cases it is not.

As an example, suppose a soft-drink company is filling 12-ounce cans with cola. Recognizing that only under perfect conditions would all cans have exactly 12 ounces of cola, the company hopes that the cans are averaging that amount. A quality controller is worried that a machine is out of control and wants to conduct a hypothesis test to help determine whether that is true. The quality controller hopes the machine is functioning properly, but is really interested in determining whether the machine is malfunctioning with the result that cans are over- or underfilled. The null hypothesis is that there is no problem and the mean value is 12 ounces. The alternative hypothesis is that there is some kind of problem and the cans are not averaging 12 ounces of cola. The null and alternative hypothesis for this problem follow.

$$H_0: \mu = 12 \text{ ounces}$$
$$H_a: \mu \neq 12 \text{ ounces}$$

In testing these hypotheses, the null hypothesis is assumed to be true; that is, the assumption is that the average fill of the cans is 12 ounces. The quality controller randomly selects and tests cans. If there is enough evidence (a sample average fill that is too low or too high), the null hypothesis is rejected. A rejection of the null hypothesis results in the acceptance of the alternative hypothesis.

All statistical conclusions reached in the hypothesis-testing process are stated in reference to the null hypothesis. We either *reject the null hypothesis or fail to reject the null hypothesis.* When we fail to reject the null hypothesis, we never say that we "accept the null hypothesis" because we have not proven that the null hypothesis is true. The null hypothesis is assumed to be true at the beginning of the hypothesis-testing procedure. Failure to attain evidence to reject the null hypothesis in favor of the alternative hypothesis does not equate to "proof" that the null hypothesis is true. It merely means that we did not have enough evidence to reject the null hypothesis and thereby failed to reject the null hypothesis.

Recall the scenario of an international business analyst who wants to test the hypothesis that manufacturing workers in Germany no longer receive an average of 30 vacation days per year. The null hypothesis is that the mean is 30 days (as before, nothing new). The alternative hypothesis is that the mean is not still 30 days.

$$H_0: \mu = 30 \text{ days}$$
$$H_a: \mu \neq 30 \text{ days}$$

The assumption is that the mean number of vacation days is 30 days. Data are gathered. If enough evidence shows that the average is less than or more than 30 days, the conclusion is that the null hypothesis is rejected. If there is not enough evidence, the conclusion is that the null hypothesis cannot be rejected. In such a case, we have not proven that the mean number of vacation days is still 30; rather, we have failed to disprove it.

Note that in establishing the null and alternative hypotheses, the equality point *must* be assigned to the null hypothesis. In addition, one of the two hypotheses must *exactly* match the research question being stated in the problem. In both of the preceding examples, the equality is in the null hypothesis. If the research question for the soft-drink company example is whether the process is out of control, the alternative hypothesis contains this question. In the vacation days problem, the alternative hypothesis states the research question that the mean number of vacation days is not 30.

Acceptance and Rejection Regions

After establishing the null and alternative hypotheses, the business analyst can set up decision rules to determine whether the null hypothesis is going to be rejected or not. In the can-filling problem, suppose the business analyst decides to test the null hypothesis by randomly sampling cans and measuring the fills. How much fill would the business analyst have to find in the 12-ounce cans to reject the null hypothesis? Common sense says that expecting all cans to be filled with exactly 12 ounces of fluid is unrealistic. More reasonable is to expect that the cans *average* 12 ounces. In this way the hypotheses are structured around the mean value, not individual fills. Would the null hypothesis be rejected if a sample mean of 11.99 ounces is obtained? In testing a hypothesis, the business analyst should establish a **rejection region** after determining the null and alternative hypotheses. Figure 9.1 shows a normal distribution from the soft-drink can example with the mean in the middle and rejection regions in the tails. In this case, the mean of the distribution is 12 ounces. The rejection regions are established in the tails of the distribution because the only way to reject a null hypothesis of $H_0: \mu = 12$ ounces is to get a result in the region of $\mu \neq 12$ ounces.

Rejection region
If a computed statistic lies in this portion of a distribution, the null hypothesis will be rejected.

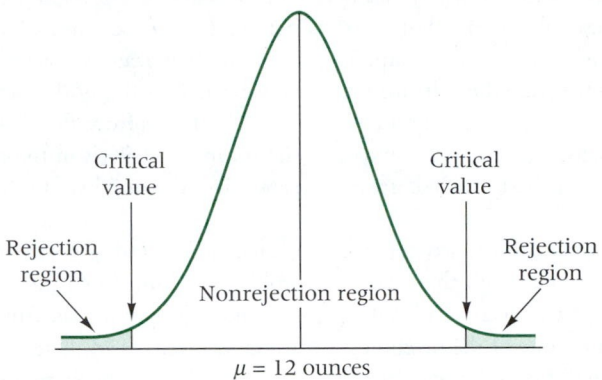

Figure 9.1

Rejection and
nonrejection regions

Each rejection region is divided from the rest of the distribution by a point called the **criti-cal value.** If results obtained from the data yield a computed value in a rejection region beyond the critical value, the null hypothesis is rejected. The *rest of the distribution, which is not in a rejection region, is called the* **nonrejection region.**

Type I and Type II Errors

Occasionally, the sample data gathered in the research process lead to a decision to reject a null hypothesis when actually it is true. This is called **Type I error**—an error committed *when a true null hypothesis is rejected.* In drawing random samples from a population, there is always a possibility of selecting a sample from the fringe of the distribution by chance. In the soft-drink can example, the business analyst could randomly select 50 of the cans with the smallest amounts of fluid or 50 with the largest amounts, causing rejection of the null hypothesis even when the company actually is filling the cans with an average of 12 ounces. In that case, the business analyst would incorrectly conclude that the company is not filling the cans with an average of 12 ounces and commit a Type I error.

An analogy of the Type I error can be found in manufacturing. Suppose a production line worker hears some unusual noises and on that evidence decides to push the red button that shuts down the production line. If the null hypothesis is that there is no problem on the production line, the worker has just rejected that null hypothesis by taking action. Suppose an investigation shows there really is no problem with the production line (perhaps the noise was coming from outside). The worker has just committed a Type I error. In manufacturing, a significant cost usually is involved in shutting down the production line (production ceases, but labor costs and fixed costs are still incurred). As another example, suppose a manager suspects a worker is cheating the company. The manager starts accumulating evidence until he is convinced the worker is guilty. The manager fires the worker for cheating. But suppose the worker was not really cheating the company and the data gathered were merely coincidental occurrences that cast the worker in a bad light. The manager has committed a Type I error. The cost to the company is the loss of a qualified and trained worker along with the potential for a lawsuit. In a court of law, a Type I error is committed when an innocent person is convicted.

Alpha (α) or the **level of significance** is *the probability of committing a Type I error.* Alpha is the proportion of the area of the curve occupied by the rejection region. The most commonly used values of alpha are .001, .01, .05, and .10. Recall that determination of alpha is step 3 in the hypothesis-testing procedure. It is sometimes referred to as the amount of *risk* taken in an experiment. The larger the area of the rejection region, the greater is the risk of committing a Type I error.

Critical value
The value that divides the nonrejection region from the rejection region.

Nonrejection region
Any portion of a distribution that is not in the rejection region. If the observed statistic falls in this region, the decision is to fail to reject the null hypothesis.

Type I error
An error committed by rejecting a true null hypothesis.

alpha (α) or Level of significance
The probability of committing a Type I error.

Type II error
An error committed by failing to reject a false null hypothesis.

A **Type II error** is *committed by failing to reject a false null hypothesis.* In some instances the null hypothesis is not true, but the data gathered yield a computed value that is the nonrejection region. For example, suppose a consumer advocate is testing a null hypothesis of $\mu = 12$ ounces and the soft-drink company actually is filling the cans with an average of 11.85 ounces. She could select a random batch of cans from this distribution of fills and get a sample average of 11.95 ounces. If the mean of 11.95 is in the nonrejection region of the null hypothesis, she incorrectly fails to reject the null hypothesis.

We can view the Type II error in light of the preceding manufacturing example as well. Suppose the production line worker hears unusual noises and is concerned about them but is not convinced enough that there is a problem to shut down the line. He abstains from pushing the red button and fails to reject the null hypothesis. Suppose, however, that a major belt on one of the machines is coming unraveled and there really is a production line problem. The worker has committed a Type II error. Such an error in manufacturing might result in poor-quality products being shipped, which can lead to loss of sales, increased cost of warranty judgments, and/or increased cost of rework or scrap. What about the worker who was suspected of cheating? Suppose the manager gathers evidence but feels it is not sufficient to justify firing the person. If the worker really is cheating the company, the manager has committed a Type II error. The cost to the company is continued cheating by the worker, which can result in low morale and lost revenue. In a court of law, a Type II error is committed when a guilty person is declared not guilty and set free.

beta (β)
The probability of committing a Type II error.

The *probability of committing a Type II error* is represented by **beta (β).** The value of beta varies within an experiment, depending on various alternative values of the parameter (in this case, the mean). Whereas alpha is determined before the experiment, beta is computed by using alpha, the hypothesized parameter, and various theoretical alternatives to the null hypothesis.

There is an inverse relationship between α and β. That is, for a given sample size, β increases as the business analyst decreases α and vice versa. Thus, there are trade-offs in the two errors. Suppose a plant manager really wants to protect against having the production line shut down for no reason (Type I error). The manager will either create a climate that makes shutting down the line difficult or create test standards for rejection that are so difficult to attain that the line is rarely shut down without real reason. The result is likely to be a decrease in Type I errors (a reduction in α). However, when there really is a production line problem, workers will be less apt to act on data collected and less inclined to shut down the line. The result is likely to be an increase in Type II errors (β). Consider the case of the potentially cheating worker. Suppose company policy dictates that managers act to fire potential cheaters more quickly and on less evidence. The result could be a reduction in Type II errors. However, there is more potential for an innocent worker to be fired (a Type I error). In a court of law, making it harder to convict innocent people (reducing Type I error) increases the possibility of not convicting guilty people (increases Type II error).

Ideally, managers and decision makers want to reduce both types of error simultaneously. One way to accomplish that statistically is to use larger sample sizes. If larger samples are not possible, one strategy is for the business analyst to establish an alpha value that is the largest possible value he or she is willing to tolerate. Thus, the probability of beta will be minimized for the business analyst's hypothesis-testing process.

Power
The probability of rejecting a false null hypothesis.

Power, which is equal to $1 - \beta$, is *the probability of a test rejecting the null hypothesis when the null hypothesis is false.* Figure 9.2 shows the relationship between α, β, and power.

Two-tailed test
A statistical test wherein the business analyst is interested in testing both sides of the distribution.

Two-Tailed and One-Tailed Tests

Statistical hypothesis testing can be done with either two-tailed or one-tailed tests. The two preceding problems are examples of **two-tailed tests.** Recall that in testing to determine whether cola cans were filled with an average of 12 ounces, the alternative hypothe-

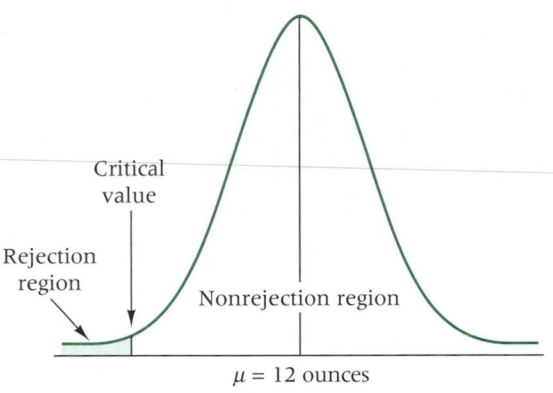

Figure 9.2

Alpha, beta, and power

	State of nature	
	Null true	Null false
Fail to reject null	Correct decision	Type II error (β)
Reject null	Type I error (α)	Correct decision (power)

Action

Figure 9.3

Rejection region for the soft-drink example with a one-tailed test

Critical value

Rejection region

Nonrejection region

$\mu = 12$ ounces

---

sis was $\mu \neq 12$ ounces. The alternative hypothesis does not state whether the business analyst believes the cans are overfilled or underfilled. Figure 9.1 showed rejection regions in both tails that cover both possibilities. In two-tailed tests, the alternative hypothesis is always stated with a does-not-equal sign ($\neq$), and there is a rejection region in *both* tails of the distribution.

At times the business analyst is interested in only one direction of the test. For example, if a consumer group is concerned that the soft-drink company is "short-changing" purchasers of the drink by underfilling the cans, the consumer group may be interested in testing only the alternative hypothesis

$$H_a: \mu < 12 \text{ ounces.}$$

That is, they are interested only in determining whether the consumer is being cheated by the company (by underfill). They are not interested in or worried about whether the company overfills the cans. This is an example of a **one-tailed test.** Figure 9.3 shows the rejection region, the critical value, and the nonrejection region for this one-tailed test. Notice that only the lower tail is shaded. Any time hypotheses are established so that the alternative hypothesis is directional (less than or greater than), the test is one-tailed. With a one-tailed test, α is concentrated at one end of the sampling distribution. Figure 9.4 shows the sampling distribution for the soft-drink can example.

Setting up the null and alternative hypotheses is more difficult with one-tailed tests than with two-tailed tests because the direction of the inequality must be determined. To use a one-tailed test, the business analyst must have some knowledge of the subject matter being studied to determine the direction of the hypotheses.

One-tailed test
A statistical test wherein the business analyst is interested only in testing one side of the distribution.

Figure 9.4

Sampling distribution with alpha for the soft-drink example

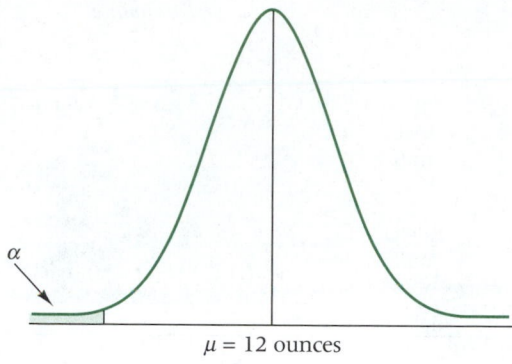

$\mu = 12$ ounces

Figure 9.5

Rejection regions for a two-tailed test for the soft-drink example

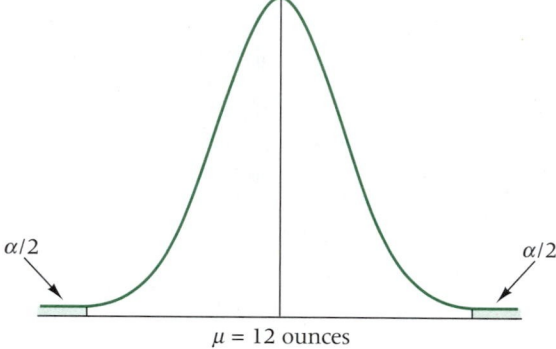

$\mu = 12$ ounces

A two-tailed test is nondirectional. In a two-tailed test, the null hypothesis has an equals sign and the alternative hypothesis has a does-not-equal sign. A two-tailed test is used when the business analyst either has no idea which direction the study will go or is interested in testing both directions. If a new technique, theory, or product is being developed, a two-tailed test might be used because the business analyst does not have enough knowledge of the phenomena to perform a directional test. At other times, a two-tailed test might be used because the business analyst is interested in both ends of the distribution.

In the case of the two-tailed test, α is divided in half, and $\alpha/2$ is the probability of the mean being in the rejection region at either end of the distribution by chance, as depicted in Figure 9.5. For any value of alpha, a two-tailed test causes the critical value to be farther from the center of the distribution than a one-tailed test because the two-tailed test splits the α and results in a smaller area in the tail.

In most situations, a two-tailed test is recommended. One-tailed tests are appropriate only when the outcome of the opposite tail is of no interest to and is completely meaningless to the business analyst. Even when the business analyst is fairly certain of the direction the research will take, unexpected and surprising results can occur. If the business analyst chooses to conduct a study with a one-tailed test, the null hypothesis is still the equality and the alternate hypothesis contains the direction of interest.

For example, one-tailed hypotheses for the soft-drink problem from the consumer group's perspective are

$$H_0: \mu = 12 \text{ ounces}$$
$$H_a: \mu < 12 \text{ ounces}.$$

Notice that the greater-than sign ($>$) is not in the hypotheses. If the data justify rejection of the null hypothesis in favor of the alternative hypothesis, then certainly any values greater than 12 ounces are rejected. It is generally standard practice to state the null hypothesis as the equality. However, because a rejection of such a null hypothesis in favor of the alternative would include a rejection of values greater than 12 (which are farther away), some business analysts would write the null hypothesis as

$$H_0: \mu \geq 12 \text{ ounces.}$$

In this text, we will use only the equality sign in the null hypothesis.

One of the most basic hypothesis tests is a test about a population mean. A business analyst might be interested in testing to determine whether an established or accepted mean value for an industry is still true. Or a business analyst might be interested in testing a mean value for a new theory or product. The test of a single population mean can be used to accomplish either objective. Formula 9.1 can be used to test hypotheses about a single population mean if the sample size is large ($n \geq 30$). The same formula can also be used for small samples ($n < 30$) if X is normally distributed *and* σ is known.

9.2
Testing Hypotheses about a Single Mean Using Large Samples

(9.1)	$$Z = \dfrac{\overline{X} - \mu}{\dfrac{\sigma}{\sqrt{n}}}$$	Z TEST FOR A SINGLE MEAN

Using the Observed Value to Reach a Decision

A survey of CPAs across the United States found that the average net income for sole proprietor CPAs is \$74,914.[*] Since this survey is now over 5 years old, suppose an accounting business analyst wants to test this figure by taking a random sample of 112 sole proprietor accountants in the United States to determine whether the net income figure has changed since the survey was taken. The business analyst could use the eight steps of hypothesis testing to do so. Assume the population standard deviation of net incomes for sole proprietor CPAs is \$14,530.

At step 1, the hypotheses must be established. As the business analyst is testing to determine whether the figure has changed, the alternative hypothesis is that the mean net income is not \$74,914. The null hypothesis is that the mean still equals \$74,914. These hypotheses follow.

$$H_0: \mu = \$74,914$$
$$H_a: \mu \neq \$74,914$$

Step 2 is to determine the appropriate statistical test and sampling distribution. Because sample size is large ($n = 112$) and the business analyst is using the sample mean as the statistic, the Z test in Formula 9.1 is the appropriate test statistic.

$$Z = \dfrac{\overline{X} - \mu}{\dfrac{\sigma}{\sqrt{n}}}$$

---

[*]ADAPTED FROM: Daniel J. Flaherty, Raymond A. Zimmerman, and Mary Ann Murray, "Benchmarking Against the Best," *Journal of Accountancy*, July 1995, 85–88.

Figure 9.6

CPA net income
example

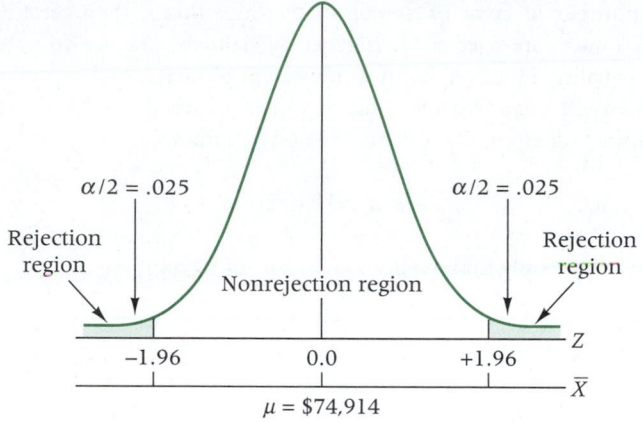

Step 3 is to specify the Type I error rate, or alpha, which is .05 in this problem. Step 4 is to state the decision rule. Because the test is two-tailed and alpha is .05, there is an $\alpha/2$ or .025 area in each of the tails of the distribution. Thus, the rejection region is the two ends of the distribution with 2.5% of the area in each. There is a .4750 area between the mean and each of the critical values that separate the tails of the distribution (the rejection region) from the nonrejection region. By using this .4750 area and Table A.5, the critical Z value can be obtained.

$$Z_{\alpha/2} = \pm 1.96$$

Figure 9.6 displays the problem with the rejection regions and the critical values of Z. The decision rule is that if the data gathered produce a Z value greater than 1.96 or less than −1.96, the test statistic is in one of the rejection regions and the decision is to reject the null hypothesis. If the Z value calculated from the data is between −1.96 and +1.96, the decision is to not reject the null hypothesis because the calculated Z value is in the nonrejection region.

Step 5 is to gather the data. Suppose the 112 CPAs who respond produce a sample mean of $78,695. At step 6, the value of the test statistic is calculated by using $\bar{X}$ = $78,695, $n = 112$, σ = $14,530, and a hypothesized μ = $74,914.

$$Z = \frac{78,695 - 74,914}{\dfrac{14,530}{\sqrt{112}}} = 2.75$$

Because this test statistic, $Z = 2.75$, is greater than the critical value of Z in the upper tail of the distribution, $Z = +1.96$, the statistical conclusion reached at step 7 of the hypothesis-testing process is the reject the null hypothesis.

Step 8 is to make a managerial decision. What does this result mean? Statistically, the business analyst has enough evidence to reject the figure of $74,914 as the true national average net income for sole proprietor CPAs. Although the business analyst conducted a two-tailed test, the evidence gathered indicates that the national average may have increased. The sample mean of $78,695 is $3781 higher than the national mean being tested. The business analyst can conclude that the national average is more than before, but because the $78,695 is only a sample mean, there is no guarantee that the national average for all sole proprietor CPAs is $3781 more. If a confidence interval were constructed with the sample data, $78,695 would be the point estimate. Other samples might produce different sample means. Managerially, this statistical finding may mean that CPAs will be more expensive to hire either as full-time employees or as consultants. It may mean that consulting services have gone up in price. For new accountants, it may mean the potential for greater earning power. Such a finding could serve as a motivation for CPAs in multiowner firms to strike out on their own.

Using a Sample Standard Deviation

In many real-life situations, the population value for the standard deviation is unavailable. With large sample sizes ($n \geq 30$), use of the sample standard deviation as a good approximate substitute for the population standard deviation, σ, is permitted.

(9.2)	$$Z = \dfrac{\bar{X} - \mu}{\dfrac{S}{\sqrt{n}}}$$	*Z* FORMULA TO TEST A MEAN WITH σ UNKNOWN—LARGE SAMPLES ONLY

Formula 9.2 can be used only for *large* sample sizes, regardless of the shape of the distribution of *X*. In the example involving CPA sole proprietorship net incomes, the sample standard deviation was 14,543. This approximation of σ could have been used to calculate the *Z* value.

Testing the Mean with a Finite Population

If the hypothesis test for the population mean is being conducted with a known finite population, the population information can be incorporated into the hypothesis-testing formula. Doing so can increase the potential for rejecting the null hypothesis. Formula 9.1 can be amended to include the population information.

(9.3)	$$Z = \dfrac{\bar{X} - \mu}{\dfrac{\sigma}{\sqrt{n}}\sqrt{\dfrac{N-n}{N-1}}}$$	FORMULA TO TEST HYPOTHESES ABOUT μ WITH A FINITE POPULATION

In the CPA net income example, suppose there are only 600 sole proprietor CPAs in the United States. A sample of 112 CPAs taken from a population of only 600 CPAs is 18.67% of the population and therefore is much more likely to be representative of the population than a sample of 112 CPAs taken from a population of 20,000 CPAs (.56% of the population). The finite correction factor takes this into consideration and allows for an increase in the calculated value of *Z*. The calculated *Z* value would change to

$$Z = \frac{\bar{X} - \mu}{\dfrac{\sigma}{\sqrt{n}}\sqrt{\dfrac{N-n}{N-1}}} = \frac{78,695 - 74,914}{\dfrac{14,530}{\sqrt{112}}\sqrt{\dfrac{600-112}{600-1}}} = \frac{3781}{1,239.2} = 3.05$$

Use of the finite correction factor increased the calculated Z value from 2.75 to 3.05. The decision to reject the null hypothesis does not change with this new information. However, on occasion, the finite correction factor can make the difference between rejecting and failing to reject the null hypothesis.

Using the Critical Value Method to Test Hypotheses

Critical value method
A method of testing hypotheses in which the sample statistic is compared to a critical value in order to reach a conclusion about rejecting or failing to reject the null hypothesis.

One alternative method of testing hypotheses is the **critical value method.** In the preceding example, the null hypothesis was rejected because the computed value of Z was in the rejection zone. What mean income would it take to cause the calculated Z value to be in the rejection zone? The critical value method determines the critical mean value required for Z to be in the rejection region and uses it to test the hypotheses.

This method also uses Formula 9.1. However, instead of a calculated Z, a critical $\bar{X}$ value, $\bar{X}_c$, is determined. The critical table value of Z_c is inserted into the formula, along with μ and σ. Thus,

$$Z_c = \frac{\bar{X}_c - \mu}{\frac{\sigma}{\sqrt{n}}}$$

Substituting values from the preceding example gives

$$\pm 1.96 = \frac{\bar{X}_c - 74,914}{\frac{14,530}{\sqrt{112}}}$$

or

$$\bar{X}_c = 74,914 \pm 1.96 \frac{14,530}{\sqrt{112}} = 74,914 \pm 2691$$

$$\text{lower } \bar{X}_c = 72,223 \quad \text{and} \quad \text{upper } \bar{X}_c = 77,605.$$

Figure 9.7 depicts graphically the rejection and nonrejection regions in terms of means instead of Z scores.

Figure 9.7

Rejection and nonrejection regions for critical value method

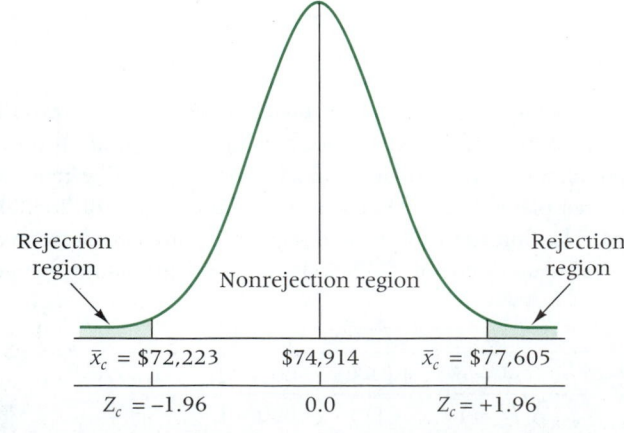

With the critical value method, most of the computational work is done ahead of time unless the sample standard deviation is being used as an estimate for σ. In this problem, before the sample means are computed, the analyst knows that a sample mean value of greater than \$77,605 or less than \$72,223 must be attained to reject the hypothesized population mean. Because the sample mean for this problem was \$78,695, which is greater than \$77,605, the analyst rejects the null hypothesis. This method is particularly attractive in industrial settings where standards can be set ahead of time and then quality control technicians can gather data and compare actual measurements of products to specifications.

Using the p-Value Method to Test Hypotheses

Another way to reach statistical conclusions in hypothesis testing problems is by the **p-value method,** sometimes referred to as **observed significance level.** The p-value method is growing in importance with the increasing use of statistical computer packages to test hypotheses. There is no preset value of α in the p-value method. Instead, the probability of getting a test statistic at least as extreme as the observed test statistic (computed from the data) is computed under the assumption that the null hypothesis is true. Virtually every statistical computer program yields this probability (p value). The p value defines the smallest value of alpha for which the null hypothesis can be rejected. For example, if the p value of a test is .038, the null hypothesis cannot be rejected at $\alpha = .01$ because .038 is the smallest value of alpha for which the null hypothesis can be rejected. However, the null hypothesis can be rejected for $\alpha = .05$.

Suppose a business analyst is conducting a one-tailed test with a rejection region in the upper tail and obtains an observed test statistic of $Z = 2.04$ from the sample data. Using the standard normal table, Table A.5, we find that the probability of randomly obtaining a Z value this great or greater by chance is $.5000 - .4793 = .0207$. The p value is .0207. Using this information, the business analyst would reject the null hypothesis for $\alpha = .05$ or .10 or any value more than .0207. The business analyst would not reject the null hypothesis for any alpha value less than or equal to .0207 (in particular, $\alpha = .01$, .001, etc.).

For a two-tailed test, recall that we have been splitting alpha to determine the critical value of the test statistic. With the p-value method and a two-tailed test, the probability of getting a test statistic at least as extreme as the observed value is computed, doubled, and then reported as the p value.

As an example of using p values with a two-tailed test, consider the CPA net income problem. The observed test statistic for this problem is $Z = 2.75$. Using Table A.5, we know that the probability of obtaining a test statistic at least this extreme if the null hypothesis is true is $.5000 - .4970 = .0030$. To reach a statistical conclusion from this p value, the business analyst must either double it to .0060 and compare it to α or compare it (.0030) to $\alpha/2$.

> **p-value method, or Observed significance level**
> A method of testing hypotheses in which there is no preset level of α. The probability of getting a test statistic at least as extreme as the observed test statistic is computed under the assumption that the null hypothesis is true. This probability is called the p value, and it is the smallest value of α for which the null hypothesis can be rejected.

In an attempt to determine why customer service is important to managers in the United Kingdom, business analysts surveyed managing directors of manufacturing plants in Scotland.* One of the reasons proposed was that customer service is a means of retaining customers. On a scale from 1 to 5, with 1 being low and 5 being high, the survey respondents rated this reason more highly than any of the others, with a mean response of 4.30. Suppose U.S. business analysts believe American manufacturing managers would not rate this reason as highly and conduct a hypothesis test to prove their theory. Alpha is set at .05. Data are gathered and the following results are obtained. Use these

> **DEMONSTRATION PROBLEM 9.1**

---

*ADAPTED FROM: William G. Donaldson, "Manufacturers Need to Show Greater Commitment to Customer Service," *Industrial Marketing Management,* 24, October 1995, 421–430. The 1-to-5 scale has been reversed here for clarity of presentation.

data and the eight steps of hypothesis testing to determine whether U.S. managers rate this reason significantly lower than the 4.30 mean ascertained in the United Kingdom.

3	4	5	5	4	5	5	4	4	4	4
4	4	4	4	5	4	4	4	3	4	4
4	3	5	4	4	5	4	4	4	5	

SOLUTION

STEP **1.** Establish hypotheses. Because the U.S. business analysts are interested only in "proving" that the mean figure is lower in the United States, the test is one-tailed. The alternative hypothesis is that the population mean is lower than 4.30. The null hypothesis states the equality case.

$$H_0: \mu = 4.30$$
$$H_a: \mu < 4.30$$

STEP **2.** Determine the appropriate statistical test. The test statistic is

$$Z = \frac{\bar{X} - \mu}{\frac{S}{\sqrt{n}}}$$

STEP **3.** Specify the Type I error rate.

$$\alpha = .05$$

STEP **4.** State the decision rule. As this is a one-tailed test, the critical Z value is found by looking up $.5000 - .0500 = .4500$ as the area in Table A.5. The critical value of the test statistic is $Z_{.05} = -1.645$. An observed test statistic must be less than -1.645 to reject the null hypothesis. The rejection region and critical value can be depicted as in the following diagram.

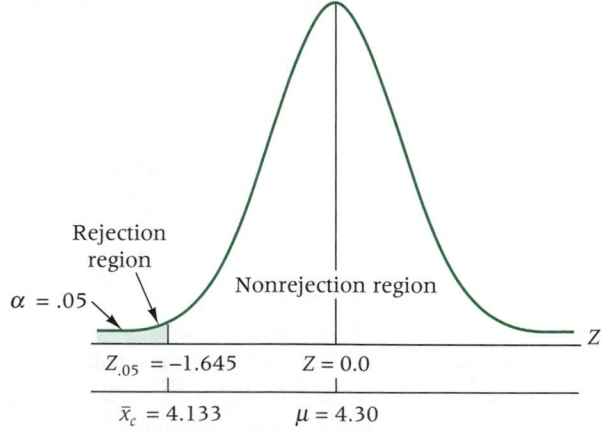

STEP **5.** Gather the sample data. The data are shown above.

STEP **6.** Calculate the value of the test statistic.

$$\bar{X} = 4.156 \qquad S = .574$$

$$Z = \frac{4.156 - 4.30}{\frac{.574}{\sqrt{32}}} = -1.42$$

STEP **7.** State the statistical conclusion. Because the observed test statistic is not less than the critical value and is not in the rejection region, the statistical conclusion is that the null hypothesis cannot be rejected.

STEP **8.** Make a managerial decision. There is not enough evidence to conclude that U.S. managers think it is less important to use customer service as a means of retaining customers than do United Kingdom managers. Customer service is an important tool for retaining customers in both countries according to managers.

Using the critical value method: For what sample mean (or more extreme) value would the null hypothesis be rejected? This critical sample mean can be determined by using the critical *Z* value associated with alpha, $Z_{.05} = -1.645$. Because the decision rule must be set before data are gathered, a population standard deviation or an estimate of one from previous studies is needed. For illustration purposes, we will use the sample standard deviation computed above in this calculation (let $\sigma = .574$).

$$Z_c = \frac{\bar{X}_c - \mu}{\dfrac{\sigma}{\sqrt{n}}}$$

$$-1.645 = \frac{\bar{X}_c - 4.30}{\dfrac{.574}{\sqrt{32}}}$$

$$\bar{X}_c = 4.133$$

The decision rule is that a sample mean less than 4.133 would be necessary to reject the null hypothesis. As the mean obtained from the sample data is 4.156, the business analysts fail to reject the null hypothesis. The diagram above includes a scale with the critical sample mean and the rejection region for the critical value method.

Using the p-value method: The calculated observed test statistic is $Z = -1.42$. From Table A.5, the probability of getting a *Z* value at least this extreme when the null hypothesis is true is $.5000 - .4222 = .0778$. Hence, the null hypothesis cannot be rejected at $\alpha = .05$ because the smallest value of alpha for which the null hypothesis can be rejected is .0778.

Analysis Using Excel

In Section 8.1 of the prior chapter, we introduced the **FAST \\ STAT** macro procedure called **1 Mean Using Z Dist.** In that chapter we used the procedure to compute confidence intervals for one mean with the normal distribution. This procedure also computes the statistics for conducting a hypothesis test for one mean with the normal distribution. The input dialog box for this procedure is shown in Figure 9.8.

To conduct a hypothesis test, you first enter the sample values into an Excel worksheet. Next click on **FAST \\ STAT** on the menu bar and select **1 Mean Using Z Dist.** from the subsequent pull-down menu. Now drag through the sample values to enter the cell range in the first box of the dialog box. If the confidence interval is not of interest, the entry for the confidence level may be set equal to any value between 0 and 100. Then enter values for the hypothesized mean and the significance level. When the data for Demonstration Problem 9.1 are entered as in Figure 9.8, a click on the **OK** command button yields the results of Figure 9.9.

In this figure in Column A you will find the sample values. In Columns C through G, the four dialog box inputs are repeated. The output values that follow include four parts. The first part displays the values for the sample mean, standard deviation, and sample size. The second and third parts provide the point estimate and confidence interval values as discussed

Figure 9.8

FAST＼STAT dialog box for **1 Mean Using Z Dist.**

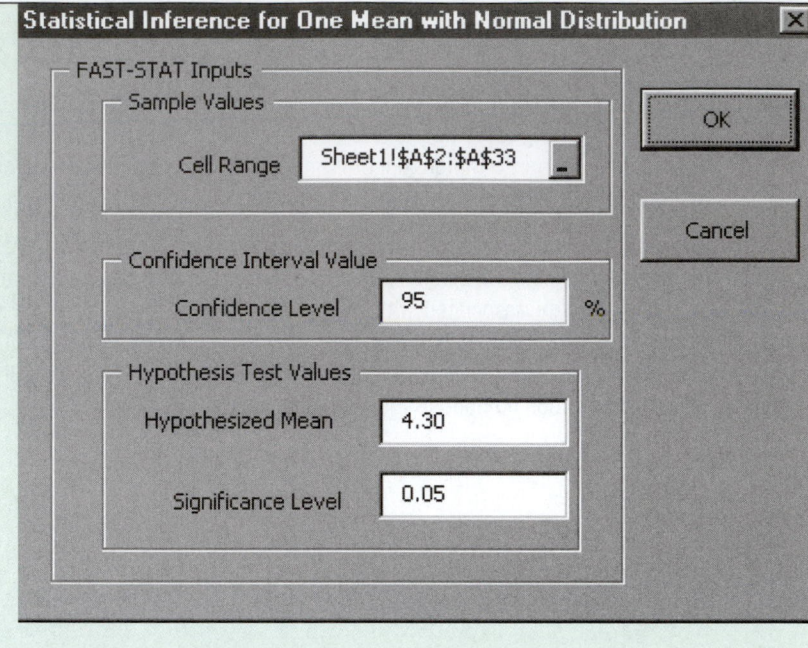

Statistical Inference for One Mean with Normal Distribution

FAST-STAT Inputs
Sample Values
Cell Range   Sheet1!$A$2:$A$33

Confidence Interval Value
Confidence Level   95   %

Hypothesis Test Values
Hypothesized Mean   4.30

Significance Level   0.05

OK   Cancel

Figure 9.9

FAST＼STAT output for Demonstration Problem 9.1

	A	B	C	D	E	F	G	H
1	**Sample Values**		**Statistical Inference for One Mean-Normal Distribution**					
2	3		*INPUTS*					
3	4			Cell Range for Sample Values			Sheet1!$A$2:$A$33	
4	4							
5	4			Level of Confidence			95	%
6	4							
7	3			Hypothesized Mean			4.3	
8	5			Significance Level of Test			0.050	
9	4		*OUTPUTS*					
10	5		*Sample Statistics*					
11	5			Sample Mean			4.15625	
12	4			Sample Standard Deviation			0.57414	
13	4			Sample Size			32	
14	4							
15	5		*Point Estimate*				4.15625	
16	4							
17	5		*Confidence Interval*					
18	4							
19	5			Z Value for Level of Confidence			1.960	
20	5			Standard Error of the Mean			0.101495	
21	4			Error of Estimation			0.198926	
22	4			Interval Lower Limit			3.957324	
23	4			Interval Upper Limit			4.355176	
24	4			Confidence Interval		3.957324481	to	4.355176
25	4		*Hypothesis Test*					
26	4			Test Statistic			-1.416	
27	3			Critical Z Value for Two-Tailed Test			1.960	
28	4			Critical Z Value for Upper-Tailed Test			1.645	
29	4			Critical Z Value for Lower-Tailed Test			-1.645	
30	4			p-Value for Two-Tailed Test			0.157	
31	5			p-Value for Upper-Tailed Test			0.922	
32	4			p-Value for Lower-Tailed Test			0.078	
33	4							

in Chapter 8. The fourth part presents the hypothesis test values. Here you will note the calculated Z test statistic value and the three critical Z values for conducting the three possible forms of the hypothesis test. In addition, three p-values are given, one for each possible form of the hypothesis test.

Consider the results of Figure 9.9. Since this is a lower-tailed test, a comparison of the calculated Z value of -1.416 to the critical Z of -1.645 results in failure to reject the null hypothesis. In like manner, a comparison of the p-value of 0.078 to the significance level of 0.05 results in the same conclusion as it must.

9.2 Problems

9.1 The Environmental Protection Agency releases figures on urban air soot in selected cities in the United States. For the city of St. Louis, the EPA claims that the average number of micrograms of suspended particles per cubic meter of air is 82. Suppose St. Louis officials have been working with businesses, commuters, and industries to reduce this figure. These city officials hire an environmental company to take random measures of air soot over a period of several weeks. The resulting data follow. Use these data to determine whether the urban air soot in St. Louis is significantly lower than it was when the EPA conducted its measurements. Let $\alpha = .01$.

81.6	66.6	70.9	82.5	58.3	71.6	72.4
96.6	78.6	76.1	80.0	73.2	85.5	73.2
68.6	74.0	68.7	83.0	86.9	94.9	75.6
77.3	86.6	71.7	88.5	87.0	72.5	83.0
85.8	74.9	61.7	92.2			

9.2 According to the U.S. Bureau of Labor Statistics, the average weekly earnings of a production worker in 1997 were $424.20. Suppose a labor business analyst wants to test to determine whether this figure is still accurate today. The business analyst randomly selects 54 production workers from across the United States and obtains a representative earnings statement for one week from each. The resulting sample average is $432.69, with a standard deviation of $33.90. Use these data and hypothesis-testing techniques along with a 5% level of significance to determine whether the mean weekly earnings of a production worker have changed.

9.3 Thirty-eight percent of all U.S. households own a wireless phone according to Personal Communications Industry Association. They report that the average wireless phone user earns $62,600 per year. Suppose a business analyst believes that the average annual earnings of a wireless phone user are higher than that now, and he sets up a study in an attempt to prove his theory. He randomly samples 48 wireless phone users and finds out that the average annual salary for this sample is $64,820, with a standard deviation of $7810. Use $\alpha = .01$ to test the business analyst's theory.

9.4 A manufacturing company produces valves in various sizes and shapes. One particular valve plate is supposed to have a tensile strength of 5 lb/mm. The company tests a random sample of 42 such valve plates from a lot of 650 valve plates. The sample mean is a tensile strength of 5.0611 lb/mm, with a standard deviation of .2803 lb/mm. Use $\alpha = .10$ and test to determine whether the lot of valve plates has an average tensile strength of 5 lb/mm.

9.5 A manufacturing firm has been averaging 18.2 orders per week for several years. However, during a recession, orders appear to have slowed. Suppose the firm's production manager randomly samples 32 weeks and finds a sample mean of 15.6 orders, with a sample standard deviation of 2.3 orders. Test to determine whether the average number of orders is down by using $\alpha = .10$.

9.6 A study conducted by Runzheimer International showed that Paris is the most expensive place to live of the 12 European Community cities. Paris ranks second in housing expense, with a rental unit of six to nine rooms costing an average of $4292 a month. Suppose a company's CEO believes this figure is too high and decides to conduct her own survey. Her assistant contacts the owners of 55 randomly selected rental units of six to nine rooms and finds that the sample average cost is $4008, with a standard deviation of $386. Using the sample results and $\alpha = .01$, test to determine whether the figure published by Runzheimer International is too high.

9.7 The American Water Works Association estimates that the average person in the United States uses 123 gallons of water per day. Suppose some business analysts believe that the actual figure is lower than this and want to test to determine whether this is so. They randomly select a sample of Americans and carefully keep track of the water used by each sample member for a day. The data from this sample are shown below. Use $\alpha = .05$ to test this hypothesis.

105	80	119	146	144	158	56	107
141	85	95	69	65	88	93	58
98	102	117	103	104	103	148	111
108	100	164	136	175	96	156	71
127	136	163	111	99	131	103	133

9.3
Testing Hypotheses about a Single Mean Using Small Samples: σ Unknown

There are times when a business analyst is testing hypotheses about a single population mean and, for reasons such as time, money, convenience, or availability, is able to gather only a small random sample ($n < 30$) of data. In such cases, if the data are normally distributed in the population and σ is known, the Z test can be used. However, in reality the sample standard deviation is often used as an estimate for the population standard deviation in hypothesis testing about the population mean because the population standard deviation is unknown. Thus, the Z test has limited usage for small-sample analysis of single population means.

Chapter 8 presented the t distribution, which can be used to analyze hypotheses about a single population mean for small sample sizes when σ is unknown *if* the population is normally distributed for the measurement being studied. In this section, we will examine the t test for a single population mean. In general, this t test is applicable whenever the business analyst is drawing a single random sample to test the value of a population mean (μ) when using small samples, when the population standard deviation is unknown, and when the population is normally distributed for the measurement of interest. The formula for testing such hypotheses follows.

---

t TEST FOR μ

$$t = \frac{\bar{X} - \mu}{\frac{S}{\sqrt{n}}}$$

(9.4)

$$\text{df} = n - 1$$

The U.S. Farmers' Production Company builds large harvesters. For a harvester to be properly balanced when operating, a 25-pound plate is installed on its side. The machine that produces these plates is set to yield plates that average 25 pounds. The distribution of plates produced from the machine is normal. However, the shop supervisor is worried that the machine is out of adjustment and is producing plates that do not average 25 pounds. To test this concern, he randomly selects 20 of the plates produced the day

22.6	22.2	23.2	27.4	24.5
27.0	26.6	28.1	26.9	24.9
26.2	25.3	23.1	24.2	26.1
25.8	30.4	28.6	23.5	23.6

$\bar{X} = 25.51,\ S = 2.1933,\ n = 20$

TABLE 9.1
Weights in Pounds
of a Sample of 20 Plates

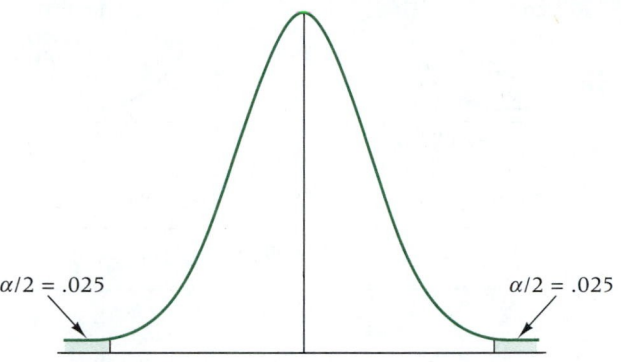

Figure 9.10

Rejection regions
for the machine
plate example

$\alpha/2 = .025$ $\alpha/2 = .025$

before and weighs them. Table 9.1 shows the weights obtained, along with the computed sample mean and sample standard deviation.

The test is to determine whether the machine is out of control, and the shop supervisor has not specified whether he believes the machine is producing plates that are too heavy or too light. Thus a two-tailed test is appropriate. The following hypotheses are tested.

$$H_0:\ \mu = 25 \text{ pounds}$$
$$H_a:\ \mu \neq 25 \text{ pounds}$$

An α of .05 is used. Figure 9.10 shows the rejection regions.

Because $n = 20$, the degrees of freedom for this test are 19 (20 − 1). The t distribution table is a one-tailed table but the test for this problem is two-tailed, so alpha must be split, which yields $\alpha/2 = .025$, the value in each tail. (To obtain the table t value when conducting a two-tailed test, always split alpha and use $\alpha/2$.) The table t value for this example is 2.093. Table values such as this one are often written in the following form:

$$t_{.025,19} = 2.093.$$

Figure 9.11 depicts the t distribution for this example, along with the critical values, the calculated t value, and the rejection regions. In this case, the decision rule is to reject the null hypothesis if the calculated value of t is less than −2.093 or greater than +2.093 (in the tails of the distribution). Computation of the test statistic yields

$$Z = \frac{\bar{X} - \mu}{\dfrac{S}{\sqrt{n}}} = \frac{25.51 - 25.0}{\dfrac{2.1933}{\sqrt{20}}} = 1.04 \text{ (Calculated } t)$$

Because the calculated value is +1.04, the null hypothesis is not rejected. There is not enough evidence in this sample to reject the hypothesis that the population mean is 25 pounds.

Figure 9.12 presents the FAST⟍STAT output for the machine plate example. The FAST⟍STAT results provide the calculated t value, the critical t values, and the p-values

Figure 9.11

Graph of calculated
and critical t values
for the machine plate
example

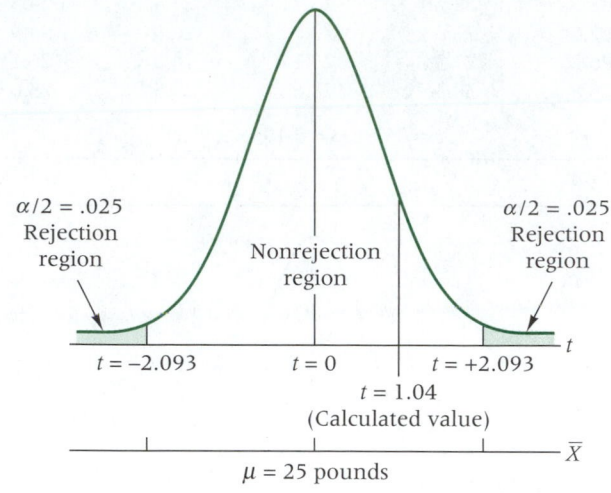

$\alpha/2 = .025$
Rejection
region

Nonrejection
region

$\alpha/2 = .025$
Rejection
region

$t = -2.093$ $t = 0$ $t = +2.093$

$t = 1.04$
(Calculated value)

$\mu = 25$ pounds

| Figure 9.12 | FAST☆STAT output for the machine plate example |

Microsoft Excel - Book1

File Edit View Insert Format Tools Data Window FAST-STAT Help

100% Arial 10 B I U $ %

R34 =

	A	B	C	D	E	F	G	H	I	J	K	L	M	N	O
1	**Sample Values**		**Statistical Inference for One Mean-t Distribution**												
2	22.6		*INPUTS*												
3	27			Cell Range for Sample Values			Sheet2!$A$2:$A$21								
4	26.2														
5	25.8			Level of Confidence			95 %								
6	22.2														
7	26.6			Hypothesized Mean			25								
8	25.3			Significance Level of Test			0.050								
9	30.4		*OUTPUTS*												
10	23.2		*Sample Statistics*												
11	28.1			Sample Mean			25.51								
12	23.1			Sample Standard Deviation			2.193267								
13	28.6			Sample Size			20								
14	27.4														
15	26.9		*Point Estimate*				25.51								
16	24.2														
17	23.5		*Confidence Interval*												
18	24.5			Degrees of Freedom			19								
19	24.9			t Value for Level of Confidence			2.093								
20	26.1			Standard Error of the Mean			0.490429								
21	23.6			Error of Estimation			1.026481								
22				Interval Lower Limit			24.48352								
23				Interval Upper Limit			26.53648								
24				Confidence Interval		24.48351903	to	26.53648							
25			*Hypothesis Test*												
26				Test Statistic			1.040								
27				Critical t Value for Two-Tailed Test			2.093								
28				Critical t Value for Upper-Tailed Test			1.729								
29				Critical t Value for Lower-Tailed Test			-1.729								
30				p-Value for Two-Tailed Test			0.311								
31				p-Value for Upper-Tailed Test			0.156								
32				p-Value for Lower-Tailed Test			0.844								
33															
34															
35															

CI-HT / Sheet1 / Sheet2 / Sheet3 /

for all three possible forms of the hypothesis test. Since this is a two-tailed test, a comparison of the calculated t value of 1.040 to the critical t value of 2.093 results in a failure to reject the null hypothesis. In like manner, a comparison of the p-value of 0.311 (p-value for a two-tailed test) to the significance level of 0.05 results in the same conclusion.

DEMONSTRATION PROBLEM 9.2

Figures released by the U.S. Department of Agriculture show that the average size of farms has been increasing since 1940. In 1940, the mean size of a farm was 174 acres; by 1997, the average size was 471 acres. Between those years the number of farms decreased but the amount of tillable land remained relatively constant, so there are now bigger farms. This trend might be explained, in part, by the inability of small farms to compete with the prices and costs of large-scale operations and to produce a level of income necessary to support the farmers' desired standard of living. Suppose an agri-business analyst believes the average size of farms has increased from the 1997 mean figure of 471 acres. To test this notion, she randomly sampled 23 farms across the United States and ascertained the size of each farm from county records. The data she gathered follow. Use a 5% level of significance to test her hypothesis.

445	489	474	505	553	477	454	463	466
557	502	449	438	500	466	477	557	433
545	511	590	561	560				

SOLUTION

STEP **1.** The business analyst's hypothesis is that the average size of a U.S. farm is more than 471 acres. Because this is an unproven theory, it is the alternate hypothesis. The null hypothesis is that the mean is still 471 acres.

$$H_0: \mu = 471$$
$$H_a: \mu > 471$$

STEP **2.** The statistical test to be used is

$$t = \frac{\bar{X} - \mu}{\frac{S}{\sqrt{n}}}$$

STEP **3.** The value of alpha is .05.

STEP **4.** With 23 data points, df $= n - 1 = 23 - 1 = 22$. This test is one-tailed, and the critical table t value is

$$t_{.05,22} = 1.717.$$

The decision rule is to reject the null hypothesis if the observed test statistic is greater than 1.717.

STEP **5.** The gathered data are shown above.

STEP **6.** The sample mean is 498.78 and the sample standard deviation is 46.94. The computed t value is

$$t = \frac{\bar{X} - \mu}{\frac{S}{\sqrt{n}}} = \frac{498.78 - 471}{\frac{46.94}{\sqrt{23}}} = 2.84$$

STEP **7.** The computed t value of 2.84 is greater than the table t value of 1.717, so the business analyst rejects the null hypothesis. She accepts the alternative hypothesis and concludes that the average size of a U.S. farm is now more than 471 acres. The following graph represents this analysis pictorially.

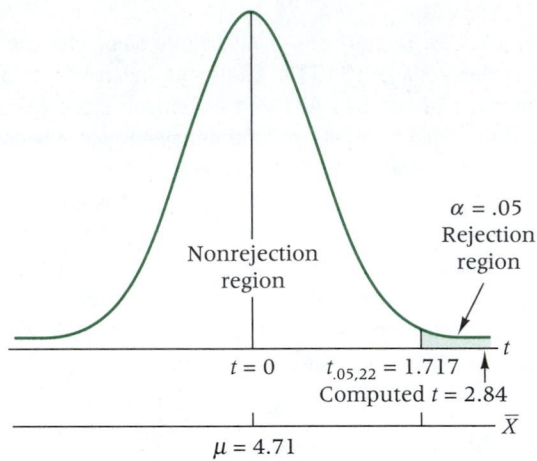

STEP **8.** Agri-business analysts can speculate about what it means to have larger farms. It could mean that small farms are not financially viable. It might mean that corporations are buying out small farms and that large company farms are on the increase. Such a trend might spark legislative movements to protect the small farm. Larger farm sizes might also affect commodity trading.

Analysis Using Excel

In Section 8.2 of the prior chapter, we introduced the **FAST ⟍ STAT** macro procedure called **1 Mean Using t Dist.** In that chapter we used the procedure to compute confidence intervals for one mean with the t distribution. This procedure also computes the statistics for conducting a hypothesis test for one mean with the t distribution. The input dialog box for this procedure is shown in Figure 9.13.

To conduct the hypothesis test, first enter the sample values into an Excel worksheet. Next click on **FAST ⟍ STAT** on the menu bar and select **1 Mean Using t Dist.** from the subsequent pull-down menu. Then drag through the sample values to enter the cell range into the first box of the dialog box. If the confidence interval is not of interest, the entry for the confidence level may be set equal to any value between 0 and 100. Now enter the hypothesized mean and the significance level. When the data for Demonstration Problem 9.2 are entered as in Figure 9.13, a click on the **OK** command button yields the results of Figure 9.14.

Figure 9.14 for the t distribution looks just like Figure 9.9 for the normal distribution with one exception. Row 18 of the worksheet of Figure 9.14 includes the additional entry for degrees of freedom for the hypothesis test.

The FAST ⟍ STAT results provide the calculated t value, the critical t values and the p-values for all three possible forms of the hypothesis test. Since this is an upper-tailed test, a comparison of the calculated t value of 2.838 to the critical t value of 1.717 results in the rejection of the null hypothesis. Similarly, a comparison of the p-value of 0.0096 to the significance level of 0.05 results in the same conclusion.

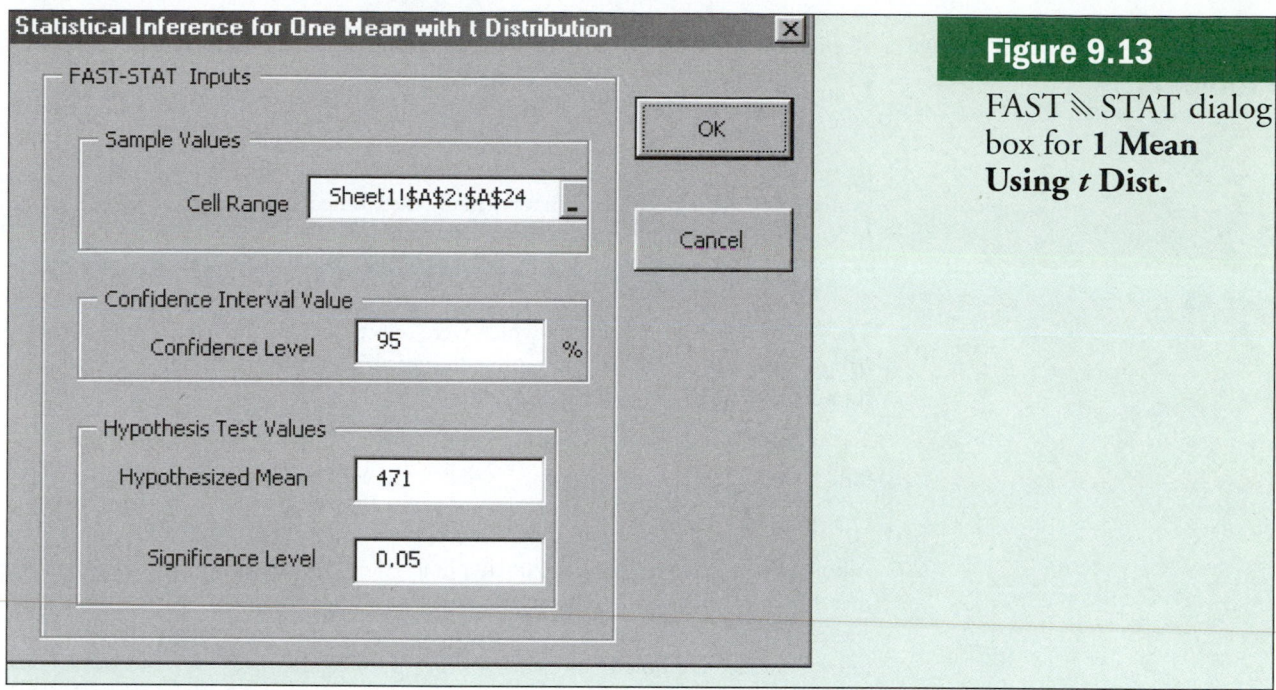

Figure 9.13

FAST⟍STAT dialog box for **1 Mean Using *t* Dist.**

Statistical Inference for One Mean with t Distribution

FAST-STAT Inputs

Sample Values

Cell Range   Sheet1!$A$2:$A$24

OK

Cancel

Confidence Interval Value

Confidence Level   95   %

Hypothesis Test Values

Hypothesized Mean   471

Significance Level   0.05

Figure 9.14   FAST⟍STAT output for Demonstration Problem 9.2

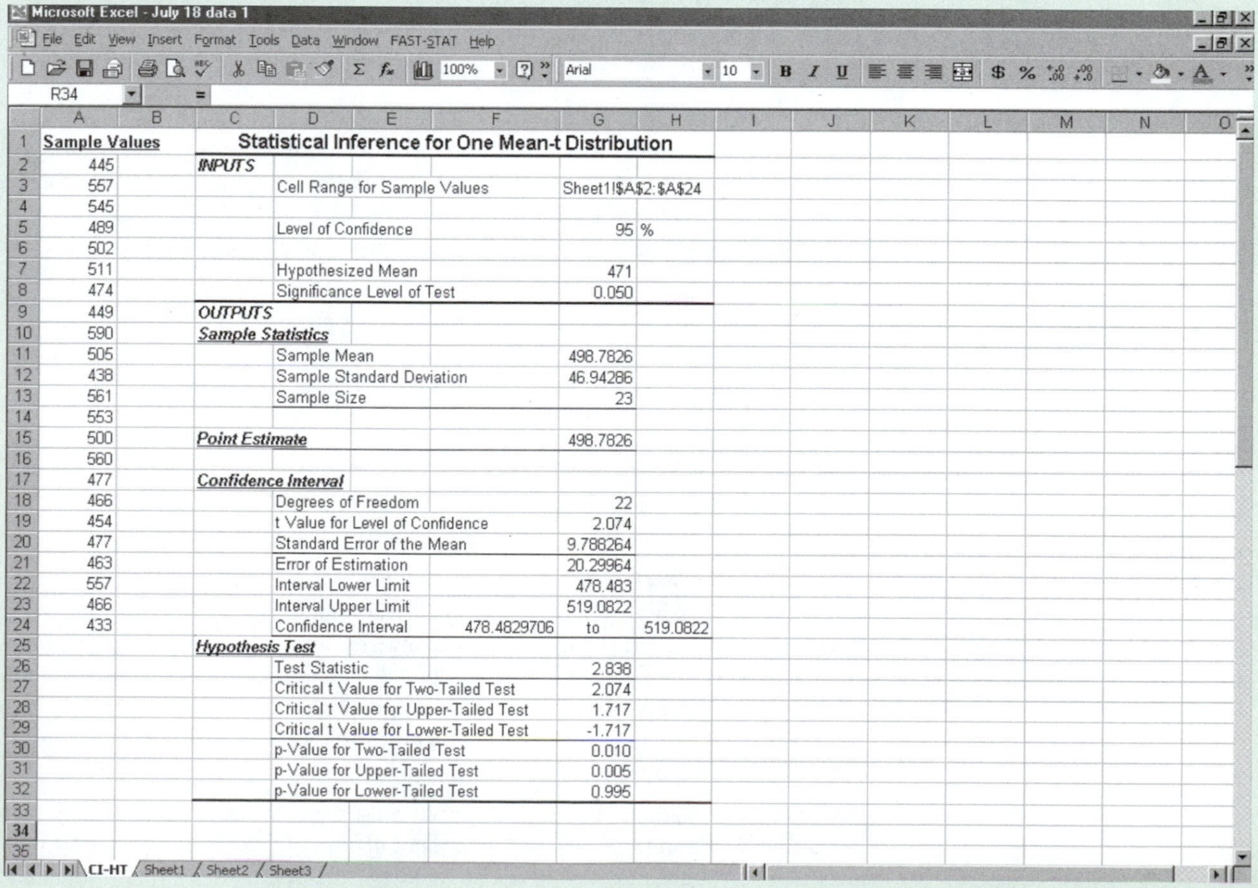

Microsoft Excel - July 18 data 1

File Edit View Insert Format Tools Data Window FAST-STAT Help

100% Arial 10 B I U $ %

R34   =

	A	B	C	D	E	F	G	H	I	J	K	L	M	N	O
1	**Sample Values**		**Statistical Inference for One Mean-t Distribution**												
2	445		*INPUTS*												
3	557			Cell Range for Sample Values			Sheet1!$A$2:$A$24								
4	545														
5	489			Level of Confidence			95 %								
6	502														
7	511			Hypothesized Mean			471								
8	474			Significance Level of Test			0.050								
9	449		*OUTPUTS*												
10	590		*Sample Statistics*												
11	505			Sample Mean			498.7826								
12	438			Sample Standard Deviation			46.94286								
13	561			Sample Size			23								
14	553														
15	500		*Point Estimate*				498.7826								
16	560														
17	477		*Confidence Interval*												
18	466			Degrees of Freedom			22								
19	454			t Value for Level of Confidence			2.074								
20	477			Standard Error of the Mean			9.788264								
21	463			Error of Estimation			20.29964								
22	557			Interval Lower Limit			478.483								
23	466			Interval Upper Limit			519.0822								
24	433			Confidence Interval		478.4829706	to	519.0822							
25			*Hypothesis Test*												
26				Test Statistic			2.838								
27				Critical t Value for Two-Tailed Test			2.074								
28				Critical t Value for Upper-Tailed Test			1.717								
29				Critical t Value for Lower-Tailed Test			-1.717								
30				p-Value for Two-Tailed Test			0.010								
31				p-Value for Upper-Tailed Test			0.005								
32				p-Value for Lower-Tailed Test			0.995								
33															
34															
35															

CI-HT / Sheet1 / Sheet2 / Sheet3 /

9.8   The following data were gathered from a random sample of 11 items.

| 1200 | 1175 | 1080 | 1275 | 1201 | 1387 |
| 1090 | 1280 | 1400 | 1287 | 1225 | |

Use these data and a 5% level of significance to test the following hypotheses, assuming that the data come from a normally distributed population.

$$H_0: \mu = 1160 \qquad H_a: \mu > 1160$$

9.9   The following data (in pounds), which were selected randomly from a normally distributed population of values, represent measurements of a machine part that is supposed to weigh, on average, 8.3 pounds.

| 8.1 | 8.4 | 8.3 | 8.2 | 8.5 | 8.6 | 8.4 | 8.3 | 8.4 | 8.2 |
| 8.8 | 8.2 | 8.2 | 8.3 | 8.1 | 8.3 | 8.4 | 8.5 | 8.5 | 8.7 |

Use these data and $\alpha = .01$ to test the hypothesis that the parts average 8.3 pounds.

9.10   A hole-punch machine is set to punch a hole 1.84 cm in diameter in a strip of sheet metal in a manufacturing process. The strip of metal is then creased and sent on to the next phase of production, where a metal rod is slipped through the hole. It is important that the hole be punched to the specified diameter of 1.84 cm. To test punching accuracy, technicians have randomly sampled 12 punched holes and measured the diameters. The data (in cm) follow. Use an alpha of .10 to determine whether the holes are being punched an average of 1.84 cm. Assume the punched holes are normally distributed in the population.

1.81	1.89	1.86	1.83
1.85	1.82	1.87	1.85
1.84	1.86	1.88	1.85

9.11   Suppose a study reports that the average price for a gallon of self-serve regular unleaded gasoline is $1.16. You believe that the figure is higher in your area of the country. You decide to test this claim for your part of the United States by randomly calling gasoline stations. Your random survey of 25 stations produces the following prices.

$1.27	$1.29	$1.16	$1.20	$1.37
1.20	1.23	1.19	1.20	1.24
1.16	1.07	1.27	1.09	1.35
1.15	1.23	1.14	1.05	1.35
1.21	1.14	1.14	1.07	1.10

Assume gasoline prices for a region are normally distributed. Do the data you obtained provide enough evidence to reject the claim? Use a 1% level of significance.

9.12   Suppose that in past years the average price per square foot for warehouses in the United States has been $32.28. A national real estate investor wants to determine whether that figure has changed now. The investor hires a business analyst who randomly samples 19 warehouses that are for sale across the United States and finds that the mean price per square foot is $31.67, with a standard deviation of $1.29. If the business analyst uses a 5% level of significance, what statistical conclusion can be reached? What are the hypotheses?

9.13   According to a National Public Transportation survey, the average commuting time for people who commute to a city with a population of 1 to 3 million is 19.0 minutes. Suppose a business analyst lives in a city with a population of 2.4 million and wants to test to determine if commuting time has increased. She takes a random sample of 26 commuters and gathers the data shown below. Using an alpha of .05 and assuming that commuting time is normally distributed, what did she find?

19	16	20	23	23
24	13	19	23	16
17	15	14	27	17
23	18	18	20	18
18	18	23	19	19
28				

9.4

Testing Hypotheses about a Proportion

The formula for proportions based on the central limit theorem makes possible the testing of hypotheses about the population proportion in a manner similar to that of the formula used to test sample means. A proportion is a value between 0 and 1 that expresses the part of the whole that has a given characteristic. For example, according to Forrester Research, Inc., .41 of all companies offer payment confirmation on their Website. Whereas means are computed by averaging measurements, proportions are calculated by counting or tallying the number of items in a population that have a characteristic and then dividing that number by the total. Recall that $\hat{p}$ denotes a sample proportion and P denotes the population proportion.

The central limit theorem applied to sample proportions states that $\hat{p}$ values are approximately normally distributed. It has been shown that the mean of the distribution of a set of $\hat{p}$ values is P and the standard deviation is $\sqrt{(P \cdot Q)/n}$ when $n \cdot P \geq 5$ and $n \cdot Q \geq 5$. A Z test is used to test hypotheses about P.

(9.5)

$$Z = \frac{\hat{p} - P}{\sqrt{\dfrac{P \cdot Q}{n}}}$$

Z TEST OF A POPULATION PROPORTION

where:
$\hat{p}$ = sample proportion
P = population proportion
$Q = 1 - P$

A manufacturer believes exactly 8% of its products contain at least one minor flaw. Suppose a company business analyst wants to test this belief. The null and alternative hypotheses are

$$H_0: P = .08$$
$$H_a: P \neq .08.$$

This is a two-tailed test because the hypothesis being tested is whether or not the proportion of products with at least one minor flaw is .08. Alpha is selected to be .10. Figure 9.15 shows the distribution, with the rejection regions and $Z_{.05}$. Because α is divided for a two-tailed test, the table value for an area of $(1/2)(.10) = .05$ is $Z_{.05} = \pm 1.645$.

For the business analyst to reject the null hypothesis, the calculated Z value must be greater than 1.645 or less than −1.645. The business analyst randomly selects a sample of

Figure 9.15

Distribution with
rejection regions
for flawed-product
example

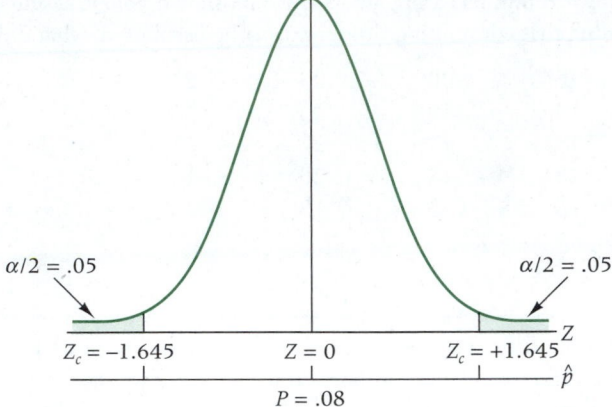

200 products, inspects each item for flaws, and determines that 33 items have at least one minor flaw. Calculating the sample proportion gives

$$\hat{p} = \frac{33}{200} = .165.$$

The calculated Z value is

$$Z = \frac{\hat{p} - P}{\sqrt{\dfrac{P \cdot Q}{n}}} = \frac{.165 - .080}{\sqrt{\dfrac{(.08)(.92)}{200}}} = \frac{.085}{.019} = 4.43.$$

Note that the denominator of the Z formula contains the population proportion. Although the business analyst does not actually know the population proportion, he is testing a population proportion value. Hence he uses the hypothesized population value in the denominator of the formula as well as in the numerator. This method contrasts with the confidence interval formula, where the sample proportion is used in the denominator.

The calculated value of Z is in the rejection region (calculated $Z = 4.43 >$ table $Z_{.05} = +1.645$), so the business analyst rejects the null hypothesis. He concludes that the proportion of items with at least one minor flaw in the population from which the sample of 200 was drawn is not .08. With $\alpha = .10$, the risk of committing a Type I error in this example is .10.

The calculated value of $Z = 4.43$ is outside the range of most values in virtually all Z tables. Thus if the business analyst were using the p-value method to arrive at a decision about the null hypothesis, the probability would be .0000, and he would reject the null hypothesis.

Suppose the business analyst wanted to use the critical value method. He would enter the table value of $Z_{.05} = 1.645$ in the Z formula for single sample proportions, along with the hypothesized population proportion and n, and solve for the critical value of $\hat{p}$, $\hat{p}_c$. The result is

$$Z_{\alpha/2} = \frac{\hat{p}_c - P}{\sqrt{\dfrac{P \cdot Q}{n}}}$$

$$\pm 1.645 = \frac{\hat{p}_c - .08}{\sqrt{\dfrac{(.08)(.92)}{200}}}$$

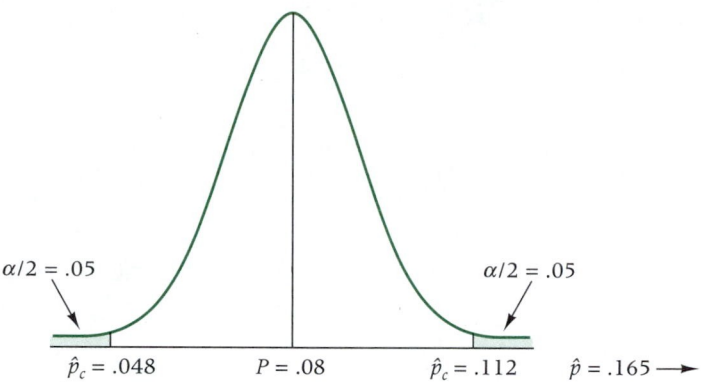

Figure 9.16

Distribution using critical value method for the flawed-product example

and

$$\hat{p}_c = .08 \pm 1.645 \sqrt{\frac{(.08)(.92)}{200}} = .08 \pm .032$$

$$= .048 \text{ and } .112.$$

Examination of the sample proportion, $\hat{p} = .165$, and Figure 9.16 clearly shows that the sample proportion is in the rejection region. The statistical conclusion is to reject the null hypothesis. The proportion of products with at least one flaw is not .08.

DEMONSTRATION PROBLEM 9.3

A survey of the morning beverage market has shown that the primary breakfast beverage for 17% of Americans is milk. A milk producer in Wisconsin, where milk is plentiful, believes the figure is higher for Wisconsin. To test this idea, she contacts a random sample of 550 Wisconsin residents and asks which primary beverage they consumed for breakfast that day. Suppose 115 replied that milk was the primary beverage. Using a level of significance of .05, test the idea that the milk figure is higher for Wisconsin.

SOLUTION

STEP **1.** The milk producer's theory is that the proportion of Wisconsin residents who drink milk for breakfast is higher than the national proportion, which is the alternative hypothesis. The null hypothesis is that the proportion in Wisconsin does not differ from the national average. That is, the hypotheses for this problem are

$$H_0: P = .17$$
$$H_a: P > .17.$$

STEP **2.** The test statistic is

$$Z = \frac{\hat{p} - P}{\sqrt{\dfrac{P \cdot Q}{n}}}.$$

STEP **3.** The Type I error rate is .05.

STEP **4.** This is a one-tailed test, and the table value is $Z_{.05} = +1.645$. The sample results must yield an observed Z value greater than 1.645 for the milk producer to reject the null hypothesis. The following diagram shows $Z_{.05}$ and the rejection region for this problem.

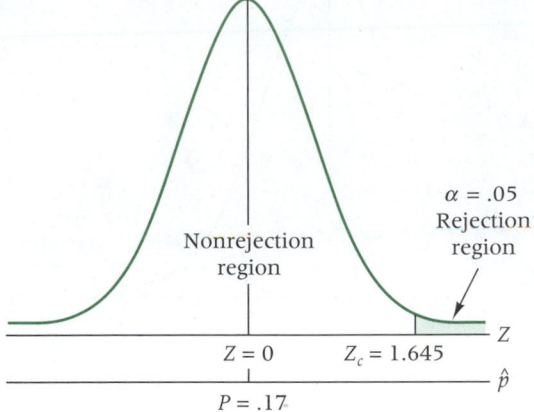

STEP **5.** $n = 550$ and $X = 115$

$$\hat{p} = \frac{115}{550} = .209$$

STEP **6.**

$$Z = \frac{\hat{p} - P}{\sqrt{\dfrac{P \cdot Q}{n}}} = \frac{.209 - .17}{\sqrt{\dfrac{(.17)(.83)}{550}}} = \frac{.039}{.016} = 2.44$$

STEP **7.** As $Z = 2.44$ is beyond $Z_{.05} = 1.645$ in the rejection region, the milk producer rejects the null hypothesis. On the basis of the random sample, the producer is ready to conclude that the proportion of Wisconsin residents who drink milk as the primary beverage for breakfast is higher than the national proportion.

STEP **8.** If the proportion of residents who drink milk for breakfast is higher in Wisconsin than in other parts of the United States, milk producers might have a market opportunity in Wisconsin that is not available in other parts of the country. Perhaps Wisconsin residents are being loyal to home-state products, in which case marketers of other Wisconsin products might be successful in appealing to residents to support their products. The fact that more milk is sold in Wisconsin might mean that if Wisconsin milk producers appealed to markets outside Wisconsin in the same way they do their own, they might increase their market share of the breakfast beverage market in other states.

The probability of obtaining a $Z \geq 2.44$ by chance is .0073. As this probability is less than $\alpha = .05$, the null hypothesis is also rejected with the p-value method.

A critical proportion can be solved for by

$$Z_{.05} = \frac{\hat{p} - P}{\sqrt{\dfrac{P \cdot Q}{n}}};$$

$$1.645 = \frac{\hat{p}_c - .17}{\sqrt{\dfrac{(.17)(.83)}{550}}};$$

$$\hat{p}_c = .17 + .026 = .196.$$

With the critical value method, a sample proportion greater than .196 must be obtained to reject the null hypothesis. The sample proportion for this problem is .209, so the null hypothesis is also rejected with the critical value method.

Analysis Using Excel

In Section 8.3 of the prior chapter, we introduced the **FAST ⟍ STAT** macro procedure called **1 Proportion C.I. and H.T.** In that chapter we used the procedure to compute confidence intervals for one proportion with the normal distribution. This procedure also computes the statistics for conducting a hypothesis test for one proportion with the normal distribution. The input dialog box for this procedure is shown in Figure 9.17.

To conduct a hypothesis test, first enter the sample size and number of items in the sample having the characteristic of interest in the first grouping of FAST ⟍ STAT inputs. If the confidence interval is not of interest, the entry for the confidence level may be set equal to any value between 0 and 100. Next enter the hypothesized proportion and the significance level of the test in the third grouping of inputs. When the data for Demonstration Problem 9.3 are entered as in Figure 9.17, a click on the **OK** command button yields the results of Figure 9.18.

Figure 9.18 for the 1 proportion is quite similar to the prior figures demonstrating the statistical inference computations for 1 mean: Figures 9.9, 9.12, and 9.14. The hypothesis test output of Figure 9.18 includes the calculated Z value, the critical Z values and the p-values for all three possible forms of the hypothesis test. Since this is an upper-tailed test, a comparison of the calculated Z value of 2.441 to the critical Z value of 1.645 results in the rejection of the null hypothesis. Similarly, a comparison of the p-value of 0.007 to the significance level of 0.05 results in the same conclusion.

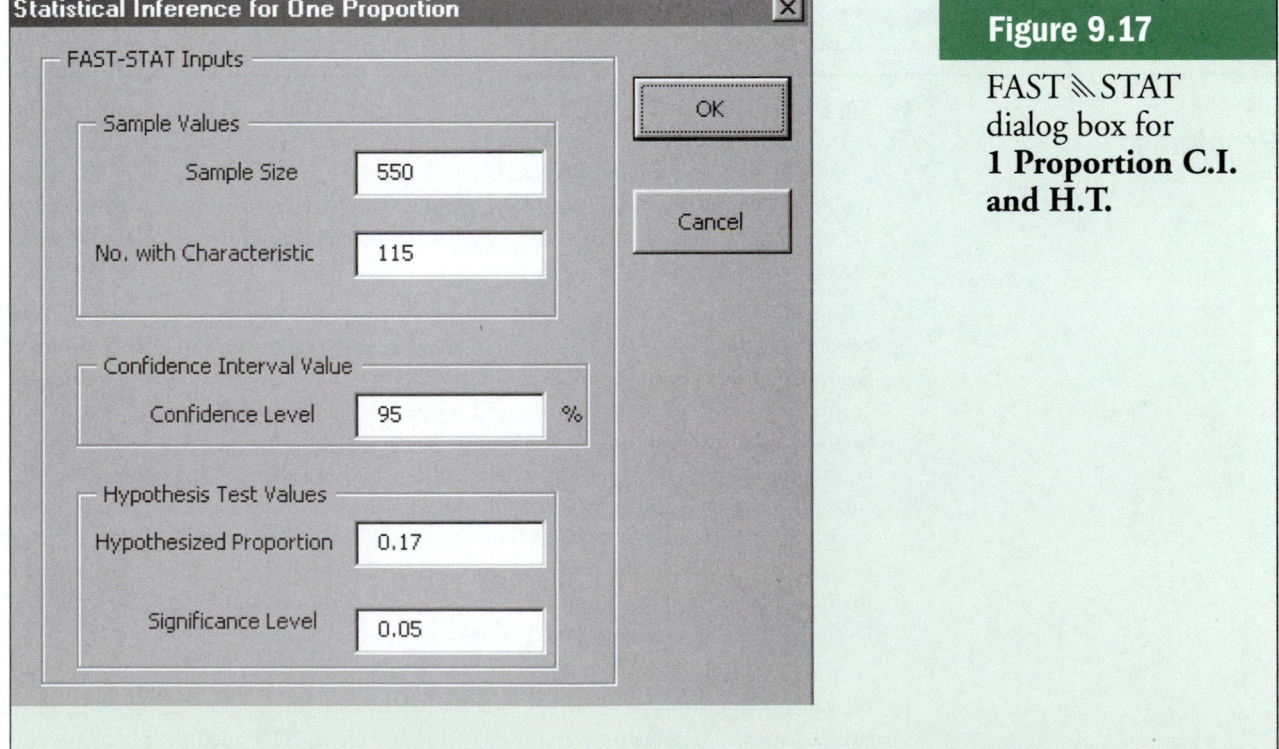

Figure 9.17

FAST ⟍ STAT dialog box for **1 Proportion C.I. and H.T.**

| Figure 9.18 | FAST⧵STAT output for Demonstration Problem 9.3 |

Microsoft Excel - Book1

File Edit View Insert Format Tools Data Window FAST-STAT Help

A1   = Statistical Inference for One Proportion

	A	B	C	D	E	F	G
1	**Statistical Inference for One Proportion**						
2	*INPUTS*						
3				Sample Size		550	
4		Number of Items with Characteristic				115	
5							
6			Level of Confidence			95 %	
7							
8			Hypothesized Proportion			0.17	
9			Level of Significance			0.05	
10	*OUTPUTS*						
11	*Point Estimate*					0.209091	
12							
13	*Confedence Interval*						
14		Z Value for Level of Confidence				1.960	
15		Standard Error of the Proportion				0.0173	
16		Error of Estimation				0.0340	
17		Interval Lower Limit				0.1751	
18		Interval Upper Limit				0.2431	
19		Confidence Interval			0.1751	to	0.2431
20	*Hypothesis Test*						
21		Test Statistic				2.4406	
22		Critical Z Value for Two-Tailed Test				1.960	
23		Critical Z Value for Upper-Tailed Test				1.645	
24		Critical Z Value for Lower-Tailed Test				-1.645	
25		p-Value for Two-Tailed Test				0.015	
26		p-Value for Upper-Tailed Test				0.007	
27		p-Value for Lower-Tailed Test				0.993	

1 Proportion / Sheet1 / Sheet2 / Sheet3 /

9.4 Problems

9.14 Suppose you are testing H_0: $P = .29$ versus H_a: $P \neq .29$. A random sample of 740 items shows that 207 have this characteristic. With a .05 probability of committing a Type I error, test the hypothesis. If you had used the critical value method, what would the two critical values be? How do the sample results compare with the critical values? For the p-value method, what is the probability of the calculated Z value for this problem?

9.15 The Independent Insurance Agents of America conducted a survey of insurance consumers and discovered that 48% of them always reread their insurance policies, 29% sometimes do, 16% rarely do, and 7% never do. Suppose a large insurance company invests considerable time and money in rewriting policies so that they will be more attractive and easy to read and understand. After using the new policies for a year, company managers want to determine whether rewriting the policies has significantly changed the proportion of policyholders who always reread their insurance policy. They contact 380 of the company's insurance consumers who have purchased a policy in the past year and ask them whether they always reread their insurance policies. One hundred and sixty-four respond that they do. Use a 1% level of significance to test the hypothesis.

9.16 A study by Hewitt Associates showed that 79% of companies offer employees flexible scheduling. Suppose a business analyst believes that in accounting firms this figure is lower. The business analyst randomly selects 415 accounting firms and through interviews determines that 303 of these firms have flexible scheduling. With a 1% level of significance, is there enough evidence to conclude that a significantly lower proportion of accounting firms offer employees flexible scheduling?

9.17 A survey was undertaken by Bruskin/Goldring Research for Quicken to determine how people plan to meet their financial goals in the next year. Respondents were allowed to select more than one way to meet their goals. Thirty-one percent said that they were using a financial planner to help them meet their goals. Twenty-four percent were using family/friends to help them meet their financial goals followed by broker/accountant (19%), computer software (17%), and books (14%). Suppose another business analyst takes a similar survey of 600 people to test these results. If 200 people respond that they are going to use a financial planner to help them meet their goals, is this enough evidence to reject the 31% figure generated in the Bruskin/Goldring survey using $\alpha = .10$? If 130 respond that they are going to use family/friends to help them meet their financial goals, is this enough evidence to declare that the proportion is significantly lower than Bruskin/Goldring's figure of .24 if $\alpha = .05$?

9.18 Eighteen percent of U.S.-based multinational companies provide an allowance for personal long-distance calls for executives living overseas, according to the Institute for International Human Resources and the National Foreign Trade Council. Suppose a business analyst thinks that U.S.-based multinational companies are having a more difficult time recruiting executives to live overseas and that an increasing number of these companies are providing an allowance for personal long-distance calls to these executives to ease the burden of living away from home. To test this, a new study is conducted by contacting 376 multinational companies. Twenty-two percent of these surveyed companies are providing an allowance for personal long-distance calls to executives living overseas. Is this enough evidence to declare that a significantly higher proportion of multinational companies are doing this? Let $\alpha = .01$.

9.19 A large manufacturing company investigated the service it has received from suppliers and discovered that, in the past, 32% of all materials shipments have been received late. However, the company recently installed a just-in-time system in which suppliers are linked more closely to the manufacturing process. A random sample of 118 deliveries since the just-in-time system was installed reveals that 22 deliveries were late. Use this sample information to test whether the proportion of late deliveries has been reduced significantly. Let $\alpha = .05$.

9.20 Where do CFOs get their money news? According to Robert Half International, 47% get their money news from newspapers, 15% get it from communication/colleagues, 12% get it from television, 11% from the Internet, 9% from magazines, 5% from radio, and 1% don't know. Suppose a business analyst wants to test these results. She randomly samples 67 CFOs and finds that 40 of them get their money news from newspapers. Is this enough evidence to reject the findings of Robert Half International? Use $\alpha = .05$.

Summary

Hypothesis testing is a very important tool for business statisticians and is used to test statistical hypotheses. The process begins with the selection of a null hypothesis and an alternative hypothesis. The null and alternative hypotheses are structured so that either one or the other is true but not both. In testing hypotheses, the business analyst assumes that the null hypothesis is true. By examining the sampled data, the business analyst either rejects or does nor reject the null hypothesis. If the sample data are significantly in opposition to the null hypothesis, the business analyst rejects the null hypothesis and accepts the alternative hypothesis by default.

Hypothesis tests can be one-tailed or two-tailed. Two-tailed tests always utilize = and ≠ in the null and alternative hypotheses. These tests are nondirectional in that significant deviations from the hypothesized value that are either greater than or less than the value are in rejection regions. The one-tailed test is directional, and the alternative hypothesis contains < or > signs. In these tests, only one end or tail of the distribution contains a rejection region. In a one-tailed test, the business analyst is interested only in deviations from the hypothesized value that are either greater than or less than the value but not both.

When a business analyst makes a decision about the null hypothesis, it can involve an error. If the null hypothesis is true, the business analyst can make a Type I error by rejecting the null hypothesis. The probability of making a Type I error is alpha (α). Alpha is usually set by the business analyst when establishing the hypotheses. Another expression sometimes used for the value of α is level of significance.

If the null hypothesis is false and the business analyst fails to reject it, a Type II error has been committed. Beta (β) is the probability of committing a Type II error. Type II errors must be computed from the hypothesized value of the parameter, α, and a specific alternative value of the parameter being examined. There are as many possible Type II errors in a problem as there are possible alternative statistical values.

If a null hypothesis is true and the business analyst fails to reject it, no error has been committed, and the business analyst has made a correct decision. Similarly, if a null hypothesis is false and it is rejected, no error has been committed. Power ($1 - \beta$) is the probability of a statistical test rejecting the null hypothesis when the null hypothesis is false.

Included in this chapter were hypothesis tests for a single mean for both large and small sample sizes and a test of a single population proportion. Three different analytic approaches were presented: (1) standard method, (2) critical value method, and (3) the *p*-value method.

Key Terms

alpha (α)	observed significance level
alternative hypothesis	one-tailed test
beta (β)	*p*-value method
critical value	power
critical value method	rejection region
hypothesis testing	two-tailed test
level of significance	Type I error
nonrejection region	Type II error
null hypothesis	

SUPPLEMENTARY PROBLEMS

9.21 According to a survey by ICR for Vienna Systems, a majority of American households have tried to cut long-distance phone bills. Of those who have tried to cut the bills, 32% have done so by switching long-distance companies. Suppose that lately there has been a frenzy of "slamming" (where the customer's long-distance provider is switched without the customer's knowledge or approval) and long-distance company solicitation and advertising. Because of this, ICR conducts another survey by randomly contacting 80 American households who have tried to cut long-distance phone bills. If 39% of the contacted households say they have tried to cut their long-distance phone bills by switching long-distance companies, is this enough evidence to state that a significantly higher proportion of American households are trying to cut long-distance phone bills by switching companies? Let $\alpha = .01$.

9.22 According to Zero Population Growth, the average urban U.S. resident uses 3.3 pounds of food per day. Is this figure accurate for rural U.S. residents? Suppose 64 rural U.S. residents are identified by a random procedure and their average consumption per day is 3.45 pounds of food. Assume a population variance of 1.31 pounds of food per day. Use a 5% level of significance to determine whether the Zero Population Growth figure for urban U.S. residents also is true for rural U.S. residents on the basis of the sample data.

9.23 Brokers generally agree that bonds are a better investment during times of low interest rates than during times of high interest rates. A survey of executives during a time of low interest rates showed that 57% of them had some retirement funds invested in bonds. Assume this percentage is constant for bond market investment by executives with retirement funds. Suppose interest rates have risen lately and the proportion of executives with retirement investment money in the bond market may have dropped. To test this idea, a business analyst randomly samples 210 executives who have retirement funds. Of these, 93 now have retirement funds invested in bonds. For $\alpha = .10$, is this enough evidence to declare that the proportion of executives with retirement fund investments in the bond market is significantly lower than .57?

9.24 Highway engineers in Ohio are painting white stripes on a highway. The stripes are supposed to be approximately 10 feet long. However, because of the machine, the operator, and the motion of the vehicle carrying the equipment, there is considerable variation among the stripe lengths. Shown below are a sample of 12 measured stripe lengths from a recent highway painting. Using these data, test to determine if stripe lengths are averaging 10 feet. Assume stripe length is normally distributed. Let $\alpha = .05$.

STRIPE LENGTHS IN FEET

10.3	9.4	9.8	10.1
9.2	10.4	10.7	9.9
9.3	9.8	10.5	10.4

9.25 A computer manufacturer estimates that its line of minicomputers has, on average, 8.4 days of downtime per year. To test this claim, a business analyst contacts seven companies that own one of these computers and is allowed to access company computer records. It is determined that, for the sample, the average number of downtime days is 5.6, with a sample standard deviation of 1.3 days. Assuming that number of downtime days is normally distributed, test to determine whether these minicomputers actually average 8.4 days of downtime in the entire population. Let $\alpha = .01$.

9.26 A life insurance salesperson claims the average worker in the city of Cincinnati has no more than $25,000 of personal life insurance. To test this claim, you randomly sample 100 workers in Cincinnati. You find that this sample of workers has an average of $26,650 of personal life insurance and that the standard deviation is $12,000.

Determine whether there is enough evidence to reject the null hypothesis posed by the salesperson. Assume the probability of committing a Type I error is .05.

9.27 A study of MBA graduates by Universum for The American Graduate Survey 1999 revealed that MBA graduates have several expectations from prospective employers beyond their base pay. In particular, according to the study 46% expect a performance-related bonus, 46% expect stock options, 42% expect a signing bonus, 28% expect profit sharing, 27% expect extra vacation/personal days, 25% expect tuition reimbursement, 24% expect health benefits, and 19% expect guaranteed annual bonuses. Suppose a study is conducted in an ensuing year to see if these expectations have changed. If 125 MBA graduates are randomly selected and if 66 expect stock options, is this enough evidence to declare that a significantly higher proportion of MBAs expect stock options? Let $\alpha = .05$.

9.28 Suppose the number of beds filled per day in a medium-size hospital is normally distributed. A hospital administrator has been quoted as having told the board of directors that, on the average, at least 185 beds are filled on any given day. One of the board members believes this figure is inflated, and she manages to secure a random sample of figures for 16 days. The data are shown here. Use $\alpha = .05$ and the sample data to test whether the hospital administrator's statement is false. Assume the number of filled beds per day is normally distributed in the population.

NUMBER OF BEDS OCCUPIED PER DAY

173	149	166	180
189	170	152	194
177	169	188	160
199	175	172	187

9.29 According to the International Data Corporation, Compaq Computers holds a 16% share of the personal computer market in the United States and a 12.7% share of the worldwide market. Suppose a market business analyst believes that Compaq holds a higher share of the market in the southwestern region of the United States. To verify this theory, he randomly selects 428 people who have purchased a personal computer in the last month in the southwestern region of the United States. Eighty-four of these purchases were Compaq Computers. Using a 1% level of significance, test the market business analyst's theory. What is the probability of making a Type I error?

9.30 A national publication reported that a college student living away from home spends, on average, no more than $15 per month on laundry. You believe this figure is too low and want to disprove this claim. To conduct the test, you randomly select 35 college students and ask them to keep track of the amount of money they spend during a given month for laundry. The sample produces an average expenditure on laundry of $19.34, with a standard deviation of $4.52. Use these sample data to conduct the hypothesis test. Assume you are willing to take a 10% risk of making a Type I error.

9.31 A study of pollutants showed that certain industrial emissions should not exceed 2.5 parts per million. You believe a particular company may be exceeding this average. To test this supposition, you randomly take a sample of nine air tests. The sample average is 3.4 parts per million, with a sample standard deviation of .6. Is this enough evidence for you to conclude that the company has been exceeding the safe limit? Use $\alpha = .01$. Assume emissions are normally distributed.

9.32 The average cost per square foot for office rental space in the central business district of Philadelphia is $23.58, according to Cushman & Wakefield, Inc. A large real estate company wants to confirm this figure. The firm conducts a telephone survey of 95 offices in the central business district of Philadelphia and asks the office managers how much they pay in rent per square foot. Suppose the sample average is $22.83 per square foot, with a standard deviation of $5.11.

Conduct a hypothesis test using $\alpha = .05$ to determine whether the cost per square foot reported by Cushman & Wakefield, Inc. should be rejected.

9.33 The American Water Works Association reports that, on average, men use between 10 and 15 gallons of water daily to shave when they leave the water running. Suppose the following data are the numbers of gallons of water used in a day to shave by 12 randomly selected men and the data come from a normal distribution of data. Use these data and a 5% level of significance to test to determine whether the population mean for such water usage is less than 12.5 gallons.

10	8	13	17	13	9
12	13	5	8	9	7

ANALYZING THE DATABASES

1. Suppose the average number of employees per industry group in the manufacturing database is believed to be less than 150 (1000s). Test this belief as the alternative hypothesis by using the 140 SIC code industries given in the database as the sample. Let $\alpha = .01$. What did you decide and why?

2. Examine the hospital database. Suppose you want to "prove" that the average hospital in the United States has an average of more than 700 births per year. Use the hospital database as your sample and test this hypothesis. Let alpha be .01. On average, do hospitals in the United States have fewer than 800 personnel? Use the hospital database as your sample and an alpha of .10 to test this as the alternative hypothesis.

3. Consider the financial database. Is the average earnings per share for companies in the stock market more than $2.50? Use the sample of companies represented by this database to test that hypothesis. Let $\alpha = .05$. Test to determine whether the average return on equity for all companies is equal to 21. Use this database as the sample and $\alpha = .10$.

4. Fifteen years ago, the average production in the United States for green beans was 166,770 per month. Use the 12 months in 1997 (the last 12 months in the database) in the agriculture database as a sample to test to determine whether the mean monthly production figure for green beans in the United States is now different from the old figure. Let $\alpha = .01$.

FRITO-LAY TARGETS THE HISPANIC MARKET

Frito Company was founded in 1932 in San Antonio, Texas, by Elmer Doolin. H. W. Lay & Company was founded in Atlanta, Georgia, by Herman W. Lay in 1938. In 1961, the two companies merged to form Frito-Lay, Inc., with headquarters in Texas. Frito-Lay, Inc. produced, distributed, and marketed snack foods with particular emphasis on various types of chips. In 1965, the company merged with Pepsi-Cola to form PepsiCo, Inc. Three decades later, Pepsi-Cola combined its domestic and international snack food operations into one business unit called Frito-Lay Company. According to data released by Information Resources, Inc., Frito-Lay brands account for over 60% of the share of the snack chip market.

One problem facing Frito-Lay has been its general lack of appeal to the Hispanic market, which is a growing segment of the U.S. population. In an effort to better penetrate that market, Frito-Lay hired various market analysts to determine why Hispanics do not purchase their products as often as company officials had hoped and what could be done about the problem.

Driving giant RVs through Hispanic neighborhoods and targeting Hispanic women (who tend to buy most of the groceries for their families), the business analysts tested various brands and discovered several things. Hispanics thought Frito-Lay products were too bland, not spicy enough. Hispanics also were relatively unaware of Frito-Lay advertising. In addition, they tended to purchase snacks in small bags rather than in large family-style bags and at small local grocery stores rather than at large supermarkets.

After the "road test," focus groups composed of male teens and male young adults—a group that tends to consume a lot of chips—were formed. The business analysts determined that while many of the teens spoke English at school, they spoke Spanish at home with their family. From this, it was concluded that Spanish advertisements would be needed to reach Hispanics. In addition, it was discovered that the use of Spanish rock music, a growing movement in the Hispanic youth culture, would be effective in some ads.

Business analysts also found that using a "Happy Face" logo, which is an icon of Frito-Lay's sister company in Mexico, was effective. Because it reminded the 63% of all Hispanics in the United States who are Mexican-American of snack foods from home, the logo increased product familiarity.

As a result of this research, Frito-Lay launched its first Hispanic products in San Antonio in 1997. Since that time, sales of the Doritos brand improved 32% in Hispanic areas and Doritos Salsa Verde sales have grown to represent 15% of all sales. Frito-Lay since has expanded its line of products into other areas of the United States with large Hispanic populations.

DISCUSSION

In the research process for Frito-Lay Company, many different numerical questions were raised regarding Frito-Lay products, advertising techniques, and purchase patterns among Hispanics. In each of these areas, statistics—in particular, hypothesis testing—plays a central role. Using the case information and the concepts of statistical hypothesis testing, discuss the following:

1. Many proportions were generated in the focus groups and market research that were conducted for this project, including the proportion of the market that is Hispanic, the proportion of Hispanic grocery shoppers that are women, the proportion of chip purchasers that are teens, and so on. Use techniques presented in this chapter to analyze each of the following and discuss how the results might affect marketing decision makers about the Hispanic market.

   a. The case information stated that 63% of all U.S. Hispanics are Mexican-American. How might we test that figure? Suppose 850 U.S. Hispanics are randomly selected using U.S. Bureau of the Census information. Suppose 575 state that they are Mexican-Americans. Test the 63% percentage using an alpha of .05.

   b. Suppose that in the past 94% of all Hispanic grocery shoppers are women. Perhaps due to changing cultural values, we believe that more Hispanic men are now grocery shopping. We randomly sample 689 Hispanic grocery shoppers from around the United States and 606 are women. Is this enough evidence to conclude that a lower proportion of Hispanic grocery shoppers now are women?

   c. What proportion of Hispanics listen primarily to advertisements in Spanish? Suppose one source says that in the past the proportion has been about .83. We want to test to determine whether this figure is true. A random sample of 438 Hispanics is selected, and 347 listened primarily in Spanish. Use $\alpha = .05$ and any appropriate calculations to reach some conclusions from these data.

2. The statistical mean can be used to measure various aspects of the Hispanic culture and the Hispanic market, including size of purchase, frequency of purchase, age of consumer, size of store, and so on. Use techniques presented in this chapter to analyze each of the following and discuss how the results might affect marketing decisions.

   a. What is the average age of a purchaser of Doritos Salsa Verde? Suppose initial tests indicate that the mean age is 31. Is this really correct? To test this, a business analyst randomly contacts 24 purchasers of Doritos Salsa Verde and asks their age with results shown below. Discuss the output in terms of a hypothesis test to determine whether the mean age is actually 31. Let α be .01.

$$\bar{X} = 27.61415$$
$$t = -1.8557$$
$$p\text{-value (two-tailed)} = .07635$$

   b. What is the average expenditure of a Hispanic customer on chips per year? Suppose it is hypothesized that the figure is $45 per year. A business analyst who knows the Hispanic market believes that this figure is too high and wants to prove her case. She randomly selects 18 Hispanics, has them keep a log of grocery purchases for one year, and obtains the following figures. Analyze the data using techniques from this chapter and an alpha of .05.

$55	37	59	57	27	28
16	46	34	62	9	34
4	25	38	58	3	50

ADAPTED FROM: "From Bland to Brand," *American Demographics,* March 1999, 57; the Frito-Lay website at http://www.fritolay.com; and Alsop, Ronald J., Editor. *The Wall Street Journal Almanac 1999.* New York: Ballantine Books, 1998 by Dow Jones & Company, Inc., 202.

10
Hypothesis Testing with Two Samples

Learning Objectives

Chapter 10 focuses on hypothesis testing about parameters from two populations, thereby enabling you to:

1. Test hypotheses about the difference in two population means using data from large independent samples.

2. Test hypotheses about the difference in two population means using data from small independent samples when the populations are normally distributed.

3. Test hypotheses about the population mean difference in two related populations when the populations are normally distributed.

4. Test hypotheses about the difference in two population proportions.

5. Test hypotheses about two population variances when the populations are normally distributed.

The presentation of hypothesis testing in Chapter 9 was centered on drawing a single sample from one population. Included in Chapter 9 were hypothesis tests about a population mean and a population proportion. In Chapter 10, we examine a number of hypothesis testing techniques that utilize two samples in an effort to test inferences about two populations. Some examples of such tests are:

- Testing the hypothesis that there was an increase in annual consumer expenditures on automobile insurance from the year 1991 to the year 2001 using a random sample from each year.
- Testing the hypothesis that there is a difference in the mean tensile strength of metal rods produced on two machines by taking a random sample of rods from each machine.
- Testing the hypothesis that female managers score higher on sensitivity ratings than males by taking a random sample of male and female managers matched on age, experience, and level of responsibility from each of several companies.
- Testing the hypothesis that there is a difference in the market shares of a given product in two markets using a sample from each market.
- Testing the hypothesis that there is a greater variance in the thickness of plastic bottles produced at manufacturing plant A than there is at manufacturing plant B by taking a random sample of bottles at each plant.

Independent samples

Two or more samples in which the selected items are related only by chance.

Dependent samples, or Related samples

Two or more samples selected in such a way as to be dependent or related; each item or person in one sample has a corresponding matched or related item in the other samples.

In this chapter, we will consider several different techniques for analyzing data that come from two samples. One technique is used with proportion, one is used with variances, and the others are used with means. The techniques for analyzing means are separated into those used with large samples and those used with small samples. In four of the five techniques presented in this chapter, the two samples are assumed to be **independent samples.** The samples are independent because *the items or people sampled in each group are in no way related to those in the other group.* Any similarity between items or people in the two samples is coincidental and due to chance. One of the techniques presented in the chapter is for analyzing data from two **dependent** or **related samples,** which are *samples selected in such a way as to be dependent or related.* In this case, *items or persons in one sample are matched in some way with items or persons in the other sample.* We begin with techniques for analyzing the difference in two independent large samples by using means.

10.1
Hypothesis Testing about the Difference in Two Means: Large, Independent Samples

In certain research designs, the sampling plan calls for selecting two different, independent samples. If the business analyst selects the mean as the statistic and if *two* samples are randomly chosen, the business analyst has *two* sample means to compare. This type of analysis is particularly useful in business when the business analyst is attempting to determine, for example, whether there is a difference in the effectiveness of two brands of toothpaste or the difference in wear of two brands of tires. Research might be conducted to study the difference in the productivity of men and women on an assembly line under certain conditions. An engineer might want to determine differences in the strength of aluminum produced under two different temperatures. Is there a difference in the average cost of a two-bedroom, one-story house between Boston and Seattle? If so, how much is the difference? These and many other interesting questions can be researched by comparing results obtained from two random samples.

How does a business analyst approach the analysis of the difference of two samples by using sample means? The central limit theorem states that the difference in two sample means, $\overline{X}_1 - \overline{X}_2$, is normally distributed for large sample sizes (both n_1 and $n_2 \geq 30$) regardless of the shape of the populations. It can also be shown that

$$\mu_{\bar{X}_1 - \bar{X}_2} = \mu_1 - \mu_2$$

$$\sigma_{\bar{X}_1 - \bar{X}_2} = \sqrt{\frac{\sigma_1^2}{n_1} + \frac{\sigma_2^2}{n_2}}$$

These expressions lead to a test statistic for the difference in two sample means.

(10.1)

$$Z = \frac{(\bar{X}_1 - \bar{X}_2) - (\mu_1 - \mu_2)}{\sqrt{\dfrac{\sigma_1^2}{n_1} + \dfrac{\sigma_2^2}{n_2}}}$$

Z FORMULA FOR THE DIFFERENCE IN TWO SAMPLE MEANS FOR n_1 AND $n_2 \geq 30$ (INDEPENDENT SAMPLES)

where:

μ_1 = the mean of population 1
μ_2 = the mean of population 2
n_1 = size of sample 1
n_2 = size of sample 2

σ_1^2 = the variance of population 1
σ_2^2 = the variance of population 2
$\bar{X}_1$ = the mean of sample 1
$\bar{X}_2$ = the mean of sample 2

This formula makes possible the solution of problems involving two random independent samples and their means.

Note: If two populations are known to be normally distributed on the measurement being studied and if the population variances are known, Formula 10.1 can be used for small sample sizes (n_1 or $n_2 < 30$).

In addition, sample variances can be used in Formula 10.1 in place of population variances when population variances are unknown and sample sizes are large (n_1, $n_2 \geq 30$) because the sample variances are good approximations of the population variances for large sample sizes.

Hypothesis Testing

In many instances, a business analyst wants to test the differences in the mean values of two populations. One example might be to test the difference between the mean values of men and women for achievement, intelligence, or other characteristics. A consumer organization might want to test two brands of light bulbs to determine whether one burns longer than the other. A company wanting to relocate might want to determine whether there is a significant difference in the average price of a home between Newark, New Jersey, and Cleveland, Ohio. Formula 10.1 can be used to test the difference between two population means.

As a specific example, suppose we want to conduct a hypothesis test to determine if the average annual wage for an advertising manager is different from the average annual wage of an auditing manager. Because we are testing to determine if the means are different, it might seem logical that the null and alternative hypotheses would be:

$$H_0: \mu_1 = \mu_2$$

$$H_a: \mu_1 \neq \mu_2$$

where: advertising managers are population 1
auditing managers are population 2

However, statisticians generally construct these hypotheses as:

$$H_0: \mu_1 - \mu_2 = \delta$$
$$H_a: \mu_1 - \mu_2 \neq \delta$$

This allows the business analyst not only to test if the population means are equal, but also affords her the opportunity to hypothesize about a particular difference in the means (δ). Generally speaking, most business analysts are only interested in testing whether or not there is a zero difference in the means. Thus, δ is set equal to zero resulting in the following hypotheses which we will use for this problem (and many others):

$$H_0: \mu_1 - \mu_2 = 0$$
$$H_a: \mu_1 - \mu_2 \neq 0$$

A random sample of 32 advertising managers from across the United States is taken. The advertising managers are contacted by telephone and asked what is their annual salary. A similar random sample is taken of 34 auditing managers. The resulting salary data are listed in Table 10.1, along with the sample means, the sample standard deviations, and the sample variances.

In this problem, the business analyst is testing to determine whether there is a difference in the average wage of an advertising manager and an auditing manager; and therefore, the test is two-tailed. If the business analyst had hypothesized that one type of manager was paid more than the other (or less than the other), the test would have been one-tailed.

Suppose $\alpha = .05$. Because this is a two-tailed test, each of the two rejection regions has an area of .025, leaving .475 of the area in the distribution between each critical value and the mean of the distribution. The associated critical table $Z_{\alpha/2}$ value for this area is $Z_{.025} = \pm 1.96$. Figure 10.1 shows the critical table Z value along with the rejection regions.

Note that the population variances are not available in this problem. So long as the sample size is large, S^2 is a good approximation of σ^2. Hence, the following formula is equivalent to Formula 10.1 for large samples.

Z FORMULA TO TEST THE DIFFERENCE IN POPULATION MEANS WITH σ_1^2, σ_2^2 UNKNOWN AND n_1, n_2 LARGE; INDEPENDENT SAMPLES	$$Z = \dfrac{(\bar{X}_1 - \bar{X}_2) - (\mu_1 - \mu_2)}{\sqrt{\dfrac{S_1^2}{n_1} + \dfrac{S_2^2}{n_2}}}$$	(10.2)

Formula 10.2 and the data in Table 10.1 yield a Z value to complete the hypothesis test.

$$Z = \frac{(70.700 - 62.187) - (0)}{\sqrt{\dfrac{264.164}{32} + \dfrac{166.411}{34}}} = 2.35$$

The observed value of 2.35 is greater than the critical value obtained from the Z table, 1.96. The business analyst rejects the null hypothesis and can say that there is a significant difference between the average annual wage of an advertising manager and the average an-

ADVERTISING MANAGER	AUDITING MANAGER
74.256	69.962
96.234	55.052
89.807	57.828
93.261	63.362
103.030	37.194
74.195	99.198
75.932	61.254
80.742	73.065
39.672	48.036
45.652	60.053
93.083	66.359
63.384	61.261
57.791	77.136
65.145	66.035
96.767	54.335
77.242	42.494
67.056	83.849
64.276	67.160
74.194	37.386
65.360	59.505
73.904	72.790
54.270	71.351
59.045	58.653
68.508	63.508
71.115	43.649
67.574	63.369
59.621	59.676
62.483	54.449
69.319	46.394
35.394	71.804
86.741	72.401
57.351	56.470
	67.814
	71.492
$n_1 = 32$	$n_2 = 34$
$\overline{X}_1 = 70.700$	$\overline{X}_2 = 62.187$
$S_1 = 16.253$	$S_2 = 12.900$
$S_1^2 = 264.164$	$S_2^2 = 166.411$

TABLE 10.1
Wages for Advertising Managers and Auditing Managers ($1,000)

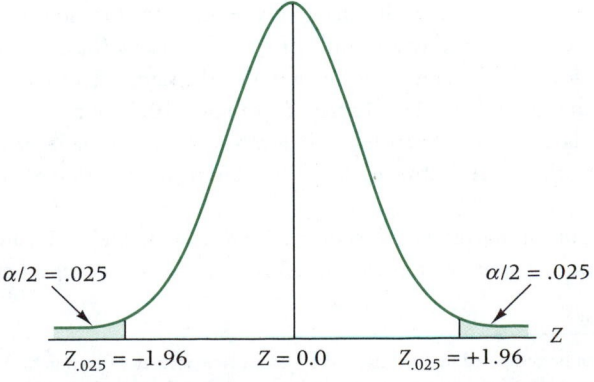

Figure 10.1

Critical values and rejection regions for the wage example

Figure 10.2

Location of
calculated Z value
for the wage example

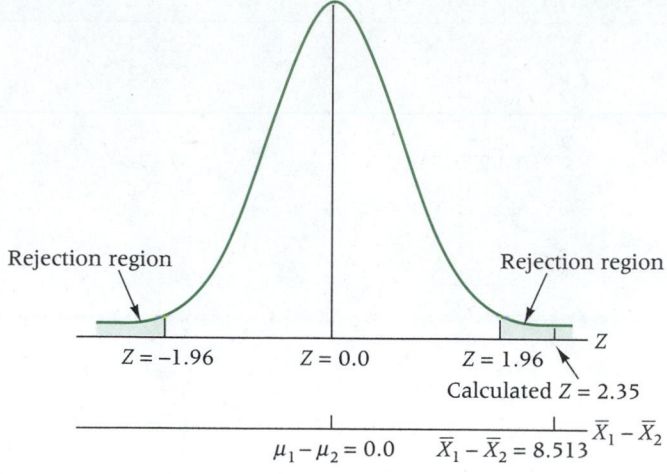

nual wage of an auditing manager. The business analyst then examines the sample means (70.700 for advertising managers and 62.187 for auditing managers) and uses common sense to conclude that advertising managers earn more, on the average, than do auditing managers. Figure 10.2 shows the relationship between the observed Z (2.35) and the critical Z (1.96).

This conclusion could have been reached by using the p-value method. Looking up the probability of $Z \geq 2.35$ in the Z distribution table in Appendix A.5 yields an area of $.5000 - .4906 = .0094$. This value is less than $\alpha/2 = .025$. The decision is to reject the null hypothesis.

Analysis Using Excel

Excel has the capability of testing hypotheses between two means from large samples using the Z test if the population variances are known. In most cases, business analysts do not know the values of the population variances. However, as pointed out earlier, for large sample sizes the sample variances are close enough approximations of the population variances to suffice. Thus, we can use the Excel **z-Test: Two Sample for Means** data analysis tool by first computing the sample variances and substituting these for the unknown population variances*. After obtaining the sample variances, the Z test for two sample means can be accomplished by selecting **Tools** from the menu bar and **Data Analysis** from the subsequent drop-down menu. **z-Test: Two Sample for Means** is one of the options in the **Data Analysis** scrolling list box. The subsequent dialog box for **z-Test: Two Sample for Means** is shown in Figure 10.3. Note that the dialog box requires the locations of the data for each of the two variables, the hypothesized difference of the means (in most cases this is zero), the variances of each variable, and the value of alpha, the significance level of the test.

Observe that the Excel dialog box displayed in Figure 10.3 contains the locations of the two columns of data for the advertising manager and auditing manager problem, the hypothesized mean difference (zero), each of the values of the sample variances, and the value of alpha.

The Excel output for this problem is shown in Figure 10.4. The results include the means, variances, and sample sizes of the two samples along with the hypothesized difference.

---

*Sample variances can be computed in Excel using the **Descriptive Statistics** function of the Data Analysis tool or by using selecting the **VAR** function from the **Statistics** category of the Paste Function

Figure 10.3 *z*-Test: Two Sample for Means Excel dialog box

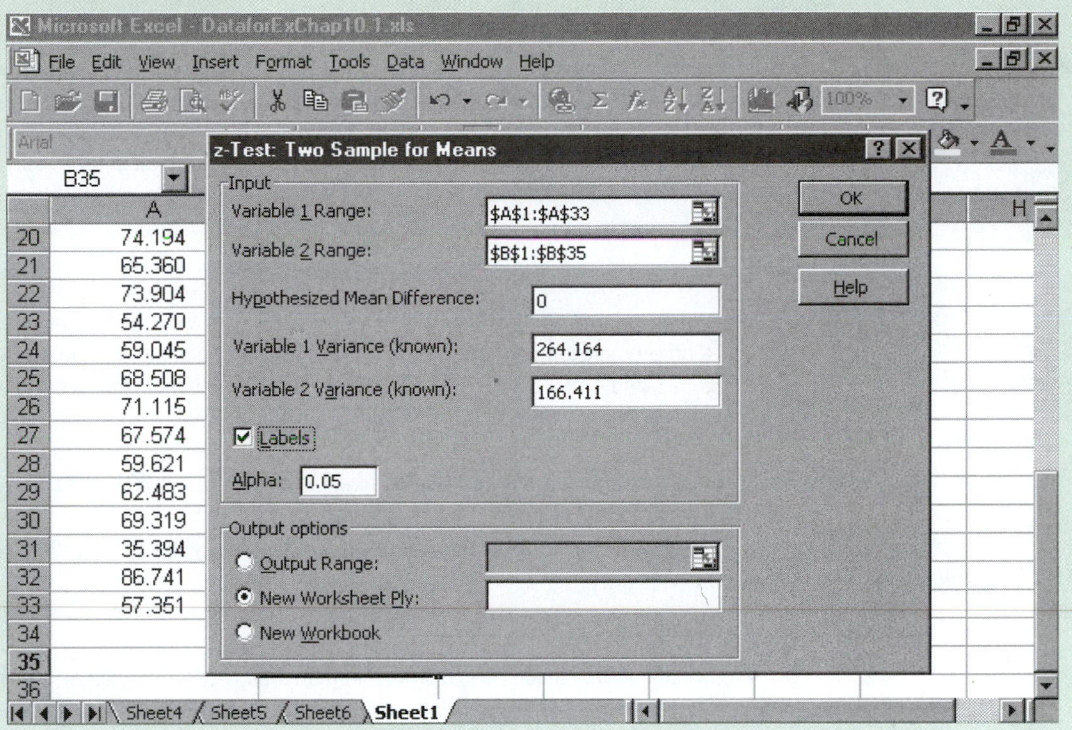

Figure 10.4 Excel output for wage problem

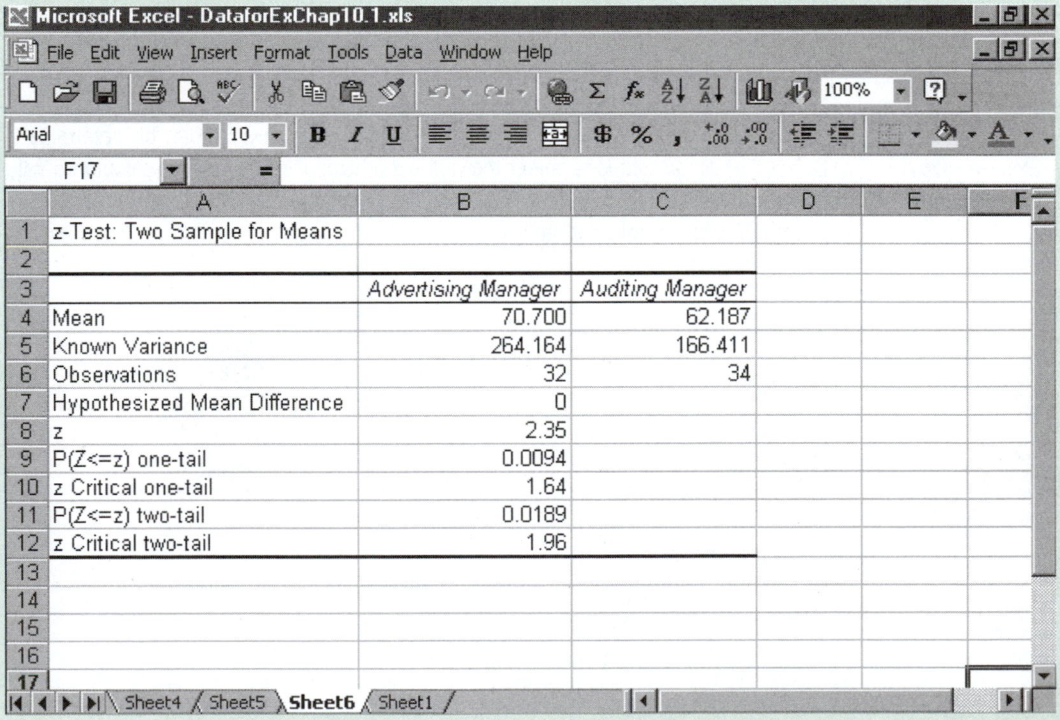

z-Test: Two Sample for Means		
	Advertising Manager	*Auditing Manager*
Mean	70.700	62.187
Known Variance	264.164	166.411
Observations	32	34
Hypothesized Mean Difference	0	
z	2.35	
P(Z<=z) one-tail	0.0094	
z Critical one-tail	1.64	
P(Z<=z) two-tail	0.0189	
z Critical two-tail	1.96	

In addition, the output includes the observed Z value, $Z = 2.35$, and its associated p-value of .0189 for a two-tail test indicating statistical significance at $\alpha = .05$. The p-value for a one-tail test, .0094, is also given and can be used in the wage problem to compare against $\alpha/2$ (one-tail) = .025. The Excel output also includes critical table values for business analysts who prefer using the table rather than using p-values. The critical table Z value is 1.96. Since the observed Z value of 2.35 is greater than the critical table Z value of 1.96, the decision is to reject the null hypothesis, which is consistent with the conclusion reached using p-values.

DEMONSTRATION PROBLEM 10.1	A sample of 87 professional working women showed that the average amount paid annually into a private pension fund per person was $3343, with a sample standard deviation of $1226. A sample of 76 professional working men showed that the average amount paid annually into a private pension fund per person was $5568, with a sample standard deviation of $1716. A women's activist group wants to "prove" that women do not pay as much per year as men into private pension funds. If they use $\alpha = .001$ and these sample data, will they be able to reject a null hypothesis that women annually pay the same as or more than men into private pension funds? Use the eight-step hypothesis-testing process.

SOLUTION

STEP **1.** This test is one-tailed. Because the women's activist group wants to prove that women pay less than men into private pension funds annually, the alternative hypothesis should be $\mu_w - \mu_m < 0$, and the null hypothesis is that women pay the same as or more than men, $\mu_w - \mu_m = 0$.

STEP **2.** The test statistic is

$$Z = \frac{(\bar{X}_1 - \bar{X}_2) - (\mu_1 - \mu_2)}{\sqrt{\dfrac{S_1^2}{n_1} + \dfrac{S_2^2}{n_2}}}.$$

STEP **3.** Alpha has been specified as .001.

STEP **4.** By using this value of alpha, a critical $Z_{.001} = -3.08$ can be determined. The decision rule is to reject the null hypothesis if the observed calculated value of the test statistic, Z, is less than -3.08.

STEP **5.** The sample data follow.

WOMEN	MEN
$X_1 = 3343$	$\bar{X}_2 = 5568$
$S_1 = 1226$	$S_2 = 1716$
$n_1 = 87$	$n_2 = 76$

STEP **6.** Solving for Z gives

$$\frac{(3343 - 5568) - (0)}{\sqrt{\dfrac{1226^2}{87} + \dfrac{1716^2}{76}}} = -9.40.$$

STEP **7.** The observed calculated Z value of -9.40 is deep in the rejection region, well past the table value of $Z_c = -3.08$. Even with the small $\alpha = .001$, the null hypothesis is rejected.

STEP **8.** The evidence is substantial that women, on average, pay less than men into private pension funds annually. The following diagram displays these results.

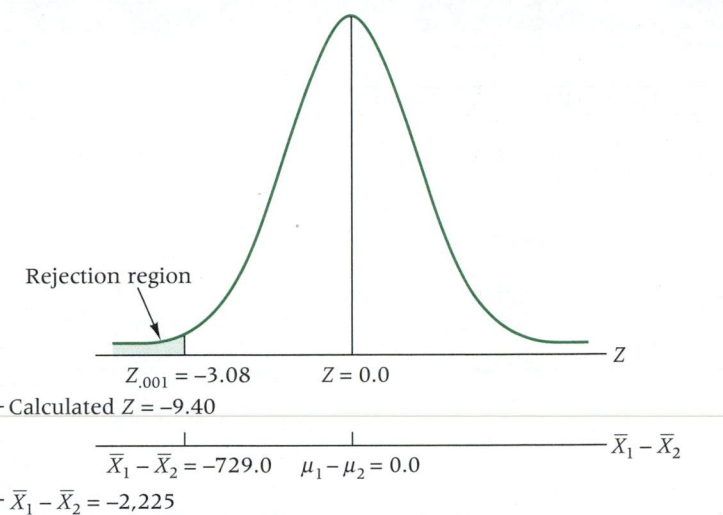

The probability of obtaining a calculated Z value of -9.40 by chance is virtually zero, because the value is beyond the limits of the Z table. By the p-value method, the null hypothesis is rejected because the probability is .0000, or less than $\alpha = .001$.

If this problem were worked by the critical value method, what critical value of the difference in the two means would have to be surpassed to reject the null hypothesis for a table Z value of -3.08? The answer is

$$(\bar{X}_1 - \bar{X}_2)_c = (\mu_1 - \mu_2) - Z\sqrt{\frac{S_1^2}{n_1} + \frac{S_2^2}{n_2}}$$

$$= 0 - 3.08(236.7) = -729.0.$$

The difference in sample means would need to be at least 729.0 to reject the null hypothesis. The actual sample difference in this problem was -2225, which is considerably larger than the critical value of difference. Thus, with the critical value method also, the null hypothesis is rejected.

Shown on the next page is the Excel output for Demonstration Problem 10.1. Note that the observed $Z = -9.40$ has an associated p-value of 0 confirming the conclusion reached above to reject the null hypothesis. The one-tail critical table Z value is reported by Excel to be -3.09. Since the observed $Z = -9.40$ is less than the critical table Z value of -3.09, the decision is to reject the null hypothesis.

```
Microsoft Excel - DataforExChap10.1.xls                                    _ 8 X
File  Edit  View  Insert  Format  Tools  Data  Window  Help                _ 8 X

[toolbar]  100%

Arial        10   B  I  U   ≡ ≡ ≡ 国   $ % ,   .0 .00  ≡ ≡    ▪ ▪ A ▪ ▪

   G16           =
        A                    B              C            D       E       F
1  z-Test: Two Sample for Means
2
3                      Working Women   Working Men
4  Mean                        3343          5568
5  Known Variance           1502818       2944276
6  Observations                  87            76
7  Hypothesized Mean Difference   0
8  z                           -9.40
9  P(Z<=z) one-tail                0
10 z Critical one-tail          3.09
11 P(Z<=z) two-tail                0
12 z Critical two-tail          3.29
13
14
15
16
17
|◄ ◄ ► ►|\ Sheet6 / Sheet2 / Sheet3 / Sheet7 \ Sheet8 / Sheet1 / |◄|
```

10.1
Problems

10.1 Consider the data given below.

SAMPLE 1						SAMPLE 2				
58	50	47	43	53		47	47	63	55	52
60	46	60	55	52		50	45	48	52	53
60	46	59	49	57		47	48	58	61	57
67	50	49	55	59		47	59	62	54	63
46	46	42	61	53		52	62	45	49	57
48	54	52	46	50		55	50	49	57	55
65						61	57			

a. Test the following hypotheses of the difference in population means on these data with $\alpha = .10$

$$H_0: \mu_1 - \mu_2 = 0 \quad H_a: \mu_1 - \mu_2 < 0$$

b. Use the critical value method to find the critical difference in the mean values required to reject the null hypothesis.

c. What is the p-value for this problem?

10.2 Examine the following data. Use the data to test the following hypotheses ($\alpha = .02$).

$$H_0: \mu_1 - \mu_2 = 0 \quad H_a: \mu_1 - \mu_2 \neq 0$$

SAMPLE 1			SAMPLE 2		
90	88	80	78	85	82
88	87	91	90	80	76
81	84	84	77	75	79
88	90	91	82	83	88
89	95	97	80	90	74
88	83	94	81	75	76
81	83	88	83	88	77
87	87	93	86	90	75
88	84	83	80	80	74
95	93	97	89	84	79

10.3 The Trade Show Bureau conducted a survey to determine why people go to trade shows. The respondents were asked to rate a series of reasons on a scale from 1 to 5, with 1 representing little importance and 5 representing great importance. Shown below are the responses for 50 people from the computers/electronics industry and 50 people from the food/beverage industry. Use these data and $\alpha = .01$ to determine whether there is a significant difference between people in these two industries on this question.

COMPUTERS/ELECTRONICS											FOOD/BEVERAGE									
1	2	1	3	2	0	3	3	2	1		3	3	2	4	3	4	5	2	4	3
3	3	1	2	2	3	2	2	2	2		3	2	3	2	3	4	3	3	3	3
1	2	3	2	1	1	1	3	3	2		2	4	2	3	3	2	4	4	4	4
2	1	4	1	4	2	3	0	1	0		3	5	3	3	2	2	0	2	2	5
3	3	2	2	3	2	1	0	2	3		4	3	3	2	3	4	3	3	3	2

10.4 A company's auditor believes the per diem cost in Nashville, Tennessee, rose significantly between 1995 and 2001. To test this belief, the auditor samples 33 business trips from the company's records for 1995 and 36 business trips for 2001. The data are shown below. Test to determine if the average per diem cost rose significantly from 1995 to 2001. Let the risk of committing a Type I error be .01.

1995						2001					
176	164	206	183	184	193	188	181	194	198	189	205
217	182	180	177	181	196	207	189	213	208	167	208
194	209	192	206	177	156	213	209	182	200	212	197
199	169	186	203	202	189	158	217	188	191	191	182
185	214	185	172	207	193	204	190	194	224	186	202
220	168	189				159	189	192	178	201	196

10.5 Suppose a market analyst wants to determine the difference in the average price of a gallon of whole milk in Seattle and Atlanta. To do so, he takes a telephone survey of 31 randomly selected consumers in Seattle. He first asks whether they have purchased a gallon of milk during the past 2 weeks. If they say no, he continues to select consumers until he selects $n = 31$ people who say yes. If they say yes, he asks them how much they paid for the milk. The analyst undertakes a similar survey in Atlanta with 31 respondents. Using the resulting sample information that follows, use $\alpha = .05$ to test to determine if there is a significant difference in the mean price of a gallon of milk between the two cities.

	SEATTLE			ATLANTA	
$2.55	$2.36	$2.43	$2.25	$2.40	$2.39
2.67	2.54	2.43	2.30	2.33	2.40
2.50	2.54	2.38	2.19	2.29	2.23
2.61	2.80	2.49	2.41	2.18	2.29
3.10	2.61	2.57	2.39	2.59	2.53
2.86	2.56	2.71	2.26	2.38	2.19
2.50	2.64	2.97	2.19	2.25	2.45
2.47	2.72	2.65	2.42	2.61	2.33
2.76	2.73	2.80	2.60	2.25	2.51
2.65	2.83	2.69	2.38	2.29	2.36
		2.71			2.44

10.6 Employee suggestions can provide useful and insightful ideas for management. Some companies solicit and receive employee suggestions more than others, and company culture influences the use of employee suggestions. Suppose a study is conducted to determine whether there is a significantly higher mean number of suggestions annually per employee at the Johnson Corporation than at the Davison Corporation. Random samples of employees are selected from each Corporation and the resulting data are given below. Use these data and $\alpha = .10$ to test the hypothesis.

		JOHNSON											DAVISON								
6	6	7	6	9	4	8	4	6	6		7	5	4	6	4	6	6	7	5	6
8	8	7	7	5	7	4	5	5	3		6	6	7	6	3	4	5	6	3	5
6	5	5	3	5	5	5	9	6	4		4	5	4	5	4	7	5	3	5	4
3	2	8	4	6	9						4	3	8	4	5	5	5	6	4	6
											6	8	4	3	4					

10.2
Hypothesis Testing about the Difference in Two Population Means: Small Independent Samples

The techniques presented in Section 10.1 are for use whenever sample sizes are large or the population variances are known. There are many occasions when business analysts wish to test hypotheses about the difference in two population means and the population variances are not known. If the sample sizes are small, the Z methodology is not appropriate. This section presents methodology for handling the small-sample situation when the population variances are unknown.

Hypothesis Testing

The hypothesis test presented in this section is a test that compares the means of two samples to determine whether there is a difference in the two population means from which the samples come. An assumption underlying this technique is that the measurement or characteristic being studied is normally distributed for both populations. This technique is used whenever sample size is small (n_1, $n_2 < 30$), the population variances are unknown (and hence the sample variances must be used), and the samples are independent (not related an any way). In Section 10.1, the difference in large sample means was analyzed by Formula 10.1:

$$Z = \frac{(\bar{X}_1 - \bar{X}_2) - (\mu_1 - \mu_2)}{\sqrt{\dfrac{\sigma_1^2}{n_1} + \dfrac{\sigma_2^2}{n_2}}}.$$

If $\sigma_1^2 = \sigma_2^2$, Formula 10.1 algebraically reduces to

$$Z = \frac{(\bar{X}_1 - \bar{X}_2) - (\mu_1 - \mu_2)}{\sigma \sqrt{\dfrac{1}{n_1} + \dfrac{1}{n_2}}}.$$

If σ is unknown, it can be estimated by *pooling* the two sample variances and computing a pooled sample standard deviation.

$$\sigma \approx S = \sqrt{\frac{S_1^2(n_1 - 1) + S_2^2(n_2 - 1)}{n_1 + n_2 - 2}}$$

Substituting this expression for σ and changing Z to t produces a formula to test the difference in means.

(10.3) $\quad t = \dfrac{(\bar{X}_1 - \bar{X}_2) - (\mu_1 - \mu_2)}{\sqrt{\dfrac{S_1^2(n_1 - 1) + S_2^2(n_2 - 1)}{n_1 + n_2 - 2}}\sqrt{\dfrac{1}{n_1} + \dfrac{1}{n_2}}}$ $\mathrm{df} = n_1 + n_2 - 2$	*t* FORMULA TO TEST THE DIFFERENCE IN MEANS ASSUMING $\sigma_1^2 = \sigma_2^2$

Formula 10.3 is constructed by assuming that the two population variances, σ_1^2 and σ_2^2, are equal. Thus, when using Formula 10.3 to test hypotheses about the difference in two means for small independent samples when the population variances are unknown, we must assume that the two samples come from populations in which the variances are essentially equal. If that is not possible, the following formula should be used.

(10.4) $\quad t = \dfrac{\bar{X}_1 - \bar{X}_2}{\sqrt{\dfrac{S_1^2}{n_1} + \dfrac{S_2^2}{n_2}}}$ $\mathrm{df} = \dfrac{\left[\dfrac{S_1^2}{n_1} + \dfrac{S_2^2}{n_2}\right]^2}{\dfrac{\left(\dfrac{S_1^2}{n_1}\right)^2}{n_1 - 1} + \dfrac{\left(\dfrac{S_2^2}{n_2}\right)^2}{n_2 - 1}}$	*t* FORMULA TO TEST THE DIFFERENCE IN MEANS ASSUMING $\sigma_1^2 \neq \sigma_2^2$

In Formula 10.4, the population variances are not assumed to be equal. Since this formula has a more complex degrees-of-freedom component, it may be unattractive to some users. Excel offers the user a choice of the "equal variances" formula or the "unequal variances" formula. The "equal variances" formula in Excel is Formula 10.3, in which equal population variances are assumed. The "unequal variances" formula is Formula 10.4 and is used when population variances cannot be assumed to be equal. Again, in each of these

TABLE 10.2
Test Scores for New
Employees After Training

TRAINING METHOD A					TRAINING METHOD B			
56	50	52	44	52	59	54	55	65
47	47	53	45	48	52	57	64	53
42	51	42	43	44	53	56	53	57

formulas, the populations from which the two samples are drawn are assumed to be normally distributed for the phenomenon being measured.

At the Hernandez Manufacturing Company, an application of the test of the difference in small sample means arises. New employees are expected to attend a 3-day seminar to learn about the company. At the end of the seminar, they are tested to measure their knowledge about the company. The traditional training method has been lecture and a question and answer session. Management has decided to experiment with a different training procedure, which processes new employees in 2 days by using videocassettes and having no question and answer session. If this procedure works, it could save the company thousands of dollars over a period of several years. However, there is some concern about the effectiveness of the 2-day method, and company managers would like to know whether there is any difference in the effectiveness of the two training methods.

To test the difference in the two methods, the managers randomly select one group of 15 newly hired employees to take the 3-day seminar (method A) and a second group of 12 new employees for the 2-day videocassette method (method B). Table 10.2 shows the test scores of the two groups. Using $\alpha = .05$, the managers want to determine whether there is a significant difference in the mean scores of the two groups. They assume that the scores for this test are normally distributed and that the population variances are approximately equal.

STEP **1.**   The hypotheses for this test follow.

$$H_0: \mu_1 - \mu_2 = 0$$
$$H_a: \mu_1 - \mu_2 \neq 0$$

STEP **2.**   The statistical test to be used is Formula 10.3.

STEP **3.**   The value of alpha is .05.

STEP **4.**   Because the hypotheses are = and ≠, this test is two-tailed. The degrees of freedom are 25 (15 + 12 − 2 = 25) and alpha is .05. The t table requires an alpha value for one tail only, and, as this test is two-tailed, alpha is split from .05 and .025 to obtain the table t value: $t_{.025,25} = \pm 2.060$.

The null hypothesis will be rejected if the observed t value is less than −2.060 or greater than +2.060.

STEP **5.**   The sample data are given in Table 10.2. From these data, we can calculate the sample statistics. The sample means and variances follow.

METHOD A	METHOD B
$\overline{X}_1 = 47.73$	$\overline{X}_2 = 56.50$
$S_1^2 = 19.495$	$S_2^2 = 18.273$
$n_1 = 15$	$n_2 = 12$

STEP **6.**   The calculated value of t is

$$t = \frac{(47.73 - 56.50) - 0}{\sqrt{\dfrac{(19.495)(14) + (18.273)(11)}{(15 + 12 - 2)}} \sqrt{\dfrac{1}{15} + \dfrac{1}{12}}} = -5.20.$$

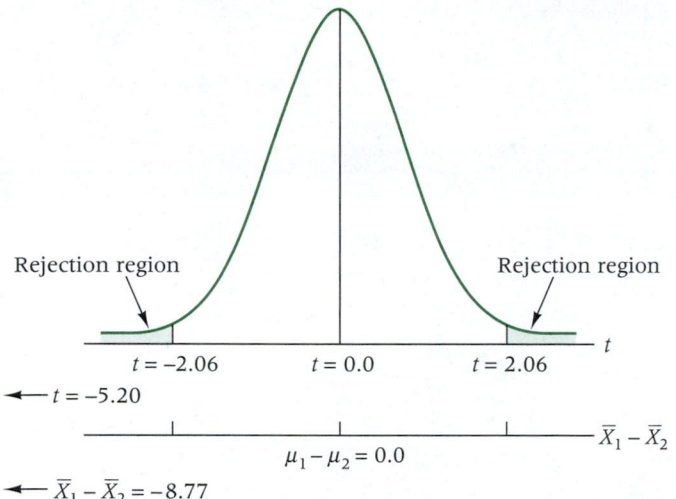

Figure 10.5

t Values for the training methods example

STEP **7.** Because the calculated value, $t = -5.20$, is less than the lower critical table value, $t = -2.06$, the calculated value of t is in the rejection region. The null hypothesis is rejected. There *is* a significant difference in the mean scores of the two tests.

STEP **8.** Figure 10.5 shows the critical areas, the calculated t value, and the decision for this test. Note that the computed t value is -5.20, which is enough to cause the managers of the Hernandez Manufacturing Company to reject the null hypothesis. Their conclusion is that there is a significant difference in the effectiveness of the training methods. Upon examining the sample means, they realize that method B (the 2-day videocassette method) actually produced an average score that was more than eight points higher than that for the group trained with method A.

In a test of this sort, which group is group 1 and which is group 2 is an arbitrary decision. If the two samples had been designated in reverse, the calculated t value would have been $t = +5.20$ (same magnitude but different sign), and the decision would have been the same.

Analysis Using Excel

Excel can test hypotheses about the difference in two means for independent samples using the t test for both the equal and the unequal variance cases using two separate **Data Analysis** tools. The equal variance t test of two means is **t-Test: Two-Sample Assuming Equal Variances,** and the dialog box for this test is shown in Figure 10.6. The t test of two means assuming unequal variances is **t-Test: Two-Sample Assuming Unequal Variances,** and the dialog box for this test is shown in Figure 10.7. Each of these dialog boxes requires the locations of the data for the two variables being analyzed, along with the hypothesized difference in means (usually zero) and the value of alpha.

The Hernandez Company training method problem can be analyzed using Excel, and the results are displayed in Figure 10.8. Note that the two-sample t-test assuming equal variances was used for this analysis. The output includes the mean, variance, and size of each sample, along with the hypothesized mean difference and the degrees of freedom for the test. The observed t value is -5.20, which is the same as that obtained previously using Formula 10.3. Since this problem involves a two-tail test, the critical table values are ± 2.06, even though Excel also lists the critical table value for a one-tail test (1.71). Since

Figure 10.6 — Excel dialog box for the **Two-Sample *t*-Test Assuming Equal Variances**

	N					V	
1	Method A	Meth					
2	56						
3	50						
4	52						
5	44						
6	52						
7	47						
8	47						
9	53						
10	45						
11	48						
12	42						
13	51						
14	42						
15	43						
16	44						

t-Test: Two-Sample Assuming Equal Variances

Input
Variable 1 Range: N1:N16
Variable 2 Range: O1:O13
Hypothesized Mean Difference: 0
☑ Labels
Alpha: 0.05

Output options
○ Output Range:
⊙ New Worksheet Ply:
○ New Workbook

OK   Cancel   Help

Sheet6 / Sheet2 / Sheet3 / Sheet7 / Sheet8 \ Sheet1

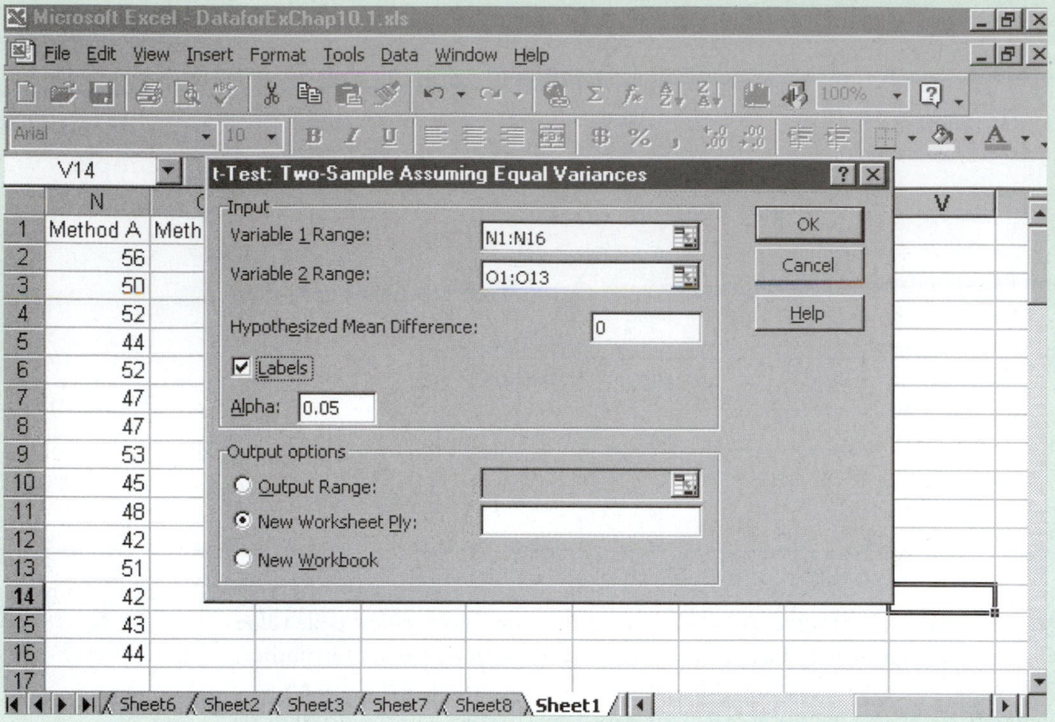

Figure 10.7 — Excel dialog box for the **Two-Sample *t*-Test Assuming Unequal Variances**

	N					V	
1	Method A	Meth					
2	56						
3	50						
4	52						
5	44						
6	52						
7	47						
8	47						
9	53						
10	45						
11	48						
12	42						
13	51						
14	42						
15	43						
16	44						

t-Test: Two-Sample Assuming Unequal Variances

Input
Variable 1 Range: $N$1:$N$16
Variable 2 Range: $O$1:$O$13
Hypothesized Mean Difference: 0
☑ Labels
Alpha: 0.05

Output options
○ Output Range:
⊙ New Worksheet Ply:
○ New Workbook

OK   Cancel   Help

Sheet2 / Sheet3 / Sheet7 / Sheet8 / Sheet9 \ Sheet1

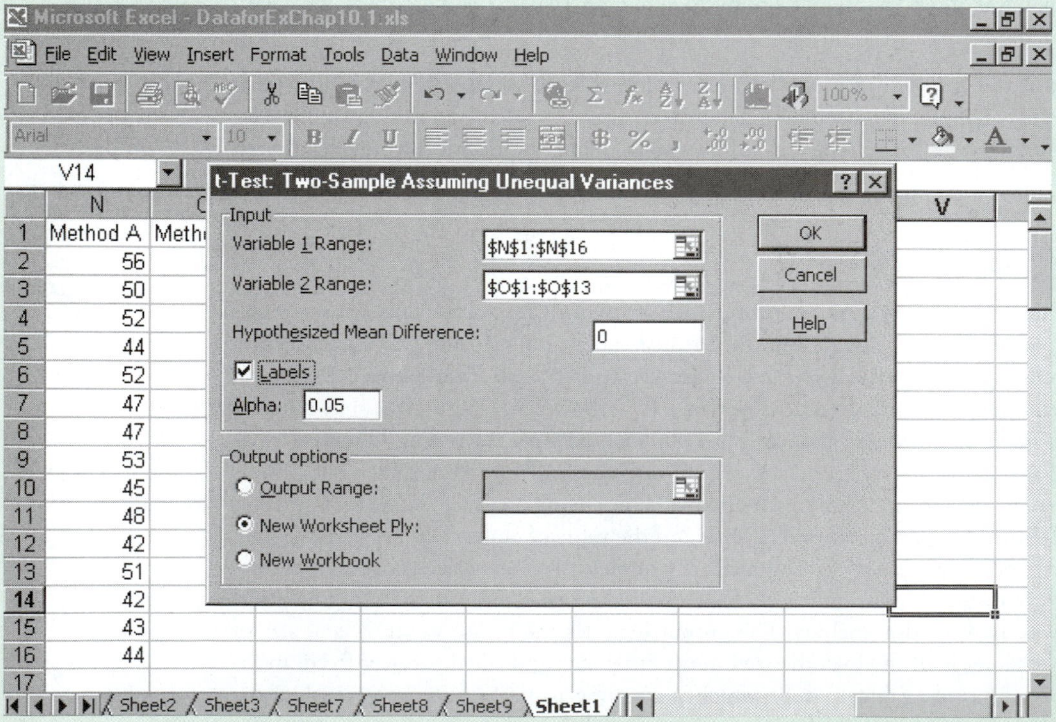

Figure 10.8 — Excel output for the Hernandez Company problem

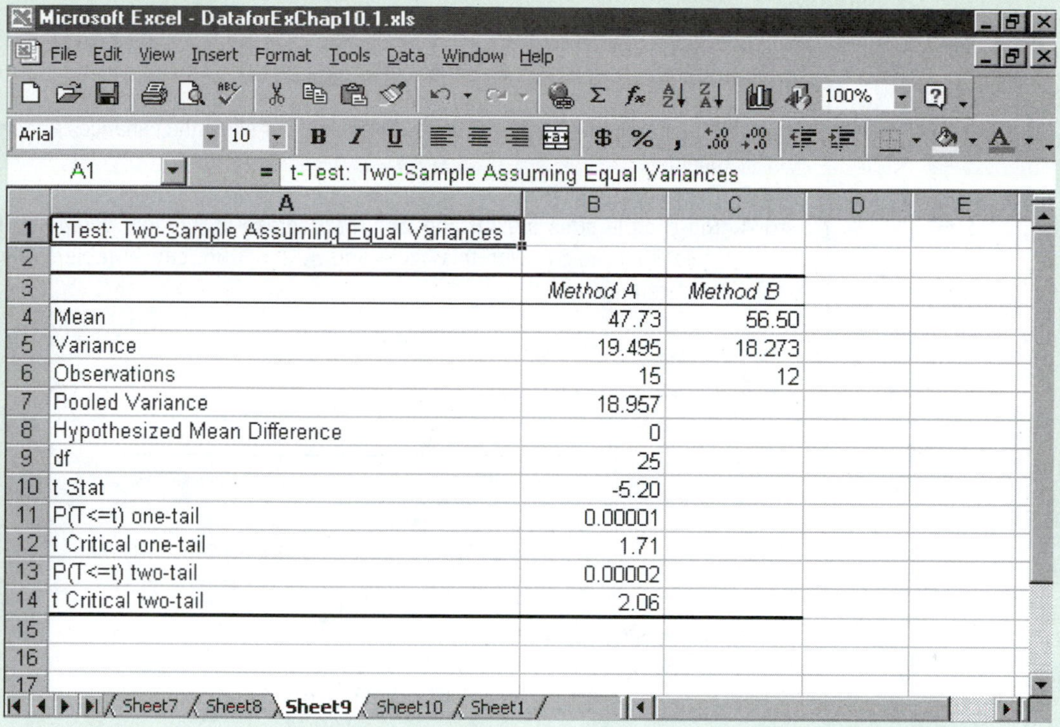

Microsoft Excel - DataforExChap10.1.xls

A1 = t-Test: Two-Sample Assuming Equal Variances

	A	B	C	D	E
1	t-Test: Two-Sample Assuming Equal Variances				
2					
3		Method A	Method B		
4	Mean	47.73	56.50		
5	Variance	19.495	18.273		
6	Observations	15	12		
7	Pooled Variance	18.957			
8	Hypothesized Mean Difference	0			
9	df	25			
10	t Stat	-5.20			
11	P(T<=t) one-tail	0.00001			
12	t Critical one-tail	1.71			
13	P(T<=t) two-tail	0.00002			
14	t Critical two-tail	2.06			

Sheet7 / Sheet8 / **Sheet9** / Sheet10 / Sheet1

the observed t value of -5.20 is less than the critical table value of -2.06, the decision is to reject the null hypothesis and conclude that the means for the populations are not equal. In addition, Excel reports the p-value for both a one-tail and a two-tail test. The two-tail p-value is .00002, which is less than $\alpha = .05$ and underscores the decision to reject the null hypothesis. This conclusion is consistent with the results obtained before using Formula 10.3. While not shown here, an analysis of this problem using the t-test for two-samples assuming unequal variances has been done. The only difference in the results is that the unequal variance-based t test yields an observed t value of -5.22 which is .02 different from that obtained using the equal variance-based test. Thus, for this problem, there is a negligible difference in the results.

DEMONSTRATION PROBLEM 10.2

Is there a difference in the way Chinese cultural values affect the purchasing strategies of industrial buyers in Taiwan and mainland China? A study by business analysts at the National Chiao-Tung University in Taiwan attempted to determine whether there is a significant difference in the purchasing strategies of industrial buyers in Taiwan and mainland China on the cultural dimension labeled "integration." Integration is being in harmony with one's self, family, and associates. For the study, 46 Taiwanese buyers and 26 mainland Chinese buyers were contacted and interviewed. Buyers were asked to respond to 35 items using a 9-point scale with possible answers ranging from no importance (1) to extreme importance (9). The resulting statistics for the two groups are shown in step 5. Using $\alpha = .01$, test to determine whether there is a significant difference between buyers of Taiwan and mainland China on integration.

SOLUTION

STEP **1.** If a two-tailed test is undertaken, the hypotheses and the table t value are as follows.

$$H_0: \mu_1 - \mu_2 = 0$$

$$H_a: \mu_1 - \mu_2 \neq 0$$

STEP **2.** The appropriate statistical test is Formula 10.4 since the variances appear to be unequal.

STEP **3.** The value of alpha is .01.

STEP **4.** The sample sizes are 46 and 26. The degrees of freedom computed by Formula 10.4 are 67. With this figure and $\alpha/2 = .005$, critical table t values can be determined.

$$t_{.005,67} = 2.70$$

STEP **5.** The sample data follow.

INTEGRATION

Taiwanese Buyers	Mainland Chinese Buyers
$n_1 = 46$	$n_2 = 26$
$X_1 = 5.47$	$X_2 = 5.03$
$S_1^2 = (.6958)^2 = .4842$	$S_2^2 = (.4838)^2 = .2341$

STEP **6.** The calculated t value is

$$t = \frac{(5.47 - 5.03) - (0)}{\sqrt{\dfrac{(.4842)}{46} + \dfrac{(.2341)}{26}}} = 3.15.$$

STEP **7.** Because the calculated value of $t = 3.15$ is greater than the critical table value of $t = 2.70$, the decision is to reject the null hypothesis.

STEP **8.** The Taiwan industrial buyers scored significantly higher than the mainland China industrial buyers on integration. Managers should keep in mind in dealing with Taiwanese buyers that they may be more likely to place worth on personal virtue and social hierarchy than do the mainland Chinese buyers.

The following graph shows the critical t values, the rejection regions, the observed t value, and the difference in the raw means.

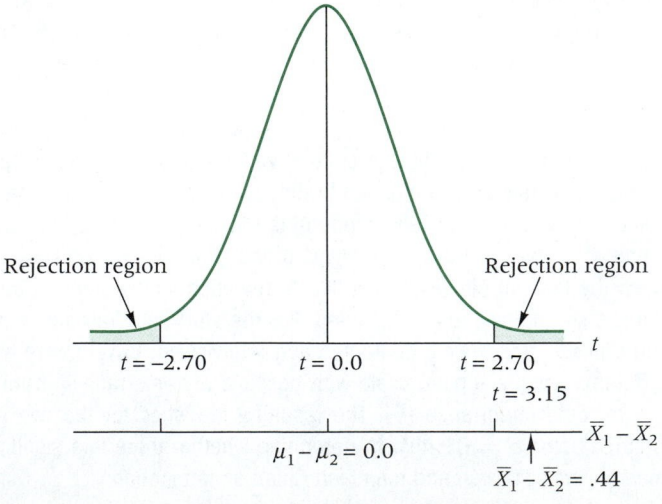

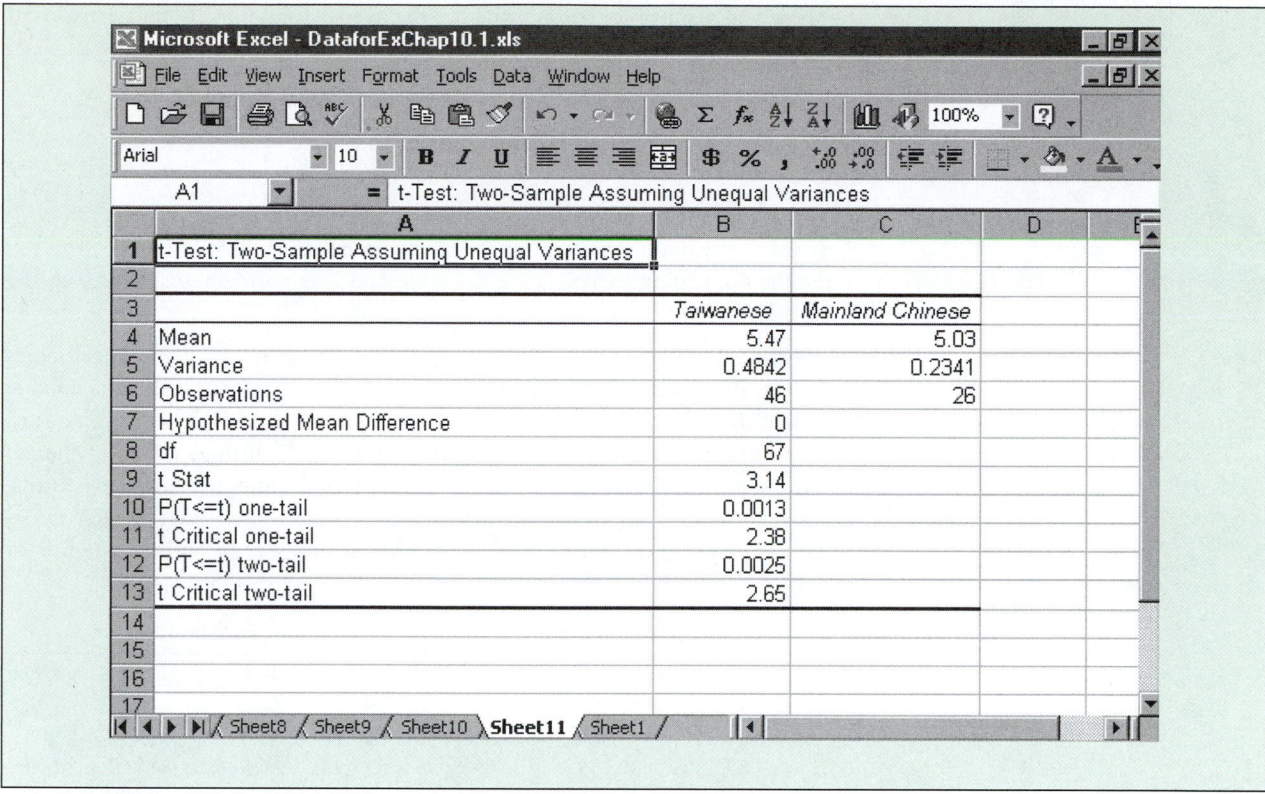

Shown above is the Excel output for Demonstration Problem 10.2 using the **t-Test: Two-Sample Assuming Unequal Variances** analysis. Note that the observed t value is 3.14, which is essentially the same as the value obtained above using the formula. The p-value associated with this observed t is .0025, which is less than the value of α, which is .01, indicating that the decision should be to reject the null hypothesis. The Excel-reported critical table t value for a two-tail test with these degrees of freedom is 2.65.

10.7 Use the data given below to test the following hypotheses.

$$H_0: \mu_1 - \mu_2 = 0 \qquad H_a: \mu_1 - \mu_2 < 0$$

SAMPLE 1					SAMPLE 2			
24.43	23.79	23.19	28.91		27.76	33.28	26.08	28.17
21.39	20.90	23.87	24.63		30.18	28.19	29.76	21.54
					27.23	18.60	25.86	

10.8 Suppose that for years the mean of population 1 has been accepted to be the same as the mean of population 2, but that now population 1 is believed to have a greater mean than population 2. Letting $\alpha = .05$ and assuming the populations have equal variances and X is approximately normally distributed, use the following data to test this belief.

SAMPLE 1		SAMPLE 2	
43.6	45.7	40.1	36.4
44.0	49.1	42.2	42.3
45.2	45.6	43.1	38.8
40.8	46.5	37.5	43.3
48.3	45.0	41.0	40.2

10.2
Problems

10.9   Suppose you want to determine whether the average values for populations 1 and 2 are different, and you randomly gather the following data.

SAMPLE 1						SAMPLE 2					
2	10	7	8	2	5	10	12	8	7	9	11
9	1	8	0	2	8	9	8	9	10	11	10
11	2	4	5	3	9	11	10	7	8	10	10

Test your conjecture, using a probability of committing a Type I error of .01. Assume the population variances are the same and X is normally distributed in the populations.

10.10   Suppose a realtor is interested in comparing the asking prices of midrange homes in Peoria, Illinois and Evansville, Indiana. The realtor conducts a small telephone survey in the two cities, asking the prices of midrange homes. The resulting data from a random sample of 21 listings in Peoria and 26 listings in Evansville are shown below. Assuming that the prices of midrange homes are normally distributed, but that the variances of the prices in the two cities are different, test to determine whether there is any difference in the mean prices of midrange homes in the two cities. Let $\alpha = .05$.

PEORIA				EVANSVILLE			
87,743	84,144	85,545	89,022	87,741	85,959	80,911	86,724
85,579	87,251	86,486	84,721	85,742	83,099	88,744	83,797
83,101	84,527	91,343	87,658	84,517	84,392	84,477	83,091
85,550	82,480	87,159	87,706	85,428	84,534	84,769	82,046
84,213	88,841	87,598	85,908	85,403	84,214	79,190	82,311
91,012				83,566	81,600	84,742	82,336
				80,805	83,781		

10.11   There is some indication that mean daily car rental rates may be higher for Boston than for Dallas. Suppose a survey of eight car rental companies in Boston is taken, and the resulting data are those shown below. Further, suppose a survey of nine car rental companies in Dallas is taken, and the data are also shown below. Use a .01 level of significance to test to determine whether the average daily car rental rates in Boston are significantly higher than those in Dallas. Assume that car rental rates are normally distributed and the population variances are approximately equal.

BOSTON				DALLAS			
45.73	44.56	45.50	45.79	42.90	48.27	49.42	39.73
44.97	51.91	49.19	47.30	40.81	47.46	49.10	41.33
				45.47			

10.12   A study was made to compare the costs of supporting a family of four Americans for a year in different foreign cities. The lifestyle of living in the United States on an annual income of $75,000 was the standard against which living in foreign cities was compared. A comparable living standard in Toronto and Mexico City was attained for about $64,000. Suppose an executive wants to determine whether there is any difference in the average annual cost of supporting her family of four in the manner to which they are accustomed between Toronto and Mexico City. She uses the following data, randomly gathered from 11 families in each city, and an alpha of .01 to test this difference. She assumes the annual cost is normally distributed and the population variances are equal. What does the executive find?

TORONTO	MEXICO CITY
$69,000	$65,000
64,500	64,000
67,500	66,000
64,500	64,900
66,700	62,000
68,000	60,500
65,000	62,500
69,000	63,000
71,000	64,500
68,500	63,500
67,500	62,400

In the preceding section, hypotheses were tested about the difference in two population means when the samples are *independent*. In this section, a method is presented to analyze dependent samples or related samples. Some business analysts refer to this test as the **matched-pairs test.** Others call it the *t test for related measures* or the *correlated t test.*

What are some types of situations in which the two samples being studied are related or dependent? Let's begin with the before-and-after study. Sometimes as an experimental control mechanism, the same person or object is measured both before and after a treatment. Certainly, the after measurement is *not* independent of the before measurement because the measurements are taken on the same person or object in both cases. Another example of dependent or related samples is when pairs of people such as twins, spouses, siblings, coworkers, or others are used as matched pairs from which the two measurements are taken (one from each member of the pair). Data gathered from research designs such as these from dependent samples are analyzed differently than data gathered from two independent samples. Table 10.3 gives data from a hypothetical study in which people were asked to rate a company before and after 1 week of viewing a 4-minute videocassette of the company twice a day. The before scores are one sample and the after scores are a second sample, but each pair of scores is related because the two measurements apply to the same person. The before scores and the after scores are not likely to vary from each other as much as scores gathered from independent samples because individuals bring their biases about businesses and the company to the study. These individual biases affect both the before scores and the after scores in the same way because each pair of scores is measured on the same person.

Other examples of related measures samples include studies in which twins, siblings, or spouses are matched and placed in two different groups. For example, a fashion merchandiser might be interested in comparing men's and women's perceptions of women's clothing. If the men and women selected for the study are spouses or siblings, a built-in

10.3
Hypothesis Testing about the Mean Difference in Two Related Populations

Matched-pairs test
A *t* test to test the differences in two related or matched samples; sometimes called the *t* test for related measures or the correlated *t* test.

INDIVIDUAL	BEFORE	AFTER
1	32	39
2	11	15
3	21	35
4	17	13
5	30	41
6	38	39
7	14	22

TABLE 10.3
Rating of a Company
(on a Scale from 0 to 50)

relatedness to the measurements of the two groups in the study is likely. Their scores are more apt to be alike or related than those of randomly chosen independent groups of men and women because of similar backgrounds or tastes.

Hypothesis Testing

To ensure the use of the proper hypothesis-testing techniques, the business analyst must determine whether the two samples being studied are dependent or independent. The approach to analyzing two *related* samples is different from the techniques used to analyze independent samples. Use of the techniques in Section 10.2 to analyze related group data can result in a loss of power and an increase in Type II errors.

The matched-pairs test for related samples requires that the two samples be the same size and that the individual related scores be matched. Formula 10.5 is used to test hypotheses about dependent populations.

---

t FORMULA TO TEST THE DIFFERENCE IN TWO DEPENDENT POPULATIONS

$$t = \frac{\bar{d} - D}{\frac{S_d}{\sqrt{n}}}$$

$$\text{df} = n - 1$$

(10.5)

where:

    n = number of pairs
    d = sample difference in pairs
    D = mean population difference
    S_d = standard deviation of sample difference
    $\bar{d}$ = mean sample difference

This t test for dependent measures uses the sample difference, d, between individual matched sample values as the basic measurement of analysis instead of individual sample values. Analysis of the d values effectively converts the problem from a two-sample problem to a single sample of differences, which is an adaptation of the single-sample means formula. This test utilizes the sample mean of differences, $\bar{d}$, and the standard deviation of differences, S_d, which can be computed by using Formulas (10.6) and (10.7).

FORMULAS FOR $\bar{d}$ AND S_d

$$\bar{d} = \frac{\Sigma d}{n}$$

(10.6)

$$S_d = \sqrt{\frac{\Sigma(d - \bar{d})^2}{n - 1}} = \sqrt{\frac{\Sigma d^2 - \frac{(\Sigma d)^2}{n}}{n - 1}}$$

(10.7)

An assumption for this test in the analysis of small samples is that the differences of the two populations are normally distributed.

Analyzing data by this method involves calculating a t value with Formula 10.5 and comparing it with a critical t value obtained from the table. The critical t value is obtained from the t distribution table in the usual way, with the exception that, in the degrees of freedom $(n - 1)$, n is the number of matched pairs of scores.

COMPANY	1998 P/E RATIO	1999 P/E RATIO
1	8.9	12.7
2	38.1	45.4
3	43.0	10.0
4	34.0	27.2
5	34.5	22.8
6	15.2	24.1
7	20.3	32.3
8	19.9	40.1
9	61.9	106.5

TABLE 10.4
P/E Ratios for Nine Randomly Selected Companies

Suppose a stock market investor is interested in determining if there is a significant different in the P/E (price to earnings) ratio for companies from one year to the next. In an effort to study this, the investor randomly samples nine companies from the *Handbook of Common Stocks* and records the P/E ratios for each of these companies at the end of the 1998 year and at the end of the 1999 year. The data are shown in Table 10.4.

These data are related data because each P/E value for 1999 has a corresponding 1998 measurement on the same company. Since there is no prior indication as to whether or not it is believed that P/E ratios have gone up or down, the hypothesis tested is two-tailed. Assume $\alpha = .01$.

STEP **1.**
$$H_0: D = 0$$
$$H_a: D \neq 0$$

STEP **2.** The appropriate statistical test is

$$t = \frac{\bar{d} - D}{\frac{S_d}{\sqrt{n}}}$$

STEP **3.** $\alpha = .01$.

STEP **4.** Because $\alpha = .01$ and this test is two-tailed, $\alpha/2 = .005$ is used to obtain the table t value. With nine pairs of data, $n = 9$, df $= n - 1 = 8$. The table t value is $t_{.005,8} = \pm 3.355$. If the observed test statistic is greater than 3.355 or less than -3.355, the null hypothesis will be rejected.

STEP **5.** The sample data are given in Table 10.4.

STEP **6.** Table 10.5 shows the calculations to obtain the observed value of the test statistic, which is $t = -0.70$.

STEP **7.** Because the observed t value is greater than the critical table t value in the lower tail ($t = -0.70 > t = -3.355$), it is in the non rejection region.

STEP **8.** There is not enough evidence from the data to declare a significant difference in the average P/E ratio between 1998 and 1999. The graph in Figure 10.9 depicts the rejection regions, the critical values of t, and the computed (observed) value of t for this example.

Analysis Using Excel

Excel contains a *t*-**Test: Paired Two Sample for Means** data analysis tool that can be selected and used to test hypotheses about the difference in two related sample means. The

TABLE 10.5
Analysis of P/E Ratio Data

COMPANY	1998 P/E	1999 P/E	d
1	8.9	12.7	-3.8
2	38.1	45.4	-7.3
3	43.0	10.0	33.0
4	34.0	27.2	6.8
5	34.5	22.8	11.7
6	15.2	24.1	-8.9
7	20.3	32.3	-12.0
8	19.9	40.1	-20.2
9	61.9	106.5	-44.6

$$\bar{d} = -5.033, \quad S_d = 21.599, \quad n = 9$$

$$\text{Calculated (observed) } t = \frac{-5.033 - 0}{\dfrac{21.599}{\sqrt{9}}} = -0.70$$

Figure 10.9

Graphical depiction
of P/E ratio analysis

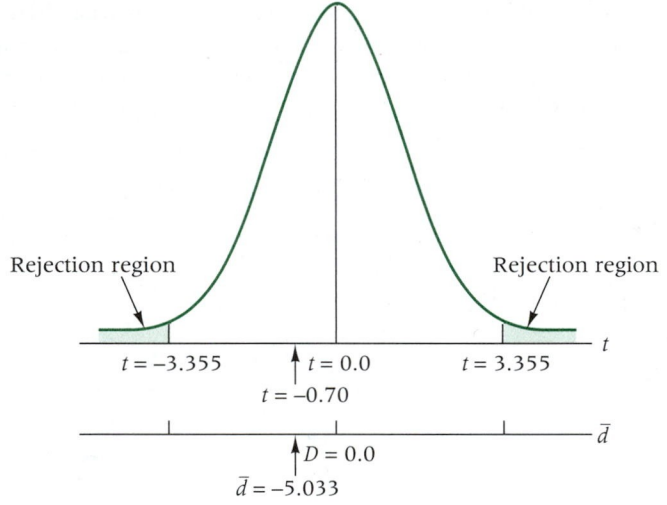

dialog box for this test, shown in Figure 10.10, requests the location of data for the first group (variable one), the location of data for the second group (variable two), the hypothesized mean difference (usually zero), and the value of alpha.

The P/E ratio data are analyzed using Excel, and the resulting output is shown in Figure 10.11. The output includes the mean, the variance, and the number of observations in each sample. In addition, the hypothesized mean difference is given along with the degrees of freedom, the observed t value, the p-value, and the critical table t values. The observed t value shown here is the same as the value computed earlier ($t = -0.70$). Since this is a two-tailed test, we use the critical table t values of ± 3.355 shown in Figure 10.11. A comparison of the observed value to the critical table value reveals that the observed value ($t = -0.70$) is in the non-rejection region, and the decision is to fail to reject the null hypothesis. In addition, the two-tailed p-value of this observed t statistic (p-value = .5043) is greater than $\alpha = .01$ underscoring the decision to fail to reject the null hypothesis. There appears to be no evidence of significant change in P/E ratios from 1998 to 1999.

Figure 10.10 Excel dialog box for the **Paired Two Sample *t*-Test for Means**

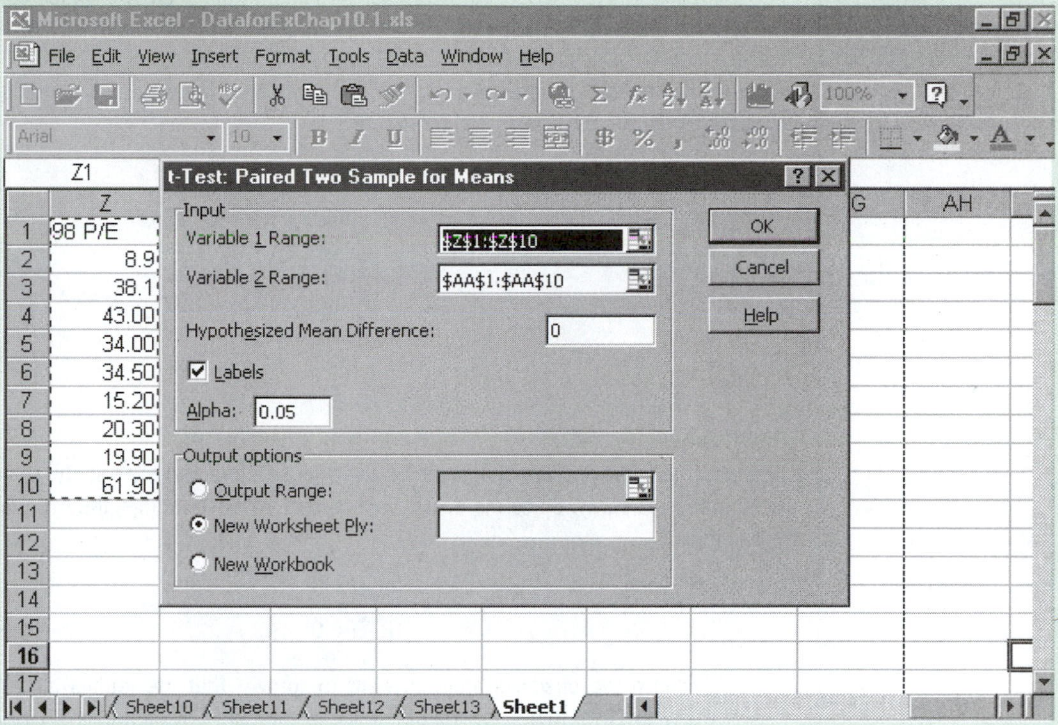

Figure 10.11 Excel output for the P/E ratio problem

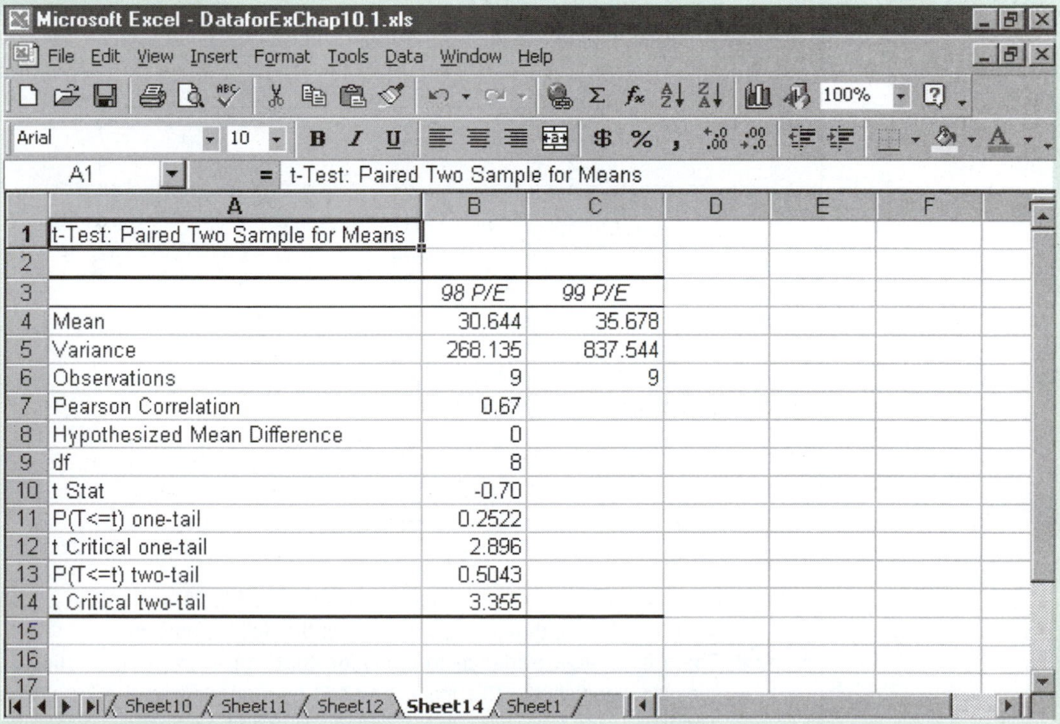

Let us revisit the hypothetical study discussed earlier in the section in which consumers are asked to rate a company both before and after of viewing a video on the company twice a day for a week. The data from Table 10.4 are displayed again here. Use an alpha of .05 to test to determine if there is a significant increase in the ratings of the company after the one-week video treatment.

INDIVIDUAL	BEFORE	AFTER
1	32	39
2	11	15
3	21	35
4	17	13
5	30	41
6	38	39
7	14	22

SOLUTION

Because the same individuals are being used in a before-and-after study, this is a related measures study. Since the desired effect is to increase ratings, the hypothesis test is one-tailed.

STEP **1.**
$$H_0: D = 0$$
$$H_a: D < 0$$

Because the business analysts want to "prove" that the ratings increase from 1998 to 1999 and since the difference is computed by subtracting 1999 ratings from 1998, the desired alternative hypothesis is $D < 0$.

STEP **2.** The appropriate test statistic is Formula 10.5.

STEP **3.** The Type I error rate is .05.

STEP **4.** The degrees of freedom are $n - 1 = 7-1 = 6$. For $\alpha = .05$, the table t value is $t_{.05,6} = -1.943$. The decision rule is to reject the null hypothesis if the observed value is less than -1.943.

STEP **5.** The sample data and some calculations follow.

INDIVIDUAL	BEFORE	AFTER	d
1	32	39	−7
2	11	15	−4
3	21	35	−14
4	17	13	4
5	30	41	−11
6	38	39	−1
7	14	22	−8

$$\bar{d} = -5.857 \qquad S_d = 6.0945$$

STEP **6.** The calculated (observed) t value is:

$$t = \frac{-5.857 - 0}{\dfrac{6.0945}{\sqrt{7}}} = -2.54$$

STEP **7.** Since the calculated (observed) value of −2.54 is less than the critical, table value of −1.943, the decision is to reject the null hypothesis.

STEP **8.** There is enough evidence to conclude that, on average, the ratings have increased significantly. This might be used by managers to support a decision to continue using the videos or to expand the use of such videos in an effort to increase public support for their company.

The following graph depicts the calculated value, the rejection region, and the critical *t* value for the problem.

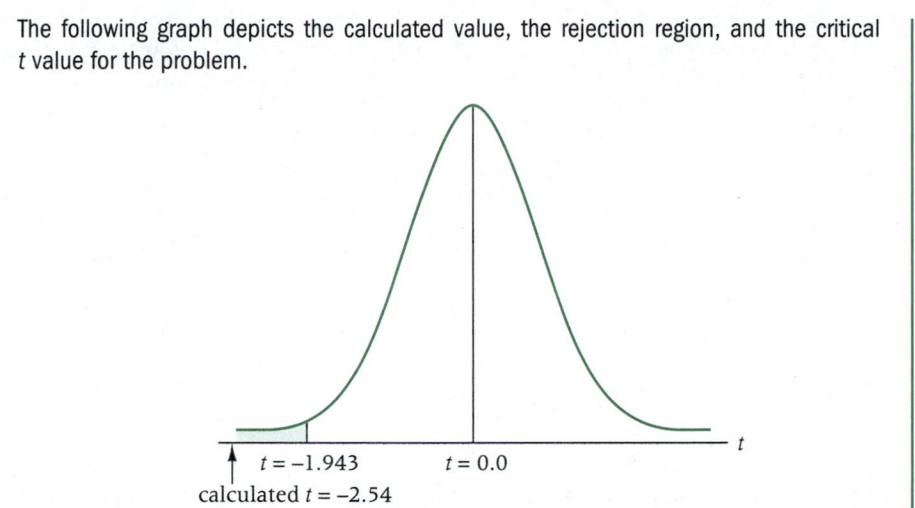

Shown below is Excel output for Demonstration Problem 10.3 using the **t-Test: Paired Two Sample for Means.** The observed value of −2.54 (same as computed above) is less than the one-tailed critical table value of −1.94, leading the business analyst to reject the null hypothesis. The one-tailed *p*-value of .0220, which is less than .05, also underscores the decision to reject the null hypothesis.

	A	B	C	D	E	F
1	t-Test: Paired Two Sample for Means					
2						
3		*Before*	*After*			
4	Mean	23.286	29.143			
5	Variance	103.238	146.810			
6	Observations	7	7			
7	Pearson Correlation	0.86				
8	Hypothesized Mean Difference	0				
9	df	6				
10	t Stat	-2.54				
11	P(T<=t) one-tail	0.0220				
12	t Critical one-tail	1.94				
13	P(T<=t) two-tail	0.0439				
14	t Critical two-tail	2.45				

10.13 Use the data given and a 1% level of significance to test the following hypotheses. Assume the differences are normally distributed in the population.

$$H_0: D = 0 \qquad H_a: D > 0$$

PAIR	SAMPLE 1	SAMPLE 2
1	38	22
2	27	28
3	30	21
4	41	38
5	36	38
6	38	26
7	33	19
8	35	31
9	44	35

10.14 Use the data given to test the following hypotheses ($\alpha = .05$). Assume the differences are normally distributed in the population.

$$H_0: D = 0 \qquad H_a: D \neq 0$$

INDIVIDUAL	BEFORE	AFTER
1	107	102
2	99	98
3	110	100
4	113	108
5	96	89
6	98	101
7	100	99
8	102	102
9	107	105
10	109	110
11	104	102
12	99	96
13	101	100

10.15 Because of uncertainty in real estate markets, many homeowners are considering remodeling and constructing additions rather than selling. Probably the most expensive room in the house to remodel is the kitchen, with an average cost of around $23,400. In terms of resale value, is remodeling the kitchen worth the cost? Below are cost and added resale figures published by *Remodeling* magazine for 11 cities. Use these data and $\alpha = .01$ to test to determine if the added resale value of a kitchen is significantly more than the cost. Assume the differences are normally distributed in the population.

CITY	COST	RESALE
Atlanta	$20,427	$25,163
Boston	27,255	24,625
Des Moines	22,115	12,600
Kansas City, MO	23,256	24,588
Louisville	21,887	19,267
Portland, OR	24,255	20,150
Raleigh-Durham	19,852	22,500
Reno	23,624	16,667
Ridgewood, NJ	25,885	26,875
San Francisco	28,999	35,333
Tulsa	20,836	16,292

10.16 The vice president of marketing brought to the attention of sales managers that most of the company's manufacturer representatives contacted clients and maintained client relationships in a disorganized, haphazard way. The sales managers brought the reps in for a 3-day seminar and training session on how to use an organizer to schedule visits and recall pertinent information about each client more effectively. Sales reps were taught how to schedule visits most efficiently to maximize their efforts. Sales managers were given data on the number of site visits by sales reps on a randomly selected day both before and after the seminar. Use the following data to test whether there were significantly more site visits after the seminar ($\alpha = .05$). Assume the differences in the number of site visits are normally distributed.

REP	BEFORE	AFTER
1	2	4
2	4	5
3	1	3
4	3	3
5	4	3
6	2	5
7	2	6
8	3	4
9	1	5

10.17 Eleven employees were put under the care of the company nurse because of high cholesterol readings. The nurse lectured them on the dangers of this condition and put them on a new diet. Shown are the cholesterol readings of the 11 employees both before the new diet and 1 month after use of the diet began. Using the data given below, test to determine if the cholesterol readings have been significantly lowered after the 1-month diet. Let $\alpha = .05$. Assume that differences in cholesterol readings are normally distributed in the population.

EMPLOYEE	BEFORE	AFTER
1	255	197
2	230	225
3	290	215
4	242	215
5	300	240
6	250	235
7	215	190
8	230	240
9	225	200
10	219	203
11	236	223

10.18 Lawrence and Glover published the results of a study in the *Journal of Managerial Issues* in which they examined the effects of accounting firm mergers on auditing delay. Auditing delay is the time between a company's fiscal year-end and the date of the auditor's report. The hypothesis is that with the efficiencies gained through mergers, the length of the audit delay would decrease. Suppose to test their hypothesis they examined the audit delays on 27 clients of Big Six firms from both before and after the Big Six firm merger (a span of 5 years), and the resulting data are those given on the next page. Use these data and $\alpha = .01$ to test to determine whether the audit delays after the merger were significantly lower than before the merger. Assume that the differences in before and after delays are normally distributed in the population.

COMPANY	BEFORE	AFTER	COMPANY	BEFORE	AFTER
1	21	16	15	26	22
2	30	27	16	20	17
3	22	14	17	22	23
4	15	16	18	29	21
5	29	23	19	16	15
6	27	26	20	21	17
7	19	19	21	26	28
8	25	20	22	19	17
9	37	26	23	15	14
10	31	24	24	26	19
11	24	25	25	33	28
12	19	16	26	25	25
13	23	17	27	21	20
14	21	20			

10.4
Hypothesis Testing about the Difference in Two Population Proportions

Sometimes a business analyst wishes to make inferences about the difference in two population proportions. This type of analysis has many applications in business, such as comparing the market share of a product for two different markets, studying the difference in the proportion of female customers in two different geographic regions, or comparing the proportion of defective products from one period to another. In making inferences about the difference in two population proportions, the statistic normally used is the difference in the sample proportions: $\hat{p}_1 - \hat{p}_2$. This statistic is computed by taking random samples and determining $\hat{p}$ for each sample for a given characteristic, then calculating the difference in these sample proportions.

The central limit theorem states that for large samples (each of $n_1 \cdot \hat{p}_1$, $n_1 \cdot \hat{q}_1$, $n_2 \cdot \hat{p}_2$, and $n_2 \cdot \hat{q}_2 > 5$, where $\hat{q} = 1 - \hat{p}$), the difference in sample proportions is normally distributed with a mean difference of

$$\mu_{\hat{p}_1 - \hat{p}_2} = P_1 - P_2$$

and a standard deviation of the difference of sample proportions of

$$\sigma_{\hat{p}_1 - \hat{p}_2} = \sqrt{\frac{P_1 \cdot Q_1}{n_1} + \frac{P_2 \cdot Q_2}{n_2}}.$$

From this information, a Z formula for the difference in sample proportions can be developed.

Z FORMULA FOR THE DIFFERENCE IN TWO POPULATION PROPORTIONS	$$Z = \frac{(\hat{p}_1 - \hat{p}_2) - (P_1 - P_2)}{\sqrt{\frac{P_1 \cdot Q_1}{n_1} + \frac{P_2 \cdot Q_2}{n_2}}}$$	(10.8)

where:

$\hat{p}_1$ = proportion from sample 1 P_1 = proportion from population 1
$\hat{p}_2$ = proportion from sample 2 P_2 = proportion from population 2
n_1 = size of sample 1 $Q_1 = 1 - P_1$
n_2 = size of sample 2 $Q_2 = 1 - P_2$

Hypothesis Testing

Formula 10.8 is the formula that can be used to determine the probability of getting a particular difference in two sample proportions when given the values of the population proportions. In testing hypotheses about the difference in two population proportions, particular values of the population proportions are not usually known or assumed. Rather, the hypotheses are about the difference in the two population proportions $(P_1 - P_2)$. Note that Formula 10.8 requires knowledge of the values of P_1 and P_2. Hence, a modified version of Formula 10.8 is used when testing hypotheses about $P_1 - P_2$. This formula utilizes a pooled value obtained from the sample proportions to replace the population proportions in the denominator of Formula 10.8.

The denominator of Formula 10.8 is the standard deviation of the difference in two sample proportions and uses the population proportions in its calculations. However, the population proportions are unknown, so an estimate of the standard deviation of the difference in two sample proportions is made by using sample proportions as point estimates of the population proportions. The sample proportions are combined by using a weighted average to produce $\overline{P}$, which in conjunction with $\overline{Q}$ and the sample sizes produces a point estimate of the standard deviation of the difference in sample proportions. The result is Formula 10.9 which we shall use to test hypotheses about the difference in two population proportions.

(10.9)

$$Z = \frac{(\hat{p}_1 - \hat{p}_2) - (P_1 - P_2)}{\sqrt{(\overline{P} \cdot \overline{Q})\left(\dfrac{1}{n_1} + \dfrac{1}{n_2}\right)}}$$

Z FORMULA TO TEST THE DIFFERENCE IN POPULATION PROPORTIONS

where:

$$\overline{P} = \frac{X_1 + X_2}{n_1 + n_2} = \frac{n_1 \hat{p}_1 + n_2 \hat{p}_2}{n_1 + n_2}$$

$$\overline{Q} = 1 - \overline{P}$$

Testing the difference in two population proportions is useful whenever the business analyst is interested in comparing the proportion of one population that has a certain characteristic with the proportion of a second population that has the same characteristic. For example, a business analyst might be interested in determining whether the proportion of people driving new cars (less than 1 year old) in Houston is different from the proportion in Denver. A study could be conducted with a random sample of Houston drivers and a random sample of Denver drivers to test this idea. The results could be used to compare the new-car potential of the two markets and the propensity of drivers in these areas to buy new cars.

Do consumers and CEOs have different perceptions of ethics in business? A group of business analysts attempted to determine whether there was a difference in the proportion of consumers and the proportion of CEOs who believe that fear of getting caught or losing one's job is a strong influence of ethical behavior. In their study, they found that 57% of consumers said that fear of getting caught or losing one's job was a strong influence on ethical behavior but only 50% of CEOs felt the same way.

Suppose these data were determined from a sample of 755 consumers and 616 CEOs. Is this enough evidence to declare that a significantly higher proportion of consumers than of CEOs believe fear of getting caught or losing one's job is a strong influence on ethical behavior?

STEP **1.** Suppose sample 1 is the consumer sample and sample 2 is the CEO sample. Because we are trying to prove that a higher proportion of consumers than of CEOs believe this, the alternative hypothesis should be $P_1 - P_2 > 0$. The following hypotheses are being tested.

$$H_0: P_1 - P_2 = 0$$
$$H_a: P_1 - P_2 > 0$$

where:

P_1 is the proportion of consumers who select the factor
P_2 is the proportion of CEOs who select the factor

STEP **2.** The appropriate statistical test is Formula 10.9.
STEP **3.** Let $\alpha = .10$.
STEP **4.** As this is a one-tailed test, the critical table Z value is $Z_c = 1.28$. If an observed value of Z of more than 1.28 is obtained, the null hypothesis will be rejected. Figure 10.12 shows the rejection region and the critical value for this problem.
STEP **5.** The sample information follows.

CONSUMERS	CEOs
$n_1 = 755$	$n_2 = 616$
$\hat{p}_1 = .57$	$\hat{p}_2 = .50$

STEP **6.**

$$\overline{P} = \frac{n_1 \hat{p}_1 + n_2 \hat{p}_2}{n_1 + n_2} = \frac{(755)(.57) + (616)(.50)}{755 + 616} = .539$$

If the statistics had been given as raw data instead of sample proportions, we would have used the following formula.

$$\overline{P} = \frac{X_1 + X_2}{n_1 + n_2}$$

Figure 10.12

Rejection region for the ethics example

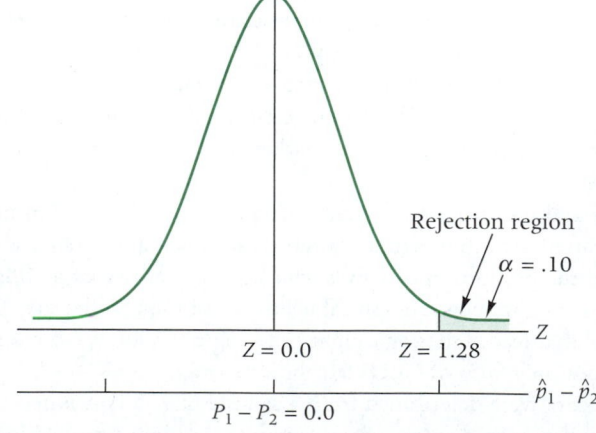

The calculated Z value is

$$Z = \frac{(.57 - .50) - (0)}{\sqrt{(.539)(.461)\left(\frac{1}{755} + \frac{1}{616}\right)}} = 2.59.$$

STEP **7.** Because $Z = 2.59$ is greater than the critical table Z value of 1.28 and is in the rejection region, the null hypothesis is rejected.

STEP **8.** A significantly higher proportion of consumers than of CEOs believe fear of getting caught or losing one's job is a strong influence on ethical behavior. CEOs might want to take another look at ways to influence ethical behavior. If employees are more like consumers than CEOs, CEOs might be able to use fear of getting caught or losing one's job as a means of ensuring ethical behavior on the job. By transferring the idea of ethical behavior to the consumer, retailers might use fear of being caught and prosecuted to retard shoplifting in the retail trade.

Analysis Using Excel

Excel does not have the built-in capability of testing hypotheses about the difference in two proportions. However, **FAST** ⑃ **STAT**, can perform such tests. The two proportion hypothesis test is **2 Proportions Hyp. Test** selected from the FAST ⑃ STAT pull-down menu. The dialog box for this test is shown in Figure 10.13. Along with the sample sizes and the value of alpha, this dialog box requires the raw number of "successes" from each sample and the hypothesized difference in population proportions (usually zero).

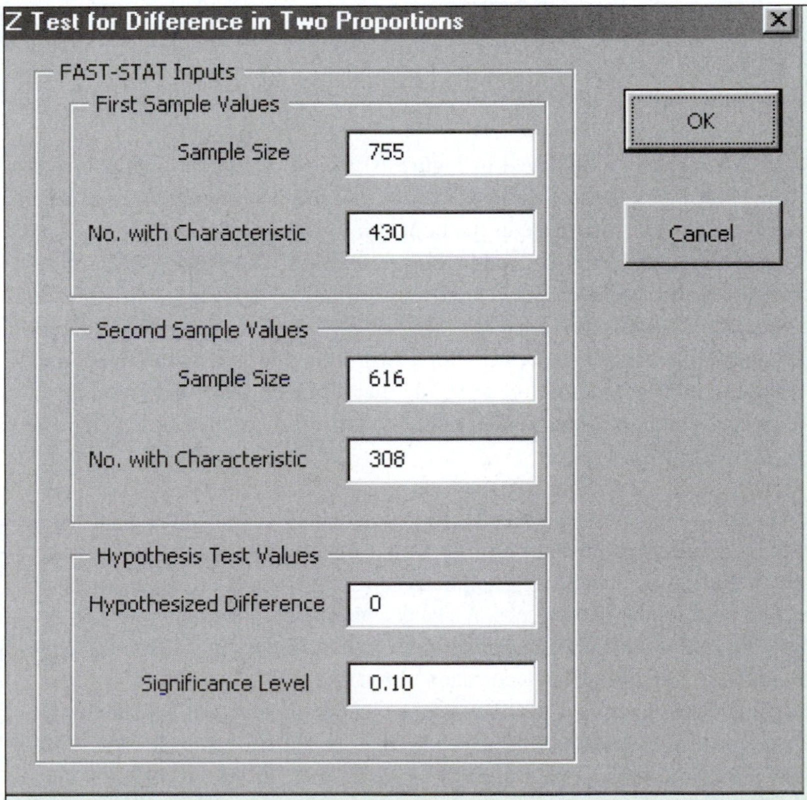

Figure 10.13

FAST ⑃ STAT dialog box for the **2 Proportions Hyp. Test**

Figure 10.14　FAST \\ STAT output for the consumer/CEO ethics problem

	A	B	C	D	E	F	G	H	I	J
1			Z Test for Difference in Two Proportions							
2	INPUTS									
3		First Sample			Sample Size =	755				
4				Number with Characteristic =		430				
5		Second Sample			Sample Size =	616				
6				Number with Characteristic =		308				
7			Hypothesized Proportion Difference =			0				
8				Significance Level of Test :		0.1				
9	OUTPUTS									
10	Sample Statistics									
11			First Sample Proportion			0.5695				
12			Second Sample Proportion			0.5000				
13			Difference Between 2 Proportions			0.0695				
14			Proportion for Combined Samples			0.5383				
15			Pooled Standard Error			0.0271				
16			Z Test Statistic Value			2.5690				
17	Critical z Values		Two Tailed Test Upper Critical Value			1.6449				
18			Two Tailed Test Lower Critical Value			-1.6449				
19			Upper-Tailed Test Critical Value			1.2816				
20			Lower-Tailed Test Critical Value			-1.2816				
21	p-Values		Two-Tailed Test p-Value			0.0102				
22			Upper-Tail Test p-Value			0.0051				
23			Lower-Tail Test p-Value			0.9949				
24										

The **FAST \\ STAT** results are displayed in Figure 10.14. Note that the results include a summary of the input data including the sample sizes, number of successes in each sample, the hypothesized value of the difference in the population proportions, and alpha. In addition, the outputs include the observed Z value (Z Test Statistic Value) along with its associated p-value. Note that the observed (calculated) value of Z shown in Figure 10.14 is 2.57, similar to the observed Z previously calculated in Step 6 (with slight rounding difference). Since the displayed p-value of .0051 (Upper-Tail Test-p-Value) is less than $\alpha = .10$, the decision is to reject the null hypothesis and conclude that a higher proportion of consumers believe fear of getting caught or losing one's job is a strong influence on ethical behavior.

DEMONSTRATION PROBLEM 10.4

A study of female entrepreneurs was conducted to determine their definition of success. The women were offered optional choices such as happiness/self-fulfillment, sales/profit, and achievement/challenge. The women were divided into groups according to the gross sales of their businesses. A significantly higher proportion of female entrepreneurs in the $100,000 to $500,000 category than in the less than $100,000 category seemed to rate sales/profit as a definition of success.

Suppose you decide to test this result by taking a survey of your own and identify female entrepreneurs by gross sales. You interview 100 female entrepreneurs with gross sales of less than $100,000, and 24 of them define sales/profit as success. You then interview 95 female entrepreneurs with gross sales of $100,000 to $500,000, and 39 cite sales/profit as a definition of success. Use this information to test to determine whether there is a significant difference in the proportions of the two groups that define success as sales/profit. Use $\alpha = .01$.

SOLUTION

STEP **1.** You are testing to determine whether there is a difference between two groups of entrepreneurs, so a two-tailed test is required. The hypotheses follow.

$$H_0: P_1 - P_2 = 0$$

$$H_a: P_1 - P_2 \neq 0$$

STEP **2.** The appropriate statistical test is Formula 10.9.

STEP **3.** Alpha has been specified as .01.

STEP **4.** With $\alpha = .01$, you obtain a critical Z value from Table A.5 for $\alpha/2 = .005$, $Z_{.005} = \pm 2.575$. If the observed Z value is more than 2.575 or less than -2.575, the null hypothesis is rejected.

STEP **5.** The sample information follows.

LESS THAN $100,000	$100,000 TO $500,000
$n_1 = 100$	$n_2 = 95$
$X_1 = 24$	$X_2 = 39$
$\hat{p}_1 = \dfrac{24}{100} = .24$	$\hat{p} = \dfrac{39}{95} = .41$

where:

$$\bar{P} = \frac{X_1 + X_2}{n_1 + n_2} = \frac{24 + 39}{100 + 95} = \frac{63}{195} = .323$$

$X =$ the number of entrepreneurs who define sales/profits as success

STEP **6.** The calculated Z value is

$$Z = \frac{(\hat{p}_1 - \hat{p}_2) - (P_1 - P_2)}{\sqrt{(\bar{P} \cdot \bar{Q})\left(\dfrac{1}{n_1} + \dfrac{1}{n_2}\right)}} = \frac{(.24 - .41) - 0}{\sqrt{(.323)(.677)\left(\dfrac{1}{100} + \dfrac{1}{95}\right)}}$$

$$= \frac{-.17}{.067} = -2.54.$$

STEP **7.** Although this calculated value is near the rejection region, it is in the nonrejection region. The null hypothesis is not rejected. That is, there is not enough evidence here to reject the null hypothesis and declare that the responses to the question by the two groups are different statistically. Note that alpha was small and that a two-tailed test was conducted. If a one-tailed test had been used, Z_c would have been $Z_{.01} = 2.33$, and the null hypothesis would have been rejected. If alpha had been .05, Z_c would have been $Z_{.025} = 1.96$, and the null hypothesis would have been rejected. This result underscores the crucial importance of selecting alpha and determining whether to use a one-tailed or two-tailed test in hypothesis testing.

The following diagram shows the critical values, the rejection regions, and the observed value for this problem.

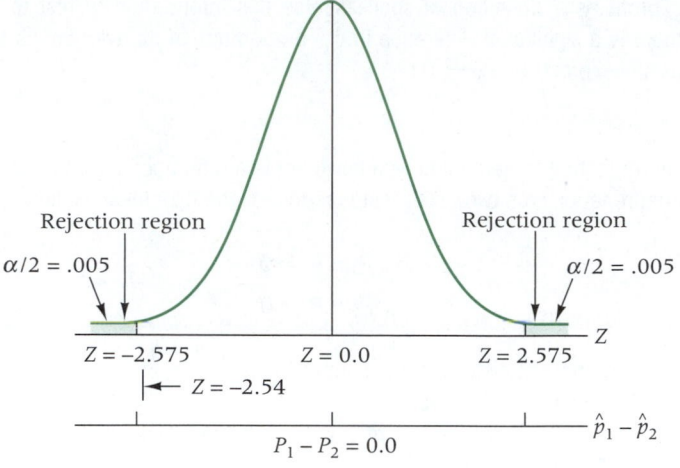

STEP **8.** We cannot statistically conclude that a greater proportion of female entrepreneurs in the higher gross sales category define success as sales/profit. One of the payoffs of such a determination is to find out what motivates the people with whom we do business. If sales/profits motivate people, offers or premises of greater sales and profits can be a means of attracting their services, their interest, or their business. If sales/profits do not motivate people, such offers would not generate the kind of response wanted and we would need to look for other ways to motivate them.

Displayed below are the FAST $\diagdown$ STAT results for Demonstration Problem 10.4 using the Two Proportion Test. The observed value of 2.5451 has a *p*-value of 0.0109 for the two-tailed test being done here. Since the *p*-value is greater than alpha = 0.01, the decision is to fail to reject the null hypothesis.

	Z Test for Difference in Two Proportions				
INPUTS					
	First Sample		Sample Size =	100	
		Number with Characteristic =		24	
	Second Sample		Sample Size =	95	
		Number with Characteristic =		39	
		Hypothesized Proportion Difference =		0	
		Significance Level of Test :		0.01	
OUTPUTS					
Sample Statistics					
		First Sample Proportion		0.2400	
		Second Sample Proportion		0.4105	
		Difference Between 2 Proportions		-0.1705	
		Proportion for Combined Samples		0.3231	
		Pooled Standard Error		0.0670	
		Z Test Statistic Value		-2.5451	
Critical z Values		Two Tailed Test Upper Critical Value		2.5758	
		Two Tailed Test Lower Critical Value		-2.5758	
		Upper-Tailed Test Critical Value		2.3263	
		Lower-Tailed Test Critical Value		-2.3263	
p-Values		Two-Tailed Test p-Value		0.0109	
		Upper-Tail Test p-Value		0.9945	
		Lower-Tail Test p-Value		0.0055	

10.4

Problems

10.19 Using the given sample information, test the following hypotheses.

a. $H_0: P_1 - P_2 = 0$ $H_a: P_1 - P_2 \neq 0$

SAMPLE 1	SAMPLE 2	
$n_1 = 368$	$n_2 = 405$	
$X_1 = 175$	$X_2 = 182$	Let $\alpha = .05$.

Note that X is the number in the sample having the characteristic of interest.

b. $H_0: P_1 - P_2 = 0$ $H_a: P_1 - P_2 > 0$

SAMPLE 1	SAMPLE 2	
$n_1 = 649$	$n_2 = 558$	
$\hat{p}_1 = .38$	$\hat{p}_2 = .25$	Let $\alpha = .10$.

10.20 According to a study conducted for Gateway Computers, 59% of men and 70% of women say that weight is an extremely/very important factor in purchasing a laptop computer. Suppose this survey was conducted using 374 men and 481 women. Is there enough evidence in these data to declare that a significantly higher proportion of women than men believe that weight is an extremely/very important factor in purchasing a laptop computer? Use a 5% level of significance.

10.21 Does age make a difference in the amount of savings a worker feels is needed to be secure at retirement? A study by CommSciences for Transamerica Asset Management found that .24 of workers in the 25–33 age category feel that $250,000 to $500,000 is enough to be secure at retirement. However, .35 of the workers in the 34–52 age category feel that this is enough. Suppose 210 workers in the 25–33 age category and 176 workers in the 34–52 age category were involved in this study. Use these data to determine if there is a significant difference in population proportions by age. Let $\alpha = .01$.

10.22 Companies that had recently developed new products were asked to rate which activities are most difficult to accomplish with new products. Options included such activities as assessing market potential, market testing, finalizing the design, developing a business plan, and the like. A business analyst wants to conduct a similar study to compare the results between two industries: the computer hardware industry and the banking industry. He takes a random sample of 56 computer firms and 89 banks. The business analyst asks whether market testing is the most difficult activity to accomplish in developing a new product. Some 48% of the sampled computer companies and 56% of the sampled banks respond that it is the most difficult activity. Use a level of significance of .20 to test whether there is a significant difference in the responses to the question from these two industries.

10.23 A large production facility uses two machines to produce a key part for its main product. Inspectors have expressed concern about the quality of the finished product. Quality control investigation has revealed that the key part made by the two machines is defective at times. The inspectors randomly sampled 35 units of the key part from each machine. Of those produced by machine A, five were defective. Seven of the 35 sampled parts from machine B were defective. The production manager is interested in estimating the difference in proportions of the populations of parts that are defective between machine A and machine B. Test to determine if there is a significant difference in the proportions of defective parts between machine A and machine B. Assume $\alpha = .10$.

10.24 According to a CCH Unscheduled Absence survey, 9% of small businesses use telecommuting of workers in an effort to reduce unscheduled absenteeism. This compares to 6% for all businesses. Is there really a significant difference between small businesses and all businesses on this issue? Use these data and an alpha of .05 to test this question. Assume that there were 780 small businesses and 915 other businesses in this survey.

10.5
Hypotheses Testing about the Difference in Two Population Variances

Sometimes we are interested in the variability of a population of data rather than a measure of central tendency such as the mean or proportion. Recall from Chapter 3 that a variance is a measure of dispersion or variability. There are occasions when business analysts are interested in testing hypotheses about two population variances. This section discusses how to do such tests. When would a business analyst be interested in the variances from two populations?

In quality control, statisticians often examine both a measure of central tendency (mean or proportion) and a measure of variability. Suppose a manufacturing plant has made two batches of an item, produced items on two different machines, or produced items on two different shifts. It might be of interest to compare the variances from the two batches or groups in an effort to determine whether there is more variability in one than another. For example, if one machine or shift has more variability than another, managers might want to investigate why that machine or shift is not as consistent as the other.

Variance is sometimes used as a measure of the risk of a stock in the stock market. The greater the variance, the greater the risk. By using techniques discussed here, a financial business analyst could determine whether the variances (or risk) of two stocks are the same.

In testing hypotheses about two population variances, the sample variances are used. It makes sense that if two samples come from the same population (or populations with equal variances), the ratio of the sample variances, S_1^2/S_2^2, should be about 1. However, because of sampling error, sample variances even from the same population (or from two populations with equal variances) will vary. This *ratio of two sample variances* formulates what is called an **F value.**

F value
The ratio of two sample variances, used to reach statistical conclusions regarding the null hypothesis; in ANOVA, the ratio of the treatment variance to the error variance.

$$F = \frac{S_1^2}{S_2^2}$$

These ratios, if computed repeatedly for pairs of sample variances taken from a population, are distributed as an **F distribution.** The F distribution will vary by the sizes of the samples, which are converted to degrees of freedom.

With the F distribution, there are degrees of freedom associated with the numerator (of the ratio) and the denominator. An assumption underlying the F distribution is that the populations from which the samples are drawn are normally distributed for X. The F test of two population variances is extremely sensitive to violations of the assumption that the populations are normally distributed. The business analyst should carefully investigate the shape of the distributions of the populations from which the samples are drawn to be certain the populations are normally distributed. The formula used to test hypotheses comparing two population variances follows.

F distribution
A distribution based on the ratio of two random variances; used in testing two variances and in analysis of variance.

F TEST FOR TWO POPULATION VARIANCES

$$F = \frac{S_1^2}{S_2^2} \qquad (10.10)$$

$\text{df}_{\text{numerator}} = v_1 = n_1 - 1$
$\text{df}_{\text{denominator}} = v_2 = n_2 - 1$

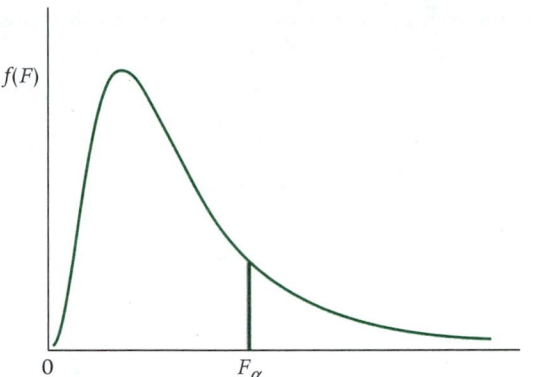

Figure 10.15

An F distribution for $v_1 = 6$, $v_2 = 30$

Table A.7 contains F distribution table values for $\alpha = .10, .05, .025, .01,$ and $.005.$ Figure 10.15 shows an F distribution for $v_1 = 6$ and $v_2 = 30$. Notice that the distribution is nonsymmetric. This can be a problem when we are conducting a two-tailed test and want to determine the critical value for the lower tail. Table A.7 contains only F values for the upper tail. However, the F distribution is not symmetric nor does it have a mean of zero as do the Z and t distributions; therefore, we cannot merely place a minus sign on the upper-tail critical value and obtain the lower-tail critical value (in addition, the F ratio is always positive—it is the ratio of two variances). This dilemma can be solved by using Formula 10.11, which essentially states that the critical F value for the lower tail $(1 - \alpha)$ can be solved for by taking the inverse of the F value for the upper tail (α). The degrees of freedom numerator for the upper-tail critical value is the degrees of freedom denominator for the lower-tail critical value, and the degrees of freedom denominator for the upper-tail critical value is the degrees of freedom numerator for the lower-tail critical value.

(10.11)	$$F_{1-\alpha, v_2, v_1} = \frac{1}{F_{\alpha, v_1, v_2}}$$	FORMULA FOR DETERMINING THE CRITICAL VALUE FOR THE LOWER-TAIL F

A hypothesis test can be done by using two sample variances and Formula 10.10. The following example illustrates this process.

Suppose a machine produces metal sheets that are specified to be 22 mm thick. Because of the machine, the operator, the raw material, the manufacturing environment, and other factors, there is variability in the thickness. Two machines produce these sheets. Operators are concerned about the consistency of the two machines. To test consistency, they randomly sample 10 sheets produced by machine 1 and 12 sheets produced by machine 2. The thickness measurements of sheets from each machine are given in the accompanying table. Assume sheet thickness is normally distributed in the population. How can we test to determine whether the variance from each sample comes from the same population variance (population variances are equal) or from different population variances (population variances are not equal)?

STEP **1.** Determine the null and alternative hypotheses. In this case, we are conducting a two-tailed test (variances are the same or not), and the following hypotheses are used.

$$H_0: \sigma_1^2 = \sigma_2^2$$
$$H_a: \sigma_1^2 \neq \sigma_2^2$$

STEP **2.** The appropriate statistical test is

$$F = \frac{S_1^2}{S_2^2}.$$

STEP **3.** Let $\alpha = .05$.

STEP **4.** As we are conducting a two-tailed test, $\alpha/2 = .025$. Because $n_1 = 10$ and $n_2 = 12$, the degrees of freedom numerator for the upper-tail critical value is $v_1 = n_1 - 1 = 10 - 1 = 9$ and the degrees of freedom denominator for the upper-tail critical value is $v_2 = n_2 - 1 = 12 - 1 = 11$. The critical F value for the upper tail obtained from Table A.7 is

$$F_{.025,9,11} = 3.59.$$

Table 10.6 is a copy of the F distribution for a one-tailed $\alpha = .025$ (which yields equivalent values for two-tailed $\alpha = .05$ where the upper tail contains .025 of the area). Locate $F_{.025,9,11} = 3.59$ in the table by finding the numerator degrees of freedom (9) across the top of the table, the denominator degrees of freedom (11) down the left side of the table, and determining where the two degrees of freedom meet in the table as shown in Table 10.6. The lower-tail critical value can be calculated from the upper-tail value by using Formula 10.11.

$$F_{.975,11,9} = \frac{1}{F_{.025,9,11}} = \frac{1}{3.59} = .28$$

The decision rule is to reject the null hypothesis if the observed F value is greater than 3.59 or less than .28.

STEP **5.** Next we computed the sample variances. The data are shown here.

MACHINE 1		MACHINE 2	
22.3	21.9	22.0	21.7
21.8	22.4	22.1	21.9
22.3	22.5	21.8	22.0
21.6	22.2	21.9	22.1
21.8	21.6	22.2	21.9
		22.0	22.1

$$S_1^2 = .1138 \qquad\qquad S_2^2 = .0202$$
$$n_1 = 10 \qquad\qquad\quad n_2 = 12$$

TABLE 10.6   A Portion of the F Distribution Table

PERCENTAGE POINTS OF THE F DISTRIBUTION

$f(F)$

$0 \qquad F_\alpha$

				$\alpha = 0.025$					
v_2 \ v_1	\multicolumn{9}{c}{Numerator Degrees of Freedrom}
	1	*2*	*3*	*4*	*5*	*6*	*7*	*8*	*9*
1	647.80	799.5	864.2	899.6	921.8	937.1	948.2	956.7	963.3
2	38.51	39.00	39.17	39.25	39.30	39.33	39.36	39.37	39.39
3	17.44	16.04	15.44	15.10	14.88	14.73	14.62	14.54	14.47
4	12.22	10.65	9.98	9.60	9.36	9.20	9.07	8.98	8.90
5	10.01	8.43	7.76	7.39	7.15	6.98	6.85	6.76	6.68
6	8.81	7.26	6.60	6.23	5.99	5.82	5.70	5.60	5.52
7	8.07	6.54	5.89	5.52	5.29	5.12	4.99	4.90	4.82
8	7.57	6.06	5.42	5.05	4.82	4.65	4.53	4.43	4.36
9	7.21	5.71	5.08	4.72	4.48	4.32	4.20	4.10	4.03
10	6.94	5.46	4.83	4.47	4.24	4.07	3.95	3.85	3.78
11	6.72	5.26	4.63	4.28	4.04	3.88	3.76	3.66	(3.59)
12	6.55	5.10	4.47	4.12	3.89	3.73	3.61	3.51	3.44
13	6.41	4.97	4.35	4.00	3.77	3.60	3.48	3.39	3.31
14	6.30	4.86	4.24	3.89	3.66	3.50	3.38	3.29	3.21
15	6.20	4.77	4.15	3.80	3.58	3.41	3.29	3.20	3.12
16	6.12	4.69	4.08	3.73	3.50	3.34	3.22	3.12	3.05
17	6.04	4.62	4.01	3.66	3.44	3.28	3.16	3.06	2.98
18	5.98	4.56	3.95	3.61	3.38	3.22	3.10	3.01	2.93
19	5.92	4.51	3.90	3.56	3.33	3.17	3.05	2.96	2.88
20	5.87	4.46	3.86	3.51	3.29	3.13	3.01	2.91	2.84
21	5.83	4.42	3.82	3.48	3.25	3.09	2.97	2.87	2.80
22	5.79	4.38	3.78	3.44	3.22	3.05	2.93	2.84	2.76
23	5.75	4.35	3.75	3.41	3.18	3.02	2.90	2.81	2.73
24	5.72	4.32	3.72	3.38	3.15	2.99	2.87	2.78	2.70
25	5.69	4.29	3.69	3.35	3.13	2.97	2.85	2.75	2.68
26	5.66	4.27	3.67	3.33	3.10	2.94	2.82	2.73	2.65
27	5.63	4.24	3.65	3.31	3.08	2.92	2.80	2.71	2.63
28	5.61	4.22	3.63	3.29	3.06	2.90	2.78	2.69	2.61
29	5.59	4.20	3.61	3.27	3.04	2.88	2.76	2.67	2.59
30	5.57	4.18	3.59	3.25	3.03	2.87	2.75	2.65	2.57
40	5.42	4.05	3.46	3.13	2.90	2.74	2.62	2.53	2.45
60	5.29	3.93	3.34	3.01	2.79	2.63	2.51	2.41	2.33
120	5.15	3.80	3.23	2.89	2.67	2.52	2.39	2.30	2.22
∞	5.02	3.69	3.12	2.79	2.57	2.41	2.29	2.19	2.11

Denominator Degrees of Freedom

$F_{.025,9,11}$

Figure 10.16

Graph of F values
and rejection region
for the sheet metal
example

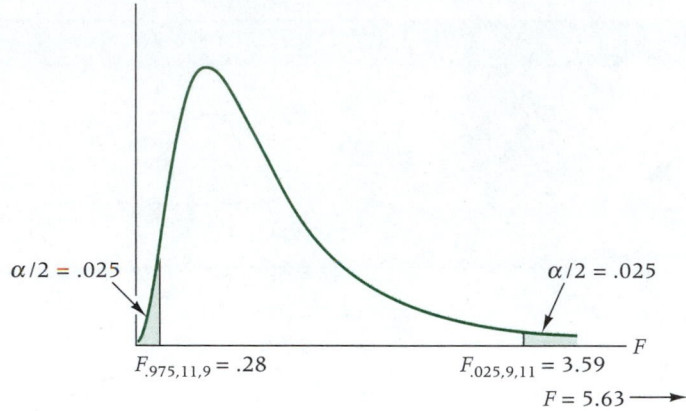

$\alpha/2 = .025$ $\alpha/2 = .025$

$F_{.975,11,9} = .28$ $F_{.025,9,11} = 3.59$

$F = 5.63 \longrightarrow$

STEP **6.**
$$F = \frac{S_1^2}{S_2^2} = \frac{.1138}{.0202} = 5.63$$

The ratio of sample variances is 5.63.

STEP **7.** The calculated F value is 5.63, which is greater than the upper-tail critical value of 3.59. As Figure 10.16 shows, this F value is in the rejection region. Thus, the decision is to reject the null hypotheses. The population variances are not equal.

STEP **8.** An examination of the sample variances reveals that the variance from machine 1 measurements is greater than that from machine 2 measurements. The operators and process managers might want to examine machine 1 further; an adjustment may be needed or there may be some other reason for the seemingly greater variations on that machine.

Analysis Using Excel

Excel can test hypotheses about two population variances using the **F-Test Two-Sample for Variances** selection under **Data Analysis** from the **Tools** menu. The dialog box for this feature is presented in Figure 10.17. Note that the dialog box requires the location of the data for each of the two variables, along with the value of alpha. Excel always does a one-tailed test for two-sample variances. If the business analyst is only interested in the lower tail, place the location of the variable with the smaller variance in box 1. If the analysis is for the upper tail, then place the location of the variable with the larger variance in box 1. For a two-tailed test, place the location of the variable with the larger variance in box 1 and place the value of $\alpha/2$ in the box for alpha. Note that in Figure 10.17 the value of α entered is .025, which is one-half the value of alpha for the machine problem since it is a two-tailed test.

The Excel output shown in Figure 10.18 for the machine variance problem includes the means, the variances, the number of observations, and the degrees of freedom for each sample. In addition, the output includes the observed value of F (5.63), which is the same as the value computed in Step 6. The Excel output also includes the critical table F value of 3.59 and the p-value associated with the observed F. Since the observed F of 5.63 is greater than the critical table F of 3.59, the decision is to reject the null hypothesis. This decision is underscored using a p-value criterion since the p-value of .0047 is less than $\alpha/2 = .025$.

Figure 10.17
Excel dialog box for the **F-Test Two-Sample for Variances**

Microsoft Excel - DataforExChap10.1.xls

File Edit View Insert Format Tools Data Window Help

AN15

	AF	AG	AH	AI	AJ	AK	AL	AM	AN
1	Machine 1	Machine 2							
2	22.3	22							
3	21.8	22.1							
4	22.30	21.80							
5	21.60	21.90							
6	21.80	22.20							
7	21.90	22.00							
8	22.40	21.70							
9	22.50	21.90							
10	22.20	22.00							
11	21.60	22.10							
12		21.90							
13		22.10							
14									
15									
16									
17									

F-Test Two-Sample for Variances

Input
Variable 1 Range: $AF$1:$AF$11
Variable 2 Range: $AG$1:$AG$13
☑ Labels
Alpha: .025

Output options
○ Output Range:
● New Worksheet Ply:
○ New Workbook

OK
Cancel
Help

Sheet14 / Sheet15 / Sheet16 / Sheet13 \ Sheet1

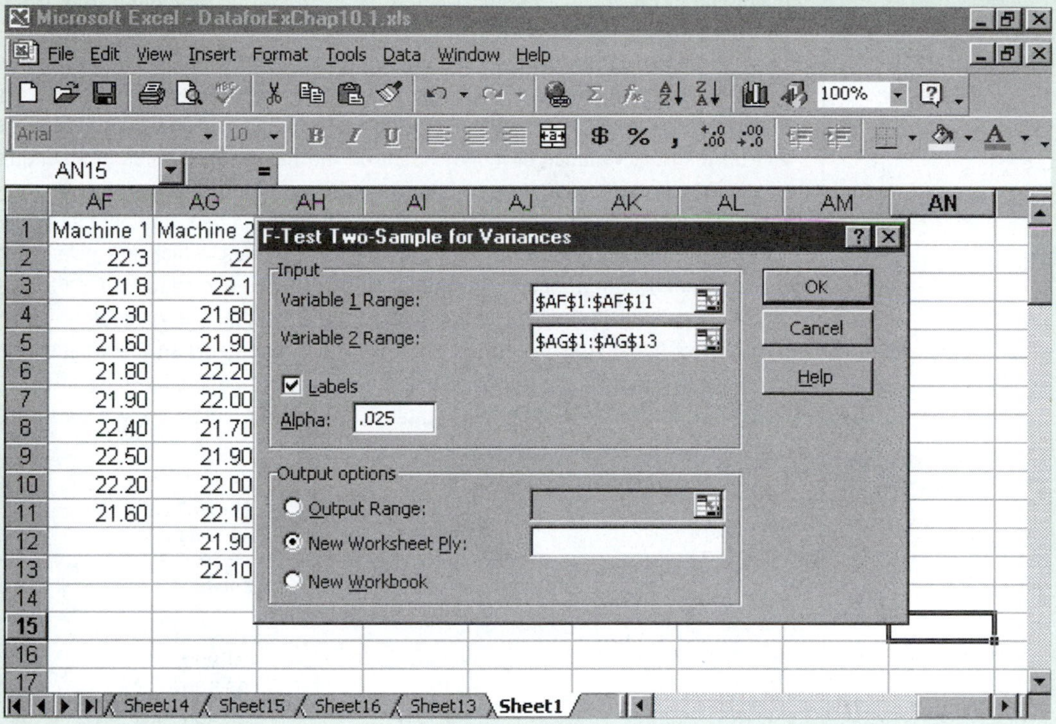

Figure 10.18
Excel output for the **F-Test Two-Sample for Variances**

Microsoft Excel - DataforExChap10.1.xls

File Edit View Insert Format Tools Data Window Help

A1 = F-Test Two-Sample for Variances

	A	B	C	D	E	F
1	F-Test Two-Sample for Variances					
2						
3		Machine 1	Machine 2			
4	Mean	22.040	21.975			
5	Variance	0.1138	0.0202			
6	Observations	10	12			
7	df	9	11			
8	F	5.63				
9	P(F<=f) one-tail	0.0047				
10	F Critical one-tail	3.59				
11						
12						
13						
14						
15						
16						
17						

Sheet15 / Sheet16 / Sheet13 \ Sheet17 / Sheet1

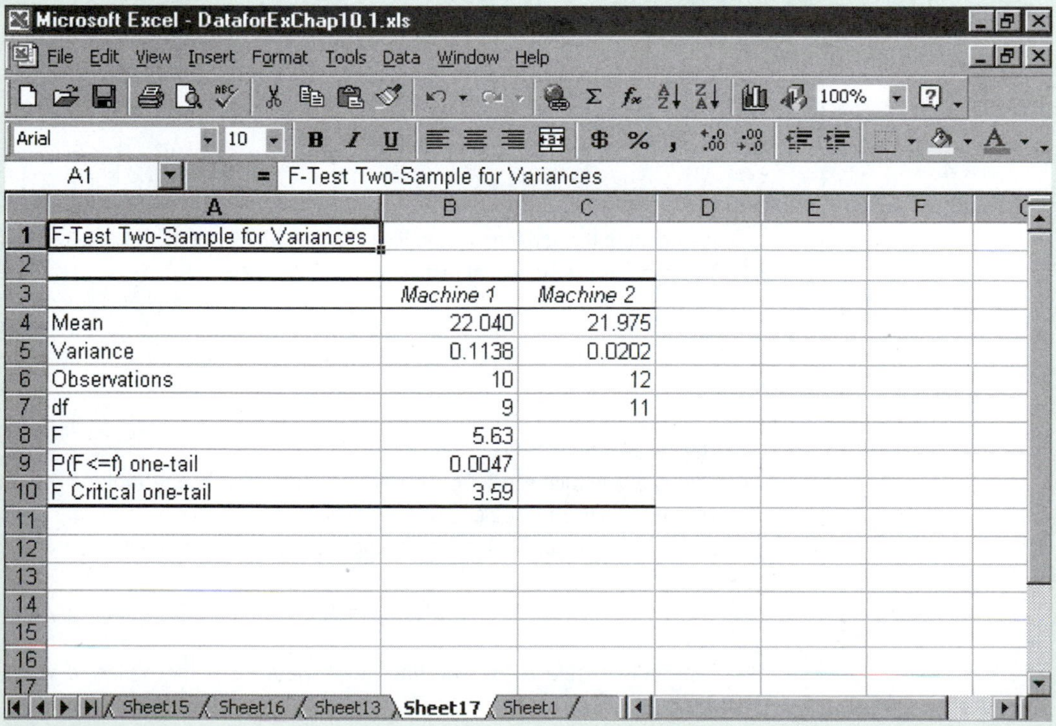

<table>
<tr><td>

DEMONSTRATION PROBLEM 10.5

</td><td>

According to Runzheimer International, a family of four in Manhattan with $60,000 annual income spends more than $22,000 a year on basic goods and services. In contrast, a family of four in San Antonio with the same annual income spends only $15,460 on the same items. Suppose we want to test to determine whether the variance of money spent per year on the basics by families across the United States is greater than the variance of money spent on the basics by families in Manhattan—that is, whether the amounts spent by families of four in Manhattan are more homogeneous than the amounts spent by such families nationally. Suppose a random sample of eight Manhattan families produces the accompanying figures, which are given along with those reported from a random sample of seven families across the United States. Complete a hypothesis-testing procedure to determine whether the variance of values taken from across the United States can be shown to be greater than the variance of values obtained from families in Manhattan. Let $\alpha = .01$. Let population 1 be "Across United States" and population 2 be "Manhattan." Assume the amount spent on the basics is normally distributed in the population.

</td></tr>
</table>

AMOUNT SPENT ON BASICS BY FAMILY OF FOUR
WITH $60,000 ANNUAL INCOME

Across United States	Manhattan
$18,500	$23,000
19,250	21,900
16,400	22,500
20,750	21,200
17,600	21,000
21,800	22,800
14,750	23,100
	21,300

SOLUTION

STEP **1.** This is a one-tailed test with the following hypotheses.

$$H_0: \sigma_1^2 = \sigma_2^2$$
$$H_a: \sigma_1^2 > \sigma_2^2$$

Note that what we are trying to prove—that the variance for the U.S. population is greater than the variance for families in Manhattan—is in the alternative hypothesis.

STEP **2.** The appropriate statistical test is

$$F = \frac{S_1^2}{S_2^2}.$$

STEP **3.** The Type I error rate is .01.

STEP **4.** This is a one-tailed test, so we will use the F distribution table in Appendix A.7 with $\alpha = .01$. The degrees of freedom for $n_1 = 7$ and $n_2 = 8$ are $v_1 = 6$ and $v_2 = 7$. The critical F value for the upper tail of the distribution is

$$F_{.01,6,7} = 7.19.$$

The decision rule is to reject the null hypothesis if the observed value of F is greater than 7.19.

STEP **5.** The following sample variances are computed from the data.

$$S_1^2 = 5{,}961{,}428.6$$

$$n_1 = 7$$

$$S_2^2 = 737{,}142.9$$

$$n_2 = 8$$

STEP **6.** The calculated F value can be determined by

$$F = \frac{S_1^2}{S_2^2} = \frac{5{,}961{,}428.6}{737{,}142.9} = 8.09.$$

STEP **7.** Because the calculated value of $F = 8.09$ is greater than the table critical F value of 7.19, the decision is to reject the null hypothesis.

STEP **8.** The variance for families in the United States is greater than the variance of families in Manhattan. Families in Manhattan are more homogeneous in amount spent on basics than families across the United States. Marketing managers need to understand this as they attempt to find niches in the Manhattan population. There may not be as many different subgroups in Manhattan as there are across the United States. The task of locating market niches may be easier in Manhattan than in the rest of the country because there are likely to be fewer possibilities. The following graph shows the rejection region as well as the critical and calculated values of F.

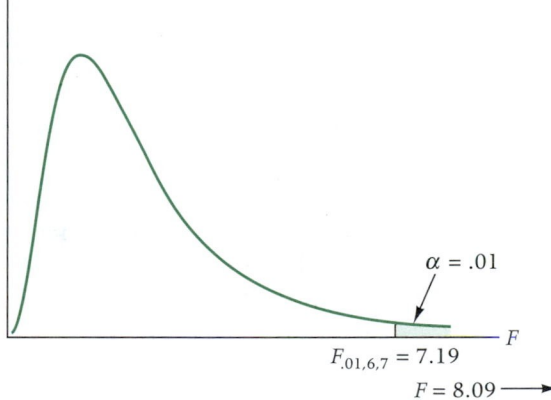

$\alpha = .01$

$F_{.01, 6, 7} = 7.19$

$F = 8.09 \longrightarrow$

Note: Some authors recommend the use of this F test to determine whether the data being analyzed by a t test for two population means are meeting the assumption of equal population variances. However, some statistical analysts suggest that for equal sample sizes, the t test is insensitive to the equal variance assumption, and therefore the F test is not needed in that situation. For unequal sample sizes, the F test of variances is "not generally capable of detecting assumption violations that lead to poor performance" with the t test.* This text does not present the application of the F test to determine whether variance assumptions for the t test have been met.

---

*Carol A. Markowski and Edward P. Markowski, "Conditions for the Effectiveness of a Preliminary Test of Variance," *The American Statistician,* 44, November 1990, 322–326.

	A	B	C	D	E	F	G
	F-Test Two-Sample for Variances						
1	F-Test Two-Sample for Variances						
2							
3		U.S.	Manhattan				
4	Mean	18435.71	22100.00				
5	Variance	5961428.57	737142.86				
6	Observations	7	8				
7	df	6	7				
8	F	8.09					
9	P(F<=f) one-tail	0.0072					
10	F Critical one-tail	7.19					

Shown above is Excel output for Demonstration Problem 10.5. Note that the observed value of $F = 8.09$ is greater than the critical table F value of 7.19 underscoring the decision to reject the null hypothesis. In addition, the p-value of .0072 is less than $\alpha = .01$.

10.5
Problems

10.25 Test the following hypotheses by using the given sample information and $\alpha = .01$. Assume the populations are normally distributed.

$$H_0: \sigma_1^2 = \sigma_2^2 \qquad H_a: \sigma_1^2 < \sigma_2^2$$

SAMPLE 1	SAMPLE 2
130	160
146	92
124	164
152	166
147	176
102	197
173	104
177	149
122	162
130	136
	157
	110

10.26 Test the following hypotheses by using the given sample information and $\alpha = .05$. Assume the populations are normally distributed.

$$H_0: \sigma_1^2 = \sigma_2^2 \qquad H_a: \sigma_1^2 \neq \sigma_2^2$$

SAMPLE 1		SAMPLE 2		
19	20	22	19	20
23	21	20	23	24
27	20	25	20	22
19	19	21	20	19
15	23	22	18	

10.27 Suppose the data shown here are the results of a survey to investigate gasoline prices. Ten service stations were selected randomly in each of two cities and the figures represent the prices of a gallon of unleaded regular gasoline on a given day. Use the F test to determine whether there is a significant difference in the variances of the prices of unleaded regular gasoline between these two cities. Let $\alpha = .10$. Assume gasoline prices are normally distributed.

CITY 1					CITY 2				
1.58	1.47	1.53	1.55	1.54	1.48	1.45	1.59	1.57	1.61
1.53	1.54	1.53	1.43	1.49	1.52	1.54	1.54	1.53	1.51

10.28 How long are resale houses on the market? One survey by the Houston Association of Realtors reported that in Houston, resale houses are on the market an average of 112 days. Of course, the length of time will vary by market. Suppose random samples of 13 houses in Houston and 11 houses in Chicago that are for resale are traced. The data shown here represent the number of days each house was on the market before being sold. Use the given data and a 1% level of significance to determine whether the population variances for the number of days until resale are different in Houston than in Chicago. Assume the numbers of days resale houses are on the market are normally distributed.

HOUSTON	CHICAGO
132	118
138	85
131	113
127	81
99	94
126	93
134	56
126	69
94	67
161	54
133	137
119	
88	

10.29 One recent study showed that the average annual amount spent by an East Coast household on frankfurters was $23.84 compared with an average of $19.83 for West Coast households. Suppose a random sample of 12 East Coast households showed that the standard deviation of these purchases (frankfurters) was $7.52, whereas a random sample of 15 West Coast households resulted in a standard deviation of $6.08. Is there enough evidence from these samples to conclude that the variance of annual frankfurter purchases for East Coast households is greater than the variance of annual frankfurter purchases for West Coast households? Let alpha be .05. Assume amounts spent per year on frankfurters are normally distributed. Suppose the data did show that the variance among East Coast households is greater than that among West Coast households. What might this mean to decision makers in the frankfurter industry?

EAST COAST				WEST COAST			
28.18	22.83	23.52	20.91	18.08	20.62	17.58	19.33
25.02	21.64	19.54	20.02	19.40	21.43	21.12	18.76
25.98	23.64	21.83	22.97	20.70	21.19	21.36	20.50
				19.07	20.55	20.06	

10.30 According to the General Accounting Office of the U.S. government, the average age of a male federal worker is 43.6 and that of a male worker in the nonfederal sector is 37.3 years. Is there any difference in the variation of ages of men in the federal sector and men in the nonfederal sector? Suppose a random sample of 15 male federal workers is taken and a random sample of 15 male nonfederal workers is taken, each is asked their age, and the data below are the results. Use these data and $\alpha = .01$ to answer the question. Assume ages of workers in each of the populations are normally distributed.

FEDERAL SECTOR					NONFEDERAL SECTOR				
46	53	45	27	54	26	27	40	31	44
27	41	59	45	42	27	47	36	30	31
40	32	53	46	58	40	42	40	30	30

Summary

Business analysts sometimes want to study the differences in two populations. This chapter examines the differences in two populations using three different parameters (means, proportions, and variances) through the use of hypothesis testing.

The population means are analyzed by comparing two sample means. When sample sizes are large ($n \geq 30$), a Z test is used. When sample sizes are small, the population variances are known, and the populations are normally distributed, the Z test is used to analyze the population means. If sample size is small, the population variances are unknown, and the populations are normally distributed, the t test of means for independent samples is used. For populations that are related on some measure, such as twins or before-and-after, a t test for dependent measures (matched pairs) is used.

The difference in population proportions can be tested by using the difference in two sample proportions, one from each population, as the sample statistic. The Z distribution is used to test this difference in sample proportions.

The population variances are analyzed by an F test when the assumption that the populations are normally distributed is met. The F value is a ratio of the two variances. The F distribution is a distribution of possible ratios of two sample variances taken from one population or from two populations containing the same variance.

Key Terms

dependent samples	matched-pairs test
F distribution	paired data
F value	related measures
independent samples	related samples
matched-pairs data	

SUPPLEMENTARY PROBLEMS

10.31 Test the following hypotheses with the data given. Let $\alpha = .10$.

$$H_0: \mu_1 - \mu_2 = 0 \qquad H_a: \mu_1 - \mu_2 \neq 0$$

SAMPLE 1

150	140	138	132	131	143	134
140	138	143	139	149	133	136
139	141	135	140	149	156	136
129	137	152	120	144	144	139
130	136	135	140	156	137	145
142	148	132	147	141	144	146
138	122	144	143	142	137	

SAMPLE 2

138	143	136	149	144	146	147
151	141	134	144	135	151	143
125	129	127	144	127	155	141
148	136	123	146	131	145	138
121	137	147	140	142	138	144
131	134	144	148			

10.32 The following data come from independent samples drawn from normally distributed populations. Use these data to test the following hypotheses. Let the Type I error rate by .05. Assume the population variances are unequal.

$$H_0: \mu_1 - \mu_2 = 0 \qquad H_0: \mu_1 - \mu_2 > 0$$

SAMPLE 1

1.82	1.85	1.91	2.01	2.53
2.18	1.68	2.00	1.86	1.86
2.08	2.33			

SAMPLE 2

1.83	1.93	1.89	1.86	1.86	2.11
1.80	1.84	2.03	2.00	1.89	1.90
1.87	1.89	1.78			

10.33 The following data have been gathered from two related samples. The differences are assumed to be normally distributed in the population. Use these data and alpha of .01 to test the following hypotheses.

$$H_0: D = 0$$
$$H_a: D \neq 0$$

RESPONDENT	BEFORE	AFTER
1	47	63
2	33	35
3	38	36
4	50	56
5	39	44
6	27	29
7	35	32
8	46	54
9	41	47

10.34 Test the following hypotheses by using the given data and alpha equal to .05.

$$H_0: P_1 - P_2 = 0 \qquad H_a: P_1 - P_2 \neq 0$$

SAMPLE 1	SAMPLE 2
$n_1 = 783$	$n_2 = 896$
$X_1 = 345$	$X_2 = 421$

10.35 Test the following hypotheses by using the given data. Let alpha = .05.

$$H_0: \sigma_1^2 = \sigma_2^2 \qquad H_a: \sigma_1^2 \neq \sigma_2^2$$

SAMPLE 1	SAMPLE 2
75.3 82.6 70.4 82.5	82.5 75.4 77.7 83.8
84.8 79.6 77.3 64.8	72.9 77.8 87.6 83.7
	89.7 80.1

10.36 A study is conducted to estimate the average difference in bus ridership for a large city during the morning and afternoon rush hours. The transit authority's business analyst randomly selects nine buses because of the variety of routes they represent. On a given day the number of riders on each bus is counted at 7:45 A.M. and at 4:45 P.M., with the following results.

BUS	MORNING	AFTERNOON
1	43	41
2	51	49
3	37	44
4	24	32
5	47	46
6	44	42
7	50	47
8	55	51
9	46	49

Use these data and a 5% level of significance to test to determine if there is a significant difference between the average morning ridership and the average afternoon ridership.

10.37 A study was conducted to compare the salaries of accounting clerks and data entry operators. One of the hypotheses to be tested is that the variability of salaries among accounting clerks is the same as the variability of salaries of data entry operators. To test this, a random sample of 11 accounting clerks was taken and a random sample of 12 data entry operators was taken. The salary data are shown below. Use these data and $\alpha = .05$ to test to determine whether the population variance of salaries is the same for accounting clerks as it is for data operators. Assume the salaries are normally distributed in the population.

ACCOUNTING CLERKS			DATA ENTRY OPERATORS		
28,431	23,536	26,765	24,183	25,467	25,120
23,703	28,833	28,163	24,756	25,047	26,978
25,307	26,895	23,051	25,141	24,136	26,252
26,772	26,323		26,105	26,504	25,769

10.38 A national grocery store chain wants to estimate the difference in the average weight of turkeys sold in Detroit and the average weight of turkeys sold in Charlotte. The chain's business analyst has a random sample of 20 turkeys selected in Detroit and 24 turkeys selected in Charlotte. The weights of the turkeys are recorded and displayed below. Use a 1% level of significance to determine whether there is a difference in the mean weight of turkeys sold in these two cities. Assume the population variances are approximately the same and that the weights of turkeys sold in the stores are normally distributed.

DETROIT

11.9	21.0	14.6	17.1	18.8
16.5	18.9	17.5	15.4	19.6
9.7	16.5	13.7	21.2	18.9
18.9	14.7	17.1	I7.5	18.1

CHARLOTTE

16.7	13.8	12.0	12.9	12.2
12.9	15.7	15.4	21.2	14.5
18.4	11.3	13.4	14.6	10.1
14.5	20.4	10.3	15.3	11.6
19.0	12.7	12.0	14.9	

10.39 A tree nursery has been experimenting with fertilizer to increase the growth of seedlings. A sample of 35 two-year-old pine trees is grown for three more years with a cake of fertilizer buried in the soil near the trees' roots. A second sample of 35 two-year-old pine trees is grown for three more years under identical conditions (soil, temperature, water) as the first group, but not fertilized. Tree growth is measured over the 3-year period with the following results.

TREES WITH FERTILIZER

31.2	46.9	33.3	26.3	35.8
28.4	34.0	21.2	54.6	34.6
35.9	25.1	32.3	31.0	43.9
25.3	26.0	28.4	42.9	45.3
48.1	20.5	39.8	51.8	28.2
45.6	42.8	59.2	40.5	48.9
31.7	41.3	51.8	33.0	35.4

TREES WITHOUT FERTILIZER

21.9	26.1	19.7	23.0	22.2
26.1	27.8	14.6	21.4	28.0
14.5	22.8	22.9	12.2	20.6
20.0	22.5	24.7	19.7	22.6
23.3	27.0	23.0	26.5	31.0
21.3	19.3	13.4	20.5	28.3
15.6	14.8	21.7	18.0	14.7

Do the data support the theory that the population of trees with the fertilizer grew significantly larger during the period in which they were fertilized than the nonfertilized trees? Use $\alpha = .01$.

10.40 One of the most important aspects of a store's image is the perceived quality of its merchandise. Other factors include merchandise pricing, assortment of products, convenience of location, and service. Suppose image perceptions of shoppers of specialty stores and shoppers of discount stores are being compared. A random sample of shoppers is taken at each type of store, and the shoppers are asked whether the quality of merchandise is a determining factor in their perception of the store's image. Some 75% of the 350 shoppers at the specialty stores say yes, but only 52% of the 500 shoppers at the discount store say yes.

Use these data and test to determine if there is a significant difference between specialty stores and discount stores in the proportion of shoppers who say that the quality of merchandise is a determining factor in their perceptions of the store's image. Use an alpha of .05.

10.41 Is the average price of a name-brand soup greater than the average price of a store-brand soup? To test this, a business analyst randomly samples eight stores. Each store sells its own brand and a national name brand. The prices of a can of name-brand tomato soup and a can of store-brand tomato soup follow. Use these data and $\alpha = .10$ to test this theory. Assume that prices are normally distributed in the population.

STORE	NAME BRAND	STORE BRAND
1	54¢	49¢
2	55	50
3	59	52
4	53	51
5	54	50
6	61	56
7	51	47
8	53	49

10.42 As the prices of heating oil and natural gas increase, consumers become more careful about heating their homes. Business analysts want to know how warm homeowners keep their houses in January and how the results from Wisconsin and Tennessee compare. The business analysts randomly call 23 Wisconsin households between 7 P.M. and 9 P.M. on January 15 and ask the respondent how warm the house is according to the thermostat. The business analysts then call 19 households in Tennessee the same night and ask the same question. The results follow.

WISCONSIN				TENNESSEE			
71	71	65	68	73	75	74	71
70	61	67	69	74	73	74	70
75	68	71	73	72	71	69	72
74	68	67	69	74	73	70	72
69	72	67	72	69	70	67	
70	73	72					

For $\alpha = .01$, is the average temperature of a house in Tennessee significantly higher than that of a house in Wisconsin on the evening of January 15? Assume the population variances are equal and the house temperatures are normally distributed in each population.

10.43 A manufacturer has two machines that drill holes in pieces of sheet metal used in engine construction. The workers who attach the sheet metal to the engine become inspectors in that they reject sheets that have been so poorly drilled that they cannot be attached. The production manager is interested in knowing whether one machine produces more defective drillings than the other machine. As an experiment, employees mark the sheets so that the manager can determine which machine was used to drill the holes. A random

sample of 191 sheets of metal drilled by machine 1 is taken, and 38 of the sheets are defective. A random sample of 202 sheets of metal drilled by machine 2 is taken, and 21 of the sheets are defective. Use $\alpha = .05$ to determine whether there is a significant difference in the proportion of sheets drilled with defective holes between machine 1 and machine 2.

10.44 Executives often spend so many hours in meetings that they have relatively little time to manage their individual areas of operation. What is the difference in mean time spent in meetings by executives of the aerospace industry and executives of the automobile industry? Suppose random samples of 33 aerospace executives and 35 automobile executives are monitored for a week to determine how much time they spend in meetings. The results follow.

AEROSPACE						
15	9	6	11	11	12	15
16	14	12	11	10	17	9
11	10	13	14	13	14	9
12	16	13	12	12	10	12
12	15	10	13	13		

AUTOMOBILE						
7	4	6	4	4	8	7
3	6	8	5	8	2	6
2	1	4	2	2	10	7
4	6	2	4	6	7	3
5	4	8	5	3	6	4

Use these data to test to determine if there is a significant difference between aerospace executives and automobile executives in the average number of hours spent per week in meetings. Let alpha be .01.

10.45 Is there more variation in output of one shift in a manufacturing plant than in another shift? In an effort to study this question, plant managers gathered productivity reports from the 8 A.M. to 4 P.M. shift for 8 days. The reports indicated that the following numbers of units were produced on each day for this shift.

5528 4779 5112 5380 4918 4763 5055 5106

Productivity information was also gathered from 7 days for the 4 P.M. to midnight shift, resulting in the following data.

4325 4016 4872 4559 3982 4754 4116

Use these data and $\alpha = .01$ to test to determine whether the variances of productivity for the two shifts are the same. Assume productivity is normally distributed in the population.

10.46 Various types of retail outlets sell toys during the holiday season. Among them are specialty toy stores, the large discount toy stores, and other retailers that carry toys as only one part of their stock of goods. Is there any difference in the dollar amount of a customer purchase between a large discount toy store and a specialty toy store if they carry relatively comparable types of toys? Suppose in December a random sample of 40 sales slips is selected from a large discount toy outlet and a random sample of 36 sales slips is selected from a specialty toy store. The data gathered from these samples follow.

DISCOUNT

34.60	31.52	32.89	52.47	33.63
38.85	43.65	57.90	57.94	42.20
54.19	42.13	53.41	31.27	47.41
61.41	26.51	13.76	45.74	47.68
47.88	67.62	55.61	41.38	47.89
36.73	56.09	45.34	42.36	35.79
43.06	37.12	49.96	62.36	57.56
56.58	58.40	21.04	36.86	46.14

SPECIALTY

22.00	24.30	46.89	49.47	39.41
34.52	33.35	28.94	40.91	33.34
37.53	35.64	28.53	24.57	27.52
13.54	27.68	25.63	27.48	39.66
14.34	20.63	26.77	25.68	51.17
25.66	6.45	21.19	39.21	21.44
36.83	12.69	40.15	25.25	24.66
34.05				

Use $\alpha = .01$ and the data to determine whether there is a significant difference in the average size of purchases at these stores.

10.47 One of the new thrusts of quality control management is to examine the process by which a product is produced. This approach also applies to paperwork. In industries where large long-term projects are undertaken, days and even weeks may elapse as a change order makes its way through a maze of approvals before receiving final approval. This process can result in long delays and stretch schedules to the breaking point. Suppose a quality control consulting group claims that it can significantly reduce the number of days required for such paperwork to receive approval. In an attempt to "prove" its case, the group selects five jobs for which it revises the paperwork system. The following data show the number of days required for a change order to be approved before the group intervened and the number of days required for a change order to be approved after the group instituted a new paperwork system.

BEFORE	AFTER
12	8
7	3
10	8
16	9
8	5

Use $\alpha = .01$ to determine whether there was a significant drop in the number of days required to process paperwork to approve change orders. Assume that in each case the number of days of paperwork is normally distributed.

10.48 There are two large newspapers in your city. You are interested in knowing whether there is a significant difference in the average number of pages in each newspaper dedicated solely to advertising. You randomly select 10 editions of newspaper A and six editions of newspaper B (excluding weekend editions). The data follow. Use $\alpha = .01$ to test whether there is a significant difference in averages. Assume the number of pages of advertising per edition is normally distributed and the population variances are approximately equal.

A		B	
17	17	8	14
21	15	11	10
11	19	9	6
19	22		
26	16		

10.49 Is there a difference in the proportion of construction workers who are under 35 years of age and the proportion of telephone repair people who are under 35 years of age? Suppose a study is conducted in Calgary, Alberta, using random samples of 338 construction workers and 281 telephone repair people. The sample of construction workers includes 297 people under 35 years of age and the sample of telephone repair people includes 192 people under that age. Use these data to test to determine if there is a significant difference in the proportions of workers who are under 35 years of age in construction and in telephone repair. Let alpha be .10.

10.50 Suppose a large insurance company wants to test to determine if the average amount of term life insurance purchased per family is greater than the average amount of whole life insurance purchased per family. To test this hypothesis, one of the company's actuaries randomly selects 27 families who have term life insurance only and 29 families who have whole life insurance only. Each sample is taken from families in which the leading provider is younger than 45 years of age. Suppose the data from this study are those given below. Use these data to test the hypothesis. Use a 5% level of significance, assume that the amount of insurance is normally distributed, and assume that the variances are unequal.

TERM LIFE ($1,000)

80	65	65	70	60	100	75
115	120	60	75	60	45	75
95	30	35	25	60	45	95
75	60	100	80	110	115	

WHOLE LIFE ($1,000)

35	60	40	55	30	45	20
15	60	60	55	60	65	35
65	40	40	35	30	45	35
25	50	30	15	45	70	50
55						

ANALYZING THE DATABASES

1. Test to determine whether there is a significant difference between mean value added by the manufacturer and the cost of materials in manufacturing. Use the manufacturing database as the sample data and let alpha be .01.
2. Use the manufacturing database to test to determine whether there is a significantly greater variance among the values of end-of-year inventories than among cost of materials. Let $\alpha = .05$.
3. Is there a difference between the average number of admissions at a general medical hospital and a psychiatric

hospital? Use the hospital database to test this hypothesis with $\alpha = .10$. The variable Service in the hospital database differentiates general medical hospitals (coded 1) and psychiatric hospitals (coded 2). Now test to determine whether there is a difference between these two types of hospitals on the variables Beds and Total Expenses.
4. Use the financial database to test to determine whether there is a significant difference in the proportion of companies whose earnings per share are more than $2.00 and the proportion of companies whose dividends per share are more than $1.00. Let $\alpha = .05$.

CASE

SEITZ CORPORATION: PRODUCING QUALITY GEAR-DRIVEN AND LINEAR-MOTION PRODUCTS

The Seitz Corporation is a QS 9000 certified company that designs and manufactures thermoplastic mechanical drives, such as gears and pulleys, and pin-feed tractors for printers. They specialize in complete gear train design and converting drive systems from metals to plastics for cost reductions and higher performance. Founded in 1949 by the late Karl F. Seitz, this family-owned company based in Torrington, Connecticut has plants in Connecticut and Illinois and is in the process of opening a plant in Loveland, Colorado. Currently, Seitz's plants employ around 300 people; the new Loveland facility is expected to add around 200 employees to that number in the next two years.

Seitz began as a small toolmaking business and grew slowly. In the late 1960s, the company expanded its services to include custom injection molding. As their customer base grew to include leading printer manufacturers, Seitz developed and patented a proprietary line of perforated-form-handling tractors. Utilizing its injection-molding technology, the company engineered an all-plastic tractor called Data Motion, which replaced the costly metal version. By the late 1970s, business was booming, and Data Motion had become the worldwide industry leader.

In the 1980s, foreign competition entered the business-equipment market, and many of Seitz's customers relocated or closed shop. The ripple effect hit Seitz as sales declined and profits eroded. Employment at the company dropped from a high of 313 in 1985 to only 125 in 1987. Drastic changes had to be made at Seitz.

To meet the challenge in 1987, Seitz made a crucial decision to change the way it did business. The company implemented a formal 5-year plan with measurable goals called "World-Class Excellence Through Total Quality." Many hours were devoted by senior managers to improving employee training and involvement. New concepts were explored and integrated into the business plan. Teams and programs were put into place to immediately correct deficiencies in Seitz's systems that were revealed in customer-satisfaction surveys. All employees from machine operators to accountants were taught that quality means understanding customers' needs and fulfilling them correctly the first time.

Once the program started, thousands of dollars in cost savings and two new products generating almost $1 million in sales resulted. Annual sales grew from $10.8 million in 1987 to $19 million in 1990. Seitz's customer base expanded from 312 in 1987 to 550 at the end of 1990.

In the decade of the 1990s, Seitz continued its steady growth. By 1999, Seitz was shipping products to 28 countries, and customers included Xerox, Hewlett Packard, Canon, U.S. Tsubaki, and many more worldwide. By 1998, sales had topped the $30 million mark and were expected to reach $45 million by the year 2000. CEO and President Alan F. Seitz stated that the "Seitz Corporation is dedicated to providing products and services that consistently meet or exceed our customers' requirements."

DISCUSSION

1. Seitz has a growing list of several hundred business-to-business customers. Managers would like to know whether the average dollar amount of sales per transaction per customer has changed from last year to this year. Suppose company accountants sampled 20 customers randomly from last year's records and 25 customers randomly from this year's files and the resulting dollar amount of each sales transaction is shown below. Analyze these data and summarize your findings for managers. Explain how this information can be used by decision makers. Assume that sales per customer are normally distributed and that the variances are likely to be unequal.

SALES LAST YEAR						SALES THIS YEAR				
1739	2482	2163	2938	2710		2199	1766	2425	1374	2188
2047	2096	2324	2462	2192		2227	1866	3100	2281	2768
1733	1380	2816	2045	1964		2736	2913	2613	2530	2977
2929	2618	2575	2067	2114		2420	2236	2046	1360	1866
						2206	2462	1895	3121	2337

2. One common approach to measuring a company's quality is through the use of customer-satisfaction surveys. Suppose a random sample of Seitz's customers are asked whether the plastic tractor produced by Seitz has outstanding quality (yes or no). Assume Seitz produces these tractors at two different plant locations and that the tractor customers can be divided according to where their tractors were manufactured. Suppose a random sample of 45 customers who bought tractors made at plant 1 results in 18 saying the tractors have excellent quality and a random sample of 51 customers who bought tractors made at plant 2 results in 12 saying the tractors have excellent quality. Test the difference in population proportions of excellent ratings between the two groups of customers using $\alpha = .05$. Does it seem to matter which plant produces the tractors in terms of the quality rating received from customers? What would you report from these data?

3. Suppose the customer-satisfaction survey included a question on the overall quality of Seitz measured on a 5-point scale, with 1 indicating low quality and 5 indicating high quality. Company managers monitor the figures from year to year to help determine whether Seitz is improving customers' perception of its quality. Suppose random samples of the responses from 2000 and 2001 customers are taken and analyzed on this question, and the following Excel analysis of the data results. Help managers interpret this analysis so that comparisons can be made between 2000 and 2001. Discuss the samples, the statistics, and the conclusions.

```
Microsoft Excel - DataforExChap10.1.xls
File  Edit  View  Insert  Format  Tools  Data  Window  Help
```

A1 = z-Test: Two Sample for Means

	A	B	C	D	E	F	G
1	z-Test: Two Sample for Means						
2							
3		2000	2001				
4	Mean	3.282	3.312				
5	Known Variance	0.0424	0.0337				
6	Observations	75	93				
7	Hypothesized Mean Difference	0					
8	z	-1.01					
9	P(Z<=z) one-tail	0.1561					
10	z Critical one-tail	1.64					
11	P(Z<=z) two-tail	0.3123					
12	z Critical two-tail	1.96					
13							
14							
15							
16							
17							

Sheet54 / Sheet55 / Sheet56 \ **Sheet57** / Sheet1 /

4. Suppose Seitz produces pulleys that are specified to be 50 mm in diameter. A large batch of pulleys is made in week 1 and another is made in week 5. Quality control people want to determine whether there is a difference in the variance of the diameters of the two batches. Assume that a sample of six pulleys from the week-1 batch results in the following diameter measurements (in mm): 51, 50, 48, 50, 49, 51. Assume that a sample of seven pulleys from the week-5 batch results in the following diameter measurements (in mm): 50, 48, 48, 51, 52, 50, 52. Conduct a test to determine whether the variance in diameters differs between these two populations. Why would the quality control people be interested in such a test? What results of this test would you relate to them? What about the means of these two batches? Analyze these data in terms of the means and report on the results.

ADAPTED FROM: "Seitz Corporation," *Strengthening America's Competitiveness: Resource Management Insights for Small Business Success.* Published by Warner Books on behalf of Connecticut Mutual Life Insurance Company and the U.S. Chamber of Commerce in association with the Blue Chip Enterprise Initiative, 1991. See also the Seitz Corporation's website located at http://www.seitzcorp.com.

11

Analysis of Variance and Chi-Square Applications

Learning Objectives

The focus of this chapter is the analysis of variance and applications of the chi-square statistic, thereby enabling you to:

1. Understand the differences between various experimental designs and when to use them.

2. Compute and interpret the results of a one-way ANOVA.

3. Compute and interpret the results of a random block design.

4. Compute and interpret the results of a two-way ANOVA.

5. Understand and interpret interaction.

6. Understand the chi-square goodness-of-fit test and how to use it.

7. Analyze data by using the chi-square test of independence.

Chapter 11 explores two very important and widely used types of statistical tests: analysis of variance (ANOVA) and chi-square. Analysis of variance tests are used to extend the hypothesis tests of means from two independent populations presented in Chapter 10 to include more than two populations and can be used to statistically test hypotheses from more complex research designs. The chi-square techniques presented in this chapter are appropriate for testing hypotheses using categorical data. The first part of the chapter includes three analysis of variance techniques presented in a design of experiments setting. The last two sections of the chapter present two of the more widely used chi-square techniques.

11.1 Introduction to Design of Experiments

Sometimes business research entails more complicated hypothesis-testing scenarios than those presented to this point in the text. Instead of comparing the wear of tire tread for two brands of tires to determine whether there is a significant difference between the brands, as we could have done by using Chapter 10 techniques, a tire researcher may choose to compare three, four, or even more brands of tires at the same time. In addition, the researcher may want to include different levels of quality of tires in the experiment, such as low-quality, medium-quality, and high-quality tires. Tests may be conducted under varying conditions of temperature, precipitation, or road surface.

How does a business analyst set up designs for such experiments as these? How can the data be analyzed? These questions can be answered, in part, through the use of analysis of variance and the design of experiments.

An **experimental design** is *a plan and a structure to test hypotheses in which the researcher either controls or manipulates one or more variables.* It contains *independent* and *dependent* variables. In an experimental design, an **independent variable** may be either a treatment viable or a classification variable. A **treatment variable** is one *the experimenter controls or modifies in the experiment.* A **classification variable** is some characteristic of the experimental subjects that was *present prior to the experiment and is not a result of the experimenter's manipulations or control.* Independent variables are sometimes also referred to as **factors.** Wal-Mart executives might sanction an in-house study to compare daily sales volumes for a given size store in four different demographic settings: (1) Inner-city stores (large city), (2) Suburban stores (large city), (3) Stores in a medium-size city, and (4) Stores in a small town. Managers might also decide to compare sales on the five different weekdays (Monday through Friday). In this study, the independent variables are store demographics and day of the week. A finance researcher might conduct a study to determine whether there is a significant difference in application fees for home loans in five geographic regions of the United States and might include three different types of lending organizations. In this study, the independent variables are geographic region and types of lending organizations. Or suppose a manufacturing organization produces a valve that is specified to have an opening of 6.37 cm. Quality controllers within the company might decide to test to determine how the openings for produced valves vary among four different machines on three different shifts. This experiment includes the independent variables of type of machine and work shift.

Whether an independent variable can be manipulated by the business analyst depends on the concept being studied. Independent variables such as work shift, gender of employee, geographic region, type of machine, and quality of tire are classification variables with conditions that existed prior to the study. The business analyst cannot change the characteristic of the variable, so he or she studies the phenomenon being explored under several conditions of the various aspects of the variable. As an example, the valve experiment is conducted under the conditions of all three work shifts.

Experimental design
A plan and a structure to test hypotheses in which the researcher either controls or manipulates one or more variables.

Independent variable
In an analysis of variance, the treatment or factor being analyzed. In regression analysis, the predictor variable.

Treatment variable
The independent variable of an experimental design that the researcher either controls or modifies.

Classification variable
The independent variable of an experimental design that was present prior to the experiment and is not the result of the researcher's manipulations or control.

Factors
Another name for the independent variables of an experimental design.

However, some independent variables can be manipulated by the researcher. For example, in the well-known Hawthorne studies of the Western Electric Company in the 1920s in Illinois, the amount of light in production areas was varied to determine the effect of light on productivity. In theory, this independent variable could be manipulated by the researcher to allow any level of lighting. Other examples of independent variables that can be manipulated include the amount of bonuses offered workers, level of humidity, and temperature.

Each independent variable has two or more levels, or classifications. **Levels, or classifications,** of independent variables are *the subcategories of the independent variable used by the researcher in the experimental design.* For example, the different demographic settings listed for the Wal-Mart study are four levels, or classifications, of the independent variable store demographics: (1) inner-city store, (2) suburban store, (3) store in a medium-size city, and (4) store in small town. In the valve experiment, there are four levels, or classifications of machines within the independent variable machine type: machine 1, machine 2, machine 3, and machine 4.

The other type of variable in an experimental design is a dependent variable. A **dependent variable** is *the response to the different levels of the independent variables.* It is the measurement taken under the conditions of the experimental design that reflect the effects of the independent variable(s). In the Wal-Mart study, the dependent variable is probably the dollar amount of daily total sales. For the study on loan application fees, the fee charged for a loan application is probably the dependent variable. In the valve experiment, the dependent variable is the size of the opening of the valve.

Experimental designs in this chapter are analyzed statistically by a group of techniques referred to as **analysis of variance** or **ANOVA.** The analysis of variance concept begins with the notion that individual items being studied, such as employees, machine-produced products, district offices, hospitals, and so on, are not all the same. Note the measurements for the openings of 24 valves randomly selected from an assembly line that are given in Table 11.1. The mean opening is 6.34 cm. Only one of the 24 valve openings is actually the mean. Why do the valve openings vary? Notice that the total sum of squares of deviation of these valve openings around the mean is .3915 cm². Why is this value not zero? Using various types of experimental designs, we can explore some possible reasons for this variance with analysis of variance techniques. As we explore each of the experimental designs and their associated analysis, note that the statistical technique is attempting to "break down" the total variance among the objects being studied into possible causes. In the case of the valve openings, this variance of measurements might be due to such variables as machine, operator, shift, supplier, and production conditions, among others.

Many different types of experimental designs are available to business analysts. In this chapter, we will present and discuss three specific types of experimental designs: completely randomized design, randomized block design, and factorial experiments.

Levels, or Classifications
The subcategories of the independent variable used by the researcher in the experimental design.

Dependent variable
In analysis of variance, the measurement that is being analyzed; the response to the different levels of the independent variables. In regression analysis, the variable that is being predicted.

Analysis of variance (ANOVA)
A technique for statistically analyzing the data from a completely randomized design; uses the F test to determine whether there is a significant difference in two or more independent groups.

6.26	6.19	6.33	6.26	6.50
6.19	6.44	6.22	6.54	6.23
6.29	6.40	6.23	6.29	6.58
6.27	6.38	6.58	6.31	6.34
6.21	6.19	6.36	6.56	

$\overline{X} = 6.34$

Total Sum of Squares Deviation = SST = $\Sigma(X_i - \overline{X})^2 = .3915$

TABLE 11.1
Valve Opening Measurements (in cm) for 24 Valves Produced on an Assembly Line

11.2
The Completely Randomized Design (One-Way ANOVA)

Completely randomized design

An experimental design wherein there is one treatment or independent variable with two or more treatment levels and one dependent variable. This design is analyzed by analysis of variance.

One of the simplest experimental designs is the completely randomized design. In the **completely randomized design,** subjects are assigned randomly to treatments. The completely randomized design contains only one independent variable, with two or more treatment levels, or classifications. If there are only two treatment levels, or classifications, of the independent variable, the design is the same one used to test the difference in means of two independent populations presented in Chapter 10, which used the *t* test to analyze the data.

In this section, we will focus on completely randomized designs with three or more classification levels. Analysis of variance, or ANOVA, will be used to analyze the data that result from the treatments. Completely randomized design experiments contain only one dependent variable.

A completely randomized design could be structured for a tire-quality study in which tire quality is the independent variable and the treatment levels are low, medium, and high quality. The dependent variable might be the number of miles driven before the tread fails state inspection. A study of daily sales volumes for Wal-Mart stores could be undertaken by using a completely randomized design with demographic setting as the independent variable. The treatment levels, or classifications, would be inner-city stores, suburban stores, stores in medium-size cities and stores in small towns. The dependent variable would be sales dollars.

Suppose a business analyst decides to analyze the effects of the machine operator on the valve opening measurements of valves produced in a manufacturing plant, like those shown in Table 11.1. The independent variable in this design is machine operator. Suppose further that there are three different operators (one for each shift). These three machine operators are the levels of treatment, or classification, of the independent variable. The dependent variable is the opening measurement of the valve. Figure 11.1 shows the structure of this completely randomized design. Is there a significant difference in the mean valve openings between the three machine operators? The data from Table 11.1 have been organized by machine operator as shown in Figure 11.1 and are displayed in Table 11.2.

Figure 11.1

Completely randomized design

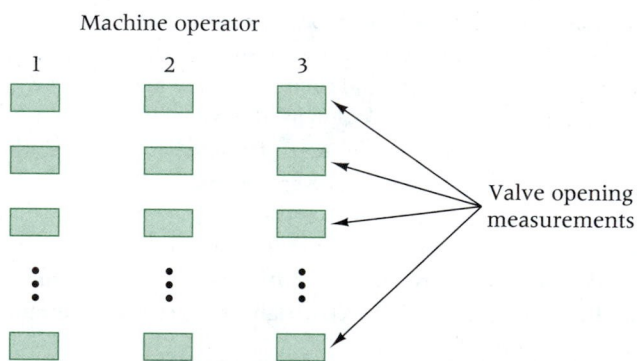

Machine operator

Valve opening measurements

TABLE 11.2
Valve Openings by Machine Operator

OPERATOR 1	OPERATOR 2	OPERATOR 3
6.56	6.38	6.39
6.40	6.19	6.33
6.54	6.26	6.29
6.34	6.23	6.43
6.58	6.22	6.36
6.44	6.27	6.41
6.36	6.29	6.31
6.50	6.19	6.50

One-Way Analysis of Variance

In the machine operator example, is it possible to analyze the three samples by using a t test for the difference in two sample means? These three samples would require $_3C_2 = 3$ individual t tests to accomplish the analysis of two groups at a time. Recall that if $\alpha = .05$ for a particular test, there is a 5% chance of rejecting a null hypothesis that is true (i.e., committing a Type I error). If enough tests are done, eventually one or more null hypotheses will be falsely rejected by chance. Hence, $\alpha = .05$ is valid only for one t test. In this problem, with three t tests, the error rate compounds, so when the analyst is finished with the problem there is a much greater than .05 chance of committing a Type I error. Fortunately, a technique has been developed that analyzes all the sample means at one time and thus precludes the buildup of error rate: analysis of variance (ANOVA). A completely randomized design is analyzed by a **one-way analysis of variance.**

In general, if k samples are being analyzed, the following hypotheses are being tested in a one-way ANOVA.

$$H_0: \mu_1 = \mu_2 = \mu_3 = \cdots = \mu_k$$

$$H_a: \text{At least one of the means is different from the others.}$$

The null hypothesis states that the population means for all treatment levels are equal. Because of the way the alternative hypothesis is stated, if even one of the population means is different from the others, the null hypothesis is rejected.

Testing these hypotheses by using one-way ANOVA is accomplished by partitioning the total variance of the data into the following two variances.

1. The variance resulting from the treatment (columns)
2. The error variance, or that portion of the total variance unexplained by the treatment

As part of this process, the total sum of squares of deviation of values around the mean can be divided into two additive and independent parts.

$$\text{SST} \quad = \quad \text{SSC} \quad + \quad \text{SSE}$$

$$\sum_{i=1}^{n_j}\sum_{j=1}^{C}(X_{ij} - \overline{X})^2 = \sum_{j=1}^{C} n_j(\overline{X}_j - \overline{X})^2 + \sum_{i=1}^{n_j}\sum_{j=1}^{C}(X_{ij} - \overline{X}_j)^2$$

where:
   i = particular member of a treatment level
   j = a treatment level
   C = number of treatment levels
   n_j = number of observations in a given treatment level
   $\overline{X}$ = grand mean
   $\overline{X}_j$ = mean of a treatment group or level
   X_{ij} = individual value

This relationship is shown in Figure 11.2. Observe that the total sum of squares of variation is partitioned into the sum of squares of treatment (columns) and the sum of squares of error.

The formulas used to accomplish one-way analysis of variance are developed from this relationship. The double summation sign indicates that the values are summed within a treatment level and across treatment levels. Basically, ANOVA compares the relative sizes of the treatment variation and the *error* variation (within-group variation). The error variation is unaccounted-for variation and can be viewed at this point as variation due to individual differences within treatment groups. If there is a significant difference in treatments, the treatment variation should be large relative to the error variation.

One-way analysis of variance
The process used to analyze a completely randomized experimental design. This process involves computing a ratio of the variance between treatment levels of the independent variable to the error variance. This ratio is an F value, which is then used to determine whether there are any significant differences between the means of the treatment levels.

Figure 11.2

Partitioning total
sum of squares
of variation

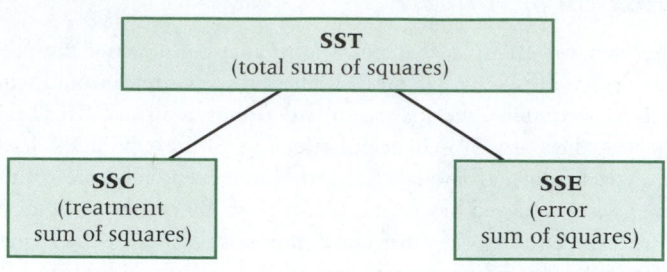

Figure 11.3

Valve openings
by operator

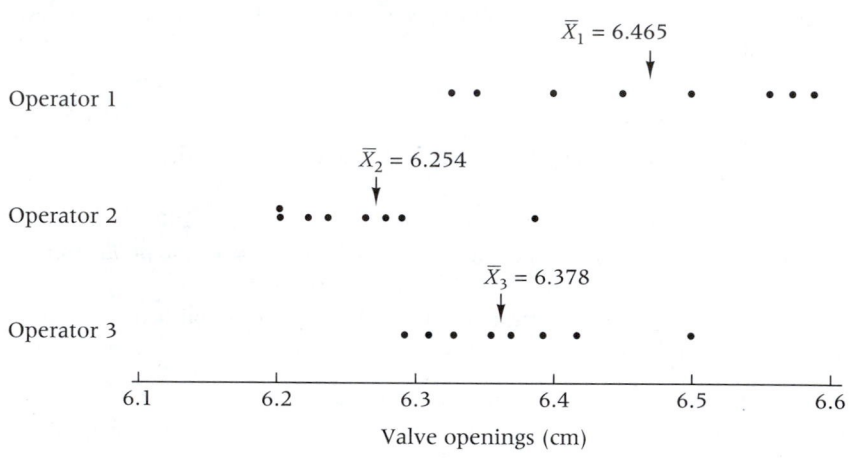

Figure 11.3 displays the data from the machine operator example in terms of treatment level. Note the variation of values (*X*) *within* each treatment level. Now examine the variation *between* levels 1 through 3 (the difference in the machine operator groupings). Note that the means are located somewhat apart from each other. This difference also is underscored by the mean values for each treatment level:

$$\bar{X}_1 = 6.465 \quad \bar{X}_2 = 6.254 \quad \bar{X}_3 = 6.378.$$

Analysis of variance is used to determine statistically whether the variance between the treatment level means is greater than the variances within levels (error variance). There are several important assumptions underlying analysis of variance:

1. Observations are drawn from normally distributed populations.
2. Observations represent random samples from the populations.
3. Variances of the populations are equal.

These assumptions are similar to those for using the *t* test for small independent samples in Chapter 10. It is assumed that the populations are normally distributed and that the population variances are equal. These techniques should be used only with random samples.

An ANOVA is computed with the three sums of squares: total, treatment (columns), and error. Shown here are the formulas to compute a one-way analysis of variance. The term SS represents sum of squares, and the term MS represents mean square. SSC is the sum of squares columns, which yields the sum of squares between treatments. This measures the variation between columns or between treatments since the independent variable treatment levels are presented as columns. SSE is the sum of squares of error, which yields the variation within treatments (or columns). Some say that this is a measure of the individual differences unaccounted for by the treatments. SST is the total sum of squares and is a measure of all variation in the dependent variable. As shown previously, SST contains both SSC and SSE and can be partitioned into SSC and SSE. MSC, MSE, and MST are the mean squares of column, error, and total, respectively. Mean square is an average and is computed by dividing the sum of squares by the degrees of freedom. Finally, the **F value** is determined by dividing the treatment variance (MSC) by the error variance (MSE). As discussed in Chapter 10, the F is a ratio of two variances. In the ANOVA situation, the F is *a ratio of the treatment variance to the error variance.*

F value
The ratio of two variances, used to reach statistical conclusions regarding the null hypothesis; in ANOVA, the ratio of the treatment variance to the error variance.

$$\text{SSC} = \sum_{j=1}^{C} n_j (\overline{X}_j - \overline{X})^2 \qquad \text{df}_C = C - 1 \qquad \text{MSC} = \frac{\text{SSC}}{\text{df}_C}$$

$$\text{SSE} = \sum_{i=1}^{n_j} \sum_{j=1}^{C} (X_{ij} - \overline{X}_j)^2 \qquad \text{df}_E = N - C \qquad \text{MSE} = \frac{\text{SSE}}{\text{df}_E}$$

$$\text{SST} = \sum_{i=1}^{n_j} \sum_{j=1}^{C} (X_{ij} - \overline{X})^2 \qquad \text{df}_T = N - 1 \qquad F = \frac{\text{MSC}}{\text{MSE}}$$

FORMULAS FOR COMPUTING A ONE-WAY ANOVA

where:
 i = a particular member of a treatment level
 j = a treatment level
 C = number of treatment levels
 n_j = number of observations in a given treatment level
 $\overline{X}$ = grand mean
 $\overline{X}_j$ = column mean
 X_{ij} = individual value

Performing these calculations for the machine operator example yields the following.

Valve Openings by Machine Operator

OPERATOR 1	OPERATOR 2	OPERATOR 3
6.56	6.38	6.39
6.40	6.19	6.33
6.54	6.26	6.29
6.34	6.23	6.43
6.58	6.22	6.36
6.44	6.27	6.41
6.36	6.29	6.31
6.50	6.19	6.50

T_j:	$T_1 = 51.72$	$T_2 = 50.03$	$T_3 = 51.02$	$T = 152.77$
n_j:	$n_1 = 8$	$n_2 = 8$	$n_3 = 8$	$n = 24$
$\overline{X}_j$:	$\overline{X}_1 = 6.465$	$\overline{X}_2 = 6.25375$	$\overline{X}_3 = 6.3775$	$\overline{X} = 6.36542$

$$\text{SSC} = \sum_{j=1}^{C} n_j (\overline{X}_j - \overline{X})^2$$

$$= [8(6.465 - 6.36542)^2 + 8(6.25375 - 6.36542)^2 + 8(6.3775 - 6.36542)^2]$$

$$= 0.07933 + 0.09976 + 0.00117$$

$$= 0.18026$$

$$\text{SSE} = \sum_{i=1}^{n_j} \sum_{j=1}^{C} (X_{ij} - \overline{X}_j)^2$$

$$= [(6.56 - 6.465)^2 + (6.40 - 6.465)^2 + (6.54 - 6.465)^2 + (6.34 - 6.465)^2$$

$$+ (6.58 - 6.465)^2 + (6.44 - 6.465)^2 + (6.36 - 6.465)^2 + (6.50 - 6.465)^2$$

$$+ (6.38 - 6.25375)^2 + (6.19 - 6.25375)^2 + \cdots + (6.31 - 6.3775)^2$$

$$+ (6.50 - 6.3775)^2]$$

$$= 0.121738$$

$$\text{SST} = \sum_{i=1}^{n_j} \sum_{j=1}^{C} (X_{ij} - \overline{X})^2$$

$$= [(6.56 - 6.36542)^2 + (6.40 - 6.36542)^2 + (6.54 - 6.36542)^2$$

$$+ \cdots + (6.31 - 6.36542)^2 + (6.50 - 6.36542)^2]$$

$$= 0.301997$$

$$\text{df}_C = C - 1 = 3 - 1 = 2$$

$$\text{df}_E = N - C = 24 - 3 = 21$$

$$\text{df}_T = N - 1 = 24 - 1 = 23$$

$$\text{MSC} = \frac{\text{SSC}}{\text{df}_C} = \frac{0.18026}{2} = 0.090130$$

$$\text{MSE} = \frac{\text{SSE}}{\text{df}_E} = \frac{0.121738}{21} = 0.005797$$

$$F = \frac{\text{MSC}}{\text{MSE}} = \frac{0.090130}{0.005797} = 15.55$$

From these computations, an analysis of variance chart can be constructed, as shown in Table 11.3. The observed F value is 15.55. It is compared to a critical value from the F table to determine whether there is a significant difference in treatment or classification.

F distribution

A distribution based on the ratio of two random variances; used in testing two variances and in analysis of variance.

Reading the F Distribution Table

The **F distribution** table is in Table A.7. Associated with every F value in the table are two unique df values: degrees of freedom in the numerator (df_C) and degrees of freedom in the denominator (df_E). To look up a value in the F distribution table, the researcher must know both degrees of freedom. Because each F distribution is determined by a

TABLE 11.3
Analysis of Variance Table for the Machine Operator Example

SOURCE OF VARIANCE	df	SS	MS	F
Between (columns)	2	0.180259	0.090130	15.55
Error	21	0.121738	0.005797	
Total	23	0.301997		

unique pair of degrees of freedom, there are many F distributions. Space constraints limit Table A.7 to F values for only $\alpha = .005, .01, .025, .05,$ and $.10$. However, statistical computer software packages for computing ANOVAs usually give a probability for the F value, which allows a hypothesis-testing decision for any alpha based on the p-value method.

In the one-way ANOVA, the df_C values are the treatment (column) degrees of freedom, $C - 1$. The df_E values are the error degrees of freedom, $N - C$. Table 11.4 contains an abbreviated F distribution table for $\alpha = .05$. For the machine operator example, $df_C = 2$ and $df_E = 21$. $F_{.05,2,21}$ from Table 11.4 is 3.47. This value is the critical value of the F test. Analysis of variance tests are *always* one-tailed tests with the rejection region in the upper tail. The decision rule is to reject the null hypothesis if the observed F value is greater than the critical F value ($F_{.05,2,21} = 3.47$). In this case, the observed F value of 15.55 is larger than the table F value of 3.47, so the null hypothesis is rejected. Not all means are equal, so there is a significant difference in the mean number of valve openings by machine operator. Figure 11.4 is a graph of an F distribution showing the critical F value for this example and the rejection region. Note that the F distribution begins at zero and contains no negative values. The reason is that an F value is the ratio of two variances, and variances are always positive.

Multiple Comparisons

Analysis of variance techniques are particularly useful in testing hypotheses about the differences of means in multiple groups because ANOVA utilizes only one single overall test. The advantage of this is that the probability of committing a Type I error, α, is controlled. As noted in Section 11.2, if three groups are tested two at a time, it takes three tests ($_3C_2$) to analyze hypotheses between all possible pairs. In general, if k groups are tested two at a time, there are $_kC_2 = k(k - 1)/2$ possible paired comparisons.

TABLE 11.4
An Abbreviated F Table for $\alpha = .05$

| | NUMERATOR DEGREES OF FREEDOM | | | | | | | | |
	1	2	3	4	5	6	7	8	9
DENOMINATOR DEGREES OF FREEDOM · · · 20	4.35	3.49	3.10	2.87	2.71	2.60	2.51	2.45	2.39
21	4.32	3.47	3.07	2.84	2.68	2.57	2.49	2.42	2.37
22	4.30	3.44	3.05	2.82	2.66	2.55	2.46	2.40	2.34

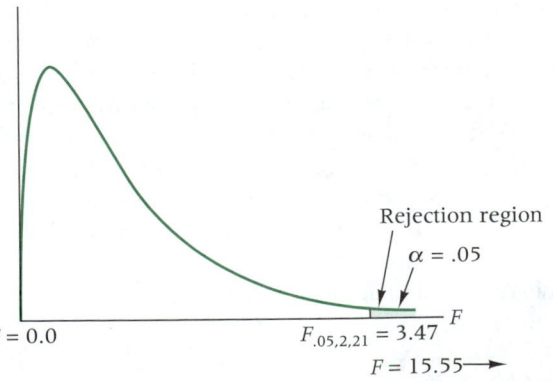

Figure 11.4

Graph of F values for the machine operator example

Suppose alpha for an experiment is to be .05. If two different pairs of comparisons are made in the experiment using alpha of .05 in each, there is a .95 probability of not making a Type I error in each comparison. This results in a .9025 probability of not making a Type I error in either comparison (.95 × .95), and a .0975 probability of committing a Type I error in at least one comparison (1 − .9025). Thus, the probability of committing a Type I error for this experiment is not .05 but .0975. In an experiment where the means of three groups are being tested two at a time (like the machine operator problem), three different tests would need to be conducted. If each is analyzed using α = .05, the probability that no Type I error would be committed in any of the three tests is .95 × .95 × .95 = .857 and the probability of committing at least one Type I error in the three tests is 1 − .857 = .143. However, computing one ANOVA on all three groups simultaneously using α = .05 maintains the value of alpha for the experiment.

Sometimes the researcher is satisfied with conducting an overall test of differences in groups such as the one ANOVA provides. However, when it is determined that there is an overall difference in population means, it is often desirable to go back to the groups and determine from the data which pairs of means are significantly different, if any. Such pairwise analyses can lead to the buildup of the Type I experimental error rate, as mentioned. Fortunately, several techniques, referred to as **multiple comparisons,** have been developed to handle this problem.

Multiple comparisons
Statistical techniques used to compare pairs of treatment means when the analysis of variance yields an overall significant difference in the treatment means.

Multiple comparisons are to be used only when an overall significant difference between groups has been obtained by using the F value of the analysis of variance. Some of these techniques protect more for Type I errors and others protect more for Type II errors. Some multiple comparison techniques require equal sample sizes. There seems to be some difference of opinion in the literature about which techniques are more appropriate. The variety of multiple comparison tests available include the following:* Dunn's multiple comparison procedure, Fisher's LSD test, Tukey's HSD test, Scheffe's S method, Newman-Keuls test, and Duncan's new multiple range test. These tests differ in the way they analyze the means and the manner in which they control error. The use of these techniques is beyond the scope of this text, and they will not be presented in detail here.

Analysis Using Excel

It is relatively easy to compute a one-way ANOVA in Excel. From **Tools** and then **Data Analysis,** select **ANOVA: Single Factor.** The dialog box for **ANOVA: Single Factor** is displayed in Figure 11.5. For the **Input Range,** insert the location of all the data. Check whether the data are arranged in columns or rows. The default is columns. Check whether or not there are labels in the first row, and insert the value of alpha for the test.

The Excel output for the machine operator problem is shown in Figure 11.6. This output contains a summary table and an ANOVA table. The summary table displays descriptive data for the samples including size, sum, mean, and variance. The ANOVA table is a standard ANOVA table, which contains the sum of squares (SS) for the between groups, the sum of squares for the within groups (error), and the total sum of squares along with degrees of freedom, the mean squares, and the observed F value. Note that these computed values are essentially the same as those computed by hand in this section (with slight rounding error). The p-value, reported for this problem in scientific notation as 7.19E-05 (7.19×10^{-5}), is .0000719. Since this is less than the assumed alpha (.05), the decision is to reject the null hypothesis and declare that there is a significant difference between machine operators. In addition, the Excel output contains the table value, which for this problem is F = 3.47. This Excel feature is a convenience for those business analysts who prefer using a critical value rather than a p-value for decision making.

---

*Kirk, Roger, *Experimental Design: Procedures for the Behavioral Sciences.* (Belmont, CA: Brooks/Cole, 1968).

Figure 11.5

Excel dialog box for **ANOVA: Single Factor**

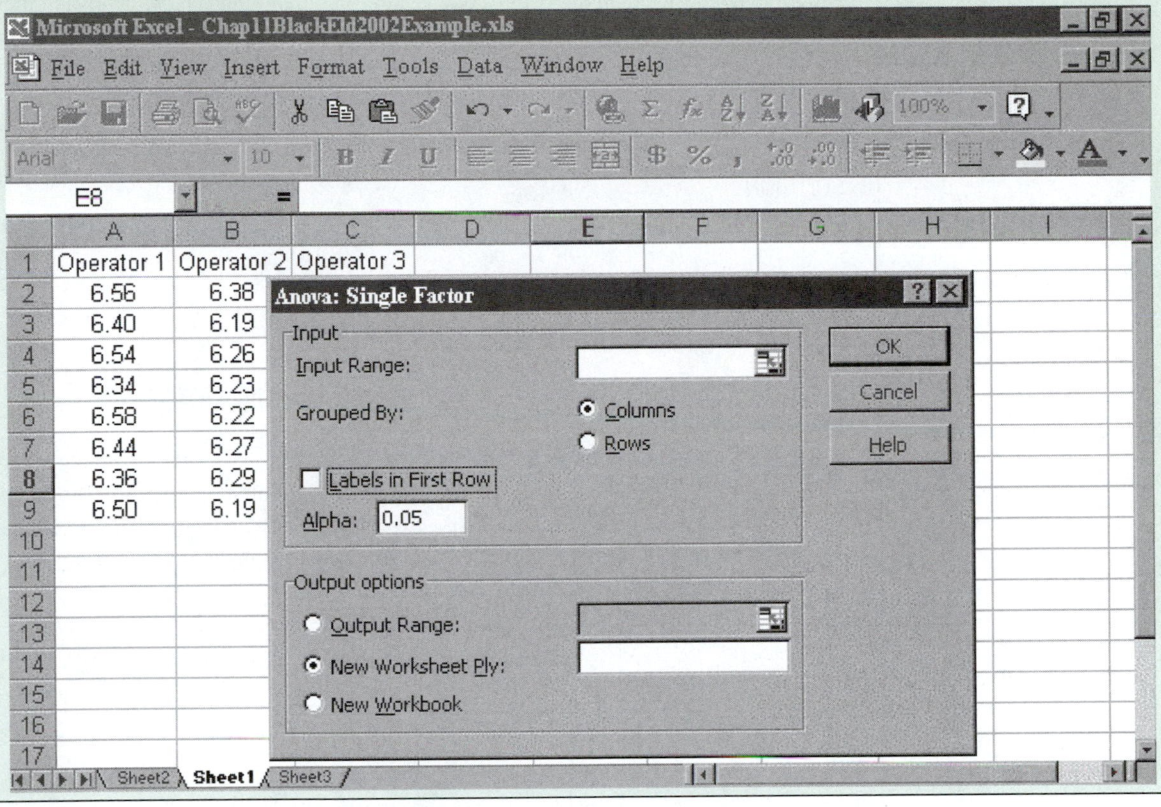

Figure 11.6

Excel output for the machine operator problem

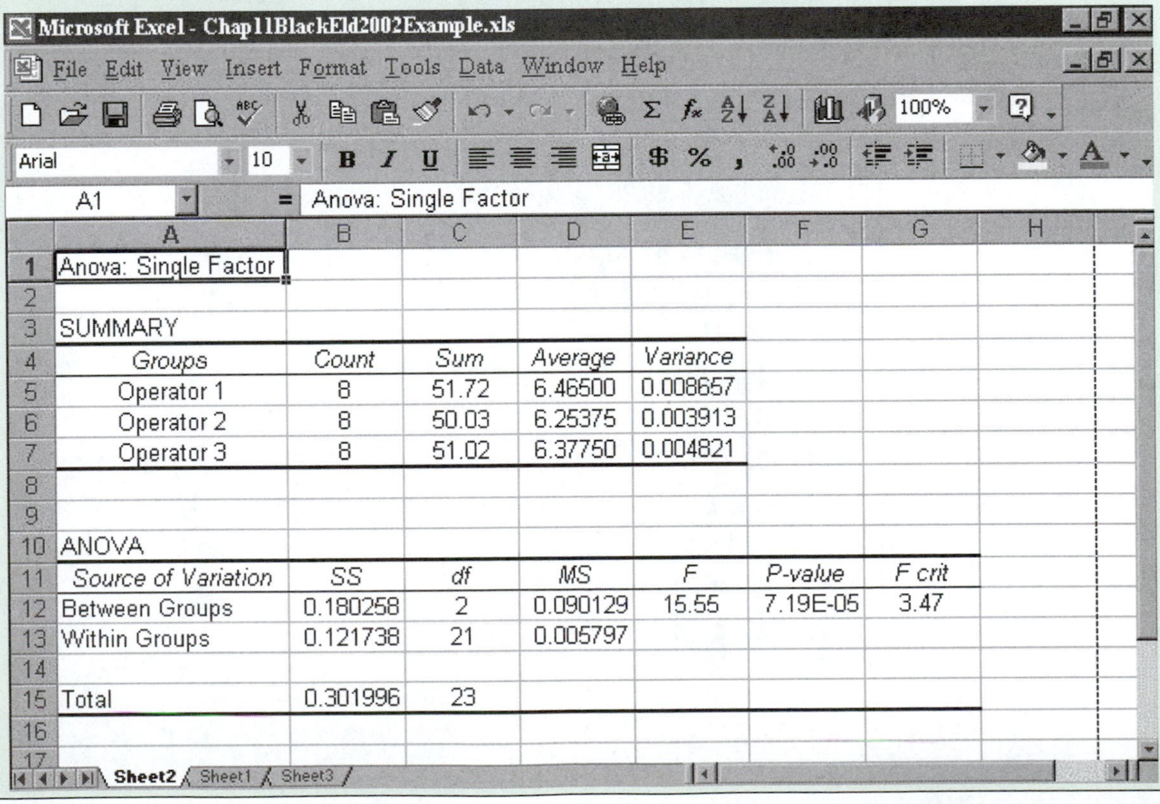

Anova: Single Factor

SUMMARY

Groups	Count	Sum	Average	Variance
Operator 1	8	51.72	6.46500	0.008657
Operator 2	8	50.03	6.25375	0.003913
Operator 3	8	51.02	6.37750	0.004821

ANOVA

Source of Variation	SS	df	MS	F	P-value	F crit
Between Groups	0.180258	2	0.090129	15.55	7.19E-05	3.47
Within Groups	0.121738	21	0.005797			
Total	0.301996	23				

DEMONSTRATION PROBLEM 11.1

A company has four manufacturing plants, and company officials want to determine whether there is a difference in the average age of workers at the four locations. The following data are the ages of randomly selected workers at each plant. Perform a one-way ANOVA to determine whether there is a significant difference in the mean ages of the workers at the four plants. Use $\alpha = .01$.

SOLUTION

STEP **1.** The hypotheses follow.

$$H_0: \mu_1 = \mu_2 = \mu_3 = \mu_4$$

H_a: At least one of the means is different from the others.

STEP **2.** The appropriate test statistic is the F test calculated from ANOVA.

STEP **3.** The value of α is .01.

STEP **4.** The degrees of freedom for this problem are $4 - 1 = 3$ for the numerator (treatments − columns) and $18 - 4 = 14$ for the denominator (error). The critical F value is $F_{.01,3,14} = 5.56$. Because ANOVAs are always one-tailed with the rejection region in the upper tail, the decision rule is to reject the null hypothesis if the observed value of F is greater than 5.56.

STEP **5.**

PLANT (EMPLOYEE AGES)

1	2	3	4
29	32	25	27
27	33	24	24
30	31	24	26
27	34	25	
	30	26	
	28		

STEP **6.**

T_j:	$T_1 = 113$	$T_2 = 188$	$T_3 = 124$	$T_4 = 77$	$T = 502$
n_j:	$n_1 = 4$	$n_2 = 6$	$n_3 = 5$	$n_4 = 3$	$N = 18$
$\bar{X}_j$:	$\bar{X}_1 = 28.25$	$\bar{X}_2 = 31.33$	$\bar{X}_3 = 24.8$	$\bar{X}_4 = 25.67$	$\bar{X} = 27.89$

$$SSC = 4(28.25 - 27.89)^2 + 6(31.33 - 27.89)^2 + 5(24.8 - 27.89)^2$$
$$+ 3(25.67 - 27.89)^2$$
$$= 134.23$$

$$SSE = (29 - 28.25)^2 + (27 - 28.25)^2 + \cdots + (24 - 25.67)^2 + (26 - 25.67)^2$$
$$= 37.55$$

$$SST = (29 - 27.89)^2 + (27 - 27.89)^2 + \cdots + (24 - 27.89)^2 + (26 - 27.89)^2$$
$$= 171.78$$

$$df_C = 4 - 1 = 3$$
$$df_E = 18 - 4 = 14$$
$$df_T = 18 - 1 = 17$$

SOURCE OF VARIANCE	SS	df	MS	F
Between	134.23	3	44.74	16.7
Error	37.55	14	2.68	
Total	171.78	17		

STEP **7.** The decision is to reject the null hypothesis because the observed *F* value of 16.7 is greater than the critical table *F* value of 5.56.

STEP **8.** There is a significant difference in the mean ages of workers at the three plants.

Following is the Excel output for Demonstration Problem 11.1.

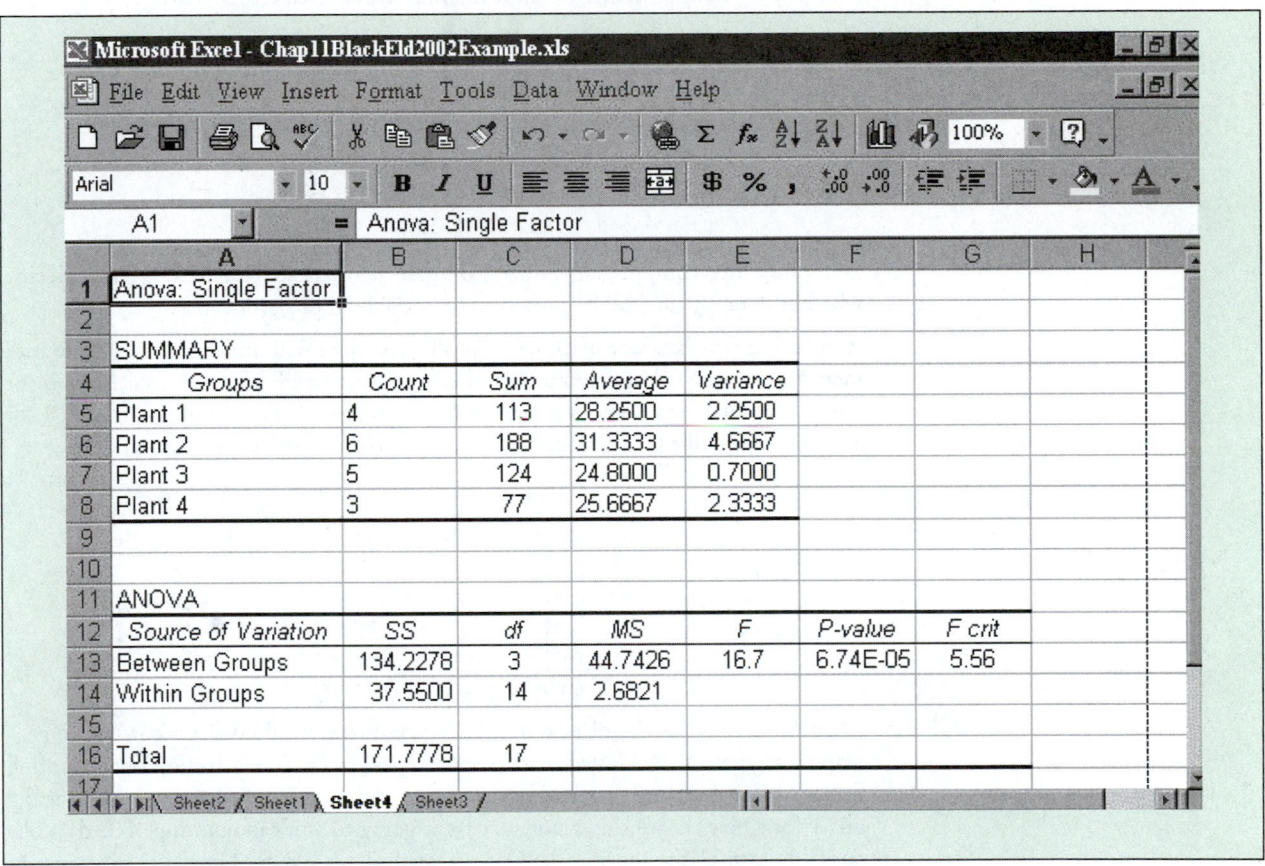

11.1 Compute a one-way ANOVA on the following data.

1	2	3
2	5	3
1	3	4
3	6	5
3	4	5
2	5	3
1		5

Determine the computed *F* value. Compare the *F* value with the critical table *F* value and decide whether to reject the null hypothesis. Use $\alpha = .05$.

11.2

Problems

11.2 Compute a one-way ANOVA on the following data.

1	2	3	4	5
14	10	11	16	14
13	9	12	17	12
10	12	13	14	13
	9	12	16	13
	10		17	12
				14

Determine the computed F value. Compare the F value with the critical table F value and decide whether to reject the null hypothesis. Use $\alpha = .01$.

11.3 Develop a one-way ANOVA on the following data.

1	2	3	4
113	120	132	122
121	127	130	118
117	125	129	125
110	129	135	125

Determine the computed F value. Compare it to the critical F value and decide whether to reject the null hypothesis. Use a 1% level of significance.

11.4 A milk company has four machines that fill gallon jugs with milk. The quality control manager is interested in determining whether the average fill for these machines is the same. The following data represent random samples of fill measures (in quarts) for 19 jugs of milk filled by the different machines. Use $\alpha = .01$ to test the hypotheses.

MACHINE 1	MACHINE 2	MACHINE 3	MACHINE 4
4.05	3.99	3.97	4.00
4.01	4.02	3.98	4.02
4.02	4.01	3.97	3.99
4.04	3.99	3.95	4.01
	4.00	4.00	
	4.00		

11.5 That the starting salaries of new accounting graduates would differ according to geographic regions of the United States seems logical. A random selection of accounting firms is taken from three geographic regions, and each is asked to state the starting salary for a new accounting graduate who is going to work in auditing. The data obtained follow. Use a one-way ANOVA to analyze these data. Note that the data can be restated to make the computations more reasonable (example: $32,500 = 3.25$). Use a 1% level of significance.

SOUTH	NORTHEAST	WEST
$30,500	$41,000	$35,500
31,500	39,500	33,500
30,000	39,000	35,000
31,000	38,000	36,500
31,500	39,500	36,000

11.6 A management consulting company presents a 3-day seminar on project management to various clients. The seminar is basically the same each time it is given. However, sometimes it is presented to high-level managers, sometimes to midlevel managers, and sometimes to low-level managers. The seminar facilitators believe evaluations of the seminar may vary with the audience. Suppose the following data are some randomly selected evaluation scores from different levels of managers after they have attended the seminar. The ratings are on a scale from 1 to 10, with 10 being the highest. Use a one-way ANOVA to determine whether there is a significant difference in the evaluations according to manager level. Assume $\alpha = .05$.

HIGH LEVEL	MIDLEVEL	LOW LEVEL
7	8	5
7	9	6
8	8	5
7	10	7
9	9	4
	10	8
	8	

11.7 Family transportation costs are usually higher than most people believe because those costs include car payments, insurance, fuel costs, repairs, parking, and public transportation. Twenty randomly selected families in four major cities are asked to use their records to estimate a monthly figure for transportation cost. Use the data obtained and ANOVA to test whether there is a significant difference in monthly transportation costs for families living in these cities. Assume that $\alpha = .05$.

ATLANTA	NEW YORK	LOS ANGELES	CHICAGO
$650	$250	$850	$540
480	525	700	450
550	300	950	675
600	175	780	550
675	500	600	600

A second research design is the **randomized block design.** The randomized block design is similar to the completely randomized design in that there is one independent variable (treatment variable) of interest. However, the randomized block design also includes a second variable, referred to as a blocking variable, that can be used to control for confounding or concomitant variables.

Confounding or **concomitant** variables are *variables that are not being controlled by the researcher in the experiment but can have an effect on the outcome of the treatment being studied.* For example, suppose a completely randomized design is used to analyze the effects of temperature on the tensile strengths of metal. Other variables not being controlled by the business analyst in this experiment may affect the tensile strength of metal, such as humidity, raw materials, machine, and shift. One way to control for these variables is to include them in the experimental design. The randomized block design has the capability of adding one of these variables into the analysis as a blocking variable. A **blocking variable** is *a variable that the researcher wants to control but is not the treatment variable of interest.*

One of the first people to use the randomized block design was Sir Ronald A. Fisher. He applied the design to the field of agriculture, where he was interested in studying the growth patterns of various varieties of seeds for a given type of plant. The seed variety was his independent variable. However, he realized that as he experimented on different plots of ground, the "block" of ground might make some difference in the experiment. Fisher designated several different plots of ground as blocks, which he controlled as a second variable. Each of the seed varieties was planted on each of the blocks. The main thrust of his study was to compare the seed varieties (independent variable). He merely wanted to control for the difference in plots of ground (blocking variable).

In the example of the problem of analyzing the effects of temperature on the tensile strengths of metal, blocking variables might be machine number (if several machines are used to make the metal), worker, shift, or day of the week. The business analyst probably

11.3
The Randomized Block Design

Randomized block design
An experimental design in which there is one independent variable of interest and a second variable, known as a blocking variable, that is used to control for confounding or concomitant variables.

Confounding variables, or Concomitant variables
Variables that are not being controlled by the researcher in the experiment but can have an effect on the outcome of the treatment being studied.

Blocking variable
A variable that the researcher wants to control but is not the treatment variable of interest.

already knows that different workers or different machines will produce at least slightly different metal tensile strengths because of individual differences. However, designating the variable (machine or worker) as the blocking variable and computing a randomized block design affords the potential for a more powerful analysis. In other experiments, some other possible variables that might be used as blocking variables include gender of subject, age of subject, intelligence of subject, economic level of subject, brand, supplier, or vehicle.

A special case of the randomized block design is the repeated measures design. The **repeated measures design** is *a randomized block design in which each block level is an individual item or person, and that person or item is measured across all treatments.* Thus, where a block level in a randomized block design is night shift and items produced under different treatment levels on the night shift are measured, in a repeated measures design, a block level might be an individual machine or person; items produced by that person or machine are then randomly chosen across all treatments. Thus, there is a *repeated measure* of the person or machine across all treatments. This repeated measures design is an extension of the *t* test for dependent samples presented in Section 10.3.

Repeated measures design
A randomized block design in which each block level is an individual item or person, and that person or item is measured across all treatments.

The sum of squares in a completely randomized design is

$$SST = SSC + SSE.$$

In a randomized block design, the sum of squares is

$$SST = SSC + SSR + SSE$$

where:
SST = sum of squares total,
SSC = sum of squares columns (treatment),
SSR = sum of squares rows (blocking), and
SSE = sum of squares error.

SST and SSC are the same for a given analysis whether a completely randomized design or a randomized block design is used. Because of this, the SSR (blocking effects) must come out of the SSE. That is, some of the error variation in the completely randomized design is accounted for in the blocking effects of the randomized block design, as shown in Figure 11.7. By reducing the error term, it is possible that the value of F for treatment will increase (the denominator of the F value is decreased). However, if there is not sufficient difference between levels of the blocking variable, the use of a randomized block design can lead to a less powerful result than would a completely randomized design computed on the same problem. Thus, the researcher should seek out blocking variables that he or she believes are significant contributors to variation among measurements of the dependent variable. Figure 11.8 shows the layout of a randomized block design.

In each of the intersections of independent variable and blocking variable in Figure 11.8, one measurement is taken. In the randomized block design, there is one measurement for each treatment level under each blocking level.

The null and alternate hypotheses for the treatment effects in the randomized block design are

$$H_0: \mu_{.1} = \mu_{.2} = \mu_{.3} = \cdots = \mu_{.C}$$

H_a: At least one of the treatment means is different from the others.

For the blocking effects, they are

$$H_0: \mu_{1.} = \mu_{2.} = \mu_{3.} = \cdots = \mu_{R.}$$

H_a: At least one of the blocking means is different from the others.

Figure 11.7

Partitioning the total sum of squares in a randomized block design

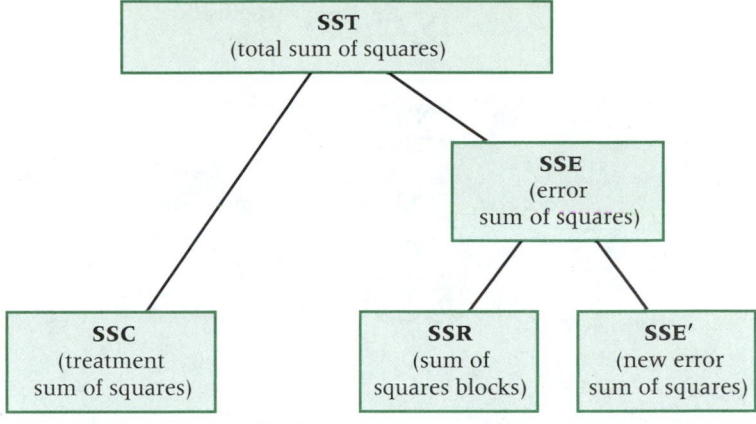

Figure 11.8

A randomized block design

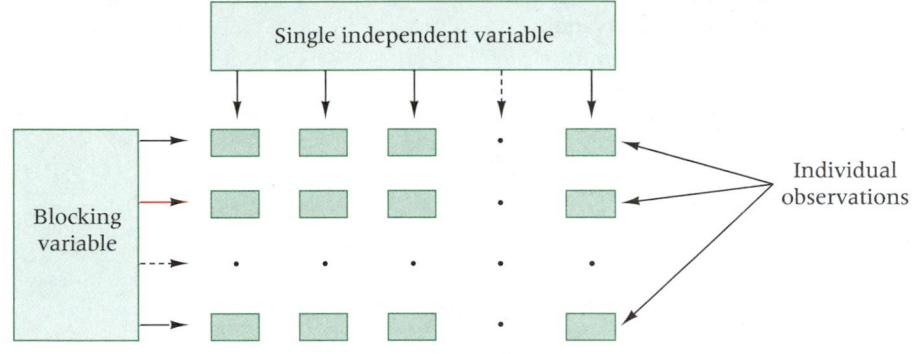

Essentially, we are testing the null hypothesis that the population means of the treatment groups are equal. If the null hypothesis is rejected, at least one of the population means does not equal the others.

The formulas for computing a randomized block design are on the following page.

The F value for treatments shown with the randomized block design design formulas is compared to a table F value, which is ascertained from Appendix A.7 by using α, df_C (treatment), and df_E (error). If the observed F value is greater than the table value, the null hypothesis is rejected for that alpha value. In addition, Excel output provides a p-value for the observed F. As usual, if the p-value is less than α, the decision is to reject the null hypothesis. Such a result would indicate that not all population treatment means are equal. At this point, the business analyst has the option of computing multiple comparisons if the null hypothesis has been rejected.

Some business analysts also compute an F value for blocks even though the main emphasis in the experiment is on the treatments. This value is compared to a critical table F value determined from Appendix A.7 by using α, df_R (blocks), and df_E (error). If the F value for blocks is greater than the critical F value, the null hypothesis that all block population means are equal is rejected. This tells the researcher that including the blocking in the design was probably worthwhile and that a significant amount of variance was drawn off from the error term, thus increasing the power of the treatment test. In this text, we have omitted F_{blocks} from the normal presentation and problem solving. We leave the use of this F value to the discretion of the reader.

$$SSC = n\sum_{j=1}^{C}(\bar{X}_j - \bar{X})^2$$

$$SSR = C\sum_{i=1}^{n}(\bar{X}_i - \bar{X})^2$$

$$SSE = \sum_{i=1}^{n}\sum_{j=1}^{C}(X_{ij} - \bar{X}_j - \bar{X}_i + \bar{X})^2$$

$$SST = \sum_{i=1}^{n}\sum_{j=1}^{C}(X_{ij} - \bar{X})^2$$

where:

i = block group (row)
j = treatment level (column)
C = number of treatment levels (columns)
n = number of observations in each treatment level (number of blocks or rows)
X_{ij} = individual observation
$\bar{X}_j$ = treatment (column) mean
$\bar{X}_i$ = block (row) mean
$\bar{X}$ = grand mean
N = total number of observations

$$df_C = C - 1$$

$$df_R = n - 1$$

$$df_E = (C - 1)(n - 1) = N - n - C + 1$$

$$MSC = \frac{SSC}{C - 1}$$

$$MSR = \frac{SSR}{n - 1}$$

$$MSE = \frac{SSE}{N - n - C + 1}$$

$$F_{treatments} = \frac{MSC}{MSE}$$

$$F_{blocks} = \frac{MSR}{MSE}$$

As an example of the application of the randomized block design, consider a tire company that has developed a new tire. The company has conducted tread-wear tests on the tire to determine whether there is a significant difference in tread wear if the average speed with which the automobile is driven varies. The company set up an experiment in which the independent variable was speed of automobile. There were three treatment levels: slow speed (car is driven 20 mph), medium speed (car is driven 40 mph), and high speed (car is driven 60 mph). The company researchers realized that several possible variables could confound the study. One of these variables was supplier. The company uses five suppliers to provide a major component of the rubber from which the tires are made. To control for this variable experimentally, the researchers used supplier as a blocking variable. Fifteen tires were randomly selected for the study, three from each supplier. Each of the three was assigned to be tested under a different speed condition. The data are given here, along with treatment and block totals. These figures represent tire wear in units of 10,000 miles.

| | SPEED | | | BLOCK MEANS |
SUPPLIER	Slow	Medium	Fast	$\overline{X}_i$
1	3.7	4.5	3.1	3.77
2	3.4	3.9	2.8	3.37
3	3.5	4.1	3.0	3.53
4	3.2	3.5	2.6	3.10
5	3.9	4.8	3.4	4.03
Treatment Means ($\overline{X}_j$)	3.54	4.16	2.98	$\overline{X} = 3.56$

To analyze this randomized block design using $\alpha = .01$, the computations are as follows.

$$C = 3$$
$$n = 5$$
$$N = 15$$

$$SSC = n\sum_{j=1}^{C}(\overline{X}_j - \overline{X})^2$$

$$= 5[(3.54 - 3.56)^2 + (4.16 - 3.56)^2 + (2.98 - 3.56)^2]$$

$$= 3.484$$

$$SSR = C\sum_{i=1}^{n}(\overline{X}_i - \overline{X})^2$$

$$= 3[(3.77 - 3.56)^2 + (3.37 - 3.56)^2 + (3.53 - 3.56)^2 + (3.10 - 3.56)^2$$

$$+ (4.03 - 3.56)^2]$$

$$= 1.549$$

$$SSE = \sum_{i=1}^{n}\sum_{j=1}^{C}(X_{ij} - \overline{X}_j - \overline{X}_i + \overline{X})^2$$

$$= (3.7 - 3.54 - 3.77 + 3.56)^2 + (3.4 - 3.54 - 3.37 + 3.56)^2 + \cdots$$

$$+ (2.6 - 2.98 - 3.10 + 3.56)^2 + (3.4 - 2.98 - 4.03 + 3.56)^2$$

$$= .143$$

$$SST = \sum_{i=1}^{n}\sum_{j=1}^{C}(X_{ij} - \overline{X})^2$$

$$= (3.7 - 3.56)^2 + (3.4 - 3.56)^2 + \cdots + (2.6 - 3.56)^2 + (3.4 - 3.56)^2$$

$$= 5.176$$

$$MSC = \frac{SSC}{C-1} = \frac{3.484}{2} = 1.742$$

$$MSR = \frac{SSR}{n-1} = \frac{1.549}{4} = .387$$

$$MSE = \frac{SSE}{N-n-C+1} = \frac{.143}{8} = .018$$

$$F = \frac{MSC}{MSE} = \frac{1.742}{0.018} = 96.78$$

SOURCE OF VARIANCE	SS	df	MS	F
Treatment	3.484	2	1.742	96.78
Block	1.549	4	.387	
Error	.143	8	.018	
Total	5.176	14		

For alpha of .01, the critical F value is

$$F_{.01,2,8} = 8.65.$$

Because the observed value of F for treatment (96.78) is greater than this critical F value, the null hypothesis is rejected. At least one of the population means of the treatment levels is not the same as the others. That is, there is a significant difference in tread wear for cars driven at different speeds. If this problem had been set up as a completely randomized design, the SSR would have been a part of the SSE. The degrees of freedom for the blocking effects would have been combined with degrees of freedom of error. Thus, the value of SSE would have been $1.549 + .143 = 1.692$, and df_E would have been $4 + 8 = 12$. These would then have been used to recompute MSE = $1.692/12 = .141$. The value of F for treatments would have been

$$F = \frac{\text{MSC}}{\text{MSE}} = \frac{1.742}{0.141} = 12.35.$$

Thus, the F value for treatment *with* the blocking was 96.78 and *without* the blocking was 12.35. By using the random block design, a much larger observed F value was determined.

Analysis Using Excel

Excel can analyze a randomized block design using its **Anova: Two-Factor Without Replication** feature. This feature is located in the **Data Analysis** drop-down menu. **Data Analysis** is selected from the drop-down menu provided by **Tools** on the menu bar. The dialog box for **Anova: Two-Factor Without Replication** is displayed in Figure 11.9.

After loading the data in the worksheet, you have the option of placing labels on the rows and the columns. Give the location of the data under **Input Range** including all labels. If you have labels, then check the **Labels'** box. If you place labels on rows, then you must place labels on columns. Do not place an overall label for rows or columns in addition to the individual row and column labels. As an example, in the tire tread problem, you may place the numbers 1, 2, 3, 4, 5 beside each row to designate the supplier. However, do not place the label "supplier" in another column beside these numbers because Excel will only accept one column of labels and one row of labels. Insert the value of alpha in the box, **Alpha.**

The Excel output for the tire tread wear problem is shown in Figure 11.10. The summary table shows the number in each row and each column, the sum total for each row and column, the row and column means, and the row and column variances. The ANOVA table contains the sum of squares (SS) and the degrees of freedom (df) for the rows (blocks), the columns (treatments), error, and total. In addition, the mean squares (MS) are given for rows (blocks), columns (treatments), and error. An observed F value is given for rows (blocks) and columns (treatments) along with its associated p-value and critical table F value.

Note the observed F value of 97.68 for columns (treatments). The difference between this value and the hand calculated value is due to rounding error. The p-value for this observed F is .00000, which is less than $\alpha = .01$ indicating that the decision is to reject the

Figure 11.9 — Dialog box for **Anova: Two-Factor Without Replication**

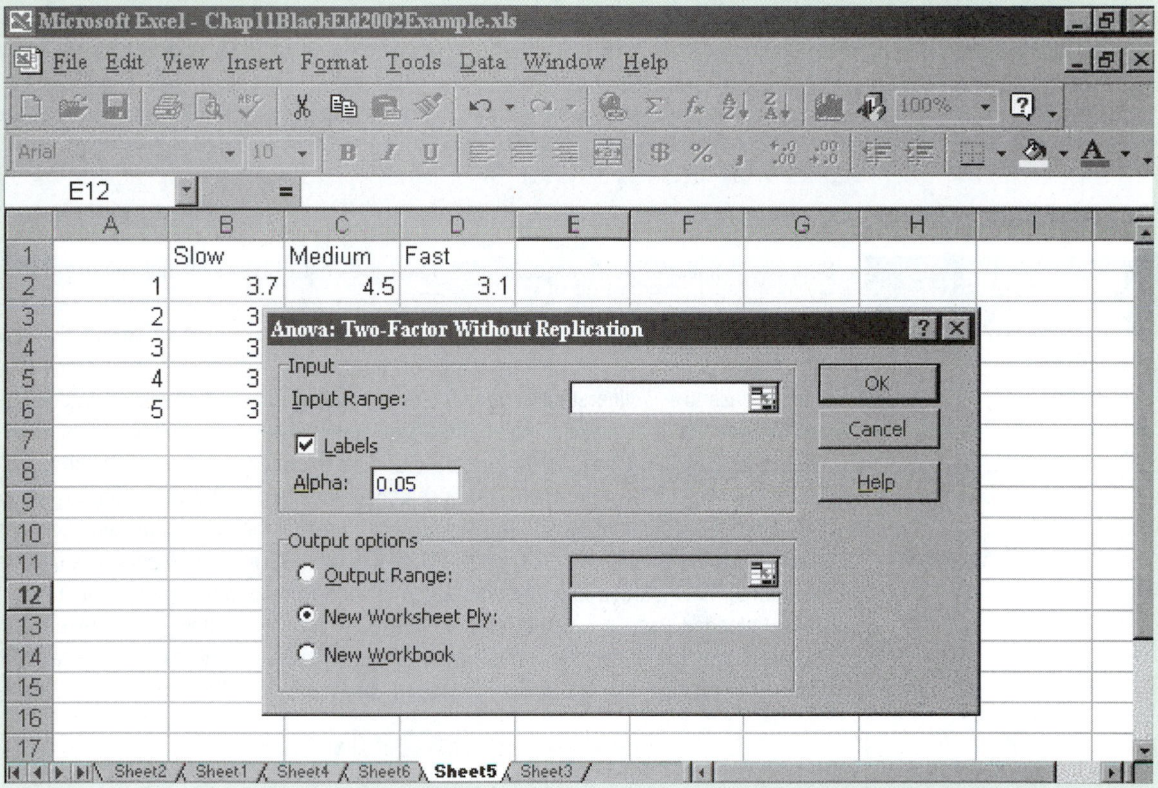

Figure 11.10 — Output for the tire tread wear problem

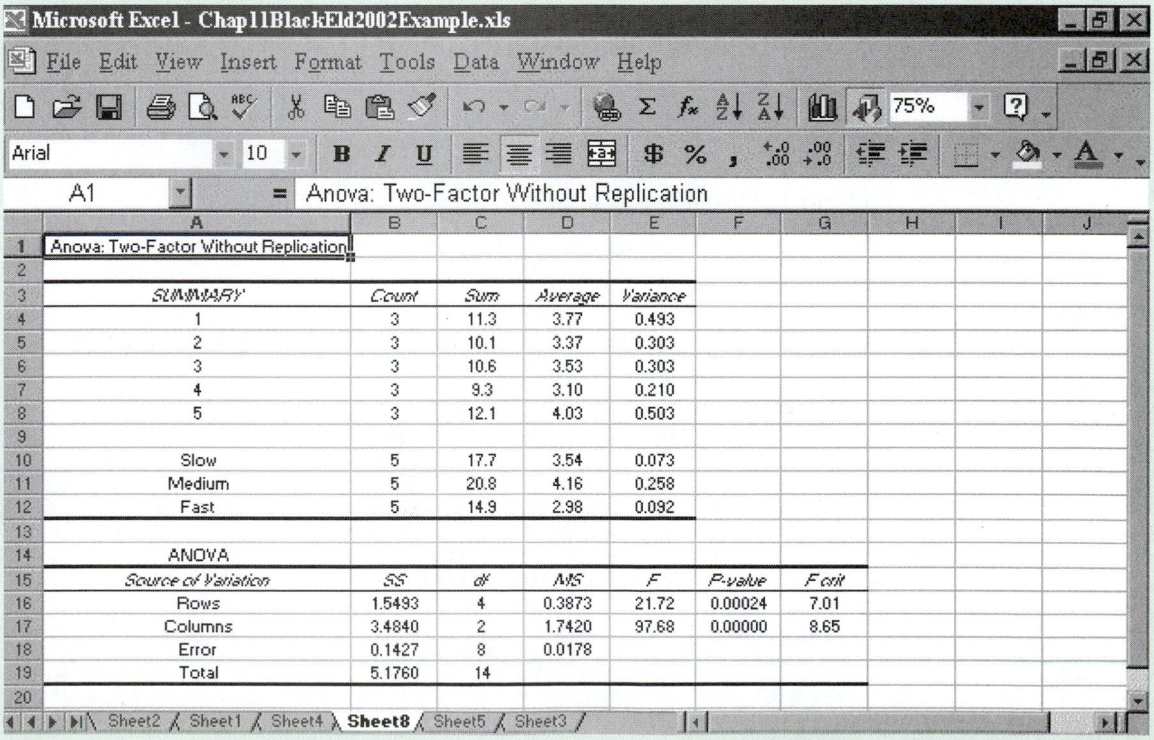

Anova: Two-Factor Without Replication

SUMMARY	Count	Sum	Average	Variance
1	3	11.3	3.77	0.493
2	3	10.1	3.37	0.303
3	3	10.6	3.53	0.303
4	3	9.3	3.10	0.210
5	3	12.1	4.03	0.503
Slow	5	17.7	3.54	0.073
Medium	5	20.8	4.16	0.258
Fast	5	14.9	2.98	0.092

ANOVA

Source of Variation	SS	df	MS	F	P-value	F crit
Rows	1.5493	4	0.3873	21.72	0.00024	7.01
Columns	3.4840	2	1.7420	97.68	0.00000	8.65
Error	0.1427	8	0.0178			
Total	5.1760	14				

null hypothesis. The Excel-produced critical F of 8.65 for columns (treatments) is identical to the table value obtained earlier during hand calculations. Since the observed $F = 97.68$ is greater than this critical $F = 8.65$, the decision is to reject the null hypothesis consistent with the results obtained using the p-value method. While the main emphasis in these problems is not on the F value for blocking effects, the Excel output shows that the observed F for blocking (21.72) is significant at $\alpha = .01$ with a p-value of .00024 and a critical F of 7.01.

DEMONSTRATION PROBLEM 11.2

Suppose a national travel association studied the cost of premium unleaded gasoline in the United States during the summer of 2001. From experience, association directors believed there was a significant difference in the average cost of a gallon of premium gasoline among urban areas in different parts of the country. To test this belief, they placed random calls to gasoline stations in five different cities. In addition, the researchers realized that the brand of gasoline might make a difference. They were mostly interested in the differences between cities, so they made city their treatment variable. To control for the fact that pricing varies with brand, the researchers included brand as a blocking variable and selected six different brands to participate. The researchers randomly telephoned one gasoline station for each brand in each city, resulting in 30 measurements (five cities and six brands). Each station operator was asked to report the current cost of a gallon of premium unleaded gasoline at that station. The data are shown here. Test these data by using a randomized block design analysis to determine whether there is a significant difference in the average cost of premium unleaded gasoline by city. Let $\alpha = .01$.

	GEOGRAPHIC REGION					
BRAND	Miami	Philadelphia	Minneapolis	San Antonio	Oakland	$\overline{X}_i$
A	1.47	1.40	1.38	1.32	1.50	1.414
B	1.43	1.41	1.42	1.35	1.44	1.410
C	1.44	1.41	1.43	1.36	1.45	1.418
D	1.46	1.45	1.40	1.30	1.45	1.412
E	1.46	1.40	1.39	1.39	1.48	1.424
F	1.44	1.43	1.42	1.39	1.49	1.434
$\overline{X}_j$	1.450	1.417	1.407	1.352	1.468	$\overline{X} = 1.419$

SOLUTION

STEP **1.** The hypotheses follow.
For treatments,

$$H_0: \mu_{.1} = \mu_{.2} = \mu_{.3} = \mu_{.4} = \mu_{.5}$$

H_a: At least one of the treatment means is different from the others.

For blocks,

$$H_0: \mu_{1.} = \mu_{2.} = \mu_{3.} = \mu_{4.} = \mu_{5.} = \mu_{6.}$$

H_a: At least one of the blocking means is different from the others.

STEP **2.** The appropriate statistical test is the F test in the ANOVA for randomized block designs.

STEP **3.** Let $\alpha = .01$.

STEP **4.** There are four degrees of freedom for the treatment ($C - 1 = 5 - 1 = 4$), five degrees of freedom for the blocks ($n - 1 = 6 - 1 = 5$), and 20 degrees of freedom for error [$(C - 1)(n - 1) = (4)(5) = 20$]. Using these, $\alpha = .01$, and Table A.7, we find the critical F values.

$$F_{.01,4,20} = 4.43 \text{ for treatments}$$

$$F_{.01,5,20} = 4.10 \text{ for blocks}$$

The decision rule is to reject the null hypothesis for treatments if the observed F value for treatments is greater than 4.43 and to reject the null hypothesis for blocking effects if the observed F value for blocks is greater than 4.10.

STEP **5.** The sample data including row and column means and the grand mean are given in the preceding table.

STEP **6.**

$$SSC = n \sum_{j=1}^{C} (\overline{X}_j - \overline{X})^2$$

$$= 6[(1.450 - 1.419)^2 + (1.417 - 1.419)^2 + (1.407 - 1.419)^2$$

$$+ (1.352 - 1.419)^2 + (1.468 - 1.419)^2]$$

$$= .04851$$

$$SSR = C \sum_{i=1}^{n} (\overline{X}_i - \overline{X})^2$$

$$= 5[(1.414 - 1.419)^2 + (1.410 - 1.419)^2 + (1.418 - 1.419)^2$$

$$+ (1.412 - 1.419)^2 + (1.424 - 1.419)^2 + (1.434 - 1.419)^2]$$

$$= .00203$$

$$SSE = \sum_{i=1}^{n} \sum_{j=1}^{C} (X_{ij} - \overline{X}_j - \overline{X}_i + \overline{X})^2$$

$$= (1.47 - 1.450 - 1.414 + 1.419)^2 + (1.43 - 1.450 - 1.410 + 1.419)^2$$

$$+ \cdots + (1.48 - 1.468 - 1.424 + 1.419)^2 + (1.49 - 1.468 - 1.434 + 1.419)^2$$

$$= .01281$$

$$SST = \sum_{i=1}^{n} \sum_{j=1}^{C} (X_{ij} - \overline{X})^2$$

$$= (1.47 - 1.419)^2 + (1.43 - 1.419)^2 + \cdots + (1.48 - 1.419)^2 + (1.49 - 1.419)^2$$

$$= .06335$$

$$MSC = \frac{SSC}{C - 1} = \frac{.04851}{4} = .01213$$

$$MSR = \frac{SSR}{n - 1} = \frac{0.00203}{5} = .00041$$

$$MSE = \frac{SSE}{(C - 1)(n - 1)} = \frac{.01281}{20} = .00064$$

$$F = \frac{MSC}{MSE} = \frac{.01213}{.00064} = 18.95$$

SOURCE OF VARIANCE	SS	df	MS	F
Treatment	.04851	4	.01213	18.95
Block	.00203	5	.00041	
Error	.01281	20	.00064	
Total	.06335	29		

STEP **7.** Because $F_{treat} = 18.95 > F_{.01,4,20} = 4.43$, the null hypothesis is rejected for the treatment effects. There is a significant difference in the average price of a gallon of premium unleaded gasoline in various cities.

A glance at the MSR reveals that there appears to be relatively little blocking variance. The result of determining an F value for the blocking effects is

$$F = \frac{MSR}{MSE} = \frac{.00041}{.00064} = 0.64.$$

The value of F for blocks is *not* significant at $\alpha = .01$ ($F_{.01,5,20} = 4.10$). This indicates that the blocking portion of the experimental design did not contribute significantly to the analysis. If the blocking effects (SSR) are added back into SSE and the df_R are included with df_E, the MSE becomes .00059 instead of .00064. Using the value .00059 in the denominator for the treatment F increases the observed treatment F value to 20.56. Thus, including nonsignificant blocking effects in the original analysis caused a loss of power.

Shown below is the Excel output for this problem.

Microsoft Excel - Chap11BlackEld2002Example.xls

File   Edit   View   Insert   Format   Tools   Data   Window   Help

Arial   10   **B** *I* <u>U</u> $ %   75%

J24

	A	B	C	D	E	F	G	H	I	J
1	Anova: Two-Factor Without Replication									
2										
3	*SUMMARY*	*Count*	*Sum*	*Average*	*Variance*					
4	A	5	7.07	1.414	0.0052					
5	B	5	7.05	1.410	0.0013					
6	C	5	7.09	1.418	0.0013					
7	D	5	7.06	1.412	0.0045					
8	E	5	7.12	1.424	0.0018					
9	F	5	7.17	1.434	0.0013					
10										
11	Miami	6	8.70	1.450	0.0002					
12	Philadelphia	6	8.50	1.417	0.0004					
13	Minneapolis	6	8.44	1.407	0.0004					
14	San Antonio	6	8.11	1.352	0.0013					
15	Oakland	6	8.81	1.468	0.0006					
16										
17	ANOVA									
18	*Source of Variation*	*SS*	*df*	*MS*	*F*	*P-value*	*F crit*			
19	Rows	0.00203	5	0.00041	0.63	0.6769	4.10			
20	Columns	0.04851	4	0.01213	18.94	0.0000	4.43			
21	Error	0.01281	20	0.00064						
22	Total	0.06335	29							
23										
24										

Sheet2 / Sheet1 / Sheet4 \ **Sheet10** / Sheet9 / Sheet8 / Sheet5

The ANOVA table in the Excel output summarizes the results of this problem. The F value for treatments shown as Columns yields a p value of .0000, which indicates that the treatment F is significant at an alpha of .0001. The F value for blocks shown as Rows is less than 1. The p value is .6769, which means that this F would not be significant even at $\alpha = .10$. The critical table F values are given for $\alpha = .01$.

STEP **8.** The fact that there is a significant difference in the price of gasoline in different parts of the country can be useful information to decision makers. For example, companies in the ground transportation business are impacted greatly by increases in the cost of fuel. Knowledge of price differences in fuel can help these companies plan strategies and routes. Fuel price differences can sometimes be indications of cost-of-living differences and/or distribution problems, which can impact a company's relocation decision or cost-of-living increases given to employees who transfer to the higher-priced locations. Knowing that the price of gasoline varies around the country can generate interest among market researchers who might want to study why the differences are there and what drives them. This can sometimes result in a better understanding of the marketplace.

11.8 Use ANOVA to analyze the data from the randomized block design given here. Let $\alpha = .05$. State the null and alternative hypotheses and determine whether the null hypothesis is rejected.

11.3
Problems

TREATMENT LEVEL

		1	2	3	4
	1	23	26	24	24
	2	31	35	32	33
BLOCK	3	27	29	26	27
	4	21	28	27	22
	5	18	25	27	20

11.9 The following data have been gathered from a randomized block design. Use $\alpha = .01$ to test for a significant difference in the treatment levels. Establish the hypotheses and reach a conclusion about the null hypothesis.

TREATMENT LEVEL

		1	2	3
	1	1.28	1.29	1.29
	2	1.40	1.36	1.35
BLOCK	3	1.15	1.13	1.19
	4	1.22	1.18	1.24

11.10 Safety in motels and hotels is a growing concern among travelers. Suppose a survey was conducted by the National Motel and Hotel Association to determine the U.S. travelers' perception of safety in various motel chains. The association chose four different national chains from the economy lodging sector and randomly selected 10 people who had stayed overnight in a motel in each of the four chains in the past 2 years. Each selected traveler was asked to rate each motel chain on a scale from 0 to 100 to indicate how safe he or she felt at that motel. A score of 0 indicates completely unsafe and a score of 100 indicates perfectly safe. The scores follow. Test this randomized block design to determine whether there is a significant difference in the safety ratings of the four motels. Use $\alpha = .05$.

TRAVELER	MOTEL 1	MOTEL 2	MOTEL 3	MOTEL 4
1	40	30	55	45
2	65	50	80	70
3	60	55	60	60
4	20	40	55	50
5	50	35	65	60
6	30	30	50	50
7	55	30	60	55
8	70	70	70	70
9	65	60	80	75
10	45	25	45	50

11.11 In recent years, there has been constant debate over the U.S. economy. The electorate seems somewhat divided as to whether the economy is in a recovery or not. Suppose a survey was undertaken to ascertain whether the perception of economic recovery differs according to political affiliation. People were selected for the survey from the Democratic party, the Republican party, and those classifying themselves as independents. A 25-point scale was developed in which respondents gave a score of 25 if they felt the economy was definitely in complete recovery, a 0 if the economy was definitely not in a recovery, and some value in between for more uncertain responses. To control for differences in socioeconomic class, a blocking variable was maintained using five different socioeconomic categories. The data are given here in the form of a randomized block design. Use $\alpha = .01$ to determine whether there is a significant difference in mean responses according to political affiliation.

SOCIOECONOMIC CLASS	POLITICAL AFFILIATION		
	Democrat	*Republican*	*Independent*
Upper	11	5	8
Upper middle	15	9	8
Middle	19	14	15
Lower middle	16	12	10
Lower	9	8	7

11.12 As part of a manufacturing process, a plastic container is supposed to be filled with 46 ounces of saltwater solution. The plant has three machines that fill the containers. Managers are concerned that the machines might not be filling the containers with the same amount of saltwater solution, so they set up a randomized block design to test this concern. There is a pool of five machine operators who at different times operate each of the three machines. Company technicians randomly select five containers filled by each machine (one container for each of the five operators). The measurements are shown below. Use $\alpha = .05$ to analyze the design.

		MACHINE		
		1	2	3
	1	46.05	45.99	46.02
	2	45.97	46.08	45.98
OPERATOR	3	45.91	46.05	45.95
	4	46.01	46.03	46.12
	5	45.96	46.04	45.99

There are times when an experiment is designed so that *two or more treatments* (independent variables) *are explored simultaneously.* Such experimental designs are referred to as **factorial designs.** In factorial designs, *every level of each treatment is studied under the conditions of every level of all other treatments.* Factorial designs can be arranged such that three, four, or *n* treatments or independent variables are studied simultaneously in the same experiment. As an example, consider the valve opening data in Table 11.1. The mean valve opening for the 24 measurements is 6.34 cm. However, every valve but one in the sample measures something other than the mean. Why? Company workers realize that valves at this firm are made on different machines, by different operators, on different shifts, on different days, with raw materials from different suppliers. Business analysts who are interested in finding the sources of variation might decide to set up a factorial design that incorporates all five of these independent variables in one study. In this text, we explore the factorial designs with two treatments only.

Advantages of the Factorial Design

If two variables are analyzed by using a completely randomized design, the effects of each variable are explored separately (one per design). Thus, it takes two completely randomized designs to analyze the effects of the two variables. By using a factorial design, the researcher can analyze both variables at the same time in one design, saving the time and effort of doing two different analyses and minimizing the experimentwise error rate.

Some researchers use the factorial design as a way to control confounding or concomitant variables in a study. By building variables into the design, the researcher is attempting to control for the effects of multiple variables *in* the experiment. With the completely randomized design, the variables are studied in isolation. With the factorial design, there is potential for increased power over the completely randomized design because the additional effects of the second variable are removed from the error sum of squares.

The business analyst can explore the possibility of interaction between the two treatment variables in a two-factor factorial design if multiple measurements are taken under every combination of levels of the two treatments. Interaction will be discussed later.

Factorial designs with two treatments are similar to randomized block designs. However, whereas randomized block designs focus on one treatment variable and *control* for a blocking effect, a two-treatment factorial design focuses on the effects of both variables. Because the randomized block design contains only one measure for each (treatment–block) combination, interaction cannot be analyzed in randomized block designs.

Factorial Designs with Two Treatments

The structure of a two-treatment factorial design is featured in Figure 11.11. Note that there are two independent variables (two treatments) and that there is an intersection of each level of each treatment. These intersections are referred to as *cells.* One treatment is arbitrarily designated as *row* treatment (forming the rows of the design) and the other treatment is designated as *column* treatment (forming the columns of the design). Although it is possible to analyze factorial designs with unequal numbers of items in the cells, the analysis of unequal cell designs is beyond the scope of this text. All factorial designs discussed here have cells of equal size.

Treatments (independent variables) of factorial designs must have at least two levels each. The simplest factorial design is a 2 × 2 factorial design, where each treatment has two levels. If such a factorial design were diagrammed in the manner of Figure 11.11, there would be two rows and two columns, forming four cells.

Factorial design
An experimental design in which two or more independent variables are studied simultaneously and every level of each treatment is studied under the conditions of every level of all other treatments. Also called a factorial experiment.

Figure 11.11

Two-way factorial
design

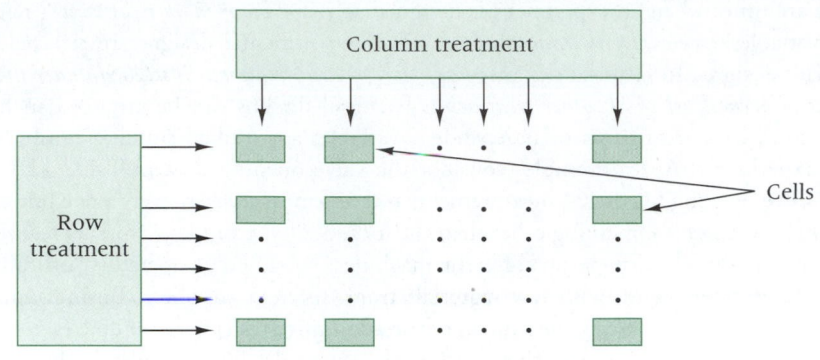

In this section, we study only factorial designs with $n > 1$ measurements for each combination of treatment levels (cells). This allows us to attempt to measure the interaction of the treatment variables. As with the completely randomized design and the randomized block design, there is only *one* dependent variable in a factorial design.

Applications

There are many possible applications of the factorial design in business research. For example, the natural gas industry can design an experiment to study usage rates and how they are affected by temperature and precipitation. Theorizing that the outside temperature and type of precipitation make a difference in natural gas usage, industry researchers can gather usage measurements for a given community over a variety of temperature and precipitation conditions. At the same time, they can make an effort to determine whether certain types of precipitation combined with certain temperature levels affect usage rates differently than other combinations of temperature and precipitation (interaction effects).

Stock market analysts can select a company from an industry such as the construction industry and observe the behavior of its stock under different conditions. A factorial design can be set up by using volume of the market and prime interest as two independent variables. For volume of the market, researchers can select some days when the volume is up from the day before, some days when the volume is down from the day before, and some other days when the volume is essentially the same as on the preceding day. These groups of days would constitute three levels of the independent variable, volume. The business analysts can do the same thing with prime rate. Levels can be selected such that the prime rate is (1) up, (2) down, and (3) essentially the same. For the dependent variable, the researchers would measure how much the company's stock rises or falls on those randomly selected days. Using the factorial design, the researcher can determine whether stock rates are different under various levels of volume, whether stock rates are different under various levels of the prime interest rate, and whether stock rates react differently under various combinations of volume and prime rate (interaction effects).

Two-way analysis of variance (two-way ANOVA)
The process used to statistically test the effects of variables in factorial designs with two independent variables.

Statistically Testing the Factorial Design

Analysis of variance is used to analyze data gathered from factorial designs. For factorial designs with two factors (independent variables), a **two-way analysis of variance (two-way ANOVA)** is used to test hypotheses statistically. The following hypotheses are tested by a two-way ANOVA.

Row effects: H_0: Row means all are equal.

H_a: At least one row mean is different from the others.

Column effects: H_0: Column means are all equal.

H_a: At least one column mean is different from the others.

Interaction effects: H_0: The interaction effects are zero.

H_a: There is an interaction effect.

Formulas for computing a two-way ANOVA are given in the following box. These formulas are computed in a manner similar to computations for the completely randomized design and the randomized block design. F values are determined for three effects:

1. row effects,
2. column effects, and
3. interaction effects.

The row effects and the column effects are sometimes referred to as the *main* effects. Although F values are determined for these main effects, an F value is also computed for interaction effects. Using these F values, the researcher can make a decision about the null hypotheses for each effect.

Each of these observed F values is compared to a table F value. The table F value is determined by α, df_{num}, and df_{denom}. The degrees of freedom for the numerator (df_{num}) are determined by the effect being studied. If the observed F value is for columns, the degrees of freedom for the numerator are $C - 1$. If the observed F value is for rows, the degrees of freedom for the numerator are $R - 1$. If the observed F value is for interaction, the degrees of freedom for the numerator are $(R - 1)(C - 1)$. The number of degrees of freedom for the denominator of the table value for each of the three effects is the same, the error degrees of freedom, $RC(n - 1)$. The table F values (critical F) for a two-way ANOVA follow.

Row effects: $F_{\alpha, R-1, RC(n-1)}$ Column effects: $F_{\alpha, C-1, RC(n-1)}$ Interaction effects: $F_{\alpha, (R-1)(C-1), RC(n-1)}$	TABLE F VALUES FOR A TWO-WAY ANOVA

$$SSR = nC\sum_{i=1}^{R}(\bar{X}_i - \bar{X})^2$$ $$SSC = nR\sum_{j=1}^{C}(\bar{X}_j - \bar{X})^2$$ $$SSI = n\sum_{i=1}^{R}\sum_{j=1}^{C}(\bar{X}_{ij} - \bar{X}_i - \bar{X}_j + \bar{X})^2$$ $$SSE = \sum_{i=1}^{R}\sum_{j=1}^{C}\sum_{k=1}^{n}(X_{ijk} - \bar{X}_{ij})^2$$ $$SST = \sum_{i=1}^{R}\sum_{j=1}^{C}\sum_{k=1}^{n}(X_{ijk} - \bar{X})^2$$	FORMULAS FOR COMPUTING A TWO-WAY ANOVA

continued

FORMULAS FOR
COMPUTING A
TWO-WAY ANOVA
(continued)

where:

n = number of observations per cell
C = number of column treatments
R = number of row treatments
i = row treatment level
j = column treatment level
k = cell member
X_{ijk} = individual observation
$\overline{X}_{ij}$ = cell mean
$\overline{X}_i$ = row mean
$\overline{X}_j$ = column mean
$\overline{\overline{X}}$ = grand mean

$$\text{df}_R = R - 1 \qquad \text{MSR} = \frac{\text{SSR}}{R-1} \qquad F_R = \frac{\text{MSR}}{\text{MSE}}$$

$$\text{df}_C = C - 1 \qquad \text{MSC} = \frac{\text{SSC}}{C-1} \qquad F_C = \frac{\text{MSC}}{\text{MSE}}$$

$$\text{df}_I = (R-1)(C-1)$$

$$\text{df}_E = RC(n-1) \qquad \text{MSI} = \frac{\text{SSI}}{(R-1)(C-1)} \qquad F_I = \frac{\text{MSI}}{\text{MSE}}$$

$$\text{df}_T = N - 1$$

$$\text{MSE} = \frac{\text{SSE}}{RC(n-1)}$$

Interaction

As noted before, along with testing the effects of the two treatments in a factorial design, it is possible to test for the interaction effects of the two treatments whenever multiple measures are taken in each cell of the design. **Interaction** occurs *when the effects of one treatment vary according to the levels of treatment of the other effect.* For example, in a study examining the impact of temperature and humidity on a manufacturing process, it is possible that temperature and humidity will interact in such a way that the effect of temperature on the process varies with the humidity. Low temperatures might not be a significant manufacturing factor when humidity is low but might be a factor when humidity is high. Similarly, high temperatures might be a factor with low humidity but not with high humidity.

As another example, suppose a business analyst is studying the amount of red meat consumed by families per month and is examining economic class and religion as two independent variables. Class and religion might interact in such a way that with certain religions, economic class does not matter in the consumption of red meat, but with other religions, class does make a difference.

In terms of the factorial design, interaction occurs when the pattern of cell means in one row (going across columns) varies from the pattern of cell means in other rows. This indicates that the differences in column effects depend on which row is being examined. Hence, there is an interaction of the rows and columns. The same thing can happen when the pattern of cell means within a column is different from the pattern of cell means in other columns.

Interaction can be depicted graphically by plotting the cell means within each row (and can also be done by plotting the cell means within each column). The means within each row (or column) are then connected by a line. If the broken lines for each row (or column) are parallel, there is no interaction.

Figure 11.12 is a graph of the means for each cell in each row in a 2 × 3 (2 rows, 3 columns) factorial design with interaction. Note that the lines connecting the means in

Interaction
When the effects of one treatment in an experimental design vary according to the levels of treatment of the other effect(s).

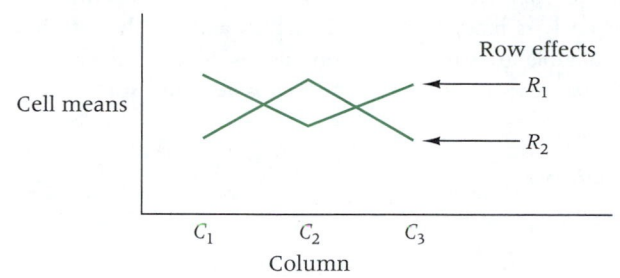

Figure 11.12

A 2 × 3 factorial design with interaction

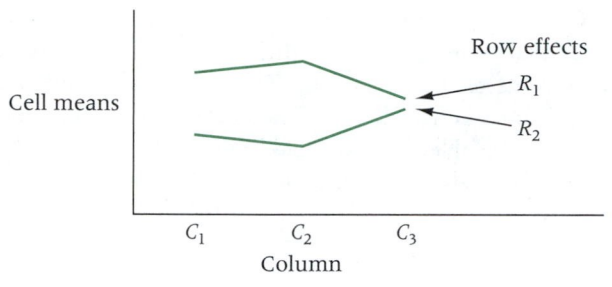

Figure 11.13

A 2 × 3 factorial design with some interaction

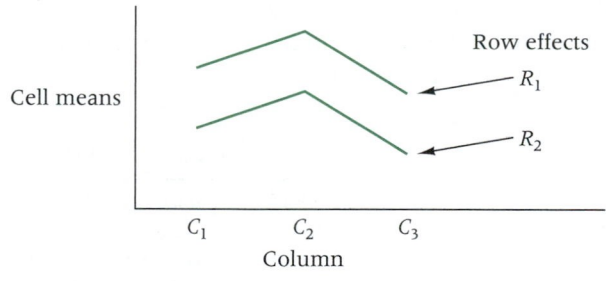

Figure 11.14

A 2 × 3 factorial design with no interaction

each row cross each other. In Figure 11.13 the lines converge, indicating the likely presence of some interaction. Figure 11.14 depicts a 2 × 3 factorial design with no interaction.

When the interaction effects are significant, the main effects (row and column) are confounded and should not be analyzed in the usual manner. In this case, it is not possible to state unequivocally that the row effects or the column effects are significantly different because the difference in means of one main effect varies according to the level of the other main effect (interaction is present). There are some recommended ways of examining main effects when significant interaction is present. However, these techniques are beyond the scope of material presented here. Hence, in this text, whenever interaction effects are present (F_{inter} is significant), the business analyst should *not* attempt to interpret the main effects (F_{row} and F_{col}).

As an example of a factorial design, consider the fact that at the end of a financially successful fiscal year, CEOs often must decide whether to award a dividend to stockholders or to make a company investment. One factor in this decision would seem to be whether attractive investment opportunities are available.* To determine whether this factor is important, business analysts randomly select 24 CEOs and ask them to rate how important "availability of profitable investment opportunities" is in deciding whether to pay dividends or invest. The CEOs are requested to respond to this item on a scale from 0 to 4, where

  0 = no importance,

  1 = slight importance,

  2 = moderate importance,

  3 = great importance, and

  4 = maximum importance.

The 0–4 response is the dependent variable in the experimental design.

The business analysts are concerned that where the company's stock is traded (New York Stock Exchange, American Stock Exchange, and over the counter) might make a difference in the CEOs' response to the question. In addition, the business analysts believe that how stockholders are informed of dividends (annual reports vs. presentations) might affect the outcome of the experiment. Thus, a two-way ANOVA is set up with "where the company's stock is traded" and "how stockholders are informed of dividends" as the two independent variables. The variable "how stockholders are informed of dividends" has two treatment levels, or classifications.

1. Annual/quarterly reports
2. Presentations to analysts

The variable "where company stock is traded" has three treatment levels, or classifications.

1. New York Stock Exchange
2. American Stock Exchange
3. Over the counter

This factorial design is a 2 × 3 design (2 rows, 3 columns) with four measurements (ratings) per cell, as shown in the following table.

		Where Company Stock Is Traded			
		New York Stock Exchange	*American Stock Exchange*	*Over the Counter*	$\overline{X}_i$
	Annual/Quarterly Reports	2 1 2 1 $\overline{X}_{11} = 1.5$	2 3 3 2 $\overline{X}_{12} = 2.5$	4 3 4 3 $\overline{X}_{13} = 3.5$	2.5
How Stockholders Are Informed of Dividends	*Presentations to Analysts*	2 3 1 2 $\overline{X}_{21} = 2.0$	3 3 2 4 $\overline{X}_{22} = 3.0$	4 4 3 4 $\overline{X}_{23} = 3.75$	2.9167
	$\overline{X}_j =$	1.75	2.75	3.625	
			$\overline{X} = 2.7083$		

*ADAPTED FROM: H. Kent Baker, "Why Companies Pay No Dividends," *Akron Business and Economic Review,* 20. Summer 1989, 48–61.

These data are analyzed by using a two-way analysis of variance and $\alpha = .05$.

$$SSR = nC \sum_{i=1}^{R} (\bar{X}_i - \bar{X})^2$$

$$= 4(3)[(2.5 - 2.7083)^2 + (2.9167 - 2.7083)^2]$$

$$= 1.0418$$

$$SSC = nR \sum_{j=1}^{C} (\bar{X}_j - \bar{X})^2$$

$$= 4(2)[(1.75 - 2.7083)^2 + (2.75 - 2.7083)^2 + (3.625 - 2.7083)^2]$$

$$= 14.0833$$

$$SSI = n \sum_{i=1}^{R} \sum_{j=1}^{C} (\bar{X}_{ij} - \bar{X}_i - \bar{X}_j + \bar{X})^2$$

$$= 4[(1.5 - 2.5 - 1.75 + 2.7083)^2 + (2.5 - 2.5 - 2.75 + 2.7083)^2$$

$$+ (3.5 - 2.5 - 3.625 + 2.7083)^2 + (2.0 - 2.9167 - 1.75 + 2.7083)^2$$

$$+ (3.0 - 2.9167 - 2.75 + 2.7083)^2 + (3.75 - 2.9167 - 3.625 + 2.7083)^2]$$

$$= .0833$$

$$SSE = \sum_{i=1}^{R} \sum_{j=1}^{C} \sum_{k=1}^{n} (X_{ijk} - \bar{X}_{ij})^2$$

$$= (2 - 1.5)^2 + (1 - 1.5)^2 + \cdots + (3 - 3.75)^2 + (4 - 3.75)^2$$

$$= 7.7500$$

$$SST = \sum_{i=1}^{R} \sum_{j=1}^{C} \sum_{k=1}^{n} (X_{ijk} - \bar{X})^2$$

$$= (2 - 2.7083)^2 + (1 - 2.7083)^2 + \cdots + (3 - 2.7083)^2 + (4 - 2.7083)^2$$

$$= 22.9583$$

$$MSR = \frac{SSR}{R - 1} = \frac{1.0418}{1} = 1.0418$$

$$MSC = \frac{SSC}{C - 1} = \frac{14.0833}{2} = 7.0417$$

$$MSI = \frac{SSI}{(R - 1)(C - 1)} = \frac{.0833}{2} = .0417$$

$$MSE = \frac{SSE}{RC(n - 1)} = \frac{7.7500}{18} = .4306$$

$$F_R = \frac{MSR}{MSE} = \frac{1.0418}{.4306} = 2.42$$

$$F_C = \frac{MSC}{MSE} = \frac{7.0417}{.4306} = 16.35$$

$$F_I = \frac{MSI}{MSE} = \frac{.0417}{.4306} = 0.10$$

SOURCE OF VARIANCE	SS	df	MS	F
Row	1.0418	1	1.0418	2.42
Column	14.0833	2	7.0417	16.35*
Interaction	.0833	2	.0417	0.10
Error	7.7500	18	.4306	
Total	22.9583	23		

*Denotes significance at $\alpha = .01$.

The critical F value for the interaction effects at $\alpha = .05$ is

$$F_{.05,2,18} = 3.55.$$

The observed F value for interaction effects is .10. As this value is less than the critical table value (3.55), there are no significant interaction effects. Because there are no significant interaction effects, it is possible to examine the main effects. The critical F value of the row effects at $\alpha = .05$ is

$$F_{.05,1,18} = 4.41.$$

The calculated F value of 2.42 is less than the table value. Hence, there are no significant row effects. The critical F value of the column effects at $\alpha = .05$ is

$$F_{.05,2,18} = 3.55.$$

This value is coincidently the same as the critical table value for interaction because in this problem the degrees of freedom are the same for interaction and column effects. The observed F value for columns (16.35) is greater than this critical value. Hence, there is a significant difference in row effects at $\alpha = .05$.

There is a significant difference in the CEOs' mean ratings of the item "availability of profitable investment opportunities" according to where the company's stock is traded. A cursory examination of the means for the three levels of the column effects (where stock is traded) reveals that the lowest mean was from CEOs whose company traded stock on the New York Stock Exchange. The highest mean rating was from CEOs whose company traded stock over the counter. Using multiple comparison techniques, the business analysts could statistically test for differences in the means of these three groups.

Analysis Using Excel

A two-way ANOVA with more than one observation per cell can be computed using Excel. Begin by selecting **Tools** from the menu bar. From the **Tools'** pull-down menu, select **Data Analysis.** From the **Data Analysis'** list box, select **Anova: Two-Factor With Replication.** This feature will compute a two-way ANOVA with interaction. For two-way ANOVA designs with only one observation per cell, select **Anova: Two-Factor Without Replication.** The dialog box for **Anova: Two-Factor With Replication** is shown is Figure 11.15.

Load the data into a worksheet. The worksheet entries for the CEO/Stock problem are displayed in Figure 11.16. Excel assumes there will be labels and will not execute without a label for each "treatment level" row and column. Place the number of observations per sample in the **Rows** per sample box. Because Excel is told how many row lines there are per sample (number of observations per sample), there need not be a label for each row line (each row of the worksheet). As an example, note that in Figure 11.16 each column

Figure 11.15 Dialog box for **Anova: Two-Factor With Replication**

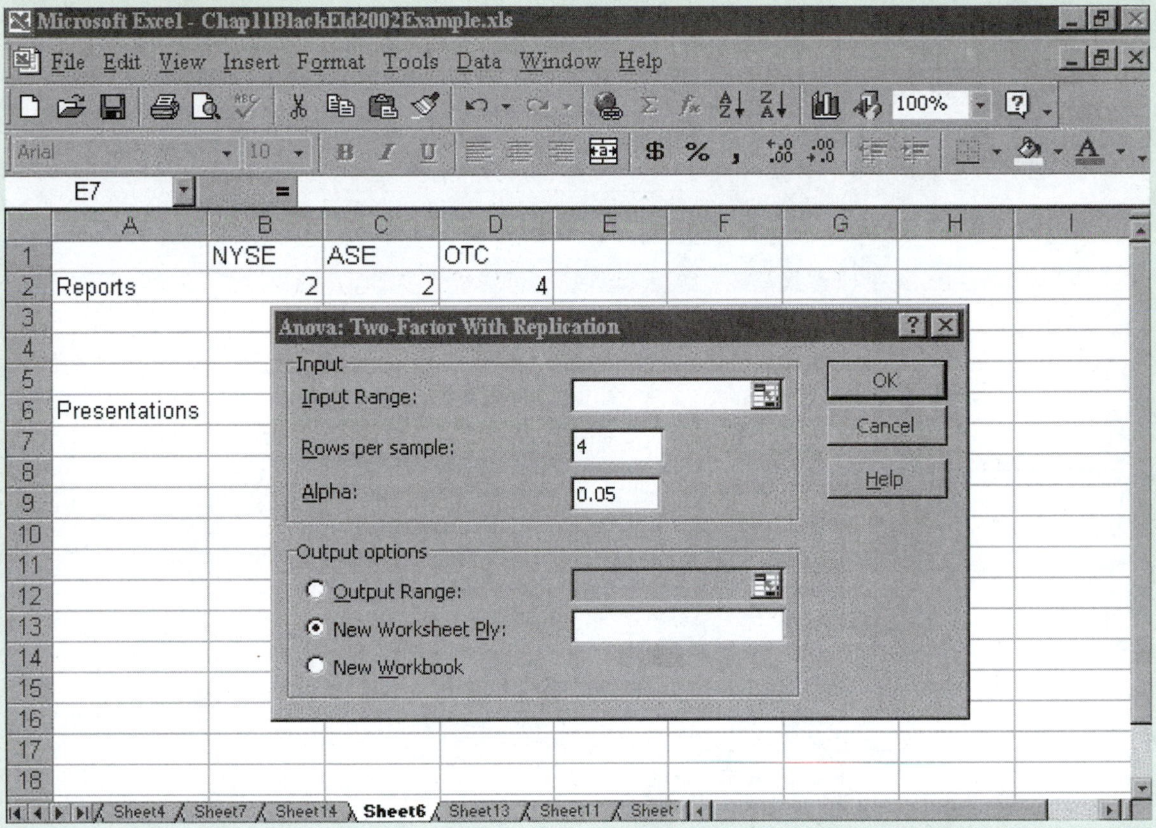

Figure 11.16 Spreadsheet entries for CEO/stock problem

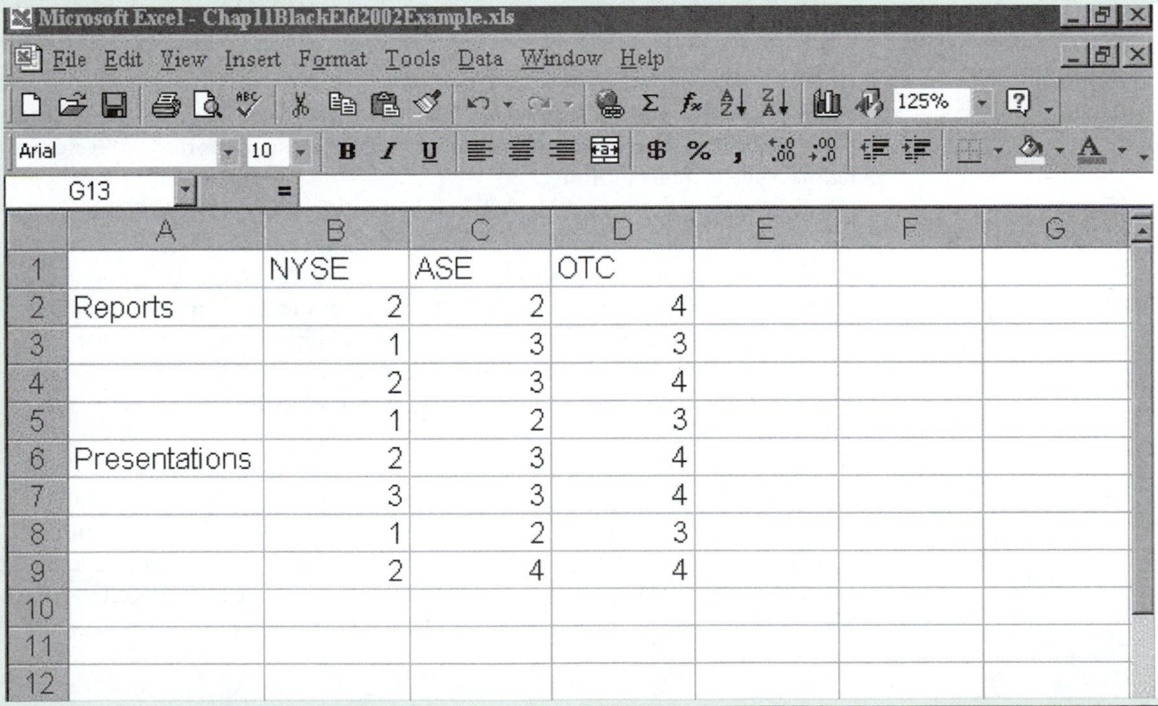

	A	B	C	D
1		NYSE	ASE	OTC
2	Reports	2	2	4
3		1	3	3
4		2	3	4
5		1	2	3
6	Presentations	2	3	4
7		3	3	4
8		1	2	3
9		2	4	4

Figure 11.17   Excel output for CEO/stock problem

Microsoft Excel - Chap11BlackEld2002Example.xls

File Edit View Insert Format Tools Data Window Help

Arial   10   **B** *I* U

M28   =

	A	B	C	D	E	F	G	H	I	J	K	L	M
1	Anova: Two-Factor With Replication												
2													
3	SUMMARY	NYSE	ASE	OTC	Total								
4	*Reports*												
5	Count	4	4	4	12								
6	Sum	6	10	14	30								
7	Average	1.5	2.5	3.5	2.5								
8	Variance	0.3333	0.3333	0.3333	1								
9	*Presentations*												
10	Count	4	4	4	12								
11	Sum	8	12	15	35								
12	Average	2	3	3.75	2.917								
13	Variance	0.6667	0.6667	0.2500	0.9924								
14	*Total*												
15	Count	8	8	8									
16	Sum	14	22	29									
17	Average	1.75	2.75	3.625									
18	Variance	0.5	0.5	0.2679									
19													
20	ANOVA												
21	*Source of Variation*	*SS*	*df*	*MS*	*F*	*P-value*	*F crit*						
22	Sample	1.042	1	1.042	2.42	0.1373	4.41						
23	Columns	14.083	2	7.042	16.35	0.0001	3.55						
24	Interaction	0.083	2	0.042	0.10	0.9082	3.55						
25	Within	7.750	18	0.431									
26	Total	22.958	23										
27													
28													
29													

Sheet4 **Sheet7** Sheet6 Sheet13 Sheet11 Sheet10 Sheet8

has a label and each row treatment <u>level</u> has a label but not each row <u>line</u> because there are four row lines (observations) per sample (row treatment level). Each sample constitutes a row treatment <u>level</u>; and there are two row treatment levels: **Reports** and **Presentations.** Enter the value of alpha in **Alpha.**

The Excel output for the CEO/stock problem is shown in Figure 11.17. Note that the output contains a Summary table with counts, sums, averages, and variances of each sample and of each treatment level. In addition, there is an ANOVA table containing the sum of squares (SS) and degrees of freedom (df) for rows (denoted in the table as Sample), columns, interaction, error (denoted in the table as Within), and total. There are mean squares (MS) for rows, columns, interaction, and error. There are observed F values with associated p-values and critical (table) F values for rows (Sample), columns, and interaction. Note that the F values obtained here are the same as those values calculated by hand in the chapter. In addition, the provided p-values show that the observed F value for columns is significant at $\alpha = .05$ because the p-value is .0001 while the p-values for rows (Sample) and interaction are not (.1373 and .9062 respectively). For those business analysts who are more comfortable using the F tables, critical F values are provided. These values are the same as those displayed earlier in the chapter when the problem was worked by hand.

Some theorists believe that training warehouse workers can reduce absenteeism.* Suppose an experimental design is structured to test this belief. Warehouses in which training sessions have been held for workers are selected for the study. The four types of warehouses are (1) general merchandise, (2) commodity, (3) bulk storage, and (4) cold storage. The training sessions are differentiated by length. Researchers identify three levels of training sessions according to the length of sessions: (1) 1–20 days, (2) 21–50 days, and (3) more than 50 days. Three warehouse workers are selected randomly for each particular combination of type of warehouse and session length. The workers are monitored for the next year to determine how many days they are absent. The resulting data are in the following 4×3 design (4 rows, 3 columns) structure. Using this information, calculate a two-way ANOVA to determine whether there are any significant differences in effects. Use $\alpha = .05$.

SOLUTION

STEP **1.** The following hypotheses are being tested.
For row effects:

H_0: $\mu_{1.} = \mu_{2.} = \mu_{3.} = \mu_{4.}$

H_a: At least one of the row means is different from the others.

For column effects:

H_0: $\mu_{.1} = \mu_{.2} = \mu_{.3}$

H_a: At least one of the column means is different from the others.

For interaction effects:

H_0: The interaction effects are zero.

H_a: There is an interaction effect.

STEP **2.** The two-way ANOVA with the F test is the appropriate statistical test.
STEP **3.** $\alpha = .05$
STEP **4.**

$$df_{rows} = 4 - 1 = 3$$
$$df_{columns} = 3 - 1 = 2$$
$$df_{interaction} = (3)(2) = 6$$
$$df_{error} = (4)(3)(2) = 24$$

For row effects, $F_{.05,3,24} = 3.01$; for column effects, $F_{.05,2,24} = 3.40$; and for interaction effects, $F_{.05,6,24} = 2.51$. For each of these effects, if any observed F value is greater than its associated critical F value, the respective null hypothesis will be rejected.

---

*ADAPTED FROM: Paul R. Murphy and Richard F. Poist, "Managing the Human Side of Public Warehousing: An Overview of Modern Practices," *Transportation Journal,* 31, Spring 1992, 54–63.

STEP **5.**

	Length of Training Session (Days)			
	1–20	21–50	More than 50	$\overline{X}_r$
General Merchandise	3 4.5 4	2 2.5 2	2.5 1 1.5	2.5556
Commodity	5 4.5 4	1 3 2.5	0 1.5 2	2.6111
Bulk Storage	2.5 3 3.5	1 3 1.5	3.5 3.5 4	2.8333
Cold Storage	2 2 3	5 4.5 2.5	4 4.5 5	3.6111
$\overline{X}_c$	3.4167	2.5417	2.75	

Types of Warehouses (row label on left side)

$$\overline{X} = 2.9028$$

STEP **6.** The Excel output for this problem follows.

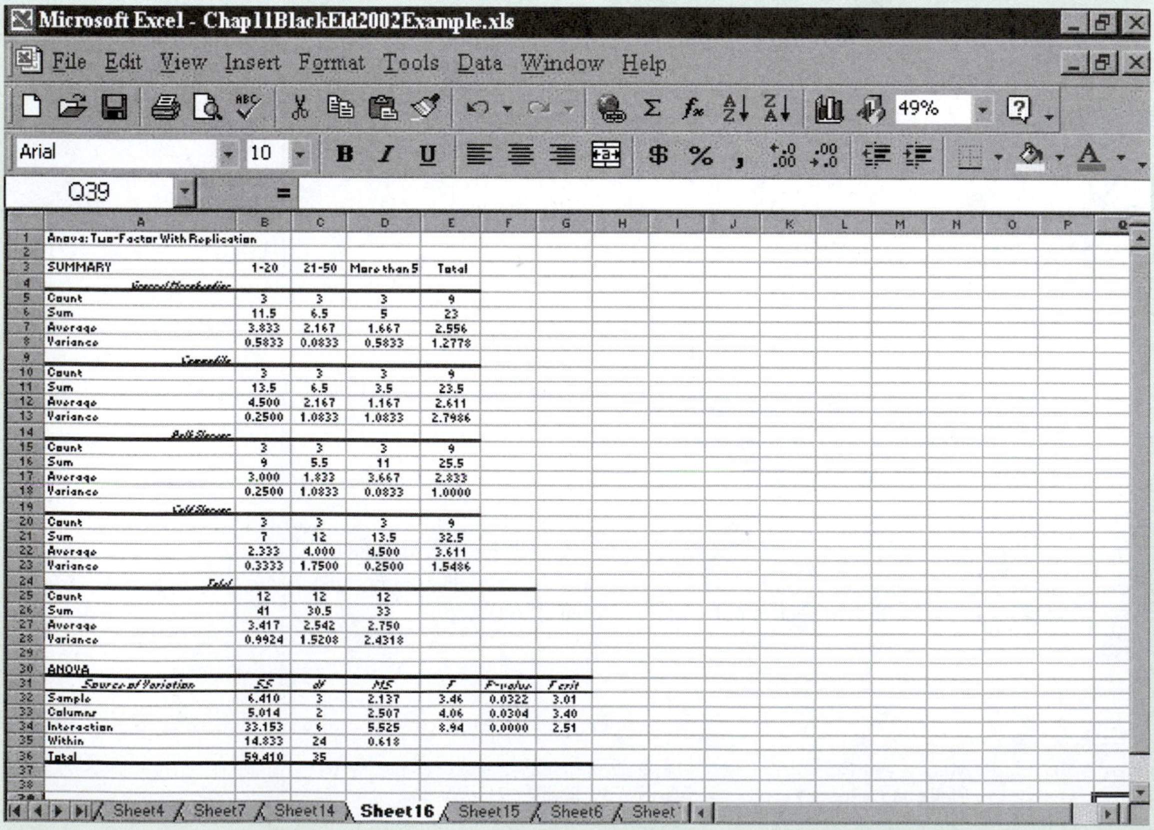

Microsoft Excel - Chap11BlackEld2002Example.xls

File Edit View Insert Format Tools Data Window Help

Q39

	A	B	C	D	E	F	G
1	Anova: Two-Factor With Replication						
2							
3	SUMMARY	1-20	21-50	More than 5	Total		
4	*General Merchandise*						
5	Count	3	3	3	9		
6	Sum	11.5	6.5	5	23		
7	Average	3.833	2.167	1.667	2.556		
8	Variance	0.5833	0.0833	0.5833	1.2778		
9	*Commodity*						
10	Count	3	3	3	9		
11	Sum	13.5	6.5	3.5	23.5		
12	Average	4.500	2.167	1.167	2.611		
13	Variance	0.2500	1.0833	1.0833	2.7986		
14	*Bulk Storage*						
15	Count	3	3	3	9		
16	Sum	9	5.5	11	25.5		
17	Average	3.000	1.833	3.667	2.833		
18	Variance	0.2500	1.0833	0.0833	1.0000		
19	*Cold Storage*						
20	Count	3	3	3	9		
21	Sum	7	12	13.5	32.5		
22	Average	2.333	4.000	4.500	3.611		
23	Variance	0.3333	1.7500	0.2500	1.5486		
24	*Total*						
25	Count	12	12	12			
26	Sum	41	30.5	33			
27	Average	3.417	2.542	2.750			
28	Variance	0.9924	1.5208	2.4318			
29							
30	ANOVA						
31	*Source of Variation*	SS	df	MS	F	P-value	F crit
32	Sample	6.410	3	2.137	3.46	0.0322	3.01
33	Columns	5.014	2	2.507	4.06	0.0304	3.40
34	Interaction	33.153	6	5.525	8.94	0.0000	2.51
35	Within	14.833	24	0.618			
36	Total	59.410	35				
37							
38							

Sheet4 / Sheet7 / Sheet14 / **Sheet16** / Sheet15 / Sheet6 / Sheet

STEP **7.** Looking at the source of variation table, we must first examine the interaction effects. The observed F value for interaction is 8.94. The observed F value for interaction is greater than the critical F value. The interaction effects are statistically significant at $\alpha = .05$. The p value for interaction is .0000. The interaction effects are significant at $\alpha = .0001$. The researcher should not bother to examine the main effects because the significant interaction confounds the main effects.

STEP **8.** The significant interaction effects indicate that certain warehouse types in combination with certain lengths of training session result in different absenteeism rates than do other combinations of levels for these two variables. Using the cell means shown here and Excel, we can depict the interactions graphically.

		Length of Training Session (Days)		
		1–20	*21–50*	*More than 50*
	General Merchandise	3.8	2.2	1.7
Types of Warehouses	*Commodity*	4.5	2.2	1.2
	Bulk Storage	3.0	1.8	3.7
	Cold Storage	2.3	4.0	4.5

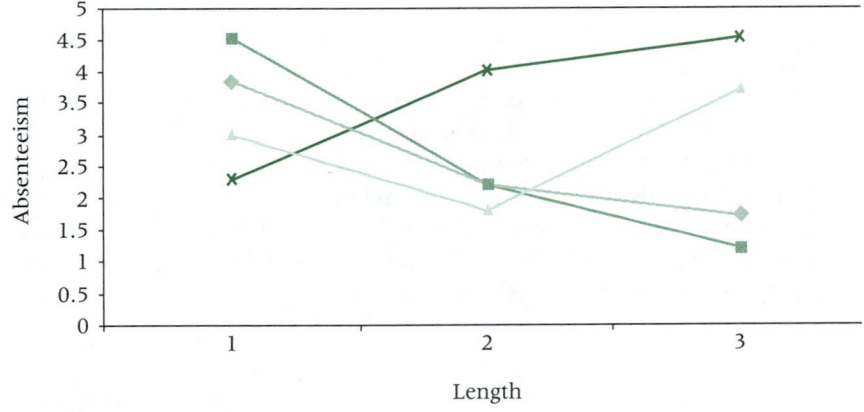

Note the intersecting and crossing lines, which indicate interaction. Under the short-length training sessions, 1, cold-storage workers had the lowest rate of absenteeism and workers at commodity warehouses had the highest. However, for medium-length sessions, 2, cold-storage workers had the highest rate of absenteeism and bulk-storage had the lowest. For the longest training sessions, 3, commodity warehouse workers had the lowest rate of absenteeism, even though these workers had the highest rate of absenteeism for short-length sessions. Thus, the rate of absenteeism for workers at a particular type of warehouse depended on length of session. There was an interaction between type of warehouse and length of session. This graph could be constructed with the row levels along the bottom axis instead of column levels.

11.4
Problems

11.13 The data gathered from a two-way factorial design follow. Use the two-way ANOVA to analyze these data. Let $\alpha = .01$.

		TREATMENT 1		
		A	B	C
TREATMENT 2	A	23	21	20
		25	21	22
	B	27	24	26
		28	27	27

11.14 Suppose the following data have been gathered from a study with a two-way factorial design. Use $\alpha = .05$ and a two-way ANOVA to analyze the data. State your conclusions.

		TREATMENT 1				TREATMENT 1	
		A	B			A	B
TREATMENT 2	A	1.2	1.9	C		1.7	1.9
		1.3	1.6			1.8	2.2
		1.3	1.7			1.7	1.9
		1.5	2.0			1.6	2.0
	B	2.2	2.7	D		2.4	2.8
		2.1	2.5			2.3	2.6
		2.0	2.8			2.5	2.4
		2.3	2.8			2.4	2.8

11.15 Children are generally believed to have considerable influence over their parents in the purchase of certain items, particularly food and beverage items. To study this notion further, a study is conducted in which parents are asked to report how many food and beverage items purchased by the family per week are purchased mainly because of the influence of their children. Because the age of the child may have an effect on the study, parents are asked to focus on one particular child in the family for the week, and to report the age of the child. Four age categories are selected for the children: 4–5 years, 6–7 years, 8–9 years, and 10–12 years. Also, because the number of children in the family might make a difference, three different sizes of families are chosen for the study: families with one child, families with two children, and families with three or more children. Suppose the following data represent the reported number of child-influenced buying incidents per week. Use the data to compute a two-way ANOVA. Let $\alpha = .05$.

		NUMBER OF CHILDREN IN FAMILY		
		1	2	3 or more
AGE OF CHILD (YEARS)	4–5	2	1	1
		4	2	1
	6–7	5	3	2
		4	1	1
	8–9	8	4	2
		6	5	3
	10–12	7	3	4
		8	5	3

11.16 A shoe retailer has conducted a study to determine whether there is a difference in the number of pairs of shoes sold per day by stores according to the number of competitors within a 1-mile radius and the location of the store. The company researchers have selected three types of stores for consideration in the study: stand-alone suburban stores, mall stores, and downtown stores. These stores have varied numbers of competing stores within a 1-mile radius, which have been reduced to four categories: 0 competitors, 1 competitor, 2 competitors, and 3 or more competitors. Suppose the following data represent the number of pairs of shoes sold per day for each of these types of stores with the given number of competitors. Use $\alpha = .05$ and a two-way ANOVA to analyze the data.

| | | NUMBER OF COMPETITORS | | | |
		0	1	2	3 or more
	Stand-Alone	41	38	59	47
		30	31	48	40
		45	39	51	39
	Mall	25	29	44	43
STORE		31	35	48	42
LOCATION		22	30	50	53
	Downtown	18	22	29	24
		29	17	28	27
		33	25	26	32

11.17 In Section 11.2, we tested to determine if there were any significant differences in the mean valve openings among three different operators. Suppose four different machines are used to make the valves by the three operators and that the quality controllers want to know whether there is any difference in the mean measurements of valve openings by operator or by machine. The data are given below, organized by machine and operator. Analyze these data using a two-way ANOVA and $\alpha = .01$.

| | | VALVE OPENINGS (cm) | | |
		Operator 1	Operator 2	Operator 3
	1	6.56	6.38	6.29
		6.40	6.19	6.23
	2	6.54	6.26	6.19
MACHINE		6.34	6.23	6.33
	3	6.58	6.22	6.26
		6.44	6.27	6.31
	4	6.36	6.29	6.21
		6.50	6.19	6.58

In Chapter 5, we studied the binomial distribution in which only two possible outcomes can occur on a single trial in an experiment. An extension of the binomial distribution is a multinomial distribution in which more than two possible outcomes can occur in a single trial. The **chi-square goodness-of-fit test** is *used to analyze probabilities of multinomial distribution trials along a single dimension.* For example, if the variable being studied is type of industry with the three possible outcomes of 1) information technology, 2) financial, and 3) transportation, the single dimension is industry and the three possible outcomes are information technology, financial, and transportation. On each trial, one and only one of the outcomes can occur; that is, the company or respondent must be in one and only one of these three industries. The chi-square goodness-of-fit test uses the chi-square distribution in the analysis of data as does many other statistical techniques including the chi-square test of independence, which is introduced in Section 11.6.

11.5

Chi-Square Goodness-of-Fit Test

Chi-square goodness-of-fit test
A statistical test that compares expected or theoretical frequencies of categories from a population distribution to the actual or observed frequencies from a distribution.

The chi-square goodness-of-fit test compares the *expected*, or theoretical, *frequencies* of categories from a population distribution to the *observed*, or actual, *frequencies* from a distribution to determine whether there is a difference between what was expected and what was observed. For example, airline industry officials might theorize that the ages of airline ticket purchasers are distributed in a particular way. To validate or reject this expected distribution, an actual sample of ticket purchaser ages can be gathered randomly, and the observed results can be compared to the expected results with the chi-square goodness-of-fit test. This test also can be used to determine whether the observed arrivals at teller windows at a bank are Poisson distributed, as might be expected. In the paper industry, manufacturers can use the chi-square goodness-of-fit test to determine whether the demand for paper follows a uniform distribution throughout the year.

Formula 11.1 is used to compute a chi-square goodness-of-fit test.

CHI-SQUARE
GOODNESS-OF-FIT
TEST

$$\chi^2 = \sum \frac{(f_o - f_e)^2}{f_e} \qquad (11.1)$$

$$df = k - 1 - c$$

where:
f_o = frequency of observed values
f_e = frequency of expected values
k = number of categories
c = number of parameters being estimated from the sample data

---

This formula compares the frequency of observed values to the frequency of the expected values across the distribution. The test loses one degree of freedom because the total number of expected frequencies must equal the number of observed frequencies. That is, the observed total taken from the sample is used as the total for the expected frequencies. In addition, in some instances a population parameter, such as λ, μ, or σ, is estimated from the sample data to determine the frequency distribution of expected values. Each time this estimation occurs, an additional degree of freedom is lost. As a rule, if a uniform distribution is being used as the expected distribution or if an expected distribution of values is *given*, $k - 1$ degrees of freedom are used in the test. In testing to determine whether an observed distribution is Poisson, the degrees of freedom are $k - 2$ because an additional degree of freedom is lost in estimating λ. In testing to determine whether an observed distribution is normal, the degrees of freedom are $k - 3$ because two additional degrees of freedom are lost in estimating μ and σ from the observed sample data.

Chi-square distribution
A continuous distribution determined by the sum of the squares of k independent random variables.

Karl Pearson introduced the chi-square test in 1900. The **chi-square distribution** is *the sum of the squares of k independent random variables* and therefore can never be less than zero; it extends indefinitely in the positive direction. Actually the chi-square distributions constitute a family, with each distribution defined by the degrees of freedom (df) associated with it. For small df values the chi-square distribution is skewed considerably to the right (positive values). As the df increase, the chi-square distribution begins to approach the normal curve. Table values for the chi-square distribution are given in Appendix A. Because of space limitations, chi-square values are listed only for certain probabilities.

How can the chi-square goodness-of-fit test be applied to business situations? One survey of U.S. consumers conducted by *The Wall Street Journal* and NBC News asked the question: "In general, how would you rate the level of service that American businesses provide?" The distribution of responses to this question was as follows:

Excellent	8%
Pretty good	47%
Only fair	34%
Poor	11%

Suppose a store manager wants to find out whether the results of this consumer survey apply to customers of supermarkets in her city. To do so, she interviews 207 randomly selected consumers as they leave supermarkets in various parts of the city. She asks the customers how they would rate the level of service at the supermarket from which they had just exited. The response categories are excellent, pretty good, only fair, and poor. The observed responses from this study are given in Table 11.5. Now the manager can use a chi-square goodness-of-fit test to determine whether the observed frequencies of responses from this survey are the same as the frequencies that would be expected on the basis of the national survey.

STEP **1.**   The hypotheses for this example follow.

H_0: The observed distribution is the same as the expected distribution.
H_a: The observed distribution is not the same as the expected distribution.

STEP **2.**   The statistical test being used is

$$\chi^2 = \sum \frac{(f_o - f_e)^2}{f_e}$$

STEP **3.**   Let $\alpha = .05$.

STEP **4.**   Chi-square goodness-of-fit tests are one-tailed because a chi-square of zero indicates perfect agreement between distributions. Any deviation from zero difference occurs in the positive direction only because chi-square is determined by a sum of squared values and can never be negative. With four categories in this example (excellent, pretty good, only fair, and poor), $k = 4$. The degrees of freedom are $k - 1$ because the expected distribution is given: $k - 1 = 4 - 1 = 3$. For $\alpha = .05$ and df $= 3$, the critical chi-square value is

$$\chi^2_{.05,3} = 7.815.$$

After the data have been analyzed, an observed chi-square greater than 7.815 must be computed in order to reject the null hypothesis.

STEP **5.**   The observed values gathered in the sample data from Table 11.5 sum to 207. Thus $n = 207$. The expected *proportions* are given, but the expected *frequencies* must be calculated by multiplying the expected proportions by the sample total of the observed frequencies, as shown in Table 11.6.

STEP **6.**   The chi-square goodness-of-fit can then be calculated, as shown in Table 11.7.

STEP **7.**   The observed value of chi-square is 6.25 versus a critical table value of 7.815. Because the observed chi-square is not greater than the critical chi-square, the store manager will not reject the null hypothesis.

RESPONSE	FREQUENCY (f_o)
Excellent	21
Pretty good	109
Only fair	62
Poor	15

TABLE 11.5
Results of a Local Survey of Consumer Satisfaction

TABLE 11.6
Construction of Expected Values for Service Satisfaction Study

RESPONSE	EXPECTED PROPORTION	EXPECTED FREQUENCY (f_e) (PROPORTION × SAMPLE TOTAL)
Excellent	.08	(.08)(207) = 16.56
Pretty good	.47	(.47)(207) = 97.29
Only fair	.34	(.34)(207) = 70.38
Poor	.11	(.11)(207) = 22.77
		207.00

TABLE 11.7
Calculation of Chi-Square for Service Satisfaction Example

RESPONSE	f_o	f_e	$\dfrac{(f_o - f_e)^2}{f_e}$
Excellent	21	16.56	1.19
Pretty good	109	97.29	1.41
Only fair	62	70.38	1.00
Poor	15	22.77	2.65
	207	207.00	6.25

$$\chi^2 = \sum \frac{(f_o - f_e)^2}{f_e} = 6.25$$

Figure 11.18

Graph of chi-square distribution for service satisfaction example

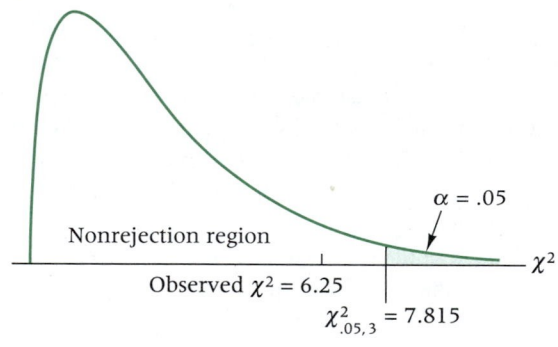

Nonrejection region

$\alpha = .05$

Observed $\chi^2 = 6.25$

$\chi^2_{.05,3} = 7.815$

STEP **8.** Thus the data gathered in the sample of 207 supermarket shoppers indicates that the distribution of responses of supermarket shoppers in the manager's city is not significantly different from the distribution of responses to the national survey.

The store manager may conclude that her customers do not appear to have attitudes different than those people who took the survey. Figure 11.18 depicts the chi-square distribution for this example, along with the observed and critical values.

Analysis Using Excel

By using two different Excel statistical functions, Excel has the capability of computing an observed χ^2 along with its associated p-value and a critical value of χ^2 when conducting a chi-square goodness-of-fit test. Start with the **CHITEST** from the paste function. The dialog box for this is shown in Figure 11.19. The dialog box requires only the location of the actual values and the location of the expected values. The output for **CHITEST** is limited to the p-value of the observed chi-square only.

Using the p-value obtained from **CHITEST**, it is possible to determine the observed χ^2 by using the **CHIINV** feature of the paste function. The dialog box for **CHIINV** is displayed in Figure 11.20. Also, by using **CHIINV**, it is possible to determine the critical

Figure 11.19   Excel dialog box for **CHITEST**

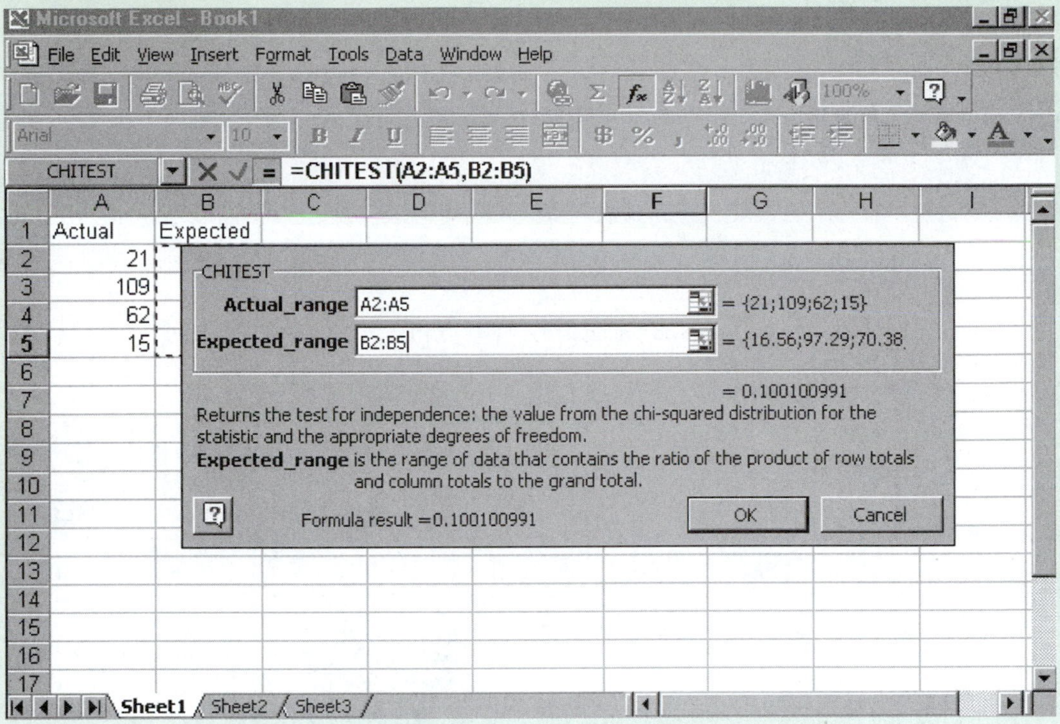

Figure 11.20   Excel dialog box for **CHIINV**

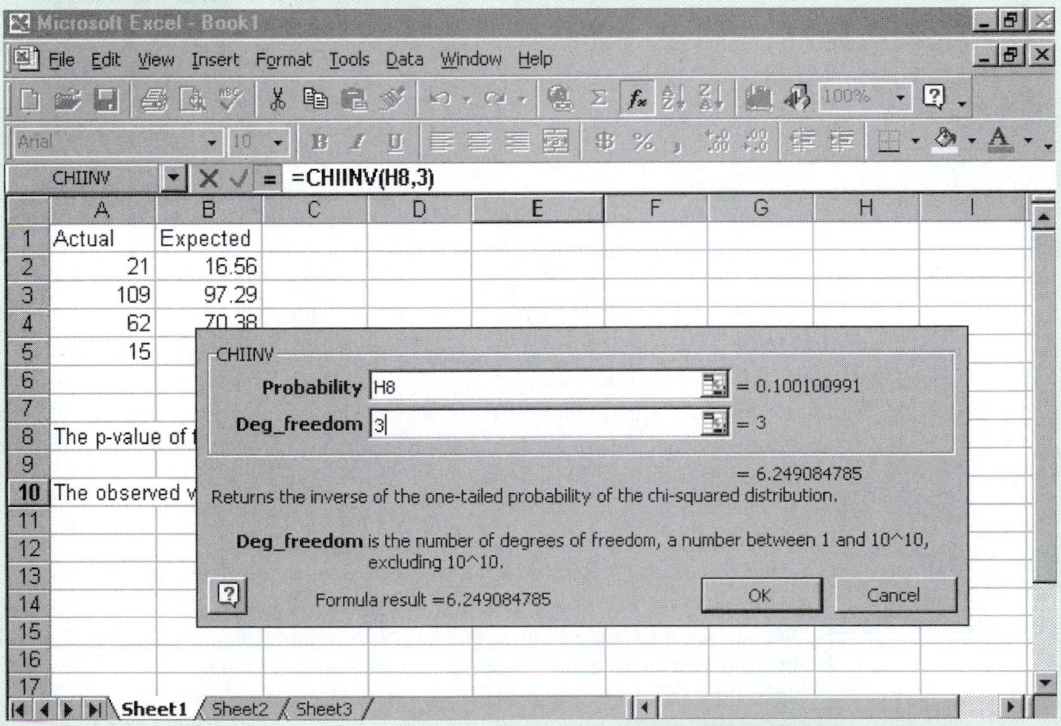

Figure 11.21

Excel output of the chi-square goodness-of-fit test for the consumer satisfaction problem

```
Microsoft Excel - Book1
File Edit View Insert Format Tools Data Window Help

N33         =
      A     B     C     D     E     F         G           H        I     J     K     L     M     N
1
2
3
4
5
6                 The p-value for the observed Chi-square is        0.1001
7                 The observed Chi-square is                        6.2491
8                 The critical Chi-square is                        7.8147
9
10
...
33
Sheet1 / Sheet2 / Sheet3 /
```

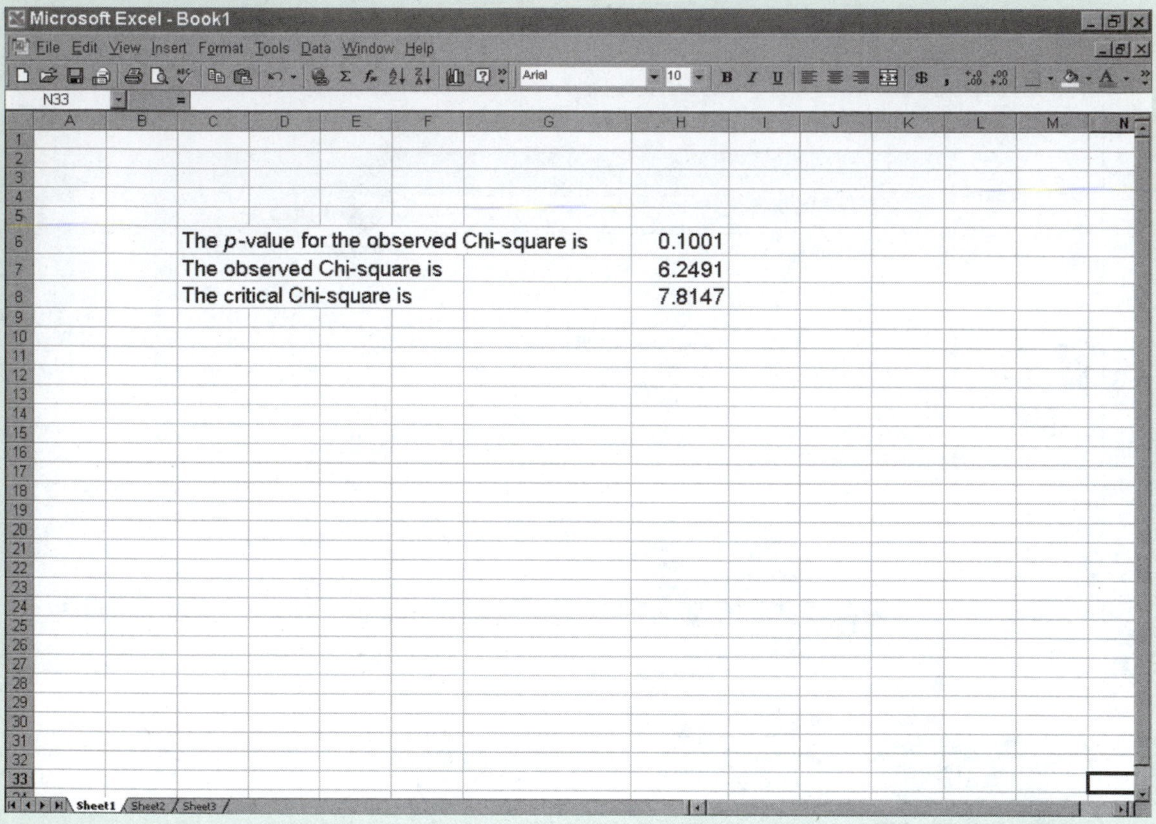

table value. In both cases, **CHIINV** requires the probability which for the observed value is the *p*-value and for the critical value is the value of alpha. In addition, **CHIINV** requires the degrees of freedom for the test.

The output for **CHITEST** is the *p*-value for the observed chi-square. The output for **CHIINV** is the value of the chi-square. Figure 11.21 displays the output for the consumer satisfaction problem including the *p*-value, the observed value of chi-square, and the critical value of chi-square. Note that the *p*-value is .1001, which is greater than $\alpha = .05$ and that the observed chi-square is less than the critical chi-square underscoring the decision to fail to reject the null hypothesis.

DEMONSTRATION PROBLEM 11.4

Dairies would like to know whether the sales of milk are distributed uniformly over a year so they can plan for milk production and storage. A uniform distribution means that the frequencies are the same in all categories. In this situation, the producers are attempting to determine whether the amounts of milk sold are the same for each month of the year. They ascertain the number of gallons of milk sold by sampling one large supermarket each month during a year, obtaining the following data. Use $\alpha = .01$ to test whether the data fit a uniform distribution.

MONTH	GALLONS	MONTH	GALLONS
January	1,553	August	1,450
February	1,585	September	1,495
March	1,649	October	1,564
April	1,590	November	1,602
May	1,497	December	1,609
June	1,443	Total	18,447
July	1,410		

SOLUTION

STEP **1.** The hypotheses follow.

H_0: The monthly figures for milk sales are uniformly distributed.

H_a: The monthly figures for milk sales are not uniformly distributed.

STEP **2.** The statistical test used is

$$\chi^2 = \sum \frac{(f_o - f_e)^2}{f_e}.$$

STEP **3.** Alpha is .01.

STEP **4.** There are 12 categories and a uniform distribution is the expected distribution, so the degrees of freedom are $k - 1 = 12 - 1 = 11$. For $\alpha = .01$, the critical value is $\chi^2_{.01,11} = 24.725$. An observed chi-square value of more than 24.725 must be obtained to reject the null hypothesis.

STEP **5.** The data are given above.

STEP **6.** The first step in calculating the test statistic is to determine the expected frequencies. The total for the expected frequencies must equal the total for the observed frequencies (18,447). If the frequencies are uniformly distributed, the same number of gallons of milk are expected to be sold each month. The expected monthly figure is

$$\frac{18,447}{12} = 1537.25 \text{ gallons.}$$

The following table shows the observed frequencies, the expected frequencies, and the chi-square calculations for this problem.

MONTH	f_o	f_e	$\frac{(f_o - f_e)^2}{f_e}$
January	1,553	1,537.25	0.16
February	1,585	1,537.25	1.48
March	1,649	1,537.25	8.12
April	1,590	1,537.25	1.81
May	1,497	1,537.25	1.05
June	1,443	1,537.25	5.78
July	1,410	1,537.25	10.53
August	1,450	1,537.25	4.95
September	1,495	1,537.25	1.16
October	1,564	1,537.25	.47
November	1,602	1,537.25	2.73
December	1,609	1,537.25	3.35
Total	18,447	18,447.00	$\chi^2 = 41.59$

step **7.** The observed χ^2 value of 41.59 is greater than the critical table value of $\chi^2_{01,11} = 24.725$, so the decision is to reject the null hypothesis. There is enough evidence in this problem to indicate that the distribution of milk sales is not uniform.

step **8.** As retail milk demand is not uniformly distributed, sales and production managers need to generate a production plan to cope with uneven demand. In times of heavy demand, more milk will need to be processed or on reserve; in times of less demand, provision for milk storage or for a reduction in the purchase of milk from dairy farmers will be necessary.

The following graph depicts the chi-square distribution, critical chi-square value, and observed chi-square value.

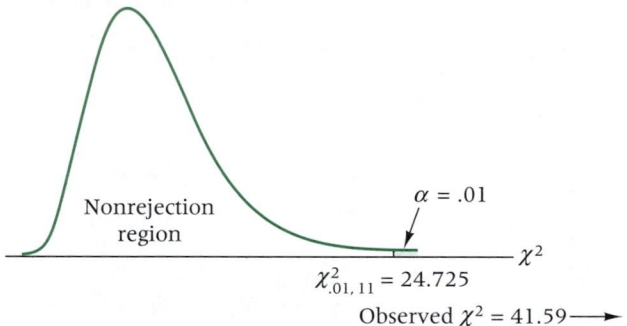

Shown below is the Excel output for Demonstration Problem 11.4.

The *p*-value for the observed Chi-square is	0.000019	
The observed Chi-square is	41.623	
The critical Chi-square is	24.725	

Testing a Population Proportion by Using the Chi-Square Goodness-of-Fit Test as an Alternative Technique to the Z Test

In Chapter 9 we discussed a technique for testing the value of a population proportion. When sample size is large enough ($n \cdot P \geq 5$ and $n \cdot Q \geq 5$), sample proportions are normally distributed and the following formula can be used to test hypotheses about P.

$$Z = \frac{\hat{p} - P}{\sqrt{\dfrac{P \cdot Q}{n}}}$$

The chi-square goodness-of-fit test can also be used to conduct tests about P; this situation can be viewed as a special case of the chi-square goodness-of-fit test where the number of classifications equals two (binomial distribution situation). The observed chi-square is computed in the same way as in any other chi-square goodness-of-fit test but, as there are only two classifications (success/failure), $k = 2$ and the degrees of freedom are $k - 1 = 2 - 1 = 1$.

As an example, we will work two problems from Section 9.4 by using chi-square methodology. The first example in Section 9.4 tests the hypothesis that exactly 8% of a manufacturer's products are defective. The following hypotheses are being tested.

$$H_0: P = .08$$

$$H_a: P \neq .08$$

The value of alpha was given to be .10. To test this claim, a researcher randomly selected a sample of 200 items and determined that 33 of the items had at least one flaw.

Working this problem by the chi-square goodness-of-fit test, we view this as a two-category expected distribution in which we expect .08 defects and .92 nondefects. The observed categories are 33 defects and $200 - 33 = 167$ nondefects. Using the total observed items (200), we can determine an expected distribution as $.08(200) = 16$ and $.92(200) = 184$. Shown here are the observed and expected frequencies.

	f_o	f_e
Defects	33	16
Nondefects	167	184

Alpha is .10 and this is a two-tailed test, so $\alpha/2 = .05$. The degrees of freedom are 1. The critical table chi-square value is

$$\chi^2_{.05,1} = 3.841.$$

An observed chi-square value greater than this value must be obtained to reject the null hypothesis. The chi-square for this problem is calculated as follows.

$$\chi^2 = \sum \frac{(f_o - f_e)^2}{f_e} = \frac{(33 - 16)^2}{16} + \frac{(167 - 184)^2}{184} = 18.06 + 1.57 = 19.63$$

Notice that this observed value of chi-square, 19.63, is greater than the critical table value, 3.841. The decision is to reject the null hypotheses. The manufacturer does not produce 8% defects according to this analysis. Observing the actual sample result, in which .165 of the sample was defective, indicates that the proportion of the population that is defective might be greater than 8%.

| Figure 11.22 | Excel output for the defects problem |

The results obtained are approximately the same as those computed in Chapter 9, in which an observed Z value of 4.43 was determined and compared to a critical Z value of 1.645, causing us to reject the null hypothesis. This is not surprising to business analysts, who understand that when the degrees of freedom equal 1, the value of χ^2 equals Z^2.

Figure 11.22 shows the Excel output for this problem. Note the observed value of chi-square and its associated p-value. Since the p-value of .0000094 is smaller than .05, the decision based on a p-value analysis is to reject the null hypothesis, which is consistent with the findings previously computed by hand.

11.5
Problems

11.18 Use the following data and $\alpha = .01$ to determine whether the observed frequencies represent a uniform distribution.

CATEGORY	f_o
1	19
2	17
3	14
4	18
5	19
6	21
7	18
8	18

11.19 In one survey, successful female entrepreneurs were asked to state their personal definition of success in terms of several categories from which they could select. Thirty-nine percent responded that happiness was their definition of success, 12% said that sales/profit was their definition, 18% responded that helping others was their definition, and 31% responded that achievements/challenge was their definition. Suppose you wanted to determine whether male entrepreneurs felt the same way and took a random sample of men, resulting in the data shown. Use the chi-square goodness-of-fit test to determine whether the observed frequency distribution of data for men is the same as the distribution for women. Let $\alpha = .05$.

DEFINITION	f_o
Happiness	42
Sales/profit	95
Helping others	27
Achievements/challenge	63

11.20 The following percentages are from a national survey of the ages of prerecorded-music shoppers. A local survey produced the observed values. Is there evidence in the observed data to reject the national survey distribution for local prerecorded-music shoppers? Use $\alpha = .01$.

AGE	PERCENT FROM SURVEY	f_o
10–14	9	22
15–19	23	50
20–24	22	43
25–29	14	29
30–34	10	19
≥ 35	22	49

11.21 The Springfield Emergency Medical Service keeps records of emergency telephone calls. A study of 150 five-minute time intervals resulted in the distribution of number of calls shown. For example, there were 18 five-minute intervals in which there were no calls. Use the chi-square goodness-of-fit test and $\alpha = .01$ to determine whether this distribution is Poisson.

NUMBER OF CALLS (PER 5-MIN INTERVAL)	FREQUENCY
0	18
1	28
2	47
3	21
4	16
5	11
6 or more	9

11.22 According to an extensive survey conducted for *Business Marketing* by Leo J. Shapiro & Associates, 66% of all computer companies are going to spend more on marketing this year than in previous years. Only 33% of other information technology companies and 28% of non-information technology companies are going to spend more. Suppose a researcher wanted to conduct a survey of her own to test the claim that 28% of all non-information technology companies are spending more on marketing next year than this year. She randomly selects 270 companies and determines that 62 of the companies do plan to spend more on marketing next year. Use $\alpha = .05$, the chi-square goodness-of-fit test, and the sample data to test to determine whether the 28% figure holds for all non-information technology companies.

11.23   Cross-cultural training is rapidly becoming a popular way to prepare executives for foreign management positions within their company. This training includes such things as foreign language, previsit orientations, meetings with former expatriates, and cultural background information on the country. According to Runzheimer International, 30% of all major companies provide formal cross-cultural programs to their executives being relocated in foreign countries. Suppose a researcher wants to test this figure for companies in the communications industry to determine whether the figure is too high for that industry. A random sample of 180 communications firms are contacted; 42 provide such a program. Let $\alpha = .05$ and use the chi-square goodness-of-fit test to determine whether the .30 proportion for all major companies is too high for this industry.

11.6
Contingency Analysis: Chi-Square Test of Independence

Chi-square test of independence
A statistical test used to analyze the frequencies of two variables with multiple categories to determine whether the two variables are independent.

Contingency table
A two-way table that contains the frequencies of responses to two questions; also called a raw values matrix.

Contingency analysis
A chi-square test of independence.

The chi-square goodness-of-fit test is used to analyze the distribution of frequencies for categories of *one* variable, such as age or number of bank arrivals, to determine whether the distribution of these frequencies is the same as some hypothesized or expected distribution. However, the goodness-of-fit test cannot be used to analyze *two* variables simultaneously. A different chi-square test, the **chi-square test of independence,** can be *used to analyze the frequencies of two variables with multiple categories to determine whether the two variables are independent.* Many times this type of analysis is desirable. For example, a market researcher might want to determine whether the type of soft drink preferred by a consumer is independent of the consumer's age. An organizational behaviorist might want to know whether absenteeism is independent of job classification. Financial investors might want to determine whether type of preferred stock investment is independent of the region where the investor resides.

The chi-square test of independence can be used to analyze any level of data measurement, but it is particularly useful in analyzing nominal data. Suppose a researcher is interested in determining whether geographic region is independent of type of financial investment. On a questionnaire, the following two questions might be used to measure geographic region and type of financial investment.

In which region of the country do you reside?

A. Northeast   **B.** Midwest   **C.** South   **D.** West

Which type of financial investment are you most likely to make today?

E. Stocks   **F.** Bonds   **G.** Treasury Bills

The researcher would tally the frequencies of responses to these two questions into a two-way table called a **contingency table.** Because the chi-square test of independence uses a contingency table, this test is sometimes referred to as **contingency analysis.**

Depicted in Table 11.8 is a contingency table for these two variables. Variable 1, geographic region, has four categories: A, B, C, and D. Variable 2, type of financial investment, has three categories: E, F, and G. The observed frequency for each cell is denoted as o_{ij}, where i is the row and j is the column. Thus, o_{13} is the observed frequency for the cell in the first row and third column. The expected frequencies are denoted in a similar manner.

If the two variables are independent, they are not related. In a sense, the chi-square test of independence is a test of whether the variables are related. The null hypothesis for a chi-square test of independence is that the two variables are independent. If the null hypothesis is rejected, the conclusion is that the two variables are not independent and are related.

Assume at the beginning that variable 1 and variable 2 are independent. The probability of the intersection of two of their respective categories, A and F, can be found by using the multiplicative law for independent events presented in Chapter 4:

$$P(A \cap F) = P(A) \cdot P(F),$$

TABLE 11.8
Contingency Table
for the Investment Example

Type of
Financial
Investment

	E	F	G	
A			o_{13}	n_A
B				n_B
C				n_C
D				n_D
	n_E	n_F	n_G	N

Geographic Region

if A and F are independent. Then

$$P(A) = \frac{n_A}{N}, \quad P(F) = \frac{n_F}{N}, \quad \text{and} \quad P(A \cap F) = \frac{n_A}{N} \cdot \frac{n_F}{N}.$$

If $P(A \cap F)$ is multiplied by the total number of frequencies, N, the expected frequency for the cell of A and F can be determined.

$$e_{AF} = \frac{n_A}{N} \cdot \frac{n_F}{N}(N) = \frac{n_A \cdot n_F}{N}$$

In general, if the two variables are independent, the expected frequency values of each cell can be determined by

$$e_{ij} = \frac{(n_i)(n_j)}{N},$$

where:
  i = the row,
  j = the column,
  n_i = the total of row i,
  n_j = the total of column j, and
  N = the total of all frequencies.

Using these expected frequency values and the observed frequency values, we can compute a chi-square test of independence to determine whether the variables are independent. Formula 11.2 is the formula for accomplishing this.

(11.2)

$$\chi^2 = \sum \sum \frac{(f_o - f_e)^2}{f_e}$$

CHI-SQUARE TEST
OF INDEPENDENCE

where:
  df = $(r - 1)(c - 1)$
  r = number of rows
  c = number of columns

The null hypothesis for a chi-square test of independence is that the two variables are independent. The alternative hypothesis is that the variables are not independent. This test is one-tailed. The degrees of freedom are $(r-1)(c-1)$. Note that Equation 11.2 is similar to Equation 11.1, with the exception that the values are summed across both rows and columns and the degrees of freedom are different.

Suppose a business analyst wants to determine whether type of gasoline preferred is independent of a person's income. He takes a random survey of gasoline purchasers, asking them one question about gasoline preference and a second question about income. The respondent is to check whether he or she prefers (1) regular gasoline, (2) premium gasoline, or (3) extra premium gasoline. The respondent also is to check his or her income brackets as being (1) less than \$30,000, (2) \$30,000 to \$49,999, (3) \$50,000 to \$99,999, or (4) more than \$100,000. The business analyst tallies the responses and obtains the results in Table 11.9. Using $\alpha = .01$, he can use the chi-square test of independence to determine whether type of gasoline preferred is independent of income level.

STEP **1.**   The hypotheses follow.

$$H_0: \text{Type of gasoline is independent of income.}$$

$$H_a: \text{Type of gasoline is not independent of income.}$$

STEP **2.**   The appropriate statistical test is

$$\chi^2 = \sum\sum \frac{(f_o - f_e)^2}{f_e}$$

STEP **3.**   Alpha is .01.

STEP **4.**   Here, there are four rows $(r = 4)$ and three columns $(c = 3)$. The degrees of freedom are $(4-1)(3-1) = 6$. The critical value of chi-square for $\alpha = .01$ is $\chi^2_{.01,6} = 16.812$. The decision rule is to reject the null hypothesis if the observed chi-square is greater than 16.812.

STEP **5.**   The observed data are in Table 11.9.

STEP **6.**   To determine the observed value of chi-square, the business analyst must compute the expected frequencies. The expected values for this example are calculated as follows, with the first term in the subscript (and numerator) representing the row and the second term in the subscript (and numerator) representing the column.

TABLE 11.9

Contingency Table for the Gasoline Consumer Example

		Type of Gasoline		
		Regular	Premium	Extra Premium
	Less than \$30,000	41	16	6	63
Income	\$30,000 to \$49,999	70	27	13	110
	\$50,000 to \$99,999	52	18	15	85
	More than \$100,000	19	17	25	61
		182	78	59	319

$$e_{11} = \frac{(n_1.)(n._1)}{N} = \frac{(63)(182)}{319} = 35.94$$

$$e_{12} = \frac{(n_1.)(n._2)}{N} = \frac{(63)(78)}{319} = 15.40$$

$$e_{13} = \frac{(n_1.)(n._3)}{N} = \frac{(63)(59)}{319} = 11.65$$

$$e_{21} = \frac{(n_2.)(n._1)}{N} = \frac{(110)(182)}{319} = 62.76$$

$$e_{22} = \frac{(n_2.)(n._2)}{N} = \frac{(110)(78)}{319} = 26.90$$

$$e_{23} = \frac{(n_2.)(n._3)}{N} = \frac{(110)(59)}{319} = 20.34$$

$$e_{31} = \frac{(n_3.)(n._1)}{N} = \frac{(85)(182)}{319} = 48.50$$

$$e_{32} = \frac{(n_3.)(n._2)}{N} = \frac{(85)(78)}{319} = 20.78$$

$$e_{33} = \frac{(n_3.)(n._3)}{N} = \frac{(85)(59)}{319} = 15.72$$

$$e_{41} = \frac{(n_4.)(n._1)}{N} = \frac{(61)(182)}{319} = 34.80$$

$$e_{42} = \frac{(n_4.)(n._2)}{N} = \frac{(61)(78)}{319} = 14.92$$

$$e_{43} = \frac{(n_4.)(n._3)}{N} = \frac{(61)(59)}{319} = 11.28$$

The business analyst then lists the expected frequencies in the cells of the contingency tables along with observed frequencies. In this text, expected frequencies are enclosed in parentheses. Table 11.10 is the contingency table for this example.

		Type of Gasoline			
		Regular	*Premium*	*Extra Premium*	
	Less than $30,000	(35.94) 41	(15.40) 16	(11.65) 6	63
	$30,000 to $49,999	(62.76) 70	(26.90) 27	(20.34) 13	110
Income	*$50,000 to $99,999*	(48.50) 52	(20.78) 18	(15.72) 15	85
	More than $100,000	(34.80) 19	(14.92) 17	(11.28) 25	61
		182	78	59	319

TABLE 11.10

Contingency Table of Observed and Expected Frequencies for Gasoline Consumer Example

Figure 11.23

Graph of chi-square distribution for gasoline consumer example

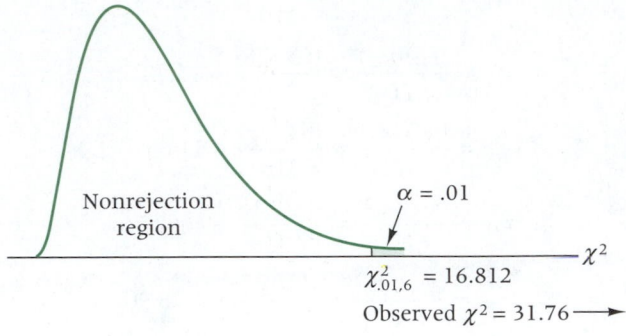

Next, the business analyst computes the chi-square value by summing $(f_o - f_e)^2/f_e$ for all cells.

$$\chi^2 = \frac{(41 - 35.94)^2}{35.94} + \frac{(16 - 15.40)^2}{15.40} + \frac{(6 - 11.65)^2}{11.65} + \frac{(70 - 62.76)^2}{62.76}$$

$$+ \frac{(27 - 26.90)^2}{26.90} + \frac{(13 - 20.34)^2}{20.34} + \frac{(52 - 48.50)^2}{48.50} + \frac{(18 - 20.78)^2}{20.78}$$

$$+ \frac{(15 - 15.72)^2}{15.72} + \frac{(19 - 34.80)^2}{34.80} + \frac{(17 - 14.92)^2}{14.92} + \frac{(25 - 11.28)^2}{11.28}$$

$$= 0.71 + 0.02 + 2.74 + 0.84 + 0.00 + 2.65 + 0.25 + 0.37 + 0.03 + 7.17$$

$$+ 0.29 + 16.69$$

$$= 31.76$$

STEP **7.**   The observed value of chi-square, 31.76, is greater than the critical value of chi-square, 16.812, obtained from Table A.8. The business analyst's decision is to reject the null hypothesis. That is, type of gasoline preferred is not independent of income.

STEP **8.**   Having established that conclusion, the business analyst can then examine the outcome to determine which people, by income brackets, tend to purchase which type of gasoline and use this information in market decisions.

Figure 11.23 is the chi-square graph with the critical value, the rejection region, and the observed χ^2.

Analysis Using Excel

While Excel can compare observed values to expected values and compute p-values for the resulting chi-square, it does not have the capability of computing chi-square values for tests of independence. For this then, we turn to **FAST ⟋ STAT,** which can analyze contingency tables testing for independence with the chi-square statistic. From the pull-down menu provided by **FAST ⟋ STAT**, select **Chi-Square Indep. Test.** Observe the dialog box for this test in Figure 11.24. Note that this dialog box requires the location of the observed frequencies and the value of alpha (significance level).

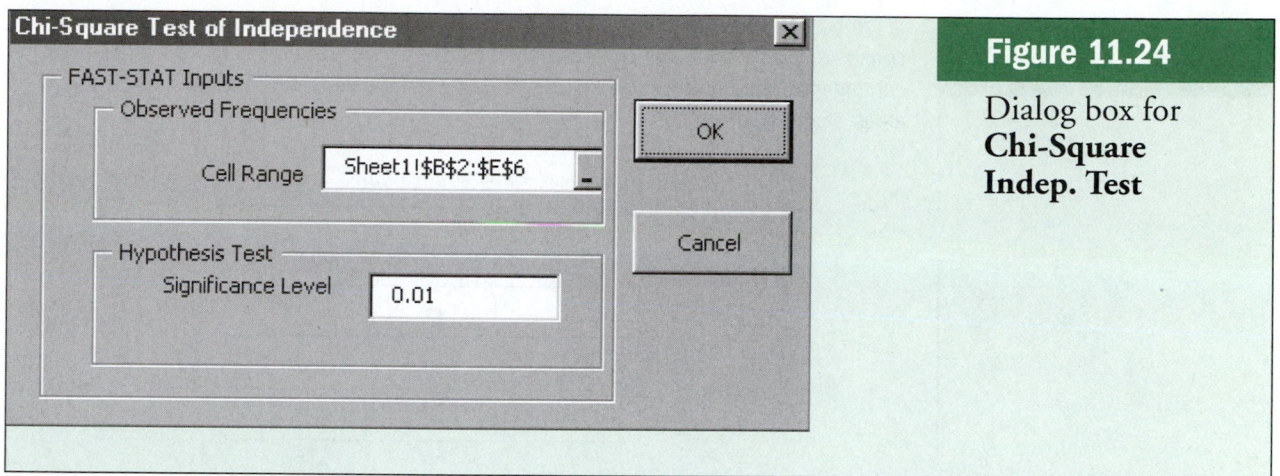

Figure 11.24

Dialog box for **Chi-Square Indep. Test**

Figure 11.25 FAST ＼ STAT output for the gasoline problem

	A	B	C	D	E	F	G	H	I	J	K	L	M	N	O
1				Chi-Square Test of Independence											
2	INPUTS														
3		Significance Level of the Test			0.01										
4															
5		Observed Frequencies													
6	First Variable		Second Variable												
7		Regular	Premium	Extra Prem.											
8	<$30,000	41	16	6											
9	$30-49,999	70	27	13											
10	$50-99,999	52	18	15											
11	>=$100,00	19	17	25											
12															
13															
14															
15															
16															
17															
18	OUTPUTS														
19		Sample Statistics													
20		Total of All Observed Frequencie:			319										
21															
22		No. of Categories-First Variable			4										
23		No. of Categories-Second Variab			3										
24		Degrees of Freedom			6										
25															
26		Chi-Square Test Statistic			31.7689										
27		Chi-Square Critical Value			16.8119										
28		p-Value for Test			0.0000										
29															
30		Expected Frequencies													
31	First Variable		Second Variable												
32		Regular	Premium	Extra Pren											
33	<$30,000	35.94357	15.40439	11.65204											
34	$30-49,999	62.75862	26.89655	20.34483											
35	$50-99,999	48.4953	20.7837	15.721											
36	>=$100,00	34.80251	14.91536	11.28213											

Shown in Figure 11.25 are the results for **FAST ＼ STAT's Chi-Square Indep. Test** analysis on the gasoline problem. Note the calculated value of chi-square, its associated p-value, and the critical value of chi-square. Since the p-value of .00002 is less than $\alpha = .01$, the decision is to reject the null hypothesis. This conclusion is consistent with that previously computed by hand.

DEMONSTRATION PROBLEM 11.5

Is the type of beverage ordered with lunch at a restaurant independent of the age of the consumer? A random poll of 259 lunch customers is taken, resulting in the following contingency table of observed values. Use $\alpha = .01$ to determine whether the two variables are independent.

		Preferred Beverage		
		Coffee/Tea	Soft Drink	Other (Milk, etc.)	
	21-34	26	45	18	89
Age	35-55	41	40	20	101
	>55	24	13	32	69
		91	98	70	259

SOLUTION

STEP **1.** The hypotheses follow.

H_0: Type of beverage preferred is independent of age.

H_a: Type of beverage preferred is not independent of age.

STEP **2.** The appropriate statistical test is

$$\chi^2 = \sum\sum \frac{(f_o - f_e)^2}{f_e}$$

STEP **3.** Alpha is .01.

STEP **4.** The degrees of freedom are $(3 - 1)(3 - 1) = 4$, and the critical value is $\chi^2_{.01,4} = 13.277$. The decision rule is to reject the null hypothesis if the observed value of chi-square is greater than 13.277.

STEP **5.** The sample data are shown above.

STEP **6.** The expected frequencies are the product of the row and column totals divided by the grand total. The contingency table, with expected frequencies, follows.

		Preferred Beverage		
		Coffee/Tea	Soft Drink	Other (Milk, etc.)	
	21-34	(31.27) 26	(33.68) 45	(24.05) 18	89
Age	35-55	(35.49) 41	(38.22) 40	(27.30) 20	101
	>55	(24.24) 24	(26.11) 13	(18.65) 32	69
		91	98	70	259

For these values, the observed χ^2 is

$$\chi^2 = \frac{(26 - 31.27)^2}{31.27} + \frac{(45 - 33.68)^2}{33.68} + \frac{(18 - 24.05)^2}{24.05} + \frac{(41 - 35.49)^2}{35.49}$$
$$+ \frac{(40 - 38.22)^2}{38.22} + \frac{(20 - 27.30)^2}{27.30} + \frac{(24 - 24.24)^2}{24.24} + \frac{(13 - 26.11)^2}{26.11}$$
$$+ \frac{(32 - 18.65)^2}{18.65}$$

$$= 0.89 + 3.80 + 1.52 + 0.86 + 0.08 + 1.95 + 0.00 + 6.58 + 9.56$$

$$= 25.24$$

STEP **7.** The observed value of chi-square, 25.24, is greater than the critical value, 13.277, so the null hypothesis is rejected.

STEP **8.** The two variables—preferred beverage and age—are not independent. The type of beverage that a customer orders with lunch is related to or dependent on age. Examination of the categories reveals that younger people tend to prefer soft drinks and older people prefer other types of beverages. Managers of eating establishments and marketers of beverage products can utilize such information in targeting their market and in providing appropriate products.

Shown below is FAST $\diagdown$ STAT output for Demonstration Problem 11.5. Note the *p*-value of .000045, which is less than $\alpha = .01$, indicating a decision to reject the null hypothesis.

Caution: As with the chi-square goodness-of-fit test, small expected frequencies can lead to inordinately large chi-square values with the chi-square test of independence. Hence contingency tables should not be used with expected cell values of less than 5. One way to avoid small expected values is to collapse (combine) columns or rows whenever possible and whenever doing so makes sense.

11.6
Problems

11.24 Use the following contingency table to test whether variable 1 is independent of variable 2. Let $\alpha = .01$.

	Variable 2			
Variable 1	24	13	47	58
	93	59	187	244

11.25 Use the following contingency table and the chi-square test of independence to determine whether social class is independent of number of children in a family. Let $\alpha = .05$.

		Social Class		
		Lower	*Middle*	*Upper*
Number of Children	*0*	7	18	6
	1	9	38	23
	2 or 3	34	97	58
	More than 3	47	31	30

11.26 A group of 30-year-olds is interviewed to determine whether the type of music most listened to by people in their age category is independent of the geographic location of their residence. Use the chi-square test of independence, $\alpha = .01$, and the following contingency table to determine whether music preference is independent of geographic location.

		Type of Music Preferred			
		Rock	*R & B*	*Country*	*Classical*
Geographic Region	*Northeast*	140	32	5	18
	South	134	41	52	8
	West	154	27	8	13

11.27 Is the transportation mode used to ship goods independent of type of industry? Suppose the following contingency table represents frequency counts of types of transportation used by the publishing and the computer hardware industries. Analyze the data by using the chi-square test of independence to determine whether type of industry is independent of transportation mode. Let $\alpha = .05$.

Transportation Mode

		Air	Train	Truck
Industry	*Publishing*	32	12	41
	Computer Hardware	5	6	24

11.28 According to data released by the U.S. Department of Housing and Urban Development about new homes built in the United States, there is an almost 50–50 split between one-story and two-story homes. In addition, over half of all new homes have three bedrooms. Suppose a study is done to determine whether the number of bedrooms in a new home is independent of the number of stories. Use $\alpha = .10$ and the following contingency table to conduct a chi-square test of independence to determine whether, in fact, the number of bedrooms is independent of the number of stories.

Number of Bedrooms

		≤ 2	3	≥ 4
Number of Stories	*1*	116	101	57
	2	90	325	160

11.29 A study was conducted to determine the impact of a major Mexican peso devaluation on U.S. border retailers. As a part of the study, data were gathered on the magnitude of business that U.S. border retailers were doing with Mexican citizens. Forty-one shoppers of border city department stores were interviewed; 24 were Mexican citizens, and the rest were U.S. citizens. Thirty-five discount store shoppers were interviewed, as were 30 hardware store and 60 shoe store customers. In these three groups, 20, 11, and 32 were Mexican citizens, and the remaining shoppers were U.S. citizens. Use a chi-square contingency analysis to determine whether the shoppers' citizenship (Mexican versus U.S.) is independent of type of border city retailer (department, discount, hardware, shoe) for these data. Let $\alpha = .05$.

Summary

In this chapter, two types of statistical tests are presented: analysis of variance (ANOVA) and chi-square. Analysis of variance tests are used to test the differences in the means of two or more populations and are often used to test statistically hypotheses from complex experimental designs. The chi-square techniques presented in this chapter are used for testing hypotheses using categorical data.

Sound business research requires that the business analyst plan and establish a design for the experiment before a study is undertaken. The design of experiment should encompass the treatment variables to be studied, manipulated, and controlled. These variables are often referred to as the independent variables. It is possible to study several independent variables and several levels, or classifications, of each of those variables in one design. In addition, the business analyst selects one measurement to be taken from sample items under the conditions of the experiment. This measurement is referred to as the dependent variable because if the treatment effect is significant, the measurement of the dependent variable will "depend" on the independent variable(s) selected. This chapter explored three types of experimental designs: completely randomized design, randomized block design, and the factorial experimental designs.

The completely randomized design is the simplest of the experimental designs presented in this chapter. It has only one independent, or treatment, variable. With the completely randomized design, subjects are assigned randomly to treatments. If the treatment variable has only two levels, the design becomes identical to the one used to test the difference in means of independent populations presented in Chapter 10. The data from a completely randomized design are analyzed by a one-way analysis of variance (ANOVA). A one-way ANOVA produces an F value that can be compared to table F values in Appendix A.7 to determine whether the ANOVA F value is statistically significant. If it is, the null hypothesis that all population means are equal is rejected and at least one of the means is different from the others. Analysis of variance does not tell the business analyst which means, if any, are significantly different from others. Although the business analyst can visually examine means to determine which ones are greater and lesser, statistical techniques called multiple comparisons must be used to determine statistically whether pairs of means are significantly different.

A second experimental design is the randomized block design. This design contains a treatment variable (independent variable) and a blocking variable. The independent variable is the main variable of interest in this design. The blocking variable is a variable the business analyst is interested in controlling rather than studying. A special case of randomized block design is the repeated measures design, in which the blocking variable represents subjects or items for which repeated measures are taken across the full range of treatment levels.

In randomized block designs, the variation of the blocking variable is removed from the error variance. This can potentially make the test of treatment effects more powerful. If the blocking variable contains no significant differences, the blocking can make the treatment effects test less powerful. Usually an F is computed only for the treatment effects in a randomized block design. Sometimes an F value is computed for blocking effects to determine whether the blocking was useful in the experiment.

A third experimental design is the factorial design. A factorial design enables the business analyst to test the effects of two or more independent variables simultaneously, In complete factorial designs, every treatment level of each independent variable is studied under the conditions of every other treatment level for all independent variables. This chapter focused only on factorial designs with two independent variables. Each independent variable can have two or more treatment levels. These two-way factorial designs are analyzed by two-way analysis of variance (ANOVA). This analysis produces an F value for each of the two treatment effects and for interaction. Interaction is present when the results of one treatment vary significantly according to the levels of the other treatment. There must be at least two measurements per cell in order to compute interaction. If the F value for interaction is statistically significant, the main effects of the experiment are confounded and should not be examined in the usual manner.

The chi-square goodness-of-fit test is used to compare a theoretical or expected distribution of measurements for several categories of variable with the actual or observed distribution of measurements. It can be used to determine whether a distribution of values fits a given distribution, such as the Poisson or normal distribution. If there are only two categories, the test can be used as the equivalent of a Z test for a single proportion.

The chi-square test of independence is used to analyze frequencies for categories of two variables to determine whether the two variables are independent. The data used in analysis by a chi-square test of independence are arranged in a two-dimensional table called a contingency table. For this reason, the test is sometimes referred to as contingency analysis. A chi-square test of independence is computed in a manner similar to that used with the chi-square goodness-of-fit test. Expected values are computed for each cell of the contingency table and then compared to observed values with the chi-square statistic. Both the chi-square test of independence and the chi-square goodness-of-fit test require that expected values be greater than or equal to 5.

Key Terms

analysis of variance (ANOVA)
blocking variable
chi-square distribution
chi-square goodness-of-fit test
chi-square test of independence
classification variable
classifications
completely randomized design
concomitant variables
confounding variables
contingency analysis
contingency table
dependent variable
experimental design

F distribution
F value
factorial design
factors
independent variable
interaction
levels
multiple comparisons
one-way analysis of variance
randomized block design
repeated measures design
treatment variable
two-way analysis of variance

SUPPLEMENTARY PROBLEMS

11.30 Compute a one-way ANOVA on the following data. Use $\alpha = .05$.

TREATMENT

1	2	3	4
10	9	12	10
12	7	13	10
15	9	14	13
11	6	14	12

11.31 Compute a one-way ANOVA on the following data. Let $\alpha = .01$.

TREATMENT

1	2	3
7	11	8
12	17	6
9	16	10
11	13	9
8	10	11
9	15	7
11	14	10
10	18	
7		
8		

11.32 Analyze the following data, gathered from a randomized block design using $\alpha = .05$.

		TREATMENT			
		A	B	C	D
	1	17	10	9	21
	2	13	9	8	16
BLOCKING	3	20	17	18	22
VARIABLE	4	11	6	5	10
	5	16	13	14	22
	6	23	19	20	28

11.33 Compute a two-way ANOVA on the following data ($\alpha = .01$).

		TREATMENT 1		
		A	B	C
		5	2	2
	A	3	4	3
		6	4	5
		11	9	13
	B	8	10	12
		12	8	10
TREATMENT 2		6	7	4
	C	4	6	6
		5	7	8
		9	8	8
	D	11	12	9
		9	9	11

11.34 Use a chi-square goodness-of-fit test to determine whether the following observed frequencies are distributed the same as the expected frequencies. Let $\alpha = .01$.

CATEGORY	f_o	f_e
1	214	206
2	235	232
3	279	268
4	281	284
5	264	268
6	254	232
7	211	206

11.35 Use the chi-square contingency analysis to test to determine whether variable 1 is independent of variable 2. Use a 5% level of significance.

	VARIABLE 2		
	12	23	21
VARIABLE 1	8	17	20
	7	11	18

11.36 A company has conducted a consumer research project to ascertain customer service ratings from its customers. The customers were asked to rate the company on a scale from 1 to 7 on various quality characteristics. One question was the promptness of company response to a repair problem. The following data represent customer responses to this question. The customers were divided by geographic region and by age. Use analysis of variance to analyze the responses. Let $\alpha = .05$. Graph the cell means and observe any interaction.

		GEOGRAPHIC REGION			
		Southeast	West	Midwest	Northeast
	21–35	3	2	3	2
		2	4	3	3
		3	3	2	2
AGE	36–50	5	4	5	6
		5	4	6	4
		4	6	5	5
	Over 50	3	2	3	3
		1	2	2	2
		2	3	3	1

11.37 A major automobile manufacturer wants to know whether there is any difference in the average mileage of four different brands of tires (A, B, C, and D) because the manufacturer is trying to select the best supplier in terms of tire durability. The manufacturer selects comparable levels of tires from each company and tests some on comparable cars. The mileage results follow.

A	B	C	D
31,000	24,000	30,500	24,500
25,000	25,500	28,000	27,000
28,500	27,000	32,500	26,000
29,000	26,500	28,000	21,000
32,000	25,000	31,000	25,500
27,500	28,000		26,000
	27,500		

Use $\alpha = .05$ to test whether there is a significant difference in the mean mileage of these four brands. Assume tire mileage is normally distributed.

11.38 Is a manufacturer's geographic location independent of type of customer? Use the following data for companies with primarily industrial customers and companies with primarily retail customers to test this question. Let $\alpha = .10$.

		Geographic Location		
		Northeast	West	South
Customer Type	Industrial Customer	230	115	68
	Retail Customer	185	143	89

11.39 Agricultural researchers are studying three different ways of planting peanuts to determine whether significantly different levels of production yield will result. The researchers have access to a very large peanut farm on which to conduct their tests. They identify six blocks of land. In each block of land, peanuts are planted in each of the three different ways. At the end of the growing season, the peanuts are harvested and the average number of pounds per acre is determined for peanuts planted under each method in each block. Using the following data and $\alpha = .01$, test to determine whether there is a significant difference in yields among the planting methods.

BLOCK	METHOD 1	METHOD 2	METHOD 3
1	1310	1080	850
2	1275	1100	1020
3	1280	1050	780
4	1225	1020	870
5	1190	990	805
6	1300	1030	910

11.40 According to *Beverage Digest/Maxwell Report,* the distribution of market share for the top six soft drinks in the United States was Coca-Cola Classic 20.6%, Pepsi 14.5%, Diet Coke 8.5%, Mountain

Dew 6.3%, Sprite 6.2%, Dr. Pepper 5.9%, and others 38%. Suppose a marketing analyst wants to determine whether this distribution fits that of her geographic region. She randomly surveys 1726 local people and asks them to name their favorite soft drink. The responses are: Classic Coke 361, Pepsi 272, Diet Coke 192, Mountain Dew 121, Sprite 102, Dr. Pepper 94, and others 584. She then tests to determine whether the local distribution of soft-drink preferences is the same or different from the national figures, using $\alpha = .05$. What does she find?

11.41 The Construction Labor Research Council lists a number of construction labor jobs that seem to pay approximately the same wages per hour. Some of these are bricklaying, iron working, and crane operation. Suppose a labor researcher takes a random sample of workers from each of these types of construction jobs and from across the country and asks what their hourly wages are. If this survey yields the following data, is there a significant difference in mean hourly wages for these three jobs? Let $\alpha = .05$.

JOB TYPE

Bricklaying	Iron Working	Crane Operation
19.25	26.45	16.20
17.80	21.10	23.30
20.50	16.40	22.90
24.33	22.86	19.50
19.81	25.55	27.00
22.29	18.50	22.95
21.20		25.52
		21.20

11.42 Are the types of professional jobs held in the computing industry independent of the number of years a person has worked in the industry? Suppose 246 workers are interviewed. Use the results obtained to determine whether type of professional job held in the computer industry is independent of years worked in the industry. Let $\alpha = .01$.

Professional Position

	Manager	Programmer	Operator	Systems Analyst
0-3	6	37	11	13
4-8	28	16	23	24
More than 8	47	10	12	19

Years

11.43 A study by Market Facts/TeleNation for Personnel Decisions International (PDI) found that the average workweek is getting longer for U.S. full-time workers. Forty-three percent of the responding workers in the survey cited "more work, more business" as the number one reason for this increase in workweek. Suppose you want to test this figure in California to determine whether California workers feel the same way. A random sample of 315 California full-time workers whose workweek has been getting longer is chosen. They are offered a selection of possible reasons for this increase and 120 pick "more work, more business." Use techniques presented in this chapter and an alpha of .05 to test to determine whether the 43% U.S. figure for this reason holds true in California.

11.44 Why are mergers attractive to CEOs? One of the reasons might be a potential increase in market share that can come with the pooling of company markets. Suppose a random survey of CEOs is taken, and they are asked to respond on a scale from 1 to 5 (5 representing strongly agree) whether increase in market share is a good reason for considering a merger of their company with another. Suppose also that the data are as given here and that CEOs have been categorized by size of company and years they have been with their company. Use a two-way ANOVA to determine whether there are any significant differences in the responses to this question. Let $\alpha = .05$.

			COMPANY SIZE ($ MILLION PER YEAR IN SALES)			
			0-5	6-20	21-100	>100
		0-2	2	2	3	3
			3	1	4	4
			2	2	4	4
			2	3	5	3
YEARS WITH THE COMPANY		3-5	2	2	3	3
			1	3	2	3
			2	2	4	3
			3	3	4	4
		Over 5	2	2	3	2
			1	3	2	3
			1	1	3	2
			2	2	3	3

11.45 Are some unskilled office jobs viewed as having more status than others? Suppose a study is conducted in which eight unskilled, unemployed people are interviewed. The people are asked to rate each of five positions on a scale from 1 to 10 to indicate the status of the position, with 10 denoting most status and 1 denoting least status. The resulting data are given here. Use $\alpha = .05$ to analyze these repeated measures randomized block design data.

JOB

RESPONDENT		Mail Clerk	Typist	Reception-ist	Secretary	Telephone Operator
	1	4	5	3	7	6
	2	2	4	4	5	4
	3	3	3	2	6	7
	4	4	4	4	5	4
	5	3	5	1	3	5
	6	3	4	2	7	7
	7	2	2	2	4	4
	8	3	4	3	6	6

11.46 Following is Excel output for an ANOVA problem. Describe the experimental design. The given value of alpha was .05. Discuss the output in terms of significant findings.

ANOVA: Two-Factor Without Replication

SUMMARY	Count	Sum	Average	Variance
1	3	72	24	1
2	3	80	26.67	0.333
3	3	80	26.67	4.333
4	3	87	29	9
5	3	86	28.67	4.333
6	3	82	27.33	1.333
1	6	165	27.5	5.9
2	6	166	27.67	6.667
3	6	156	26	3.2

ANOVA

Source of Variation	SS	df	MS	F	P-value	F crit
Rows	48.278	5	9.656	3.16	0.057	3.33
Columns	10.111	2	5.056	1.65	0.239	4.10
Error	30.556	10	3.056			
Total	88.944	17				

11.47 Interpret the following Excel output. Discuss the structure of the experimental design and any significant effects. Alpha is .01.

ANOVA: Two-Factor With Replication

SUMMARY	Column 1	Column 2	Column 3	Total
Row 1				
Count	3	3	3	9
Sum	611	645	559	1815
Average	203.67	215	186.33	201.67
Variance	6.333	1	2640.333	818.25
Row 2				
Count	3	3	3	9
Sum	657	681	698	2036
Average	219	227	232.67	226.22
Variance	13	13	9.333	44.194
Row 3				
Count	3	3	3	9
Sum	618	626	635	1879
Average	206	208.67	211.67	208.78
Variance	9	6.333	2.333	10.444
Row 4				
Count	3	3	3	9
Sum	628	631	629	1888
Average	209.33	210.33	209.67	209.78
Variance	2.333	2.333	4.333	2.444
Total				
Count	12	12	12	
Sum	2514	2583	2521	
Average	209.5	215.25	210.08	
Variance	42.818	60.205	776.627	

ANOVA

Source of Variation	SS	df	MS	F	P-value	F crit
Sample	2913.889	3	971.296	4.30	0.0146	3.01
Columns	240.389	2	120.194	0.53	0.5940	3.40
Interaction	1342.944	6	223.824	0.99	0.4533	2.51
Within	5419.333	24	225.806			
Total	9916.556	35				

ANALYZING THE DATABASES

1. Do various financial indicators differ significantly according to type of company? Use a one-way ANOVA and the financial database to answer this question. Let Type of Company be the independent variable with seven levels (Apparel, Chemical, Electric Power, Grocery, Healthcare Products, Insurance, and Petroleum). Compute three one-way ANOVAs, one for each of the following dependent variables: Earnings Per Share, Dividend, and Average P/E Ratio.

2. Use the stock market database to determine whether there is any difference in stock market statistics for different parts of the month. Use a one-way ANOVA with Composite Index as the dependent variable and Part of the Month (1 = 10th, 2 = 20th, and 3 = 30th) as the independent variable with three levels. Compute a second ANOVA with Stock Volume as the dependent variable and Part of the Month as the independent variable. Is there a significant difference in Part of the Month on either of these variables?

3. In the manufacturing database, the Value of Industrial Shipments has been precoded into four classifications (1–4) according to magnitude of value. Let this be the independent variable with four levels of classifications. Compute a one-way ANOVA to determine whether there is any significant difference in classification of the Value of Industrial Shipments on the Number of Production Workers (dependent variable). Perform the same analysis using End-of-Year Inventory as the dependent variable. Now change the independent variable to Industry Group, of which there are 20, and perform first a one-way ANOVA using Number of Production Workers as the dependent variable and then a one-way ANOVA using End-of-Year Inventory as the dependent variable.

4. The hospital database contains data on hospitals from seven different geographic regions. Let this variable be the independent variable. Determine whether there is a significant difference in Admissions for these geographic regions using a one-way ANOVA. Perform the same analysis using Births as the dependent variable. Control is a variable with four levels of classification denoting the type of control the hospital is under (such as federal government or for-profit). Use this variable as the independent variable and test to determine whether there is a significant difference in the Admissions of a hospital by Control. Perform the same test using Births as the dependent variable.

5. Use a chi-square test of independence to determine whether Control is independent of Service in the hospital database.

PROLIGHT: A BUMPY PATH TO A BRIGHT FUTURE

In 1983, Boyd Berends started the Progressive Technology Lighting Company (ProLight) in Holland, Michigan. Berends' goal was to develop and market compact fluorescent lights, which he felt had a big future as energy-savers to replace incandescent bulbs. The road to success was a bumpy one.

Berends designed the specifications of the bulb that he wanted to sell. However, a manufacturer that Berends thought would produce a ballast for his bulb declined. The manufacturer did offer to sell Berends the ballast design and do some of the work if Berends would pay for tooling. Berends accepted this challenge, and he and his wife took out a second mortgage on their home as financing.

ProLight began manufacturing with several other companies and produced 10,000 units for sale. However, because of an oversight, all 10,000 proved to be defective. The second-mortgage money was gone, and there was still no product to sell. Berends was able to get a line of credit and new bulbs were produced. During this time, however, Berends was involved in a legal dispute with one manufacturer, which ended in an out-of-court settlement.

At this point, a manufacturer that was assembling the light bulbs refused to continue. ProLight ended up having to buy equipment, half-finished product, and the raw-material inventory from the manufacturer. The total cost was around $70,000. ProLight's financial base was seriously eroded. Nevertheless, a lender was persuaded to supply additional funds to the company.

ProLight was now a manufacturer and a marketer. Product and raw-material inventory continued to grow. More product and inventory existed than could be sold. There was a cash-flow crisis. Suppliers were persuaded to wait for payment. At this point, Berends bought out a manager who had been given stock as an enticement to join the company.

Two years after start-up, just when things appeared to be going well for the company, a competitor emerged with a similar product. The rival firm was well financed and marketed aggressively. ProLight's sales went flat.

ProLight reacted by designing new fluorescent products that sold well. The company worked to cut costs and improved assembly techniques so much that ProLight believes it is cheaper to assemble the bulbs in Michigan than it would be in a low-wage country. The company imports only parts it does not make itself.

ProLight's sales doubled annually for 6 years. Profits have been on the rise, and the competitor that arose a few years back filed for bankruptcy. Currently, ProLight's product line includes the widest variety of compact fluorescent products available in the industry as well as LED (light emitting diode) energy-saving lighting products.

DISCUSSION

In producing a fluorescent bulb, both mercury vapor and argon gas are used. A fluorescent powder coating on the inner surface of the tube allows for the chemical reaction to produce visible light. This coating consists of several compounds, including zinc silicate, cadmium borate, and barium silicate, which convert ultraviolet radiation into visible light. The fluorescent bulb has pins on the end, through which a source of electricity flows. Because the color of light depends on the makeup of the phosphor, more than two dozen different colors of white can be produced.

1. Suppose you work for ProLight in research and design. You want to study the effects of certain variables on the power, life, or cost of the bulbs. Use the discussion given here to describe a completely randomized design that could be developed to study these bulbs. What are some possible independent variables? Select one independent variable and name some possible levels of treatment for this variable. Name one dependent variable.

2. Suppose ProLight fluorescent bulbs are sold in packages of four bulbs per package. Suppose further that ProLight's managers believe wholesalers in different regions of the country charge different prices to businesses for their bulbs. ProLight researchers set up an experimental design to determine whether the average price at which a package of ProLight bulbs is sold to businesses by wholesalers differs by region. A random sample of wholesalers taken from three regions (Northeast, West, and South) yielded the following data on the price per package of four bulbs. Analyze these data and write a brief report on what you find. Discuss the implications of the results for ProLight in terms of management and marketing decisions.

NORTHEAST	WEST	SOUTH
$5.62	$5.93	$5.78
5.71	5.98	5.83
5.57	6.03	5.80
5.62	5.84	5.83
5.56	5.91	5.87
	5.96	5.84
		5.86

3. ProLight produces light bulbs in many shades of white. Is there any difference in cost to produce bulbs in various shades of white? Cost of bulb production may depend on which of three major suppliers of components is used. To test these ideas, suppose a two-way ANOVA is set up in which there are two independent variables, shade of white and major supplier. Four shades of white and three major suppliers are used in the study. The costs of producing each of three light bulbs under each shade of white and each major supplier are analyzed. The Excel output follows. Your job is to analyze the output and write a brief report of your findings.

ANOVA: Two-Factor With Replication

SUMMARY	Supplier A	Supplier B	Supplier C	Total
Color A				
Count	3	3	3	9
Sum	3.37	3.18	3.34	9.89
Average	1.123333	1.06	1.113333	1.098889
Variance	0.000633	0.0007	0.000233	0.001261
Color B				
Count	3	3	3	9
Sum	3.19	3	3.23	9.42
Average	1.063333	1	1.076667	1.046667
Variance	0.000233	0.0004	0.000433	0.001525
Color C				
Count	3	3	3	9
Sum	3.36	3.2	3.31	9.87
Average	1.12	1.066667	1.103333	1.096667
Variance	1E-04	0.000133	0.000233	0.000675
Color D				
Count	3	3	3	9
Sum	3.5	3.33	3.46	10.29
Average	1.166667	1.11	1.153333	1.143333
Variance	0.000233	1E-04	0.000233	0.0008
Total				
Count	12	12	12	
Sum	13.42	12.71	13.34	
Average	1.118333	1.059167	1.111667	
variance	0.001688	0.001917	0.001033	

ANOVA

Source of Variation	SS	df	MS	F	P-value	F crit
Color	0.042142	3	0.014047	45.97273	4.21E-10	3.008786
Supplier	0.025206	2	0.012603	41.24545	1.72E-08	3.402832
Interaction	0.00155	6	0.000258	0.845455	0.547739	2.508187
Within	0.007333	24	0.000306			
Total	0.076231	35				

4. Suppose ProLight strongly encourages its employees to make formal suggestions to improve the process, the product, and the working environment. Suppose a quality auditor keeps records of the suggestions and persons who have submitted them. A

possible breakdown of the number of suggestions over a 3-year period by employee gender and function follows. Is there any relationship between the function of the employee and gender in terms of number of suggestions?

		GENDER	
		Male	*Female*
	Engineering	209	32
FUNCTION	Manufacturing	483	508
	Shipping	386	185

ADAPTED FROM: "ProLight: A Bumpy Path to a Bright Future," *Real-World Lessons for America's Small Businesses: Insights from the Blue Chip Enterprise Initiative.* Published by *Nation's Business Magazine* on behalf of Connecticut Mutual Life Insurance Company and the U.S. Chamber of Commerce in association with the Blue Chip Enterprise Initiative, 1992. Also from *Encyclopedia Americana,* vol. 11 (Danbury, CT: Grolier International, 1988), 466. See also ProLight's website at http://www.prolight.com/.

12

Simple Regression and Correlation Analysis

Learning Objectives

The overall objective of this chapter is to give you an understanding of bivariate linear regression analysis and correlation, thereby enabling you to:

1. Compute the equation of a simple regression line from a sample of data and interpret the slope and intercept of the equation.

2. Understand the usefulness of residual analysis in testing the assumptions underlying regression analysis and in examining the fit of the regression line to the data.

3. Compute a standard error of the estimate and interpret its meaning.

4. Compute a coefficient of determination and interpret it.

5. Test hypotheses about the slope of the regression model and interpret the results.

6. Estimate values of Y by using the regression model.

7. Compute a coefficient of correlation and interpret it.

In many business research situations, the key to decision making lies in understanding the relationships between two or more variables. For example, in an effort to predict the value of airline stock from day to day, an analyst might find it helpful to determine whether the price of an airline stock is related to the price of West Texas intermediate (WTI) crude oil. In studying the behavior of the bond market, a broker might find it useful to know whether the interest rate of bonds is related to the prime interest rate. In studying the effect of advertising on sales, an account executive might find it useful to know whether there is a strong relationship between advertising dollars and sales dollars for a company.

What variables are related to unemployment rates? Are minimum hourly wage rates, the inflation rate, or the wholesale price index usable for predicting an unemployment rate? This chapter presents techniques that can be used to determine the strength of a relationship between variables and to construct mathematical models for predicting one variable from another.

Two main types of techniques are presented: correlation techniques and regression techniques. **Correlation** is *a measure of the degree of relatedness of two variables.* Correlation is widely used in exploratory research when the objective is to locate variables that might be related in some way to the variable of interest. **Regression** analysis is *the process of constructing a mathematical model or function that can be used to predict or determine one variable by another variable* (multiple variables in multiple regression). Whereas correlation attempts to determine the strength of a relationship between variables, regression attempts to determine the functional relationship between the variables. The first part of this chapter focuses on regression analysis and its ramifications.

Correlation
A measure of the degree of relatedness of two or more variables.

Regression
The process of constructing a mathematical model or function that can be used to predict or determine one variable by any other variable.

12.1
Introduction to Simple Regression Analysis

The most elementary regression model is called **simple regression,** which is *bivariate linear regression.* That is, simple regression involves only two variables; one variable is predicted by another variable. *The variable to be predicted* is called the **dependent variable** and is designated as *Y. The predictor* is called the **independent variable,** or *explanatory variable,* and is designated as *X.* In simple regression analysis, only a straight-line relationship between two variables is examined. Nonlinear relationships and/or regression models with more than one independent variable can be explored by using multiple regression models, which are presented in Chapter 13.

Can the cost of flying a commercial airliner be predicted using regression analysis? If so, what variables are related to such cost? A few of the many variables that can potentially contribute are type of plane, distance, number of passengers, amount of luggage/freight, weather conditions, direction of destination, and perhaps even pilot skill. Suppose a study is conducted using only Boeing 737s traveling 500 miles on comparable routes during the same season of the year in an effort to reduce the number of possible predictor variables. Can the number of passengers predict the cost of flying such routes? It seems logical that more passengers result in more weight and more baggage, which could in turn result in increased fuel consumption and other costs. Suppose the data displayed in Table 12.1 are the costs and associated number of passengers for 12 five-hundred-mile commercial airline flights using Boeing 737s during the same season of the year. We will use these data to develop a regression model to predict cost by number of passengers.

Simple regression
Bivariate, linear regression.

Dependent variable
In regression analysis, the variable that is being predicted.

Independent variable
In regression analysis, the predictor variable.

Scatter Plots

Usually, the first step in simple regression analysis is to construct a **scatter plot** (or *scatter diagram*). The independent (predictor) variable is scaled along the *X* axis and the dependent variable (variable being predicted or determined) is scaled along the *Y* axis. Graphing the data in this way yields preliminary information about the shape and spread of the data. Beware that one of the ways to "cheat" or mislead with statistics is to plot variables on axes having different scales. However, many real-life regression problems

Scatter plot
A plot or graph of the pairs of data from a simple regression analysis.

contain data stated on different scales, and caution should be exercised in interpreting scatter plots.

Figure 12.1 is an Excel scatter plot of the data in Table 12.1. Figure 12.2 is a more closeup view of the scatter plot produced by Excel. Try to imagine a line passing through

TABLE 12.1
Airline Cost Data

NUMBER OF PASSENGERS	COST ($1000)
61	4.280
63	4.080
67	4.420
69	4.170
70	4.480
74	4.300
76	4.820
81	4.700
86	5.110
91	5.130
95	5.640
97	5.560

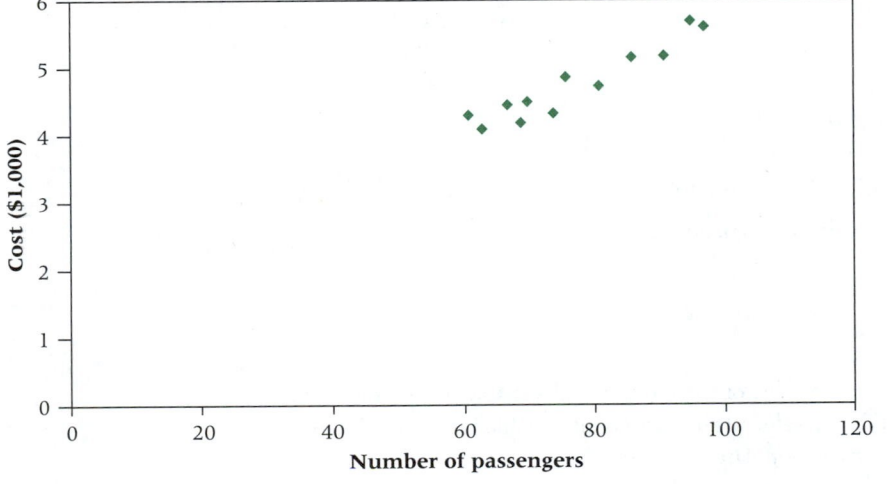

Figure 12.1

Excel scatter plot of airline cost data

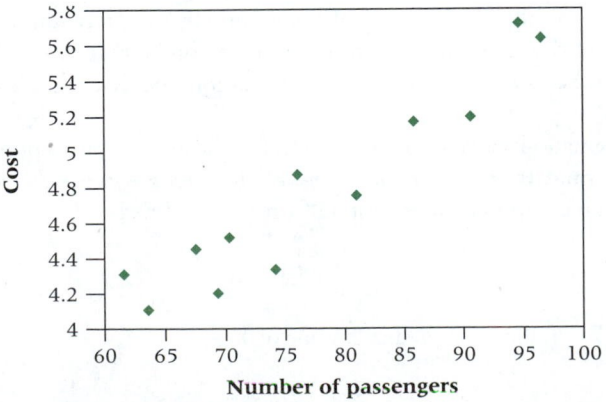

Figure 12.2

Close-up Excel scatter plot of airline cost data

the points. Is there a linear fit? Would a curve fit the data better? The scatter plot gives some idea of how well a regression line fits the data. Later in the chapter, statistical techniques are presented that can be used to determine more precisely how well a regression line fits the data.

12.2
Determining the Equation of the Regression Line

The first step in determining the equation of the regression line that passes through the sample data is to establish the equation's form. Several different types of equations of lines are discussed in algebra, finite math, or analytic geometry courses. Recall that among these equations of a line are the two-point form, the point-slope form, and the slope-intercept form. In regression analysis, business analysts use the slope-intercept equation of a line. In math courses, the slope-intercept form of the equation of a line often takes the form

$$Y = mX + b$$

where:
   m = slope of the line, and
   b = Y intercept of the line.

In statistics, the slope-intercept form of the equation of the regression line through the population points is

$$\hat{Y} = \beta_0 + \beta_1 X$$

where:
   $\hat{Y}$ = the predicted value of Y,
   β_0 = the population Y intercept, and
   β_1 = the population slope.

For any specific dependent variable value, Y_i,

$$Y_i = \beta_0 + \beta_1 X_i + \epsilon_i$$

where:
   X_i = the value of the independent variable for the ith value
   Y_i = the value of the dependent variable for the ith value,
   β_0 = the population Y intercept,
   β_1 = the population slope, and
   ϵ_i = the error of prediction for the ith value.

Unless the points being fitted by the regression equation are in perfect alignment, the regression line will miss at least some of the points. In the preceding equation, ϵ_i represents the error of the regression line in fitting these points, that is, the distances the data values are from their projected values on the regression line. If a point is on the regression line, $\epsilon_i = 0$.

These mathematical models can be either deterministic models or probabilistic models. **Deterministic models** are *mathematical models that produce an "exact" output for a given input.* For example, suppose the equation of a regression line is

Deterministic model
Mathematical models that produce an "exact" output for a given input.

$$Y = 1.68 + 2.40X.$$

For a value of $X = 5$, the exact predicted value of Y is

$$Y = 1.68 + 2.40(5) = 13.68.$$

We recognize, however, that most of the time the values of Y will not equal exactly the values yielded by the equation. Random error will occur in the prediction of the Y values for values of X because it is likely that the variable X does not explain all the variability of the variable Y. For example, suppose we are trying to predict the volume of sales (Y) for a company through regression analysis by using the annual dollar amount of advertising (X) as the predictor. Although sales are often related to advertising, there are other factors related to sales that are not accounted for by amount of advertising. Hence, a regression model to predict sales volume by amount of advertising probably involves some error. For this reason, in regression, we present the general model as a probabilistic model. A **probabilistic model** is *one that includes an error term that allows for the Y values to vary for any given value of X.*

A deterministic regression model is

$$Y = \beta_0 + \beta_1 X.$$

The probabilistic regression model is

$$Y = \beta_0 + \beta_1 X + \epsilon.$$

$\beta_0 + \beta_1 X$ is the deterministic portion of the probabilistic model, $\beta_0 + \beta_1 X + \epsilon$. In a deterministic model, all points are assumed to be on the line and in all cases ϵ is zero.

Virtually all regression analyses of business data involve sample data, not population data. As a result, β_0 and β_1 are unattainable and must be estimated by using the sample statistics, b_0 and b_1. Hence the equation of the regression line contains the sample Y intercept, b_0, and the sample slope, b_1.

$$\hat{Y} = b_0 + b_1 X$$
where:
$\quad b_0$ = the sample intercept
$\quad b_1$ = the sample slope

EQUATION OF THE SIMPLE REGRESSION LINE

To determine the equation of the regression line for a sample of data, the business analyst must determine the values for b_0 and b_1. This process is sometimes referred to as least squares analysis. **Least squares analysis** is *a process whereby a regression model is developed by producing the minimum sum of the squared error values.* On the basis of this premise and calculus, a particular set of equations has been developed to produce components of the regression model.

Examine the regression line fit through the points in Figure 12.3 using Excel. The points (X, Y) contain the actual values of Y for given values of X. Observe that the line does not actually pass through any of these points. The vertical distance from each point to the line is the error of the prediction. In theory, an infinite number of lines could be constructed to pass through these points in some manner. The least squares regression line is the regression line that results in the smallest sum of errors squared.

Fitting a Line Using Excel

Excel has the ability to fit lines and curves through data. To accomplish this, first construct a scatter chart of the regression data using the Chart Wizard (see Chapter 2), producing a scatter plot like the one displayed in Figure 12.2. To fit a line or curve through the scatter plot while in Excel, touch one of the points on the graph with the mouse

Probabilistic model
A model that includes an error term that allows for various values of output to occur for a given value of input.

Least squares analysis
The process by which a regression model is developed based on calculus techniques that attempt to produce a minimum sum of the squared error values.

Figure 12.3

Excel plot of a regression line through the airline cost data

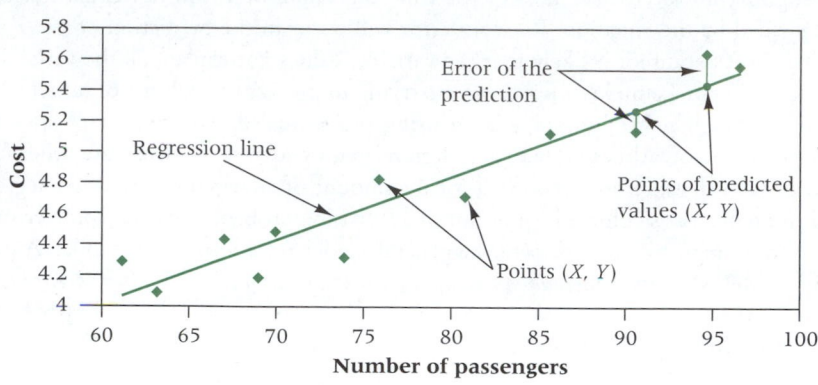

Figure 12.4   Excel dialog box for **Trendline Feature**

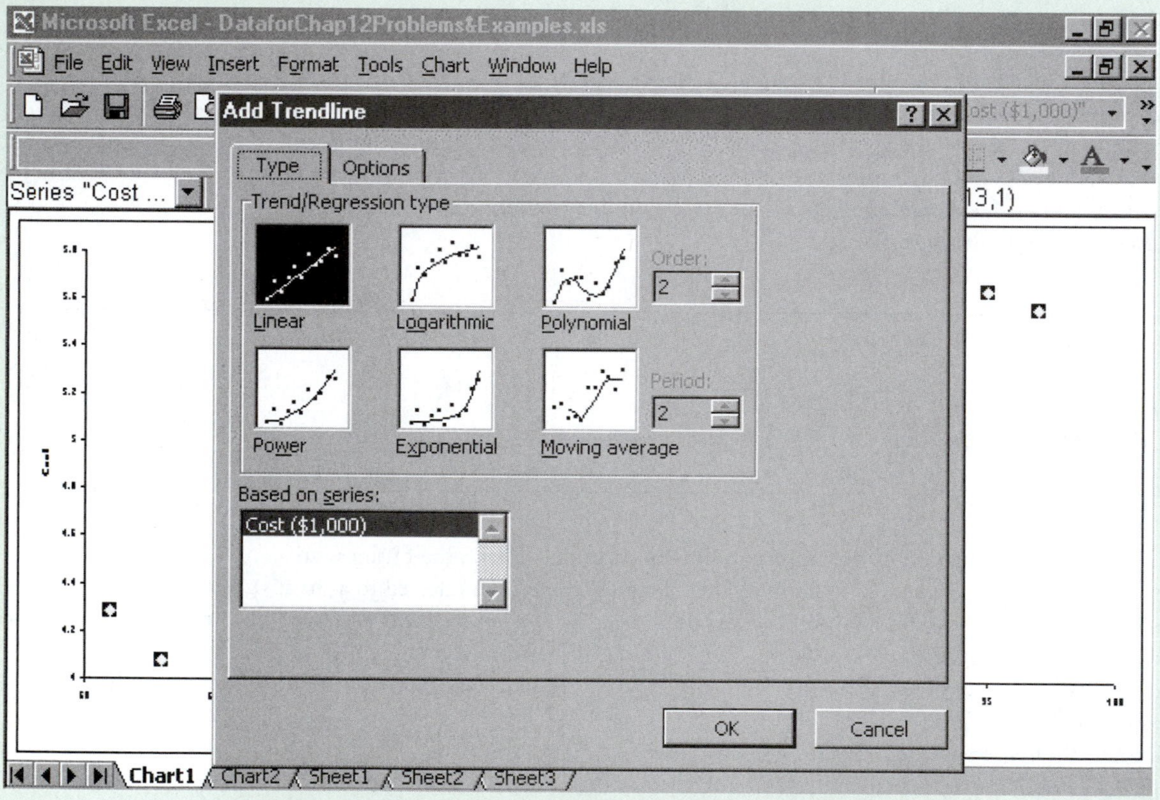

pointer and "right-click" on the mouse. A shortcut menu will appear and one of the items on it is **Add Trendline.** Clicking on this item will produce the dialog box shown in Figure 12.4. Note that there are six types of trend line/curves that can be fit through the data of the scatter plot. Since this chapter deals only with the linear trend (simple regression analysis), click on **Linear** and the result is the graph displayed in Figure 12.3. Curvilinear fits will be explored in Chapter 13 (Multiple Regression Analysis) and Chapter 14 (Time Series Forecasting).

Equation 12.1 is a formula for computing the value of the sample slope. Three versions of the equation are given to afford latitude in doing the computations.

$$(12.1) \qquad b_1 = \frac{\Sigma(X - \bar{X})(Y - \bar{Y})}{\Sigma(X - \bar{X})^2} = \frac{\Sigma XY - n\bar{X}\bar{Y}}{\Sigma X^2 - n\bar{X}^2} = \frac{\Sigma XY - \dfrac{(\Sigma X)(\Sigma Y)}{n}}{\Sigma X^2 - \dfrac{(\Sigma X)^2}{n}}$$

SLOPE OF THE
REGRESSION LINE

The expression in the numerator of the slope formula (12.1) appears frequently in this chapter and is denoted as SS_{XY}.

$$SS_{XY} = \Sigma(X - \bar{X})(Y - \bar{Y}) = \Sigma XY - \frac{(\Sigma X)(\Sigma Y)}{n}$$

The expression in the denominator of the slope formula (12.1) also appears frequently in this chapter and is denoted as SS_{XX}.

$$SS_{XX} = \Sigma(X - \bar{X})^2 = \Sigma X^2 - \frac{(\Sigma X)^2}{n}$$

With these abbreviations, the formula for the slope can be expressed as in Equation 12.2.

$$(12.2) \qquad b_1 = \frac{SS_{XY}}{SS_{XX}}$$

ALTERNATIVE
FORMULA FOR SLOPE

Equation 12.3 is used to compute the sample Y intercept. The slope must be computed before the Y intercept.

$$(12.3) \qquad b_0 = \bar{Y} - b_1\bar{X} = \frac{\Sigma Y}{n} - b_1\frac{(\Sigma X)}{n}$$

Y INTERCEPT OF
THE REGRESSION LINE

Equations 12.1, 12.2, and 12.3 show that the following data are needed from sample information to compute the slope and intercept: ΣX, ΣY, ΣX^2, and ΣXY, unless sample means are used. Table 12.2 contains the results of solving for the slope and intercept and determining the equation of the regression line for the data in Table 12.1.

The least squares equation of the regression line for this problem is

$$\hat{Y} = 1.570 + .0407X.$$

The slope of this regression line is .0407. Since the X values were recoded for the ease of computation and are actually in $1000 denominations, the slope is actually $40.70. One interpretation of the slope in this problem is that for every unit increase in X (every person added to the flight of the airplane), there is a $40.70 increase in the cost of the flight. The Y intercept is the point where the line crosses the Y axis (where X is zero). Sometimes in regression analysis, the Y intercept is meaningless in terms of the variables studied. However, in this problem, one interpretation of the Y intercept, which is 1.570 or $1570, is

TABLE 12.2
Solving for the Slope and the Y Intercept of the Regression Line for the Airline Cost Example

NUMBER OF PASSENGERS X	COST ($1,000) Y	X^2	XY
61	4.280	3,721	261.080
63	4.080	3,969	257.040
67	4.420	4,489	296.140
69	4.170	4,761	287.730
70	4.480	4,900	313.600
74	4.300	5,476	318.200
76	4.820	5,776	366.320
81	4.700	6,561	380.700
86	5.110	7,396	439.460
91	5.130	8,281	466.830
95	5.640	9,025	535.800
97	5.560	9,409	539.320
$\Sigma X = 930$	$\Sigma Y = 56.690$	$\Sigma X^2 = 73,764$	$\Sigma XY = 4462.220$

$$SS_{XY} = \Sigma XY - \frac{\Sigma X \Sigma Y}{n} = 4462.220 - \frac{(930)(56.690)}{12} = 68.745$$

$$SS_{XX} = \Sigma X^2 - \frac{(\Sigma X)^2}{n} = 73,764 - \frac{(930)^2}{12} = 1689$$

$$b_1 = \frac{SS_{XY}}{SS_{XX}} = \frac{68.745}{1689} = .0407$$

$$b_0 = \frac{\Sigma Y}{n} - b_1 \frac{\Sigma X}{n} = \frac{56.690}{12} - (.0407)\frac{930}{12} = 1.570$$

$$\hat{Y} = 1.570 + .0407X$$

that even if there were no passengers on the commercial flight, it would still cost $1570. That is, even when the plane flies empty, there are still costs associated with the flight.

Analysis Using Excel

The equation of the regression line can be obtained in Excel by using the **Regression** feature of the **Data Analysis** tool. The dialog box for **Regression** is shown in Figure 12.5. This dialog box requires input of the location of the Y variable followed by the location of the X variable. Note that there are several options for types of output given at the bottom of the dialog box, one of which is to include a fitted line plot (scatter plot of the data with a line fit to the data) like the one displayed in Figure 12.3. Residuals will be discussed in Section 12.3 of this chapter.

The resulting output is shown in Figure 12.6. Note that listed under *Coefficients* in the output are: *Intercept* with 1.5655 and *Number of Passengers* with 0.0407. Intercept is the value of the Y intercept in the regression equation. The other coefficient, in this case Number of Passengers, is the value of the slope. From these two figures, we can produce the regression equation of:

$$\hat{Y} = 1.5655 + 0.0407 \text{ Number of Passengers},$$

which is essentially the same as the equation displayed in Table 12.2 (differences due to rounding error).

Figure 12.5	Excel dialog box for **Regression**

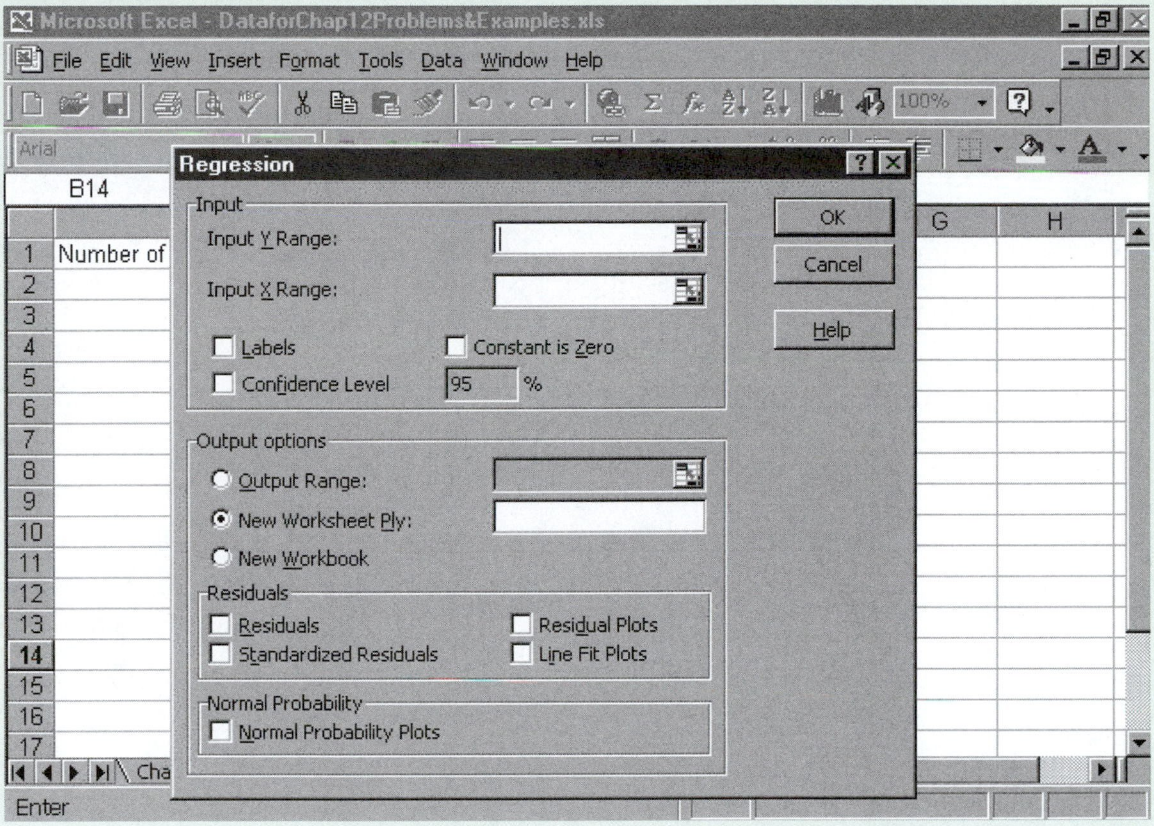

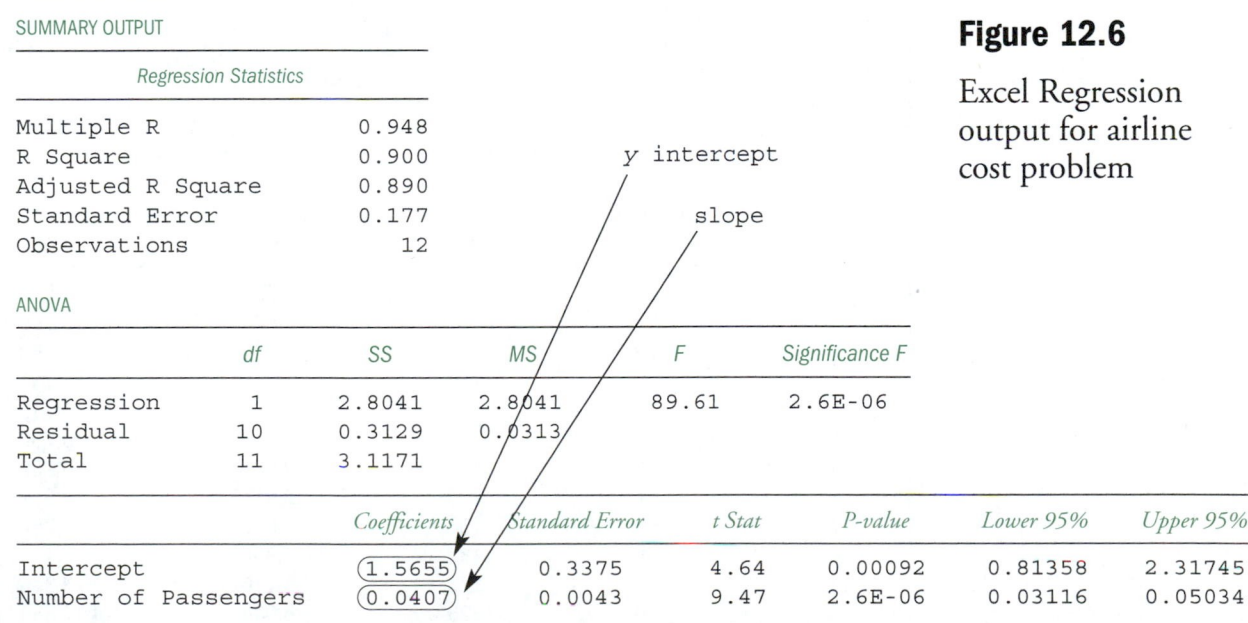

SUMMARY OUTPUT

Regression Statistics	
Multiple R	0.948
R Square	0.900
Adjusted R Square	0.890
Standard Error	0.177
Observations	12

ANOVA

	df	SS	MS	F	Significance F
Regression	1	2.8041	2.8041	89.61	2.6E-06
Residual	10	0.3129	0.0313		
Total	11	3.1171			

	Coefficients	Standard Error	t Stat	P-value	Lower 95%	Upper 95%
Intercept	1.5655	0.3375	4.64	0.00092	0.81358	2.31745
Number of Passengers	0.0407	0.0043	9.47	2.6E-06	0.03116	0.05034

y intercept

slope

Figure 12.6

Excel Regression output for airline cost problem

DEMONSTRATION PROBLEM 12.1

A specialist in hospital administration stated that the number of FTEs (full-time employees) in a hospital can be estimated by counting the number of beds in the hospital (a common measure of hospital size). A business analyst decided to develop a regression model in an attempt to predict the number of FTEs of a hospital by the number of beds. She surveyed 12 hospitals and obtained the following data. The data are presented in sequence, according to the number of beds.

NUMBER OF BEDS	FTES	NUMBER OF BEDS	FTES
23	69	50	138
29	95	54	178
29	102	64	156
35	118	66	184
42	126	76	176
46	125	78	225

SOLUTION

The following Excel graph is a scatter plot of these data. Note the linear appearance of the data.

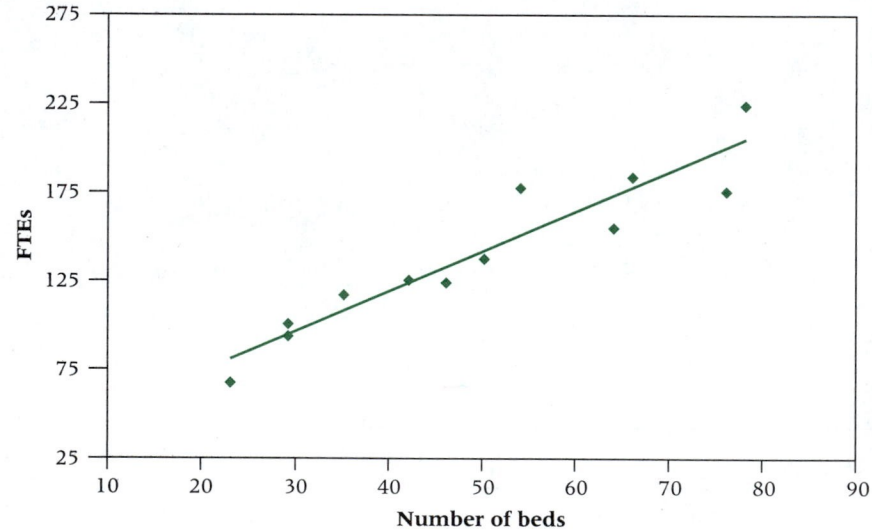

Next, the business analyst determined the values of ΣX, ΣY, ΣX^2, and ΣXY.

HOSPITAL	NUMBER OF BEDS X	FTES Y	X^2	XY
1	23	69	529	1,587
2	29	95	841	2,755
3	29	102	841	2,958
4	35	118	1,225	4,130
5	42	126	1,764	5,292
6	46	125	2,116	5,750
7	50	138	2,500	6,900
8	54	178	2,916	9,612
9	64	156	4,096	9,984
10	66	184	4,356	12,144
11	76	176	5,776	13,376
12	78	225	6,084	17,550
	$\Sigma X = 592$	$\Sigma Y = 1,692$	$\Sigma X^2 = 33,044$	$\Sigma XY = 92,038$

Using these values, the business analyst solved for the sample slope (b_1) and the sample Y intercept (b_0).

$$SS_{XY} = \Sigma XY - \frac{\Sigma X \Sigma Y}{n} = 92{,}038 - \frac{(592)(1692)}{12} = 8566$$

$$SS_{XX} = \Sigma X^2 - \frac{(\Sigma X)^2}{n} = 33{,}044 - \frac{(592)^2}{12} = 3838.667$$

$$b_1 = \frac{SS_{XY}}{SS_{XX}} = \frac{8566}{3838.667} = 2.232$$

$$b_0 = \frac{\Sigma Y}{n} - b_1 \frac{\Sigma X}{n} = \frac{(1692)}{12} - 2.232 \left(\frac{592}{12} \right) = 30.888$$

The least squares equation of the regression line is

$$\hat{Y} = 30.888 + 2.23X.$$

The slope of the line, $b_1 = 2.232$, means that for every unit increase of X (every bed), Y (number of FTEs) is predicted to increase by 2.232. Even though the Y intercept helps the business analysts sketch the graph of the line by being one of the points on the line (0, 30.888), it is meaningless information in terms of this solution because $X = 0$ is not in the range of the data.

Shown below is a portion of the Excel output for this problem. Note the coefficients denoting the Y intercept and the slope of the regression line.

	COEFFICIENTS
Intercept	30.912
Number of Beds	2.232

12.2 Problems

12.1 Create a scatter plot for the following data and observe the plot to indicate whether there seems to be a linear trend in the data. Now use Excel's trendline feature to fit a line through the data.

X	6	11	9	14	5	3
Y	5	2	3	1	7	11

12.2 Create a scatter plot for the following data and determine whether there seems to be a linear relationship between X and Y based on your observations. Use Excel's trendline feature to assist you in this decision.

X	36	45	52	20	12	40	63
Y	14	26	35	32	48	18	51

12.3 Create a scatter plot for the following data and determine the equation of the regression line.

X	12	21	28	8	20
Y	17	15	22	19	24

12.4 Create a scatter plot from the following data and determine the equation of the regression line.

X	140	119	103	91	65	29	24
Y	25	29	46	70	88	112	128

12.5 A corporation owns several companies. The strategic planner for the corporation believes dollars spent on advertising can to some extent be a predictor of total sales dollars. As an aid in long-term planning, she gathers the following sales and advertising information from several of the companies for 2001 (in $ millions).

ADVERTISING	SALES
12.5	148
3.7	55
21.6	338
60.0	994
37.6	541
6.1	89
16.8	126
41.2	379

Develop the equation of the simple regression line to predict sales from advertising expenditures using these data.

12.6 Investment analysts generally believe the interest rate on bonds is inversely related to the prime interest rate for loans. That is, bonds perform well when lending rates are down and perform poorly when interest rates are up. Can the bond rate be predicted by the prime interest rate? Use the following data to construct a least squares regression line to predict bond rates by the prime interest rate.

BOND RATE	PRIME INTEREST RATE
5%	16%
12	6
9	8
15	4
7	7

12.7 Is it possible to predict the annual number of business failures in the United States by the number of business starts the previous year? It might seem that the more business starts there are in a given year, the more potential there is for business failure the next year. The following data from Dun & Bradstreet show the number of business failures over a 10-year period and the number of business starts for each of the previous years. Use these data to develop the equation of a regression line to predict the number of business failures from the number of business starts the previous year. Discuss the slope and Y intercept of the model.

NUMBER OF BUSINESS STARTS FOR THE PREVIOUS YEAR	NUMBER OF BUSINESS FAILURES
233,710	57,097
199,091	50,361
181,645	60,747
158,930	88,140
155,672	97,069
164,086	86,133
166,154	71,558
188,387	71,128
168,158	71,931
170,475	83,384

12.8 It appears that over the past 35 years, the number of farms in the United States has declined while the average size of farms has increased. The following data provided by the U.S. Department of Agriculture show five-year interval data for U.S. farms. Use these data to develop the equation of a regression line to predict the average size of a farm by the number of farms. Discuss the slope and Y intercept of the model.

NUMBER OF FARMS (MILLIONS)	AVERAGE SIZE (ACRES)
5.65	213
4.65	258
3.96	297
3.36	340
2.95	374
2.52	420
2.44	426
2.29	441
2.15	460
2.07	469

12.9 Can the annual new orders for manufacturing in the United States be predicted by the raw steel production in the United States? Shown here are the annual new orders for 10 years according to the U.S. Bureau of the Census and the raw steel production for the same 10 years as published by the American Iron & Steel Institute. Use these data to develop a regression model to predict annual new orders by raw steel production. Construct a scatter plot and draw the regression line through the points.

RAW STEEL PRODUCTION (100,000s OF NET TONS)	NEW ORDERS ($ TRILLION)
99.9	2.74
97.9	2.87
98.9	2.93
87.9	2.87
92.9	2.98
97.9	3.09
100.6	3.36
104.9	3.61
105.3	3.75
108.6	3.95

Simple regression analysis can be performed on data even when the data are not "linear." How do we know that the linear regression analysis "fits" the data well enough to justify the analysis? We now explore three techniques for testing the "fit" of the regression line to the data. These techniques include residual analysis, coefficient of determination (r^2), and standard error of the estimate (S_e).

Residual Analysis

How does a business analyst test a regression line to determine mathematically whether the line is a *good* fit of the data? One type of information available is the *historical data* used to construct the equation of the line. In other words, there are actual Y values that correspond to the X values used in constructing the regression line. Why not insert the historical X values into the equation of the sample regression line and get predicted Y values (denoted $\hat{Y}$) and then compare these predicted values to the actual Y values to determine how much error the equation of the regression line produced? *Each difference between the actual Y values and the predicted Y values is the error of the regression line at a given point, $Y - \hat{Y}$,* and is referred to as the **residual.** It is the sum of squares of these residuals that is minimized to find the least squares line.

12.3

Analyzing the Fit of the Regression Line: Residuals, r^2, and Standard Error of the Estimate

Residual

The difference between the actual Y value and the Y value predicted by the regression model; the error of the regression model in predicting each value of the dependent variable.

TABLE 12.3
Predicted Values and
Residuals for the Airline
Cost Example

NUMBER OF PASSENGERS X	COST ($1,000) Y	PREDICTED VALUE $\hat{Y}$	RESIDUAL $Y - \hat{Y}$
61	4.280	4.053	.227
63	4.080	4.134	−.054
67	4.420	4.297	.123
69	4.170	4.378	−.208
70	4.480	4.419	.061
74	4.300	4.582	−.282
76	4.820	4.663	.157
81	4.700	4.867	−.167
86	5.110	5.070	.040
91	5.130	5.274	−.144
95	5.640	5.436	.204
97	5.560	5.518	.042
			$\Sigma(Y - \hat{Y}) = -.001$

Table 12.3 shows $\hat{Y}$ values and the residuals for each pair of data for the airline cost regression model developed in Section 12.2. The predicted values are calculated by inserting an X value into the equation of the regression line and solving for $\hat{Y}$. For example, when $X = 61$, $\hat{Y} = 1.57 + .0407(61) = 4.053$, as displayed in column 3 of the table. Each of these predicted Y values is subtracted from the actual Y value to determine the error, or residual. For example, the first Y value listed in the table is 4.280 and the first predicted value is 4.053, resulting in a residual of $4.280 - 4.053 = .227$. The residuals for this problem are given in column 4 of the table.

Note that the sum of the residuals is approximately zero. Except for rounding error, the sum of the residuals *always* is *zero*. The reason is that a residual is geometrically the vertical distance from the regression line to a data point. The equations used to solve for the slope and intercept place the line geometrically in the middle of all points. That is, *vertical* distances from the line to the points will cancel each other and sum to zero. Figure 12.7 is an Excel-produced scatter plot of the data and the residuals for the airline cost example.

An examination of the residuals may give the business analyst an idea of how well the regression line fits the historical data points. The largest residual for the airline cost example is −.282, and the smallest is .040. Since the objective of the regression analysis was to predict the cost of flight in $1000s, the regression line produces an error of $282 when there are 74 passengers and an error of only $40 when there are 86 passengers. This result presents the *best* and *worst* cases for the residuals. The business analyst must examine other residuals to determine how well the regression model fits other data points.

Outliers

Data points that lie apart
from the rest of the points.

Sometimes residuals are used to locate outliers. **Outliers** are *data points that lie apart from the rest of the points.* Outliers can produce residuals with large magnitudes and are usually easy to identify on scatter plots. Outliers can be the result of misrecorded or miscoded data, or they may simply be data points that do not conform to the general trend. The equation of the regression line is influenced by every data point used in its calculation in a manner similar to the arithmetic mean. Therefore outliers sometimes can unduly influence the regression line by "pulling" the line toward the outliers. The origin of outliers must be investigated to determine whether they should be retained or whether the regression equation should be recomputed without them.

Residuals are usually plotted against the X axis, which reveals a view of the residuals as X increases. Figure 12.8 shows the residuals plotted by Excel against the X axis for the airline cost example.

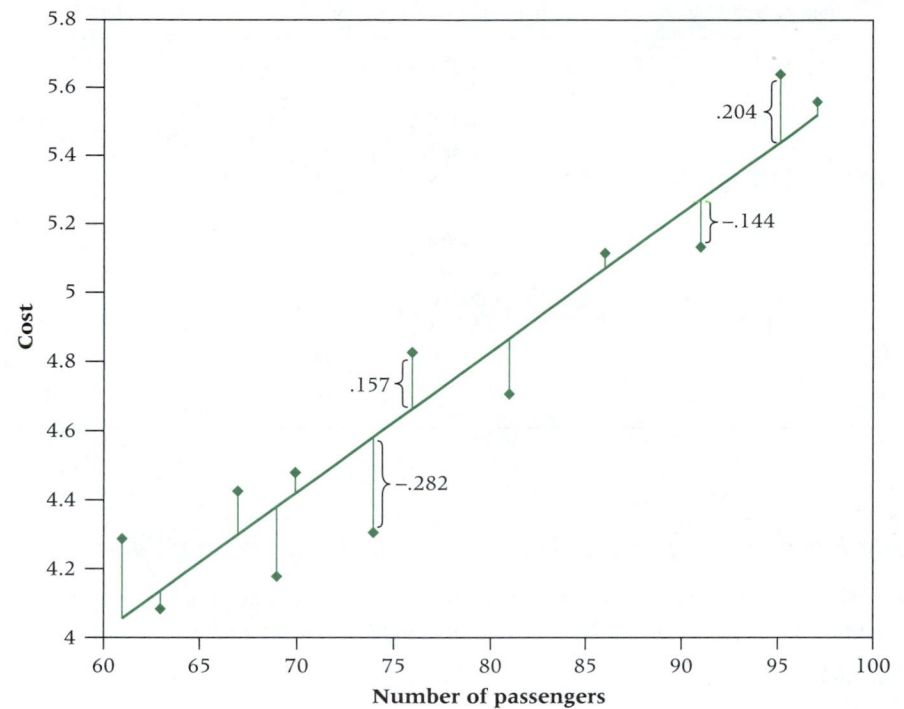

Figure 12.7

Close-up Excel scatter plot with residuals for the airline cost example

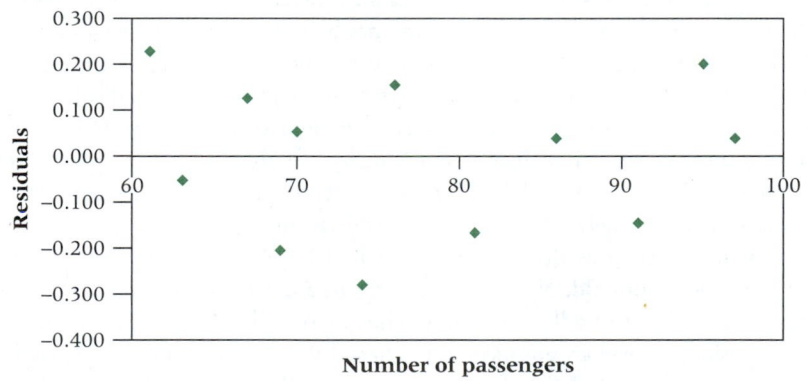

Figure 12.8

Excel graph of residuals for airline cost problem

Using Excel to Compute and Display Residuals

Excel allows you to compute residuals as an option with the **Regression** tool. Note in the Excel dialog box displayed in Figure 12.5 there is a section of **Residuals** optional outputs that can be checked. By selecting **Residuals,** you get a printout of all the residuals as shown in Figure 12.9 below; and by selecting **Residual Plots,** you get a plot of the residuals against the X axis like that shown in Figure 12.8.

Figure 12.9

Excel-produced
residuals for the
airline cost problem

OBSERVATION	PREDICTED COST ($1,000)	RESIDUALS
1	4.051	0.229
2	4.133	−0.053
3	4.296	0.124
4	4.377	−0.207
5	4.418	0.052
6	4.581	−0.281
7	4.662	0.158
8	4.866	−0.166
9	5.070	0.040
10	5.273	−0.143
11	5.436	0.204
12	5.518	0.042

Using Residuals to Test the Assumptions of the Regression Model

One of the major uses of residual analysis is to test some of the assumptions underlying regression. The following are the assumptions of simple regression analysis.

1. The model is linear.
2. The error terms have constant variances.
3. The error terms are independent.
4. The error terms are normally distributed.

Residual plot

A type of graph in which the residuals for a particular regression model are plotted along with their associated values of X.

A particular method for studying the behavior of residuals is the **residual plot.** The residual plot is *a type of graph in which the residuals for a particular regression model are plotted along with their associated value of X* as an ordered pair $(X, Y - \hat{Y})$. Information about how well the regression assumptions are met by the particular regression model can be gleaned by examining the plots. Residual plots are more meaningful with larger sample sizes. For small sample sizes, residual plot analyses can be problematic and subject to over-interpretation. Hence, because the airline cost example is constructed from only 12 pairs of data, one should be cautious in reaching conclusions from Figure 12.8. The residual plots in Figures 12.10, 12.11, and 12.12, however, represent large numbers of data points and therefore are more likely to depict overall trends accurately.

If a residual plot such as the one in Figure 12.10 appears, the assumption that the model is linear does not hold. Note that the residuals are negative for low and high values of X and are positive for middle values of X. The graph of these residuals is parabolic, not linear. The residual plot does not have to be shaped like this for a nonlinear relationship to exist. Any significant deviation from an approximately linear residual plot may mean that a nonlinear relationship exists between the two variables.

Homoscedasticity

The condition that occurs when the error variances produced by a regression model are constant.

The assumption of constant error variance sometimes is called **homoscedasticity.** If *the error variances are not constant* (called **heteroscedasticity**), the residual plots might look like one of the two plots in Figure 12.11. Note in Figure 12.11(a) that the error variance is greater for small values of X and smaller for large values of X. The situation is reversed in Figure 12.11(b).

Heteroscedasticity

The condition that occurs when the error variances produced by a regression model are not constant.

If the error terms are not independent, the residual plots could look like one of the graphs in Figure 12.12. According to these graphs, instead of each error term being independent of the one next to it, the value of the residual is a function of the residual value next to it. For example, a large positive residual is next to a large positive residual and a small negative residual is next to a small negative residual.

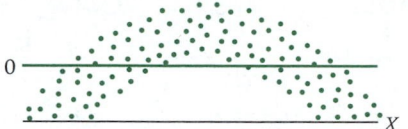

Figure 12.10

Nonlinear residual plot

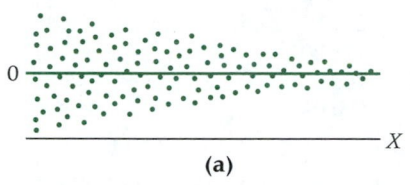
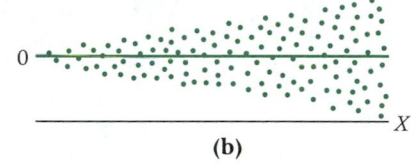

(a) (b)

Figure 12.11

Nonconstant error variance

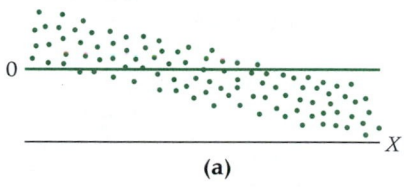
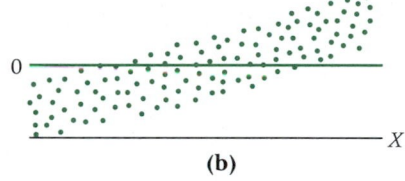

(a) (b)

Figure 12.12

Graphs of nonindependent error terms

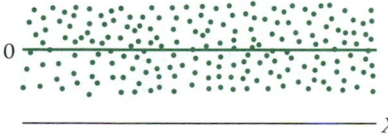

Figure 12.13

Healthy residual graph

The graph of the residuals from a regression analysis that meets the assumptions—a *healthy residual graph*—might look like the graph in Figure 12.13. The plot is relatively linear; the variance of the errors are about equal for each value of X, and the error terms do not appear to be related to adjacent terms.

Compute the residuals for Demonstration Problem 12.1 in which a regression model was developed to predict the number of full-time equivalent workers (FTEs) by the number of beds in a hospital. Analyze the residuals.

DEMONSTRATION PROBLEM 12.2

SOLUTION
The data and computed residuals are shown in the following table.

HOSPITAL	NUMBER OF BEDS X	FTES Y	PREDICTED VALUE $\hat{Y}$	RESIDUALS $Y - \hat{Y}$
1	23	69	82.22	−13.22
2	29	95	95.62	−.62
3	29	102	95.62	6.38
4	35	118	109.01	8.99
5	42	126	124.63	1.37
6	46	125	133.56	−8.56
7	50	138	142.49	−4.49
8	54	178	151.42	26.58
9	64	156	173.74	−17.74
10	66	184	178.20	5.80
11	76	176	200.52	−24.52
12	78	225	204.98	20.02

$$\Sigma(Y - \hat{Y}) = -.01$$

Note that the regression model fits these particular data well for hospitals 2 and 5, as indicated by residuals of −.62 and 1.37 FTEs, respectively. For hospitals 1, 8, 9, 11, and 12, the residuals are relatively large, indicating that the regression model does not fit the data for these hospitals well.

Shown below is the Excel-produced residual plot for this problem. An examination of the graph shows that there appears to be greater variability among the residuals towards the higher values of X indicating the potential for heteroscedasticity.

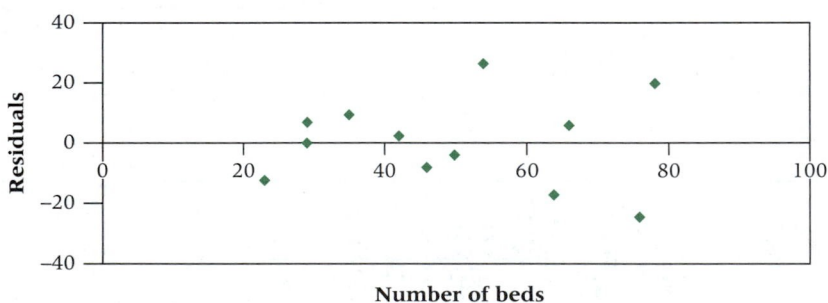

Standard Error of the Estimate

Residuals represent errors of estimation for individual points. With large samples of data, residual computations become laborious. Even with computers, a business analyst sometimes has difficulty working through pages of residuals in an effort to understand the error of the regression model. An alternative way of examining the error of the model is the standard error of the estimate, which provides a single measurement of the regression error.

Because the sum of the residuals is zero, attempting to determine the total amount of error by summing the residuals is fruitless. This zero-sum characteristic of residuals can be avoided by squaring the residuals and then summing them.

Table 12.4 contains the airline cost data from Table 12.1, along with the residuals and the residuals squared. The *total of the residuals squared* column is called the **sum of squares of error (SSE).**

Sum of squares of error (SSE)

The sum of the residuals squared for a regression model.

NUMBER OF PASSENGERS	COST ($1000)	RESIDUAL	
X	Y	$Y - \hat{Y}$	$(Y - \hat{Y})^2$
61	4.280	.227	.05153
63	4.080	−.054	.00292
67	4.420	.123	.01513
69	4.170	−.208	.04326
70	4.480	.061	.00372
74	4.300	−.282	.07952
76	4.820	.157	.02465
81	4.700	−.167	.02789
86	5.110	.040	.00160
91	5.130	−.144	.02074
95	5.640	.204	.04162
97	5.560	.042	.00176
		$\Sigma(Y - \hat{Y}) = -.001$	$\Sigma(Y - \hat{Y})^2 = .31434$

Sum of squares of error = SSE = .31434

TABLE 12.4
Determining SSE for the Airline Cost Example

$$SSE = \Sigma(Y - \hat{Y})^2$$

SUM OF SQUARES OF ERROR

In theory, infinitely many lines can be fit to a sample of points. However, equations 12.1 and 12.3 produce a line of *best fit* for which the SSE is the smallest for any line that can be fit to the sample data. This result is guaranteed, because equations 12.1 and 12.3 are derived from calculus to minimize SSE. For this reason, the regression process used in this chapter is called *least squares* regression.

There is a computational version of the equation for computing SSE. This equation is less meaningful in terms of interpretation than $\Sigma(Y - \hat{Y})^2$, but it is usually easier to compute. The computational formula for SSE follows.

$$SSE = \Sigma Y^2 - b_0 \Sigma Y - b_1 \Sigma XY$$

COMPUTATIONAL FORMULA FOR SSE

For the airline cost example,

$$\Sigma Y^2 = \Sigma[(4.280)^2 + (4.080)^2 + (4.420)^2 + (4.170)^2 + (4.480)^2 + (4.300)^2 + (4.820)^2$$
$$+ (4.700)^2 + (5.110)^2 + (5.130)^2 + (5.640)^2 + (5.560)^2 = 270.9251$$
$b_0 = 1.5697928$
$b_1 = .0407016*$
$\Sigma Y = 56.69$
$\Sigma XY = 4462.22$
$SSE = \Sigma Y^2 - b_0 \Sigma Y - b_1 \Sigma XY$
$\quad = 270.9251 - (1.5697928)(56.69) - (.0407016)(4462.22) = .31405$

The slight discrepancy between this value and the value computed in Table 12.4 is due to rounding error.

---

*Note: In previous sections, the values of the slope and intercept were rounded off for ease of computation and interpretation. They are shown here with more precision in an effort to reduce rounding error.

Standard error of the estimate (S_e)
A standard deviation of the error of a regression model.

The sum of squares error is in part a function of the number of pairs of data being used to compute the sum, which lessens the value of SSE as a measurement of error. A more useful measurement of error is the standard error of the estimate. The **standard error of the estimate,** denoted S_e, is *a standard deviation of the error of the regression model* and has a more practical use than SSE. The standard error of the estimate follows.

STANDARD ERROR OF THE ESTIMATE	$$S_e = \sqrt{\frac{SSE}{n-2}}$$

The standard error of the estimate for the airline cost example is

$$S_e = \sqrt{\frac{SSE}{n-2}} = \sqrt{\frac{.31434}{10}} = .1773.$$

How is the standard error of the estimate used? As previously mentioned, the standard error of the estimate is a standard deviation of error. Recall from Chapter 3 that if data are approximately normally distributed, the empirical rule states that about 68% of all values are within $\mu \pm 1\sigma$ and that about 95% of all values are within $\mu \pm 2\sigma$. One of the assumptions for regression states that for a given X the error terms are normally distributed. Because the error terms are normally distributed, S_e is the standard deviation of error, and the average error is zero, approximately 68% of the error values (residuals) should be within $0 \pm 1S_e$ and 95% of the error values (residuals) should be within $0 \pm 2S_e$. By having knowledge of the variables being studied and by examining the value of S_e, the business analyst can often make a judgment about the fit of the regression model to the data by using S_e. How can the S_e value for the airline cost example be interpreted?

The regression model in that example is used to predict airline cost by number of passengers. Note that the range of the airline cost data in Table 12.1 is from 4.08 to 5.64 ($4080 to $5640). The regression model for the data yields an S_e of .1773. An interpretation of S_e is that the standard deviation of error for the airline cost example is $177.30. If the error terms were normally distributed about the given values of X, approximately 68% of the error terms would be within $\pm\$177.30$ and 95% would be within $\pm 2(\$177.30) = \pm\$354.60$. Examination of the residuals reveals that 100% of the residuals are within $2S_e$. The standard error of the estimate provides a single measure of error, which, if the business analyst has enough background in the area being analyzed, can be used to understand the magnitude of errors in the model. In addition, some business analysts use the standard error of the estimate to identify outliers. They do so by looking for data that are outside $\pm 2S_e$ or $\pm 3S_e$.

DEMONSTRATION PROBLEM 12.3

Compute the sum of squares of error and the standard error of the estimate for Demonstration Problem 12.1, in which a regression model was developed to predict the number of FTEs at a hospital by the number of beds.

SOLUTION

HOSPITAL	NUMBER OF BEDS X	FTES Y	RESIDUALS $Y - \hat{Y}$	$(Y - \hat{Y})^2$
1	23	69	−13.22	174.77
2	29	95	−.62	−0.38
3	29	102	6.38	40.70
4	35	118	8.99	80.82

5	42	126	1.37	1.88
6	46	125	-8.56	73.27
7	50	138	-4.49	20.16
8	54	178	26.58	706.50
9	64	156	-17.74	314.71
10	66	184	5.80	33.64
11	76	176	-24.52	601.23
12	78	225	20.02	400.80
	$\Sigma X = 592$	$\Sigma Y = 1692$	$\Sigma(Y - \hat{Y}) = -.01$	$\Sigma(Y - \hat{Y})^2 = 2448.86$

$$\text{SSE} = 2448.86$$

$$S_e = \sqrt{\frac{\text{SSE}}{n - 2}} = \sqrt{\frac{2448.86}{10}} = 15.65$$

The standard error of the estimate is 15.65 FTEs. An examination of the residuals for this problem reveals that eight of 12 (67%) are within $\pm 1 S_e$ and 100% are within $\pm 2 S_e$. Is this an acceptable size of error? Hospital administrators probably can best answer that question.

Coefficient of Determination

A widely used measure of fit for regression models is the **coefficient of determination,** or r^2. The coefficient of determination is *the proportion of variability of the dependent variable (Y) accounted for or explained by the independent variable (X).*

The coefficient of determination ranges from 0 to 1. An r^2 of zero means that the predictor accounts for none of the variability of the dependent variable and that there is no regression prediction of Y by X. An r^2 of 1 means that there is perfect prediction of Y by X and that 100% of the variability of Y is accounted for by X. Of course, most r^2 values are between the extremes. The business analyst must interpret whether a particular r^2 is high or low, depending on the use of the model and the context within which the model was developed.

In exploratory research where the variables are less understood, low values of r^2 are likely to be more acceptable than they are in areas of research where the parameters are more developed and understood. One NASA analyst who uses vehicular weight to predict mission cost searches for regression models that have an r^2 of .90 or higher. However, a business analyst who is trying to develop a model to predict the motivation level of employees might be pleased to get an r^2 near .50 in the initial research.

The dependent variable, Y, being predicted in a regression model has a variation that is measured by the sum of squares of $Y(\text{SS}_{YY})$,

$$\text{SS}_{YY} = \Sigma(Y - \bar{Y})^2 = \Sigma Y^2 - \frac{(\Sigma Y)^2}{n},$$

and is the sum of the squared deviations of the Y values from the mean value of Y. This variation can be broken into two additive variations: the *explained variation,* measured by the sum of squares of regression (SSR), and the *unexplained variation,* measured by the sum of squares of error (SSE). This relationship can be expressed in equation form as

$$\text{SS}_{YY} = \text{SSR} + \text{SSE}.$$

If each term in the equation is divided by SS_{YY}, the resulting equation is

$$1 = \frac{\text{SSR}}{\text{SS}_{YY}} + \frac{\text{SSE}}{\text{SS}_{YY}}$$

Coefficient of determination (r^2) The proportion of variability of the dependent variable accounted for or explained by the independent variable in a regression model.

The term r^2 is the proportion of the Y variability that is explained by the regression model and represented here as

$$r^2 = \frac{SSR}{SS_{YY}}.$$

Substituting this into the preceding relationship gives

$$1 = r^2 + \frac{SSE}{SS_{YY}}.$$

Solving for r^2 yields Equation 12.4.

COEFFICIENT OF DETERMINATION	$r^2 = 1 - \dfrac{SSE}{SS_{YY}} = 1 - \dfrac{SSE}{\Sigma Y^2 - \dfrac{(\Sigma Y)^2}{n}}$	(12.4)

Note: $0 \leq r^2 \leq 1$.

The value of r^2 for the airline cost example is solved as follows.

$$SSE = .31434$$

$$SS_{YY} = \Sigma Y^2 - \frac{(\Sigma Y)^2}{n} = 270.9251 - \frac{(56.69)^2}{12} = 3.11209$$

$$r^2 = 1 - \frac{SSE}{SS_{YY}} = 1 - \frac{.31434}{3.11209} = .899$$

That is, 89.9% of the variability of the cost of flying a Boeing 737 airplane on a commercial flight is accounted for or predicted by the number of passengers. This result also means that 11.1% of the variance in airline flight cost, Y, is unaccounted for by X or unexplained by the regression model.

DEMONSTRATION PROBLEM 12.4

Compute the coefficient of determination (r^2) for Demonstration Problem 12.1, in which a regression model was developed to predict the number of FTEs of a hospital by the number of beds.

SOLUTION

$$SSE = 2448.6$$

$$SS_{YY} = 260,136 - \frac{1692^2}{12} = 21,564$$

$$r^2 = 1 - \frac{SSE}{SS_{YY}} = 1 - \frac{2448.6}{21,564} = .886$$

This regression model accounts for 88.6% of the variance in FTEs, leaving only 11.4% unexplained variance.

Shown below is the Excel output for this problem including the values of S_e and r^2.

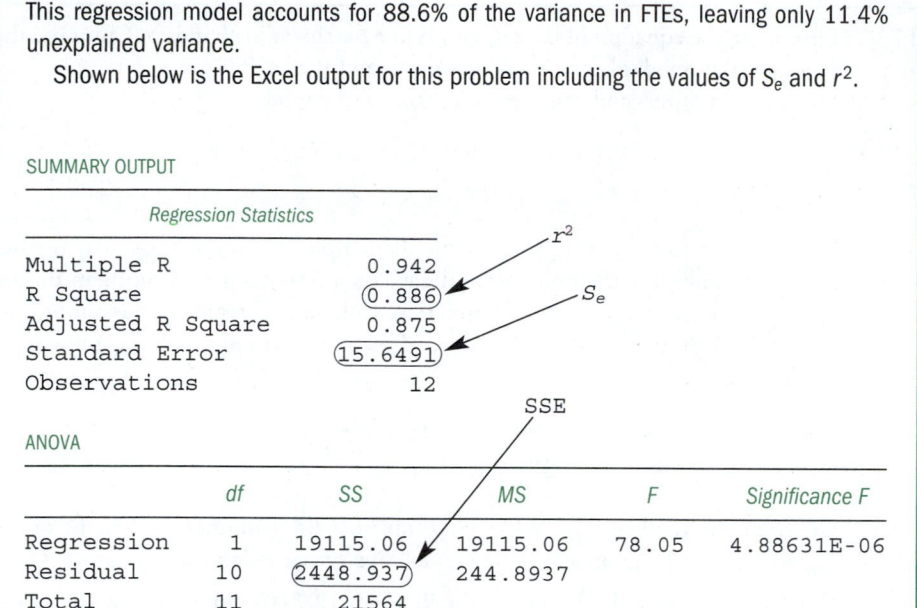

Using Excel to Solve for S_e and r^2

Excel's **Regression** tool includes the Standard Error of the Estimate and r^2 in its output. Figure 12.14 displays a portion of the Excel-produced regression output for the airline cost problem. Note the standard error of the estimate, .1769, and the value of the coefficient of determination, $r^2 = .900$ (differences between these values and those calculated before are due to rounding error). In addition, the sum of squares of error, used in the calculation of each of these, is shown in the row beside the word *Residual* under SS (0.31293).

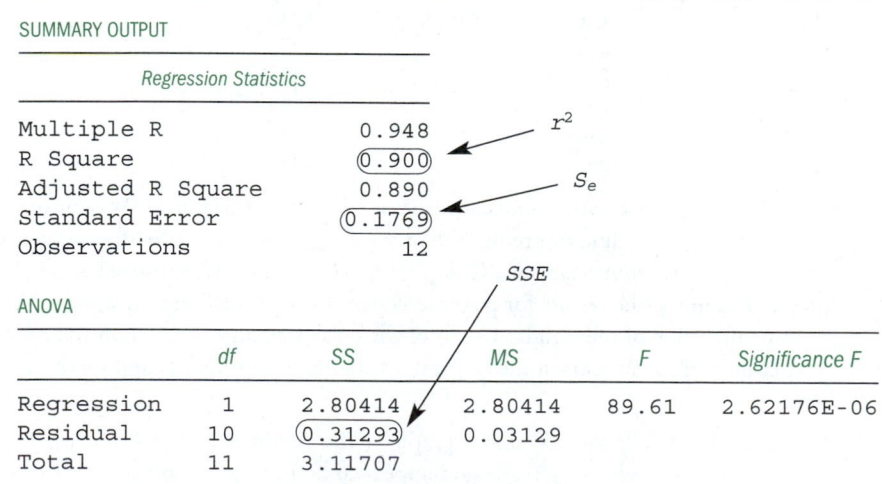

Figure 12.14

Excel Regression output for the airline cost problem including S_e and r^2

12.3

Problems

12.10 Determine the equation of the regression line for the following data. Compute the residuals, the standard error of the estimate, and the coefficient of determination. Study these outputs and comment on the fit of the model.

X	15	8	19	12	5
Y	47	36	56	44	21

12.11 Consider the following data. Develop the simple regression model to predict Y from X. Compute the residuals and develop a residual plot. Comment on the size of the residuals and on any apparent violation of regression assumptions. Compute S_e and r^2 and use these to discuss the strength of the regression model.

X	5	7	11	12	19	25
Y	47	38	32	24	22	10

12.12 Compute the residuals, S_e, and r^2 for Problem 12.5. Comment on the fit of the regression model and on any apparent violation of regression assumptions.

12.13 Compute the residuals, S_e, and r^2 for Problem 12.7. Comment on the fit of the regression model and on any apparent violation of regression assumptions.

12.14 Compute the residuals, S_e, and r^2 for Problem 12.8. Comment on the fit of the regression model and on any apparent violation of regression assumptions.

12.15 Compute the residuals, S_e, and r^2 for Problem 12.9. Comment on the fit of the regression model and on any apparent violation of regression assumptions.

12.16 Wisconsin is an important milk-producing state. Some people might argue that because of transportation costs, the cost of milk increases with the distance of markets from Wisconsin. Suppose the milk prices in eight cities are as follows.

COST OF MILK (PER GALLON)	DISTANCE FROM MADISON (MILES)
$2.64	1245
2.31	425
2.45	1346
2.52	973
2.19	255
2.55	865
2.40	1080
2.37	296

Use the prices along with the distance of each city from Madison, Wisconsin, to develop a regression line to predict the price of a gallon of milk by the number of miles the city is from Madison. Compute the residuals and construct a residual plot and examine the results for possible violations of regression assumptions. Determine the value of the standard error of the estimate and r^2. Based on the results of this analysis, comment on the strength of the regression model and its fit to the data.

12.17 For each of the following, sketch a graph of the residuals and indicate which of the assumptions underlying regression appear to be in jeopardy on the basis of the graph.

X	Y – Ŷ
213	−11
216	−5
227	−2
229	−1
237	+6
247	+10
263	+12

X	Y – Ŷ	X	Y – Ŷ
5	−21	13	−7
6	+16	14	+5
8	+14	17	−2
9	−11	18	+1
12	−8		

X	Y – Ŷ	X	Y – Ŷ
10	+6	14	−3
11	+3	15	+2
12	−1	16	+5
13	−11	17	+8

12.18 Determine the equation of the regression line to predict annual sales of a company from the yearly stock market volume of shares sold in a recent year. Compute the standard error of the estimate and r^2 for this model. Does volume of shares sold appear to be a good predictor of a company's sales? Why or why not?

COMPANY	ANNUAL SALES (BILLIONS)	ANNUAL VOLUME (MILLION SHARES)
Merck	10.5	728.6
Philip Morris	48.1	497.9
IBM	64.8	439.1
Eastman Kodak	20.1	377.9
Bristol-Myers Squibb	11.4	375.5
General Motors	123.8	363.8
Ford Motors	89.0	276.3

12.19 The Conference Board produces a Consumer Confidence Index (CCI) that reflects people's feelings about general business conditions, employment opportunities, and their own income prospects. Some business analysts may feel that consumer confidence is a function of the median household income. Shown here are the CCIs for 9 years and the median household incomes for the same 9 years published by the U.S. Bureau of the Census. Determine the equation of the regression line to predict the CCI from the median household income. Compute the standard error of the estimate for this model. Compute the value of r^2. Does median household income appear to be a good predictor of the CCI? Why or why not?

CCI	MEDIAN HOUSEHOLD INCOME ($1000)
116.8	37.415
91.5	36.770
68.5	35.501
61.6	35.047
65.9	34.700
90.6	34.942
100.0	35.887
104.6	36.306
125.4	37.005

12.4
Hypothesis Tests for the Slope of the Regression Model and Testing the Overall Model

Testing the Slope

Another way to determine how well a regression model fits the data (besides using residual analysis, the standard error of the estimate, and the coefficient of determination) is by conducting a hypothesis test using the sample slope of the regression model to see if the population slope is significantly different from zero.

Suppose a business analyst decided that it is not worth the effort to develop a linear regression model to predict Y from X. An alternative approach might be to average the Y values and use $\overline{Y}$ as the predictor of Y for all values of X. For the airline cost example, instead of using population as a predictor, the business analyst would use the average value of airline cost, $\overline{Y}$, for the sample as the predictor. In this case the average value of Y is

$$\overline{Y} = \frac{56.69}{12} = 4.7242, \text{ or } \$4724.20.$$

Using this result as a model to predict Y, if the number of passengers is 61, 70, or 95— or any other number—the predicted value of Y is still 4.7242. Essentially, this approach fits the line of $\overline{Y} = 4.7242$ through the data, which is a horizontal line with a slope of zero. Would a regression analysis offer anything more than the $\overline{Y}$ model? Using this non-regression model (the $\overline{Y}$ model) as a worst case, the business analyst can analyze the regression line to determine whether it adds a more significant amount of predictability of Y than does the $\overline{Y}$ model. Because the slope of the $\overline{Y}$ line is zero, one way to determine whether the regression line adds significant predictability is to test the *population* slope of the regression line to find out if the slope is different from zero. As the slope of the regression line diverges from zero, the regression model is adding predictability that the $\overline{Y}$ line is not generating. For this reason, testing the slope of the regression line to determine whether the slope is different from zero is important. If the slope is not different from zero, the regression line is doing nothing more than the $\overline{Y}$ line in predicting Y.

How does the business analyst go about testing the slope of the regression line? Why not just examine the calculated slope? For example, the slope of the regression line for the airline cost data is .0407. This value obviously is not zero. The problem is that this slope is a *sample* slope obtained from a sample of 12 data points. If another sample of airline cost and number of passengers were used, a different slope likely would be obtained. Thus the sample slope is a function of the particular sample from which it is obtained. What has to be tested here is the *population* slope. If all the pairs of data points for the population were available, would the slope of that regression line be different from zero? Here the sample slope, b_1, is used as evidence to test whether the population slope is different from zero. The hypotheses for this test follow.

$$H_0: \beta_1 = 0$$
$$H_a: \beta_1 \neq 0$$

Note that this test is two-tailed. The null hypothesis can be rejected if the slope is either negative or positive. A negative slope indicates an inverse relationship between X and Y. That is, larger values of X are related to smaller values of Y and vice versa. Both negative and positive slopes can be different from zero. To determine whether there is a significant positive relationship between two variables, the hypotheses would be one-tailed, or

$$H_0: \beta_1 = 0$$
$$H_a: \beta_1 > 0.$$

To test for a significant negative relationship between two variables, the hypotheses also would be one-tailed, or

$$H_0: \beta_1 = 0$$
$$H_a: \beta_1 < 0.$$

In each case, testing the null hypothesis involves a t test of the slope.

$$t = \frac{b_1 - \beta_1}{S_b}$$

t TEST OF SLOPE

where:

$$S_b = \frac{S_e}{\sqrt{SS_{XX}}}$$

$$S_e = \sqrt{\frac{SSE}{n-2}}$$

$$SS_{XX} = \Sigma X^2 - \frac{(\Sigma X)^2}{n}$$

β_1 = the hypothesized shape

df = $n - 2$

The test of the slope of the regression line for the airline cost regression model using $\alpha = .05$ follows. The regression line derived for the data is

$$\hat{Y} = 1.57 + .0407X.$$

The sample slope is $.0407 = b_1$. The value of S_e is .1773, $\Sigma X = 930$, $\Sigma X^2 = 73,764$, and $n = 12$. The hypotheses are

$$H_0: \beta_1 = 0$$
$$H_a: \beta_1 \neq 0.$$

The df $= n - 2 = 12 - 2 = 10$. As this test is two-tailed, $\alpha/2 = .025$. The table t value is $t_{.025,10} = \pm 2.228$. The calculated t value for this sample slope is

$$t = \frac{.0407 - 0}{\dfrac{.1773}{\sqrt{73,764 - \dfrac{(930)^2}{12}}}} = 9.43$$

As shown in Figure 12.15, the t value calculated from the sample slope is in the rejection region and the observed $t = 9.43$ is greater than the critical $t = 2.228$. The null hypothesis that the population slope is zero is rejected. This linear regression model is adding significantly more predictive information to the $\overline{Y}$ model (no regression).

It is desirable to reject the null hypothesis in testing the slope of the regression model. In rejecting the null hypothesis of a zero population slope, we are stating that the regression

Figure 12.15

t Test of slope from airline cost example

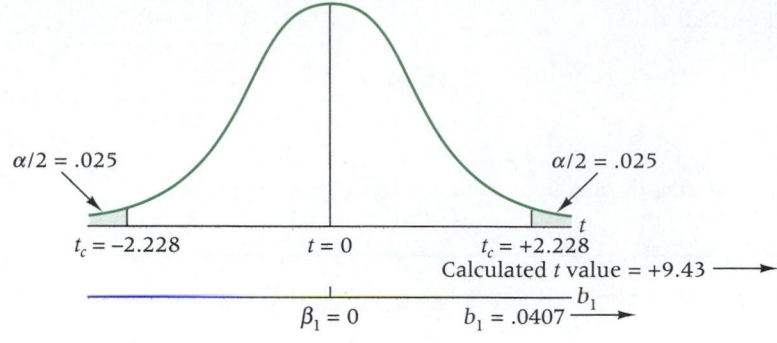

Figure 12.16

Excel Regression output for the airline cost problem including a test of the slope

	COEFFICIENTS	STANDARD ERROR	*t* STAT	*P*-VALUE
Intercept	1.5655	0.3375	4.64	0.00092
Number of Beds	0.0407	0.0043	9.47	.0000026

Observed *t* value

p-value of observed *t*

model is adding something to the explanation of the variation of the dependent variable that the average value of Y model does not. Failure to reject the null hypothesis in this test causes the business analyst to conclude that the regression model has no predictability of the dependent variable, and the model therefore has little or no use.

Using Excel to Test the Slope of the Regression Model

Excel's **Regression** tool includes a test for the slope of the regression model in its standard output. Figure 12.16 shows a portion of the Excel-produced regression output for the airline cost problem. Observe the value of *t* Stat in the *Number of Beds* row. This value, $t = 9.47$, is the observed (calculated) value of *t* for the test of the slope, which is essentially the same as the $t = 9.43$ calculated in the section above (with slight differences due to rounding error). The associated *p*-value is .0000026. Since this is less than $\alpha/2 = .025$ for the two-tailed test of the slope being conducted, the decision is to reject the null hypothesis and conclude that the population slope is significantly different from zero. Most business analysts are not interested in conducting *t* tests about the Y intercept, and we will not discuss such tests in this text.

<table>
<tr><td rowspan="2">**DEMONSTRATION PROBLEM 12.5**</td><td rowspan="2">Test the slope of the regression model developed in Demonstration Problem 12.1 to predict the number of FTEs in a hospital from the number of beds to determine whether there is a significant positive slope. Use $\alpha = .01$.

SOLUTION
The hypotheses for this problem are

$$H_0: \beta_1 = 0$$
$$H_a: \beta_1 > 0.$$</td></tr>
</table>

BEDS	FTEs
23	69
29	95
29	102
35	118
42	126
46	125
50	138
54	178
64	156
66	184
76	176
78	225

The level of significance is .01. There are 12 pairs of data, so df = 10. The critical table t value is $t_{.01,10} = 2.764$. The regression line equation for this problem is

$$\hat{Y} = 30.888 + 2.232X.$$

The sample slope, b_1, is 2.232, and $S_e = 15.65$, $\Sigma X = 592$, $\Sigma X^2 = 33{,}044$, and $n = 12$. The calculated t value for the sample slope is

$$t = \frac{2.232 - 0}{\dfrac{15.65}{\sqrt{33{,}044 - \dfrac{(592)^2}{12}}}} = 8.84.$$

The calculated t value (8.84) is in the rejection region because it is greater than the critical table t value of 2.764. The null hypothesis is rejected. The population slope for this regression line is significantly different from zero in the positive direction. This regression model is adding significant predictability over the $\bar{Y}$ model.

Shown below is the portion of the Excel regression output for this problem containing the t test of the slope. Note that the t Stat for number of beds (test of the slope) is essentially the same as the value computed above. The p-value of .00000489 is less than $\alpha = .01$ supporting the decision to reject the null hypothesis of the slope.

	COEFFICIENTS	STANDARD ERROR	t STAT	P-VALUE
Intercept	30.912	13.2542	2.33	0.041888
Number of Beds	2.232	0.2526	8.83	.00000489

Testing the Overall Model

It is common in regression analysis to compute an F test to determine the overall significance of the model. Most computer software packages include the F text and its associated ANOVA table as standard regression output. In multiple regression (Chapter 13), this is a test to determine whether at least one of the regression coefficients (from multiple predictors) is different from zero. In simple regression, there is only one predictor and only one regression coefficient to test. Because the regression coefficient is the slope of the regression line, the F test for overall significance is testing the same thing as the t test in simple regression. The hypotheses being tested in simple regression by the F test for overall significance are

$$H_0: \beta_1 = 0$$
$$H_a: \beta_1 \neq 0.$$

In the case of simple regression analysis, $F = t^2$. Thus, for the airline cost example, the F value is

$$F = t^2 = (9.43)^2 = 88.92$$

The F value is computed directly by

$$F = \frac{\dfrac{SS_{reg}}{df_{reg}}}{\dfrac{SS_{err}}{df_{err}}} = \frac{MS_{reg}}{MS_{err}},$$

where:

$df_{reg} = k,$

$df_{err} = n - k - 1,$ and

k = the number of independent variables.

The values of the sum of squares (SS), degrees of freedom (df), and mean squares (MS) are obtained from the analysis of variance table, which is produced with other regression statistics as standard output from statistical software packages. Shown here is the analysis of variance table produced by Excel for the airline cost example.

ANOVA

	df	SS	MS	F	Significance F
Regression	1	2.80414	2.80414	89.61	.00000262
Residual	10	0.31293	0.03129		
Total	11	3.11707			

The F value for the airline cost example is calculated from this ANOVA table as:

$$F = \frac{\dfrac{2.80414}{1}}{\dfrac{0.31293}{10}} = \frac{2.80414}{0.03129} = 89.61$$

The difference between this value (89.61) and the value obtained by squaring the t statistic (88.92) is due to rounding error. The probability of obtaining an F value this large or larger by chance if there is no regression prediction in this model is .00000262 according to the ANOVA output (the p value). This means it is highly unlikely that the population slope is zero and it is highly unlikely that there is no prediction due to regression from this model given the sample statistics obtained. Hence, it is highly likely that this regression model adds significant predictability of the dependent variable.

Note from the ANOVA table that the degrees of freedom due to regression are equal to 1. Simple regression models have only one independent variable; therefore, $k = 1$. The degrees of freedom error in simple regression analysis is always $n - k - 1 = n - 1 - 1 = n - 2$. With the degrees of freedom due to regression (1) as the numerator degrees of freedom and the degrees of freedom due to error ($n - 2$) as the denominator degrees of freedom, Table A.7 can be used to obtain the critical F value ($F_{\alpha,1,n-2}$) to help make the hypothesis-testing decision about the overall regression model if the p value of F is not given in the computer output. This critical F value is always found in the right tail of the distribution. In simple regression, the relationship between the critical t value to test the slope and the critical F value of overall significance is

$$t_{\alpha/2,n-2}^2 = F_{\alpha,1,n-2}$$

For the airline cost example with a two-tailed test and $\alpha = .05$, the critical value of $t_{.025,10}$ is ± 2.228 and the critical value of $F_{.05,1,10}$ is 4.96.

$$t_{.025,10}^2 = (\pm 2.228)^2 = 4.96 = F_{.05,1,10}$$

12.4

Problems

12.20   Test the slope of the regression line developed in Problem 12.5. Let $\alpha = .05$.

ADVERTISING	SALES
12.5	148
3.7	55
21.6	338
60.0	994
37.6	541
6.1	89
16.8	126
41.2	379

12.21   Test the slope of the regression line developed in Problem 12.7. Let $\alpha = .01$.

STARTS	FAILURES
233,710	57,097
199,091	50,361
181,645	60,747
158,930	88,140
155,672	97,069
164,086	86,133
166,154	71,558
188,387	71,128
168,158	71,931
170,475	83,384

12.22   Test the slope of the regression line developed in Problem 12.8. Let $\alpha = .10$.

FARMS	SIZE
5.65	213
4.65	258
3.96	297
3.36	340
2.95	374
2.52	420
2.44	426
2.29	441
2.15	460
2.07	469

12.23   Test the slope of the regression line developed in Problem 12.9. Let $\alpha = .05$.

PRODUCTION	NEW ORDERS
99.9	2.74
97.9	2.87
98.9	2.93
87.9	2.87
92.9	2.98
97.9	3.09
100.6	3.36
104.9	3.61
105.3	3.75
108.6	3.95

12.24   Test the slope of the regression line developed in Problem 12.16. Let $\alpha = .05$.

COST ($)	DISTANCE
2.64	1245
2.31	425
2.45	1346
2.52	973
2.19	255
2.55	865
2.40	1080
2.37	296

12.25 Test the slope of the regression line developed in Problem 12.18. Let $\alpha = .01$.

SALES	VOLUME
10.5	728.6
48.1	497.9
64.8	439.1
20.1	377.9
11.4	375.5
123.8	363.8
89.0	276.3

12.26 Test the slope of the regression line developed in Problem 12.19. Let $\alpha = .05$.

CCI	INCOME
116.8	37.415
91.5	36.770
68.5	35.501
61.6	35.047
65.9	34.700
90.6	34.942
100.0	35.887
104.6	36.306
125.4	37.005

12.5 Estimation

One of the main uses of regression analysis is as a prediction tool. If the regression function is a good model, the business analyst can use the regression equation to determine values of the dependent variable from various values of the independent variable. For example, financial brokers would like to have a model with which they could predict the selling price of a particular stock on a certain day by some variable, such as the unemployment rate or the producer price index. Marketing managers would like to have a site location model with which they could predict the sales volume of a new location by variables such as population density or number of competitors. The airline cost example presents a regression model that has the potential to predict the cost of flying an airplane by the number of passengers.

A point estimate prediction can be made by taking a particular value of X that is of interest, substituting the value of X into the regression equation, and solving for Y. For example, if the number of passengers is 73, what is the predicted cost of the airline flight? The regression equation for this example was

$$\hat{Y} = 1.57 + .0407X.$$

Substituting $X = 73$ into this equation yields a predicted cost of 4.5411 or $4,541.10.

Confidence Intervals to Estimate the Conditional Mean of $Y: \mu_{Y|X}$

Although a point estimate is often of interest to the business analyst, the regression line is determined by a sample set of points. If a different sample is taken, a different line will result, yielding a different point estimate. Hence computing a confidence interval for the estimating often is useful. Because for any value of X (independent variable) there can be many values of Y (dependent variable), one type of confidence interval is an estimate of the *average* value of Y for a given X. This average value of Y is denoted $E(Y_X)$—the expected value of Y.

(12.5)	$\hat{Y} \pm t_{\alpha/2, n-2} S_e \sqrt{\dfrac{1}{n} + \dfrac{(X_0 - \bar{X})^2}{SS_{XX}}}$	CONFIDENCE INTERVAL TO ESTIMATE $E(Y_X)$ FOR A GIVEN VALUE OF X

where:

X_0 = a particular value of X

$$SS_{XX} = \Sigma X^2 - \frac{(\Sigma X)^2}{n}$$

Use of this formula can be illustrated with construction of a 95% confidence interval to estimate the average value of Y (airline cost) for the airline cost example when X (number of passengers) is 73. This confidence interval utilizes a t value obtained through the degrees of freedom and $\alpha/2$. For a 95% confidence interval, $\alpha = .05$ and $\alpha/2 = .025$. The df $= n - 2 = 12 - 2 = 10$. The table t value is $t_{.025,10} = 2.228$. In addition, other needed values for this problem, which were solved for previously, are

$$S_e = .1773, \quad \Sigma X = 930, \quad \bar{X} = 77.5, \quad \Sigma X^2 = 73,764.$$

For $X_0 = 73$, the value of $\hat{Y}$ is 4.5411. The computed confidence interval for the average value of Y, $E(Y_{73})$, is

$$4.5411 \pm (2.228)(.1773) \sqrt{\frac{1}{12} + \frac{(73 - 77.5)^2}{73,764 - \dfrac{(930)^2}{12}}} = 4.5411 \pm .1220,$$

$$4.4191 \leq E(Y_{73}) \leq 4.6631.$$

That is, the statement can be made with 95% confidence that the average value of Y for $X = 73$ is between 4.4191 and 4.6631.

Table 12.5 shows confidence intervals computed for the airline cost example for several values of X to estimate the average value of Y. Note that as X values get farther from the mean X value (77.5), the confidence intervals get wider; as the X values get closer to the mean, the confidence intervals get narrower. The reason is that the numerator of the second term under the radical sign approaches zero as the value of X nears the mean and increases as X departs from the mean.

X	CONFIDENCE INTERVAL	
62	$4.0934 \pm .1876$	3.9058 to 4.2810
68	$4.3376 \pm .1461$	4.1915 to 4.4837
73	$4.5411 \pm .1220$	4.4191 to 4.6631
85	$5.0295 \pm .1349$	4.8946 to 5.1644
90	$5.2230 \pm .1656$	5.0674 to 5.3986

TABLE 12.5

Confidence Intervals to Estimate the Average Value of Y for Some X Values in the Airline Cost Example

Prediction Intervals to Estimate a Single Value of Y

A second type of interval in regression estimation is a prediction interval to estimate a single value of Y for a given value of X.

PREDICTION INTERVAL TO ESTIMATE Y FOR A GIVEN VALUE OF X	$$\hat{Y} \pm t_{\alpha/2,n-2}S_e\sqrt{1 + \frac{1}{n} + \frac{(X_0 - \bar{X})^2}{SS_{XX}}} \qquad (12.6)$$ where: X_0 = a particular value of X $SS_{XX} = \Sigma X^2 - \dfrac{(\Sigma X)^2}{n}$

Equation 12.6 is virtually the same as Equation 12.5, except for the additional value of 1 under the radical. This additional value widens the prediction interval to estimate a single value of Y from the confidence interval to estimate the average value of Y. This result seems logical because the average value of Y is toward the middle of a group of Y values. Thus the confidence interval to estimate the average need not be as wide as the prediction interval produced by Equation 12.6, which takes into account all the Y values for a given X.

A 95% prediction interval can be computed to estimate the single value of Y for $X = 73$ from the airline cost example by using Equation 12.6. The same values used to construct the confidence interval to estimate the average value of Y are used here.

$$t_{.025,10} = 2.228, \quad S_e = .1773, \quad \Sigma X = 930, \quad \bar{X} = 77.5, \quad \Sigma X^2 = 73,764$$

For $X_0 = 73$, the value of $\hat{Y} = 4.5411$. The computed prediction interval for the single value of Y is

$$4.5411 \pm (2.228)(.1773)\sqrt{1 + \frac{1}{12} + \frac{(73 - 77.5)^2}{73,764 - \dfrac{(930)^2}{12}}} = 4.5411 \pm .4134,$$

$$4.1277 \leq Y \leq 4.9545.$$

Caution: A regression line is determined from a sample of points. The line, the r^2, the S_e, and the confidence intervals change for different sets of sample points. That is, the linear relationship developed for a set of points does not necessarily hold for values of X outside the domain of those used to establish the model. In the airline cost example, the domain of X values (number of passengers) varied from 61 to 97. The regression model developed from these points may not be valid for flights of say 40, 50, or 100 because the regression model was not constructed with X values of those magnitudes. However, decision makers sometimes extrapolate regression results to values of X beyond the domain of those used to develop the formulas (often in time-series sales forecasting). Understanding the limitations of this type of use of regression analysis is essential.

Construct a 95% confidence interval to estimate the average value of Y (FTEs) for Demonstration Problem 12.1 when $X = 40$ beds. Then construct a 95% prediction interval to estimate the single value of Y for $X = 40$ beds.

SOLUTION

For a 95% confidence interval, $\alpha = .05$, $n = 12$, and df $= 10$. The table t value is $t_{.025,10} = 2.228$; $S_e = 15.65$, $\Sigma X = 592$, $\overline{X} = 49.33$, and $\Sigma X^2 = 33{,}044$. For $X_0 = 40$, $\hat{Y} = 120.17$. The computed confidence interval for the average value of Y is

$$120.17 \pm (2.228)(15.65)\sqrt{\frac{1}{12} + \frac{(40 - 49.33)^2}{33{,}044 - \dfrac{(592)^2}{12}}} = 120.17 \pm 11.35$$

$$108.82 \leq E(Y_{40}) \leq 131.52.$$

With 95% confidence, the statement can be made that the average number of FTEs for a hospital with 40 beds is between 108.82 and 131.52.

The computed prediction interval for the single value of Y is

$$120.17 \pm (2.228)(15.65)\sqrt{1 + \frac{1}{12} + \frac{(40 - 49.33)^2}{33{,}044 - \dfrac{(592)^2}{12}}} = 120.17 \pm 36.67$$

$$83.5 \leq Y \leq 156.84.$$

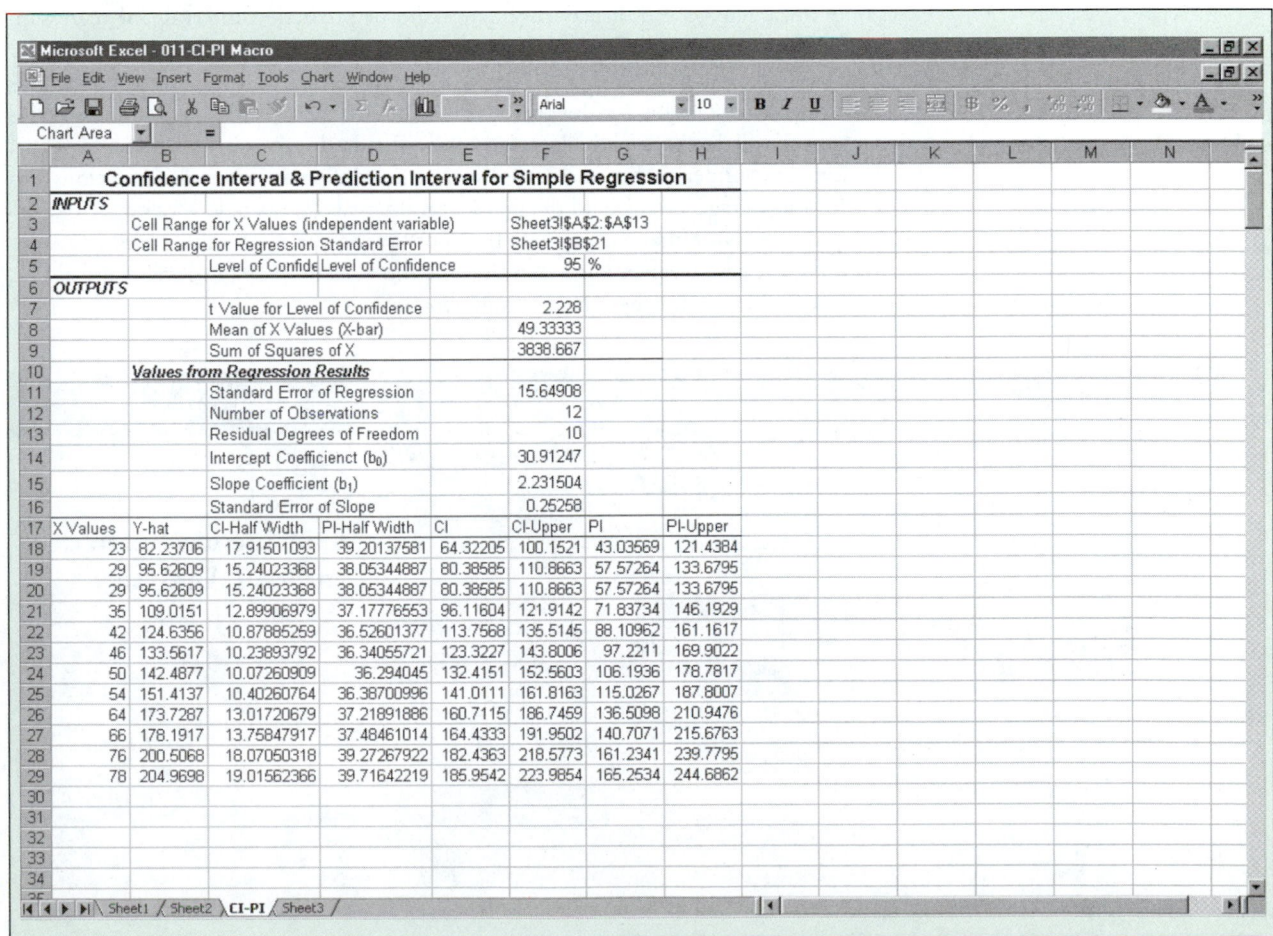

With 95% confidence, the statement can be made that a single number of FTEs for a hospital with 40 beds is between 83.5 and 156.84. Obviously this interval is much wider than the 95% confidence interval for the average value of Y for X = 40.

Shown on the previous page is a FAST ⟍ STAT-produced table showing the computed data for confidence intervals and prediction intervals for this problem, and below is a graph of the 95% interval bands for both the average Y value and the single Y values for all 12 X values in this problem. Note once again the flaring out of the bands near the extreme values of X.

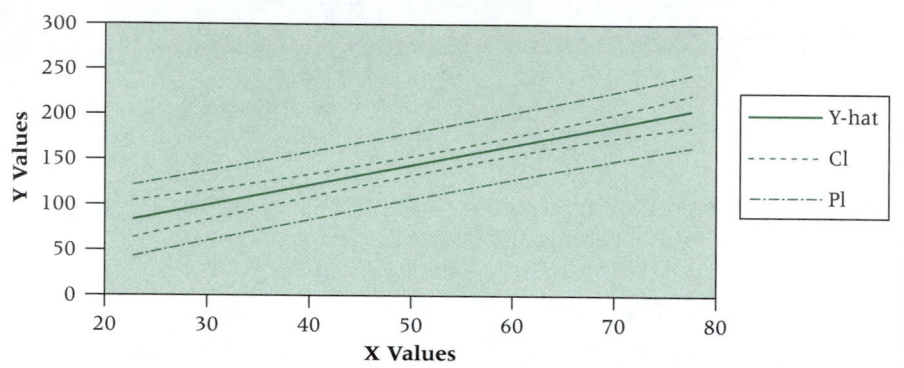

Analysis Using Excel

Excel does not calculate and display confidence and/or prediction intervals for regression analysis. However, **FAST ⟍ STAT** can produce confidence and prediction intervals for simple regression analysis utilizing its **Regression C.I. and P.I.** macro procedure. Figure 12.17 displays the dialog box for this macro. Note that this dialog box requires the location of the X variable values, the location of the standard error of the regression model and the level of confidence. Figure 12.18 displays the confidence intervals for various values of X for the average Y value and the prediction intervals for a single Y value for the airline cost problem. Note that the intervals flare out toward the ends, as the values of X depart from the average X value. Note also that the intervals for a single Y value are always wider than the intervals for the average Y value for any given value of X.

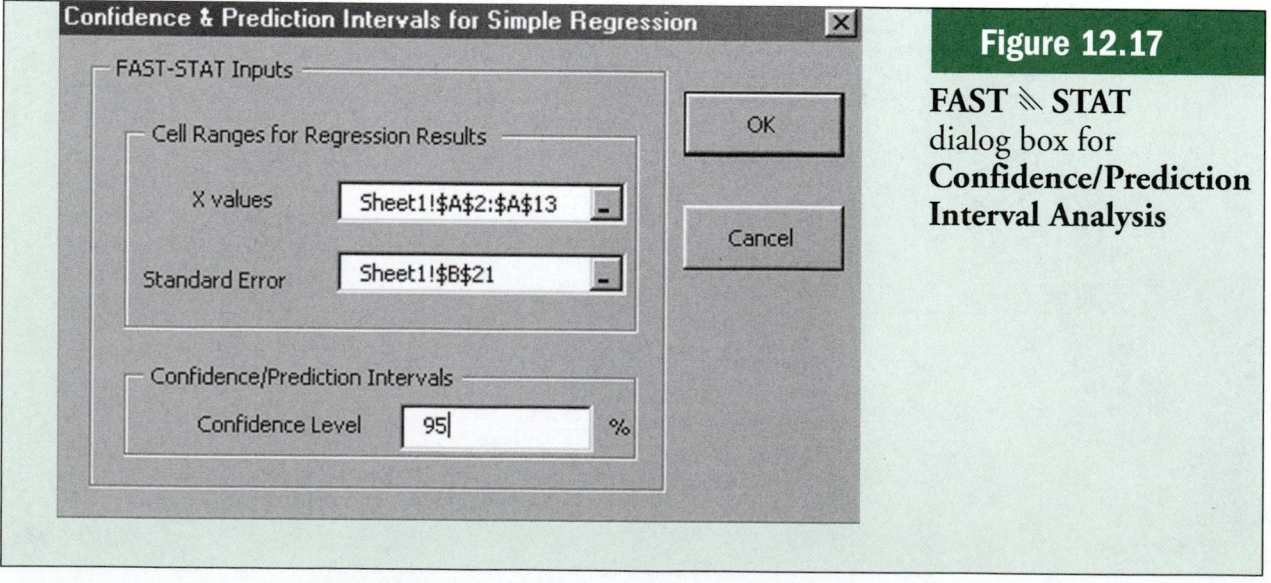

Figure 12.17

FAST ⟍ STAT dialog box for **Confidence/Prediction Interval Analysis**

An examination of the prediction interval formula to estimate Y for a given value of X explains why the intervals flare out.

$$\hat{Y} \pm t_{\alpha/2,n-2} S_e \sqrt{1 + \frac{1}{n} + \frac{(X_0 - \bar{X})^2}{\sum X^2 - \frac{(\sum X)^2}{n}}}$$

Figure 12.18	FAST STAT results for the **Confidence/Prediction Interval Analysis**

Microsoft Excel - 011-CI-PI Macro

File Edit View Insert Format Tools Chart Window Help

Arial | 10 | B I U

Chart 2

	A	B	C	D	E	F	G	H	I	J	K	L	M	N	O
1	**Confidence Interval & Prediction Interval for Simple Regression**														
2	*INPUTS*														
3		Cell Range for X Values (independent variabl			Sheet1!$A$2:$A$13										
4		Cell Range for Regression Standard Error			Sheet1!$B$21										
5		Level of Cc Level of Confidence			95	%									
6	*OUTPUTS*														
7		t Value for Level of Confidence			2.228										
8		Mean of X Values (X-bar)			77.5										
9		Sum of Squares of X			1689										
10		*Values from Regression Results*													
11		Standard Error of Regression			0.177217										
12		Number of Observations			12										
13		Residual Degrees of Freedom			10										
14		Intercept Coefficienct (b_0)			1.569793										
15		Slope Coefficient (b_1)			0.040702										
16		Standard Error of Slope			0.004312										
17	X Values	Y-hat	CI-Half Wic	PI-Half Wic	CI		CI-Upper	PI	PI-Upper						
18	61	4.05259	0.195258	0.440505	3.857332		4.247848	3.612086	4.493095						
19	63	4.133993	0.180006	0.433959	3.953987		4.314	3.700034	4.567953						
20	67	4.2968	0.15222	0.423189	4.14458		4.44902	3.87361	4.719989						
21	69	4.378203	0.140224	0.419024	4.237979		4.518428	3.959179	4.797227						
22	70	4.418905	0.134855	0.417258	4.28405		4.55376	4.001647	4.836163						
23	74	4.581711	0.118845	0.412362	4.462866		4.700556	4.169349	4.994073						
24	76	4.663114	0.114895	0.411241	4.548219		4.77801	4.251873	5.074356						
25	81	4.866622	0.118845	0.412362	4.747778		4.985467	4.45426	5.278984						
26	86	5.07013	0.140224	0.419024	4.929906		5.210355	4.651106	5.489155						
27	91	5.273638	0.172677	0.430971	5.100961		5.446316	4.842667	5.704609						
28	95	5.436445	0.203136	0.444053	5.233308		5.639581	4.992392	5.880497						
29	97	5.517848	0.219307	0.451679	5.298541		5.737155	5.066169	5.969527						

Sheet1 Sheet2 **CI-PI** Sheet3

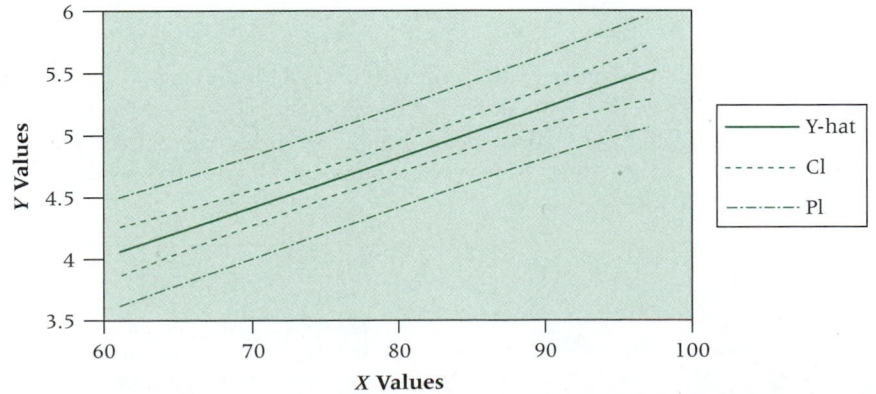

As we enter different values of X_0 from the regression analysis into the equation, the only thing that changes in the equation is $(X_0 - \bar{X})^2$. This expression increases as individual values of X_0 get farther from the mean, resulting in an increase in the width of the interval. The interval is narrower for values of X_0 nearer $\bar{X}$ and wider for values of X_0 farther from $\bar{X}$. A comparison of Equations 12.5 and 12.6 reveals them to be identical except that Equation 12.6 (to compute a prediction interval to estimate Y for a given value of X) contains a 1 under the radical sign and Equation 12.5 does not. This ensures that Equation 12.6 will yield wider intervals than Equation 12.5 for otherwise identical data.

For the optimum use of the **Regression C.I. and P.I.** macro procedure, the X variable values should be sorted either in ascending or descending order. These can either be sorted on the original data worksheet before the macro is initiated, or they can be sorted on the **CI-PI** worksheet after the macro has been run. This requirement stems from Excel's charting feature not from the FAST ⧵ STAT macro.

Two other suggestions should be considered for the best use of **Regression C.I. and P.I.** If the resulting graph is too small, you may need to adjust the scale for the two axes. To do so place the mouse pointer on the horizontal axis until the words *Value (X) Axis* appears below the pointer. Right click and then click on **Format Axis** from the resulting shortcut menu. Click on the **Scale** tab and change the value for the Minimum, for the Maximum or for both. Click on **OK.** The same process can be used for the vertical axis. If the resulting graph is still too small, you can repeat this process.

The second suggestion is appropriate if the macro is to be run a second time. If so, delete the graph from the first run. Otherwise it will be left on the worksheet and updated during the second run. This slows the computations considerably. You may not be aware the first graph is still on the worksheet since the second graph will cover it exactly. However, if you click and drag the second chart on top, you will see that the first chart has also been re-plotted. To delete the first chart before the second run, click just inside the chart and press the **Del** key. If you are to run a third time, delete the second plot and so on.

12.5

Problems

12.27 Construct a 95% confidence interval for the average value of Y for $X = 40$ in Problem 12.5 shown again below.

ADVERTISING	SALES
12.5	148
3.7	55
21.6	338
60.0	994
37.6	541
6.1	89
16.8	126
41.2	379

12.28 Construct a 90% prediction interval for the single value of Y for 200,000 business starts for Problem 12.7 shown again below. Construct a 90% prediction interval for the single value of Y for 175,000 business starts for Problem 12.7. Compare the results. Which prediction interval is greater? Why do you think this is so?

STARTS	FAILURES
233,710	57,097
199,091	50,361
181,645	60,747
158,930	88,140
155,672	97,069

continued

12.28 *continued*

164,086	86,133
166,154	71,558
188,387	71,128
168,158	71,931
170,475	83,384

12.29 Using FAST ⑆ STAT, construct the confidence interval and prediction interval bands for Problem 12.8 shown below. Discuss any "flaring out" that occurs; and if it does occur, explain why. Compare the confidence interval bands to the prediction interval bands. Which is wider and why?

FARMS	SIZE
5.65	213
4.65	258
3.96	297
3.36	340
2.95	374
2.52	420
2.44	426
2.29	441
2.15	460
2.07	469

12.30 Construct a 99% confidence interval for $X = 100$ in Problem 12.9 shown below. Construct a 99% prediction interval for $X = 100$ in Problem 12.9. Explain the difference between the two intervals and why the difference exists. Use FAST ⑆ STAT to construct a graph displaying both the confidence interval bands and the prediction interval bands for Problem 12.9. Comment on the graph.

PRODUCTION	NEW ORDERS
99.9	2.74
97.9	2.87
98.9	2.93
87.9	2.87
92.9	2.98
97.9	3.09
100.6	3.36
104.9	3.61
105.3	3.75
108.6	3.95

12.6
Measures of Association

Measures of association are numerical values that yield information about the relatedness of variables. The measures of association discussed in this chapter apply to only two variables. One measure of association is correlation.

Correlation

Whereas regression analysis involves developing a functional relationship between variables, **correlation** is *the process of determining a measure of the strength of relatedness of variables*. For example, do the stocks of two airlines rise and fall in any related manner? Logically, the prices of two stocks in the same industry should be related. For a sample of pairs of data, correlation analysis can yield a numerical value that represents the degree of relatedness of the two stock prices over time. Another example comes from the transportation

Correlation
A measure of the degree of relatedness of two or more variables.

industry. Is there a correlation between the price of transportation and the weight of the object being shipped? Is there a correlation between price and distance? How strong are the correlations? Pricing decisions can be based in part on shipment costs that are correlated with other variables. In economics and finance, how strong is the correlation between the producer price index and the unemployment rate? In retail sales, what variables are related to a particular store's sales? Is sales related to population density, number of competitors, size of the store, amount of advertising, or other variables?

Several measures of correlation are available, the selection of which depends mostly on the level of data being analyzed. Ideally, business analysts would like to solve for ρ, the population coefficient of correlation. However, because business analysts virtually always deal with sample data, this section introduces a widely used sample coefficient of correlation, r. This measure is applicable only if both variables being analyzed have at least an interval level of data.

Pearson product–moment correlation coefficient (r)
A correlation measure used to determine the degree of relatedness of two variables that are at least of interval level.

The term r is called the **Pearson product–moment correlation coefficient,** named after Karl Pearson (1857–1936), an English statistician who developed several coefficients of correlation along with other significant statistical concepts. The term r is *a measure of the linear correlation of two variables.* It is a number that ranges from –1 to 0 to +1, representing the strength of the relationship between the variables. An r value of +1 denotes a perfect positive relationship between two sets of numbers. An r value of –1 denotes a perfect negative correlation, which indicates an inverse relationship between two variables: as one variable gets larger, the other gets smaller. An r value of 0 means that there is no linear relationship between the two variables.

PEARSON PRODUCT–MOMENT CORRELATION COEFFICIENT

$$r = \frac{SS_{XY}}{\sqrt{(SS_{XX})(SS_{YY})}} = \frac{\Sigma(X - \bar{X})(Y - \bar{Y})}{\sqrt{\Sigma(X - \bar{X})^2 \Sigma(Y - \bar{Y})^2}} \qquad (12.7)$$

$$= \frac{\Sigma XY - \dfrac{(\Sigma X)(\Sigma Y)}{n}}{\sqrt{\left[\Sigma X^2 - \dfrac{(\Sigma X)^2}{n}\right]\left[\Sigma Y^2 - \dfrac{(\Sigma Y)^2}{n}\right]}}$$

Figure 12.19 depicts five different degrees of correlation: (a) represents strong negative correlation, (b) represents moderate negative correlation, (c) represents moderate positive correlation, (d) represents strong positive correlation, and (e) contains no correlation.

What is the measure of correlation between the interest rate of federal funds and the commodities futures index? With data such as those shown in Table 12.6, which represent the values for interest rates of federal funds and commodities futures indexes for a sample of 12 days, a correlation coefficient, r, can be computed. Examination of Equation 12.7 reveals that the following values must be obtained to compute r: ΣX, ΣX^2, ΣY, ΣY^2, ΣXY, and n. In contrast to regression analysis, in correlation analysis it does not matter which variable is designated X and which is designated Y. For this example, the correlation coefficient is computed as shown in Table 12.7. The r value obtained represents a relatively strong positive relationship between interest rates and commodities futures index over this 12-day period.

Relationship between r and r²

Is r, the coefficient of correlation, related to r^2, the coefficient of determination in linear regression? The answer is yes: r^2 equals $(r)^2$. That is, the coefficient of determination is the square of the coefficient of correlation. In the preceding economics example, the ob-

Figure 12.19   Five correlations

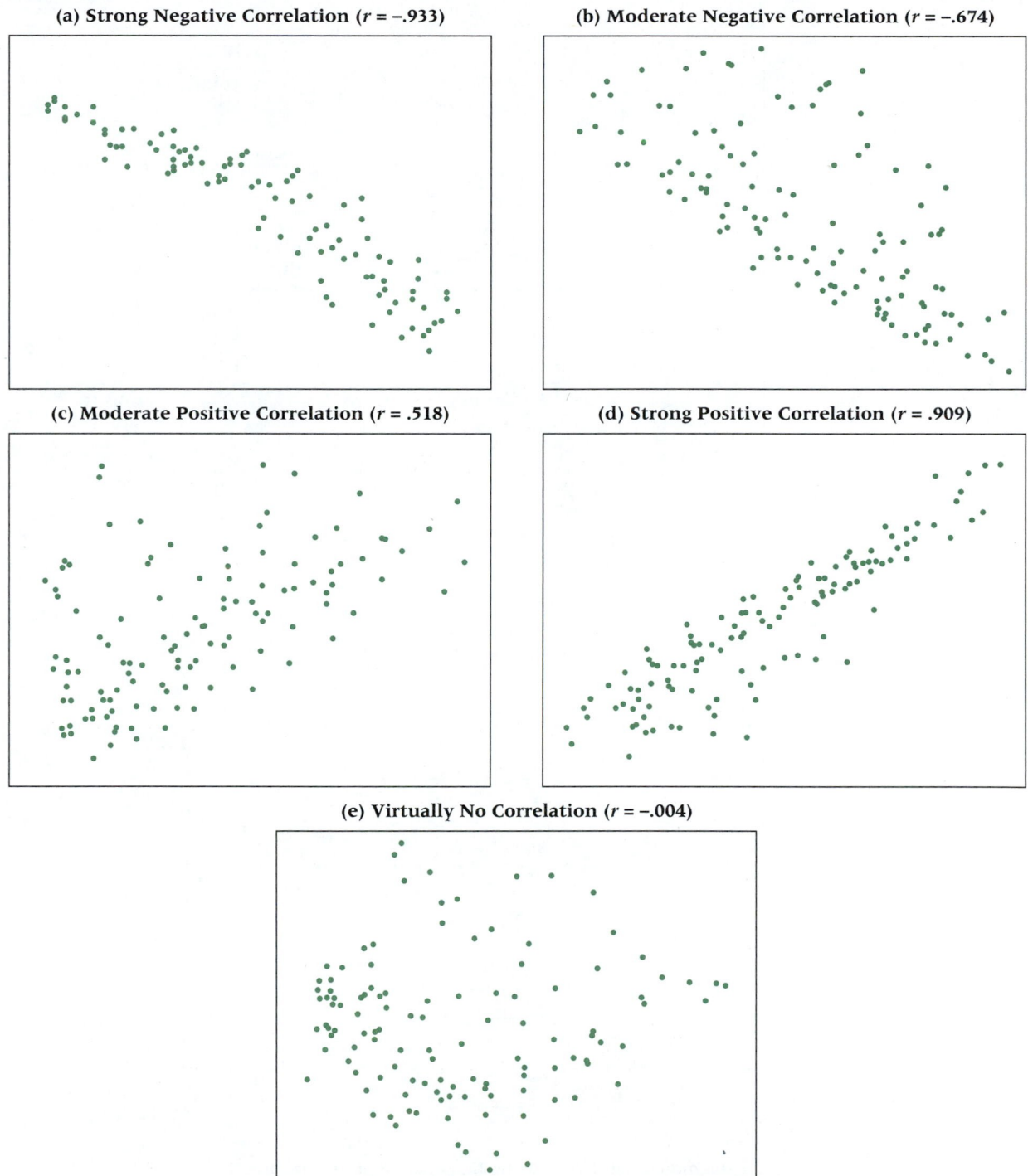

(a) Strong Negative Correlation ($r = -.933$)

(b) Moderate Negative Correlation ($r = -.674$)

(c) Moderate Positive Correlation ($r = .518$)

(d) Strong Positive Correlation ($r = .909$)

(e) Virtually No Correlation ($r = -.004$)

jective was to determine the correlation between interest rates and commodities futures indexes. If a regression model had been developed to predict commodities futures indexes by interest rates, r^2 would be $(r)^2 = (.815)^2 = .66$. In Demonstration Problem 12.1, a regression model was developed to predict FTEs by number of hospital beds. The r^2 value

TABLE 12.6
Data for the Economics
Example

DAY	INTEREST RATE	FUTURES INDEX
1	7.43	221
2	7.48	222
3	8.00	226
4	7.75	225
5	7.60	224
6	7.63	223
7	7.68	223
8	7.67	226
9	7.59	226
10	8.07	235
11	8.03	233
12	8.00	241

TABLE 12.7
Computation of r for the
Economics Example

DAY	INTEREST X	FUTURES INDEX Y	X^2	Y^2	XY
1	7.43	221	55.205	48,841	1,642.03
2	7.48	222	55.950	49,284	1,660.56
3	8.00	226	64.000	51,076	1,808.00
4	7.75	225	60.063	50,625	1,743.75
5	7.60	224	57.760	50,176	1,702.40
6	7.63	223	58.217	49,729	1,701.49
7	7.68	223	58.982	49,729	1,712.64
8	7.67	226	58.829	51,076	1,733.42
9	7.59	226	57.608	51,076	1,715.34
10	8.07	235	65.125	55,225	1,896.45
11	8.03	233	64.481	54,289	1,870.99
12	8.00	241	64.000	58,081	1,928.00
	$\Sigma X = 92.93$	$\Sigma Y = 2,725$	$\Sigma X^2 = 720.220$	$\Sigma Y^2 = 619,207$	$\Sigma XY = 21,115.07$

$$r = \frac{SS_{XY}}{\sqrt{SS_{XX} \cdot SS_{YY}}} = \frac{\Sigma XY - \frac{(\Sigma X)(\Sigma Y)}{n}}{\sqrt{\left[\Sigma X^2 - \frac{(\Sigma X)^2}{n}\right]\left[\Sigma Y^2 - \frac{(\Sigma Y)^2}{n}\right]}}$$

$$= \frac{(21,115.07) - \frac{(92.93)(2725)}{12}}{\sqrt{\left[(720.22) - \frac{(92.93)^2}{12}\right]\left[(619,207) - \frac{(2725)^2}{12}\right]}} = .815$$

for the model was .886. Taking the square root of this value yields $r = .941$, which is the correlation between the sample number of beds and FTEs. A word of caution here: Because r^2 is always positive, solving for r by taking $\sqrt{r^2}$ gives the correct magnitude of r but may give the wrong sign. The business analyst must examine the sign of the slope of the regression line to determine whether there is a positive or negative relationship between the variables and then assign the appropriate sign to the correlation value.

Analysis Using Excel

Excel has the capability of computing correlation coefficients. Figure 12.20 shows the Excel dialog box for correlation analysis. Note the *one* box for **Input Range.** Place in this box the location of all variables to be correlated. Excel's correlation feature allows for analyzing two or more variables and producing a correlation *matrix* containing multiple correlation coefficients when there are more than two variables.

The Excel output from **Correlation** is in matrix form. Examine Figure 12.21 and find the correlation between interest rates and futures indexes computed in the previous example ($r = .815$).

Figure 12.22 is a correlation matrix produced by Excel for seven variables from the financial database (on the CD-ROM). Observe that the highest correlation ($r = .647$) is between Average Yield and Dividends per Share. The lowest correlation is between Total Revenues and Return on Equity ($r = -.005$). The 1's on the diagonal of the matrix represent the correlation of each variable with itself, which of course is 1.00. The upper right portion (triangle) of the matrix is blank because r is symmetric with X and Y. That is, the correlation of return on equity and gross revenues is the same as the correlation of gross revenues and return on equity.

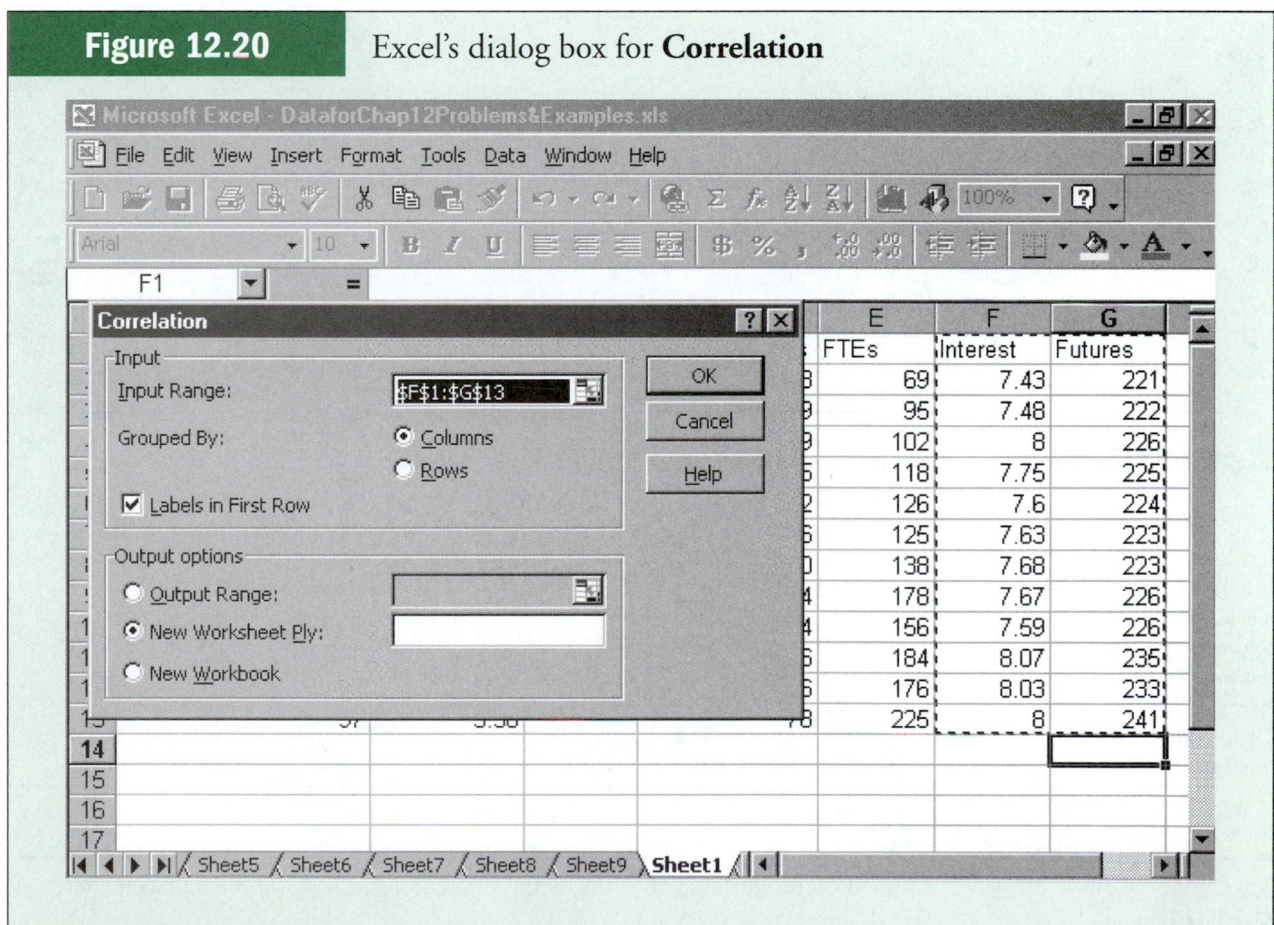

Figure 12.20   Excel's dialog box for **Correlation**

Figure 12.21   Excel computed correlation between interest and futures index

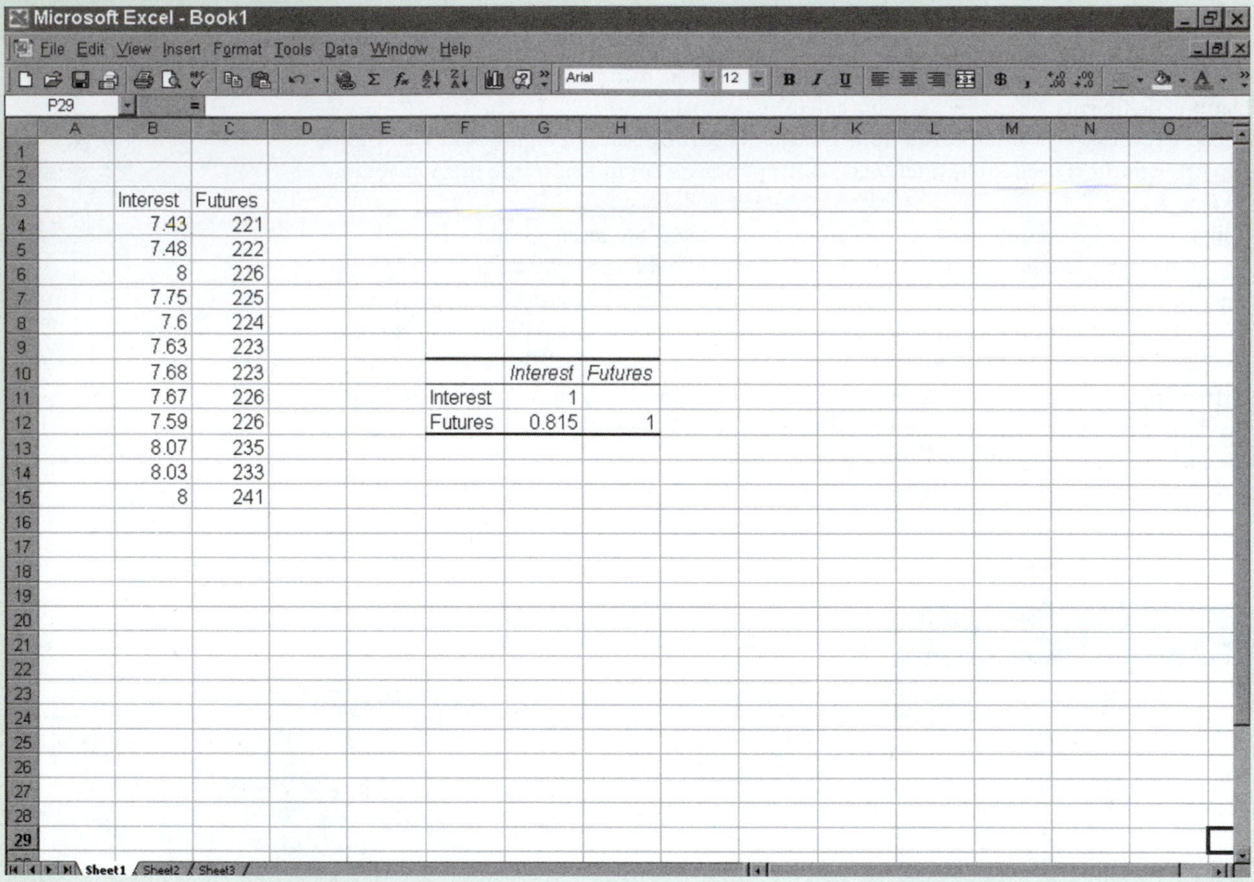

Figure 12.22   Excel correlation matrix for the financial database

	TOTAL REVENUES	TOTAL ASSETS	RETURN ON EQUITY	EARNINGS PER SHARE	AVERAGE YIELD	DIVIDENDS PER SHARE	AVERAGE P/E RATIO
Total Revenues	1						
Total Assets	.389	1					
Return on Equity	−.005	−.053	1				
Earnings per Share	.306	.168	.161	1			
Average Yield	−.034	−.105	−.144	−.077	1		
Dividends per Share	.286	.033	−.102	.435	.647	1	
Average P/E Ratio	−.039	.012	−.130	−.294	−.057	−.009	1

12.6
Problems

12.31   Determine the value of the coefficient of correlation, r, for the following data.

X	4	6	7	11	14	17	21
Y	18	12	13	8	7	7	4

12.32 Determine the value of r for the following data.

X	2	6	8	15	19	20	23
Y	5	6	13	16	22	20	21

12.33 Determine the value of r for the following data.

X	158	296	87	110	436
Y	349	510	301	322	550

12.34 In an effort to determine whether there is any correlation between the price of stocks of airlines, an analyst sampled 6 days of activity of the stock market. Using the following prices of Delta stock and Southwest Air stock, compute the coefficient of correlation. Stock prices have been rounded off to the nearest tenth for ease of computation.

DELTA	SOUTHWEST
47.6	15.1
46.3	15.4
50.6	15.9
52.6	15.6
52.4	16.4
52.7	18.1

12.35 During recent years the Dow Jones industrial average (DJIA) has been quite volatile, and in the late 1990s it rose dramatically. Is the DJIA correlated with Treasury Bill rates? Shown here are DJIA for 9 years and the rates of 6-month Treasury Bills for those corresponding years. Compute r for these data. How strong is the correlation? If an analyst wanted to develop a regression model to predict the DJIA by the Treasury Bill rates, how strong would be the model's prediction? Compute r^2 for this model.

DJIA	TREASURY BILL RATE
2200	8.12
2700	7.51
2900	5.42
3200	3.45
3700	3.02
4500	5.51
5600	5.02
7000	5.07

12.36 The following data are the claims (in $ millions) for Blue Cross & Blue Shield benefits for nine states, along with the surplus (in $ millions) that the company had in assets in those states.

STATE	CLAIMS	SURPLUS
Alabama	$1425	$277
Colorado	273	100
Florida	915	120
Illinois	1687	259
Maine	234	40
Montana	142	25
North Dakota	259	57
Oklahoma	258	31
Texas	894	141

Use the data to compute a correlation coefficient, r, to determine the correlation between claims and surplus.

12.37 The National Safety Council released the following data on the incidence rates for fatal or lost-work-time injuries per 100 employees for several industries in three recent years.

INDUSTRY	YEAR 1	YEAR 2	YEAR 3
Textile	.46	.48	.69
Chemical	.52	.62	.63
Communication	.90	.72	.81
Machinery	1.50	1.74	2.10
Services	2.89	2.03	2.46
Nonferrous metals	1.80	1.92	2.00
Food	3.29	3.18	3.17
Government	5.73	4.43	4.00

Compute r for each pair of years and determine which years are most highly correlated.

Summary

Regression is a procedure that produces a mathematical model (function) that can be used to predict one variable by other variables. Correlation is a measure of the relatedness of variables. Simple regression and correlation are bivariate in that the analysis is of two variables.

Simple regression analysis produces a model in which a Y variable, referred to as the dependent variable, is predicted by an X variable, referred to as the independent variable. The resulting regression model is linear. The general form of the equation of the simple regression line is the slope-intercept equation of a line. The equation of the simple regression model consists of a slope of the line as a coefficient of X and a Y intercept value as a constant.

After the equation of the line has been developed, several statistics are available that can be used to determine how well the line fits the data. Using the historical data values of X, predicted values of Y (denoted $\hat{Y}$) can be calculated by inserting values of X into the regression equation. The predicted values can then be compared to the actual values of Y to determine how well the regression equation fits the known data. The difference between a specific Y value and its associated predicted Y value is called the residual or error of prediction. Examination of the residuals can offer insight into the magnitude of the errors produced by a model. In addition, residual analysis can be used to help determine whether the assumptions underlying the regression analysis have been met. Specifically, graphs of the residuals can reveal (1) lack of linearity, (2) lack of homogeneity of error variance, and (3) independence of error terms. Geometrically, the residuals are the vertical distances from the Y values to the regression line. Because the equation that yields the regression line is derived in such a way that the line is in the geometric middle of the points, the sum of the residuals is zero.

A single value of error measurement called the standard error of the estimate, S_e, can be computed. It is the standard deviation of error of a model. The value of S_e can be used as a single guide to the magnitude of the error produced by the regression model.

Another widely used statistic for testing the strength of a regression model is r^2, or the coefficient of determination. The coefficient of determination is the proportion of total variance of the Y variable accounted for or predicted by X. It ranges from 0 to 1. The higher the r^2 is, the stronger the prediction by the model becomes.

Testing to determine whether the slope of the regression line is different from zero is another way to judge the fit of the regression model to the data. If the population slope of the regression line is not different from zero, the regression model is not adding significant predictability to the dependent variable. A t statistic is used to test the significance of the slope. The overall significance of the regression model can be tested by using an F statistic. In simple regression, because there is only one predictor, this test accomplishes the same thing as the t test of the slope and $F = t^2$.

One of the most prevalent uses of a regression model is to predict the values of Y for given values of X. Recognizing that the predicted value is often not the same as the actual value, a confidence interval has been developed to yield a range within which the mean Y value for a given X should be. A prediction interval for a single Y value for a given X value also is given. This second interval is wider because it allows for the wide diversity of individual values, whereas the confidence interval for the mean Y value reflects only the range of average Y values for a given X.

Bivariate correlation can be accomplished with several different measures. In this chapter, only one coefficient of correlation is presented: the Pearson product–moment coefficient of correlation, r. This value ranges from -1 to 0 to $+1$. An r value of $+1$ is perfect positive correlation, and an r value of -1 is a perfect negative correlation. Negative correlation means that as one variable increases in value, the other variable tends to decrease. For r values near zero, there is little or no correlation.

Key Terms

coefficient of determination (r^2)
correlation
dependent variable
deterministic model
heteroscedasticity
homoscedasticity
independent variable
least squares analysis
outliers
Pearson product–moment correlation
  coefficient (r)

probabilistic model
regression
residual
residual plot
scatter plot
simple regression
standard error of the estimate (S_e)
sum of squares of error (SSE)

SUPPLEMENTARY PROBLEMS

12.38 Use the following data for parts (a) through (e).

X	5	7	3	16	12	9
Y	8	9	11	27	15	13

  a. Determine the equation of the least squares regression line to predict Y by X.
  b. Using the X values, solve for the predicted values of Y and the residuals.
  c. Solve for S_e and r^2.
  d. Test the slope of the regression line. Use $\alpha = .01$.
  e. Based on the results obtained in parts (b) through (d) make a statement about the fit of the line.

12.39 Use the following data for parts (a) through (h).

X	53	47	41	50	58	62	45	60
Y	5	5	7	4	10	12	3	11

  a. Determine the equation of the simple regression line to predict Y from X.
  b. Using the X values, solve for the predicted values of Y and the residuals.
  c. Solve for SSE.
  d. Calculate the standard error of the estimate.
  e. Determine the coefficient of determination.
  f. Calculate the coefficient of correlation.
  g. Test the slope of the regression line. Assume $\alpha = .05$. What do you conclude about the slope?
  h. Comment on parts (d) through (f).

12.40 Solve for the value of r for the following data.

X	213	196	184	202	221	247
Y	76	65	62	68	71	75

If you were to develop a regression line to predict Y by X, what value would the coefficient of determination have?

12.41 Determine the equation of the least squares regression line to predict Y from the following data.

X	47	94	68	73	80	49	52	61
Y	14	40	34	31	36	19	20	21

a. Construct a 95% confidence interval to estimate the mean Y value for $X = 60$.
b. Construct a 95% prediction interval to estimate an individual Y value for $X = 70$.
c. Use FAST ⟍ STAT to produce a graph of the confidence interval and prediction interval bands.
d. Interpret the results obtained in parts (a)–(c).

12.42 Determine the Pearson product–moment correlation coefficient for the following data.

X	1	10	9	6	5	3	2
Y	8	4	4	5	7	7	9

12.43 A manager of a car dealership believes there is a relationship between the number of salespeople on duty and the number of cars sold. Use the sample data collected for five different weeks at the dealership to calculate r. Is there much of a relationship? Solve for r^2. Explain what r^2 means in this problem.

WEEK	NUMBER OF CARS SOLD	NUMBER OF SALESPEOPLE
1	79	6
2	64	6
3	49	4
4	23	2
5	52	3

12.44 Executives of a video rental chain want to predict the success of a potential new store. The company's business analyst begins by gathering information on number of rentals and average family income from several of the chain's present outlets.

RENTALS	AVERAGE FAMILY INCOME ($1000)
710	65
529	43
314	29
504	47
619	52
428	50
317	46
205	29
468	31
545	43
607	49
694	64

Use a computer to develop a regression model to predict the number of rentals per day by the average family income. By examining various statistics in the output, comment on the strength of the model.

12.45 It seems logical that restaurant chains with more units (restaurants) would have greater sales. This is mitigated, however, by several possibilities: some units may be more profitable than others, some units may be larger, some units may serve more meals, some units may serve more expensive meals, and so on. The data shown here was published by Technomic, Inc. Use these data to determine whether there is a correlation between a restaurant chain's sales and its number of units. How strong is the relationship?

CHAIN	SALES ($ BILLIONS)	NUMBER OF UNITS (1000)
McDonald's	17.1	12.4
Burger King	7.9	7.5
Taco Bell	4.8	6.8
Pizza Hut	4.7	8.7
Wendy's	4.6	4.6
KFC	4.0	5.1
Subway	2.9	11.2
Dairy Queen	2.7	5.1
Hardee's	2.7	2.9

12.46 According to the National Marine Fisheries Service, the current landings in millions of pounds of fish by U.S. fleets are almost double what they were in the 1970s. In other words, fishing has not faded as an industry. However, the growth of this industry has varied by region as shown in the following data. Some regions have remained relatively constant, the South Atlantic region has dropped in pounds caught, and the Pacific-Alaska region has grown more than threefold.

FISHERIES	1977	1997
New England	581	642
Mid-Atlantic	213	242
Chesapeake	668	729
South Atlantic	345	269
Gulf of Mexico	1476	1497
Pacific-Alaska	1776	6129

a. Compute the correlation coefficient between the 1977 and 1997 data. What does this tell, if anything, about the industry?
b. Develop a simple regression model to predict the 1997 landings by the 1977 landings. According to the model, if a region had 700 landings in 1977, what would the predicted number be for 1997? Construct a confidence interval for the average Y value for the 700

landings. Use the t statistic to test to determine whether the slope is significantly different from zero. Use $\alpha = .05$. Compute S_e and r^2 and use these along with the t test of the slope to comment on the strength of the regression model.

12.47 People in the aerospace industry believe the cost of a space project is a function of the weight of the major object being sent into space. Use the following data to develop a regression model to predict the cost of a space project by the weight of the space object. Determine r^2 and S_e.

WEIGHT (TONS)	COST (MILLIONS)
1.897	$ 53.6
3.019	184.9
0.453	6.4
0.988	23.5
1.058	33.4
2.100	110.4
2.387	104.6

12.48 The following data represent a breakdown of state banks and all savings organizations in the United States according to the Federal Reserve System.

YEAR	STATE BANKS	ALL SAVINGS
1940	1342	2330
1945	1864	2667
1950	1912	3054
1955	1847	3764
1960	1641	4423
1965	1405	4837
1970	1147	4694
1975	1046	4407
1980	997	4328
1985	1070	3626
1990	1009	2815
1995	1042	2030
1997	992	1779

a. Develop a regression model to predict the total number of state banks by the number of all savings organizations. Compute r^2, S_e, and test the slope of the regression model.
b. Determine the correlation between the number of state banks and the number of all savings organizations.
c. Based on (a) and (b) comment on the relationship between the number of state banks and the number of all savings organizations.

12.49 How strong is the correlation between the inflation rate and 30-year treasury yields? The following data published by Fuji Securities, Inc., are given as pairs of inflation rates and treasury yields for selected years over a 35-year period.

INFLATION RATE	30-YEAR TREASURY YIELD
1.57%	3.05%
2.23	3.93
2.17	4.68
4.53	6.57
7.25	8.27
9.25	12.01
5.00	10.27
4.62	8.45

Compute the Pearson product–moment correlation coefficient to determine the strength of the correlation between these two variables. Comment on the strength and direction of the correlation.

12.50 Is the amount of money spent by companies on advertising a function of the total sales of the company? Shown are sales income and advertising cost data for seven companies, published by *Advertising Age*.

COMPANY	ADVERTISING (MILLIONS)	SALES (BILLIONS)
Procter & Gamble	$1703.1	37.1
Philip Morris	1319.0	56.1
Ford Motor	973.1	153.6
PepsiCo	797.4	20.9
Time Warner	779.1	13.3
Johnson & Johnson	738.7	22.6
MCI	455.4	19.7

Use the data to develop a regression line to predict the amount of advertising by sales. Compute S_e and r^2. Assuming $\alpha = .05$, test the slope of the regression line. Comment on the strength of the regression model.

12.51 Can the consumption of water in a city be predicted by temperature? The following data represent a sample of a day's water consumption and the high temperature for that day.

WATER USE (MILLION GAL)	TEMPERATURE
219	103°
56	39
107	77
129	78
68	50
184	96
150	90
112	75

Develop a least squares regression line to predict the amount of water used in a day in a city by the high temperature for that day. What would be the predicted water usage for a temperature of 100°? Evaluate the regression model by calculating S_e, by calculating r^2, and by testing the slope. Let $\alpha = .01$.

ANALYZING THE DATABASES

1. Use the manufacturing database to correlate the Number of Employees with the Number of Production Workers. Is this a high correlation? What would you expect the correlation to be? Is there a strong correlation between Cost of Materials and Value Added by Manufacture? Why do you think it is this way?

2. Develop a regression model from the manufacturing database to predict New Capital Expenditures from Value Added by Manufacture. Discuss the model and its strength on the basis of indicators presented in this chapter. Does it seem logical that the dollars spent on New Capital Expenditure could be predicted by Value Added by Manufacture?

3. Using the hospital database, develop a regression model to predict the number of Personnel by the number of Births. Now develop a regression model to predict number of Personnel by number of Beds. Examine the regression output. Which model is stronger in predicting number of Personnel? Explain why, using techniques presented in this chapter. Use the second regression model to predict the number of Personnel in a hospital that has 110 beds. Construct a 95% confidence interval around this prediction for the average value of Y.

4. Produce a correlation matrix for the variables Beds, Admissions, Census, Outpatient Visits, Births, Total Expenditures, Payroll Expenditures, and Personnel for the hospital database. Which variables are most highly correlated? Which variables are least correlated?

5. Analyze all the variables except Type in the financial database by using a correlation matrix. The seven variables in this database are capable of producing 21 pairs of correlations. Which are most highly correlated? Select the variable that is most highly correlated with P/E ratio and use it as a predictor to develop a regression model to predict P/E ratio. How did the model do?

6. Use the stock market database to develop a regression model to predict the Utility Index by the Stock Volume. How well did the model perform? Did it perform as you expected? Why or why not? Construct a correlation matrix for the variables of this database (excluding Part of Month) so that you can explore the stock market. Did you discover any apparent relationships between variables?

CASE

DELTA WIRE USES TRAINING AS A WEAPON

The Delta Wire Corporation was founded in 1978 in Clarksdale, Mississippi. The company manufactures high-carbon specialty steel wire for global markets and at present employs around 100 people. For the past few years, sales have increased each year.

A few years ago, however, things did not look as bright for Delta Wire because it was caught in a potentially disastrous bind. With the dollar declining in value, foreign competition was becoming a growing threat to Delta's market position. In addition to the growing foreign competition, industry quality requirements were becoming increasingly tough each year.

Delta officials realized that some conditions, such as the value of the dollar, were beyond their control. However, one area that they could improve upon was employee education. The company worked with training programs developed by the state of Mississippi and a local community college to set up its own school. Delta employees were introduced to statistical process control and other quality assurance techniques. Delta reassured its customers that the company was working hard on improving quality and staying competitive. Customers were invited to sit in on the educational sessions. Because of this effort, Delta has been able to weather the storm and continues to sustain a leadership position in the highly competitive steel wire industry.

DISCUSSION

1. Delta Wire prides itself on its efforts in the area of employee education. Employee education can pay off in many ways. Discuss some of them. One payoff can be the

renewed interest and excitement generated toward the job and the company. Some people theorize that because of a more positive outlook and interest in implementing things learned, the more education received by a worker, the less likely he or she is to miss work days. Suppose the following data represent the number of days of sick leave taken by 20 workers last year along with the number of contact hours of employee education they each received in the past year. Use the techniques learned in this chapter to analyze the data. Include both regression and correlation techniques. Discuss the strength of the relationship and any models that are developed.

EMPLOYEE	HOURS OF EDUCATION	SICK DAYS	EMPLOYEE	HOURS OF EDUCATION	SICK DAYS
1	24	5	11	8	8
2	16	4	12	60	1
3	48	0	13	0	9
4	120	1	14	28	3
5	36	5	15	15	8
6	10	7	16	88	2
7	65	0	17	120	1
8	36	3	18	15	8
9	0	12	19	48	0
10	12	8	20	5	10

2. Many companies have found that the implementation of total quality management has eventually resulted in improved sales. Companies that have failed to adopt quality efforts have lost market share in many cases or have gone out of business. One measure of the effect of a company's quality improvement efforts is customer satisfaction. Suppose Delta Wire hired a research firm to measure customer satisfaction each year. The research firm developed a customer satisfaction scale in which totally satisfied customers can award a score as high as 50 and totally unsatisfied customers can award scores as low as 0. The scores are measured across many different industrial customers and averaged for a yearly mean customer score. Do sales increase with increases in customer satisfaction scores? To study this notion, suppose the average customer satisfaction score each year for Delta Wire is paired with the company's total sales of that year for the last 15 years, and a regression analysis is run on the data. Assume the following Excel output is the result. Suppose you were asked by Delta Wire to analyze the data and summarize the results. What would you find?

Excel Output

SUMMARY OUTPUT

Regression Statistics	
Multiple R	0.949
R Square	0.901
Adjusted R Square	0.894
Standard Error	0.411
Observations	15

ANOVA

	df	SS	MS	F	Significance F
Regression	1	20.098	20.098	118.80	0.0000
Residual	13	2.199	0.169		
Total	14	22.297			

	Coefficients	Standard Error	t Stat	P-value
Intercept	1.733	0.436	3.97	0.0016
CustSat	0.162	0.015	10.90	0.0000

3. One of Delta Wire's main concerns over the years has been the value of the U.S. dollar. Foreign exchange rates affect Delta's ability to sell competitively in international markets, and they also help determine the extent to which foreign firms can compete against Delta in the United States. Below is an Excel correlation matrix constructed by using the foreign exchange rates (against the U.S. dollar) for selected years from 1970 through 1997 for five countries: France, Germany, Japan, South Korea, and the United Kingdom. Over this time frame, the Japanese yen has mostly increased in strength in comparison with the dollar (fewer yen to equal a dollar). Examine the correlation matrix and discuss the exchange rates of these countries in relation to each other. If Delta were selling products in all five of these countries, would an unfavorable exchange rate in one country necessarily mean an unfavorable exchange rate in all international markets? Which countries seem to have related exchange rates? What might this mean?

	France	Germany	Japan	S.Korea	U.K.
France	1				
Germany	0.341	1			
Japan	0.005	0.906	1		
S. Korea	0.403	−0.627	−0.792	1	
U.K.	−0.612	0.426	0.710	−0.8699	1

ADAPTED FROM: "Delta Wire Corporation," *Strengthening America's Competitiveness: Resource Management Insights for Small Business Success.* Published by Warner Books on behalf of Connecticut Mutual Life Insurance Company and the U.S. Chamber of Commerce in association with The Blue Chip Enterprise Initiative, 1991; International Monetary Fund.

13
Multiple Regression Analysis

Learning Objectives

This chapter presents the potential of multiple regression analysis as a tool in business decision making and its applications, thereby enabling you to:

1. Develop a multiple regression model.

2. Understand and apply techniques that can be used to determine how well a regression model fits data.

3. Analyze and interpret nonlinear variables in multiple regression analysis.

4. Understand the role of qualitative variables and how to use them in multiple regression analysis.

5. Learn how to build and evaluate multiple regression models.

Dependent variable
In analysis of variance, the measurement that is being analyzed; the response to the different levels of the independent variables. In regression analysis, the variable that is being predicted.

Independent variable
In an analysis of variance, the treatment or factor being analyzed. In regression analysis, the predictor variable.

Simple regression analysis is bivariate linear regression in which one **dependent variable,** Y, is predicted by one **independent variable,** X. Examples of simple regression applications include models to predict retail sales by population density, Dow Jones averages by prime interest rates, crude oil production by energy consumption, and CEO compensation by quarterly sales. However, there are usually other independent variables that, taken in conjunction with these variables, can make the regression model a better fit in predicting the dependent variable. For example, sales could be predicted by the size of store and number of competitors in addition to population density. A model to predict the Dow Jones average of 30 industrials could include, in addition to the prime interest rate, such predictors as yesterday's volume, the bond interest rate, and the producer price index. A model to predict CEO compensation could be developed by using variables such as company earnings per share, age of CEO, and size of company in addition to quarterly sales. A model could perhaps be developed to predict the cost of outsourcing by such variables as unit price, export taxes, cost of money, damage in transit, and other factors. In each of these examples, there is only one dependent variable, Y, as there is with simple regression analysis. However, there are multiple independent variables, X (predictors). *Regression analysis with two or more independent variables or with at least one nonlinear predictor* is called **multiple regression** analysis.

13.1
The Multiple Regression Model

Multiple regression analysis is similar in principle to simple regression analysis. However, it is more complex conceptually and computationally. Recall from Chapter 12 that the equation of the probabilistic simple regression model is

$$Y = \beta_0 + \beta_1 X + \epsilon$$

Multiple regression
Regression analysis with one dependent variable and two or more independent variables or at least one nonlinear independent variable.

where:

Y = the value of the dependent variable,
β_0 = the population Y intercept,
β_1 = the population slope, and
ϵ = the error of prediction.

Extending this notion to multiple regression gives the general equation for the probabilistic multiple regression model.

$$Y = \beta_0 + \beta_1 X_1 + \beta_2 X_2 + \beta_3 X_3 + \ldots + \beta_k X_k + \epsilon$$

Response variable
The dependent variable in a multiple regression model; the variable that the business analyst is trying to predict.

Partial regression coefficient
The coefficient of an independent variable in a multiple regression model that represents the increase that will occur in the value of the dependent variable from a 1-unit increase in the independent variable if all other variables are held constant.

where:

Y = the value of the dependent variable
β_0 = the regression constant
β_1 = the partial regression coefficient for independent variable 1
β_2 = the partial regression coefficient for independent variable 2
β_3 = the partial regression coefficient for independent variable 3
β_k = the partial regression coefficient for independent variable k
k = the number of independent variables

In multiple regression analysis, the dependent variable, Y, is sometimes referred to as the **response variable.** The **partial regression coefficient** of an independent variable, β_i, *represents the increase that will occur in the value of Y from a 1-unit increase in the independent variable if all other variables are held constant.* The "full" (versus partial) regression coefficient of an independent variable is a coefficient obtained from the bivariate model (simple regression) in which the independent variable is the sole predictor of Y, as was the case with the slope in Chapter 12. The partial regression coefficients occur because more than one predictor is included in a model. The partial regression coefficients are analogous to β_1, the slope of the simple regression model in Chapter 12.

In actuality, the partial regression coefficients and the regression constant of a multiple regression model are population values and are unknown. These values are estimated by using sample information. Shown here is the form of the equation for estimating Y with sample information.

$$\hat{Y} = b_0 + b_1X_1 + b_2X_2 + b_3X_3 + \cdots + b_kX_k$$

where:

$\hat{Y}$ = the predicted value of Y
b_0 = the estimate of the regression constant
b_1 = the estimate of regression coefficient 1
b_2 = the estimate of regression coefficient 2
b_3 = the estimate of regression coefficient 3
b_k = the estimate of regression coefficient k
k = the number of independent variables

Multiple Regression Model with Two Independent Variables (First-Order)

The simplest multiple regression model is one constructed with two independent variables, where the highest power of either variable is 1 (first-order regression model). The regression model is

$$Y = \beta_0 + \beta_1X_1 + \beta_2X_2 + \epsilon.$$

The constant and coefficients are estimated from sample information, resulting in the following model.

$$\hat{Y} = b_0 + b_1X_1 + b_2X_2$$

Figure 13.1 is a three-dimensional graph of a series of points (X_1, X_2, Y) representing values from three variables used in a multiple regression model to predict the sales price of a house by the number of square feet in the house and the age of the house. Simple regression models yield a line that is fit through data points in the XY plane. In multiple regression analysis, the resulting model produces a **response surface.** In the multiple regression model shown here with two independent first-order variables, the response surface is a **response plane.** The response plane for such a model is fit in a three-dimensional space (X_1, X_2, Y).

If such a response plane is fit into the points shown in Figure 13.1, the result is the graph in Figure 13.2. Notice that most of the points are not on the plane. As in simple regression, there is usually error in the fit of the model in multiple regression. The distances shown in the graph from the points to the response plane are the errors of fit, or residuals $(Y - \hat{Y})$. Multiple regression models with three or more independent variables involve more than three dimensions and are difficult to depict geometrically.

Observe in Figure 13.2 that the regression model attempts to fit a plane into the three-dimensional plot of points. Notice that the plane intercepts the Y axis. Figure 13.2 depicts some values of Y for various values of X_1 and X_2. The error of the response plane (ϵ) in predicting or determining the Y values is the distance from the points to the plane.

Response surface
The surface defined by a multiple regression model.

Response plane
A plane fit in a three-dimensional space and that represents the response surface defined by a multiple regression model with two independent first-order variables.

Figure 13.1

Points in a sample
space

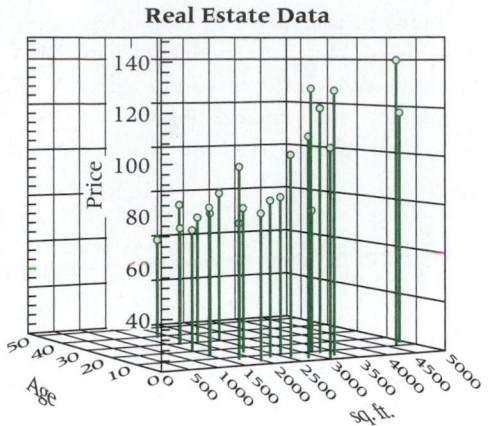

Figure 13.2

Response plane for a
first-order two-
predictor multiple
regression model

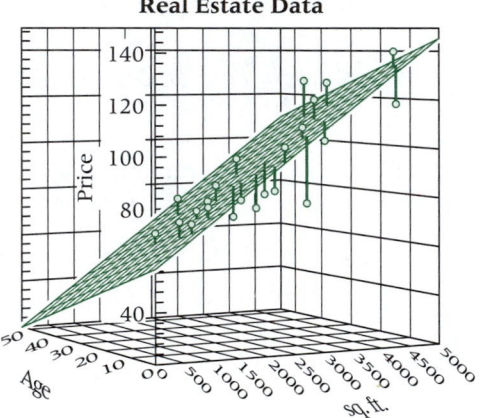

Determining the Multiple Regression Equation

The simple regression equations for determining the sample slope and intercept given in
Chapter 12 are the result of using methods of calculus to minimize the sum of squares of
error for the regression model. The procedure for developing these equations involves
solving two simultaneous equations with two unknowns, b_0 and b_1. Finding the sample
slope and intercept from these formulas requires the values of ΣX, ΣY, ΣXY, and ΣX^2.

The procedure for determining formulas to solve for multiple regression coefficients is
similar. The formulas are established to *meet an objective of minimizing the sum of squares
of error for the model.* Hence, the regression analysis shown here is referred to as **least
squares analysis.** Methods of calculus are applied, resulting in $k + 1$ equations with $k + 1$
unknowns (b_0 and k values of b_i) for multiple regression analyses with k independent variables. Thus, a regression model with six independent variables will generate seven simultaneous equations with seven unknowns ($b_0, b_1, b_2, b_3, b_4, b_5, b_6$).

For multiple regression models with two independent variables, the result is three simultaneous equations with three unknowns ($b_0, b_1,$ and b_2).

Least squares analysis
The process by which a
regression model is
developed based on
calculus techniques that
attempt to produce a
minimum sum of the
squared error values.

$$b_0 n + b_1 \Sigma X_1 + b_2 \Sigma X_2 = \Sigma Y$$
$$b_0 \Sigma X_1 + b_1 \Sigma X_1^2 + b_2 \Sigma X_1 X_2 = \Sigma X_1 Y$$
$$b_0 \Sigma X_2 + b_1 \Sigma X_1 X_2 + b_2 \Sigma X_2^2 = \Sigma X_2 Y$$

The process of solving these equations by hand is tedious and time-consuming. Solving for the regression coefficients and regression constant in a multiple regression model with two independent variables requires ΣX_1, ΣX_2, ΣY, ΣX_1^2, ΣX_2^2, $\Sigma X_1 X_2$, $\Sigma X_1 Y$, and $\Sigma X_2 Y$. In actuality, virtually all business analysts use computer statistical software packages to solve for the regression coefficients, the regression constant, and other pertinent information. In this chapter, we will discuss computer output and assume little or no hand calculation. The emphasis will be on the interpretation of the computer output.

A Multiple Regression Model

A real estate study was conducted in a small Louisiana city to determine what variables, if any, are related to the market price of a home. Several variables were explored, including the number of bedrooms, the number of bathrooms, the age of the house, the number of square feet of living space, the total number of square feet of space, and how many garages the house had. Suppose the business analyst wants to develop a regression model to predict the market price of a home by two variables, "total number of square feet in the house" and "the age of the house." Listed in Table 13.1 are the data for these three variables.

Using Excel to perform the multiple regression analysis on the real estate data results in the output given in Figure 13.3.

One of the first items that a business analyst wants to retrieve from regression output is the equation of the regression model. In Excel, the regression equation is found in the column labeled "Coefficients." From Figure 13.3, the coefficient for the number of square feet variable, X_1, is 0.0177, the coefficient for the age variable, X_2, is –0.6663, and the

MARKET PRICE ($1000) Y	TOTAL NUMBER OF SQUARE FEET X_1	AGE OF HOUSE (YEARS) X_2
63.0	1605	35
65.1	2489	45
69.9	1553	20
76.8	2404	32
73.9	1884	25
77.9	1558	14
74.9	1748	8
78.0	3105	10
79.0	1682	28
83.4	2470	30
79.5	1820	2
83.9	2143	6
79.7	2121	14
84.5	2485	9
96.0	2300	19
109.5	2714	4
102.5	2463	5
121.0	3076	7
104.9	3048	3
128.0	3267	6
129.0	3069	10
117.9	4765	11
140.0	4540	8

TABLE 13.1
Real Estate Data

| Figure 13.3 | Excel output of regression for the real estate example |

Y intercept is 57.351. From this, the equation of the regression model for the real estate data can be determined:

$$\hat{Y} = 57.351 + .0177\,X_1 - .6663\,X_2$$

The regression constant, 57.351, is the *Y* intercept. The *Y* intercept is the value of $\hat{Y}$ if both X_1 (number of square feet) and X_2 (age) are zero. In this example, a practical understanding of the *Y* intercept is meaningless. It is nonsense to say that if a house contains no square feet ($X_1 = 0$) and is new ($X_2 = 0$) it would cost $57,351. In addition, the values of $X_1 = 0$ and $X_2 = 0$ are out of the range of values for X_1, X_2 used to construct the model. Note in Figure 13.2, however, that the response plane crosses the *Y* (price) axis at 57.351.

The coefficient of X_1 (total number of square feet in the house) is .0177. This means that a 1-unit increase in square footage would result in a predicted increase of .0177 ($1000) = $17.70 in the price of the home if age were held constant. All other variables being held constant, the addition of 1 square foot of space in the house results in a predicted increase of $17.70 in the price of the home.

The coefficient of X_2 (age) is –.6663. The negative sign on the coefficient denotes an inverse relationship between the age of a house and the price of the house: the older the house, the lower the price. In this case, if the total number of square feet in the house is kept constant, a 1-unit increase in the age of the house (1 year) will result in $-.6663(\$1000) = -666.30$, a predicted $666.30 drop in the price.

In examining the regression coefficients, it is important to remember that the independent variables are often measured in different units. It is usually not wise to compare the regression coefficients of predictors in a multiple regression model and decide that the variable with the largest regression coefficient is the best predictor. In this example, the two variables are in different units, square feet and years. Just because X_2 has the larger coefficient (.6663) does not necessarily make X_2 the strongest predictor of Y.

This regression model can be used to predict the price of a house in this small Louisiana city. If the house has 2500 square feet total and is 12 years old, $X_1 = 2500$ and $X_2 = 12$. Substituting these values into the regression model yields

$$\hat{Y} = 57.351 + .0177X_1 - .6663$$
$$= 57.351 + .0177(2500) - .6663(12)$$
$$= 93.605$$

The predicted price of the house is $93,605. Figure 13.2 is a graph of these data with the response plane and the residual distances.

DEMONSTRATION PROBLEM 13.1*

Much of the freight cargo in the world is transported over roads. The volume of freight cargo shipped over roads varies from country to country depending on the size of the country, the amount of commerce, the wealth of the country, and other factors. Shown here are seven of the top 10 countries in which freight cargo is shipped over roads, along with the number of miles of roads and the number of commercial vehicles (trucks and buses) for each country. Use these data to develop a multiple regression model to predict the volume of freight cargo shipped over roads by the length of roads and the number of commercial vehicles. Determine the predicted volume of freight cargo over roads if the length of roads is 600,000 miles and the number of commercial vehicles is 3 million.

COUNTRY	FREIGHT CARGO SHIPPED BY ROAD (MILLION SHORT-TON MILES)	LENGTH OF ROADS (MILES)	NUMBER OF COMMERCIAL VEHICLES
China	278,806	673,239	5,010,000
Brazil	178,359	1,031,693	1,371,127
India	144,000	1,342,000	1,980,000
Germany	138,975	395,367	2,923,000
Italy	125,171	188,597	2,745,500
Spain	105,824	206,271	2,859,438
Mexico	96,049	157,036	3,758,034

*SOURCES: World Data; World Road Statistics; and George Thomas Kurian, *The Illustrated Book of World Rankings,* Armonk, NY, M. E. Sharpe, Inc., 1997.

SOLUTION

The following output shows the results of analyzing the data by using the regression portion of Excel.

SUMMARY OUTPUT

Regression Statistics	
Multiple R	0.812
R Square	0.659
Adjusted R Square	0.488
Standard Error	44273.867
Observations	7

ANOVA

	df	SS	MS	F	Significance F
Regression	2	15148592381	7574296191	3.86	0.116
Residual	4	7840701114	1960175278		
Total	6	22989293495			

	Coefficients	Standard Error	t Stat	P-value
Intercept	−26425.45	67624.938	−0.39	0.716
Length of Roads	0.1018	0.0435	2.34	0.079
No. of Comm. Vehicles	0.0410	0.0171	2.39	0.075

The regression equation is

$$\hat{Y} = -26{,}425.45 + .1018X_1 + .0410X_2$$

where:

Y = volume of freight cargo shipped,

X_1 = length of roads, and

X_2 = number of commercial vehicles.

The model indicates that for every 1-unit (1 mile) increase in length of roads, the predicted volume of freight cargo shipped increases by .1018 million short-ton miles, or 101,800 short-ton miles, if the number of commercial vehicles is held constant. If the number of commercial vehicles is increased by 1 unit, the predicted volume of freight cargo shipped increases by .0410 million short-ton miles, or 41,000 short-ton miles, if the length of roads is held constant.

If X_1 (length of roads) is 600,000 and X_2 (number of commercial vehicles) is 3 million, the model predicts that the volume of freight cargo shipped will be 157,655 million short-ton miles:

$$\hat{Y} = -26{,}425.45 + .1018(600{,}000) + .0410(3{,}000{,}000) = 157{,}655$$

Analysis Using Excel

To use Excel for regression analysis, begin by selecting **Data Analysis** from the **Tools** option on the menu bar. From the **Data Analysis** pull-down menu, select **Regression.** The Excel dialog box for **Regression** is shown in Figure 13.4. To use this feature, insert the location of the dependent variable (Y) in the box labeled **Input Y Range** and the location of the independent variables ($X_1, X_2, \ldots$) in box labeled **Input X Range.** If there are labels

| Figure 13.4 | Excel Regression dialog box |

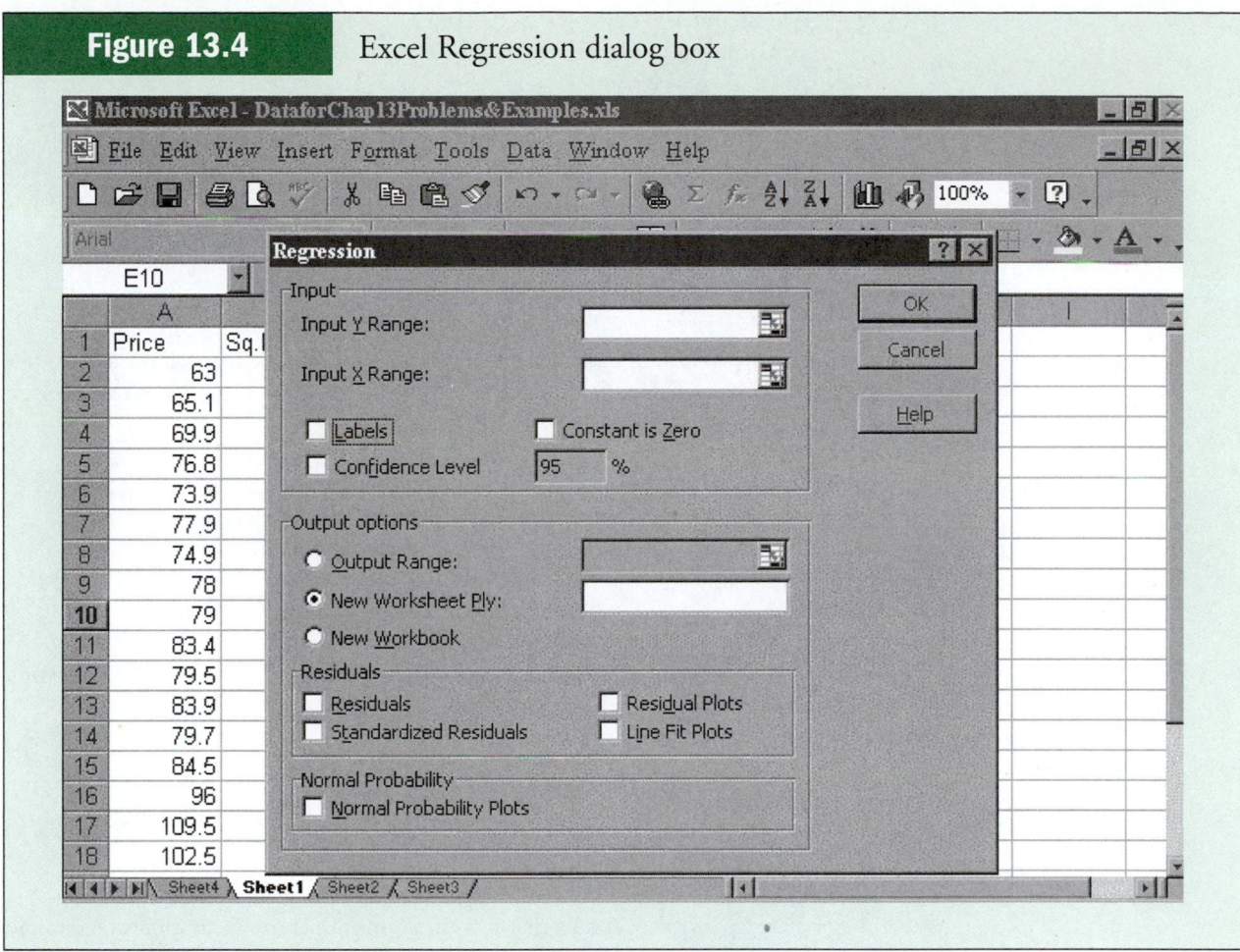

at the top of the columns, check **Labels.** Near the bottom of the dialog box are several residual options. To show the computed residuals, check **Residuals;** for a plot of the residuals, check **Residual Plots;** for a line plot, check **Line Fit Plots.** The output for the real estate data problem is shown in Figure 13.3 and will be discussed throughout the chapter.

13.1   Use a computer to develop the equation of the regression model for the following data. Comment on the regression coefficients. Determine the predicted value of Y for $X_1 = 200$ and $X_2 = 7$.

Y	X_1	X_2
12	174	3
18	281	9
31	189	4
28	202	8
52	149	9
47	188	12
38	215	5
22	150	11
36	167	8
17	135	5

13.1
Problems

13.2   Use a computer to develop the equation of the regression model for the following data. Comment on the regression coefficients. Determine the predicted value of Y for $X_1 = 33$, $X_2 = 29$, and $X_3 = 13$.

Y	X_1	X_2	X_3	Y	X_1	X_2	X_3
114	21	6	5	94	40	33	14
94	43	25	8	107	32	14	11
87	56	42	25	119	16	4	7
98	19	27	9	93	18	31	16
101	29	20	12	108	27	12	10
85	34	45	21	117	31	3	8

13.3   Using the following data, determine the equation of the regression model. How many independent variables are there? Comment on the meaning of these regression coefficients.

PREDICTOR	COEFFICIENT
Constant	121.62
X_1	−.174
X_2	6.02
X_3	.00026
X_4	.0041

13.4   Use the following data to determine the equation of the multiple regression model. Comment on the regression coefficients.

PREDICTOR	COEFFICIENT
Constant	31,409.5
X_1	.08425
X_2	289.62
X_3	−.0947

13.5   Is there a particular product that is an indicator of per capita consumption around the world? Shown below are data on per capita consumption, paper consumption, fish consumption, and gasoline consumption for nine countries. Use the data to determine the equation of the multiple regression model to predict per capita consumption by paper consumption, fish consumption, and gasoline consumption. Discuss the impact of increasing paper consumption by 1 unit on the predicted per capita consumption. Discuss the impact of a 1-unit increase in fish consumption on the predicted per capita consumption and the impact of a 1-unit increase in gasoline consumption on the predicted per capita consumption.

COUNTRY	PER CAPITA CONSUMPTION	PAPER CONSUMPTION (KG PER 1000 PEOPLE)	FISH CONSUMPTION (POUNDS)	GASOLINE CONSUMPTION (1000 BARRELS PER DAY)
Japan	$19,700	76,892	158.7	5,454
Portugal	5,570	24,126	132.7	277
United States	16,500	84,579	47.0	17,033
Venezuela	2,090	6,860	31.1	430
Greece	4,490	15,641	42.1	331
Italy	10,790	43,098	44.3	1,936
Norway	13,400	41,575	90.6	183
United Kingdom	9,040	52,335	43.9	1,803
Philippines	640	940	76.3	235

SOURCES: World Development Report; Pulp & Paper Industry; Fishery Statistic Yearbook; Energy Statistics Yearbook; and George Thomas Kurian, *The Illustrated Book of World Rankings,* Armonk, NY, M. E. Sharpe, Inc., 1997.

13.6   Jensen, Solberg, and Zorn investigated the relationship of insider ownership, debt, and dividend policies in companies. One of their findings was that firms with high insider ownership choose lower levels of both debt and dividends. Shown here is a sample of data of these three variables for 11 different industries. Use the data to develop the equation of the regression model to predict insider ownership by debt ratio and dividend payout. Comment on the regression coefficients.

INDUSTRY	INSIDER OWNERSHIP	DEBT RATIO	DIVIDEND PAYOUT
Mining	8.2	14.2	10.4
Food and beverage	18.4	20.8	14.3
Furniture	11.8	18.6	12.1
Publishing	28.0	18.5	11.8
Petroleum refining	7.4	28.2	10.6
Glass and cement	15.4	24.7	12.6
Motor vehicle	15.7	15.6	12.6
Department store	18.4	21.7	7.2
Restaurant	13.4	23.0	11.3
Amusement	18.1	46.7	4.1
Hospital	10.0	35.8	9.0

SOURCE: R. Gerald Jensen, Donald P. Solberg, and Thomas S. Zorn, "Simultaneous Determination of Insider Ownership, Debt, and Dividend Policies," *Journal of Financial and Quantitative Analysis* 27, No. 2, June 1992.

13.2 Evaluating the Multiple Regression Model

Multiple regression models can be developed to fit almost any data set if the level of measurement is adequate and there are enough data points. Once a model has been constructed, it is important to test the model to determine whether it fits the data well and whether the assumptions underlying regression analysis are met. There are several ways to examine the adequacy of the regression model, including testing the overall significance of the model, studying the significance tests of the regression coefficients, computing the residuals, examining the standard error of the estimate, and observing the coefficient of determination.

Testing the Overall Model

With simple regression, a *t* test of the slope of the regression line is used to determine whether the population slope of the regression line is different from zero—that is, whether the independent variable contributes significantly in linearly predicting the dependent variable. The hypotheses for this test, presented in Chapter 12, are

$$H_0: \beta_1 = 0$$
$$H_a: \beta_1 \neq 0.$$

For multiple regression, an analogous test makes use of the *F* statistic. The overall significance of the multiple regression model is tested with the following hypotheses.

$$H_0: \beta_1 = \beta_2 = \beta_3 = \cdots = \beta_k = 0$$
$$H_a: \text{At least one of the regression coefficients is} \neq 0.$$

If we fail to reject the null hypothesis, we are stating that the regression model has no significant predictability for the dependent variable. A rejection of the null hypothesis indicates that at least one of the independent variables is adding significant predictability for Y.

This F test of overall significance is part of the regression output of Excel and appears in the analysis of variance (ANOVA) table. Shown here is the ANOVA table for the real estate example taken from the Excel output in Figure 13.3.

ANOVA

	df	SS	MS	F	Significance F
Regression	2	8189.723	4094.862	28.63	0.0000014
Residual	20	2861.017	143.051		
Total	22	11050.740			

The F value is 28.63; because $p = .0000014$, the F value is significant at $\alpha = .001$. The null hypothesis is rejected, and there is at least one significant predictor of house price in this analysis.

The F value is calculated by the following equation.

$$F = \frac{MS_{reg}}{MS_{err}} = \frac{\dfrac{SS_{reg}}{df_{reg}}}{\dfrac{SS_{err}}{df_{err}}} = \frac{\dfrac{SSR}{k}}{\dfrac{SSE}{N-k-1}}$$

where:

MS = mean square
SS = sum of squares
df = degrees of freedom
k = number of independent variables
N = number of observations

Note that in the ANOVA table for the real estate example, $df_{reg} = 2$. The degrees of freedom formula for regression is the number of regression coefficients plus the regression constant minus 1. The net result is the number of regression coefficients, which equals the number of independent variables, k. In the real estate example, there are two independent variables and so $k = 2$. Degrees of freedom error in multiple regression equals the total number of observations minus the number of regression coefficients minus the regression constant, or $N - k - 1$. For the real estate example, $N = 23$; thus, $df_{err} = 23 - 2 - 1 = 20$.

As shown in Chapter 11, MS = SS/df. The F ratio is formed by dividing MS_{reg} by MS_{err}. In using the F distribution table to determine a critical value against which to test the calculated F value, the degrees of freedom numerator is df_{reg} and the degrees of freedom denominator is df_{err}. The table F value is obtained in the usual manner, as presented in Chapter 11. With $\alpha = .01$ for the real estate example, the table value is

$$F_{.01,2,20} = 5.85.$$

Comparing the calculated F of 28.63 to this table value shows that the decision is to reject the null hypothesis. This is the same conclusion reached using the p-value method from the computer output.

If a regression model has only one linear independent variable, it is a simple regression model. In that case, the F test for the overall model is the same as the t test for significance of the population slope. The F value displayed in the regression ANOVA table is related to the t test for the slope in the simple regression case as follows.

$$F = t^2$$

In simple regression, the F value and the t value give redundant information about the overall test of the model.

Most business analysts who use multiple regression analysis will observe the value of F and its p value rather early in the process. If F is not significant, there is no population regression coefficient that is significantly different from zero, and the regression model has no predictability for the dependent variable.

Significance Tests of the Regression Coefficients

Individual significance tests for each regression coefficient are available by using a t test. This test is analogous to the t test for the slope used in Chapter 12 for simple regression analysis. The hypotheses for testing the regression coefficient of each independent variable take the following form.

$$H_0: \beta_1 = 0$$
$$H_a: \beta_1 \neq 0$$
$$H_0: \beta_2 = 0$$
$$H_a: \beta_2 \neq 0$$
$$\vdots$$
$$H_0: \beta_k = 0$$
$$H_a: \beta_k \neq 0$$

Excel regression output includes observed t values and their associated p-values to test the individual regression coefficients as standard output. Shown here are the t values and their associated probabilities for the real estate example as displayed with the multiple regression output in Figure 13.3.

	Coefficients	Standard Error	t Stat	P-value
Intercept	57.351	10.0072	5.73	0.000013
Square Feet	0.0177	0.0031	5.63	0.000016
Age	-0.6663	0.2280	-2.92	0.008418

At $\alpha = .05$, the null hypothesis is rejected for both variables (p-value for square feet is .000016 and for age is .008418) because the probabilities (p) associated with their t values are less than .05. If the t ratios for any predictor variables are not significant (fail to reject the null hypothesis), the business analyst might decide to drop that variable(s) from the analysis as a nonsignificant predictor(s). Other factors can enter into this decision. In a later section, we will explore techniques for model-building in which there is some variable sorting.

The degrees of freedom for each of these individual tests of regression coefficients are $n - k - 1$. In this particular example, the degrees of freedom are $23 - 2 - 1 = 20$. With $\alpha = .05$ and a two-tailed test, the critical table t value is

$$t_{.025,20} = \pm 2.086.$$

Notice from the t ratios shown here that if this critical table t value had been used as the hypothesis test criterion instead of the p-value method, the results would have been the same. Testing the regression coefficients not only gives the business analyst some insight into the fit of the regression model, but it also helps in the evaluation of how worthwhile individual independent variables are in predicting Y.

Residuals, SSE, and Standard Error of the Estimate

Residual

The difference between the actual Y value and the Y value predicted by the regression model; the error of the regression model in predicting each value of the dependent variable.

The **residual,** or error, of the regression model is *the difference between the Y value and the predicted value of Y, $(Y - \hat{Y})$.* The residuals for a multiple regression model are solved for in the same manner as they are with simple regression. First, a predicted value of Y, $\hat{Y}$, is determined by entering the value for each independent variable for a given set of observations into the multiple regression equation and solving for $\hat{Y}$. Next, the value of $Y - \hat{Y}$ is computed for each set of observations. Shown here are the calculations for the residual of the first set of observations from Table 13.1. The predicted value of Y for $X_1 = 1605$ and $X_2 = 35$ is

$$\hat{Y} = 57.351 + .0177(1605) - .6663(35) = 62.44$$

Actual value of $Y = 63.0$

Residual $= Y - \hat{Y} = 63.0 - 62.44 = .56$

In Table 13.2, all residuals are shown for the real estate example.

An examination of the residuals in Table 13.2 can reveal some information about the fit of the regression model that was used to predict house prices. The business analyst can observe the residuals and decide whether the errors are small enough to support the accuracy of the model. The house price figures are in units of $1000. Two of the 23 residuals are more than 20.00, or more than $20,000 off in their prediction. On the other hand, two residuals are less than 1, or $1000 off in their prediction.

Outliers

Data points that lie apart from the rest of the points.

Residuals are also helpful in locating outliers. **Outliers** are *data points that are apart, or far, from the mainstream of the other data.* They are sometimes data points that were mistakenly recorded or measured. As every data point has an influence on the regression model, outliers can exert an overly important influence on the model because of their distance from other points. An examination of outliers is worth considering. In Table 13.2, the eighth residual listed is -27.702. This error indicates that the regression model was not nearly as successful in predicting house price on this particular house as it was with others (an error of more than $27,000). For whatever reason, this data point stands somewhat apart from other data points and may be considered an outlier.

Residuals are also useful in testing the assumptions underlying regression analysis. Figure 13.5 contains an Excel-produced plot (using the Chart Wizard and the data from Table 13.2) of the residuals against the fits ($\hat{Y}$). Notice that residual variance seems to increase in the right half of the plot, indicating potential heteroscedasticity. As discussed in Chapter 12, one of the assumptions underlying regression analysis is that the error terms have homoscedasticity or homogeneous variance. That assumption might be violated in

OBSERVATION	PREDICTED PRICE	RESIDUALS
1	62.466	0.534
2	71.465	−6.365
3	71.540	−1.640
4	78.622	−1.822
5	74.073	−0.173
6	75.627	2.273
7	82.991	−8.091
8	105.702	−27.702
9	68.495	10.505
10	81.124	2.276
11	88.265	−8.765
12	91.322	−7.422
13	85.602	−5.902
14	95.383	−10.883
15	85.442	10.558
16	102.772	6.728
17	97.659	4.841
18	107.187	13.813
19	109.356	−4.456
20	111.237	16.763
21	105.064	23.936
22	134.447	−16.547
23	132.460	7.540

TABLE 13.2

Excel-produced Residuals for the Real Estate Regression Model

this example. Figure 13.6 contains an Excel-produced histogram of the residuals (using the histogram feature of data analysis and the fits). Observe that the residuals appear to be somewhat normally distributed indicating that the assumption of normally distributed error terms probably has not been violated.

One of the properties of residuals is that for any given regression model, the residuals add to zero. This zero-sum property can be overcome by *squaring the residuals and then summing the squares.* Such an operation produces the **sum of squares of error (SSE).**

The formula for computing the sum of squares error (SSE) for multiple regression is the same as it is for simple regression.

Sum of squares of error (SSE)
The sum of the residuals squared for a regression model.

$$SSE = \Sigma(Y - \hat{Y})^2$$

For the real estate example, SSE can be computed by squaring and summing the residuals shown in Table 13.2.

$$
\begin{aligned}
SSE = {} & [(.534)^2 + (-6.365)^2 + (-1.640)^2 + (-1.822)^2 + (-0.173)^2 + (2.273)^2 \\
& + (-8.091)^2 + (-27.702)^2 + (10.505)^2 + (2.276)^2 + (-8.765)^2 + \\
& (-7.422)^2 + (-5.902)^2 + (-10.883)^2 + (10.558)^2 + (6.728)^2 + (4.841)^2 + \\
& (13.813)^2 + (-4.456)^2 + (16.763)^2 + (23.936)^2 + (-16.547)^2 + (7.540)^2] \\
= {} & 2861.0
\end{aligned}
$$

SSE can also be obtained directly from the Excel regression output by selecting the value of SS (sum of squares) listed beside "Residual." Shown here is the ANOVA portion

Figure 13.5   Excel plot of residuals against the fits

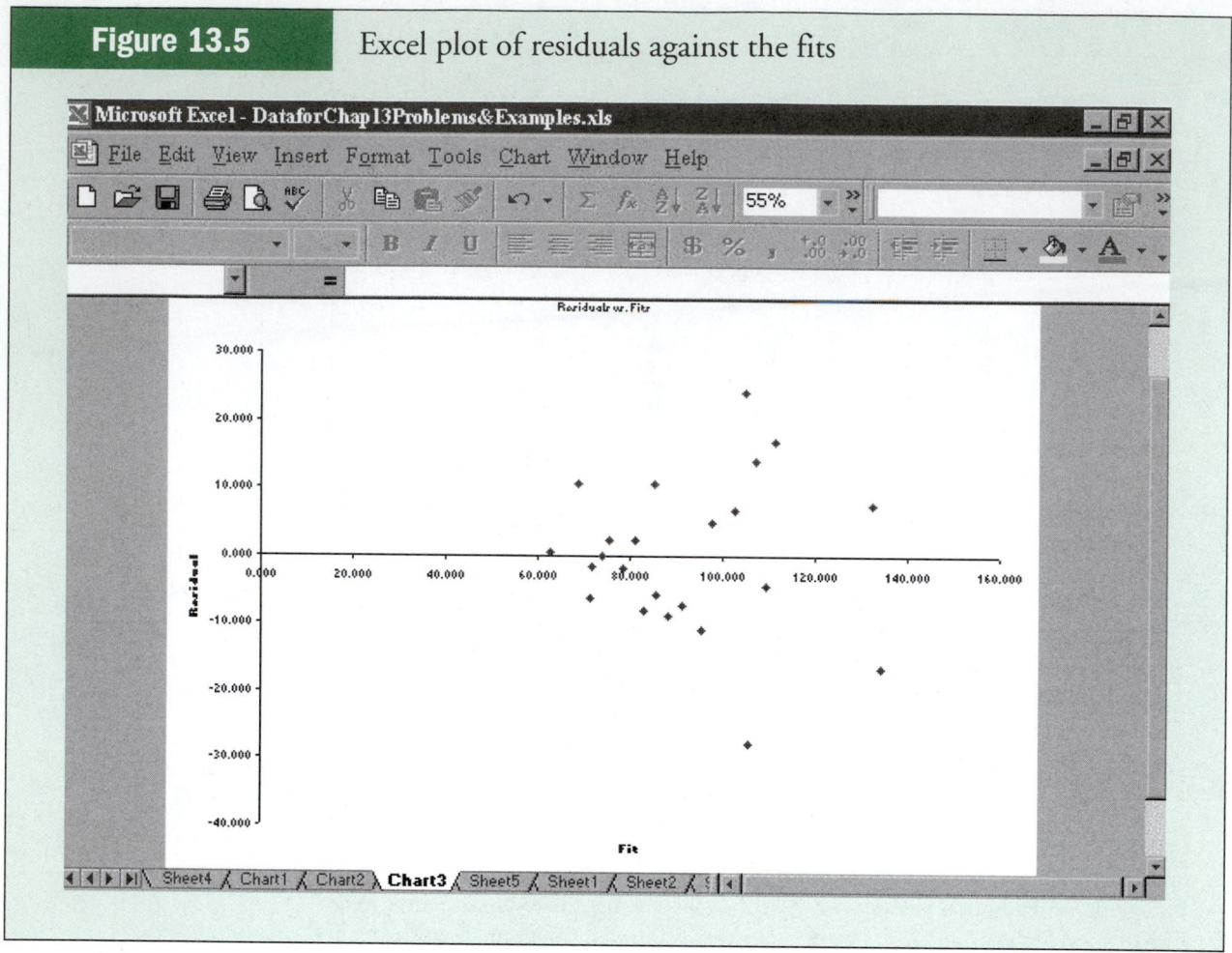

of the output displayed in Figure 13.3. Note that the SS for "Residual" shown in the ANOVA table equals the value of $\Sigma(Y - \hat{Y})^2$ just computed (2861.0).

ANOVA

	df	SS	MS	F	Significance F
Regression	2	8189.723	4094.862	28.63	0.0000014
Residual	20	(2861.017)	143.051		
Total	22	11050.740			

SSE

Standard error of the estimate (S_e)

A standard deviation of the error of a regression model.

SSE has limited usage as a measure of error. However, it is a tool that is used to solve for other, more useful measures. One of those is the **standard error of the estimate, S_e,** which is essentially *the standard deviation of residuals (error) for the regression model.* As explained in Chapter 12, an assumption underlying regression analysis is that the error terms are approximately normally distributed with a mean of zero. With this information and by the empirical rule, approximately 68% of the residuals should be within $\pm 1 S_e$ and 95% should be within $\pm 2 S_e$. This makes the standard error of the estimate a very useful tool in estimating how accurately a regression model is fitting the data.

The standard error of the estimate is computed by dividing SSE by the degrees of freedom of error for the model and taking the square root.

| Figure 13.6 | Excel-produced histogram of the residuals |

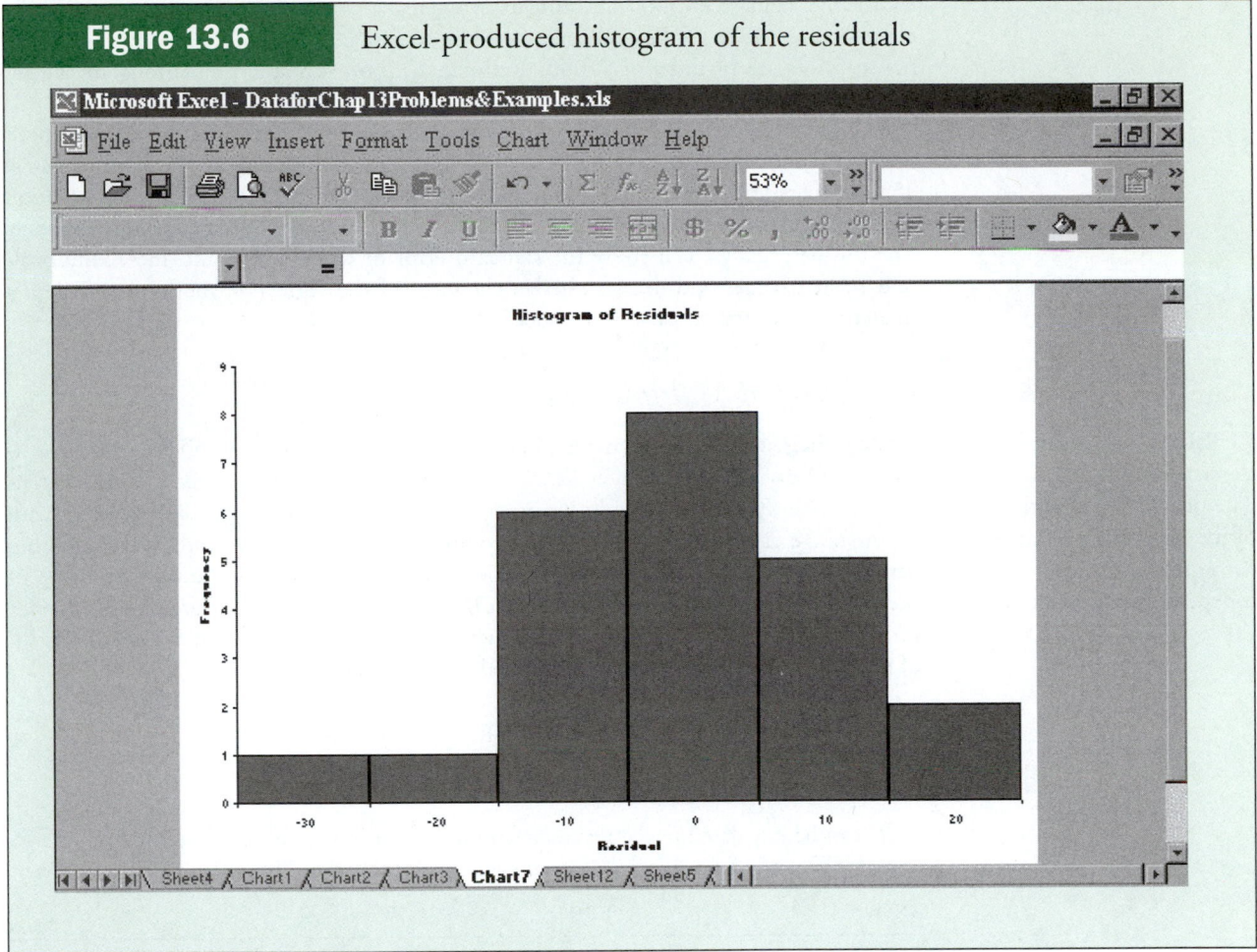

$$S_e = \sqrt{\frac{\text{SSE}}{n - k - 1}}$$

where:

n = number of observations

k = number of independent variables

The value of S_e can be computed for the real estate example as follows.

$$S_e = \sqrt{\frac{\text{SSE}}{n - k - 1}} = \sqrt{\frac{2861}{23 - 2 - 1}} = 11.96$$

The standard error of the estimate, S_e, is usually given as standard output from regression analysis by computer software packages. The Excel output displayed in Figure 13.3 contains

Standard Error = 11.960.

This is the standard error of the estimate for the real estate example. By the empirical rule, approximately 68% of the residuals should be within $\pm 1 S_e = \pm 1(11.96) = \pm 11.96$. Since house prices are in units of $1000, approximately 68% of the predictions are within $\pm 11.96(\$1000)$, or $\pm \$11,960$.

The residuals for this example, presented in Table 13.2, show that 18/23, or about 78%, of the residuals are within this span. According to the empirical rule, approximately 95% of the residuals should be within $\pm 2 S_e$, or $\pm 2(11.96) = \pm 23.92$. Further examination of the residual values in Table 13.2 shows that 21 of 23, or 91%, are within this range. The business analyst can study the standard error of the estimate and these empirical-rule–related ranges and decide whether the error of the regression model is sufficiently small to justify further use of the model.

Coefficient of Multiple Determination (R^2)

Coefficient of multiple determination (R^2)
The proportion of variation of the dependent variable accounted for by the independent variables in the regression model.

The **coefficient of multiple determination (R^2)** is analogous to coefficient of determination (r^2) discussed in Chapter 12. R^2 represents *the proportion of variation of the dependent variable, Y, accounted for by the independent variables in the regression model*. As with r^2, the range of possible values for R^2 is from 0 to 1. An R^2 of 0 indicates no relationship between the predictor variables in the model and Y. An R^2 of 1 indicates that 100% of the variability of Y has been accounted for by the predictors. Of course, it is desirable for R^2 to be high, indicating the strong predictability of a regression model. The coefficient of multiple determination can be calculated by the following formula.

$$R^2 = \frac{\text{SSR}}{\text{SS}_{YY}} = 1 - \frac{\text{SSE}}{\text{SS}_{YY}}$$

R^2 can be calculated in the real estate example by using the SS regression (SSR), the SS error (SSE), and SS total (SS_{YY}) from the ANOVA portion of Figure 13.3.

ANOVA

	df	SS	MS	F	Significance F
Regression	2	8189.723	4094.862	28.63	0.0000014
Residual	20	2861.017	143.051		
Total	22	11050.740			

$$R^2 = \frac{\text{SSR}}{\text{SS}_{YY}} = \frac{8189.723}{11050.740} = .741$$

or

$$R^2 = 1 - \frac{\text{SSE}}{\text{SS}_{YY}} = 1 - \frac{2861.017}{11050.740} = .741$$

In addition, Excel prints out R^2 as standard output with regression analysis. A reexamination of Figure 13.3 reveals that R^2 is given as

$$R \text{ Square} = 0.741$$

This result indicates that a relatively high proportion of the variation of the dependent variable, house price, is accounted for by the independent variables in this regression model.

Adjusted R^2

As additional independent variables are added to a regression model, the value of R^2 cannot decrease, and in most cases it will increase. In the formulas for determining R^2,

$$R^2 = \frac{SSR}{SS_{YY}} = 1 - \frac{SSE}{SS_{YY}}$$

The value of SS_{YY} for a given set of observations will remain the same as independent variables are added to the regression analysis because SS_{YY} is the sum of squares for the dependent variable. Because additional independent variables are likely to increase SSR at least by some amount, the value of R^2 will probably increase for any additional independent variables.

However, sometimes additional independent variables add no *significant* information to the regression model, yet R^2 increases. R^2 therefore may yield an inflated figure. Statisticians have developed an **adjusted R^2** to take into consideration both the additional information each new independent variable brings to the regression model and the changed degrees of freedom of regression. Many standard statistical computer packages now compute and report adjusted R^2 as part of the output. The formula for computing adjusted R^2 is

$$\text{Adjusted } R^2 = 1 - \frac{\dfrac{SSE}{n-k-1}}{\dfrac{SS_{YY}}{n-1}}.$$

Adjusted R^2
A modified value of R^2 in which the degrees of freedom are taken into account, thereby allowing the business analyst to determine whether the value of R^2 is inflated for a particular multiple regression model.

The value of adjusted R^2 for the real estate example can be solved by using information from the ANOVA portion of the computer output in Figure 13.3.

ANOVA	$n-k-1$	$n-1$	SSE	SS_{YY}		
	df	*SS*		*MS*	*F*	*Significance F*
Regression	2	8189.723		4094.862	28.63	0.0000014
Residual	20	2861.017		143.051		
Total	22	11050.740				

$$\text{Adj. } R^2 = 1 - \frac{\dfrac{2861.017}{20}}{\dfrac{11050.740}{22}} = 1 - .285 = .715$$

The Excel regression output in Figure 13.3 contains the value of the adjusted R^2 already computed. For the real estate example, this value is shown as

Adjusted R Square = 0.715

A comparison of R^2 (.741) with the adjusted R^2 (.715) for this example shows that the adjusted R^2 reduces the overall proportion of variation of the dependent variable accounted for by the independent variables by a factor of .026, or 2.6%. The gap between the R^2 and adjusted R^2 tends to increase as nonsignificant independent variables are added to the regression model. As n increases, the difference between R^2 and adjusted R^2 becomes less.

Figure 13.7

Annotated version of
the Excel output of
regression for the real
estate problem

SUMMARY OUTPUT

Regression Statistics

Multiple R	0.861	← Coefficient of multiple determination (R^2)
R Square	0.741	
Adjusted R Square	0.715	← Adjusted R^2
Standard Error	11.960	← Standard error of estimate (S_e)
Observations	23	

ANOVA table and *F* test for overall model

ANOVA

	df	SS	MS	F	Significance F
Regression	2	8189.723	4094.862	28.63	0.0000014
Residual	20	2861.017	143.051		
Total	22	11050.740			

	Coefficients	*Standard Error*	*t Stat*	*P-value*
Intercept	-57.351	10.0072	-5.73	0.000013
Sq. Feet	0.0177	0.0031	5.63	.0.000016
Age	-0.6663	0.2280	-2.92	0.008418

Components of regression equation

t tests of
regression
coefficients

A Reexamination of the Excel Regression Output

Figure 13.7 shows again the Excel multiple regression output for the real estate example.
Many of the concepts discussed thus far in the chapter are highlighted. Note the following items.

1. The components of equation of the regression model
2. The ANOVA table with the *F* value for the overall test of the model
3. The *t* ratios, which test the significance of the regression coefficients
4. The value of SSE
5. The value of S_e
6. The value of R^2
7. The value of adjusted R^2

**DEMONSTRATION
PROBLEM 13.2**

Discuss the Excel multiple regression output for Demonstration Problem 13.1. Comment
on the *F* test for the overall significance of the model, the *t* tests of the regression coefficients, and the values of S_e, R^2, and adjusted R^2.

SOLUTION

This regression analysis was done to predict the volume of freight cargo shipped annually in a country by road using the predictors "length of roads" and "number of commercial vehicles." The equation of the regression model was presented in the solution of

Demonstration Problem 13.1. Shown here is the complete multiple regression output from the Excel analysis of the data.

SUMMARY OUTPUT

Regression Statistics	
Multiple R	0.812
R Square	0.659
Adjusted R Square	0.488
Standard Error	44273.87
Observations	7

ANOVA

	df	SS	MS	F	Significance F
Regression	2	15148592381	7574296191	3.86	0.116
Residual	4	7840701114	1960175278		
Total	6	22989293495			

	Coefficients	Standard Error	t Stat	P-value
Intercept	-26425.45	67624.94	-0.39	0.716
Length of Roads	0.1018	0.0435	2.34	0.079
No. of Comm. Veh.	0.0409	0.0171	2.39	0.075

RESIDUAL OUTPUT

Observation	Predicted Freight Cargo	Residuals
1	247276.61	31529.39
2	134768.10	43590.90
3	191296.29	-47296.29
4	133523.80	5451.20
5	105201.93	19969.07
6	111667.11	-5843.11
7	143450.17	-47401.17

The value of F for this problem is 3.86, with a p value of .116, which is not significant at $\alpha = .05$. On the basis of this information, the null hypothesis would not be rejected for the overall test of significance. None of the regression coefficients are significantly different from zero and there is no significant predictability of the volume of freight cargo shipped by road from this regression model.

An examination of the t ratios support this conclusion using $\alpha = .05$. The t ratio for length of roads is 2.34 with an associated p value of .079, and the t ratio for number of commercial vehicles is 2.39 with an associated p value of .075. Neither p value is less than .05.

The standard error of the estimate is $S_e = 44,273.87$, indicating that approximately 68% of the residuals are within $\pm 44,273.87$. An examination of the Excel-produced residuals shows that actually five out of seven, or 71.4%, of the residuals fall in this interval. Approximately 95% of the residuals should be within $\pm 2(44,273.87) = \pm 88,547.74$, and an examination of the Excel-produced residuals shows that seven out of seven, or 100%, of the residuals are within this interval. Shipping industry analysts could examine the value of the standard error of the estimate to determine whether this model produces results with small enough error to suit their needs.

R^2 for this regression analysis is .659 or 65.9%. That is, 65.9% of the variation in the volume of freight cargo is accounted for by these two independent variables. Conversely, 34.1% of the variation is unaccounted for by this model. The adjusted R^2 is only .488 or 48.8%, indicating that the value of R^2 is considerably inflated. Thus, it could be that the two predictors of the regression model actually account for less than half of the variation of the dependent variable when R^2 is adjusted.

This problem highlights the notion that a regression model can be developed for data and not really fit the data in a significant way. By examining the values of F, t, S_e, R^2, and adjusted R^2, the business analyst can begin to understand whether the regression model is providing any significant predictability for Y.

13.2 Problems

13.7 Examine the Excel output shown here for a multiple regression analysis. How many predictors were analyzed? Determine the equation of the regression model. Comment on the overall significance of the regression model. What is the value of the standard error of the estimate? Find R^2 and compare it with the adjusted value of R^2. Discuss the t ratios of the variables and their significance.

SUMMARY OUTPUT

Regression Statistics	
Multiple R	0.664
R Square	0.440
Adjusted R Square	0.268
Standard Error	10.258
Observations	35

ANOVA

	df	SS	MS	F	Significance F
Regression	8	2153.051	269.131	2.56	0.0334
Residual	26	2736.044	105.232		
Total	34	4889.096			

	Coefficients	Standard Error	t Stat	P-value
Intercept	−56.594	19.6515	2.88	0.008
X1	0.208	0.1608	1.29	0.208
X2	−0.368	0.1434	−2.56	0.016
X3	−0.219	0.1357	−1.61	0.119
X4	0.351	0.1779	1.97	0.059
X5	0.260	0.1370	1.90	0.068
X6	−0.131	0.1308	−1.00	0.327
X7	0.044	0.1477	0.30	0.767
X8	−0.048	0.1804	−0.27	0.793

13.8 Following is Excel output for a multiple regression analysis. Study the ANOVA table, the Standard Error of the Estimate, R^2, adjusted R^2, the t ratios, and discuss the strengths and weaknesses of the regression model. Does this model appear to fit the data well? From the information here, what recommendations would you make about the predictor variables in the model?

SUMMARY OUTPUT

Regression Statistics	
Multiple R	0.788
R Square	0.621
Adjusted R Square	0.518
Standard Error	26.373
Observations	15

ANOVA

	df	SS	MS	F	Significance F
Regression	3	12550.193	4183.40	6.01	0.0111
Residual	11	7650.740	695.52		
Total	14	20200.933			

	Coefficients	Standard Error	t Stat	P-value
Intercept	116.294	37.317	3.12	0.010
X1	-1.255	0.301	-4.17	0.002
X2	0.249	0.558	0.45	0.664
X3	0.491	0.306	1.61	0.137

13.9 Using the data in Problem 13.5, develop a multiple regression model to predict per capita consumption by the consumption of paper, fish, and gasoline. Discuss the output and pay particular attention to the F test, the t tests, and the values of S_e, R^2, and the adjusted R^2.

13.10 Using the data from Problem 13.6, develop a multiple regression model to predict insider ownership from debt ratio and dividend payout. Comment on the strength of the model and the predictors by examining the ANOVA table, the t tests, and the values of S_e, R^2, and the adjusted R^2.

13.11 Develop a multiple regression model to predict Y from X_1, X_2, and X_3 using the following data. Discuss the values of F, t, S_e, R^2, and adjusted R^2. Compute the residuals. Plot a graph of the residuals against the fits. Construct a histogram of the residuals. By observing the residuals, the residual plot, and the histogram, comment on any possible violations of regression assumptions.

Y	X_1	X_2	X_1
5.3	44	11	401
3.6	24	40	219
5.1	46	13	394
4.9	38	18	362
7.0	61	3	453
6.4	58	5	468
5.2	47	14	386
4.6	36	24	357
2.9	19	52	206
4.0	31	29	301
3.8	24	37	243
3.8	27	36	228
4.8	36	21	342
5.4	50	11	421
5.8	55	9	445

13.12 Use the following data to develop a regression model to predict Y from X_1 and X_2. Comment on the output. Develop a regression model to predict Y from X_1 only. Compare the results of this model with those of the model using both predictors. How do the values of R^2 compare? What might you conclude by examining the output from both regression models? Using the residuals, a plot of the residuals against the fits, and a histogram of the residuals, discuss any possible violations of regression assumptions.

Y	X_1	X_2
28	12.6	134
43	11.4	126
45	11.5	143
49	11.1	152
57	10.4	143
68	9.6	147
74	9.8	128
81	8.4	119
82	8.8	130
86	8.9	135
101	8.1	141
112	7.6	123
114	7.8	121
119	7.4	129
124	6.4	135

13.13 Study the following Excel multiple regression output. How many predictors are there in this model? How many observations? What is the equation of the regression line? Discuss the strength of the model in terms of R^2, adjusted R^2, S_e, and F. Which predictors if any are significant? Why or why not? Comment on the overall effectiveness of the model.

SUMMARY OUTPUT

Regression Statistics	
Multiple R	0.842407116
R Square	0.709649749
Adjusted R Square	0.630463317
Standard Error	109.4295947
Observations	15

ANOVA

	df	*SS*	*MS*	*F*	*Significance F*
Regression	3	321946.8018	107315.6	8.961759	0.00272447
Residual	11	131723.1982	11974.84		
Total	14	453670			

	Coefficients	*Standard Error*	*t Stat*	*P-value*
Intercept	657.0534435	167.4595388	3.923655	0.002378
X Variable 1	5.710310868	1.791835982	3.186849	0.008655
X Variable 2	-0.416916682	0.322192459	-1.294	0.222174
X Variable 3	-3.471481072	1.442934778	-2.40585	0.03487

Some variables are referred to as **qualitative variables** (as opposed to *quantitative* variables) because qualitative variables do not yield quantifiable outcomes. Instead, qualitative variables yield nominal or ordinal level information, which is used more to categorize items. These variables have a role in multiple regression and are referred to as **indicator, or dummy, variables.** In this section, we will examine the role of indicator, or dummy, variables as predictors or independent variables in multiple regression analysis.

Indicator variables arise in many ways in business research. Mail questionnaire or personal interview demographic questions are prime candidates because they tend to generate qualitative measures on such items as gender, geographic region, occupation, marital status, level of education, economic class, political affiliation, religion, management/nonmanagement status, buying/leasing a home, method of transportation, or type of broker. In one business study, business analysts were attempting to develop a multiple regression model to predict the distances shoppers drive to malls in the greater Cleveland area. One independent variable was whether or not the mall was located on the shore of Lake Erie. In a second study, a site-location model for pizza restaurants included indicator variables for (1) whether or not the restaurant served beer and (2) whether or not the restaurant had a salad bar.

These indicator variables are qualitative in that no interval or ratio level measurement is assigned to a response. For example, if a mall is located on the shore of Lake Erie, awarding it a score of 20 or 30 or 75 because of its location makes no sense. In terms of gender, what value would you assign to a man or a woman in a regression study? Yet these types of indicator, or dummy, variables are often useful in multiple regression studies and can be included if they are coded in the proper format.

Most business analysts code indicator variables by using 0 or 1. For example, in the shopping mall study, malls located on the shore of Lake Erie could be assigned a 1, and all other malls would then be assigned a 0. The assignment of 0 or 1 is arbitrary, with the number merely holding a place for the category. For this reason, the coding is referred to as "dummy" coding; the number represents a category by holding a place and is not a measurement.

Many indicator, or dummy, variables are dichotomous, such as male/female, salad bar/no salad bar, employed/not employed, and rent/own. For these variables, a value of 1 is arbitrarily assigned to one category and a value of 0 is assigned to the other category. Some qualitative variables contain several categories, such as the variable "type of job," which might have the categories assembler, painter, and inspector. In this case, using a coding of 1, 2, and 3, respectively, is tempting. However, that type of coding creates problems for multiple regression analysis. For one thing, the category "inspector" would receive a value that is three times that of "painter." In addition, the values of 1, 2, and 3 indicate a hierarchy of job types: assembler < painter < inspector.

The proper way to code such indicator variables is with the 0, 1 coding. Two separate independent variables should be used to code the three categories of type of job. The first variable is assembler, where a 1 is recorded if the person's job is assembler and a 0 is recorded if it is not. The second variable is painter, where a 1 is recorded if the person's job is painter and a 0 is recorded if it is not. A variable should not be assigned to inspector, because all workers in the study for whom a 1 was not recorded either for the assembler variable or the painter variable must be inspectors. Thus, coding the inspector variable would result in redundant information and is not necessary. This reasoning holds for all indicator variables with more than two categories. If an indicator variable has c categories, then $c - 1$ dummy variables must be created and inserted into the regression analysis in order to include the indicator variable in the multiple regression.*

13.3 Indicator (Dummy) Variables

Qualitative variable, or Indicator variable, or Dummy variable Represents whether or not a given item or person possesses a certain characteristic and is usually coded as 0 or 1.

---

*If c indicator variables are included in the analysis, no unique estimators of the regression coefficients can be found. [J. Neter, M. H. Kutner, W. Wasserman, and C. Nachtsheim. *Applied Linear Regression Models*, 3d ed. Chicago, Richard D. Irwin, Inc., 1996.]

An example of an indicator variable with more than two categories is the result of the following question taken from a typical questionnaire.

Your office is located in which region of the country?

___ Northeast   ___ Midwest   ___ South   ___ West

Suppose a business analyst is using a multiple regression analysis to predict the cost of doing business and believes geographic location of the office is a potential predictor. How does the business analyst insert this qualitative variable into the analysis? Because $c = 4$ for this question, three dummy variables are inserted into the analysis. Table 13.3 shows one possible way this may have occurred with 13 respondents. Note that rows 2, 7, and 11 contain all zeros, which indicate that those respondents have offices in the West. Thus, a fourth dummy variable for the West region is not necessary and, indeed, should not be included because the information contained in such a fourth variable is contained in the other three variables.

A word of caution is in order. Because of degrees of freedom and interpretation considerations, it is important that a multiple regression analysis have enough observations to handle adequately the number of independent variables entered. Some business analysts recommend as a rule of thumb that there be at least three observations per independent variable. If a qualitative variable has multiple categories, resulting in several dummy independent variables, and if several qualitative variables are being included in an analysis, the number of predictors can rather quickly exceed the limit of recommended number of variables per number of observations. Nevertheless, dummy variables can be very useful and are a way in which nominal or ordinal information can be recoded and incorporated into a multiple regression model.

As an example, consider the issue of gender discrimination in the salary earnings of workers in some industries. In examining this issue, suppose a random sample of 15 workers is drawn from a pool of employed laborers in a particular industry and the workers' average monthly salaries are determined, along with their age and gender. The data are shown in Table 13.4. As gender can be only male or female, this variable is a dummy variable requiring 0, 1 coding. Suppose we arbitrarily let 1 denote male and 0 denote female. Figure 13.8 is the multiple regression model developed from the data of Table 13.4 by using Excel to predict the dependent variable, monthly salary, by two independent variables, age and gender.

The computer output in Figure 13.8 results in the regression equation for this model.

$$\text{Salary} = 0.7321 + 0.1112 \text{ Age} + 0.4587 \text{ Gender}$$

TABLE 13.3

Coding for the Indicator Variable of Geographic Location for Regression Analysis

NORTHEAST X_1	MIDWEST X_2	SOUTH X_3
1	0	0
0	0	0
1	0	0
0	0	1
0	1	0
0	1	0
0	0	0
0	0	1
1	0	0
1	0	0
0	0	0
0	1	0
0	0	1

An examination of the t ratios reveals that the dummy variable "gender" has a regression coefficient that is significant at $\alpha = .001$ ($t = 8.58$, $p = .000$). The overall model is significant at $\alpha = .001$ ($F = 48.54$, $p = .000002$). The standard error of the estimate, $S_e = .0968$, indicates that approximately 68% of the errors of prediction are within ±$96.8 (.0968 × $1000). The R^2 is relatively high at 89.0%, and the adjusted R^2 is 87.2%.

The t value for gender indicates that gender is a significant predictor of monthly salary in this model. This is apparent when one looks at the effects of this dummy variable another way. Figure 13.9 shows the graph of the regression equation when gender = 1 (male) and the graph of the regression equation when gender = 0 (female). When gender = 1 (male), the regression equation becomes

$$.7321 + .1112(\text{age}) + .4587(1) = 1.1908 + .1112(\text{age}).$$

TABLE 13.4
Data for the Monthly Salary Example

MONTHLY SALARY ($1000)	AGE (10 YEARS)	GENDER (1 = MALE, 0 = FEMALE)
1.548	3.2	1
1.629	3.8	1
1.011	2.7	0
1.229	3.4	0
1.746	3.6	1
1.528	4.1	1
1.018	3.8	0
1.190	3.4	0
1.551	3.3	1
0.985	3.2	0
1.610	3.5	1
1.432	2.9	1
1.215	3.3	0
.990	2.8	0
1.585	3.5	1

Figure 13.8

Excel regression output for the monthly salary example

Regression Statistics

Multiple R	0.943
R Square	0.890
Adjusted R Square	0.872
Standard Error	0.0968
Observations	15

ANOVA

	df	SS	MS	F	Significance F
Regression	2	0.9095	0.4547	48.54	0.000002
Residual	12	0.1124	0.0094		
Total	14	1.0219			

	Coefficients	Standard Error	t Stat	P-value
Intercept	0.7321	0.2356	3.11	0.0091
Age	0.1112	0.0721	1.54	0.1488
Gender	0.4587	0.0535	8.58	0.0000

Figure 13.9

Regression model for male and female gender

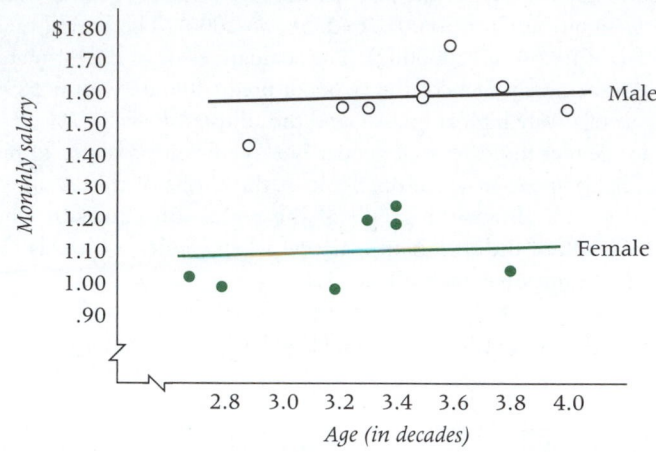

When gender = 0 (female), the regression equation becomes

$$.7321 + .1112(\text{age}) + .4587(0) = .7321 + .1112(\text{age}).$$

The full regression model (with both predictors) has a response surface that is a plane in a three-dimensional space. However, if a value of 1 is entered for gender into the full regression model, as just shown, the regression model is reduced to a line passing through the plane formed by monthly salary and age. If a value of 0 is entered for gender, as shown, the full regression model also reduces to a line passing through the plane formed by monthly salary and age. Figure 13.9 displays these two lines. Notice that the only difference in the two lines is the Y intercept. Observe the monthly salary with male gender, as depicted by $\circ$, versus the monthly salary with female gender, depicted by $\bullet$. The difference in the Y intercepts of these two lines is .4587, which is the value of the regression coefficient for gender. This signifies that, on average, men earn \$458.70 per month more than women for this population.

13.3

Problems

13.14 Analyze the following data by using a multiple regression computer software package to predict Y using X_1 and X_2. Notice that X_2 is a dummy variable. Discuss the output from the regression analysis; in particular, comment on the predictability of the dummy variable.

Y	X_1	X_2
16.8	27	1
13.2	16	0
14.7	13	0
15.4	11	1
11.1	17	0
16.2	19	1
14.9	24	1
13.3	21	0
17.8	16	1
17.1	23	1
14.3	18	0
13.9	16	0

13.15 Given here are the data from a dependent variable and two independent variables. The second independent variable is an indicator variable with several categories. Hence, this variable is represented by X_2, X_3, and X_4. How many categories are there in total for this independent variable? Use a computer to perform a multiple regression analysis on this data to predict Y from the X values. Discuss the output and pay particular attention to the dummy variables.

Y	X_1	X_2	X_3	X_4
11	1.9	1	0	0
3	1.6	0	1	0
2	2.3	0	1	0
5	2.0	0	0	1
9	1.8	0	0	0
14	1.9	1	0	0
10	2.4	1	0	0
8	2.6	0	0	0
4	2.0	0	1	0
9	1.4	0	0	0
11	1.7	1	0	0
4	2.5	0	0	1
6	1.0	1	0	0
10	1.4	0	0	0
3	1.9	0	1	0
4	2.3	0	1	0
9	2.2	0	0	0
6	1.7	0	0	1

13.16 Given here is Excel output for a multiple regression model that was developed to predict Y from two independent variables, X_1 and X_2. Variable X_2 is a dummy variable. Discuss the strength of the multiple regression model on the basis of the output. Focus on the contribution of the dummy variable. Plot X_1 and Y with X_2 as 0, and then plot X_1 and Y with X_2 as 1. Compare the two lines and discuss the differences.

Regression Statistics	
Multiple R	0.623
R Square	0.388
Adjusted R Square	0.341
Standard Error	11.744
Observations	29

ANOVA

	df	SS	MS	F	Significance F
Regression	2	2270.11	1135.05	8.23	0.0017
Residual	26	3585.75	137.91		
Total	28	5855.86			

	Coefficients	Standard Error	t Stat	P-value
Intercept	41.225	6.380	6.46	0.0000008
X_1	1.081	1.353	0.80	0.4316
X_2	−18.404	4.547	−4.05	0.0004

13.17 Falvey, Fried, and Richards* developed a multiple regression model to predict the average price of a meal at New Orleans restaurants. The variables explored included such indicator variables as the following.

Accepts reservations

Accepts credit cards

Has its own parking lot

Has a separate bar or lounge

Has a maitre d'

Has a dress code

Is candlelit

Has live entertainment

Serves alcoholic beverages

Is a steakhouse

Is in the French Quarter

Suppose a relatively simple model is developed to predict the average price of a meal at a restaurant in New Orleans from the number of hours the restaurant is open per week, the probability of being seated upon arrival, and whether the restaurant is located in the French Quarter. Use the following data and a computer to develop such a model. Comment on the output.

PRICE	HOURS	PROBABILITY OF BEING SEATED	FRENCH QUARTER
$ 8.52	65	.62	0
21.45	45	.43	1
16.18	52	.58	1
6.21	66	.74	0
12.19	53	.19	1
25.62	55	.49	1
13.90	60	.80	0
18.66	72	.75	1
5.25	70	.37	0
7.98	55	.64	0
12.57	48	.51	1
14.85	60	.32	1
8.80	52	.62	0
6.27	64	.83	0

13.18 A business analyst has gathered 155 observations on four variables: job satisfaction, occupation, industry, marital status. He or she wants to develop a multiple regression model to predict job satisfaction by the other three variables. All three predictor variables are qualitative variables with the following categories.

1. Occupation: accounting, management, marketing, finance
2. Industry: manufacturing, healthcare, transportation
3. Marital status: married, single

How many variables will be in the regression model? Delineate the number of predictors needed in each category and discuss the total number of predictors.

*ADAPTED FROM: Rodney E. Falvey, Harold O. Fried, and Bruce Richards, "An Hedonic Guide to New Orleans Restaurants," *Quarterly Review of Economics and Finance,* 32, No. 1, Spring 1992.

The regression models presented thus far, along with many others, are based on the **general linear regression model,** which has the form

$$Y = \beta_0 + \beta_1 X_1 + \beta_2 X_2 + \cdots + \beta_k X_k + \epsilon \qquad (13.1)$$

where:

β_0 = the regression constant,
$\beta_1, \beta_2, \ldots, \beta_k$ are the partial regression coefficients for the k independent variables,
$X_1, \ldots, X_k$ are the independent variables, and
k = the number of independent variables.

In this general linear model, the parameters, β_i, are linear. This does not mean, however, that the dependent variable, $Y,$ is necessarily linearly related to the predictor variables. Scatter plots sometimes reveal a curvilinear relationship between X and Y. Multiple regression response surfaces are not restricted to linear surfaces and may be curvilinear.

To this point, the variables, X_i, have represented different predictors. For example, in the real estate example presented previously, the variables, $X_1, X_2,$ represented two predictors: number of square feet in the house and the age of the house, respectively. Certainly, regression models can be developed for more than two predictors. For example, a marketing site-location model could be developed in which sales, as the response variable, is predicted by population density, number of competitors, size of the store, and number of salespeople. Such a model could take the form

$$Y = \beta_0 + \beta_1 X_1 + \beta_2 X_2 + \beta_3 X_3 + \beta_4 X_4 + \epsilon.$$

This regression model has four X_i variables, each of which represents a different predictor.

The general linear model also applies to situations in which some X_i represent recoded data from a predictor variable already represented in the model by another independent variable. In some models, X_i represents variables that have undergone a mathematical transformation to allow the model to follow the form of the general linear model.

This section explores some of these other linear models, including polynomial regression models, regression models with interaction, and models with transformed variables, along with some **nonlinear regression models.**

Polynomial Regression

Regression models in which the highest power of any predictor variable is 1 and in which there are no interaction terms—cross products $(X_i \cdot X_j)$—are referred to as *first-order models.* Simple regression models like those presented in Chapter 12 are *first-order models with one independent variable.* The general model for simple regression is

$$Y = \beta_0 + \beta_1 X_1 + \epsilon.$$

If a second independent variable is added, the model is referred to as a *first-order model with two independent variables* and appears as

$$Y = \beta_0 + \beta_1 X_1 + \beta_2 X_2 + \epsilon.$$

Polynomial regression models are regression models that are second- or higher-order models. They contain squared, cubed, or higher powers of the predictor variable(s) and contain response surfaces that are curvilinear. Yet, they are still special cases of the general linear model.

13.4

More Complex Regression Models

General linear regression model
Regression models that take the form of $Y = \beta_0 + \beta_1 X_1 + \beta_2 X_2 + \ldots + \beta_k X_k + \epsilon,$ where the parameters, $\beta_i,$ are linear.

Nonlinear regression model
Multiple regression models in which the models are nonlinear, such as polynomial models, logarithmic models, and exponential models.

Consider a regression model with one independent variable where the model includes a second predictor, which is the independent variable squared. Such a model is referred to as a *second-order model with one independent variable* because the highest power among the predictors is 2, but there is still only one independent variable. This model takes the following form.

$$Y = \beta_0 + \beta_1 X_1 + \beta_2 X_1^2 + \epsilon$$

This model can be used to explore the possible fit of a quadratic model in predicting a dependent variable. How can this be a special case of the general linear model? Let X_2 of the general linear model be equal to X_1^2; then $Y = \beta_0 + \beta_1 X_1 + \beta_2 X_1^2 + \epsilon$ becomes $Y = \beta_0 + \beta_1 X_1 + \beta_2 X_2 + \epsilon$. Through what process does a business analyst go to develop the regression constant and coefficients for a curvilinear model such as this?

The process of multiple regression analysis assumes a linear fit of the regression coefficients and regression constant, but not necessarily a linear relationship of the independent variable values (Xs). Hence, a business analyst can often accomplish curvilinear regression by recoding the data *before* the multiple regression analysis is attempted.

As an example, consider the data given in Table 13.5. This table contains sales volumes (in $1,000,000) for 13 manufacturing companies along with the number of manufacturer's representatives associated with each firm. A simple regression analysis to predict sales by the number of manufacturer's representatives results in the Excel output in Figure 13.10. This regression output shows a regression model with an r^2 of 87.0%, a standard error of the estimate equal to 51.10, a significant overall F test for the model, and a significant t ratio for the predictor number of manufacturer's representatives.

Figure 13.11(a) is a scatter plot for the data in Table 13.5. Notice that the plot of number of representatives and sales is not a straight line and is an indication that the relationship between the two variables may be curvilinear. To explore the possibility that there may be a quadratic relationship between sales and number of representatives, the business analyst creates a second predictor variable, (number of manufacturer's representatives)², to use in the regression analysis to predict sales along with number of manufacturer's representatives, as shown in Table 13.6. Thus, a variable can be created to explore second-order parabolic relationships by squaring the data from the independent variable of the linear model and entering it into the analysis. Figure 13.11(b) is a scatter plot of sales with (number of manufacturer's reps)². Note that this graph, with the squared term, more closely approaches a

TABLE 13.5

Sales Data for
13 Manufacturing Companies

MANUFACTURER	SALES ($1,000,000)	NUMBER OF MANUFACTURING REPRESENTATIVES
1	2.1	2
2	3.6	1
3	6.2	2
4	10.4	3
5	22.8	4
6	35.6	4
7	57.1	5
8	83.5	5
9	109.4	6
10	128.6	7
11	196.8	8
12	280.0	10
13	462.3	11

Regression Statistics	
Multiple R	0.933
R Square	0.870
Adjusted R Square	0.858
Standard Error	51.10
Observations	13

Figure 13.10

Excel simple regression output for manufacturing example

ANOVA

	df	SS	MS	F	Significance F
Regression	1	192395	192395	73.69	0.000
Residual	11	28721	2611		
Total	12	221117			

	Coefficients	Standard Error	t Stat	P-value
Intercept	-107.03	28.737	-3.72	0.003
MfgrRp	41.026	4.779	1.54	0.000

Figure 13.11 Scatter plots of manufacturing data

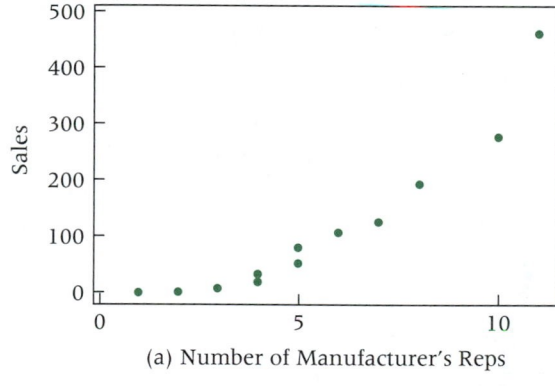

(a) Number of Manufacturer's Reps

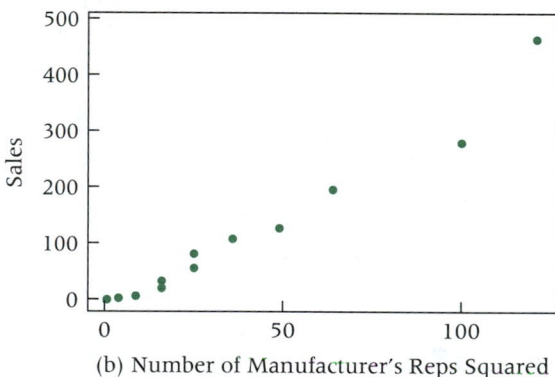

(b) Number of Manufacturer's Reps Squared

TABLE 13.6

Display of Manufacturing Data with Newly Created Variable

MANUFACTURER	SALES ($1,000,000) Y	NUMBER OF MGFR. REPS X_1	(NO. MGFR. REPS)^2 $X_2 = (X_1)^2$
1	2.1	2	4
2	3.6	1	1
3	6.2	2	4
4	10.4	3	9
5	22.8	4	16
6	35.6	4	16
7	57.1	5	25
8	83.5	5	25
9	109.4	6	36
10	128.6	7	49
11	196.8	8	64
12	280.0	10	100
13	462.3	11	121

straight line than does the graph in Figure 13.11(a). By recoding the predictor variable, the business analyst has potentially created a better regression fit.

With these data, a multiple regression model can be developed. Figure 13.12 shows the Excel output for the regression analysis to predict sales by number of manufacturer's representatives and (number of manufacturer's representatives)^2.

Examine the output in Figure 13.12 and compare it with the output in Figure 13.10 for the simple regression model. The R^2 for this model is 97.3%, which is an increase from the r^2 of 87.0% for the single linear predictor model. The standard error of the estimate for this model is 24.59, which is considerably lower than the 51.10 value obtained from the simple regression model. Remember, the sales figures were in units of $1,000,000. The quadratic model reduced the standard error of the estimate by 26.51 ($1,000,000), or $26,510,000. It appears that the quadratic model is a better model for predicting sales.

An examination of the t statistic for the squared term and its associated probability in Figure 13.12 shows that it is statistically significant at $\alpha = .001$ ($t = 6.12$ with a probability of .000). If this t statistic were not significant, the business analyst would most likely drop the squared term and revert to the first-order model (simple regression model).

In theory, third- and higher-order models can be explored. Generally, business analysts tend to utilize first- and second-order regression models more than higher-order models. Remember that most regression analysis is used in business to aid decision making. Higher-power models (third, fourth, etc.) become difficult to interpret and difficult to explain to decision makers. In addition, the business analyst is usually looking for trends and general directions. The higher the order in regression modeling, the more the model tends to follow irregular fluctuations rather than meaningful directions.

Regression Models with Interaction

Often when two different independent variables are used in a regression analysis, there is an *interaction* between the two variables. This is the same interaction discussed in Chap-

Figure 13.12

Excel output for quadratic model of manufacturing example

Regression Statistics

Multiple R	0.986
R Square	0.973
Adjusted R Square	0.967
Standard Error	24.59
Observations	13

ANOVA

	df	SS	MS	F	Significance F
Regression	2	215069	107534	177.79	0.000
Residual	10	6048	605		
Total	12	221117			

	Coefficients	Standard Error	t Stat	P-value
Intercept	18.067	24.673	0.73	0.481
MfgrRp	−15.723	9.550	−1.65	0.131
MfgrRpSq	4.750	0.776	6.12	0.000

ter 11 in two-way analysis of variance, where one variable will act differently over a given range of values for the second variable than it does over another range of values for the second variable. For example, in a manufacturing plant, temperature and humidity might interact in such a way as to have an effect on the hardness of the raw material. The air humidity may affect the raw material differently at different temperatures.

In regression analysis, interaction can be examined as a separate independent variable. An interaction predictor variable can be designed by multiplying the data values of one variable by the values of another variable, thereby creating a new variable. A model that includes an interaction variable is

$$Y = \beta_0 + \beta_1 X_1 + \beta_2 X_2 + \beta_3 X_1 X_2 + \epsilon.$$

The $X_1 X_2$ term is the interaction term. Even though this model has 1 as the highest power of any one variable, it is considered to be a second-order equation because of the $X_1 X_2$ term.

Suppose the data in Table 13.7 represent the closing stock prices for three corporations over a period of 15 months. An investment firm wants to use the prices for stocks 2 and 3 to develop a regression model to predict the price of stock 1. The form of the general linear regression equation for this model is

$$Y = \beta_0 + \beta_1 X_1 + \beta_2 X_2 + \epsilon.$$

where:

  Y = price of stock 1,
  X_1 = price of stock 2, and
  X_2 = price of stock 3.

Using Excel to develop this regression model, the firm's business analyst obtains the first output displayed in Figure 13.13(a). This regression model is a first-order model with two predictors, X_1 and X_2. This model produced a modest R^2 of .472. Both of the t ratios are small and statistically nonsignificant ($t = -.62$ with a p value of .549 and $t = -.36$ with a p value of .728). Although the overall model is statistically significant, $F = 5.37$ with probability of .022, neither predictor is significant.

TABLE 13.7
Prices of Three Stocks over a 15-month Period

STOCK 1	STOCK 2	STOCK 3
41	36	35
39	36	35
38	38	32
45	51	41
41	52	39
43	55	55
47	57	52
49	58	54
41	62	65
35	70	77
36	72	75
39	74	74
33	83	81
28	101	92
31	107	91

Sometimes the effects of two variables are not additive because there are interacting effects between the two variables. In such a case, the business analyst can use multiple regression analysis to explore the interaction effects by including an interaction term in the equation.

$$Y = \beta_0 + \beta_1 X_1 + \beta_2 X_2 + \beta_3 X_1 X_2 + \epsilon$$

Figure 13.13

Two Excel Regression outputs—without and with interaction

(a) First regression analysis with 2 predictors without interaction

SUMMARY OUTPUT

Regression Statistics	
Multiple R	0.687
R Square	0.472
Adjusted R Square	0.384
Standard Error	4.570
Observations	15

ANOVA

	df	SS	MS	F	Significance F
Regression	2	224.29	112.15	5.37	0.022
Residual	12	250.64	20.89		
Total	14	474.93			

	Coefficients	Standard Error	t Stat	P-value
Intercept	50.8555	3.791	13.41	0.000
Stock 2	−0.1190	0.193	−0.62	0.549
Stock 3	−0.0708	0.199	−0.36	0.728

(b) Second regression analysis with the interaction variable included

SUMMARY OUTPUT

Regression Statistics	
Multiple R	0.897
R Square	0.804
Adjusted R Square	0.751
Standard Error	2.909
Observations	15

ANOVA

	df	SS	MS	F	Significance F
Regression	3	381.85	127.28	15.04	0.0003
Residual	11	93.09	8.46		
Total	14	474.93			

	Coefficients	Standard Error	t Stat	P-value
Intercept	12.046	9.312	1.29	0.222
Stock 2	0.879	0.262	3.36	0.006
Stock 3	0.220	0.144	1.54	0.153
Interaction	−0.010	0.002	−4.32	0.001

The equation fits the form of the general linear model,

$$Y = \beta_0 + \beta_1 X_1 + \beta_2 X_2 + \beta_3 X_3 + \epsilon,$$

where $X_3 = X_1 X_2$. Each individual observation of X_3 is obtained through a recoding process by multiplying the associated observations of X_1 and X_2.

Applying this procedure to the stock example, the business analyst uses the interaction term and Excel to obtain the second regression output shown in Figure 13.13(b). This output contains X_1, X_2, and the interaction term, $X_1 X_2$. Observe the R^2, which equals .804 for this model. The introduction of the interaction term has caused the R^2 to increase from 47.2% to 80.4%. In addition, the standard error of the estimate has decreased from 4.57 in the first model to 2.909 in the second model. The t ratios for both the X_1 term and the interaction term are statistically significant in the second model ($t = 3.36$ with a p value of .006 for X_1 and $t = -4.32$ with a probability of .001 for $X_1 X_2$). The inclusion of the interaction term has helped the regression model account for a substantially greater amount of the dependent variable and is a significant contributor to the model.

Figure 13.14(a) is the response surface for the first regression model presented in Figure 13.13(a) (the model without interaction). As you observe the response plane with stock 3 as the point of reference, you see the plane moving upward with increasing values of stock 1 as the plane moves away from you toward smaller values of stock 2. Now examine Figure 13.14(b), the response surface for the second regression model presented in Figure 13.13(b) (the model with interaction). Note how the response plane is twisted, with its slope changing as it moves along stock 2. This pattern is caused by the interaction effects of stock 2 price and stock 3 prices. A cross section of the plane taken from left to right at any given stock 2 price produces a line that attempts to predict the price of stock 3 from the price of stock 1. As you more back through different prices of stock 2, the slope of that line changes, indicating that the relationship between stock 1 and stock 3 varies according to stock 2.

A business analyst also could develop a model using two independent variables with their squares and interaction. Such a model would be a *second-order model with two independent variables.* The model would look like this.

$$Y = \beta_0 + \beta_1 X_1 + \beta_2 X_2 + \beta_3 X_1{}^2 + \beta_4 X_2{}^2 + \beta_5 X_1 X_2 + \epsilon$$

This is sometimes referred to as the "full" quadratic model for two predictors.

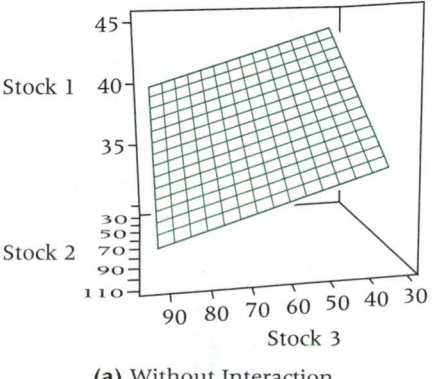

(a) Without Interaction

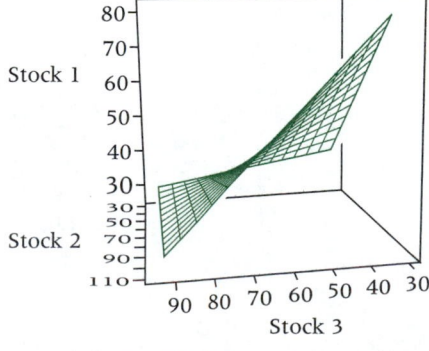

(b) With Interaction

Figure 13.14

Response surfaces for the stock example—without and with interaction

Model Transformation

To this point in examining polynomial and interaction models, the focus has been on re-coding values of X variables. Some multiple regression situations require that the dependent variable, Y, be recoded. Two such transformations include $\log Y$ and $1/Y$.

Suppose the following data represent the annual sales and annual advertising expenditures for seven companies. Can a regression model be developed from these figures that can be used to predict annual sales by annual advertising expenditures?

COMPANY	SALES ($ MILLION/YEAR)	ADVERTISING ($ MILLION/YEAR)
1	2,580	1.2
2	11,942	2.6
3	9,845	2.2
4	27,800	3.2
5	18,926	2.9
6	4,800	1.5
7	14,550	2.7

One mathematical model that is a good candidate for fitting these data is an exponential model of the form

$$Y = \beta_0 \beta_1^X \epsilon.$$

This model can be transformed (by taking the log of each side) to get it in the form of the general linear equation.

$$\log Y = \log \beta_0 + X \log \beta_1$$

This transformed model requires a recoding of the Y data through the use of logarithms. Notice that X is not recoded but that the regression constant and coefficient are in logarithmic scale. If we let $Y' = \log Y$, $\beta_0' = \log \beta_0$, and $\beta_1' = \log \beta_1$, the exponential model is in the form of the general linear model.

$$Y' = \beta_0' + \beta_1' X$$

The process begins by taking the log of the Y values. The data used to build the regression model and the Excel regression output for these data follow.

LOG SALES (Y)	ADVERTISING (X)
3.4116	1.2
4.0771	2.6
3.9932	2.2
4.4440	3.2
4.2771	2.9
3.6812	1.5
4.1629	2.7

Regression Statistics	
Multiple R	0.990
R Square	0.980
Adjusted R Square	0.977
Standard Error	0.054
Observations	7

ANOVA

	df	SS	MS	F	Significance F
Regression	1	0.7392	0.7392	250.36	0.000
Residual	5	0.0148	0.0030		
Total	6	0.7540			

	Coefficients	Standard Error	t Stat	P-value
Intercept	2.9003	0.0729	39.80	0.000
Advertising (X)	0.4751	0.0300	15.82	0.000

A simple regression model (without the log recoding of the Y variable) yields an R^2 of 87% whereas the exponential model R^2 is 98%. The t statistic for advertising is 15.82 with a p value of 0.000 in the exponential model and 5.774 with a p value of 0.002 in the simple regression model. Thus the exponential model gives a better fit than does the simple regression model. An examination of (X^2, Y) and (X^3, Y) models reveals R^2 of .930 and .969, respectively, which are quite high but still not as good as the R^2 yielded by the exponential model.

The resulting equation of the exponential regression model is

$$Y = 2.9003 + .4751X.$$

In using this regression equation to determine predicted values of Y for X, remember that the resulting predicted Y value is in logarithmic form and the antilog of the predicted Y must be taken to get the predicted Y value in raw units. For example, to get the predicted Y value (sales) for an advertising figure of 2.0 ($ million), substitute $X = 2.0$ into the regression equation.

$$Y = 2.9003 + .4751X = 2.9003 + .4751(2.0) = 3.8505$$

The log of sales is 3.8505. Taking the antilog of 3.8505 results in the predicted sales in raw units.

$$\text{antilog}(3.8505) = 7087.61 \ (\$ \ million)$$

Thus, the exponential regression model predicts that $2.0 million of advertising will result in $7,089.61 million of sales.

There are other ways to transform mathematical models so that they can be treated like the general linear model. One example is an inverse model such as

$$Y = \frac{1}{\beta_0 + \beta_1 X_1 + \beta_2 X_2 + \epsilon}.$$

Such a model can be manipulated algebraically into the form

$$\frac{1}{Y} = \beta_0 + \beta_1 X_1 + \beta_2 X_2 + \epsilon.$$

Substituting $Y' = 1/Y$ into this equation results in an equation that is in the form of the general linear model.

$$Y' = \beta_0 + \beta_1 X_1 + \beta_2 X_2 + \epsilon$$

To use this "inverse" model, recode the data values for Y by using $1/Y$. The regression analysis is done on the $1/Y$, X_1, and X_2 data. To get predicted values of Y from this model, enter the raw values of X_1 and X_2. The resulting predicted value of Y from the regression equation will be the inverse of the actual predicted Y value.

DEMONSTRATION PROBLEM 13.3

In the aerospace and defense industry, some cost estimators predict the cost of new space projects by using mathematical models that take the form

$$Y = \beta_0 X^{\beta_1} \epsilon.$$

These cost estimators often use the weight of the object being sent into space as the predictor (X) and the cost of the object as the dependent variable (Y). Quite often β_1 turns out to be a value between 0 and 1, resulting in the predicted value of Y equaling some root of X.

Use the sample cost data given here to develop a cost regression model in the form just shown to determine the equation for the predicted value of Y. Use this regression equation to predict the value of Y for $X = 3000$.

Y (COST IN BILLIONS)	X (WEIGHT IN TONS)
1.2	450
9.0	20,200
4.5	9,060
3.2	3,500
13.0	75,600
0.6	175
1.8	800
2.7	2,100

SOLUTION
The equation

$$Y = \beta_0 X^{\beta_1} \epsilon$$

is not in the form of the general linear model, but it can be transformed by using logarithms,

$$\log Y = \log \beta_0 + \beta_1 \log X + \epsilon,$$

which takes on the general linear form

$$Y' = \beta_0' + \beta_1 X',$$

where:
$Y' = \log Y$,
$\beta_0' = \log \beta_0$, and
$X' = \log X$.

This equation requires that both X and Y be recoded by taking the logarithm of each.

LOG Y	LOG X
.0792	2.6532
.9542	4.3054
.6532	3.9571
.5051	3.5441
1.1139	4.8785
−.2218	2.2430
.2553	2.9031
.4314	3.3222

Using these data, the computer produces the following regression constant and coefficient.

$$b_0' = -1.25292 \quad b_1 = .49606$$

From these values, the equation of the predicted Y value is determined to be

$$\log \hat{Y} = -1.25292 + .49606 \log X.$$

If $X = 3000$, $\log X = 3.47712$, and

$$\log \hat{Y} = -1.25292 + .49606(3.47712) = .47194,$$

then

$$\hat{Y} = \text{antilog}(\log \hat{Y}) = \text{antilog}(.47194) = 2.9644.$$

The predicted value of Y is $2.9644 billion for $X = 3000$ tons of weight.
Taking the antilog of $b_0' = -1.25292$ yields .055857. From this and $b_1 = .49606$, the model can be written in the original form:

$$Y = (.055857)X^{.49606}.$$

Substituting $X = 3000$ into this formula also yields $2.9645 billion for the predicted value of Y.

13.19 Use the following data to develop a quadratic model to predict Y from X. Develop a simple regression model from the data and compare the results of the two models. Does the quadratic model seem to provide any better predictability? Why or why not?

13.4
Problems

X	Y	X	Y
14	200	15	247
9	74	8	82
6	29	5	21
21	456	10	94
17	320		

13.20 Use the following data to develop a curvilinear model to predict Y. Include both X_1 and X_2 in the model in addition to X_1^2, X_2^2, and the interaction term X_1X_2. Comment on the overall strength of the model and the significance of each predictor. Use a computer software package to graph the response surface of the "full" quadratic model. Develop a regression model with the same independent variables as the first model but without the interaction variable. Compare this model to the "full" quadratic model. Graph the response surface of this model. How does it compare to the response surface of the "full" model?

Y	X_1	X_2
47.8	6	7.1
29.1	1	4.2
81.8	11	10.0
54.3	5	8.0
29.7	3	5.7
64.0	9	8.8
37.4	3	7.1
44.5	4	5.4
42.1	4	6.5
31.6	2	4.9
78.4	11	9.1
71.9	9	8.5
17.4	2	4.2
28.8	1	5.8
34.7	2	5.9
57.6	6	7.8
84.2	12	10.2
63.2	8	9.4
39.0	3	5.7
47.3	5	7.0

13.21 Develop a multiple regression model of the form

$$Y = b_0 b_1^X \epsilon$$

using the following data to predict Y from X.

Y	X	Y	X
2485	3.87	740	2.83
1790	3.22	4010	3.62
874	2.91	3629	3.52
2190	3.42	8010	3.92
3610	3.55	7047	3.86
2847	3.61	5680	3.75
1350	3.13	1740	3.19

13.22 The Publishers Information Bureau, Inc., in New York City released magazine advertising expenditure data compiled by leading national advertisers. The data were organized by product type over several years in time. Shown here are data on total magazine advertising expenditures and household equipment and supplies advertising expenditures. Using these data, develop a regression model to predict total magazine advertising expenditures by household equipment and supplies advertising expenditures and by (household equipment and supplies advertising expenditures)^2. Compare this model to a regression model to predict total magazine advertising expenditures by only household equipment and supplies advertising expenditures. Construct a scatter plot of the data. Does the shape of the plot suggest some alternative models. If so, develop at least one other model and compare the model to the other two previously developed.

TOTAL MAGAZINE ADVERTISING EXPENDITURES (MILLIONS $)	HOUSEHOLD EQUIPMENT AND SUPPLIES EXPENDITURES (MILLIONS $)
1193	34
2846	65
4668	98
5120	93
5943	102
6644	103

13.23 Dun & Bradstreet Corporation reports, among other things, information about new business incorporations and number of business failures over the years. Shown here are data on business failures since 1970 and current liabilities of the failing companies. Use these data and the following model to predict current liabilities of the failing companies by the number of business failures. Discuss the strength of the model.

$$Y = b_0 b_1^X \epsilon$$

Now develop a different regression model by recoding X. Compare your models.

RATE OF BUSINESS FAILURES SINCE 1970 (10,000)	CURRENT LIABILITIES OF FAILING COMPANIES (MILLIONS $)
44	1,888
43	4,380
42	4,635
61	6,955
88	15,611
110	16,073
107	29,269
115	36,937
120	44,724
102	34,724
98	39,126
65	44,261

13.24 Shown here is Excel output for two regression analyses. The first regression analysis is done with the predictor variables X_1 and X_2. The second regression analysis is done with X_1 and X_2 along with an interaction term, $X_1 X_2$. Study the output for each model and compare the models. Did the inclusion of the interaction term make much of a difference in the model?

Model without Interaction

Regression Statistics	
Multiple R	0.943
R Square	0.890
Adjusted R Square	0.882
Standard Error	4.199
Observations	30

ANOVA

	df	SS	MS	F	Significance F
Regression	2	3851.544	1925.722	109.24	0.0000
Residual	27	475.976	17.629		
Total	29	4327.521			

	Coefficients	Standard Error	t Stat	P-value
Intercept	-15.0805	9.6281	-1.57	0.1289
X1	0.5450	0.0615	8.86	0.0000
X2	0.6070	0.0556	10.92	0.0000

SUMMARY OUTPUT   Model with Interaction

Regression Statistics	
Multiple R	0.956
R Square	0.914
Adjusted R Square	0.904
Standard Error	3.778
Observations	30

ANOVA

	df	SS	MS	F	Significance F
Regression	3	3956.371	1318.79	92.38	0.0000
Residual	26	371.149	14.27		
Total	29	4327.521			

	Coefficients	Standard Error	t Stat	P-value
Intercept	103.398	44.571	2.32	0.0285
X1	-0.436	0.366	-1.19	0.2447
X2	-0.408	0.378	-1.08	0.2899
X1X2	0.008	0.003	2.71	0.0118

13.5
Model-Building: Search Procedures

To this point in the chapter, we have explored various types of multiple regression models, we have evaluated the strengths of regression models, and we have learned how to understand more about the output from Excel's regression analysis tool. In this section we examine procedures for developing several multiple regression model options to aid in the decision-making process.

Suppose a business analyst wants to develop a multiple regression model to predict the world production of crude oil. The business analyst realizes that much of the world crude oil market is driven by variables related to usage and production in the United States. She decides to use as predictors the following four independent variables.

1. U.S. energy consumption
2. Gross U.S. nuclear electricity generation
3. U.S. coal production
4. Fuel rate of U.S.-owned automobiles

The business analyst measured data for each of these variables for the year preceding each data point of world crude oil production, figuring that the world production is driven by the previous year's activities in the United States. It would seem that as the energy consumption of the United States increases, so would world production of crude oil. In addition, it makes sense that as nuclear electricity generation, coal production, and fuel rates increase, world crude oil production would decrease if energy consumption stays approximately constant.

In Table 13.8 are data for the four independent variables along with the dependent variable, world crude oil production. Using the data presented in Table 13.8, the business analyst attempted to develop a multiple regression model using four different independent variables. The result of this process was the Excel output in Figure 13.15. Examining the output, the business analyst can reach some conclusions about that particular model and its variables.

The output contains an R^2 value of 92.1%, a standard error of the estimate of 1.188, and an overall significant F value of 60.85. Notice from Figure 13.15 that the t ratios indicate that the regression coefficients of three of the predictor variables, nuclear, coal, and fuel rate, are not significant at $\alpha = .05$. If the business analyst were to drop these three variables out of the regression analysis and rerun the model with the other predictor only, what would happen to the model? What if the business analyst ran a regression model with only two or three predictors? How would these models compare to the full model with all four predictors? Are all the predictors necessary?

TABLE 13.8 Data for Multiple Regression Model to Predict Crude Oil Production

WORLD CRUDE OIL PRODUCTION (MILLION BARRELS PER DAY)	U.S. ENERGY CONSUMPTION (QUADRILLION BTUS PER YEAR)	U.S. NUCLEAR ELECTRICITY GROSS GENERATION (BILLION KILOWATT-HOURS)	U.S. COAL PRODUCTION (MILLION SHORT-TONS)	U.S. FUEL RATE FOR AUTOMOBILES (MILES PER GALLON)
55.7	74.3	83.5	598.6	13.4
55.7	72.5	114.0	610.0	13.6
52.8	70.5	172.5	654.6	14.0
57.3	74.4	191.1	684.9	13.8
59.7	76.3	250.9	697.2	14.1
60.2	78.1	276.4	670.2	14.3
62.7	78.9	255.2	781.1	14.6
59.6	76.0	251.1	829.7	16.0
56.1	74.0	272.7	823.8	16.5
53.5	70.8	282.8	838.1	16.9
53.3	70.5	293.7	782.1	17.1
54.5	74.1	327.6	895.9	17.4
54.0	74.0	383.7	883.6	17.5
56.2	74.3	414.0	890.3	17.4
56.7	76.9	455.3	918.8	18.0
58.7	80.2	527.0	950.3	18.8
59.9	81.4	529.4	980.7	19.0
60.6	81.3	576.9	1029.1	20.3
60.2	81.1	612.6	996.0	21.2
60.2	82.2	618.8	997.5	21.0
60.2	83.9	610.3	945.4	20.6
61.0	85.6	640.4	1033.5	20.8
62.3	87.2	673.4	1033.0	21.1
64.1	90.0	674.7	1063.9	21.2
66.3	90.6	628.6	1089.9	21.5
67.0	89.7	666.8	1109.8	21.6

Figure 13.15

Regression results
for oil production
example—model 15
(4 independent
variables)

SUMMARY OUTPUT

Regression Statistics	
Multiple R	0.959
R Square	0.921
Adjusted R Square	0.905
Standard Error	1.1884
Observations	26

ANOVA

	df	SS	MS	F	Significance F
Regression	4	343.877	85.941	60.85	0.00000
Residual	21	29.661	1.412		
Total	25	373.427			

	Coefficients	Standard Error	t Stat	P-value
Intercept	3.8260	8.016	0.48	0.63807
USEnCons	0.7843	0.080	9.85	0.00000
USNucGen	-0.0043	0.007	-0.64	0.52768
USCoalPr	0.0109	0.006	1.74	0.09580
FuelRate	-0.8253	0.459	-1.80	0.08629

In developing regression models for business decision making, there are at least two considerations. The first is to develop a regression model that accounts for the most variation of the dependent variable—that is, develop models that maximize the explained proportion of the deviation of the Y values. At the same time, the regression model should be as parsimonious (simple and economic) as possible. The reason is that the more complicated a quantitative model becomes, the harder it is for managers to understand and implement the model. In addition, the more variables there are in a model, the more expensive it is to gather historical data or update present data for the model. These two considerations (dependent variable explanation and parsimony of the model) are quite often in opposition to each other. Hence the business analyst, as the model builder, often needs to explore many model options.

In the world crude oil production regression model, if three variables explain the deviation of world crude oil production nearly as well as four variables, the simpler model is more attractive. Is there some way to conduct regression analysis so that the business analyst can examine several models and then choose the most attractive one? The answer is to use search procedures.

Search procedures
Processes whereby more than one multiple regression model is developed for a given database, and the models are compared and sorted by different criteria, depending on the given procedure.

Search procedures are *processes whereby more than one multiple regression model is developed for a given data base, and the models are compared and sorted by different criteria, depending on the given procedure.* Virtually all search procedures are done on a computer. Several search procedures are discussed in this section, including all possible regressions, stepwise regression, forward selection, and backward elimination.

All Possible Regressions

All possible regressions
A multiple regression search procedure in which all possible multiple linear regression models are determined from the data using all variables.

The **all possible regressions** search procedure computes all the possible linear multiple regression models from the data using all the independent variables. If a data set contains k independent variables, this approach will compute $2^k - 1$ different regression models.

ONE PREDICTOR	TWO PREDICTORS	THREE PREDICTORS	FOUR PREDICTORS
X_1	X_1, X_2	X_1, X_2, X_3	X_1, X_2, X_3, X_4
X_2	X_1, X_3	X_1, X_2, X_4	
X_3	X_1, X_4	X_1, X_3, X_4	
X_4	X_2, X_3	X_2, X_3, X_4	
	X_2, X_4		
	X_3, X_4		

TABLE 13.9
Predictors for All Possible Regressions with Four Independent Variables

For the crude oil production example, the procedure of all possible regressions would produce $2^4 - 1 = 15$ different models for the $k = 4$ independent variables. It produces all the single-predictor models. The number of these equals the combination of four items taken one at a time, $_4C_1$. As discussed in Chapter 4, $_4C_1 = 4!/1!3! = 4$. In like manner, the all possible regressions procedure produces all the models with two predictors. The number of these equals the combination of four items taken two at a time, $_4C_2$. The value of $_4C_2 = 4!/2!2! = 6$. In like manner, the procedure produces four models with three predictors and one model with four predictors for a total of 15 different models. Table 13.9 presents the 15 possible combinations of predictors.

Once all the possible models have been determined, they are evaluated according to criteria that measure the adequacy of the model to fit the data. As discussed in Section 13.2, four possible measures of fit include the coefficient of determination, R^2; the adjusted R^2; the standard error of the estimate; and a test of the overall model with the F statistic.

The first of these four criteria, R^2, will always increase as more predictor variables are added to a model. The randomness of each added variable, even irrelevant variables, will improve the fit of the model. As a consequence, the model with all k variables will have the greatest R^2 value. The second of the four criteria, the adjusted R^2, modifies R^2 to account for the relationship of the sample size to the number of variables in the model. For that reason, the adjusted R^2 is used instead of R^2.

The third of these criteria, the standard error of the estimate, always yields equivalent results to the adjusted R^2. The model with the largest adjusted R^2 will also have the smallest standard error and vice versa.

The fourth criterion, the F statistic, measures the overall fit of a model. However, it is not useful for comparing models with differing degrees of freedom. To use it for such a comparison requires the consideration of the p-values for the corresponding F statistic values.

A criterion not discussed in Section 13.2 that is frequently used for comparing alternative regression models is the C_p statistic. It provides a measure of the difference between the estimated model and the true model. It is computed through the following equation.

$$C_p = \frac{(1 - R_p^2)(n - T)}{(1 - R_T^2)} - (n - 2p) + 2$$

where:

n = sample size

p = the number of independent variables included in a specific regression model

$T = 1 +$ the number of independent variables in the full regression model ($= k + 1$)

R_p^2 = coefficient of determination for the regression model with p independent variables

R_T^2 = coefficient of determination for the full regression model

| Figure 13.16 | All possible regressions criteria values for oil production example |

Microsoft Excel - All Possible Regressions

File Edit View Insert Format Tools Data Window Help XtraStat

R29

	A	B	C	D	E	F	G	H	I	J	K	L	M	N
1	Regr.	Dependent	Independents Variables Included				R	Adjusted	Overall Model			Cp Computations		
2	No.	Variable	En Cons	Nuc Gen	Coal Pr	Fuel Rate	Squared	R Squared	F Statistic	p Value	p	Rp-sqrd.	Cp	Cp - p - 1
3	1	Cr Oil Prd	yes				0.8524	0.8462	138.6	1.86E-11	1	0.8524	17.04	15.04
4	2	Cr Oil Prd		yes			0.4500	0.4271	19.6	0.000176	1	0.4500	123.47	121.47
5	3	Cr Oil Prd			yes		0.3891	0.3637	15.3	0.000662	1	0.3891	139.57	137.57
6	4	Cr Oil Prd				yes	0.3425	0.3151	12.5	0.001686	1	0.3425	151.90	149.90
7	5	Cr Oil Prd	yes	yes			0.9055	0.8973	110.2	1.64E-12	2	0.9055	4.99	1.99
8	6	Cr Oil Prd	yes		yes		0.8828	0.8726	86.6	1.97E-11	2	0.8828	11.00	8.00
9	7	Cr Oil Prd	yes			yes	0.9083	0.9003	113.9	1.17E-12	2	0.9083	4.25	1.25
10	8	Cr Oil Prd		yes	yes		0.4524	0.4048	9.5	0.000983	2	0.4524	124.83	121.83
11	9	Cr Oil Prd		yes		yes	0.5313	0.4905	13.0	0.000164	2	0.5313	103.96	100.96
12	10	Cr Oil Prd			yes	yes	0.3946	0.3419	7.5	0.003118	2	0.3946	140.12	137.12
13	11	Cr Oil Prd	yes	yes	yes		0.9083	0.8958	72.7	1.42E-11	3	0.9083	6.25	2.25
14	12	Cr Oil Prd	yes	yes		yes	0.9091	0.8967	73.3	1.3E-11	3	0.9091	6.04	2.04
15	13	Cr Oil Prd	yes		yes	yes	0.9190	0.9080	83.2	3.66E-12	3	0.9190	3.42	-0.58
16	14	Cr Oil Prd		yes	yes	yes	0.5533	0.4924	9.1	0.00042	3	0.5533	100.14	96.14
17	15	Cr Oil Prd	yes	yes	yes	yes	0.9206	0.9054	60.8	3E-11	4	0.9206	5.00	0.00
18		Maximum or Minimum Values =					0.9206	0.9080	138.6	1.17E-12				-0.58

Sheet1 Sheet5 Sheet2 Sheet4 Sheet3

For the crude oil production example of Figure 13.15, we note that the value for n is 26, the value of T is 5 and the value for R_T^2 is 0.92057. These three values are combined with the values for R_p^2 and p for each of the 15 possible models to compute a value for C_p for each of them.

If a model only has random differences from the true model, the average value of C_p is equal to $p + 1$. Furthermore, a regression model with a good fit will have a value of C_p that is less than $p + 1$. Therefore in evaluating alternative regression models with this statistic, the objective is to identify a model such that the value of the expression $(C_p - p - 1)$ is negative.

To demonstrate the use of all these criteria, we have run all 15 of the possible regressions for the crude oil production example. The values for each of the criteria for evaluating the adequacy of the model fit for all the regressions are summarized in Figure 13.16

Figure 13.16 presents the results for the 15 models beginning with the models with only one independent variable and ending with the one with all four independent variables. The regression results for the last model, designated as Model 15 in Figure 13.16, were previously presented in Figure 13.15.

The *values shown in the R^2 column* of Figure 13.16 are as expected. In particular as independent variables are added to a model, the R^2 value increases. Accordingly, the model with all four independent variables, Model 15, has the highest R^2 value.

SUMMARY OUTPUT

Regression Statistics	
Multiple R	0.959
R Square	0.919
Adjusted R Square	0.908
Standard Error	1.1725
Observations	26

ANOVA

	df	SS	MS	F	Significance F
Regression	3	343.183	114.394	83.21	0.0000
Residual	22	30.243	1.375		
Total	25	373.427			

	Coefficients	Standard Error	t Stat	P-value
Intercept	8.454	3.463	2.44	0.0231
USEnCons	0.754	0.063	11.94	0.0000
USCoalPr	0.010	0.006	1.71	0.1022
FuelRate	−1.028	0.328	−3.14	0.0048

Figure 13.17

Regression results for oil production example—Model 13 (3 independent variables)

The *adjusted R^2 value* takes into consideration both the additional information each new independent variable brings to a regression model and the changed degrees of freedom. As indicated in Figure 13.16, the best value for it is 0.9080 for Model 13. It has the three independent variables of "energy consumption," "coal production," and "fuel rate." The regression results for this model are shown in Figure 13.17.

As shown in Figure 13.17, the *F* value for the overall model is 83.21 with a corresponding *p* value (labeled as *Significant F*) of 0.00000000000366. Accordingly, the overall model is extremely significant. However an examination of the *t* ratios (labeled as *t Stat*) and their corresponding *p* values suggests that only the two variables, "energy consumption" and "fuel rate," are significant at $\alpha = .05$. This indicates it would be best to eliminate the variable, "coal production," from the model. The removal of it yields Model 7 of Figure 13.16. The regression results for Model 7 are shown in Figure 13.18.

The *F* value for the overall model in Figure 13.18 is 113.92 with a corresponding *p* value of 0.00000000000117. This is a slight improvement over Model 13. A comparison of the *p* values for the independent variables *t* ratios shows that both of them have improved (decreased).

To summarize these considerations, the adjusted R^2 criterion identified the three-variable Model 13 as the potentially best model. However, our further consideration of the individual independent variables suggests that actually the two-variable Model 7 is preferred.

The next criterion shown in Figure 13.16 is the *F statistic* value for the overall model. The highest value for it is for Model 1. However, the *F* statistic isn't meaningful in comparing models with different numbers of variables since it doesn't account for the differing degrees of freedom for finding the critical *F* value. The corresponding *p-value* for each *F* statistic value does take into consideration the degrees of freedom. The best *p*-value as shown in Figure 13.16 is for Model 7.

The final criterion of Figure 13.16 is the C_p *statistic*. As previously indicated, the goal for it is to identify a model with a C_p value greater than $p + 1$. The last column of

Figure 13.18

Regression results for oil production example—Model 7 (2 independent variables)

SUMMARY OUTPUT

Regression Statistics	
Multiple R	0.953
R Square	0.908
Adjusted R Square	0.900
Standard Error	1.2201
Observations	26

ANOVA

	df	SS	MS	F	Significance F
Regression	2	339.186	169.593	113.92	0.0000
Residual	23	34.240	1.489		
Total	25	373.427			

	Coefficients	Standard Error	t Stat	P-value
Intercept	7.1403	3.513	2.03	0.0538
USEnCons	0.7720	0.065	11.91	0.0000
FuelRate	-0.5173	0.138	-3.75	0.0011

Figure 13.16 identifies only one model with such a condition. It is Model 13 again. The regression results for it were given in Figure 13.17. As before a further consideration of these results would lead to the conclusion that it would be best to delete the "coal production" variable. Thus, we again conclude that Model 7 is better.

In passing, we should point out that the C_p *minus p + 1* column shows a value of zero for the full model, Model 15. This will always be true for the full model. Through manipulation of the algebraic expression for C_p *minus p + 1*, it can be shown that it will always be zero for the full model.

We have considered five separate criteria in our discussion of Figure 13.16. As we have shown, two of these—R^2 and the F statistic—are not useful in comparing alternative regression models with varying number of variables. The use of the other three, the adjusted R^2, the p-value for the overall model, and the C_p statistic, have been demonstrated. For the crude oil production example, all three of these criteria resulted in identifying Model 7 as the preferred relationship. This consistency among the three criteria will not always occur. Particularly for problems with a larger number of potential independent variables, it may well be that these criteria identify different preferred models. It may then be necessary for the business analyst to use other considerations to select among these preferred models.

As previously mentioned, the all possible regressions search procedure requires considering a total of $2^k - 1$ models where k is the total number of independent variables. Accordingly, the possible number of models to be considered for various values of k is as given in column two of Table 13.10. However, this number can be reduced slightly by noting that the best *one* independent variable model will always be the model produced using the independent variable that has the highest correlation coefficient value with the dependent variable. For the oil production example, Excel provides the correlation coefficients given in Figure 13.19. From these values, we can conclude the best one variable model will be that with "energy consumption" as the independent variable. This conclusion is consistent with all five criteria for the four one independent variable models given in Figure 13.16.

TABLE 13.10
Number of Regressions to be Evaluated

k	$2^k - 1$	$2^k - k$
1	1	1
2	3	2
3	7	3
4	15	12
5	31	27
6	63	58
7	127	121

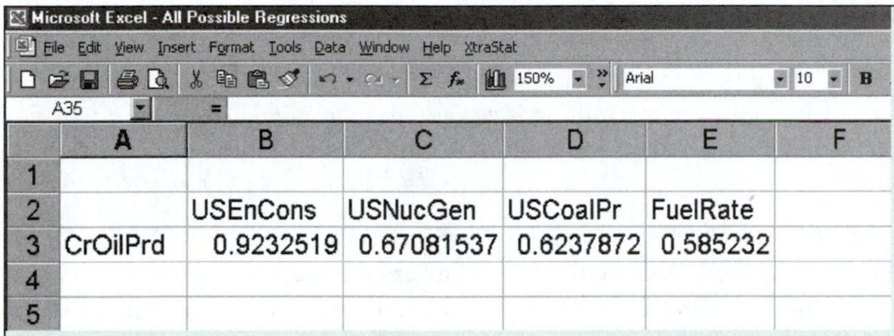

	A	B	C	D	E	F
1						
2		USEnCons	USNucGen	USCoalPr	FuelRate	
3	CrOilPrd	0.9232519	0.67081537	0.6237872	0.585232	
4						
5						

Figure 13.19
Oil production example correlation coefficients

By incorporating the use of the simple correlation coefficients for the one independent variable models into the all possible regressions procedure, the number of models to be considered is reduced by $k - 1$. Thus, the total number of models to be evaluated is computed as $(2^k - 1) - (k - 1) = (2^k - k)$. Values for this expression are given in the third column of Table 13.10.

The all possible regressions search procedure enables the business analyst to consider every possible model. In theory, this eliminates the chance that the analyst will never consider some models, as can be the case with the search procedures given below. On the other hand, the search through all the possible models may require more time and effort than is reasonable. This may particularly become an issue for problems with a large number of potential independent variables. As a result, a number of search procedures have been suggested for identifying the *best* model without the need to evaluate all possible models. Three such procedures that are frequently used are the forward selection, the backward elimination, and the stepwise regression procedures. All three of these procedures use a stepwise approach for attempting to identify the *best* model without evaluating all possible models. All three of these search procedures are oftentimes included in specialized statistical software packages.

Other Search Procedures

FORWARD SELECTION The **forward selection** search procedure starts with a model with no independent variables. It then considers all k models with one independent variable. It chooses the model with the highest simple correlation coefficient (or r^2) with the dependent variable. Suppose this variable is designated as X_1. Next the procedure examines the $k - 1$ two-variable models that include X_1 and each of the remaining $k - 1$ variables individually. It selects the two-variable model with the highest R^2 (or F) and designates this second variable as X_2. The procedure next considers the $k - 2$ three-variable models that include X_1, X_2, and each of the remaining $k - 2$ variables individually. The

Forward selection
A multiple regression search procedure that is essentially the same as stepwise regression analysis except that once a variable is entered into the process, it is never deleted.

procedure selects the three-variable model with the highest R^2 (or F). Forward selection continues adding variables in this manner until the addition of a variable results in an insignificant increase in R^2 as determined by an F test. As you will note from this description, once a variable is added to the model it doesn't leave.

Backward elimination
A step-by-step multiple regression search procedure that continues until only variables with significant t values remain in the model.

Stepwise regression
A step-by-step multiple regression search procedure that begins by developing a regression model with a single predictor variable and adds and deletes predictors one step at a time, examining the fit of the model at each step until there are no more significant predictors remaining outside the model.

BACKWARD ELIMINATION   The **backward elimination** search procedure is the opposite of the forward selection procedure. It starts with a model that includes all the independent variables. It then considers all k models in which one of the variables is removed. These models have $k - 1$ independent variables. It selects the variable that causes the smallest decrease in R^2 to be eliminated from the model. Next the procedure considers the $k - 1$ models in which one additional variable is removed. These models have $k - 2$ independent variables. The variable that causes the smallest decrease in R^2 is eliminated from the model. The procedure continues eliminating variables in this manner until the deletion of a variable results in a significant decrease in R^2 as evaluated by an F test. For this procedure once a variable leaves the model it is not considered for re-entry later.

STEPWISE REGRESSION   The **stepwise regression** search procedure is the most flexible of these three procedures. It is similar to the forward selection procedure. However at the end of each step in the process of adding variables, it considers all the variables in the model for possible removal. If the contribution of a previously added variable becomes insignificant with the addition of the latest variable, that previously added variable is removed. Thus, the procedure is a combination of the forward selection and the backward elimination procedures. The stepwise regression procedure continues adding variables and then considering the elimination of prior variables in this manner until the addition of a variable results in an insignificant increase in R^2 as determined by an F test. Each variable that is deleted from the model at some step is considering for possible re-entry at the later steps in the process.

13.5
Problems

13.25   Use Excel and the following data to develop all possible regression models to predict Y. Use adjusted R^2 and the C_p statistic to select the strongest model.

Y	X_1	X_2	X_3
21	5	108	57
17	11	135	34
14	14	113	21
13	9	160	25
19	16	122	43
15	18	142	40
24	7	93	52
17	9	128	38
22	13	105	51
20	10	111	43
16	20	140	20
13	19	150	14
18	14	126	29
12	21	175	22
23	6	98	38
18	15	129	40

13.26 Given here are the data for a dependent variable and four potential predictors. Use these data and Excel to develop all possible regression models to predict Y. Use adjusted R^2 and the C_p statistic to select the strongest model.

Y	X_1	X_2	X_3	X_4
101	2	77	1.2	42
127	4	72	1.7	26
98	9	69	2.4	47
79	5	53	2.6	65
118	3	88	2.9	37
114	1	53	2.7	28
110	3	82	2.8	29
94	2	61	2.6	22
96	8	60	2.4	48
73	6	64	2.1	42
108	2	76	1.8	34
124	5	74	2.2	11
82	6	50	1.5	61
89	9	57	1.6	53
76	1	72	2.0	72
109	3	74	2.8	36
123	2	99	2.6	17
125	6	81	2.5	48

13.27 The National Underwriter Company in Cincinnati, Ohio, publishes property and casualty insurance data. Given here is a portion of the data published. These data include information from the U.S. insurance industry about (1) net income after taxes, (2) dividends to policyholders, (3) net underwriting gain/loss, and (4) premiums earned. Use the data and all possible regressions to predict premiums earned from the other three variables. Compute the C_p statistic on each model and use it along with the adjusted R^2 to select the "best" model.

PREMIUMS EARNED	NET INCOME	DIVIDENDS	UNDERWRITING GAIN/LOSS
30.2	1.6	.6	.1
47.2	.6	.7	−3.6
92.8	8.4	1.8	−1.5
95.4	7.6	2.0	−4.9
100.4	6.3	2.2	−8.1
104.9	6.3	2.4	−10.8
113.2	2.2	2.3	−18.2
130.3	3.0	2.4	−21.4
161.9	13.5	2.3	−12.8
182.5	14.9	2.9	−5.9
193.3	11.7	2.9	−7.6

13.28 The U.S. Energy Information Administration releases figures in their publication, *Monthly Energy Review,* about the cost of various fuels and electricity. Following are the figures for four different items over a 12-year period. Use the data and all possible regressions to predict the cost of residential electricity from the cost of residential natural gas, residual fuel oil, and leaded regular gasoline. Examine the data and discuss the output in light of C_p and adjusted R^2.

RESIDENTIAL ELECTRICITY (kWh)	RESIDENTIAL NATURAL GAS (1000 FT³)	RESIDUAL FUEL OIL (GAL)	LEADED REGULAR GASOLINE (GAL)
2.54	1.29	.21	.39
3.51	1.71	.31	.57
4.64	2.98	.44	.86
5.36	3.68	.61	1.19
6.20	4.29	.76	1.31
6.86	5.17	.68	1.22
7.18	6.06	.65	1.16
7.54	6.12	.69	1.13
7.79	6.12	.61	1.12
7.41	5.83	.34	.86
7.41	5.54	.42	.90
7.49	4.49	.33	.90

13.6
Multicollinearity

Multicollinearity
A problematic condition that occurs when two or more of the independent variables of a multiple regression model are highly correlated.

One problem that can arise in multiple regression analysis is multicollinearity. **Multicollinearity** is *when two or more of the independent variables of a multiple regression model are highly correlated.* Technically, if two of the independent variables are correlated, we have *collinearity;* when three or more independent variables are correlated, we have multicollinearity, However, the two terms are frequently used interchangeably.

The reality of business research is that most of the time there is some correlation between predictors (independent variables). The problem of multicollinearity arises when the intercorrelation between predictor variables is high. This causes several other problems, particularly in the interpretation of the analysis.

1. It is difficult, if not impossible, to interpret the estimates of the regression coefficients.
2. Inordinately small t values for the regression coefficients may result.
3. The standard deviations of regression coefficients are overestimated.
4. The algebraic sign of estimated regression coefficients may be the opposite of what would be expected for a particular predictor variable.

There are many situations in business research where the problem of multicollinearity can arise in regression analysis. For example, suppose a model is being developed to predict salaries in a given industry. Independent variables such as years of education, age, years in management, experience on the job, and years of tenure with the firm might be considered as predictors. It is obvious that several of these variables are correlated and yield redundant information. Suppose a financial regression model is being developed to predict bond market rates by such independent variables as Dow Jones average, prime interest rates, GNP, producer price index, and consumer price index. Several of these predictors are likely to be intercorrelated.

In the world crude oil production example used in Section 13.5, several of the independent variables are intercorrelated, leading to the potential of multicollinearity problems. Table 13.11 gives the correlations of the predictor variables for this example. Note that r values are quite high ($r > .90$) for fuel rate and nuclear (.972), fuel rate and coal (.968), and coal and nuclear (.952).

Table 13.11 shows that fuel rate and coal production are highly correlated. Using fuel rate as a single predictor of crude oil production produces the following simple regression model.

$$\hat{Y} = 44.869 + .7838(\text{fuel rate}).$$

	ENERGY CONSUMPTION	NUCLEAR	COAL	FUEL RATE
Energy consumption	1	.856	.791	.791
Nuclear	.856	1	.952	.972
Coal	.791	.952	1	.968
Fuel rate	.796	.972	.968	1

TABLE 13.11
Correlations among Oil Production Predictor Variables

Notice that the estimate of the regression coefficient, .7838, is positive, indicating that as fuel rate increases, oil production increases. Using coal as a single predictor of crude oil production yields the following simple regression model.

$$\hat{Y} = 45.072 + .0157(\text{coal})$$

The multiple regression model developed using both fuel rate and coal to predict crude oil production is

$$\hat{Y} = 45.806 + .02278(\text{coal}) - .3934(\text{fuel rate}).$$

Observe that this regression model indicates a *negative* relationship between fuel rate and oil production (−.3934), which is in opposition to the *positive* relationship shown in the regression equation for fuel rate as a single predictor. Because of the multicollinearity between coal and fuel rate, these two independent variables interact in the regression analysis in such a way as to produce regression coefficient estimates that are difficult to interpret. Extreme caution should be exercised before interpreting these regression coefficient estimates.

The problem of multicollinearity can also affect the t values that are used to evaluate the regression coefficients. Because the problems of multicollinearity among predictors can result in an overestimation of the standard deviation of the regression coefficients, the t values tend to be underrepresentative when multicollinearity is present. There are regression models containing multicollinearity in which all t values are nonsignificant but the overall F value for the model is highly significant.

This collinearity may explain the fact that the overall model is significant but none of the predictors are significant. It also underscores one of the problems with multicollinearity: underrepresented t values. The t values test the strength of the predictor given the other variables in the model. If a predictor is highly correlated with other independent variables, it will appear not to add much to the explanation of Y and produce a low t value. However, had the predictor not been in the presence of these other variables, the predictor might have explained a high proportion of variation of Y.

Many of the problems created by multicollinearity are interpretation problems. The business analyst should be alert to and aware of multicollinearity potential with the predictors in the model and view the model outcome in light of such potential.

The problem of multicollinearity is not a simple one to overcome. However, there are several ways to make inroads into the problem. One way is to examine a correlation matrix like the one in Table 13.11 to search for possible intercorrelations among potential predictor variables. If several variables are highly correlated, the business analyst can select the variable that is most correlated to the dependent variable and use that variable to represent the others in the analysis. One problem with this idea is that correlations can be more complex than simple correlation among variables. That is, simple correlation values do not always reveal multiple correlation between variables. In some instances, variables may not appear to be correlated as pairs, but one variable is a linear combination of several other variables. This situation is also an example of multicollinearity, and a cursory observation of the correlation matrix will probably not reveal the problem.

Stepwise regression is another way to prevent the problem of multicollinearity. The search process enters the variables one at a time and compares the new variable to those in solution. If a new variable is entered and the t values on old variables become nonsignificant, the old variables are dropped out of solution. In this manner, it is more difficult for the problem of multicollinearity to affect the regression analysis. Of course, because of multicollinearity, some important predictors may not enter in to the analysis.

Other techniques are available to attempt to control for the problem of multicollinearity. One is called a **variance inflation factor,** in which a regression analysis is done to predict an independent variable by the other independent variables. In this case, the independent variable being predicted becomes the dependent variable. As this process is done for each of the independent variables, it is possible to determine whether any of the independent variables are a function of the other independent variables, yielding evidence of multicollinearity. By using the R_i^2 from such a model, a variance inflation factor (VIF) can be computed to determine whether the standard errors of the estimates are inflated:

$$\text{VIF} = \frac{1}{1 - R_i^2}$$

where R_i^2 is the coefficient of determination for any of the models to predict an independent variable by the other $k - 1$ independent variables. Some business analysts follow a guideline that if the variance inflation factor is greater than 10 or the R_i^2 value is more than .90 for the largest variance inflation factors, a severe multicollinearity problem is present.[*]

Variance inflation factor
A statistic computed using the R^2 value of a regression model developed by predicting one independent variable of a regression analysis by other independent variables; used to determine whether there is multicollinearity among the variables.

13.6
Problems

13.29   Develop a correlation matrix for the independent variables in Problem 13.25. Study the matrix and make a judgment as to whether or not substantial multicollinearity is present among the predictors. Why or why not?

13.30   Construct a correlation matrix for the four independent variables for Problem 13.26 and search for possible multicollinearity. What did you find and why?

13.31   In Problem 13.27, you were asked to compute all possible regressions to predict premiums earned by net income, dividends, and underwriting gain or loss. Study the results, including the regression coefficients, to determine whether there may be a problem with multicollinearity. Construct a correlation matrix of the three variables to aid you in this task.

13.32   Study the three predictor variables in Problem 13.28 and attempt to determine whether substantial multicollinearity is present between the predictor variables. If there is a problem of multicollinearity, how could that affect the outcome of the multiple regression analysis?

---

[*]William Mendenhall and Terry Sincich, *A Second Course in Business Statistics: Regression Analysis.* San Francisco, Dellen Publishing Company, 1989. John Neter, William Wasserman, Michael H. Kutner, *Applied Linear Regression Models,* 2d ed., Homewood, IL, Richard D. Irwin, Inc., 1989.

Multiple regression analysis is a statistical tool in which a mathematical model is developed in an attempt to predict a dependent variable by two or more independent variables or in which at least one predictor is nonlinear. Because doing multiple regression analysis by hand is extremely tedious and time-consuming, it is almost always done on a computer.

The standard output from a multiple regression analysis is similar to that of simple regression analysis. A regression equation is produced with a constant that is analogous to the Y intercept in simple regression and with estimates of the regression coefficients that are analogous to the estimate of the slope in simple regression. An F test for the overall model is computed to test to determine whether at least one of the regression coefficients is significantly different from zero. This F value is usually displayed in an ANOVA table, which is part of the regression output. The ANOVA table also contains the sum of squares of error and sum of squares of regression, which are used to compute other statistics in the model.

Computer regression output contains t values, which are used to determine the significance of the regression coefficients. Using these t values, statisticians can make decisions about including or excluding variables from the model.

Residuals, standard error of the estimate, and R^2 are also standard multiple regression computer output. The coefficient of determination for simple regression models is denoted r^2, whereas for multiple regression it is R^2. The interpretation of residuals, standard error of the estimate, and R^2 in multiple regression is very similar to that in simple regression. Because R^2 can be inflated with nonsignificant variables in the mix, an adjusted R^2 is often computed. Unlike R^2, adjusted R^2 takes into account the degrees of freedom and the number of observations.

Indicator, or dummy, variables are qualitative variables that are used to represent categorical data in the multiple regression model. These variables are coded as 0, 1 and are often used to represent nominal or ordinal classification data that the business analyst wants to use in the regression analysis. If a qualitative variable contains more than two categories, it generates multiple dummy variables. In general, if a qualitative variable contains c categories, $c - 1$ dummy variables should be created.

Multiple regression analysis can handle nonlinear independent variables. One way to do this is to recode the data and enter the variables into the analysis in the normal way.

However, some nonlinear regression models, such as exponential models, require that the entire model be transformed. Often the transformation involves the use of logarithms. In some cases, the resulting value of the regression model is in logarithmic form and the antilogarithm of the answer must be taken to determine the predicted value of Y.

Search procedures are used to help sort through the independent variables as predictors in the examination of various possible models. Several search procedures are available, including *all possible regressions, stepwise regression, forward selection,* and *backward elimination.* The *all possible regressions* procedure computes every possible regression model for a set of data. The drawbacks of this procedure include the time and energy required to compute all possible regressions and the difficulty of deciding which models are most appropriate. The *stepwise regression* procedure involves selecting and adding one independent variable at a time to the regression process after beginning with a one-predictor model. Variables are added to the model at each step if they contain the most significant t value associated with the remaining variables. If no additional t value is statistically significant at any given step, the procedure stops. With stepwise regression, at each step the process examines the variables already in the model to determine whether their t values are still significant. If not, they are dropped from the model, and the process searches for other independent variables with large, significant t values to replace the variable(s) dropped. The *forward selection procedure* is the same as stepwise regression but does not drop variables out of the model once they have been included. The *backward elimination*

Summary

procedure begins with a "full" model, a model that contains all the independent variables. The sample size must be large enough to justify a full model, which can be a limiting factor. Backward elimination starts dropping out the least important predictors one at a time until only significant predictors are left in the regression model.

One of the problems in using multiple regression is multicollinearity, or correlations among the predictor variables. This problem can cause overinflated estimates of the standard deviations of regression coefficients, misinterpretation of regression coefficients, undersized t values, and misleading signs on the regression coefficients. It can be lessened by using an intercorrelation matrix of independent variables to help recognize bivariate correlation; by using stepwise regression to sort the variables one at a time; and/or by using statistics such as a variance inflation factor.

Key Terms

adjusted R^2	outliers
all possible regressions	partial regression coefficient
backward elimination	qualitative variable
coefficient of multiple determination (R^2)	R^2
dependent variable	residual
dummy variable	response plane
forward selection	response surface
general linear regression model	response variable
independent variable	search procedures
indicator variable	standard error of the estimate (S_e)
least squares analysis	stepwise regression
multicollinearity	sum of squares of error (SSE)
multiple regression	variance inflation factor
nonlinear regression model	

SUPPLEMENTARY PROBLEMS

13.33 Use the following data to develop a multiple regression model to predict Y from X_1 and X_2. Discuss the output, including comments about the overall strength of the model, the significance of the regression coefficients, and other indicators of model fit.

X_1	X_2	Y
29	1.64	198
71	2.81	214
54	2.22	211
73	2.70	219
67	1.57	184
32	1.63	167
47	1.99	201
43	2.14	204
60	2.04	190
32	2.93	222
34	2.15	197

13.34 Given here are the data for a dependent variable, Y, and independent variables. Use these data to develop a regression model to predict Y. Discuss the output. Which variable is an indicator variable? Was it a significant predictor of Y?

X_1	X_2	X_3	Y
0	51	16.4	14
0	48	17.1	17
1	29	18.2	29
0	36	17.9	32
0	40	16.5	54
1	27	17.1	86
1	14	17.8	117
0	17	18.2	120
1	16	16.9	194
1	9	18.0	203
1	14	18.9	217
0	11	18.5	235

13.35 Use the following data and an all possible regression analysis to predict Y. In addition to the two independent variables given here, include three other predictors in your analysis: the square of each X as a predictor and an interaction predictor. Discuss the results of the process.

X_1	X_2	Y	X_1	X_2	Y
10	3	2002	5	12	1750
5	14	1747	6	8	1832
8	4	1980	5	18	1795
7	4	1902	7	4	1917
6	7	1842	8	5	1943
7	6	1883	6	9	1830
4	21	1697	5	12	1786
11	4	2021			

13.36 Use the X_1 values and the log of the X_1 values given here to predict the Y values by using an all possible regressions procedure. Discuss the output. Were either or both of the predictors significant?

Y	X_1	Y	X_1
20.4	850	13.2	204
11.6	146	17.5	487
17.8	521	12.4	192
15.3	304	10.6	98
22.4	1029	19.8	703
21.9	910	17.4	394
16.4	242	19.4	647

13.37 The U.S. Commodities Futures Trading Commission reports on the volume of trading in the U.S. commodity futures exchanges. Shown here are the figures for grain, oilseeds, and livestock products over a period of several years. Use these data to develop a multiple regression model to predict grain futures volume of trading from oilseeds volume and livestock products volume. All figures are given in units of millions. Graph each of these predictors separately with the response variable and use these to explore possible recoding schemes for nonlinear relationships. Include any of these in the regression model. Comment on the results.

GRAIN	OILSEEDS	LIVESTOCK
2.2	3.7	3.4
18.3	15.7	11.8
19.8	20.3	9.8
14.9	15.8	11.0
17.8	19.8	11.1
15.9	23.5	8.4
10.7	14.9	7.9
10.3	13.8	8.6
10.9	14.2	8.8
15.9	22.5	9.6
15.9	21.1	8.2

13.38 The U.S. Bureau of Mines produces data on the price of minerals. Shown here are the average prices per year for several minerals over a decade. Use these data and an all possible regressions procedure to produce a model to predict the average price of gold from the other variables. Comment on the results of the process.

GOLD ($ PER OZ)	COPPER (CENTS PER LB)	SILVER ($ PER OZ)	ALUMINUM (CENTS PER LB)
161.1	64.2	4.4	39.8
308.0	93.3	11.1	61.0
613.0	101.3	20.6	71.6
460.0	84.2	10.5	76.0
376.0	72.8	8.0	76.0
424.0	76.5	11.4	77.8
361.0	66.8	8.1	81.0
318.0	67.0	6.1	81.0
368.0	66.1	5.5	81.0
448.0	82.5	7.0	72.3
438.0	120.5	6.5	110.1
382.6	130.9	5.5	87.8

13.39 The Shipbuilders Council of America in Washington, DC, publishes data about private shipyards. Among the variables reported by this organization are the employment figures (per 1000), the number of naval vessels under construction, and the number of repairs or conversions done to commercial ships (in millions of dollars). Shown here are the data for these three variables over a 7-year period. Use the data to develop a regression model to predict private shipyard employment from number of naval vessels under construction and repairs or conversions of commercial ships. Graph each of these predictors separately with the response variable and use these graphs to explore possible recoding schemes for nonlinear relationships. Include any of these in the regression model. Comment on the regression model and its strengths and/or its weaknesses.

EMPLOYMENT	NAVAL VESSELS	COMMERCIAL SHIP REPAIRS OR CONVERSIONS
133.4	108	431
177.3	99	1335
143.0	105	1419
142.0	111	1631
130.3	100	852
120.6	85	847
120.4	79	806

13.40 The U.S. Bureau of Labor Statistics produces consumer price indexes for several different categories. Shown here are the percentage changes in consumer price indexes over a period of 20 years for food, shelter, apparel, and fuel oil. Also displayed are the percentage changes in consumer price indexes for all commodities. Use these data and on all possible regression procedure to develop a model that attempts to predict all commodities by the other four variables. Construct scatter plots of each of these with all commodities. Examine the graphs and use the information to develop any other appropriate predictor variables by recoding data and include them in the analysis. Comment on the result of this analysis.

ALL COMMODITIES	FOOD	SHELTER	APPAREL	FUEL OIL
.9	1.0	2.0	1.6	3.7
.6	1.3	.8	.9	2.7
.9	.7	1.6	.4	2.6
.9	1.6	1.2	1.3	2.6
1.2	1.3	1.5	.9	2.1
1.1	2.2	1.9	1.1	2.4
2.6	5.0	3.0	2.5	4.4
1.9	.9	3.6	4.1	7.2
3.5	3.5	4.5	5.3	6.0
4.7	5.1	8.3	5.8	6.7
4.5	5.7	8.9	4.2	6.6
3.6	3.1	4.2	3.2	6.2
3.0	4.2	4.6	2.0	3.3
7.4	14.5	4.7	3.7	4.0
11.9	14.3	9.6	7.4	9.3
8.8	8.5	9.9	4.5	12.0
4.3	3.0	5.5	3.7	9.5
5.8	6.3	6.6	4.5	9.6
7.2	9.9	10.2	3.6	8.4
11.3	11.0	13.9	4.3	9.2

13.41 The U.S. Department of Agriculture publishes data annually on various selected farm products. Shown here are the unit production figures for three farm products for 10 years during a 20-year period. Use these data and an all possible regression analysis to predict corn production by the production of soybeans and wheat. Comment on the results.

CORN (MILLION BUSHELS)	SOYBEANS (MILLION BUSHELS)	WHEAT (MILLION BUSHELS)
4152	1127	1352
6639	1798	2381
4175	1636	2420
7672	1861	2595
8876	2099	2424
8226	1940	2091
7131	1938	2108
4929	1549	1812
7525	1924	2037
7933	1922	2739

13.42 Cost-of-living indexes for selected metropolitan areas have been accumulated by the American Chamber of Commerce Business analysts Association. Shown here are cost-of-living indexes for 25 different cities on five different items for a recent year. Use the data to develop a regression model to predict the grocery cost-of-living index by the indexes of housing, utilities, transportation, and healthcare. Discuss the results, highlighting both the significant and nonsignificant predictors.

CITY	GROCERY ITEMS	HOUSING	UTILITIES	TRANSPORTATION	HEALTHCARE
Albany	108.3	106.8	127.4	89.1	107.5
Albuquerque	96.3	105.2	98.8	100.9	102.1
Augusta, GA	96.2	88.8	115.6	102.3	94.0
Austin	98.0	83.9	87.7	97.4	94.9
Baltimore	106.0	114.1	108.1	112.8	111.5
Buffalo	103.1	117.3	127.6	107.8	100.8
Colorado Springs	94.5	88.5	74.6	93.3	102.4
Dallas	105.4	98.9	108.9	110.0	106.8
Denver	91.5	108.3	97.2	105.9	114.3
Des Moines	94.3	95.1	111.4	105.7	96.2
El Paso	102.9	94.6	90.9	104.2	91.4
Indianapolis	96.0	99.7	92.1	102.7	97.4
Jacksonville, FL	96.1	90.4	96.0	106.0	96.1
Kansas City	89.8	92.4	96.3	95.6	93.6
Knoxville	93.2	88.0	91.7	91.6	82.3
Los Angeles	103.3	211.3	75.6	102.1	128.5
Louisville	94.6	91.0	79.4	102.4	88.4
Memphis	99.1	86.2	91.1	101.1	85.5
Miami	100.3	123.0	125.6	104.3	137.8
Minneapolis	92.8	112.3	105.2	106.0	107.5
Mobile	99.9	81.1	104.9	102.8	92.2
Nashville	95.8	107.7	91.6	98.1	90.9
New Orleans	104.0	83.4	122.2	98.2	87.0
Oklahoma City	98.2	79.4	103.4	97.3	97.1
Phoenix	95.7	98.7	96.3	104.6	115.2

13.43 Shown here is output from two Excel regression analyses on the same problem. The first output was done on a "full" model. In the second output, the variable with the smallest absolute t value has been removed, and the regression has been rerun like a second step of a backward elimination process. Examine the two outputs. Explain what happened, what the results mean, and what might happen in a third step.

Regression Statistics	
Multiple R	0.567
R Square	0.321
Adjusted R Square	0.208
Standard Error	159.681
Observations	29

ANOVA

	df	SS	MS	F	Significance F
Regression	4	289856.08	72464.02	2.84	0.046
Residual	24	611955.23	25498.13		
Total	28	901811.31			

	Coefficients	Standard Error	t Stat	P-value
Intercept	336.79	124.08	2.71	0.012
X1	1.65	1.78	0.93	0.363
X2	−5.63	13.47	−0.42	0.680
X3	0.26	1.68	0.16	0.878
X4	185.50	66.22	2.80	0.010

Regression Statistics	
Multiple R	0.566
R Square	0.321
Adjusted R Square	0.239
Standard Error	156.534
Observations	29

ANOVA

	df	SS	MS	F	Significance F
Regression	3	289238.1	96412.7	3.93	0.020
Residual	25	612573.2	24502.9		
Total	28	901811.3			

	Coefficients	Standard Error	t Stat	P-value
Intercept	342.919	115.34	2.97	0.006
X1	1.834	1.31	1.40	0.174
X2	−5.749	13.18	−0.44	0.667
X4	181.220	59.05	3.07	0.005

ANALYZING THE DATABASES

1. Use the manufacturing database to develop a multiple regression model to predict Cost of Materials by Number of Employees, New Capital Expenditures, Value Added by Manufacture, Value of Industry Shipments, and End-of-Year Inventories. Create indicator variables for values of industry shipments that have been coded from 1 to 4. Use all possible regressions procedure. Does there appear to be a problem of multicollinearity in this analysis? Discuss the results of the analysis.

2. Construct a correlation matrix for the hospital database variables. Are some of the variables highly correlated? Which ones and why? Perform all possible regressions analysis to predict Personnel by Control, Service, Beds, Admissions, Census, Outpatients, and Births. The variables Region, Control, and Service will need to be coded as indicator variables. Control has two subcategories, and Service has three.

3. Develop a regression model using the financial database. Use Total Revenues, Total Assets, Return on Equity, Earnings Per Share, Average Yield, and Dividends Per Share to predict the average P/E ratio for a company. How strong is the model? Use all possible regressions to help sort out the variables. Several of these variables may be measuring similar things. Construct a correlation matrix to explore the possibility of multicollinearity among the predictors.

4. Use the stock market database to develop a regression model to predict the composite index from Part of the Month, Stock Volume, Reported Trades, Dollar Value, and Warrants Volume. You will need to treat Part of the Month as a qualitative variable with three subcategories. Drop out the least significant variable if there is one that is not significant at $\alpha = .05$ and rerun the model. How much did R^2 drop? Continue to do this until there are only significant predictors left. Describe the final model.

CASE

VIRGINIA SEMICONDUCTOR

Virginia Semiconductor, Inc., is a producer of silicon wafers used in the manufacture of microelectronic products. The company, situated in Fredericksburg, Virginia, was founded in 1978 by two brothers, Thomas and Robert Digges. Virginia Semiconductor was growing and prospering in the early 1980s by selling a high volume of low-profit-margin wafers. However, in 1985, without notice, Virginia Semiconductor lost two major customers that represented 65% of its business. Left with only 35% of its sales base, the company desperately needed customers.

Thomas Digges, Jr., CEO of Virginia Semiconductor, decided to seek markets where his company's market share would be small but profit margin would be high because of the value of its engineering research and its expertise. This turned out to be a wise decision for the small, versatile company.

Virginia Semiconductor developed a silicon wafer that was two inches in diameter, 75 microns thick, and polished on both sides. Such wafers were needed by several customers, but had never been produced before. The company produced a number of these wafers and sold them for more than 10 times the price of conventional wafers.

Soon the company was making wafers from two to four microns thick (extremely thin), wafers with textured surfaces for infrared applications, and wafers with micromachined holes or shapes and selling them in specialized markets. It was able to deliver these products faster than competitors were able to deliver standard wafers.

Having made inroads at replacing lost sales, Virginia Semiconductor still had to streamline operations and control inventory and expenses. There were no layoffs, but the average workweek dropped to 32 hours and the president took an 80% pay reduction for a time. Expenses were cut as far as seemed possible.

The company had virtually no long-term debt and fortunately was able to make it through this period without incurring any additional significant debt. The absence of large monthly debt payments enabled the company to respond quickly to new production needs.

Virginia Semiconductor improved production quality by cross-training employees. In addition, the company participated in the State of Virginia's economic development efforts to find markets in Europe, Japan, Korea, and Israel. Exports, which were 1% of the company's business in 1985, now represent 40%.

The company continues to find new customers because of new products. One new ultramachining wafer has become a key component in auto airbags. Today the company has more than 300 active customers, whereas it had fewer than 50 in 1985.

DISCUSSION

1. It is often useful to decision makers at a company to determine what factors enter into the size of a customer's purchase. Suppose decision makers at Virginia Semiconductor want to determine from past data what variables might be predictors of size of purchase and are able to gather some data on various customer companies. Assume the following data represent information gathered on 16 companies on five variables: the total amount of purchases made during a one-year period (size of purchase), the size of the purchasing company (in total sales volume), the percentage of all purchases made by the customer company that were imports, the distance of the customer company from Virginia Semiconductor, and whether or not the customer company had a single central purchasing agent. Use these data to generate a multiple regression model to predict size of purchase by the other variables. Summarize your findings in terms of the strength of the model, significant predictor variables, and any new variables generated by recoding.

SIZE OF PURCHASE ($1000)	COMPANY SIZE ($ MILLION SALES)	% CUSTOMER IMPORTS	DISTANCE FROM VA SEMICONDUCTOR	CENTRAL PURCHASER?
27.9	25.6	41	18	1
89.6	109.8	16	75	0
12.8	39.4	29	14	0
34.9	16.7	31	117	0
408.6	278.4	14	209	1
173.5	98.4	8	114	1
105.2	101.6	20	75	0
510.6	139.3	17	50	1
382.7	207.4	53	35	1
84.6	26.8	27	15	1
101.4	13.9	31	19	0
27.6	6.8	22	7	0
234.8	84.7	5	89	1
464.3	180.3	27	306	1
309.8	132.6	18	73	1
294.6	118.9	16	11	1

2. Suppose that the next set of data is Virginia Semiconductor's sales figures for the past 11 years, along with the average number hours worked per week by a full-time employee and the number of different customers the company has for its unique wafers. How did the average workweek length and/or number of customers relate to total sales figures? Use scatter plots to examine possible relationships between sales and hours per week and sales and number of customers. Use these plots to explore possible ways to recode the data. Use an all possible regressions analysis to explore the relationships. Let the response variable be "sales" and the predictors be "average

number hours worked per week," "number of customers," and any new variables created by recoding. Explore quadratic relationships, interaction, and other relationships that seem appropriate. Summarize your findings in terms of model strength and significant predictors.

SALES ($ MILLION)	AVERAGE HOURS WORKED PER WEEK	NUMBER OF CUSTOMERS
15.6	44	54
15.7	43	52
15.4	41	55
14.3	41	55
11.8	40	39
9.7	40	28
9.6	40	37
10.2	38	58
11.3	38	67
14.3	32	186
14.8	37	226

As Virginia Semiconductor continues to grow and prosper, there is always the danger that the company will slip back into inefficient ways. Suppose that after a few years the company's sales begin to level off, but it continues hiring employees. Such figures over a 10-year period of time may look like the data given here. Graph these data, using sales as the response variable and number of employees as the predictor. Study the graph and using the information learned, develop a regression model to predict sales by the number of employees. On the basis of what you find, what would you recommend to management about the trend if it were to continue? What do you see in these data that management ought to be concerned about?

SALES ($ MILLION)	NUMBER OF EMPLOYEES
20.2	120
24.3	122
28.6	127
33.7	135
35.2	142
35.9	156
36.3	155
36.2	167
36.5	183
36.6	210

ADAPTED FROM: "Virginia Semiconductor: A New Beginning," *Real-World Lessons for America's Small Businesses: Insights from the Blue Chip Enterprise Initiative 1994.* Published by *Nation's Business* magazine on behalf of Connecticut Mutual Life Insurance Company and the U.S. Chamber of Commerce in association with The Blue Chip Enterprise Initiative, 1994.

14

Time Series Forecasting

Learning Objectives

The focus of this chapter is time series forecasting for business and industry, thereby enabling you to:

1. Gain a general understanding of the three general categories of forecasting techniques available.

2. Become aware of the four components that make up a time series.

3. Understand how to identify which components are present in a specific time series.

4. Become aware of the forecasting methods available for time series with specific components.

5. Become aware of several ways of identifying the forecasting method with the least forecasting error.

6. Understand how to forecast for time series with specific components using stationary methods, trend methods, and seasonal methods.

Forecasting
The art or science of predicting the future.

Every day, **forecasting**—*the art or science of predicting the future*—is used in the decision-making process to help business people reach conclusions about buying, selling, producing, hiring, and many other actions. As an example, consider these few items that were discussed in a recent daily newspaper:

- Investors are forecasting an increase in demand for crude oil because of a tropical storm in the Caribbean.
- Intel is considered a bellwether company for the technology stocks.
- Bank regulators see signs of an increase in bank failures due to the number of risky home equity loans being given.
- Bond investors, thinking that inflation may be on the rise, sell a week before a government report on wholesale prices and industrial production is due for publication.
- Local home sales should remain strong because there are predictions of strong job growth.
- Stock market forecasters predict a lower quarterly profit for Apple Computer because of a chip shortage.

How are these and other conclusions reached? What forecasting techniques are used? Are the forecasts accurate? In this chapter we discuss several forecasting techniques, how to measure the error of a forecast, and some of the problems that can occur in forecasting.

14.1
Introduction to Forecasting

Qualitative techniques
Forecasting techniques based on subjective estimates from informed sources.

Quantitative techniques
Forecasting techniques based on historic data.

Time series techniques
Forecasting techniques based on historic data only for the variable being forecasted.

Time series
Data values for a variable recorded over a period of time at regular intervals.

Time series plot
A graph of the values for a variable versus time.

There are three general categories of forecasting techniques. One category is called **qualitative techniques.** They are based on the judgment of experts or informed sources and are used when historical data are either scarce or non-existent. For example, these approaches might be useful for predicting the sales level for a new product. They generally use rating schemes to transform human judgment into quantitative estimates. The objective of them is to bring together in a logical, unbiased, and systematic way all information and judgments that relate to the factors being estimated. Some of these methods are the *Delphi technique, scenario writing,* and *visionary forecast.*

In contrast, **quantitative techniques** rely on historical data. This category is usually divided into two subcategories, time series techniques and causal techniques. Accordingly, the second general category of forecasting techniques is **time series techniques.** These use past data for only the variable of interest to forecast future values for that variable. The historical data consists of the numerical values for the forecasted variable recorded over a period of time at regular intervals. This set of data is called a **time series.**

Virtually all areas of business, including production, sales, marketing, finance, accounting, human relations, distribution, and inventory produce and maintain time series data. Table 14.1 presents an example time series released by the Office of Market Finance, U.S. Department of the Treasury. The table contains the bond yield rates of three-month Treasury Bills for a recent 17-year period. You will note from the data of Table 14.1, it is somewhat difficult to discern the nature of a time series from the numerical values directly. For example, the values of Table 14.1 indicate there has been a decrease in rates over the 17-year period. However, the form of the decrease isn't obvious from these values. An effective way to visualize the form of a time series is to create a **time series plot.** You may also hear it referred to as a *line plot.* It is a two-dimensional graph of the time series. The vertical axis represents the forecast variable, and the horizontal axis represents time. An example of a time series plot for the data of Table 14.1 is shown in Figure 14.1. From it, we can easily see there were large decreases in the earlier years of the 17-year period, but the decreases then became smaller, with some increases in the final years.

Time series forecasting techniques are based on the premise that the factors that influenced patterns of activity in the past will continue to do so in the same manner in the future. Accordingly, time series forecasting techniques first attempt to identify a pattern in the time series, such as that displayed in Figure 14.1. They then project the pattern into the future. Thus, these are sometimes called extrapolation methods. Some time series methods are *moving averages, exponential smoothing,* and *trend projection.* These and others are discussed at length in the remainder of this chapter.

The third category of forecasting techniques is **causal techniques.** These are the most sophisticated kind of forecasting techniques. They are based on the supposition that a relationship exists between the variable to be forecasted and other explanatory time series variables. Through regression analysis of the historic data, a mathematical relationship is estimated. For example, a company might use a causal model to forecast future sales based on values for the three explanatory variables of advertising level, sales force size, and product price. The model could then be used to forecast future sales values based on proposed values for the three explanatory variables. *Regression models, econometric models,* and *leading indicators* are examples of causal techniques.

In summary, the three general categories of forecasting techniques are qualitative, time series, and causal techniques. Table 14.2 presents a summary of some characteristics of each of these three.

Qualitative techniques based on judgment are not a subject of this quantitative text and will not be discussed further. The approach of the causal techniques based on the development of a regression relationship among variables is the subject of Chapters 12 and 13 of this text. Accordingly, these also will not be considered further in this chapter. On the other hand, time series techniques are the total content of the remainder of this chapter. To avoid confusion for the reader later, we should point out, however, that our consideration of time series techniques does include instances in which the approach of regression analysis is used. However, its use in this chapter is in the manner of time series techniques, not causal techniques. That is, it is used to forecast a variable based on historical data for only that variable, not other explanatory variables.

We continue our consideration of time series techniques in Section 14.2 with a discussion of the components that make up a time series. Section 14.3 presents a four-step procedure for analyzing and forecasting a time series. The procedure begins with an identification of the components that are present in a particular time series. This information is used to identify the potential time series methods that should be considered for forecasting the particular time series. Then the *best* of these potential methods is identified and used to make forecasts of future values.

TABLE 14.1

Bond Yields of Three-Month Treasury Bills

YEAR	AVERAGE YIELD
1	14.03%
2	10.69
3	8.63
4	9.58
5	7.48
6	5.98
7	5.82
8	6.69
9	8.12
10	7.51
11	5.42
12	3.45
13	3.02
14	4.29
15	5.51
16	5.02
17	5.07

Causal techniques

Forecasting techniques based on historic data for the variable being forecasted and one or more explanatory variables.

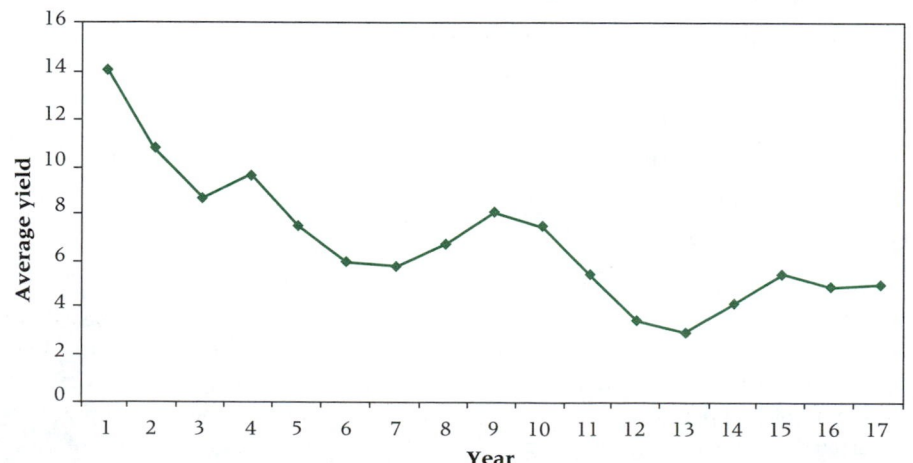

Figure 14.1

Time series plot for bond yields

TABLE 14.2
Categories of Forecasting
Techniques

GENERAL CATEGORY	SPECIFIC TECHNIQUES	EXPLANATION	USEFUL WHEN
Qualitative Techniques	Delphi Technique Scenario Writing Visionary Forecast Sales Force Estimate Historic Analogies	Uses subjective data from informed sources	Relevant historical data are scarce or nonexistent
Time Series Techniques	Moving Average Exponential Smoothing Trend Projection Autoregression Model Seasonal Regression Model	Uses historical data for only the forecast variable to find patterns	Historical data for forecast variable exhibit a pattern
Causal Techniques	Regression Models Econometric Models Leading Indicators Correlation Methods Input-Output Models	Supposes a relationship exists between forecast variable and explanatory variables	Historical data are available for forecast variable and for variables thought to influence it

Sections 14.4, 14.5, and 14.6 present potential time series methods to be considered for forecasting. These sections in turn present *stationary methods, trend methods,* and *seasonal methods.*

14.2
Time Series Components

As previously stated, a *time series* is made up of the values for a variable recorded at regular time intervals. The time interval can be years, quarters, months, weeks, days, or any other length of time that is important. For example, the marketing research department for a company might record the company's sales of a product on a daily basis. These daily time series values could then be combined for two-week periods to create a bi-weekly time series. The bi-weekly values then might be summed to provide values for a yearly, or annual, time series. The result would be a daily time series, a bi-weekly time series, and an annual time series for analysis and forecasting by the market research department.

To facilitate the analysis and forecasting of a time series, most well-used time series techniques consider the time series to be made up of four components. These are

Trend—a long-term upward or downward change in the time series

Seasonal—periodic increases and decreases that occur within a year

Cyclical—periodic increases and decreases that occur over more than a one-year period

Irregular—changes in the time series not attributable to the other three components

Trend Component

Trend component

Long-term upward or downward change in a time series.

The **trend component** represents the general tendency of a variable over a long period of time. By observing the long-term trend, we can characterize a time series as having a downward trend, having an upward trend, or having no trend at all. In addition, the trend may be observed to have either a linear relationship over time or one that is non-linear.

For example, suppose sales data have been recorded for 25 time periods and a time series plot created as shown in Figure 14.2(a). The symbols on the plot represent each of the 25 values of the time series and are connected with straight-line segments. Thus, a time series plot oftentimes is called a *line plot.* As we discussed in conjunction with Figure 14.1, it is

Figure 14.2   Stationary time series plots

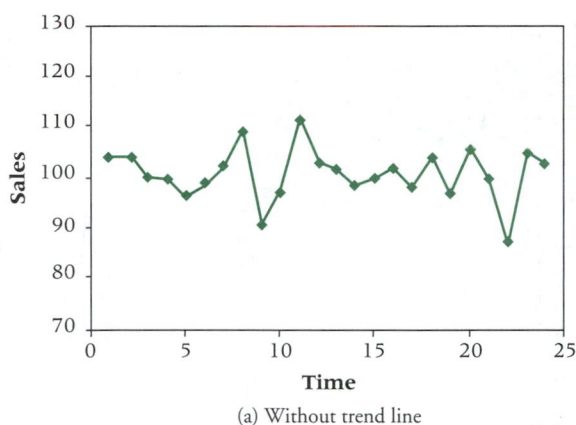

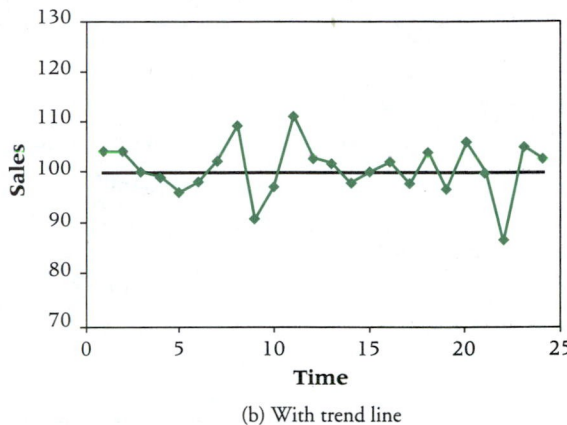

(a) Without trend line

(b) With trend line

easier to discern the nature of the time series from the plot such as Figure 14.2(a) than from the data values. However, it is also possible to go one step further by superimposing upon the plot a line of the central tendency or average values for the time series. The line can be entered by hand or by using a computerized resource such as Excel's **Trendline** feature for charts. You may recall the introduction of **Trendline** in Section 1.8 of Chapter 1 and its use for regression analysis in Section 12.2 of Chapter 12.

In Figure 14.2(b), a trend line has been added to the Figure 14.2(a) time series. As shown, it appears the time series has a trend line that is horizontal. The variation of the 25 data values about the trend line can be attributed to an irregular component. Based on this representation, it now appears this time series does not include a long-term trend. Rather it appears the average value for the variable is relatively constant or *stationary* over time. A time series that does not include a trend, seasonal, or cyclical component is called a **stationary time series.** Conversely, a stationary time series only includes an irregular component. Although we haven't demonstrated the absence of seasonal and cyclical components for this time series, they are not present so Figure 14.2 displays a stationary time series.

For a non-stationary time series with a trend component, the trend can be described either as linear or non-linear. The four line plots of Figure 14.3 and Figure 14.4 illustrate four possible forms. A trend line has been added to each of these four time series plots. Figure 14.3 shows both an increasing and a decreasing linear trend for the time series. On the other hand, Figure 14.4 demonstrates an increasing and a decreasing non-linear trend.

Seasonal Component

The **seasonal component** refers to periodic increases and decreases that occur within a year that follow the same pattern each year. Such a component is apparent in many time series data for business and industry. For example, the retail sales for many products, but particularly for such items as toys, candy, and eggnog is extremely high at the end of the year. The sale of gasoline peaks in the summer. The demand for electric power is high in the winter, lower in the spring, high in the summer, and lower in the fall.

The seasons in a year aren't necessarily the four climatic seasons of summer, fall, winter, and spring. Also, the number of seasons in the year can vary. For the electric power example referred to above, the number of seasons is four. These may in fact be the four quarters of the year that correspond to the four climatic seasons. That might also be the number of seasons for gasoline consumption example. However, it may be more meaningful to

Stationary time series
A time series that does not include a trend, seasonal or cyclical component—only an irregular component.

Seasonal component
Periodic increases and decreases that occur within one year.

Figure 14.3   Linear trend time series plots

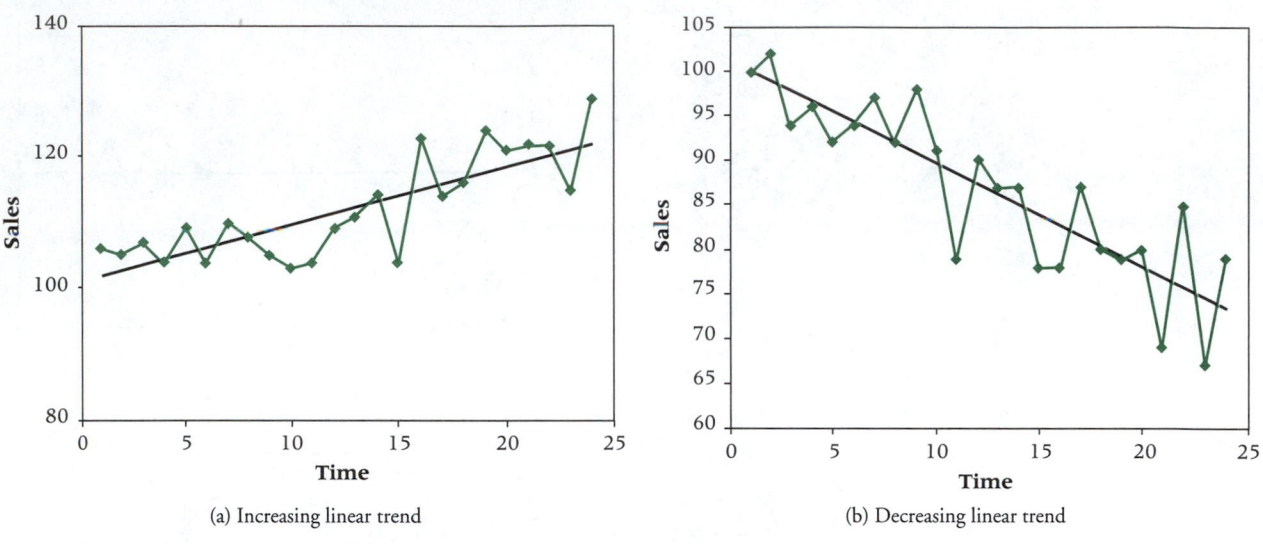

(a) Increasing linear trend

(b) Decreasing linear trend

Figure 14.4   Non-linear trend time series plots

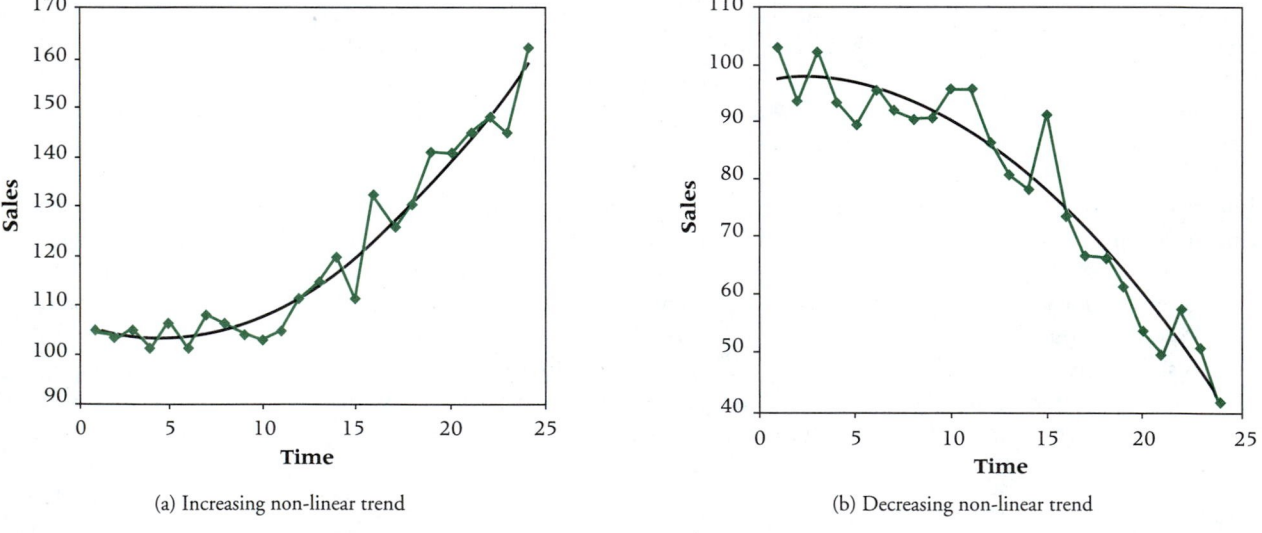

(a) Increasing non-linear trend

(b) Decreasing non-linear trend

consider gasoline consumption by one-month periods. If that were the case, the number of seasons would be 12. For items that are sold mainly at one time of the year, for example toys, perhaps the seasonal variation would best be described by two seasons, the peak season of three months and the nine months of the rest of the year. Thus the number of seasons and the length of each season in the year depend upon what is meaningful for a particular time series.

The length of the seasons also depends on the time intervals used in recording the time series data. For example, if the gasoline consumption time series is best described by

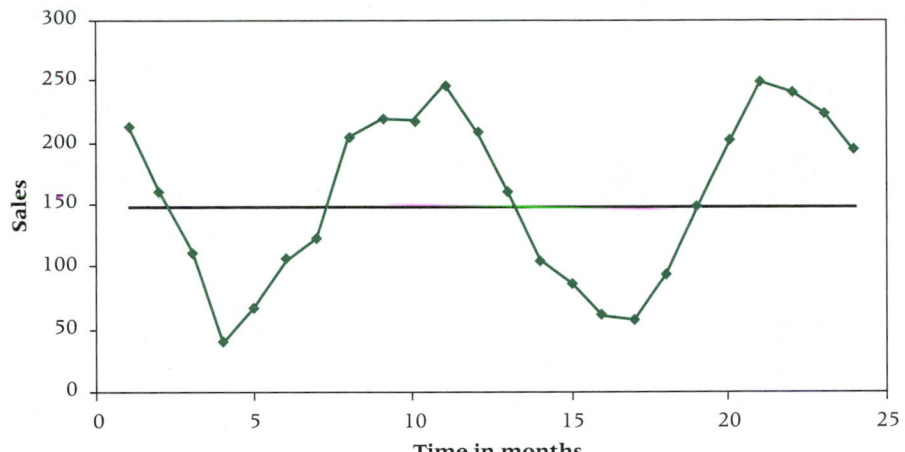

Figure 14.5

Seasonal time series plot

12 month-long seasons in a year, monthly sales data are required. If the sales data were recorded only on a quarterly basis, it would not be possible to analyze the time series on a monthly basis. When a time series is for yearly or annual data, it is not possible to determine a seasonal pattern within a year. It may be that there is a seasonal pattern within a year, but identification and analysis of it cannot be done. In order to identify the seasonal component, the time series must be recorded quarterly, monthly, weekly, daily or any other length of time that is important and is shorter than a year.

A seasonal variation can also be a regularly repeating pattern that occurs over a time period shorter than a year. For example, many persons' bank balances have a repeating pattern within a one-month period. The balance is high at the time earnings are deposited and low at another time of the month when bills are paid. Thus, the seasons within the month might correspond to the days within the month, the weeks within the month, or some other division of the month into periods.

A time series with a seasonal component can also have a trend component or not. For example, again suppose sales data have been recorded for 25 time periods. Suppose the time period is months and the time series plot is as given in Figure 14.5. We note that the high point or *peak* each of the two years occurs in months 11 and 21. In similar fashion, the low point or *trough* occurs in months 4 and 17. Since successive peaks and troughs are almost, but not exactly, 12 months apart, it suggests the presence of a seasonal component. On the other hand, Figure 14.5 also includes a superimposed trend line that suggests the lack of a trend component.

Another monthly sales time series is plotted in Figure 14.6. For it we notice the peak in the first year is month 10 and in second year is month 21. The first year trough occurs in month 4 and for the second year in month 16. Again the occurrence of a peak and a trough at about the same month in each of the two years suggests a seasonal component is present. The height of the second peak is greater than that of the first one, and the same relationship is true for the two troughs. This observation plus the slope of the trend line indicates that a trend component is also present.

Figure 14.6 shows the combination of the seasonal component of the form of Figure 14.5 with the trend component of the form of Figure 14.3(a). Of course, it is possible to have a time series that has a seasonal component of the form of Figure 14.5 combined individually with the form of the trend components shown in Figures 14.3(b), 14.4(a) or 14.4(b).

Figure 14.6

Trend-seasonal time series plot

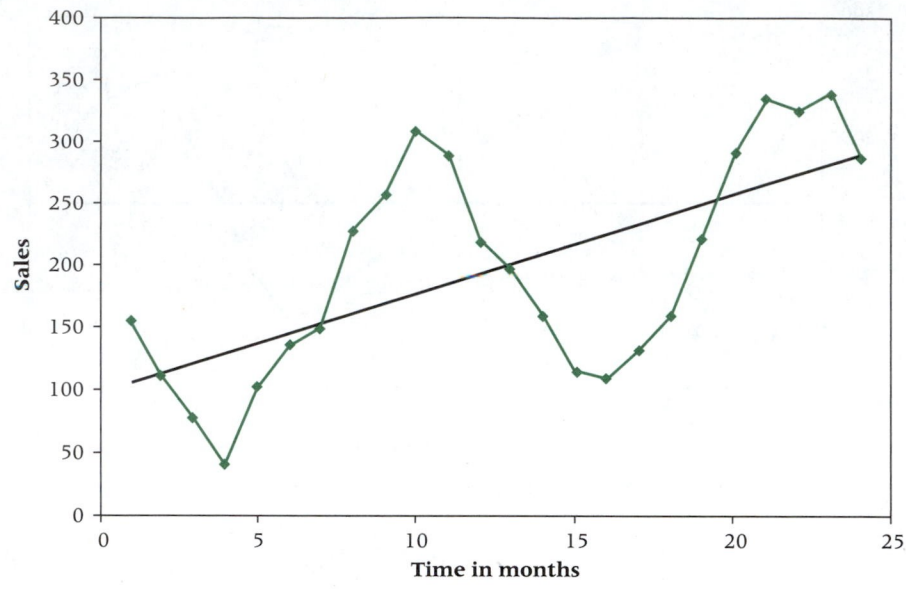

Cyclical Component

Cyclical component
Periodic increases and decreases that occur over more than a one-year period.

The **cyclical component** refers to periodic increases and decreases that occur over more than a one-year period. They follow a pattern over some multi-year period, usually of two to ten years in duration. These variations, sometimes called *business cycles,* are different from seasonal variations. They not only cover longer periods of time, but also have different causes and are less predictable. Cyclical variation generally is the result of conditions within the economy. The peaks usually occur at times of economic expansion and the troughs at times of economic contraction, that is, times of recession or depression. The troughs are then followed by a time of economic recovery. Prediction of these variations within the economy is difficult, if not impossible.

To visualize a cyclical component, again consider the time series of Figure 14.5. Suppose the time unit on the horizontal axis is quarters instead of months. Under this condition, the time series would represent six years of data. The peaks would then occur at quarter 11 and quarter 21. The time between peaks is 12 quarters or three years. In like manner, the troughs would then be at quarter 4 and quarter 17, again about three years apart. Accordingly for this situation, the time series would appear to have a three-year cyclical component. On the other hand, assuming the time unit on the horizontal axis for the time series of Figure 14.6 is quarters not months would result in a time series exhibiting both cyclical and trend components. Finally, it is possible that a time series could include all three of the components discussed so far, a trend, seasonal, and cyclical.

Irregular Component

Irregular component
Changes in the time series not attributable to the other three components.

The **irregular component** refers to changes in the time series not attributable to the other three components. This component is also called the random, residual, erratic, or error component. It represents the unpredictable or non-systematic part of the time series values caused by short-term, non-repetitive effects upon the time series values. Accordingly, this component is unpredictable and oftentimes hides the other components. To eliminate it, most time series forecasting techniques attempt to average or smooth out its effect.

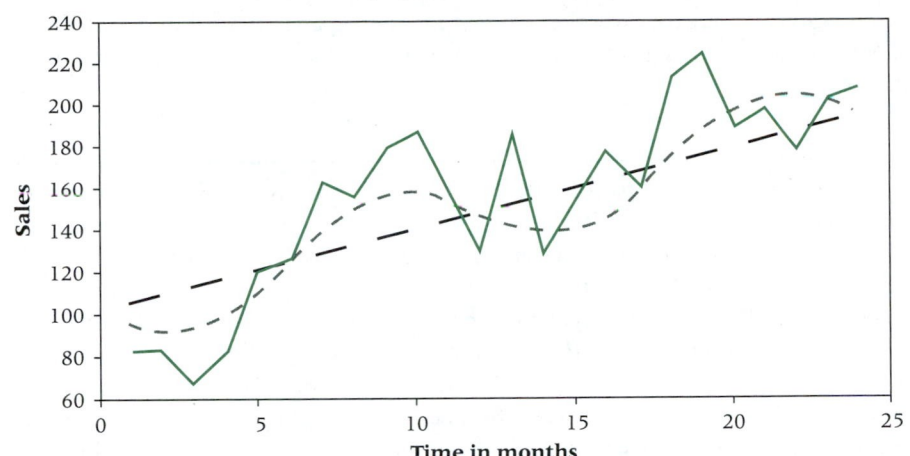

Figure 14.7

Trend, seasonal, and irregular components

Combining the Components

A time series is assumed to include all four or less of the components. For example, the monthly sales time series of Figure 14.7 includes three. It includes a trend component represented by the straight dashed line. The curved dashed line represents the addition of a seasonal component. You will note it is below the trend line during the first six months of the first year and above the trend line in the last six months of the first year. This pattern is repeated in the second year. Finally, the solid line shows the effect of including an irregular component in addition to the first two. Of course, if we were to analyze this time series, we would begin only with the solid line including all three components. The purpose then of the analysis would be to identify the trend component, identify the seasonal component, and to average or smooth out the irregular component. With this accomplished we would attempt to forecast future monthly sales values.

For the purpose of readability and understandability, Figure 14.7 does not also include the cyclical component. As before, the cyclical component is like a seasonal effect that extends over more than a one-year period. Accordingly, if the time units for the horizontal axis were quarters instead of months in Figure 14.7, the time series would represent the combined behavior of a trend, a cyclical, and an irregular effect for a three-year period of time. If, in addition, we were to include a seasonal component, it would add a pattern that would be repeated for each of the three years. An analysis of the time series would begin with the data that included all the components present.

Analysis of a time series usually supposes the four components are combined in one of two ways. One approach assumes the value for the time series variable, Y, at time period t is equal to the sum of the four components, that is,

$$Y_t = T_t + S_t + C_t + I_t$$

where Y_t is the observed value of the variable of interest at time t, and T_t, S_t, C_t, and I_t are the four component values at time t. This is called the **additive model** for a time series. It assumes that S_t, C_t, and I_t are additive deviations about T_t.

The second approach is that of the **multiplicative model.** As the name suggests, this model combines the components as

$$Y_t = T_t \cdot S_t \cdot C_t \cdot I_t$$

Additive model
Value of time series variable equals the sum of the four component effects.

Multiplicative model
Value of time series variable equals the product of the four component effects.

It assumes that S_t, C_t and I_t are multiplicative deviations about T_t.

This completes our consideration of time series components. In the next section we will see how our knowledge of them is an integral part of a four-step time series forecasting procedure.

14.3
Time Series Forecasting Procedure

Time series forecasting can be thought to consist of a sequence of four steps. Some time series forecasting methods are effective only for a time series without a trend, seasonal, or cyclical component. Such a time series has only an irregular component and is called a stationary time series. Other methods are effective for a time series that includes a trend component in addition to an irregular component. Still others perform well when, in addition, a seasonal component is present in a time series. As a result of such considerations, the first step in time series forecasting is to identify the form of the time series by determining the components that are present. The second step is to identify the potential methods suitable for the time series form. Step three is to evaluate the potential methods to determine the method that is *best*. The fourth step is to use the best method to make the required forecasts.

Step 1: Identify Time Series Form

The form of a time series is determined by the components that it includes. Our first observation is that all business time series data is considered to have an irregular component. As you may recall, this is the unpredictable component caused by random influences on the time series values. Since it cannot be predicted, forecasting methods attempt to eliminate its effect by averaging or smoothing.

A second observation considers the cyclical component. In order to identify the presence of this component, it is necessary to have a minimum of two repetitions of the cycle, although three or more would be preferred. Therefore, the discernment of a two-year business cycle would require at the very least four years of stable, relevant time series data. A ten-year cycle would require at least 20 years of data. For most time series data from business and industry, the availability of stable, relevant values for four or more years is not possible. Factors that influence such time series values do not remain unchanged over such relatively long periods of time. Accordingly, forecasts used in business operations generally do not attempt to treat the cyclical component.

This might cause you to wonder what happens in business forecasting to the cyclical component if such a component actually is present. If the cyclical component is irregular, it may be treated with the irregular component. On the other hand, if it is more regular in nature, it will be included with the trend component.

If the irregular component is always present and the cyclical component is included either with it or with the trend component, only the trend and the seasonal components remain to be identified in a time series. The discussion below suggests two methods for detecting the presence of each of these. The first method for both the trend and seasonal variation relies on a time series plot. The second method for both uses an aspect of regression and correlation analysis.

TREND COMPONENT   The determination of the presence of a trend component should always begin with a time series plot. A review of Figures 14.2, 14.3, and 14.4 shows how a time series plot may reveal the presence or not of a trend component and whether it is linear or not.

A second means for investigating the presence or absence of a trend is to use regression analysis to fit a trend line to the data. The *p value* for the slope can then be compared to an appropriate significance level, α, to determine if the trend is significantly different

Figure 14.8 Linear regression results for bond yields

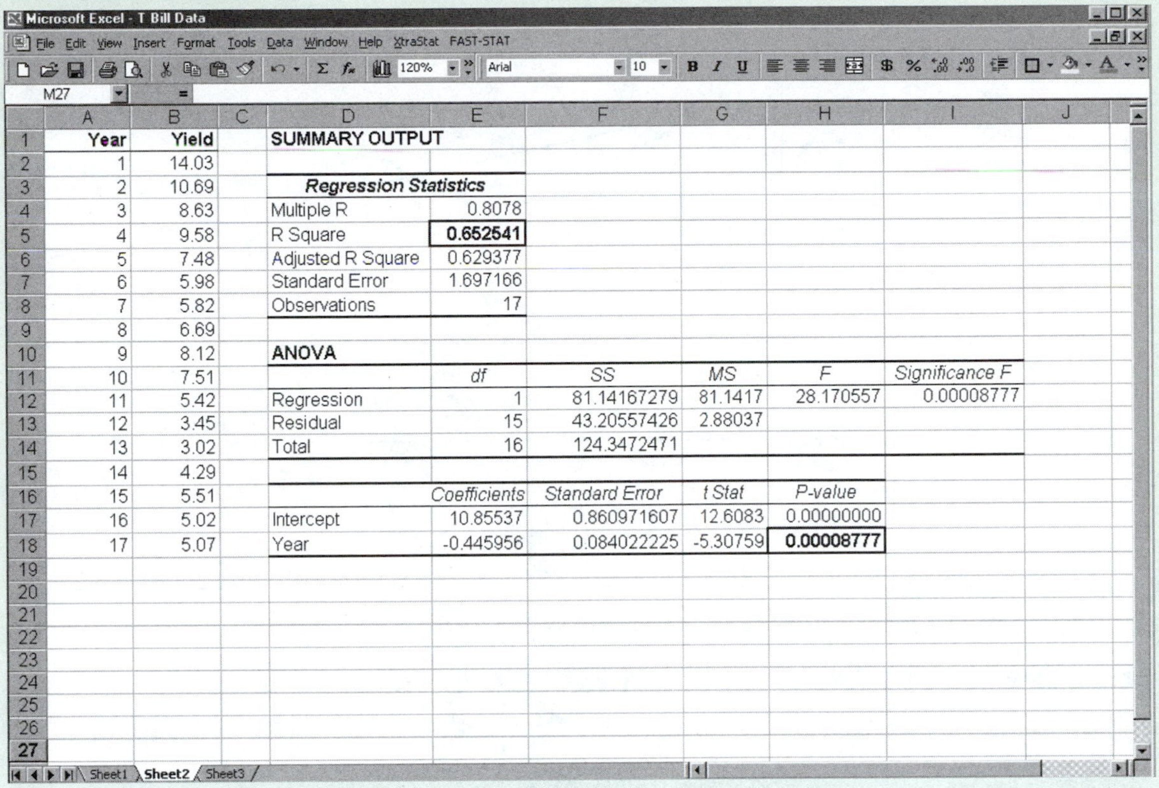

Year	Yield
1	14.03
2	10.69
3	8.63
4	9.58
5	7.48
6	5.98
7	5.82
8	6.69
9	8.12
10	7.51
11	5.42
12	3.45
13	3.02
14	4.29
15	5.51
16	5.02
17	5.07

SUMMARY OUTPUT

Regression Statistics	
Multiple R	0.8078
R Square	**0.652541**
Adjusted R Square	0.629377
Standard Error	1.697166
Observations	17

ANOVA

	df	*SS*	*MS*	*F*	*Significance F*
Regression	1	81.14167279	81.1417	28.170557	0.00008777
Residual	15	43.20557426	2.88037		
Total	16	124.3472471			

	Coefficients	*Standard Error*	*t Stat*	*P-value*
Intercept	10.85537	0.860971607	12.6083	0.00000000
Year	-0.445956	0.084022225	-5.30759	**0.00008777**

than zero. For example, the treasury bill time series of Table 14.1 and Figure 14.1 yield the regression results of Figure 14.8. As you will note, the *p value* for the slope as given in the highlighted cell at the bottom right is 0.00008777.* This value is much smaller than any reasonable value for α. Thus, we would conclude that the linear trend term is very significant. The R^2 value of 0.62541 indicates considerable predictability in the regression relationship.

Figure 14.9 presents the bond yield time series with the linear trend of Figure 14.8 superimposed upon it. You may observe that the actual data values at both ends of the time series are above the trend line. Conversely, most of the actual values in the middle are below the trend line. Such a pattern suggests that a non-linear trend may be present. A non-linear quadratic form of the regression can be found by first adding to the data a column of the time values squared. Next multiple regression can be used to determine the coefficients for the quadratic equation. The results of this process are presented in Figure 14.10.

As you will note, the *p value* for the linear term is 0.0004373 and that for the quadratic term is 0.0079776. Both are much smaller than any reasonable value for α. Thus, we would conclude that they are very significant. The R^2 value of 0.7934666 is greater than

*Strict adherence to the assumptions of this test requires that the errors about the trend line are completely independent of one another. Oftentimes this is not true for time series data. The independence of errors can be tested with the Durbin-Watson statistic discussed in Section 14.5.

Figure 14.9

Linear trend line
for bond yields

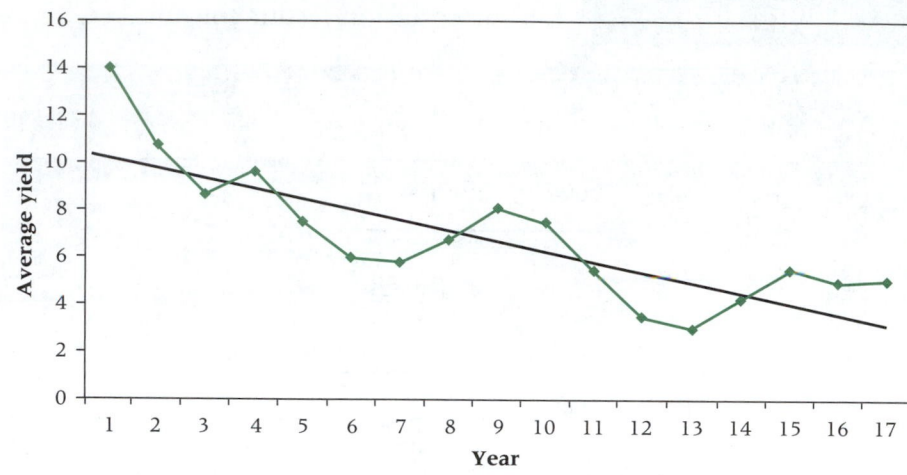

Figure 14.10   Quadratic regression results for bond yields

Microsoft Excel - T Bill Data

File Edit View Insert Format Tools Data Window Help XtraStat FAST-STAT

120%   Arial   10   B I U

R24   =

	A	B	C	D	E	F	G	H	I	J
1	Year	Year-sqrd.	Yield		SUMMARY OUTPUT					
2	1	1	14.03							
3	2	4	10.69		*Regression Statistics*					
4	3	9	8.63		Multiple R	0.8907674				
5	4	16	9.58		R Square	0.7934666				
6	5	25	7.48		Adjusted R Square	0.7639618				
7	6	36	5.98		Standard Error	1.3544071				
8	7	49	5.82		Observations	17				
9	8	64	6.69							
10	9	81	8.12		ANOVA					
11	10	100	7.51			df	SS	MS	F	Significance F
12	11	121	5.42		Regression	2	98.6653886	49.3327	26.8928248	0.0000160
13	12	144	3.45		Residual	14	25.68185846	1.83442		
14	13	169	3.02		Total	16	124.3472471			
15	14	196	4.29							
16	15	225	5.51			Coefficients	Standard Error	t Stat	P-value	
17	16	256	5.02		Intercept	13.565441	1.113970438	12.1776	0.0000000	
18	17	289	5.07		Year	-1.301769	0.284897989	-4.56924	0.0004373	
19					Year-sqrd.	0.0475451	0.015383045	3.09075	0.0079776	
20										
21										
22										
23										
24										
25										
26										
27										

Sheet1 Sheet2 Sheet3

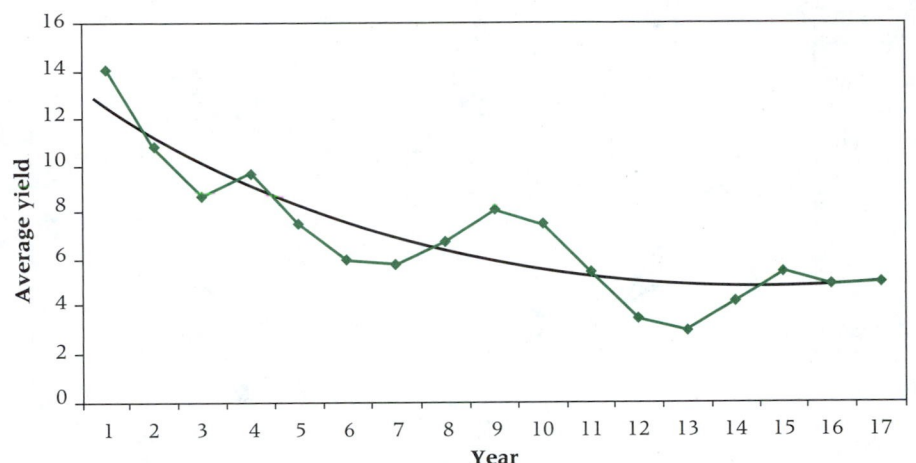

Figure 14.11

Quadratic trend line
for bond yields

that for the linear relationship. The quadratic term seems to add some predictability to the trend representation. Figure 14.11 shows the quadratic trend line superimposed on the time series plot.

SEASONAL COMPONENT   The detection of the presence of a seasonal component requires at least two years worth of data. In addition, the data must be recorded by time periods less than a year, such as quarters, months, weeks, or days. If a time series is recorded on a yearly basis, it is not possible to detect any seasonal variation. In like manner, it is not possible to detect seasonal variation with just one year of data. The data must have at least two repetitions of the seasonal pattern to ascertain whether it exists or not.

As with the trend component, the determination of the presence of a seasonal component can also begin with a time series plot. However, here we use a particular type of a time series plot. For it we will plot each year separately. That is, the horizontal axis of the plot represents only one year, and the values for each year are plotted one above, below, or on top of each other. We will call this particular type of graph a **folded annual time series plot** to indicate it is for a one-year period with the values for each year folded upon each other.

For example, consider the seasonal sales time series plot of Figure 14.5. In order to recognize the presence of a seasonal component, we would plot the first year of data and then the second. However for the second year, the first time period would be plotted as 1 instead of 13, the second as 2 instead of 14, and so on. The result would appear as shown in Figure 14.12. Note that this simple approach makes very clear the presence of a seasonal component.

Now consider the sales data of Figure 14.6 that includes both trend and seasonal components. Upon constructing a folded annual time series plot for it, the result is as shown in Figure 14.13. A comparison of Figures 14.12 and 14.13 reveals another aspect of using a folded annual time series plot. If the time series only has a seasonal component, the annual time series curves tend to be on top of each other as in Figure 14.12. However, if there also is a trend component present, the annual time series curves are above or below each other in response to the trend component. If the trend is increasing, the later annual plot will be above the earlier one, and vice versa.

A second means for investigating the presence or absence of seasonal variation is to use an aspect of regression and correlation analysis called autocorrelation. Data values for a variable over time are often correlated with prior values for that variable. There is some likelihood of this occurring in business data over time, particularly for economic variables. This condition is called **autocorrelation,** or *serial correlation*. Although autocorrelation

**Folded annual
time series plot**
A multi-year time series
plot with only one year on
the horizontal axis.

Autocorrelation
Data values for a time series
variable are correlated with
prior values for that
variable.

Figure 14.12

Folded annual time series plot for seasonal data

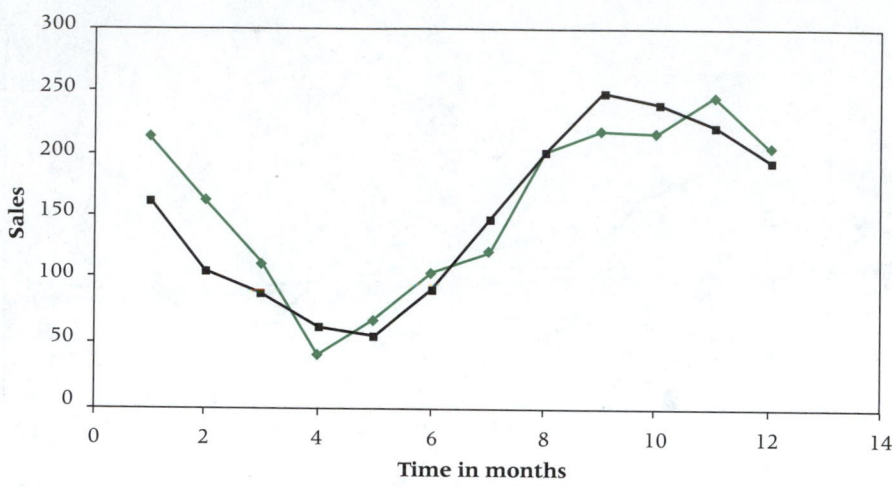

Figure 14.13

Folded annual time series plot for trend-seasonal data

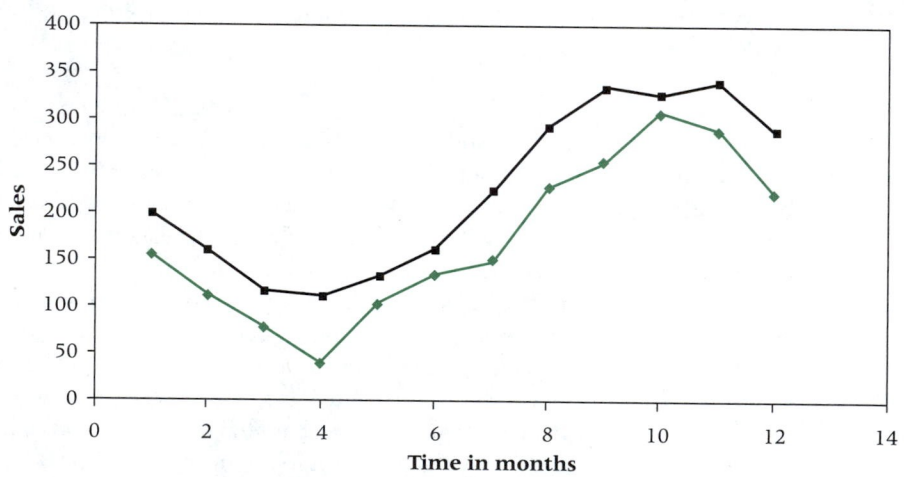

can cause problems in the use of regression analysis (see previous footnote), it also can be a benefit. One such benefit is its use in identifying the presence or absence of seasonal variation in a time series.

The correlation between time series values that are k periods apart is called the *autocorrelation of lag k*. To detect a seasonal component, we can check the autocorrelation for lags of 7 for daily data, of 52 for weekly data, of 12 for monthly data, and of 4 for quarterly data. In addition, an autocorrelation of a lag 1 is an indication of the presence of a trend component. The statistical significance of autocorrelation of lag 1 can be determined by the Durbin-Watson statistic as will be discussed in Section 14.5.

There are a number of methods of computing the autocorrelation of lag k. The usual method for computing r_k, the sample autocorrelation of lag k, is with the formula

$$r_k = \frac{\sum_{t=1}^{n-k} (Y_{t+k} - \overline{Y})(Y_t - \overline{Y})}{\sum_{t=1}^{n} (Y_t - \overline{Y})^2}$$

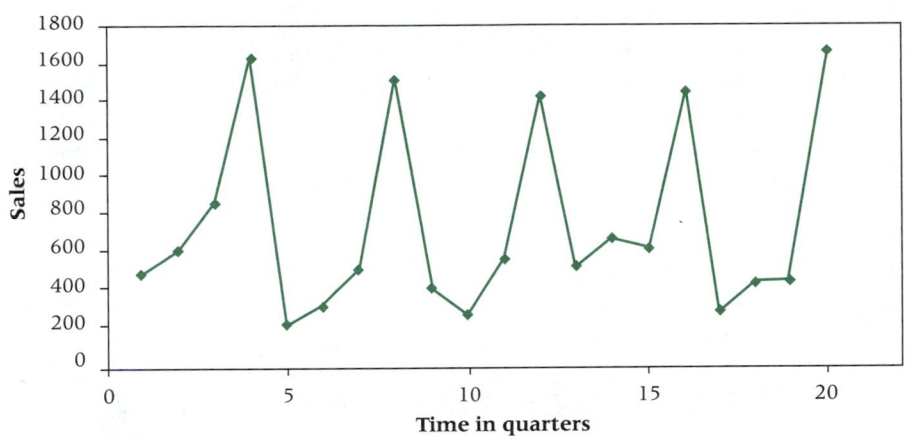

Figure 14.14

Eggnog sales time series plot

Figure 14.15 Eggnog sales and lagged sales values

			Eggnog					Sales Lagged by k =						
	Year	Quarter	Time	Sales	1	2	3	4	5	6	7	8	9	10	11	12
3	1	Jan-Mar	1	473												
4		Apr-Jun	2	599	473											
5		Jul-Sep	3	854	599	473										
6		Oct-Dec	4	1632	854	599	473									
7	2	Jan-Mar	5	211	1632	854	599	473								
8		Apr-Jun	6	313	211	1632	854	599	473							
9		Jul-Sep	7	499	313	211	1632	854	599	473						
10		Oct-Dec	8	1523	499	313	211	1632	854	599	473					
11	3	Jan-Mar	9	409	1523	499	313	211	1632	854	599	473				
12		Apr-Jun	10	265	409	1523	499	313	211	1632	854	599	473			
13		Jul-Sep	11	571	265	409	1523	499	313	211	1632	854	599	473		
14		Oct-Dec	12	1432	571	265	409	1523	499	313	211	1632	854	599	473	
15	4	Jan-Mar	13	524	1432	571	265	409	1523	499	313	211	1632	854	599	473
16		Apr-Jun	14	673	524	1432	571	265	409	1523	499	313	211	1632	854	599
17		Jul-Sep	15	621	673	524	1432	571	265	409	1523	499	313	211	1632	854
18		Oct-Dec	16	1469	621	673	524	1432	571	265	409	1523	499	313	211	1632
19	5	Jan-Mar	17	286	1469	621	673	524	1432	571	265	409	1523	499	313	211
20		Apr-Jun	18	455	286	1469	621	673	524	1432	571	265	409	1523	499	313
21		Jul-Sep	19	460	455	286	1469	621	673	524	1432	571	265	409	1523	499
22		Oct-Dec	20	1680	460	455	286	1469	621	673	524	1432	571	265	409	1523

where k is the number of periods of the lag, n is the number of values in the time series, Y_t is the time series value at time t, Y_{t+k} is the time series value at time $t + k$, and $\overline{Y}$ is the average of the time series.

To begin to illustrate these concepts, consider the sales data for the highly seasonal product eggnog. For a recent five-year period, the quarterly eggnog sales for a company is as given in Figure 14.14. The seasonality in the sales is very apparent. As an exercise, you may wish to create a *folded annual time series plot* for these data.

These sales values are given in column D of the Excel worksheet shown in Figure 14.15. In addition, this figure gives the lagged sales values for lags of 1 quarter, 2 quarters, 3 and so on up to 12 quarters. As you will note, the lagged sales for $k = 1$ are the sales values dropped down one row. Lagged sales for $k = 2$ are the sales values dropped down two rows and so on.

The computation of values for the sample autocorrelations of lag k for k of 1, 2, 3, . . . 12 are performed by the worksheet illustrated in Figure 14.16. In order to compute r_k from the equation previously given, the *sales – average sales* values are computed in column C. To accomplish this, enter the formula **= B3 – Average($B$3:$B$22)** in cell C3. Notice the absolute addresses for B3 and B22. These allow the contents of cell C3 to be copied into cell C4 through C22 to compute the remaining values of that column. The lagged values of *sales – average sales* can be entered through Excel's *Copy* and *Paste Special* commands or through the use of formulas. For example, in cell D4 enter the formula **= C3,** in E5 the formula **= D4** and so on through cell O15. These formulas can then be copied down the column to create all the lagged values needed as shown in Figure 14.16. Finally the r_k values are computed in cells D26 through O26. Enter the following formula in cell D26 and copy it into cells E26 through O26.

=SUMPRODUCT($C$3:$C$22,E3:E22)/DEVSQ($B$3:$B$22)

You must include the absolute addresses for the column C and B values, but do not include them for the column E values as indicated so they will be copied properly.

For those who wish to use this Excel worksheet without entering the formulas, it is given on the CD-ROM enclosed with this book. It is in the folder called *Chapter 14 Templates* and is entitled *Autocorrelation of Lag k.*

Let us consider the meaning of the autocorrelation values obtained in Figure 14.16. First, the seasonal variation is apparent from these 12 values. The largest r_k value, 0.6988,

Figure 14.16 Computation of autocorrelation of lag k values

Microsoft Excel - Component Plots

	A	B	C	D	E	F	G	H	I	J	K	L	M	N	O
1		Eggnog	Sales -					Sales - Average lagged by k =							
2	Time	Sales	Average	1	2	3	4	5	6	7	8	9	10	11	12
3	1	473	-274.5												
4	2	599	-148.5	-274.5											
5	3	854	106.6	-148.5	-274.5										
6	4	1632	884.6	106.6	-148.5	-274.5									
7	5	211	-536.5	884.6	106.6	-148.5	-274.5								
8	6	313	-434.5	-536.5	884.6	106.6	-148.5	-274.5							
9	7	499	-248.5	-434.5	-536.5	884.6	106.6	-148.5	-274.5						
10	8	1523	775.6	-248.5	-434.5	-536.5	884.6	106.6	-148.5	-274.5					
11	9	409	-338.5	775.6	-248.5	-434.5	-536.5	884.6	106.6	-148.5	-274.5				
12	10	265	-482.5	-338.5	775.6	-248.5	-434.5	-536.5	884.6	106.6	-148.5	-274.5			
13	11	571	-176.5	-482.5	-338.5	775.6	-248.5	-434.5	-536.5	884.6	106.6	-148.5	-274.5		
14	12	1432	684.6	-176.5	-482.5	-338.5	775.6	-248.5	-434.5	-536.5	884.6	106.6	-148.5	-274.5	
15	13	524	-223.5	684.6	-176.5	-482.5	-338.5	775.6	-248.5	-434.5	-536.5	884.6	106.6	-148.5	-274.5
16	14	673	-74.5	-223.5	684.6	-176.5	-482.5	-338.5	775.6	-248.5	-434.5	-536.5	884.6	106.6	-148.5
17	15	621	-126.5	-74.5	-223.5	684.6	-176.5	-482.5	-338.5	775.6	-248.5	-434.5	-536.5	884.6	106.6
18	16	1469	721.6	-126.5	-74.5	-223.5	684.6	-176.5	-482.5	-338.5	775.6	-248.5	-434.5	-536.5	884.6
19	17	286	-461.5	721.6	-126.5	-74.5	-223.5	684.6	-176.5	-482.5	-338.5	775.6	-248.5	-434.5	-536.5
20	18	455	-292.5	-461.5	721.6	-126.5	-74.5	-223.5	684.6	-176.5	-482.5	-338.5	775.6	-248.5	-434.5
21	19	460	-287.5	-292.5	-461.5	721.6	-126.5	-74.5	-223.5	684.6	-176.5	-482.5	-338.5	775.6	-248.5
22	20	1680	932.6	-287.5	-292.5	-461.5	721.6	-126.5	-74.5	-223.5	684.6	-176.5	-482.5	-338.5	775.6
23															
24								Autocorrelation of lag k =							
25				1	2	3	4	5	6	7	8	9	10	11	12
26				-0.1996	-0.3593	-0.3474	0.6988	-0.0967	-0.2423	-0.2319	0.5255	-0.0710	-0.1808	-0.1964	0.3954

Sheet1 / Sheet2 / Sheet4 / Sheet5 / Sheet7 / **Sheet6** / Sheet3 is for a lag of 4. A rule of thumb for testing the significance of autocorrelation coefficients of lag k with a significance level of approximately 0.05 is based on the number of values in the time series. The autocorrelation is said to be significant if $|r_k| > 2\sqrt{n}$. Since the number of eggnog sales values is 20, the rule of thumb states that values with an absolute value greater than 0.447 are significant. The second largest, 0.5255, is for a lag of 8, and the third largest, 0.3954, is for a lag of 12. The first of these is significant and the second is not. The large r_4 value suggests that seasonality is presence. The large r_8 and r_{12} values substantiate this impression.

Second the low, non-significant value for r_1 indicates a trend component is not present. A visual inspection of the time series plot of Figure 14.14 confirms this conclusion. As before, we also could have fit a trend line to the data and tested the significance of the slope to detect the absence or presence of a trend.

Finally, the number of values in the time series determines the number of viable lagged terms. Generally, it is suggested that the number of viable terms not be more than 25% of the number of values in the time series. Accordingly, the number of viable lagged terms for the time series above should be no more than 25% of 20, or 5. We computed 12 terms only for demonstration purposes.

Step 2: Select Potential Methods

Step 1 of our time series forecasting procedure identifies the form of the time series. In Step 2, this knowledge is used to select the forecasting methods that are potentially useful for forecasting the time series. Since an irregular component is always present, and since the cyclical component usually cannot be identified for business data, only the trend and seasonal components must be considered as discussed above. A summary of the possible time series forms and the forecasting methods of this chapter are given in Figure 14.17.

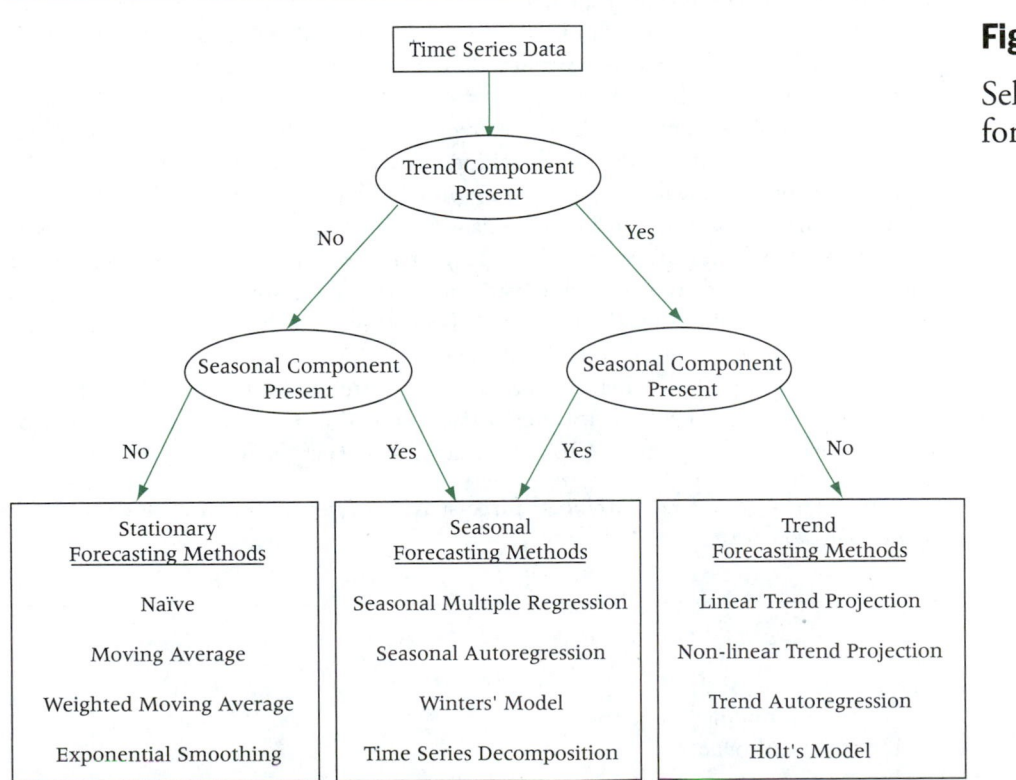

Figure 14.17

Selecting potential forecasting methods

Stationary forecasting methods are those that are effective for a stationary time series, that is, one with only an irregular component. These methods are sometimes called averaging or smoothing methods because they attempt to eliminate the irregular variation through averaging. Those stationary methods that will be discussed in Section 14.4 include the *naïve,* the *moving average,* the *weighted moving average,* and the *exponential smoothing* methods. Although these methods can be used for a non-stationary time series, the results are not satisfactory. The forecasts made based on these methods will always lag behind changes in a non-stationary times series. Accordingly, they are not effective for such time series data as is illustrated by Demonstration Problem 14.4. It is recommended that trend or seasonal methods be used for non-stationary time series.

Trend forecasting methods are effective for a time series that includes a trend component. These methods explicitly assess the trend component and utilize it in making projections into the future. Those trend methods that will be presented in Section 14.5 are the *linear trend projection,* the *non-linear trend projection, autoregression for trend,* and *Holt's model for exponential smoothing.* Trend methods are perhaps the most useful time series forecasting methods.

Seasonal forecasting methods are used for a time series that includes a trend, a seasonal, and an irregular component. They are equally useful for a time series that only includes a seasonal and an irregular component. Their use requires that the time series data be recorded at time intervals of less than a year. It is not possible to treat a seasonal variation for a time series recorded at one-year or longer time intervals. In addition, it is necessary to have at least two years of time series data to discern a seasonal variation. To be effective, it is recommended that three or more years of data be used. The seasonal methods that are introduced in Section 14.6 include the *seasonal multiple regression, autoregression for seasonality, Winters' model for exponential smoothing,* and the *time series decomposition* methods.

Step 3: Evaluate Potential Methods

Once the appropriate potential methods have been identified for a time series, the next step is to evaluate how well each method performs in forecasting the time series. This is accomplished by first using each method to forecast the historical data for the time series. Next, an evaluation of how closely these forecasts fit the actual historical data is made. This evaluation is based on finding the difference between each actual value and the forecast of that value. These are the individual forecast errors for the forecasting method. An examination of individual errors gives some insight into the accuracy of the forecasts. However, this process can be tedious, especially for large data sets, and often *a single measurement of overall* **forecasting error** is needed *for the entire set of data under consideration.* Any of several methods can be used to compute error in forecasting. The choice depends on the forecaster's objective, the forecaster's familiarity with the technique, and/or the method of error measurement used by the computer forecasting software. Several techniques have been proposed that can be used to measure overall error. These include, among others, mean error (ME), mean absolute deviation (MAD), mean square error (MSE), mean percentage error (MPE), and mean absolute percentage error (MAPE).

Forecasting error
A single measure of the overall error of a forecast for an entire set of data.

Error of an individual forecast
The difference between the actual value and the forecast of that value.

ERROR   The **error of an individual forecast** is *the difference between the actual value and the forecast of that value.*

ERROR OF AN INDIVIDUAL FORECAST	$$e_t = Y_t - F_t$$

where:
   e_t = the error of the forecast
   Y_t = the actual value
   F_t = the forecast value

MEAN ERROR (ME)   One method of computing overall error is the mean error, or ME. The **mean error** is *the average of all the errors of forecast for a group of data.* The formula for computing mean error follows.

MEAN ERROR

$$ME = \frac{\Sigma e_i}{\text{Number of Forecasts}}$$

where:

e_i = the forecast error for value i

As the formula shows, the mean error is determined by summing all the forecast errors and dividing by the number of forecasts. This, of course, produces the average error, or the mean error. One of the interesting aspects of ME is that if the forecasts are both over and under the actual values, producing positive and negative errors, ME will include some cancellation effects of the positive and negative values. In these cases, the result is that ME is less than the average of the "magnitudes" of the forecast errors. Because of this effect, ME may understate the error in terms of the measurement of overall error.

Table 14.3 contains the number of nonfarm partnership tax returns for an 11-year period (in 1000s) as reported by the U.S. Department of the Treasury. Forecasts have been made by using techniques introduced later in the chapter. Individual errors of forecast have been tallied. The forecast values along with the errors are also displayed in Table 14.3.

The mean error is computed for the data in Table 14.3 as

$$ME = (56.0 + 111.8 + 93.5 + 91.0 + 106.3 + 83.9 + 42.2 + 14.7 - 41.6 - 33.5)/10$$

$$= 52.4.$$

MEAN ABSOLUTE DEVIATION (MAD)   An examination of the data in Table 14.3 reveals that some of the forecast errors are positive and some are negative. The negative values of 41.6 and 33.5 are included in the computation of the ME, resulting in a reduction of the sum. In some situations, forecasters might prefer to examine the magnitude of the forecast errors without regard to direction.

TABLE 14.3
Nonfarm Partnership Tax Returns

YEAR	ACTUAL	FORECAST	ERROR
1	1402	—	—
2	1458	1402	56.0
3	1553	1441.2	111.8
4	1613	1519.5	93.5
5	1676	1585.0	91.0
6	1755	1648.7	106.3
7	1807	1723.1	83.9
8	1824	1781.8	42.2
9	1826	1811.3	14.7
10	1780	1821.6	−41.6
11	1759	1792.5	−33.5

The **mean absolute deviation (MAD)** is *the mean, or average, of the absolute values of the errors.*

MEAN ABSOLUTE DEVIATION	$$\text{MAD} = \frac{\Sigma|e_i|}{\text{Number of Forecasts}}$$

Of course, by taking the absolute values of the errors, we do not have to worry about the canceling effects of the positive and negative values. The mean absolute error can be computed for the forecast errors in Table 14.3 as follows.

$$\text{MAD} = (|56.0| + |111.8| + |93.5| + |91.0| + |106.3| + |83.9|$$
$$+ |42.2| + |14.7| + |-41.6| + |-33.5|)/10$$
$$= 67.5$$

Notice that this value is larger than the ME value computed on the same data (67.5 > 52.4).

Mean square error (MSE)
The average of all errors squared of a forecast for a group of data.

MEAN SQUARE ERROR (MSE)   The **mean square error (MSE)** is another way to circumvent the problem of the canceling effects of positive and negative forecast errors. The MSE is computed by *squaring each error* (thus creating a positive number) *and averaging the squared errors.* The following formula states it more formally.

MEAN SQUARE ERROR	$$\text{MSE} = \frac{\Sigma e_i^2}{\text{Number of Forecasts}}$$

The mean square error can be computed for the errors shown in Table 14.3 as follows.

$$\text{MSE} = [(56.0)^2 + (111.8)^2 + (93.5)^2 + (91.0)^2 + (106.3)^2 + (83.9)^2$$
$$+ (42.2)^2 + (14.7)^2 + (-41.6)^2 + (-33.5)^2]/10$$
$$= 55,847.1/10$$
$$= 5584.7$$

Percentage error (PE)
The ratio of the error of a forecast to the actual value being forecast multiplied by 100.

MEAN PERCENTAGE ERROR (MPE)   In decision making, it can sometimes be misleading and not particularly worthwhile to examine the raw values or even the average of raw values of error. Some forecasters like to examine errors in terms of their percentage of the actual values instead of working with the raw errors. A **percentage error (PE)** is *the ratio of the error to the actual value multiplied by 100.*

PERCENTAGE ERROR	$$\text{PE} = \frac{e_i}{Y_i}(100)$$

The average of these percentage errors is called the **mean percentage error (MPE).**

$$MPE = \frac{\Sigma\left(\frac{e_i}{Y_i} \cdot 100\right)}{\text{Number of Forecasts}}$$

Computing the MPE for the errors listed in Table 14.3 results in

$$
\begin{aligned}
MPE = [&(56.0/1458)(100) + (111.8/1553)\,(100) \\
&+ (93.5/1613)(100) + (91.0/1676)(100) \\
&+ (106.3/1755)(100) + (83.9/1807)(100) \\
&+ (42.2/1824)(100) + (14.7/1826)(100) \\
&+ (-41.6/1780)(100) + (-33.5/1759)(100)]/10 \\
= \;& 3.2\%.
\end{aligned}
$$

**Mean percentage error
(MPE)**
The average of the
percentage errors of a
forecast.

The average percentage error for the nonfarm partnership data forecasts is 3.2%.

MEAN ABSOLUTE PERCENTAGE ERROR (MAPE)   The MPE just computed includes two negative percentage errors, which cancel a portion of the sum of the positive percentage errors, resulting in a reduced value of MPE. This canceling effect can be eliminated by computing the mean absolute percentage error. The **mean absolute percentage error (MAPE)** is *the average of the absolute values of the percentage errors.*

**Mean absolute percentage
error (MAPE)**
The average of the absolute
values of the percentage
errors of a forecast.

$$MAPE = \frac{\Sigma\left(\frac{|e_i|}{Y_i} \cdot 100\right)}{\text{Number of Forecasts}}$$

The MAPE can be computed on the errors listed in Table 14.3 as

$$
\begin{aligned}
MAPE = [&(|56.0|/1458)(100) + (|111.8|/1553)(100) \\
&+ (|93.5|/1613)(100) + (|91.0|/1676)(100) \\
&+ (|106.3|/1755)(100) + (|83.9|/1807)(100) \\
&+ (|42.2|/1824)(100) + (|14.7|/1826)(100) \\
&+ (|-41.6|/1780)(100) + (|-33.5|/1759)(100)]/10 \\
= \;& 4.0\%.
\end{aligned}
$$

Notice that this value is larger than the value for MPE (3.2%). To view error as a percentage, the forecaster has the choice of using MPE or MAPE.

USE OF THE ERROR MEASURES   We will use these measures of the overall error in two ways to identify the **best** forecasting method. First, the error measures may be used to identify the *best* value for the parameters of a specific method. For example, one of the methods discussed in Section 14.4 is the simple exponential smoothing method. It utilizes a parameter called the exponential smoothing constant. Values for it can range between 0 and 1. The best value for the constant can be identified by first evaluating different values between 0 and 1. Then select the value with the lowest measure of overall error. Second the error measures may be used to identify the best method of a number considered. For example, suppose you were considering exponential smoothing and moving average methods for forecasting a stationary time series. The *best* of the two methods could be identified by comparing overall error measurements for the two methods.

The best error measures for use in these two ways are MSE and MAD. Each of these two has advantages over the other. The MSE measure tends to emphasize large errors more since it squares each value. If large errors are to be avoided, then using MSE is preferable. On the other hand, if an occasional large error isn't of prime concern, MAD may serve well to provide better overall tracking of the time series.

Forecast bias

Tendency of forecasting method to over or under predict. Also called average or mean error (ME).

The mean error, ME, is also called **forecast bias.** It is the average error so it provides a measure of the tendency of a forecasting method to over predict or under predict. If the forecast bias for a forecasting method is not near zero, it suggests the method is consistently wrong in the same way. In other words, the forecasts consistently either lag or lead the actual time series values.

Both the MPE and MAPE are relative measures. The primary use of these is in comparing the performance of *one* forecasting method on *two or more* time series. This is not the situation to be faced in this chapter. Our use involves comparing the performance of *more than one* forecasting method for *one* time series.

In summary, we will use both MSE and MAD as an overall measure of error throughout this chapter. In addition, in some instances we will examine the forecast bias of forecasting methods by computing a value for ME.

Step 4: Make Required Forecasts

The *best* forecasting method is identified as that with the smallest overall error measurement value, that is, the smallest MAD and/or MSE and also, perhaps, forecast bias. This *best* forecasting method is then used to make forecasts of one or more future values.

Suppose the historical data for a time series includes t values. The historical values would be represented by the symbols $Y_1, Y_2, Y_3, \ldots, Y_t$. The forecast for the next time period would be designated as F_{t+1}, that for the two time periods into the future as F_{t+2}, three time periods F_{t+3} and so on.

Using a stationary forecasting method will provide the forecast for one time into the future, F_{t+1}. Recall that a stationary forecasting method is based upon the premise that only irregular variation is present in the time series. Accordingly, the forecast F_{t+1} is also the forecast for all future time periods. Since stationary methods are predicated upon the lack of a trend, seasonal, or cyclical component, these methods predict only the average or smoothed value for the time series. Accordingly for these methods,

$$F_{t+1} = F_{t+2} = F_{t+3} = F_{t+4} = \ldots \ldots$$

For stationary methods, the forecasts for these future time periods are changed only when data for additional time periods are obtained.

On the other hand, the forecasts made with trend or seasonal methods will not be the same for all time periods in the future. These methods assume there is either a trend or seasonal variation or both. So for non-stationary forecasting methods,

$$F_{t+1} \neq F_{t+2} \neq F_{t+3} \neq F_{t+4} \neq \ldots \ldots$$

This completes our time series forecasting procedure. As you may recall, we began by characterizing the form of the time series. Specifically, does the data include trend and/or seasonal variation? Based on the establishment of the time series form, the diagram of Figure 14.17 identifies the forecasting methods that may be successful in forecasting the time series. Once these methods have been designated, they are evaluated using an error measure for each. For some methods, the error measure may also be used to identify the value for one or more parameters. The result is the establishment of the best parameter value(s) and the best forecasting method. The fourth and final step, then, is to make future forecasts using the best method.

The remainder of the chapter will present the twelve time series forecasting methods given in Figure 14.17. Sections 14.4, 14.5, and 14.6 respectively present the stationary methods, the trend methods, and the seasonal methods.

DEMONSTRATION PROBLEM 14.1

The following data give the average baseball salary for ten recent years as compiled by the *Major League Baseball Players Association.* Perform the four steps of our *time series forecasting procedure.* For Step 1, identify the components present in the time series. For Step 2, specify those forecasting methods that should be considered. For Step 3, evaluate one such method using the MSE measure of forecasting error. For Step 4, forecast average salaries for the next two years.

YEAR	AVG. SALARY
1	$497,254
2	$597,537
3	$851,492
4	$1,028,667
5	$1,076,089
6	$1,168,263
7	$1,110,766
8	$1,119,981
9	$1,336,609
10	$1,398,831

SOLUTION

STEP 1: *Identify Time Series Form*

Our first observation is that an irregular component is present as is assumed for all time series data. A second observation is that the cyclical component will not be treated separately within the forecasting procedure, but will be included with either the irregular and/or the trend component. A third observation for these data is that they are annual values. Accordingly, it is not possible to discern a seasonal component. That leaves only the trend component to be identified.

To determine the presence of a trend component, we first look at the time series plot for these data.

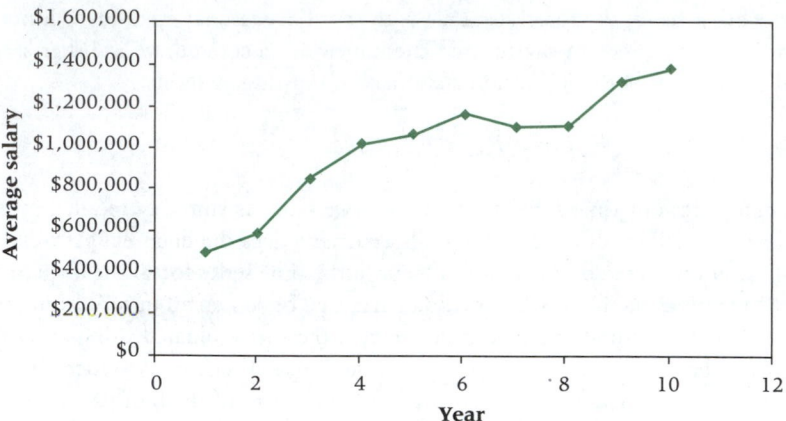

The increasing trend is quite apparent. It looks as if the trend might be either linear or slightly nonlinear. The next figure presents the linear regression fit results for the time series.

	A	B	C	D	E	F	G	H	I
1	Year	Avg. Salary		SUMMARY OUTPUT					
2	1	$497,254							
3	2	$597,537		*Regression Statistics*					
4	3	$851,492		Multiple R	0.94068				
5	4	$1,028,667		R Square	**0.88487**				
6	5	$1,076,089		Adjusted R Square	0.87048				
7	6	$1,168,263		Standard Error	105081				
8	7	$1,110,766		Observations	10				
9	8	$1,119,981							
10	9	$1,336,609		ANOVA					
11	10	$1,398,831			df	SS	MS	F	Significance F
12				Regression	1	6.8E+11	6.8E+11	61.4893	0.000050
13				Residual	8	8.8E+10	1.1E+10		
14				Total	9	7.7E+11			
15									
16					Coefficients	andard Err	t Stat	P-value	
17				Intercept	519595	71784.1	7.23831	0.000089	
18				Year	90718.9	11569	7.84151	**0.000050**	

The *p value* for the slope is extremely small, 0.000050, and the R^2 value is extremely high, 0.88487. The conclusion is that a trend component is present.

STEP 2: **Select Potential Methods**

A trend forecasting method is required for the baseball salary time series. From Figure 14.17, the potential methods to be considered include *linear trend projection, non-linear trend projection, trend autoregression,* and *Holt's model.*

STEP **3:** *Evaluate Potential Methods*

Since the information in Section 14.6 is yet to be covered, we will evaluate only two of these. *Linear trend projection* and *non-linear trend projection* are based on regression analysis as discussed in Chapters 12 and 13. In addition, the discussion above of Figures 14.8 through 14.11 relates to these two methods.

Excel's **Trendline** feature for charts (see Chapter 1) can be used to add the linear trend to the baseball salary time series plot as shown below. For the trendline dialog box, you are to select **Linear** for the *Type* of trendline. On the **Options** tab of the dialog box, click on the check box for **Display equation on chart** and on the check box for **Display R-squared value on chart**. Also on the **Options** tab under *Forecast* and *Forward*, select **2** *units*. The result is the following.

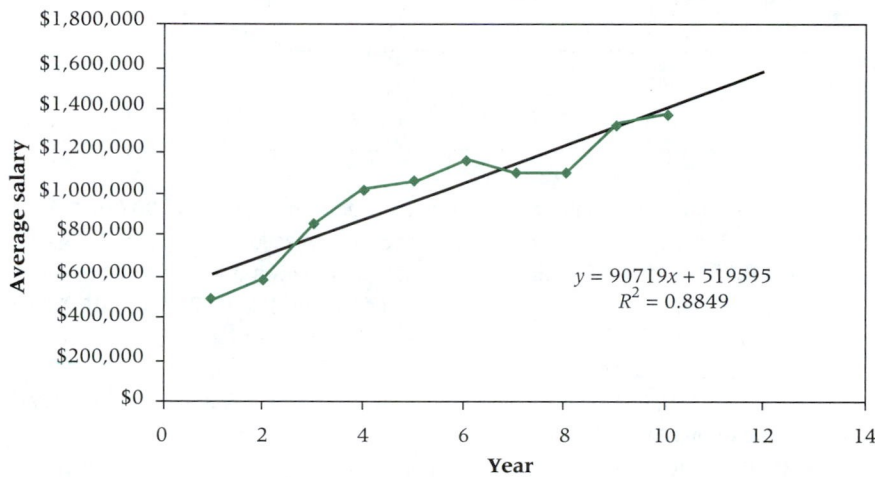

The plot shows the equation $y = 90719x + 519595$ and $R^2 = 0.8849$.

The results provided by **Trendline** include (1) the equation for the trend line, (2) a plot of the trend line projected two periods into the future and (3) the value for R^2.

On this time series plot, click on the trend line and press the delete (**Del**) key to eliminate the trend line. Next use **Trendline** to add a second-order polynomial (quadratic) trend line to the data.

The results below show an improved value for R^2.

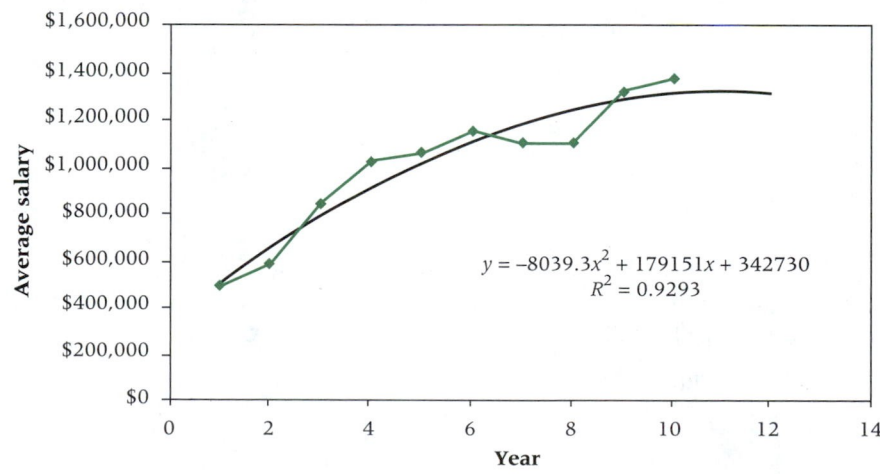

The plot shows the equation $y = -8039.3x^2 + 179151x + 342730$ and $R^2 = 0.9293$.

To compare the two forecasting methods, the following Excel worksheet shows the computation of the MSE for both methods.

	A	B	C	D	E
1	Year	Avg. Salary	Linear Forecast	Quadratic Forecast	
2	1	$497,254	610,314	513,842	
3	2	$597,537	701,033	668,875	
4	3	$851,492	791,752	807,829	
5	4	$1,028,667	882,471	930,705	
6	5	$1,076,089	973,190	1,037,503	
7	6	$1,168,263	1,063,909	1,128,221	
8	7	$1,110,766	1,154,628	1,202,861	
9	8	$1,119,981	1,245,347	1,261,423	
10	9	$1,336,609	1,336,066	1,303,906	
11	10	$1,398,831	1,426,785	1,330,310	
12		MSE =	8,833,631,511	5,421,140,320	
13	11			1,340,636	
14	12			1,334,883	
15					
16					

The formula entered into cell C2 for linear trend projection is = **90719 * A2 + 519595** and that in cell D2 for quadratic trend projection is = **−8039.3 * A2 ^2 + 179151 * A2 + 342730.** These are copied into rows 3 through 11 of their respective columns. The computation of MSE in cell C12 is = **SUMXMY2 ($B$2:$B$11,C2:C11)/10.** Note the absolute address for B2:B11 so that the formula can be correctly copied into cell D12.

Based on the two values for MSE, the quadratic trend line appears to fit best.

STEP **4: *Make Required Forecasts***

To make a forecast for the next two periods copy the formula for the quadratic trend into cells D13 and D14 for years 11 and 12. The quadratic method predicts salaries of $1,340,636 and $1,334,883 for these two years.

14.3
Problems

14.1 Use the forecast errors given here to compute ME, MAD, and MSE. Discuss the information yielded by each type of error measurement.

PERIOD	e
1	2.3
2	1.6
3	−1.4
4	1.1
5	.3
6	−.9
7	−1.9
8	−2.1
9	.7

14.2 Determine the error for each of the following forecasts. Compute ME, MAD, MSE, MPE, and MAPE.

PERIOD	VALUE	FORECAST	ERROR
1	202	—	—
2	191	202	
3	173	192	
4	169	181	
5	171	174	

continued

14.2 *continued*

6	175	172
7	182	174
8	196	179
9	204	189
10	219	198
11	227	211

14.3 Using the following data, determine the values of ME, MAD, MSE, MPE, and MAPE. Which of these measurements of error seems to yield the best information about the forecasts? Why?

PERIOD	VALUE	FORECAST
1	19.4	16.6
2	23.6	19.1
3	24.0	22.0
4	26.8	24.8
5	29.2	25.9
6	35.5	28.6

14.4 Below are figures for acres of tomatoes harvested in the United States from 1988 through 1998. The data are published by the U.S. Department of Agriculture. With these data, forecasts have been made by using techniques presented later in this chapter. Compute ME, MAD, MSE, MPE, and MAPE on these forecasts. Comment on the errors.

YEAR	NUMBER OF ACRES	FORECAST
1988	140,000	—
1989	141,730	140,000
1990	134,590	141,038
1991	131,710	137,169
1992	131,910	133,894
1993	134,250	132,704
1994	135,220	133,632
1995	131,020	134,585
1996	120,640	132,446
1997	115,190	125,362
1998	114,510	119,259

The four stationary forecasting methods presented below are effective for time series data that only include an irregular component. They are not effective in forecasting a time series that has a trend, a seasonal, or a cylical component. If these methods are used for a time series with one or more of these three components, they will produce forecasts that generally lag the actual values. The forecasts will not be able to *keep up* with the trend, seasonal, or cyclical changes in the time series data. A time series with a trend component requires a method such as those given in the next section. A time series with a seasonal component, either with or without a trend component, requires a method such as those given in Section 14.6.

We begin with the *naïve* forecasting method. It is also called the *last period method*.

14.4
Stationary Forecasting Methods

Naïve Forecasting Method

Naïve forecasting methods are simple models in which it is assumed that the more recent time periods of data represent the best predictions or forecasts for future values. Naïve methods do not take into account either trend or seasonal variation. The simplest

Naïve forecasting method
Forecasting based on the assumption that the more recent time periods of data represent the best forecast for future values.

TABLE 14.4
Number of Tracquer Tires
Sold Each Month

MONTH	NUMBER SOLD
1	96,127
2	97,654
3	101,972
4	109,301
5	91,167
6	97,403
7	110,983
8	103,258
9	101,765
10	98,231
11	99,992
12	102,111
13	97,894
14	104,267
15	91,361
16	105,901

naïve approach is that which the forecast for a given time period is the value from the previous time period.

$$F_{t+1} = Y_t$$

where:

F_{t+1} = the forecast value for the next time period $t + 1$

Y_t = the actual value for time period t

For example, if 532 pairs of shoes were sold by a retailer last week, the naïve approach would predict that the retailer will sell 532 pairs of shoes next week. With a naïve method, the actual sales for this week will be the forecast for next week.

Consider the sales data in Table 14.4 for a tire manufacturer. It shows the monthly sales of a brand of tire called the *Tracquer* for a recent 16-month period. Sales values are reported in various formats, but usually they are given in either number of units sold or sales revenue in dollars. The data of Table 14.4 are for number of tires sold. The time series plot for these data is given in Figure 14.18 and, as you will note, the sales values are given as 1000s of units sold to make the computations less cumbersome.

A naïve forecast for next month is 105,901 or, perhaps, 106,000 tires. Another version of a naïve forecast might be to use the number sold for the corresponding month of the past year as the forecast for the upcoming month. The basis behind such an approach is the belief that there is a relationship between the number sold and the month of the year. In this case, the naïve forecast for next month would be the value from 12 months back, that is, for month 5, or 91,167 or 91,000 tires. The forecaster is free to be creative with the naïve forecast method and search for other relationships or rationales within the limits of the time-series data upon which to make a valid forecast. Of course, if it is thought that relationships of this sort exist, the forecaster may wish to consider the more sophisticated trend or seasonal methods of Sections 14.5 and 14.6.

Moving Average Forecasting Method

Irregular variations
Unexplained or error variation with time-series data.

Many naïve forecasts are based on the value of one time period. Often such forecasts become a function of **irregular variations** of the data; as a result, the forecasts are "oversteered." By using *averaging models,* a forecaster can begin to "smooth" the data and enter

Figure 14.18

Monthly Tracquer tire sales in 1000s sold

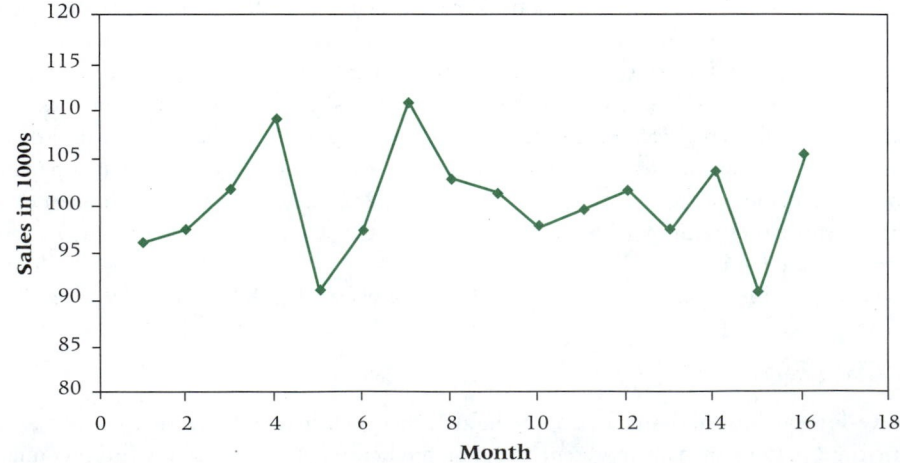

the effects of information from several time periods into the forecast. **Averaging models** are computed by *averaging data from several time periods and using the average as the forecast for the next time period.*

The most elementary of the averaging models is the **simple average model.** With this model, *the forecast for time period t + 1 is the average of the values for a given number of previous time periods,* as shown in the following equation.

$$F_{t+1} = \frac{Y_t + Y_{t-1} + Y_{t-2} + Y_{t-3} + \cdots + Y_{t-n-1}}{n}$$

A simple 12-month average could be used to forecast the sales of Tracquer tires for month 17 by averaging the values for months 5 through 16. In particular,

$$F_{17} = \frac{91.167 + 97.403 + 110.983 + \cdots + 105.901}{12}$$

$$= 100.361 \text{ thousand tires}$$

With the **simple average,** the forecast for month 17 is 100.361 thousand tires. Note that none of the previous 12 monthly sales values equal this forecast and that the average is not necessarily more closely related to values early in the period than to those late in the period. The use of the simple average over 12 months tends to smooth the variations, or fluctuations, that occur during this time.

Suppose we were to attempt to forecast tire sales for month 18 by using averages as the forecasting method. Would we still use the simple average for months 5 through 16 as we did to forecast for month 17? Instead of using the same 12 months' average used to forecast month 17, it would seem to make sense to use the 12 months *prior* to month 18 to average for the new forecast. To do this, we average the values for months 6 through 17. Suppose in month 17 the number of Tracquer tires sold is 98,763. We could forecast month 18 with a new average that includes the same months used to forecast month 17, but without the value for month 5 and with the value for month 17.

$$F_{18} = \frac{97.403 + 110.983 + 103.258 + \cdots + 98.763}{12}$$

$$= 100.994 \text{ thousand tires}$$

Computing an average of the values for months 6 through 17 produces a **moving average,** which can be used to forecast the number of Tracquer tires sold for month 18. In computing this moving average, the earliest of the previous 12 values, month 5, is dropped, and the most recent value, month 17, is included.

A moving average is an average that is updated or recomputed for every new time period being considered. The most recent information is utilized in each new moving average. This advantage is offset by the disadvantages that (1) it is difficult to choose the optimal length of time for which to compute the moving average and (2) moving averages do not usually adjust for such time-series effects as trend, cyclicality, or seasonality. To determine the more optimal lengths for which to compute the moving averages, we would need to forecast with several different average lengths and compare the errors produced by them.

Weighted Moving Average Forecasting Method

There are times when a forecaster wants to place more weight on certain periods of time than on others. For example, a forecaster might believe that the previous month's value is three times as important in forecasting as other months. *A moving average in*

Averaging models
Forecasting models in which the forecast is the average of several preceding time periods.

Simple average model
The most elementary of the forecasting averaging models, in which the forecast for the next time period is the average of values for a given number of previous time periods.

Simple average
The arithmetic mean or average for the values of a given number of time periods of data.

Moving average
In forecasting, when an average of data from previous time periods is used to forecast the value for ensuing time periods and this average is modified at each new time period by including more recent values not in the previous average and dropping out values from the more distant time periods that were in the average. It is continually updated at each new time period.

Weighted moving average
In forecasting, a moving average in which different weights are applied to the data values from different time periods.

which some time periods are weighted differently than others is called a **weighted moving average.**

As an example, suppose a 3-month weighted average is computed by weighing last month's value by 3, the value for the previous month by 2, and the value for the month before that by 1. This weighted average is computed as

$$F_{t+1} = \frac{3(Y_t) + 2(Y_{t-1}) + 1(Y_{t-2})}{6}$$

where:

Y_t = the last month's actual value
Y_{t-1} = value for the previous month
Y_{t-2} = value for the month before the previous month

Notice that the divisor is 6. With a weighted moving average, the divisor always equals the total of the weights. In this example Y_{t-1} counts three times as much as the value for Y_{t-3}.

Using this three-month moving average for the Tracquer tire sales data yields a forecast for the next month of

$$F_{17} = \frac{3(105.901) + 2(91.361) + 1(104.267)}{6}$$

$$= 100.782 \text{ thousand tires.}$$

Although not explicitly stated, note that the weights for the 13 months prior to the last three months actually are all zero. That is, they have no effect on this forecast and are not shown in the computation above.

In general, the equation for a weighted moving average is

$$F_{t+1} = w_t(Y_t) + w_{t-1}(Y_{t-1}) + w_{t-2}(Y_{t-2}) + \cdots + w_2(Y_2) + w_1(Y_1)$$

where the ws symbols represent the weights for each of the terms. The weights are numbers between zero and one, similar to probability values. In addition, the sum of all the weights should be one, similar to a probability distribution. Generally, the weights are selected to give decreasing importance to the older time series values. This means that usually

$$w_t \geq w_{t-1} \geq w_{t-2} \geq \cdots \geq w_2 \geq w_1$$

Also, many of the weights for the oldest time series values may be zero indicating they are not included in the forecast.

For the three-month moving average Tracquer tire example above, the weight for the sales value for the most recent month is 0.500 (=3/6). Similarly, that for the previous month is 0.333 (=2/6), and that for the month before the previous month is 0.167 (=1/6). The weights for all months prior to these three months are 0.000. Thus, all the weights are from zero to one; they sum to one; and they are in descending magnitude.

DEMONSTRATION PROBLEM 14.2	Compare the three prior stationary forecasting methods for forecasting the next two months for the Tracquer tire time series data of Table 14.4. Specifically, consider a *naïve* forecast, a 12-month *moving average* forecast, and a 3-month *weighted moving average* with weights of 0.500, 0.333, and 0.167 forecast. Use the MAD error measurement for selecting the *best* forecasting method.

SOLUTION

In order to compute the value of MAD for each of the three methods, it is necessary to make forecasts for all the historical time series values with each of them. The Excel worksheet given below shows all the necessary computations.

	A	B	C	D	E	F	G	H	I
			Naïve Method		12-Mo. Moving Average		3-Mo. Weighted M.A.		
1									
2	Month	Sales	Forecast	Abs. Error	Forecast	Abs. Error	Forecast	Abs. Error	
3	1	96.127							
4	2	97.654	96.127	1.527					
5	3	101.972	97.654	4.318					
6	4	109.301	101.972	7.329			99.558	9.743	
7	5	91.167	109.301	18.134			104.915	13.748	
8	6	97.403	91.167	6.236			99.010	1.607	
9	7	110.983	97.403	13.580			97.313	13.670	
10	8	103.258	110.983	7.725			103.152	0.106	
11	9	101.765	103.258	1.493			104.853	3.088	
12	10	98.231	101.765	3.534			103.802	5.571	
13	11	99.992	98.231	1.761			100.247	0.255	
14	12	102.111	99.992	2.119			99.702	2.409	
15	13	97.894	102.111	4.217	100.830	2.936	100.757	2.863	
16	14	104.267	97.894	6.373	100.978	3.289	99.649	4.618	
17	15	91.361	104.267	12.906	101.529	10.168	101.785	10.424	
18	16	105.901	91.361	14.540	100.644	5.257	96.750	9.151	
19	17		105.901		100.361		100.786		
20			MAD =	7.053		5.412		5.943	
21									

The naïve forecast in cell C4 has the formula **=B3.** Cell C4 is copied into cells C5 through C19. The 12-month moving average forecast in cell E15 has the formula **=AVERAGE(B3:B14).** It is copied into cells E16 through E19. The 3-month weighted moving average forecast in cell G6 has the formula **=0.5*B5+0.333*B4+0.167*B3.** It is copied into cells G7 through G19.

The absolute value of the error in cell D4 is computed with the formula **=ABS(B4-C4)** and is copied into cells D5 through D18. The absolute value of the error in cell F15 is computed with the formula **=ABS(B15-E15)** and is copied into F16 through F18. The absolute error in cell H6 is computed with **=ABS(B6-G6).** It is copied into cells H7 through H18.

In cell D20, the value for MAD is computed with the formula **=AVERAGE(D4:D18).** That for cell F20 is **=AVERAGE(F15:F18)** and that for cell H20 is **=AVERAGE(H6:H18).**

Note that the number of errors for each method differs. For the naïve approach, there is a forecast for all the historical values except the first. For it, there are 15 errors. For the 12-month moving average, the first forecast that can be made is for month 13 so for it there are only four error terms. For the 3-month weighted moving average, the forecasts begin for month 4, so there are 13 errors. Since the error measurement used, MAD, is the *mean* absolute deviation, the values are comparable.

The MAD values for the naïve, moving average, and weighted moving average are 7.053, 5.412, and 5.943 respectively. These values suggest that the 12-month moving average is the *best* method of these three. The 3-month weighted moving average is second best, and the naïve method is a somewhat more distant third.

The forecast for the next month, 17, using the 12-month moving average approach is 100.361 thousand tires. Since it is a stationary forecasting method, the forecast is also 100.361 for month 18 and all subsequent months.

Exponential Smoothing Forecasting Method

Exponential smoothing
A type of forecasting technique in which a weighting system is used to determine the importance of previous time periods in the forecast.

Another forecasting technique, **exponential smoothing,** is *used to weight data from previous time periods with exponentially decreasing importance in the forecast.* Exponential smoothing is accomplished by multiplying the actual value for the present time period, Y_t by a value between 0 and 1 (the exponential smoothing constant) referred to as α (not the same α used for a Type I error) and adding that result to the product of the present time period's forecast, F_t, and $(1 - \alpha)$. The following is a more formalized version.

EXPONENTIAL SMOOTHING

$$F_{t+1} = \alpha \cdot Y_t + (1 - \alpha) \cdot F_t$$

where:
   F_{t+1} = the forecast for the next time period $(t + 1)$
   F_t = the forecast for the present time period (t)
   Y_t = the actual value for the present time period
   α = a value between 0 and 1 referred to as the exponential smoothing constant

---

The value of α is determined by the forecaster. The essence of this procedure is that the new forecast is a combination of the present forecast and the present actual value. If α is chosen to be less than .5, less weight is placed on the actual value than on the forecast of that value. If α is chosen to be greater than .5, more weight is being put on the actual value than on the forecast value.

As an example, suppose the prime interest rate for a time period is 5% and the forecast of the prime interest rate for this time period was 6%. If the forecast of the prime interest rate for the next period is determined by exponential smoothing with $\alpha = .3$, the forecast is

$$F_{t+1} = (.3)(5\%) + (1.0 - .3)(6\%) = 5.7\%.$$

Notice that the forecast value of 5.7% for the next period is weighted more toward the previous forecast of 6% than toward the actual value of 5% because α is .3. Suppose we use $\alpha = .7$ as the exponential smoothing constant. Then,

$$F_{t+1} = (.7)(5\%) + (1.0 - .7)(6\%) = 5.3\%.$$

This value is closer to the actual value of 5% than the previous forecast of 6% because the exponential smoothing constant, α, is greater than .5.

To see why this is called exponential smoothing, examine the formula for exponential smoothing again.

$$F_{t+1} = \alpha \cdot Y_t + (1 - \alpha)F_t$$

If exponential smoothing has been used over a period of time, the forecast for F_t will have been obtained by

$$F_t = \alpha \cdot Y_{t-1} + (1 - \alpha)F_{t-1}.$$

Substituting this forecast value, F_t, into the preceding equation for F_{t+1} produces

$$F_{t+1} = \alpha \cdot Y_t + (1 - \alpha)[\alpha \cdot Y_{t-1} + (1 - \alpha)F_{t-1}]$$
$$= \alpha \cdot Y_t + \alpha(1 - \alpha) \cdot Y_{t-2} + (1 - \alpha)^2 F_{t-1}$$

but

$$F_{t-1} = \alpha \cdot Y_{t-2} + (1 - \alpha)F_{t-2}.$$

Substituting this value of F_{t-1} into the preceding equation for F_{t+1} produces

$$F_{t+1} = \alpha \cdot Y_t + \alpha(1 - \alpha) \cdot Y_{t-1} + (1 - \alpha)^2 F_{t-1}$$
$$= \alpha \cdot Y_t + \alpha(1 - \alpha) \cdot Y_{t-1} + (1 - \alpha)^2[\alpha \cdot Y_{t-2} + (1 - \alpha)F_{t-2}]$$
$$= \alpha \cdot Y_t + \alpha(1 - \alpha) \cdot Y_{t-1} + \alpha(1 - \alpha)^2 \cdot Y_{t-2} + (1 - \alpha)^3 F_{t-2}.$$

Continuing this process shows that the weights on previous-period values and forecasts include $(1 - \alpha)^n$ (exponential values). The following chart shows the values of α, $(1 - \alpha)$, $(1 - \alpha)^2$, and $(1 - \alpha)^3$ for three different values of α. Included is the value of $\alpha(1 - \alpha)^3$, which is the weight on the actual value for three time periods back. Notice the rapidly decreasing emphasis on values for earlier time periods. The impact of exponential smoothing on time-series data is to place much more emphasis on recent time periods. The choice of α determines the amount of emphasis.

α	$1 - \alpha$	$(1 - \alpha)^2$	$(1 - \alpha)^3$	$\alpha(1 - \alpha)^3$
.2	.8	.64	.512	.1024
.5	.5	.25	.125	.0625
.8	.2	.04	.008	.0064

Some forecasters use the computer to analyze time-series data for various values of α. By setting up criteria with which to judge the forecasting errors, forecasters can select the value of α that best fits the data.

The exponential smoothing formula

$$F_{t+1} = \alpha \cdot Y_t + (1 - \alpha) \cdot F_t$$

can be rearranged algebraically as

$$F_{t+1} = F_t + \alpha(Y_t - F_t).$$

This form of the equation shows that the new forecast, F_{t+1}, equals the old forecast, F_t, plus an adjustment based on α times the error of the old forecast $(Y_t - F_t)$. The smaller α is, the less impact the error has on the new forecast and the more the new forecast is like the old. This shows the dampening effect of α on the forecasts. If α equals 1.0, the new forecast is equal to the last actual value. This is equivalent to the naïve approach. If α equals zero, the forecast is equal to the original forecast for the first time period.

Forecast the Tracquer tire time series for the next two months using exponential smoothing. Work the problem using $\alpha = 0.05, 0.1, 0.2, 0.3$ and 0.5. Use the MSE error measurement for selecting the best value for the smoothing constant.

DEMONSTRATION PROBLEM 14.3

SOLUTION
Excel includes a **Data Analysis Tool** for performing **Exponential Smoothing** forecasts. The dialog box for it is the following.

Damping factor
One minus the exponential smoothing constant, $(1.0 - \alpha)$.

The inputs are only the historical time series values—either with or without a label above them—and the damping factor. The **damping factor** is equal to $(1.0 - \alpha)$. In using this tool, *it is extremely easy to forget and to place the value for α instead of the value for the damping factor, $(1.0 - \alpha)$.*

The results from the inputs in the dialog box above are given in cells C3 through C18 in the worksheet below.

	A	B	C	D	E	F	G	H
1			Forecast with alpha =					
2	Month	Sales	0.05	0.1	0.2	0.3	0.5	
3	1	96.127	#N/A	#N/A	#N/A	#N/A	#N/A	
4	2	97.654	96.127	96.127	96.127	96.127	96.127	
5	3	101.972	96.2034	96.2797	96.4324	96.5851	96.8905	
6	4	109.301	96.4918	96.8489	97.5403	98.2012	99.4313	
7	5	91.167	97.1322	98.0941	99.8925	101.531	104.366	
8	6	97.403	96.834	97.4014	98.1474	98.4219	97.7666	
9	7	110.983	96.8624	97.4016	97.9985	98.1162	97.5848	
10	8	103.258	97.5685	98.7597	100.595	101.976	104.284	
11	9	101.765	97.8529	99.2096	101.128	102.361	103.771	
12	10	98.231	98.0485	99.4651	101.255	102.182	102.768	
13	11	99.992	98.0577	99.3417	100.65	100.997	100.499	
14	12	102.111	98.1544	99.4067	100.519	100.695	100.246	
15	13	97.894	98.3522	99.6771	100.837	101.12	101.178	
16	14	104.267	98.3293	99.4988	100.249	100.152	99.5362	
17	15	91.361	98.6262	99.9756	101.052	101.387	101.902	
18	16	105.901	98.2629	99.1142	99.114	98.379	96.6313	
19	17		98.6448	99.0716	99.114	98.4157	96.7187	
20		MSE =	43.2452	40.296	40.0682	42.2413	49.2682	

The entry given in cell C3, **#N/A,** indicates a value is not available for that time period.

The **Exponential Smoothing Tool** inserts in cell C4 the formula **=B3.** This sets the value for the *first forecast* equal to the *first actual data value*. This is one way of satisfying the need for a first forecast in exponential smoothing. There are other less simple ways of computing a value for the first forecast.

The tool places the formula = **0.05*B4 + 0.95*C4** is cell C5 as required by the exponential smoothing equation. In cell C6 the formula is = **0.05*B5 + 0.95*C5,** and so on through cell 18.

We have copied the formula from cell C18 into C19 in order to obtain a forecast for the next time period with the exponential smoothing equation. Finally in cell C20, we have entered the formula **=SUMXMY2($B$4:$B$18,C4:C18)** to compute the MSE for the forecasts of column C. The range *B4:B18* has been designated with the absolute address symbol, **$,** so it can be properly copied into cells D20 through G20.

The exponential smoothing tool is run four more times to obtain the forecasts given in columns D though G. For these four runs, the **Damping Factor** is sequentially changed to 0.9, 0.8, 0.7, and 0.5 (not the alpha values of 0.1, 0.2, 0.3, and 0.5). The **Output Range** is sequentially changed to D3, E3, F3, and G3.

Based on the values for MSE, the best forecast model of these five is with an alpha value of 0.2.

Note that the dialog box above has a check box for **Chart Output.** If we re-run the tool for $\alpha = 0.2$ and place a check in this check box, the time series plot given below will also be given. Note this figure has been edited. The most important change has been the adjustment of the scale for the vertical axis.

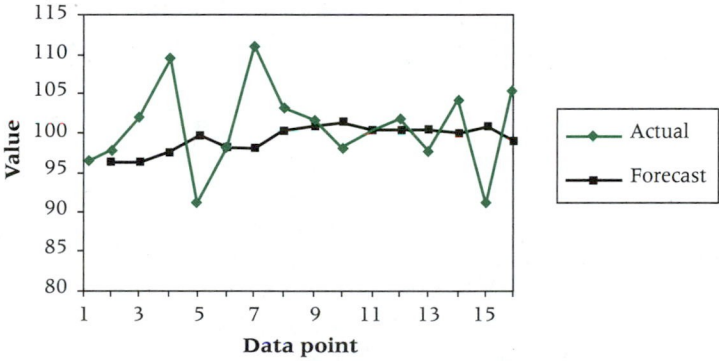

Notice that some of the forecast values are below the corresponding actual value and some are above. This is as it should be for a good application of a stationary forecasting method such as exponential smoothing. Also, note that the forecasts are less *jagged* than the actual values. The forecasts have a *smoothing* or averaging effect on the actual values. We conclude from the time series plot that the exponential smoothing model with $\alpha = 0.2$ seems to track the actual time series values quite well.

Finally, the forecast for the next two months using this *best* model is 99.114 thousand tires for month 17, and 99.114 for month 18 since this is a stationary forecasting method.

14.5 Use the following time-series data to answer the given questions.

TIME PERIOD	VALUE	TIME PERIOD	VALUE
1	27	6	66
2	31	7	71
3	58	8	86
4	63	9	101
5	59	10	97

a. Develop forecasts for periods 5 through 10 using four-month moving averages.
b. Develop forecasts for periods 5 through 10 using four-month weighted moving averages. Weight the most recent month by a factor of 4, the previous month by 2, and the other months by 1.

14.4

Problems

c. Compute the errors of the forecasts in parts (a) and (b) and observe the differences in the errors forecast by the two different techniques.

14.6 Following are time-series data for eight different periods. Use exponential smoothing to forecast the values for periods 3 through 8. Use the value for the first period as the forecast for the second period. Compute forecasts using two different values of alpha, $\alpha = .1$ and $\alpha = .8$. Compute the errors for each forecast and compare the errors produced by using the two different exponential smoothing constants.

TIME PERIOD	VALUE	TIME PERIOD	VALUE
1	211	5	242
2	228	6	227
3	236	7	217
4	241	8	203

14.7 Following are time-series data for nine time periods. Use exponential smoothing with constants of .3 and .7 to forecast time periods 3 through 9. Let the value for time period 1 be the forecast for time period 2. Compute additional forecasts for time periods 4 through 9 using a 3-month moving average. Compute the errors for the forecasts and discuss the size of errors under each method.

TIME PERIOD	VALUE	TIME PERIOD	VALUE
1	9.4	6	11.0
2	8.2	7	10.3
3	7.9	8	9.5
4	9.0	9	9.1
5	9.8		

14.8 The U.S. Bureau of the Census publishes data on factory orders for all manufacturing, durable goods, and nondurable goods industries. Shown here are factory orders in the United States from 1987 through 1997 ($billion).

a. Use these data to develop forecasts for the years 1992 through 1997 using a 5-year moving average.

b. Use these data to develop forecasts for the years 1992 through 1997 using a 5-year weighted moving average. Weight the most recent year by 6, the previous year by 4, the year before that by 2, and the other years by 1.

c. Compute the errors of the forecasts in parts (a) and (b) and observe the differences in the errors of the forecasts.

YEAR	FACTORY ORDERS ($BILLION)
1987	2512.7
1988	2739.2
1989	2874.9
1990	2934.1
1991	2865.7
1992	2978.5
1993	3092.4
1994	3356.8
1995	3607.6
1996	3749.3
1997	3952.0

14.9 The following data show the number of issues from Initial Public Offerings (IPOs) from 1985 through 1997 released by the Securities Data Company. Use these data to develop forecasts for the years 1987 through 1997 using exponential smoothing techniques with alpha values of .2 and .9. Let the forecast for 1986 be the value for 1985. Compare the results by examining the errors of the forecasts.

YEAR	NUMBER OF ISSUES
1985	332
1986	694
1987	518
1988	222
1989	209
1990	172
1991	366
1992	512
1993	667
1994	571
1995	575
1996	865
1997	609

Oftentimes a time series exhibits a trend component. A number of examples have previously been examined. The trend variation can either be accompanied by a seasonal variation, such as shown in Figures 14.6 and 14.7, or without as seen in Figures 14.3, 14.4, 14.9, and 14.11. For such time series data, the stationary forecasting methods are somewhat ineffective as is demonstrated by the following.

14.5
Trend Forecasting Methods

DEMONSTRATION PROBLEM 14.4

Consider the average baseball salary time series of Demonstration Problem 14.1. In the manner of Demonstration Problem 14.3, use Excel's **Exponential Smoothing Data Analysis Tool** to determine the *best* value for the smoothing constant, α. Also use Excel's **Moving Average Data Analysis Tool** to determine the *best* value for the number of periods included in the average. Excel terms this number the *moving average interval*. Use MSE as the overall measure of error.

SOLUTION

We will begin by using the **Exponential Smoothing Tool** to make forecasts with α values of 0.1, 0.3, 0.5, 0.7, and 0.9. The resulting MSE values for these five cases is

ALPHA	MSE
0.1	205,505,448,130
0.3	78,520,292,185
0.5	42,571,260,860
0.7	28,491,627,131
0.9	21,335,810,667

From these results, the best value for α appears to be 0.9 or higher. Since Excel does not allow alpha to be 1.0, we can try values very near it with the following results.

	A	B	C	D	E	F	G	H
1				Forecast with alpha =				
2	Year	Avg. Salary	0.9	0.99	0.999	0.9999	0.99999	
3	1	$497,254	#N/A	#N/A	#N/A	#N/A	#N/A	
4	2	$597,537	$497,254	$497,254	$497,254	$497,254	$497,254	
5	3	$851,492	587508.7	596534.17	597436.717	597526.9717	597535.9972	
6	4	$1,028,667	825093.67	848942.4217	851237.9447	851466.6035	851489.4604	
7	5	$1,076,089	1008309.667	1026869.754	1028489.571	1028649.28	1028665.228	
8	6	$1,168,263	1069311.067	1075596.808	1076041.401	1076084.256	1076088.526	
9	7	$1,110,766	1158367.807	1167336.338	1168170.778	1168253.782	1168262.078	
10	8	$1,119,981	1115526.181	1111331.703	1110823.405	1110771.749	1110766.575	
11	9	$1,336,609	1119535.518	1119894.507	1119971.842	1119980.079	1119980.908	
12	10	$1,398,831	1314901.652	1334441.855	1336392.363	1336587.337	1336606.834	
13		MSE =	21,335,810,667	19,195,200,391	19,006,864,134	18,988,262,493	18,986,404,626	
14								

It appears that the closer to 1.0 that alpha becomes, the better the forecasts fit the data. As was mentioned previously, an alpha value of 1.0 for exponential smoothing yields the results of the naïve or last period approach. The meaning of this is that there is change in the time series, and a stationary method is not able to track the changes. This is demonstrated in the following chart. It is for the best alpha value shown above, 0.99999.

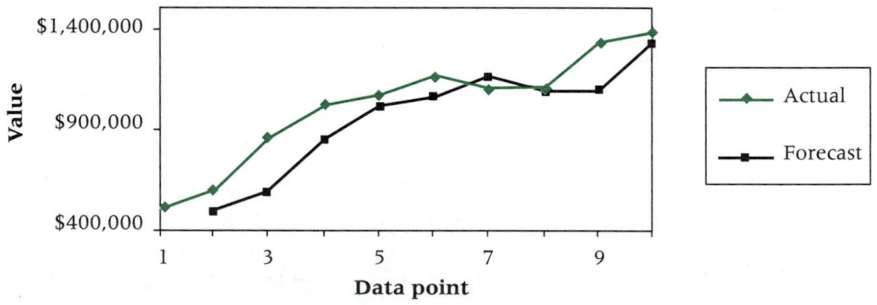

Notice that the forecast is always below the actual value except for year 7. In fact, if we compute the mean error (ME), it is 100,176. As previously mentioned, another name for ME is forecast bias. It shows the tendency of the forecasts on the average to over-predict or under-predict the actual time series values. On the average, the forecast is 100,176 below the actual value. These are typical results when using any stationary forecasting method for forecasting a non-stationary time series.

To confirm this we fit the data with a moving average. The following table shows the MSEs obtained by using Excel's **Moving Average Data Analysis Tool** with averaging intervals varying from one-to-five years. The best value is for a 1-year moving average. Again this is equivalent to the naïve or last period method.

INTERVAL	MSE
5	54,550,263,182
4	48,219,984,436
3	46,597,765,507
2	36,931,345,170
1	18,986,198,222

One final observation comes from a comparison of the MSEs computed above and those computed for Demonstration Problem 14.1. The smallest one found above is two-to-three-and-one-half times larger than those found for Demonstration Problem 14.1 using trend projection methods.

The four forecasting methods presented below are effective for time series data that include a trend and an irregular component. They are not effective if seasonal variation is present in the time series. A time series with a seasonal component, either with or without an accompanying trend component, requires a method such as those given in section 14.6.

We begin with the linear trend projection forecasting method.

Linear Trend Projection Forecasting Method

Linear trend projection is an *application of regression analysis as a time series forecasting technique.* In linear trend projection the dependent variable, Y, is the item being forecast. The independent variable, X, represents the time periods.

Many possible trend fits can be explored with time-series data. Here we examine the linear model. Seasonal effects can confound the trend analysis. Thus it is assumed here that there are no seasonal effects in the data or that they have already been removed when we determine trend.

The data in Table 14.5 represent 35 years of data on the average length of the work week in Canada for manufacturing workers. A regression line can be fit to these data by using the time periods as the independent variable and length of work week as the dependent variable. Because the time periods are consecutive, they have been renumbered from 1 to 35 and entered along with the time-series data (Y) into a regression analysis. The linear model explored in this example is

$$Y_i = \beta_0 + \beta_1 t_i + e_i$$

where:
  Y_i = data value for period i and
  t_i = ith time period.

Figure 14.19 shows the Excel regression output for this example. By using the coefficients of the t variable and intercept, the equation of the trend line can be determined to be

$$\hat{Y} = 37.4161 - .0614t.$$

The slope indicates that for every unit increase in time period, t, there is a predicted .0614 decrease in the length of the average work week in manufacturing. Thus, the trend is a .0614 decrease in the length of the average work week each year in Canada in manufacturing. The Y intercept, 37.4161, indicates that in the year prior to the first period of these data the average work week was 37.4161 hours.

The p value (.00000003) indicates that significant linear trend is present in the data. In addition, $R^2 = .611$ indicates considerable predictability in the model. Inserting the various period values (1, 2, 3, . . . , 35) into the preceding regression equation produces the predicted values of Y. For example, for period 23 the predicted value is

$$\hat{Y} = 37.4161 - .0614(23) = 36.0 \text{ hours.}$$

Linear trend projection
Forecasting by fitting a linear equation to a time series.

TABLE 14.5
Average Hours Worked per
Week in Manufacturing by
Canadian Workers over
35 Years

TIME PERIOD (YEAR)	HOURS	TIME PERIOD (YEAR)	HOURS
1	37.2	19	36.0
2	37.0	20	35.7
3	37.4	21	35.6
4	37.5	22	35.2
5	37.7	23	34.8
6	37.7	24	35.3
7	37.4	25	35.6
8	37.2	26	35.6
9	37.3	27	35.6
10	37.2	28	35.9
11	36.9	29	36.0
12	36.7	30	35.7
13	36.7	31	35.7
14	36.5	32	35.5
15	36.3	33	35.6
16	35.9	34	36.3
17	35.8	35	36.5
18	35.9		

Data prepared by the U.S. Bureau of Labor Statistics, Office of Productivity and Technology.

Figure 14.19

Excel Regression
output for hours
worked example
using linear trend

SUMMARY OUTPUT

Regression Statistics	
Multiple R	0.782
R Square	0.611
Adjusted R Square	0.600
Standard Error	0.509
Observations	35

ANOVA

	df	SS	MS	F	Significance F
Regression	1	13.4467	13.4467	51.91	.00000003
Residual	33	8.5487	0.2591		
Total	34	21.9954			

	Coefficients	Standard Error	t Stat	P-value
Intercept	37.4161	0.17582	212.81	.00000000
Year	-0.0614	0.00852	-7.20	.00000003

The model was developed with 35 periods (years). From this model, the average work week in Canada in manufacturing for period 41 can be forecast:

$$\hat{Y} = 37.4161 - .0614(41) = 34.9 \text{ hours.}$$

Excel's **Trendline** feature can be used to add a linear trend line to a time series plot created with the **Chart Wizard** as shown in Figure 14.20. For the trendline dialog box, we selected **Linear** for the *Type* of trend line. On the **Options** tab of the dialog box, we selected the check box for **Display equation on chart** and for **Display R-squared value on chart.**

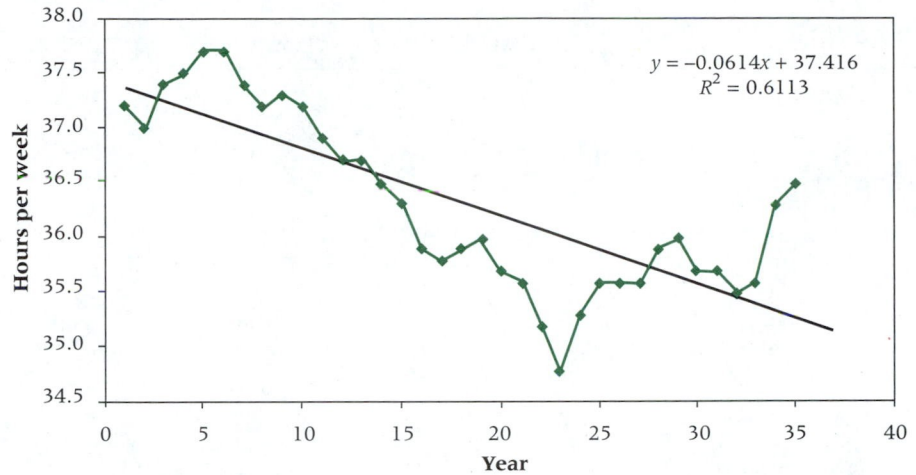

Figure 14.20

Hours worked time series with linear trend line

Also on the **Options** tab under *Forecast* and *Forward*, we selected **2** units. The regression equation and value for R^2 agrees with the results of Figure 14.19 from the Regression Data Analysis Tool. The figure shows the trend projected two time periods into the future.

Non-Linear Trend Projection Forecasting Method

For Figure 14.20, the relationship of the data values to the trend line is questionable. At the two ends the actual values are above the trend line, and in the middle they are below. This pattern suggests the trend line may be non-linear.

In addition to linear regression, forecasters can explore using **quadratic regression models** to predict data by using the time-series periods. The quadratic regression model is

$$Y_i = \beta_0 + \beta_1 \cdot t_i + \beta_2 \cdot t_i^2 + e_i$$

where:
  Y_i = the time-series data value for period i,
  t_i = the ith period, and
  t_i^2 = the square of the ith period.

This model can be implemented in time-series trend analysis by using the time periods squared as an additional predictor. Thus, in the hours worked example, besides using $t = 1, 2, 3, 4, \ldots, 35$ as a predictor, we would also use $t^2 = 1, 4, 9, 16, \ldots, 1225$ as a predictor.

Shown in Table 14.6 are the data needed to compute a quadratic regression trend model on the manufacturing work-week data. Note that the table includes the original data, the time periods, and the time periods squared.

The Excel computer output for this quadratic trend-regression analysis is shown in Figure 14.21. We see that the quadratic regression model produces an R^2 of .761 with both t and t^2 in the model. The linear model produced an R^2 of .611 with t alone. The quadratic regression seems to add some predictability to the trend model.

Trendline can add a non-linear trend to the time series plot. A quadratic equation is noted as a *polynomial trend with order of 2*. Figure 14.22 presents the results from trendline.

The **Trendline** feature makes it very simple to try other non-linear trend lines. In addition to a linear and polynomial of order 2, it includes logarithmic, exponential, and power trends. In just a few keystrokes it is possible to try each of these. For the current time

Non-linear trend projection
Forecasting by fitting a non-linear equation to a time series.

Quadratic regression model
A multiple regression model in which the predictors are a variable and the square of the variable.

TABLE 14.6   Average hours per week in manufacturing example with quadratic fit

TIME PERIOD	(TIME PERIOD)²	HOURS	TIME PERIOD	(TIME PERIOD)²	HOURS
1	1	37.2	19	361	36.0
2	4	37.0	20	400	35.7
3	9	37.4	21	441	35.6
4	16	37.5	22	484	35.2
5	25	37.7	23	529	34.8
6	36	37.7	24	576	35.3
7	49	37.4	25	625	35.6
8	64	37.2	26	676	35.6
9	81	37.3	27	729	35.6
10	100	37.2	28	784	35.9
11	121	36.9	29	841	36.0
12	144	36.7	30	900	35.7
13	169	36.7	31	961	35.7
14	196	36.5	32	1024	35.5
15	225	36.3	33	1089	35.6
16	256	35.9	34	1156	36.3
17	289	35.8	35	1225	36.5
18	324	35.9			

Data prepared by the U.S. Bureau of Labor Statistics, Office of Productivity and Technology.

Figure 14.21

Excel Regression output for hours worked example using quadratic trend

SUMMARY OUTPUT

Regression Statistics	
Multiple R	0.873
R Square	0.761
Adjusted R Square	0.747
Standard Error	0.405
Observations	35

ANOVA

	df	SS	MS	F	Significance F
Regression	2	16.7483	8.3741	51.07	1.10021E-10
Residual	32	5.2472	0.1640		
Total	34	21.9954			

	Coefficients	Standard Error	t Stat	P-value
Intercept	38.16442	0.21766	175.34	2.61329E-49
Year	-0.18272	0.02788	-6.55	2.20514E-07
YearSq	0.00337	0.00075	4.49	8.76021E-05

series, all three of these do not fit as well as the polynomial of order 2 or the linear trend. For the logarithmic, power, and exponential, the values for R^2 are 0.5969, 0.5938, and 0.6081, respectively.

The polynomial trend can have an order up to 6. As the order is increased, the fit will always be improved or stay the same. For example, a third-order polynomial for the current time series is as given in Figure 14.23, and a fourth-order in Figure 14.24.

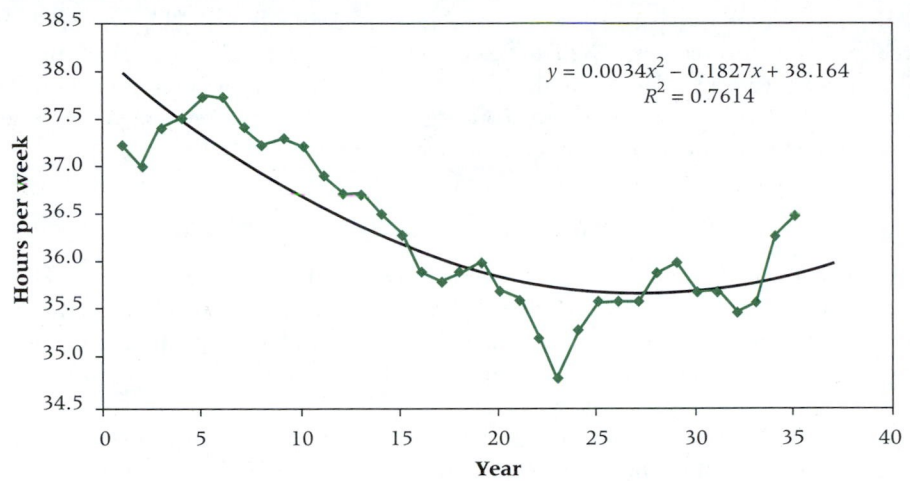

Figure 14.22

Hours worked time series with quadratic trend line

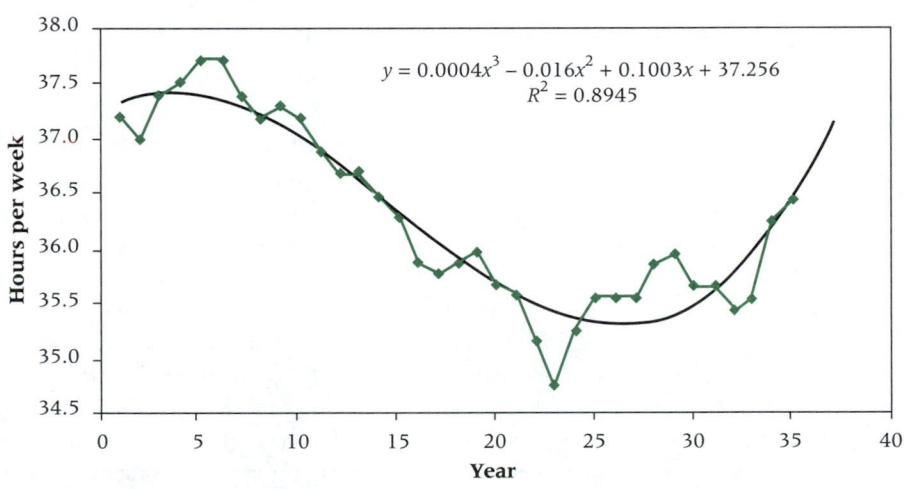

Figure 14.23

Hours worked time series with cubic trend line

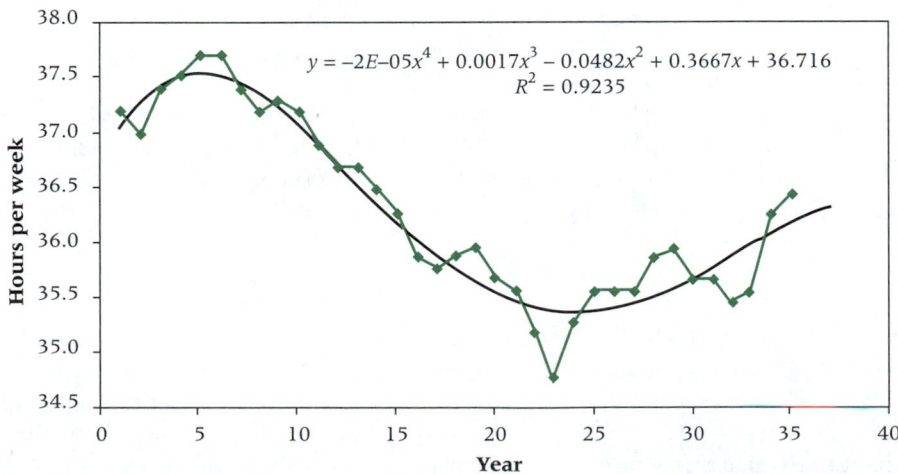

Figure 14.24

Hours worked time series with quartic trend line

The process of projecting a trend equation into the future is called *extrapolation*. Caution is necessary in extrapolating with a trend too far into the future. This is particularly true for non-linear trend models. The desire to mechanically project a trend into the future must be tempered with a consideration of whether such a projection seems to make sense or not. The forecaster must use knowledge of the time series being modeled to determine whether a particular projection seems reasonable.

Holt's Exponential Smoothing Forecasting Method

Holt's model
Exponential smoothing with a trend component.

Exponential smoothing is a simple but effective means for forecasting a time series. However, if the time series includes a trend, its forecasts consistently lag behind the actual time series values as was shown in Demonstration Problem 14.4. Holt's exponential smoothing forecasting method, commonly called **Holt's model,** rectifies this weakness by *adding the consideration of a trend to the basic exponential smoothing relation.*

Holt's model uses weights to smooth the trend in a manner similar to that which we used in exponential smoothing in the previous section. Holt's method uses the following four equations to accomplish this:

Smoothed Values:	$E_t = \alpha X_t + (1 - \alpha)(E_{t-1} + T_{t-1})$
Trend Term Update:	$T_t = \beta(E_t - E_{t-1}) + (1 - \beta)T_{t-1}$
Forecast for Next Period:	$F_{t+1} = E_t + T_t$
for k periods in the future:	$F_{t+k} = E_t + kT_t$

In simple exponential smoothing, the smoothed values are the forecasts. The next smoothed value is determined by weighting the actual value by α and the previous forecast (the last smoothed value) by $1 - \alpha$. Now we have introduced trend into consideration. Notice from the last two equations above that the forecast is now a function of both the smoothed value and the trend. Thus, we have assigned a new notation for the smoothed values, E, and reserved F for the forecast. T denotes the trend. With simple exponential smoothing there is only one smoothing constant, α. Since trend is part of the forecasting process here, we introduce a weight for trend, β. These two weights can be the same or different. As before, the weights determine how much emphasis is to be placed on recent values versus previous values. Higher values of α and β place more emphasis on recent values.

An examination of the equations can enlighten us about the process. The last two equations are for computing the forecasts. To forecast the next period, use $F_{t+1} = E_t + T_t$. To forecast more than one period in the future using current information, use $F_{t+1} = E_t + kT_t$. k is the number of periods in the future to be forecast. For example, if you want to forecast 3 periods in the future, use $F_{t+3} = E_t + 3T_t$.

Notice that the forecast for the next period is a function of the smoothed value of this period and the trend for this period. In calculating the new smoothed value using $E_t = \alpha X_t + (1 - \alpha)(E_{t-1} + T_{t-1})$, the last term, $(E_{t-1} + T_{t-1})$, is simply the forecast for this period. Substituting this notion into the smoothing formula results in:

$$E_t = \alpha X_t + (1 - \alpha)(F_t)$$

This is analogous to what is done in simple exponential smoothing to update the smoothed value. The smoothed value is a function of the actual value and the forecast.

The trend term is updated at each step of the process using the second equation, $T_t = \beta(E_t - E_{t-1}) + (1 - \beta)T_{t-1}$. Essentially, the updated trend term is a weight of the difference in this period's smoothed value and last period's smoothed value (a search for current rate of increase or decrease) and a weight of the old trend value.

As an example of the use of this technique, we will apply Holt's model to the housing starts data of Table 14.7.

We have chosen $\alpha = 0.8$ and $\beta = 0.4$ as the weights. Initializing the process can be a choice for the forecaster. Since we have no historical data given before year 1 in the problem, we will start with a trend of 0. If there is some historical trend in the process being studied, some forecasters will use that trend as the initial value. Notice also that while the first actual value is the forecast for year 2, the first smoothed value is the first actual value (1750 for year 1). Shown here are the calculations for year 2, 3, and 4.

YEAR	X_t	E_t	T_t	F_t	ERROR
1	1750	1750	0	–	–
2	1742				
3	1805				
4	1620				

Using the initial values for year 1 the forecast for year 2 can be obtained:

$$F_2 = E_1 + T_1 = 1750 + 0 = 1750$$

We can now update the smoothed value for year 2 as:

$$E_2 = 0.8\,X_2 + (1 - 0.8)(F_2) = 0.8(1742) + 0.2(1750) = 1743.6$$

The trend is now updated by:

$$T_2 = 0.4(E_2 - E_1) + (1 - 0.4)\,T_1 = 0.4(1743.6 - 1750) + 0.6(0) = -2.6$$

The forecast for year 3 is:

$$F_3 = E_2 + T_2 = 1743.6 + (-2.6) = 1741.0$$

The smoothed value for year 3 is:

$$E_3 = 0.8\,X_3 + (1 - 0.8)(F_3) = 0.8(1805) + 0.2(1741) = 1792.2$$

The trend is updated:

$$T_3 = 0.4(E_3 - E_2) + (1 - 0.4)\,T_2 = 0.4(1792.2 - 1743.6) + 0.6(-2.6) = 17.9$$

The forecast for year 4:

$$F_4 = E_3 + T_3 = 1792.2 + 17.9 = 1810.1$$

This process is repeated for the remaining years. When the equations above are entered into an Excel worksheet, such as Figure 14.25, they can be copied down the columns to compute the values for the remaining years.

The Holt's model results are a forecast of 1480.2 for year 16 and a yearly trend estimate of 15.8. Accordingly, the forecast for year 17 is $1480.2 + 15.8$, the forecast for year 18 is $1480.2 + 2 \times 15.8$, and so on.

Computed at the bottom of the worksheet are values for both MSE and MAD. The formulas used are those previously presented for examples. Four input items at the top of the worksheet can be changed in order to search for a better fitting model. The four inputs are the values for the two smoothing constants, α and β, and the two initial

TABLE 14.7
Housing Starts

YEAR	HOUSING STARTS
1	1750
2	1742
3	1805
4	1620
5	1488
6	1376
7	1193
8	1014
9	1200
10	1288
11	1457
12	1354
13	1477
14	1475
15	1450

612

Figure 14.25

Holt's model computations for housing starts: $\alpha = 0.8$, $\beta = 0.4$

	A	B	C	D	E	F	G	H
1	HOLT'S MODEL			$\alpha =$	0.8	$\beta =$	0.4	
2	Year	X_t	E_t	T_t	F_t	Error	Abs. Error	
3	1	1750	1750.0	0.0				
4	2	1742	1743.6	-2.6	1750.0	-8.0	8.0	
5	3	1805	1792.2	17.9	1741.0	64.0	64.0	
6	4	1620	1658.0	-42.9	1810.1	-190.1	190.1	
7	5	1488	1513.4	-83.6	1615.1	-127.1	127.1	
8	6	1376	1386.8	-100.8	1429.8	-53.8	53.8	
9	7	1193	1211.6	-130.6	1285.9	-92.9	92.9	
10	8	1014	1027.4	-152.0	1081.0	-67.0	67.0	
11	9	1200	1135.1	-48.1	875.4	324.6	324.6	
12	10	1288	1247.8	16.2	1086.9	201.1	201.1	
13	11	1457	1418.4	78.0	1264.0	193.0	193.0	
14	12	1354	1382.5	32.4	1496.4	-142.4	142.4	
15	13	1477	1464.6	52.3	1414.9	62.1	62.1	
16	14	1475	1483.4	38.9	1516.9	-41.9	41.9	
17	15	1450	1464.5	15.8	1522.3	-72.3	72.3	
18	16				1480.2			
19				MSE =	20473.3	MAD =	117.2	

Figure 14.26

Holt's model computations for housing starts: $\alpha = 0.8$, $\beta = 0.9$

	A	B	C	D	E	F	G	H
1	HOLT'S MODEL			$\alpha =$	0.8	$\beta =$	0.9	
2	Year	X_t	E_t	T_t	F_t	Error	Abs. Error	
3	1	1750	1750.0	0.0				
4	2	1742	1743.6	-5.8	1750.0	-8.0	8.0	
5	3	1805	1791.6	42.6	1737.8	67.2	67.2	
6	4	1620	1662.8	-111.6	1834.2	-214.2	214.2	
7	5	1488	1500.6	-157.1	1551.2	-63.2	63.2	
8	6	1376	1369.5	-133.7	1343.5	32.5	32.5	
9	7	1193	1201.6	-164.5	1235.8	-42.8	42.8	
10	8	1014	1018.6	-181.1	1037.0	-23.0	23.0	
11	9	1200	1127.5	79.9	837.5	362.5	362.5	
12	10	1288	1271.9	137.9	1207.4	80.6	80.6	
13	11	1457	1447.6	171.9	1409.8	47.2	47.2	
14	12	1354	1407.1	-19.2	1619.5	-265.5	265.5	
15	13	1477	1459.2	44.9	1387.9	89.1	89.1	
16	14	1475	1480.8	24.0	1504.1	-29.1	29.1	
17	15	1450	1461.0	-15.5	1504.8	-54.8	54.8	
18	16				1445.5			
19				MSE =	20018.2	MAD =	98.5	

values, 1750 for the initial smoothed value, and 0.0 for the initial trend. For example, changing the value of β to 0.9 reduces the value for both MSE and MAD as shown in Figure 14.26.

For those who wish to use this Excel worksheet without entering the formulas, it is given on the CD-ROM enclosed with this book. It is in the folder called *Chapter 14 Templates* and is entitled *Holt's Model.*

Trend Autoregression Forecasting Method

A forecasting method that takes advantage of the relationship of values (Y_t) to previous-period values (Y_{t-1}, Y_{t-2}, Y_{t-3}, . . .) is called **autoregression.** Autoregression is a *multiple regression technique in which the independent variables are time-lagged versions of the dependent variable,* which means we try to predict a value of Y from values of Y for previous time periods. The independent variable can be lagged for one, two, three, or more time periods. An autoregressive model containing independent variables for three time periods looks like this:

$$\hat{Y} = b_0 + b_1 Y_{t-1} + b_2 Y_{t-2} + b_3 Y_{t-3}.$$

You may recall that in Section 14.3, we introduced the closely related concept of autocorrelation. In that section we used autocorrelation as a means for identifying the presence or absence of a seasonal component in a time series. In that discussion, it was also indicated that a high autocorrelation of lag 1, that is correlation between successive values of the time series variable, indicated the presence of a trend. If so, an autoregressive model with one independent variable, Y_{t-1}, might provide a useful trend forecasting method.

We can proceed in identifying a forecasting model by first determining if a time series has a high autocorrelation of lag 1, called *first-order autocorrelation.* If it is present, we can than compute the first-order *autoregression model* to use for forecasting.

First-order correlation occurs when there is correlation between the error terms of adjacent time periods (as opposed to two or more previous periods). If first-order autocorrelation is present, the error for one time period, e_t, is a function of the error of the previous time period, e_{t-1}, as follows.

$$e_t = \rho e_{t-1} + \nu_t$$

The first-order autocorrelation coefficient, ρ, measures the correlation between the error terms. It is a value that lies between -1 and 0 and $+1$, as does the coefficient of correlation discussed in Chapter 12. ν_t is a normally distributed independent error term. If positive autocorrelation is present, the value of ρ is between 0 and $+1$. If the value of ρ is 0, $e_t = \nu_t$, which means there is no autocorrelation and e_t is just a random, independent error term.

One way to *test to determine whether autocorrelation is present in a time-series regression analysis* is by using the **Durbin-Watson test** for autocorrelation. Shown next is the formula for computing a Durbin-Watson test for autocorrelation.

$$D = \frac{\sum_{t=2}^{n} (e_t - e_{t-1})^2}{\sum_{t=1}^{n} e_t^2}$$

where:
n = the number of observations

DURBIN-WATSON
TEST

Note from the formula that the Durbin-Watson test involves finding the difference between successive values of error ($e_t - e_{t-1}$). If errors are positively correlated, this difference will be smaller than with random or independent errors. Squaring this term eliminates the cancellation effects of positive and negative terms.

Trend autoregression
Autoregression model with a one-period lagged independent variable.

Autoregression
A type of multiple regression forecasting technique in which the independent variables are time-lagged versions of the dependent variable.

Durbin-Watson test
A statistical test for determining whether significant autocorrelation is present in a time-series regression model.

The null hypothesis for this test is that there is *no* autocorrelation. For a two-tailed test, the alternative hypothesis is that there *is* autocorrelation.

$$H_0: \rho = 0$$

$$H_a: \rho \neq 0$$

As mentioned before, most business forecasting autocorrelation is positive autocorrelation. In most cases, a one-tailed test is used.

$$H_0: \rho = 0$$

$$H_a: \rho > 0$$

In the Durbin-Watson test, D is the calculated value of the Durbin-Watson statistic using the residuals from the regression analysis. A critical value for D can be obtained from the values of α, n, and k by using Table A.9 in Appendix A, where α is the level of significance, n is the number of data items, and k is the number of predictors. Two Durbin-Watson tables are given in that appendix. One table contains values for $\alpha = .01$ and the other for $\alpha = .05$. The Durbin-Watson tables in Appendix A include values for d_U and d_L. These values range from 0 to 4. If the calculated value of D is above d_U, we fail to reject the null hypothesis and there is no significant autocorrelation. If the calculated value of D is below d_L, the null hypothesis is rejected and there is autocorrelation. Sometimes the calculated statistic, D, is between the values of d_U and d_L. In this case, the Durbin-Watson test is inconclusive.

As an example, consider Table 14.8, which lists drilling data for oil wells and gas wells from 1973 through 1998 (in thousands). A regression line can be fit through these data to

TABLE 14.8
U.S. Oil and Gas Well Drilling, 1973–1998

YEAR	OIL WELLS (1000)	GAS WELLS (1000)
1973	10.167	6.933
1974	13.647	7.138
1975	16.948	8.127
1976	17.688	9.409
1977	18.745	12.122
1978	19.181	14.413
1979	20.851	15.254
1980	32.639	17.333
1981	43.598	20.166
1982	39.199	18.979
1983	37.120	14.564
1984	42.605	17.127
1985	35.118	14.168
1986	19.097	8.516
1987	16.164	8.055
1988	13.636	8.555
1989	10.204	9.539
1990	12.198	11.044
1991	11.770	9.526
1992	8.757	8.209
1993	8.407	10.017
1994	6.721	9.538
1995	7.627	8.354
1996	8.314	9.302
1997	10.436	11.327
1998	7.118	12.106

SOURCE: *Monthly Energy Review*, June 1998.

determine whether the number of oil wells drilled in a given year can be predicted by the number of gas wells drilled in a year. The resulting errors of prediction can be tested by the Durbin-Watson statistic for the presence of significant positive autocorrelation by using $\alpha = .05$. The hypotheses are

$$H_0: \rho = 0$$
$$H_a: \rho > 0.$$

The following regression equation was obtained by means of Excel results of Figure 14.27.

$$\text{Oil Wells} = -11.337 + 2.6106\ (\text{Gas Wells})$$

With the values for the number of gas wells being drilled (X) from Table 14.8 and the regression model equation shown here, predicted values of Y (number of oil wells being drilled) can be computed. From the predicted values and the actual values, the errors of prediction for each time interval, e_t can be calculated. Table 14.9 shows the values of $\hat{Y}$, e_t, e_t^2, $(e_t - e_{t-1})$, and $(e_t - e_{t-1})^2$ for this example. Note that the first predicted value of Y is

$$\hat{Y}_{1973} = -11.337 + 2.6106(6.933) = 6.7623$$

Figure 14.27 Oil and gas well regression results

Microsoft Excel - 12Durbin-Watson
File Edit View Insert Format Tools Data Window Help XtraStat FAST-STAT
A1 = OIL WELLS

	A	B	C	D	E	F	G	H	I
1	OIL WELLS	GAS WELLS		SUMMARY OUTPUT					
2	10.167	6.933							
3	13.647	7.138		Regression Statistics					
4	16.948	8.127		Multiple R	0.838917723				
5	17.688	9.409		R Square	0.703782946				
6	18.745	12.122		Adjusted R Square	0.691440568				
7	19.181	14.413		Standard Error	6.570504881				
8	20.851	15.254		Observations	26				
9	32.639	17.333							
10	43.598	20.166		ANOVA					
11	39.199	18.979			df	SS	MS	F	Significance F
12	37.120	14.564		Regression	1	2461.71293	2461.71	57.0217	8.6323E-08
13	42.605	17.127		Residual	24	1036.11683	43.1715		
14	35.118	14.168		Total	25	3497.82975			
15	19.097	8.516							
16	16.164	8.055			Coefficients	Standard Error	t Stat	P-value	
17	13.636	8.555		Intercept	-11.33706901	4.18976608	-2.7059	0.01234	
18	10.204	9.539		GAS WELLS	2.610620318	0.3457194	7.55127	8.6E-08	
19	12.198	11.044							
20	11.770	9.526		RESIDUAL OUTPUT					
21	8.757	8.209							
22	8.407	10.017		Observation	Predicted OIL WELLS	Residuals			
23	6.721	9.538		1	6.762361651	3.40463835			
24	7.627	8.354		2	7.297538816	6.34946118			
25	8.314	9.302		3	9.879442311	7.06855769			
26	10.436	11.327		4	13.22625756	4.46174244			
27	7.118	12.106		5	20.30887048	-1.56387048			
28				6	26.28980163	-7.10880163			
29				7	28.48533332	-7.63433332			
30				8	33.91281296	-1.27381296			

Sheet1 / Durbin-Watson / Sheet3

TABLE 14.9
Predicted Values and Error Terms for the Oil and Gas Well Data

YEAR	$\hat{Y}$	e_t	e_t^2	$e_t - e_{t-1}$	$(e_t - e_{t-1})^2$
1973	6.7623	3.4047	11.592	—	—
1974	7.2975	6.3495	40.317	2.9448	8.6718
1975	9.8793	7.0687	49.966	0.7191	0.5171
1976	13.2261	4.4619	19.908	−2.6068	6.7954
1977	20.3087	−1.5637	2.445	−6.0256	36.3078
1978	26.2896	−7.1086	50.532	−5.5449	30.7459
1979	28.4851	−7.6341	58.279	−0.5255	0.2762
1980	33.9125	−1.2735	1.622	6.3606	40.4572
1981	41.3084	2.2896	5.242	3.5632	12.6964
1982	38.2096	0.9894	0.979	−1.3002	1.6905
1983	26.6838	10.4362	108.915	9.4468	89.2420
1984	33.3747	9.2303	85.198	−1.2060	1.4544
1985	25.6500	9.4680	89.643	0.2378	0.0565
1986	10.8949	8.2021	67.275	−1.2659	1.6025
1987	9.6914	6.4726	41.895	−1.7295	2.9912
1988	10.9967	2.6393	6.966	−3.8333	14.6942
1989	13.5655	−3.3615	11.300	−6.0008	36.0100
1990	17.4945	−5.2965	28.053	−1.9350	3.7442
1991	13.5316	−1.7616	3.103	3.5349	12.4955
1992	10.0934	−1.3364	1.786	0.4252	0.1808
1993	14.8134	−6.4064	41.042	−5.0700	25.7049
1994	13.5629	−6.8419	46.812	−0.4355	0.1897
1995	10.4720	−2.8450	8.094	3.9970	15.9760
1996	12.9468	−4.6328	21.463	−1.7879	3.1966
1997	18.2333	−7.7973	60.797	−3.1645	10.0141
1998	20.2669	−13.1489	172.894	−5.3517	28.6407
		$\Sigma e_t = 0.0043$	$\Sigma e_t^2 = 1036.118$		$\Sigma(e_t - e_{t-1})^2 = 384.3516$

The error for 1973 is

$$\text{Actual}_{1973} - \text{Predicted}_{1973} = 10.167 - 6.7623 = 3.4047$$

The value of $e_t - e_{t-1}$ for 1973 and 1974 is computed by subtracting the error for 1973 from the error of 1974.

$$e_{1974} - e_{1973} = 6.3495 - 3.4047 = 2.9448$$

The Durbin-Watson statistic can now be computed.

$$D = \frac{\sum_{t=2}^{n}(e_t - e_{t-1})^2}{\sum_{t=1}^{n} e_t^2} = \frac{384.3516}{1036.118} = .371$$

Because we used a simple linear regression, the value of k is 1. The sample size, n, is 26, and $\alpha = .05$. The critical values in Table A.9 are

$$d_U = 1.46 \text{ and } d_L = 1.30.$$

Because the computed D statistic, .371, is less than the value of $d_L = 1.22$, the null hypothesis is rejected. There is positive autocorrelation in this example.

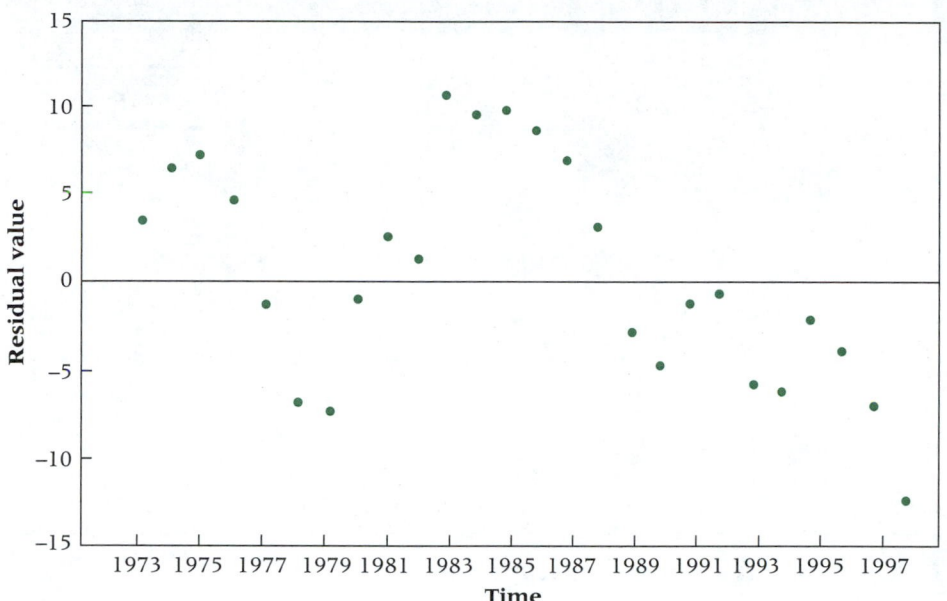

Figure 14.28

Excel graph of the residuals of the oil and gas well example

Figure 14.28 is the graph of the residuals given in Table 14.9. Note that there are several "runs" of positive and negative error terms instead of a random distribution of error terms. This is an indication of the presence of autocorrelation.

There are several ways of approaching data analysis when autocorrelation is present, two of which are the use of additional independent variables and transforming the independent variable.

ADDITION OF INDEPENDENT VARIABLES   Often the reason autocorrelation occurs in regression analyses is that one or more important predictor variables have been left out of the analysis. For example, suppose a researcher develops a regression forecasting model that attempts to predict sales of new homes by sales of used homes over some period of time. Such a model might contain significant autocorrelation. The exclusion of the variable "prime mortgage interest rate" might be a factor driving the autocorrelation between the other two variables. Adding this variable to the regression model might significantly reduce the autocorrelation.

TRANSFORMING VARIABLES   When the inclusion of additional variables is not helpful in reducing autocorrelation to an acceptable level, transforming the data in the variables may help to solve the problem. One such method is the **first-differences approach.** With the first-differences approach, *each value of X is subtracted from each succeeding time period value of X;* these "differences" become the new and transformed X variable. The same process is used to transform the Y variable. The regression analysis is then computed on the transformed X and transformed Y variables to compute a new model that is, one hopes, free of significant autocorrelation effects.

Excel doesn't have the capability to easily compute the Durbin-Watson statistic but ·FAST ⟍ STAT does. Figure 14.29 displays FAST ⟍ STAT's Durbin-Watson Statistic dialog box. Note that the only entry requirement of this feature is the location of the residual values from a prior run of Excel's **Regression Data Analysis Tool.** To obtain the residual values, you must select the first option in the **Residual** section of the **Regression**

First-differences approach

A method of transforming data in an attempt to reduce or remove autocorrelation from a time-series regression model; results in each data value being subtracted from each succeeding time period data value, producing a new and transformed value.

Figure 14.29

FAST ⧵ STAT dialog box for Durbin-Watson statistic

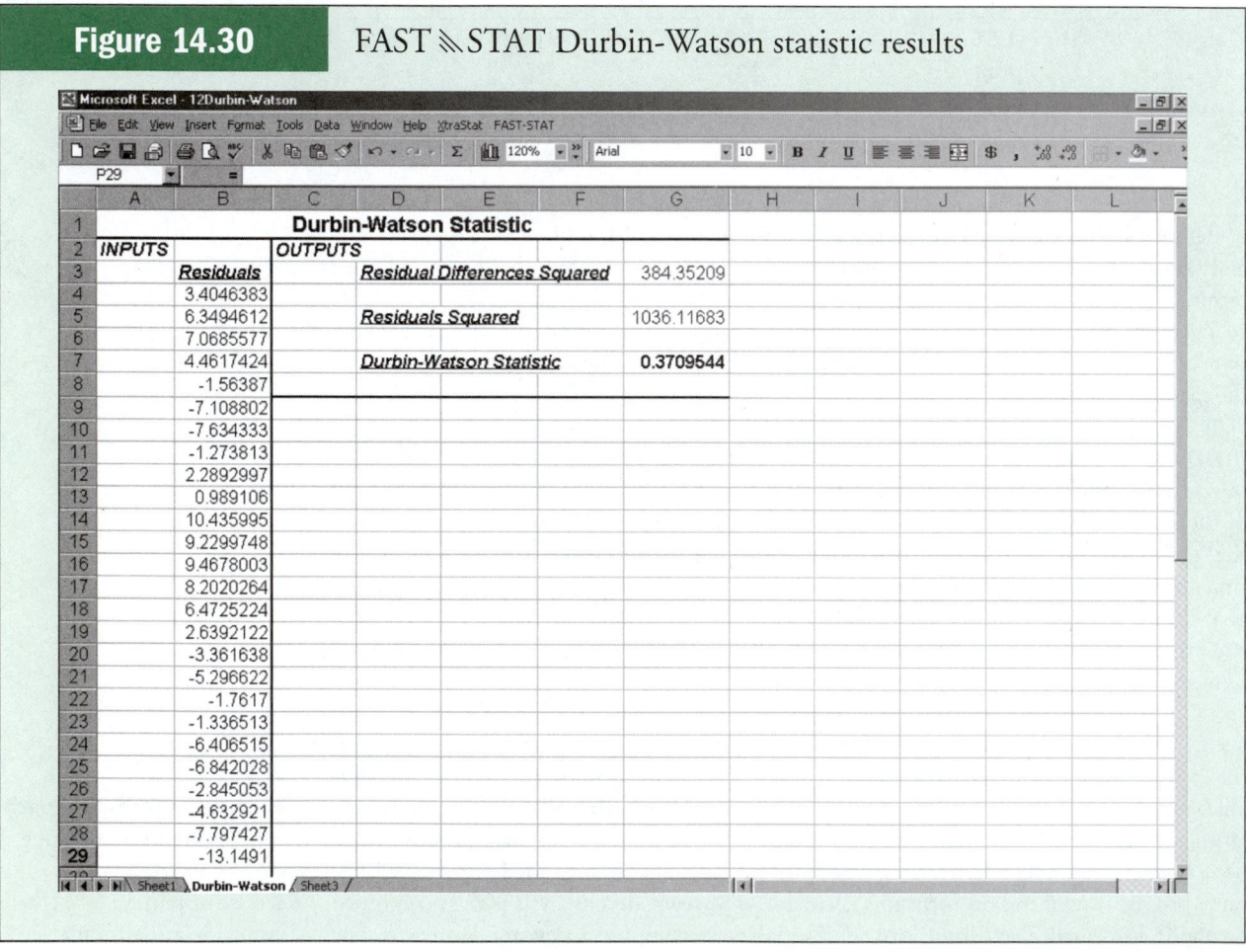

Figure 14.30

FAST ⧵ STAT Durbin-Watson statistic results

Figure 14.31 — Oil well first-order autoregression model

Figure 14.31   Oil well first-order autoregression model

	A	B	C	D	E	F	G	H	I	J
1	YEAR	OIL WELLS	Lagged 1		SUMMARY OUTPUT					
2	1973	10.167								
3	1974	13.647	10.167		Regression Statistics					
4	1975	16.948	13.647		Multiple R	0.89489943				
5	1976	17.688	16.948		R Square	0.80084499				
6	1977	18.745	17.688		Adjusted R Square	0.792186077				
7	1978	19.181	18.745		Standard Error	5.442545172				
8	1979	20.851	19.181		Observations	25				
9	1980	32.639	20.851							
10	1981	43.598	32.639		ANOVA					
11	1982	39.199	43.598			df	SS	MS	F	Significance F
12	1983	37.120	39.199		Regression	1	2739.61255	2739.61	92.48793077	0.00000000159
13	1984	42.605	37.120		Residual	23	681.289853	29.6213		
14	1985	35.118	42.605		Total	24	3420.90241			
15	1986	19.097	35.118							
16	1987	16.164	19.097			Coefficients	Standard Error	t Stat	P-value	
17	1988	13.636	16.164		Intercept	1.735640487	2.10933419	0.82284	0.419049933	
18	1989	10.204	13.636		X Variable 1	0.903418389	0.0939391	9.61706	0.00000000159	
19	1990	12.198	10.204							
20	1991	11.770	12.198							
21	1992	8.757	11.770							
22	1993	8.407	8.757		RESIDUAL OUTPUT					
23	1994	6.721	8.407							
24	1995	7.627	6.721		Observation	Predicted Y	Residuals			
25	1996	8.314	7.627		1	10.92069525	2.72630475			
26	1997	10.436	8.314		2	14.06459124	2.88340876			
27	1998	7.118	10.436		3	17.04677534	0.64122466			
28					4	17.71530495	1.02969505			
29					5	18.67021819	0.51078181			
30					6	19.0641086	1.7868914			

tool's dialog box. That option is labeled **Residuals.** For the oil and gas well example, Figure 14.27 presents these regression results. Note the residual values beginning in cell F23 and running through cell F48. The Durbin-Watson output worksheet as given in Figure 14.30 repeats the input values and provides the Durbin-Watson statistic value.

As an example of developing a trend autoregression model, consider the oil well data from Table 14.8. The data and the regression results are given in Figure 14.31 for a first-order autoregression model.

The FAST ⬅ STAT Durbin-Watson results are shown in Figure 14.32.

As shown, the Durbin-Watson statistic result is 1.22. To find the critical values for the statistic, we enter the table with $k = 1$, $n = 21$ and $\alpha = 0.05$. The table values are $d_U = 1.45$ and $d_L = 1.29$. The null hypothesis is rejected. Our conclusion is that autocorrelation is present in this time series. This is an indication that a trend component is present. Thus, it is a good candidate for a first-order autoregression model for forecasting.

The first-order autoregression model given in Figure 14.31 is

$$Y_t = 1.7356 + 0.9034 \, Y_{t-1}$$

The relatively high value for R^2 (0.8008) indicates a model with strong predictability. In addition, the p value for the slope of the relationship is essentially zero (0.00000000159).

Figure 14.32	Oil well FAST⟍STAT Durbin-Watson results

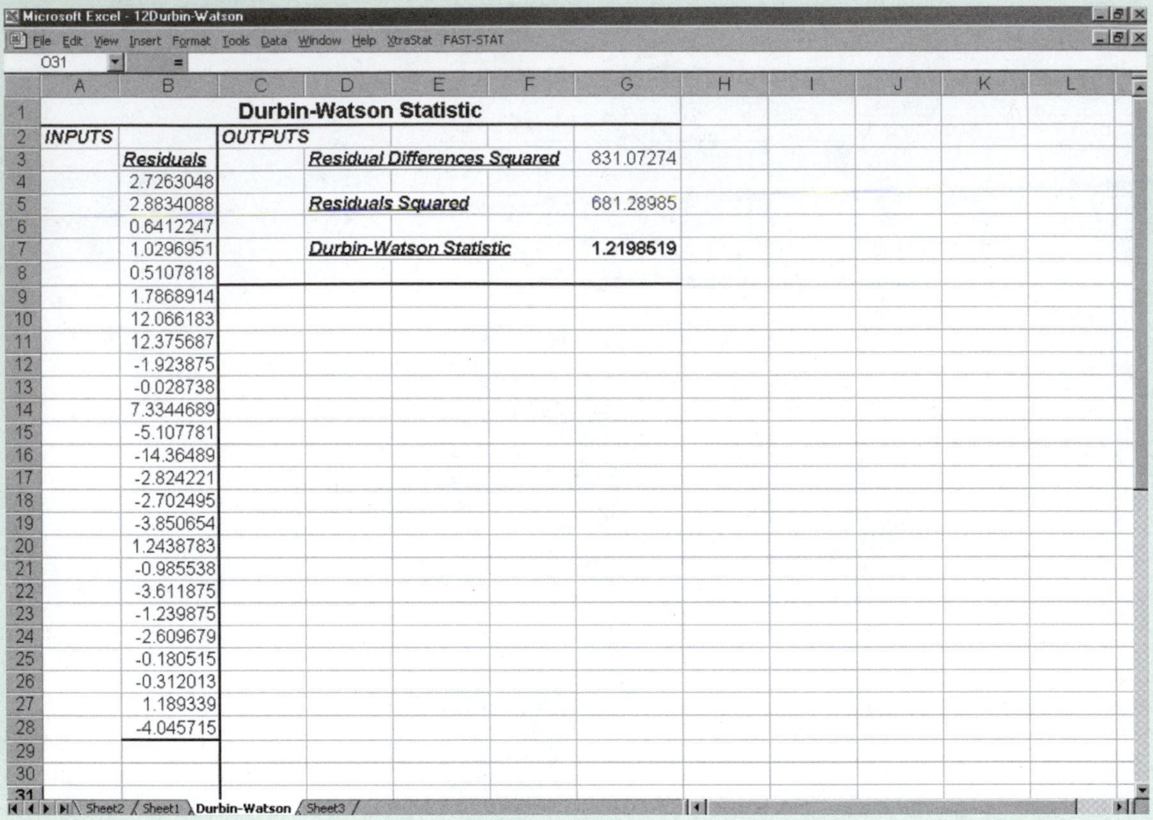

	A	B	C	D	E	F	G
1			**Durbin-Watson Statistic**				
2	INPUTS		OUTPUTS				
3		*Residuals*		*Residual Differences Squared*			831.07274
4		2.7263048					
5		2.8834088		*Residuals Squared*			681.28985
6		0.6412247					
7		1.0296951		*Durbin-Watson Statistic*			**1.2198519**
8		0.5107818					
9		1.7868914					
10		12.066183					
11		12.375687					
12		-1.923875					
13		-0.028738					
14		7.3344689					
15		-5.107781					
16		-14.36489					
17		-2.824221					
18		-2.702495					
19		-3.850654					
20		1.2438783					
21		-0.985538					
22		-3.611875					
23		-1.239875					
24		-2.609679					
25		-0.180515					
26		-0.312013					
27		1.189339					
28		-4.045715					
29							
30							

DEMONSTRATION PROBLEM 14.5	Consider the average baseball salary time series of Demonstration Problem 14.1. Perform Step 3 of the forecasting procedure of Section 14.3. Compare the four trend forecasting methods for forecasting this time series. Also, perform Step 4 of the forecasting procedure for the next two years. Compute the *best* forecast for each of the next two years.

SOLUTION

Linear Trend Projection

The following figure shows both the time series plot from Excel's **Chart Wizard** and the linear trend projection line from the **Trendline** feature. (This figure was also part of the solution to Demonstration Problem 14.1.)

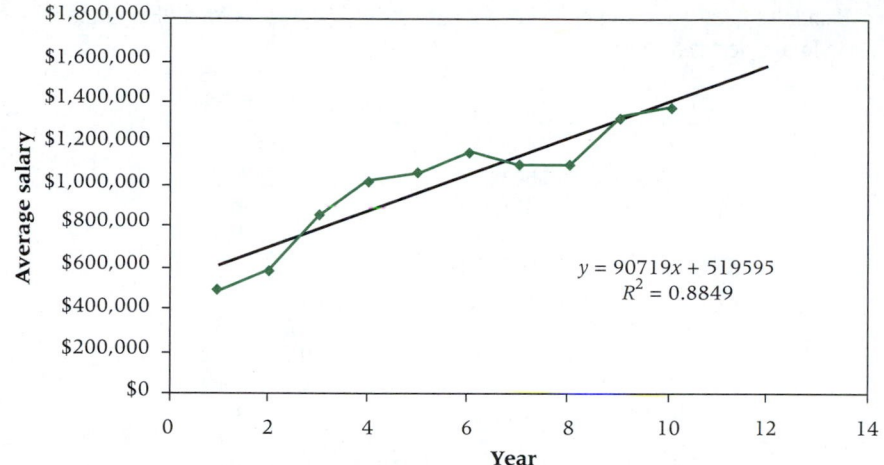

Using this linear trend equation, we can forecast the historical data in order to compute the value for MSE. In addition, we can use the equation to compute the forecasts for the next two years.

YEAR	AVG. SALARY	LINEAR FORECAST
1	$497,254	610,314
2	$597,537	701,033
3	$851,492	791,752
4	$1,028,667	882,471
5	$1,076,089	973,190
6	$1,168,263	1,063,909
7	$1,110,766	1,154,628
8	$1,119,981	1,245,347
9	$1,336,609	1,336,066
10	$1,398,831	1,426,785
		MSE = 8,833,631,511
11		1,517,504
12		1,608,223

Non-Linear Trend Projection

We can use the **Trendline** feature to explore a number of non-linear trend projection forms. These include polynomial up to an order 6, logarithmic, power, and exponential. The following figure shows the time series plot with a quadratic (polynomial of order 2) trend projection line. (This figure was also part of the solution to Demonstration Problem 14.1.)

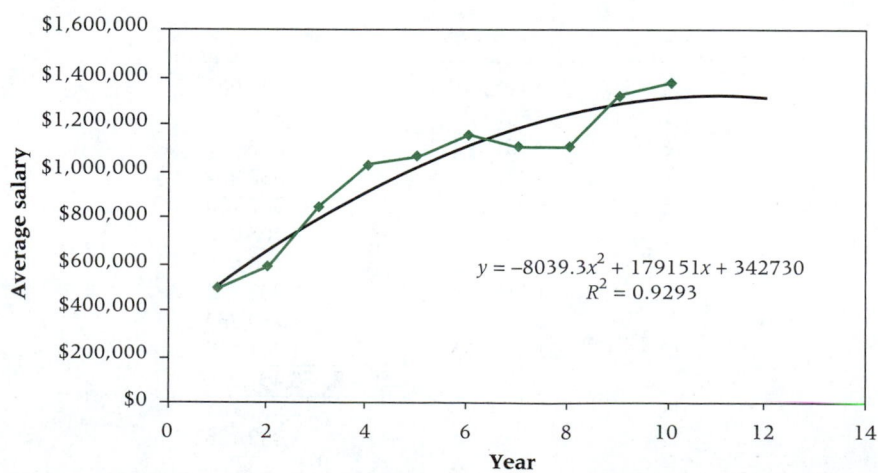

Using this quadratic trend equation, we can compute the value for MSE and the forecasts for the next two years.

YEAR	AVG. SALARY	QUADRATIC FORECAST
1	$497,254	513,842
2	$597,537	668,875
3	$851,492	807,829
4	$1,028,667	930,705
5	$1,076,089	1,037,503
6	$1,168,263	1,128,221
7	$1,110,766	1,202,861
8	$1,119,981	1,261,423
9	$1,336,609	1,303,906
10	$1,398,831	1,330,310
		MSE = 5,421,140,320
11		1,340,636
12		1,334,883

Although both the MSE and R^2 values for this equation are improved over that for the linear equation, the previous plot of the quadratic trend equation suggests that the salaries will peak for year 11 and then begin decreasing. This would not seem to be realistic and leads us to try a cubic (polynomial of order 3) trend line as shown in the next figure.

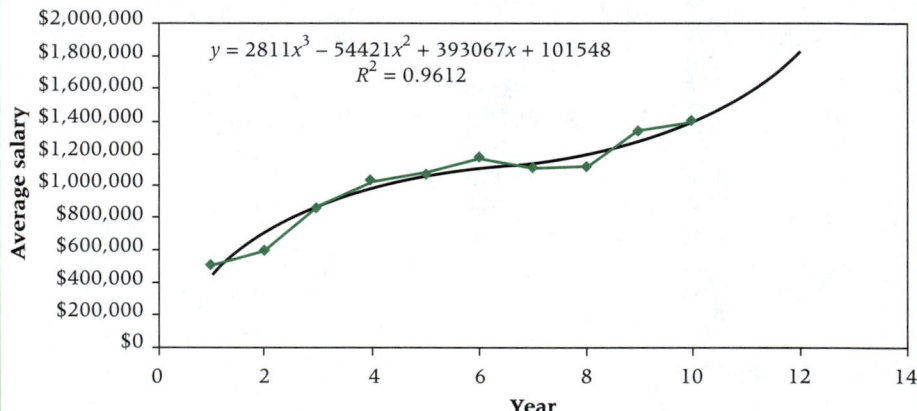

$$y = 2811x^3 - 54421x^2 + 393067x + 101548$$
$$R^2 = 0.9612$$

The R^2 value has improved further and the visual representation above looks more realistic. It suggests the average salaries will continue to increase in the future. The MSE and two-year forecast values are the following.

YEAR	AVG. SALARY	CUBIC FORECAST
1	$497,254	443,005
2	$597,537	692,486
3	$851,492	866,857
4	$1,028,667	982,984
5	$1,076,089	1,057,733
6	$1,168,263	1,107,970
7	$1,110,766	1,150,561
8	$1,119,981	1,202,372
9	$1,336,609	1,280,269
10	$1,398,831	1,401,118
		MSE = 2,980,481,978
11		1,581,785
12		1,839,136

Experimentation with other non-linear forms does not result in a non-linear trend that appears to fit the data better. For example, a quartic (polynomial of fourth order) results in an R^2 that is further improved to a value of 0.9720. However, as shown in the figure below, the forecasts for it appear to be somewhat extreme.

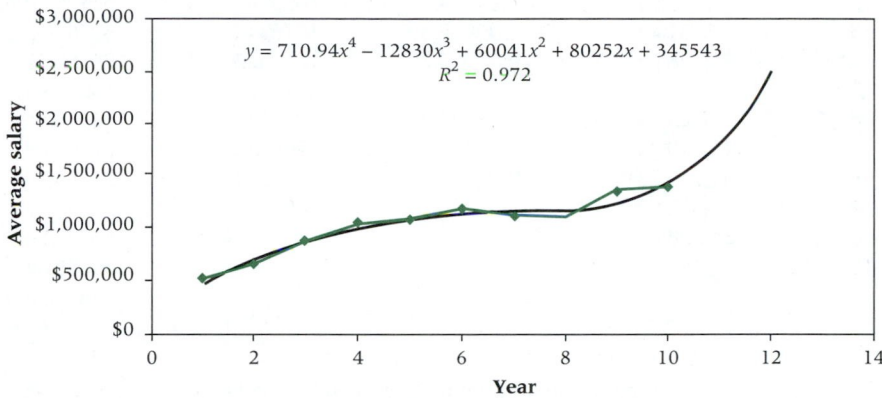

$$y = 710.94x^4 - 12830x^3 + 60041x^2 + 80252x + 345543$$
$$R^2 = 0.972$$

Holt's Model for Exponential Smoothing

The specific form of Holt's model requires four inputs. These include the values for the two smoothing constants, α and β, and the values for the initial forecast and the initial trend. For example, if we use a value of 0.2 for both the smoothing constants, the first actual time series value for the initial forecast, and a value of zero for the initial trend, we get the following results.

HOLT'S	$\alpha = 0.2$		$\beta = 0.2$	
YEAR	X_t	E_t	T_t	F_t
1	$497,254	**497254.0**	**0.0**	
2	$597,537	517310.6	4011.3	497254.0
3	$851,492	587355.9	17218.1	521321.9
4	$1,028,667	689392.6	34181.8	604574.1
5	$1,076,089	794077.4	48282.4	723574.5
6	$1,168,263	907540.4	61318.5	842359.8
7	$1,110,766	997240.4	66994.8	968859.0
8	$1,119,981	1075384.4	69224.7	1064235.2
9	$1,336,609	1183009.0	76904.7	1144609.0
15	$1,398,831	1287697.2	82461.4	1259913.7
16				*1370158.5*
17				*1452619.9*
				MSE = **67,645,593,790**

Based on the trend projection methods we have used before, we would conclude this MSE value is not very good. We can use our Excel worksheet and try different values for the four input values. Experimentation leads to the following specific form of Holt's model.

HOLT'S	$\alpha = 0.99$		$\beta = 0.05$	
YEAR	X_t	E_t	T_t	F_t
1	$497,254	**497254.0**	**100000.0**	
2	$597,537	597534.2	100014.0	597254.0
3	$851,492	849952.6	107634.2	697548.2
4	$1,028,667	1027956.2	111152.7	957586.8
5	$1,076,089	1076719.2	108033.2	1139108.9
6	$1,168,263	1168427.9	107217.0	1184752.4
7	$1,110,766	1112414.8	99055.5	1275644.9
8	$1,119,981	1120895.9	94526.8	1211470.3
9	$1,336,609	1335397.1	100525.5	1215422.7
15	$1,398,831	1399201.9	98689.5	1435922.6
16				*1497891.4*
17				*1596580.8*
				MSE = **9,401,315,045**

Although this MSE is much improved from the first Holt's model, it isn't as good as some of the previous trend models.

The following shows the time series plot and the Holt's model forecast.

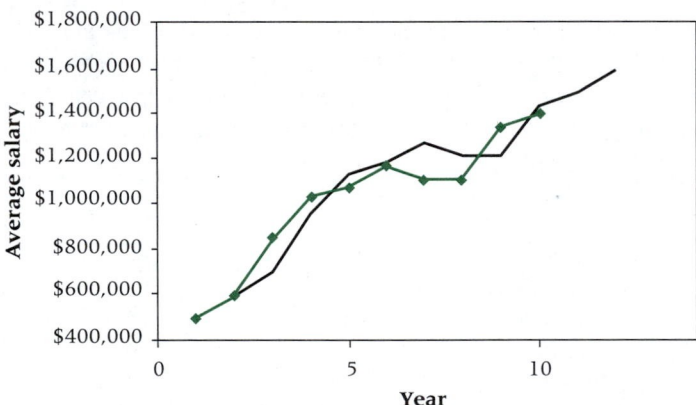

Autoregression for Trend

To begin the development of this model, we test for first order autocorrelation by computing the value for the Durbin-Watson statistic. First we run a regression on the lagged 1 values to obtain the residuals. These results are the following.

	A	B	C	D	E	F	G	H	I	J
1	Year	Avg. Salary	Lagged 1		SUMMARY OUTPUT					
2	1	$497,254								
3	2	$597,537	$497,254		Regression Statistics					
4	3	$851,492	$597,537		Multiple R	0.932985324				
5	4	$1,028,667	$851,492		R Square	0.870461615				
6	5	$1,076,089	$1,028,667		Adjusted R Square	0.851956131				
7	6	$1,168,263	$1,076,089		Standard Error	92799.45721				
8	7	$1,110,766	$1,168,263		Observations	9				
9	8	$1,119,981	$1,110,766							
10	9	$1,336,609	$1,119,981		ANOVA					
11	10	$1,398,831	$1,336,609			df	SS	MS	F	Significance F
12					Regression	1	4.05079E+11	4.1E+11	47.038	0.000240155
13					Residual	7	60282174813	8.6E+09		
14					Total	8	4.65361E+11			
15										
16						Coefficients	Standard Error	t Stat	P-value	
17					Intercept	278673.5705	120366.2641	2.31521	0.05377	
18					X Variable 1	0.817167672	0.119147953	6.85843	0.00024	
19										
20										
21										
22					RESIDUAL OUTPUT					
23										
24					Observation	Predicted Y	Residuals			
25					1	685013.4643	-87476.46425			
26					2	766961.4899	84530.51007			
27					3	974485.3062	54181.69384			
28					4	1119266.989	-43177.9885			

Using these residual values, the **FAST ＼ STAT** macro provides the following Durbin-Watson statistic value.

	A	B	C	D	E	F	G	H
1		**Durbin-Watson Statistic**						
2	INPUTS		OUTPUTS					
3		_Residuals_		_Residual Differences Squared_			120,540,277,941	
4		-87476.46						
5		84530.51		_Residuals Squared_			60,282,174,813	
6		54181.694						
7		-43177.99		_Durbin-Watson Statistic_			**1.999600683**	
8		10244.286						
9		-122574.3						
10		-66374.64						
11		142723.16						
12		27923.764						
13								
14								

The value for the statistic is about 2.0. However, it is based on a sample size of only nine. The Durbin-Watson statistic is generally inconclusive for smaller sample sizes. In fact, when we attempt to get the critical values from the table in Appendix A, we find they are not given for sample sizes of less than 15. Accordingly, we will not be able to draw a conclusion from this test. However we can still evaluate the first-order autoregressive models forecasting capability. From the prior regression results, we see the R^2 value is high (0.8705) and the *p value* for the slope of the relationship is quite small (0.000240). The MSE value is given below. We have added the **Residual Squared** column to the regression results and used Excel's **AVERAGE** function to compute the value for MSE. The forecasts are found by entering the regression equation. The forecast for the tenth time period has the actual value for the ninth time period ($1,398,831 in cell B11) as the value for the independent variable. The forecast for the eleventh time period uses the forecast in cell F35 as the value for the independent variable.

	C	D	E	F	G	H	I
21							
22			RESIDUAL OUTPUT				
23							
24			Observation	Predicted Y	Residuals	Residuals Squared	
25			1	685013.4643	-87476.46425	7652131798	
26			2	766961.4899	84530.51007	7145407132	
27			3	974485.3062	54181.69384	2935655948	
28			4	1119266.989	-43177.9885	1864338691	
29			5	1158018.714	10244.28614	104945398.6	
30			6	1233340.327	-122574.3269	15024465612	
31			7	1186355.637	-66374.63723	4405592468	
32			8	1193885.837	142723.1627	20369901162	
33			9	1370907.236	27923.76415	779736604.3	
34					MSE =	6,698,019,424	
35			10	$1,421,753.04			
36			11	$1,440,484.20			
37							
38							

The following shows the time series data and the autoregression forecasts.

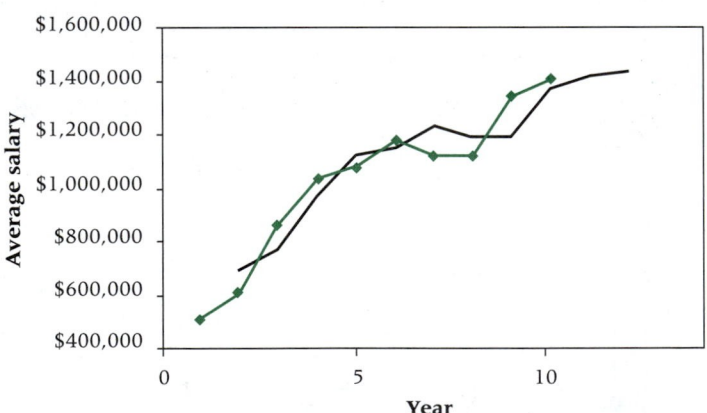

In summary the following MSE results have been obtained for the four trend forecasting methods.

METHOD	MSE
Linear Trend Projection	8,833,631,511
Non-linear Trend Projection	2,980,481,978
Holt's Model	9,401,315,045
Autoregression-Trend	6,698,019,424

Based on this evaluation, we would conclude that the non-linear projection trend method provided the *best* fit to the time series.

Finally, note that the magnitude of the MSE's in the table just above are much smaller than those computed for Demonstration Problem 14.4. In addition, the computation of the *mean error (ME)* or *forecast bias* for the trend models results in values of 0, 13, −2942, and 0, respectively for the four best models. You may recall a forecast bias of 100,176 for the best exponential smoothing model of Demonstration Problem 14.4. Again these results emphasize the unsuitability of stationary forecasting methods for forecasting time series data that include a trend component.

Forecasts for Next Two Years

Using the *best* model for the time series, the third-order polynomial, the forecasts for the next two years are $1581,781 and $1,839,136, respectively.

14.10 The Economic Report to the President of the United States included data on the amounts of manufacturers' new and unfilled orders in millions of dollars. Shown here are the figures for new orders over a 21-year period. Use a computer to develop a regression model to fit the trend effects for these data. Use a linear model and then try a quadratic model. How well does either model fit the data?

YEAR	TOTAL NUMBER OF NEW ORDERS	YEAR	TOTAL NUMBER OF NEW ORDERS
1	55,022	12	168,025
2	55,921	13	162,140
3	64,182	14	175,451
4	76,003	15	192,879
5	87,327	16	195,706
6	85,139	17	195,204
7	99,513	18	209,389
8	115,109	19	227,025
9	131,629	20	240,758
10	147,604	21	243,643
11	156,359		

14.11 The following data on the number of union members in the United States for the years 1984 through 1997 are provided by the U.S. Bureau of Labor Statistics. Using regression techniques discussed in this section, analyze the data for trend. Develop a scatter plot of the data and fit the trend line through the data. Discuss the strength of the model.

YEAR	UNION MEMBERS (1000s)	YEAR	UNION MEMBERS (1000s)
1984	17,340	1991	16,568
1985	16,996	1992	16,390
1986	16,975	1993	16,598
1987	16,913	1994	16,748
1988	17,002	1995	16,360
1989	16,960	1996	16,269
1990	16,740	1997	16,110

14.12 Following are data on the worldwide shipments of personal computers (1000) according to Dataquest. Plot the data, fit a trend line, and discuss the strength of prediction of the regression model. In addition, explore a quadratic trend. Compare the results of the two models.

YEAR	SHIPMENTS (1000)	YEAR	SHIPMENTS (1000)
1985	14,705	1992	32,411
1986	15,064	1993	38,851
1987	16,676	1994	47,894
1988	18,061	1995	60,171
1989	21,327	1996	71,065
1990	23,738	1997	82,400
1991	26,966	1998	97,321

14.13 The U.S. Bureau of Labor Statistics publishes consumer price indexes (CPIs) on many commodities. Following are the percentage changes in the CPIs for food and for shelter for 1974 through 1997. Use these data to develop a linear regression model to forecast the percentage change in food CPIs by the percentage change in shelter CPIs. Compute a Durbin-Watson statistic to determine whether there is significant autocorrelation in the model. Let $\alpha = .05$.

YEAR	FOOD	SHELTER	YEAR	FOOD	SHELTER
1974	14.3	9.6	1986	3.2	5.5
1975	8.5	9.9	1987	4.1	4.7
1976	3.0	5.5	1988	4.1	4.8
1977	6.3	6.6	1989	5.8	4.5
1978	9.9	10.2	1990	5.8	5.4
1979	11.0	13.9	1991	2.9	4.5
1980	8.6	17.6	1992	1.2	3.3
1981	7.8	11.7	1993	2.2	3.0
1982	4.1	7.1	1994	2.4	3.1
1983	2.1	2.3	1995	2.8	3.2
1984	3.8	4.9	1996	3.3	3.2
1985	2.3	5.6	1997	2.6	3.1

14.14 Use the data from Problem 14.13 to create a regression forecasting model using the first-differences data transformation. How do the results from this model differ from those obtained in Problem 14.13?

14.15 The Federal Deposit Insurance Corporation (FDIC) releases data on bank failures. Following are data on the number of U.S. bank failures in a given year and the total amount of bank deposits (in $ millions) involved in such failures for a given year. Use these data to develop a simple regression forecasting model that attempts to predict the failed bank assets involved in bank closings by the number of bank failures. Use the model to predict the failed bank assets involved if there are 150 failures. Comment on the forecasting model. Compute a Durbin-Watson statistic for this regression model and determine whether there is significant autocorrelation. Let $\alpha = .05$.

YEAR	FAILURES	FAILED BANK ASSETS	YEAR	FAILURES	FAILED BANK ASSETS
1980	11	8,189	1989	206	28,507
1981	7	104	1990	159	10,739
1982	34	1,862	1991	108	43,552
1983	45	4,137	1992	100	16,915
1984	79	36,394	1993	42	2,588
1985	118	3,034	1994	11	825
1986	144	7,609	1995	6	753
1987	201	7,538	1996	5	186
1988	221	56,620	1997	1	27

14.16 Use the data in Problem 14.15 to compute a regression model after recoding the data by the first-differences approach. Compute a Durbin-Watson statistic to determine whether there is significant autocorrelation in this first-differences model. Compare this model with the model determined in Problem 14.15 and compare the significance of the Durbin-Watson statistics for the two problems. Let $\alpha = .05$.

14.17 *Current Construction Reports* from the U.S. Bureau of the Census contains data on new privately owned housing units. Below are the data on new privately owned housing units (1000s) built in the West between 1970 and 1997. Use these time-series data to develop an autoregression model with a one-period lag. Now try an autoregression model with a two-period lag. Discuss the results and compare the two models.

YEAR	HOUSING STARTS (1000)	YEAR	HOUSING STARTS (1000)
1970	311	1984	436
1971	486	1985	468
1972	527	1986	483
1973	429	1987	420
1974	285	1988	404
1975	275	1989	396
1976	400	1990	329
1977	538	1991	254
1978	545	1992	288
1979	470	1993	302
1980	306	1994	351
1981	240	1995	331
1982	205	1996	361
1983	382	1997	364

14.18 The U.S. Department of Agriculture publishes data on the production, utilization, and value of fruits in the United States. Shown here are the amounts of noncitrus fruit processed into juice (in 1000 tons) for the years 1973 through 1997. Use these data to develop an autoregression forecasting model with a two-period lag. Discuss the results of this analysis.

YEAR	PROCESSED JUICE	YEAR	PROCESSED JUICE
1973	598	1986	1135
1974	768	1987	1893
1975	863	1988	1372
1976	818	1989	1547
1977	841	1990	1450
1978	1140	1991	1557
1979	1285	1992	1742
1980	1418	1993	1736
1981	1235	1994	1886
1982	1255	1995	1857
1983	1445	1996	1582
1984	1336	1997	1675
1985	1226		

14.19 Use Holt's model to exponentially smooth the data given in Problem 14.7. Use $\alpha = .5$, and try both $\beta = .3$ and $\beta = .7$. Calculate MAD and compare the two analyses. How did these compare to the results obtained in 14.7?

14.20. Use Holt's model to exponentially smooth the data given in Problem 14.9. Let $\alpha = .5$ and $\beta = .7$. Compare these results to those of Problem 14.9 using MSE, bias (ME) and MAD.

14.6
Seasonal Forecasting Methods

Many times a time series exhibits a seasonal component. A number of examples have previously been examined. The seasonal variation can either be accompanied by a trend variation such as shown in Figures 14.6 and 14.7, or without as shown in Figures 14.5 and 14.14. For such time series data, the stationary forecasting methods and trend forecasting methods are ineffective in treating the seasonal component.

Forecasting the effect of a seasonal component requires historical data for at least two years and preferably three or more. It is not possible to obtain a reliable estimate of the seasonal effect with less data. In addition, the data cannot be annual data. It must be recorded at time intervals of less than a year. Typically, data from business and industry are recorded by the day, the week, a two- or four-week period, the month, a quarter, or semi-annually.

The four forecasting methods presented below are effective for time series data that include a seasonal, a trend, and an irregular component. They are equally effective for a time series with only a seasonal and an irregular component.

Seasonal Multiple Regression Forecasting Method

Multiple regression analysis with the use of *dummy variables* to represent the seasons is a relatively simple approach for forecasting a seasonal time series. Dummy or indicator variables were introduced in Chapter 13 for representing qualitative variables in a multiple regression model. The number of dummy variables required here is, as it was in Chapter 13, one less than the number of seasons. So we can represent quarterly time series data with three dummy variables, monthly with 11, weekly with 51, and so on.

Seasonal multiple regression
Forecasting by using dummy variables to represent seasons.

This **seasonal multiple regression** forecasting method can model a seasonal time series either with or without a trend component. The complete model for quarterly time series is

$$\hat{Y} = b_0 + b_1 t + b_2 Q_1 + b_3 Q_2 + b_4 Q_3$$

where

b_0 is the time series average value

b_1 is the coefficient for the trend component

Q_1, Q_2, Q_3 are dummy variables for the first three quarters ($=1$ if the observation is from the specified quarter, otherwise $=0$)

b_2, b_3, b_4 are coefficients that indicate how much each quarter differs from the reference quarter, quarter 4

For example, suppose an estimated regression relationship is

$$\hat{Y} = 535 + 10t - 30Q_1 + 10Q_2 + 40Q_3$$

This indicates the average value at time period zero is 535 with a trend increase of 10 each quarter. The seasonal effect is that quarter 1 is 30 below quarter 4, quarter 2 averages 10 above quarter 4, and quarter 3 averages 40 above quarter 4.

For the above, quarter 4 is the reference quarter. Its designation as such is purely arbitrary. Any of the other three quarters could equally serve as the reference quarter. For example, if quarter 1 is designated as the reference quarter, the dummy variables would be Q_2, Q_3, and Q_4.

The model above is what was referred to at the end of Section 14.2 as an additive model for representing a time series. It is also possible to use this approach as a multiplicative model, but we will only demonstrate the more common additive structure. The additive structure assumes that the trend and seasonal effects are independent. In other words, the amount of seasonal variation is unaffected by the time series trend.

Consider the example time series of Figure 14.14 for eggnog sales by quarter for five recent years. Our analysis of Section 14.3 indicated the time series exhibited the presence of a quarterly seasonal effect and the absence of a trend effect. Accordingly, we would expect the regression analysis to have an insignificant value for b_1 and significant values for b_2, b_3, and b_4. Figure 14.33 shows the data, along with the layout for the independent variables for the multiple regression analysis. The regression model results follow in Figure 14.34.

As expected, these results indicate an insignificant trend term, *p value* of 0.5189. On the other hand, all three of the quarterly seasonal factors are significant, *p values* of 0.0000 for all three. Also, we observe that the coefficients for all three quarters are negative indicating a sales level less than that for the reference quarter, the fourth quarter. This also is just what we would expect for the time series of Figure 14.14.

One of the advantages of the seasonal multiple regression approach is its undemanding approach for developing a model to be used for forecasting. Before we use the current model to make forecasts, we would need to re-run the regression without the insignificant trend component. Such results are shown in Figure 14.35.

As shown, all of the significant levels have been improved by eliminating the non-significant trend factor. The final model is

$$\hat{Y} = 1547.2 - 1166.6Q_1 - 1086.2Q_2 - 946.2Q_3$$

Using it to make forecasts for the next year gives the results in Table 14.10.

	A	B	C	D	E	F	G	H
1			Eggnog					
2	Year	Quarter	Sales	Time	Q(1)	Q(2)	Q(3)	
3	1	Jan-Mar	473	1	1	0	0	
4		Apr-Jun	599	2	0	1	0	
5		Jul-Sep	854	3	0	0	1	
6		Oct-Dec	1632	4	0	0	0	
7	2	Jan-Mar	211	5	1	0	0	
8		Apr-Jun	313	6	0	1	0	
9		Jul-Sep	499	7	0	0	1	
10		Oct-Dec	1523	8	0	0	0	
11	3	Jan-Mar	409	9	1	0	0	
12		Apr-Jun	265	10	0	1	0	
13		Jul-Sep	571	11	0	0	1	
14		Oct-Dec	1432	12	0	0	0	
15	4	Jan-Mar	524	13	1	0	0	
16		Apr-Jun	673	14	0	1	0	
17		Jul-Sep	621	15	0	0	1	
18		Oct-Dec	1469	16	0	0	0	
19	5	Jan-Mar	286	17	1	0	0	
20		Apr-Jun	455	18	0	1	0	
21		Jul-Sep	460	19	0	0	1	
22		Oct-Dec	1680	20	0	0	0	
23								
24								

Figure 14.33

Eggnog data for seasonal multiple regression model

Figure 14.34   Eggnog seasonal multiple regression model results

	A	B	C	D	E	F	G
	M51	▾	=				
24	SUMMARY OUTPUT						
25							
26	*Regression Statistics*						
27	Multiple R	0.96519					
28	R Square	0.93159					
29	Adjusted R Squa	0.91334					
30	Standard Error	146.734					
31	Observations	20					
32							
33	ANOVA						
34		*df*	*SS*	*MS*	*F*	*Significance F*	
35	Regression	4	4397795.175	1099449	51.0637204	0.000000015	
36	Residual	15	322963.775	21530.9			
37	Total	19	4720758.95				
38							
39		*Coefficients*	*Standard Error*	*t Stat*	*P-value*		
40	Intercept	1593.18	95.65897915	16.6547	0.000000000		
41	Time	-3.83125	5.800177575	-0.66054	0.518923431		
42	Q(1)	-1178.09	94.42005016	-12.4772	0.000000003		
43	Q(2)	-1093.86	93.52505318	-11.6959	0.000000006		
44	Q(3)	-950.031	92.98392008	-10.2172	0.000000038		
45							
46							

Figure 14.35   Final eggnog seasonal multiple regression model results

	A	B	C	D	E	F	G
47	SUMMARY OUTPUT						
48							
49	*Regression Statistics*						
50	Multiple R	0.96416					
51	R Square	0.9296					
52	Adjusted R Squa	0.9164					
53	Standard Error	144.126					
54	Observations	20					
55							
56	ANOVA						
57		*df*	*SS*	*MS*	*F*	*Significance F*	
58	Regression	3	4388400.95	1462800	70.42046548	0.000000002	
59	Residual	16	332358	20772.4			
60	Total	19	4720758.95				
61							
62		*Coefficients*	*Standard Error*	*t Stat*	*P-value*		
63	Intercept	1547.2	64.45521701	24.0043	0.000000000		
64	Q(1)	-1166.6	91.15344206	-12.7982	0.000000001		
65	Q(2)	-1086.2	91.15344206	-11.9162	0.000000002		
66	Q(3)	-946.2	91.15344206	-10.3803	0.000000016		
67							
68							

YEAR	EGGNOG QUARTER	SALES	TIME	Q(1)	Q(2)	Q(3)	FORECAST
6	Jan-Mar	473	21	1	0	0	380.6
	Apr-Jun	599	22	0	1	0	461.0
	Jul-Sep	854	23	0	0	1	601.0
	Oct-Dec	1632	24	0	0	0	1547.2

TABLE 14.10
Predicted Eggnog Sales for
Seasonal Multiple Regression

Since the time series does not exhibit a trend, these forecasts would also be the current forecasts for year 7 and all subsequent years. Of course after additional historical data have been obtained, the regression model can be re-run to obtain an updated model.

Seasonal Autoregression Forecasting Method

In Section 14.5, we developed a trend autoregression model by considering the regression between the forecast variable and its values lagged by one time period. Here we will use a similar approach to forecast a variable based on its values lagged by the season length. If the data are quarterly data, we will use a lag of four. If the data are monthly, we will use a lag of 12, and so on.

As an example of developing a **seasonal autoregression** model, again consider the eggnog sales time series data. Figure 14.36 presents the data and the autoregression results. You will note the independent variables are the sales lagged one time period and sales lagged four time periods. This approach allows for both a trend and a seasonal component to be modeled.

Seasonal autoregression
Autoregression model with
an independent variable
lagged by season period.

Figure 14.36 — Eggnog first- and fourth-order autoregression model

	A	B	C	D	E	F	G	H	I	J	K	L
1			Eggnog	Lagged	Lagged		SUMMARY OUTPUT					
2	Year	Quarter	Sales	One	Four							
3	1	Jan-Mar	473				Regression Statistics					
4		Apr-Jun	599	473			Multiple R	0.90874				
5		Jul-Sep	854	599			R Square	0.8258				
6		Oct-Dec	1632	854			Adjusted R Square	0.799				
7	2	Jan-Mar	211	1632	473		Standard Error	225.934				
8		Apr-Jun	313	211	599		Observations	16				
9		Jul-Sep	499	313	854							
10		Oct-Dec	1523	499	1632		ANOVA					
11	3	Jan-Mar	409	1523	211			df	SS	MS	F	Significance F
12		Apr-Jun	265	409	313		Regression	2	3145841.05	1572921	30.813791	0.0000117
13		Jul-Sep	571	265	499		Residual	13	663597.885	51046		
14		Oct-Dec	1432	571	1523		Total	15	3809438.94			
15	4	Jan-Mar	524	1432	409							
16		Apr-Jun	673	524	265			Coefficients	Standard Error	t Stat	P-value	
17		Jul-Sep	621	673	571		Intercept	-25.1843	161.702774	-0.15574	0.8786268	
18		Oct-Dec	1469	621	1432		X Variable 1	0.0196	0.12437772	0.15761	0.8771849	
19	5	Jan-Mar	286	1469	524		X Variable 2	0.95887	0.12877054	7.44631	0.0000049	
20		Apr-Jun	455	286	673							
21		Jul-Sep	460	455	621							
22		Oct-Dec	1680	460	1469							
23												
24												

Figure 14.37   Eggnog fourth-order autoregression model

	A	B	C	D	E	F	G	H	I	J	K
			Eggnog	Lagged		SUMMARY OUTPUT					
2	Year	Quarter	Sales	Four							
3	1	Jan-Mar	473			Regression Statistics					
4		Apr-Jun	599			Multiple R	0.908553				
5		Jul-Sep	854			R Square	0.825469				
6		Oct-Dec	1632			Adjusted R Square	0.813002				
7	2	Jan-Mar	211	473		Standard Error	217.923				
8		Apr-Jun	313	599		Observations	16				
9		Jul-Sep	499	854							
10		Oct-Dec	1523	1632		ANOVA					
11	3	Jan-Mar	409	211			df	SS	MS	F	Significance F
12		Apr-Jun	265	313		Regression	1	3144573.003	3144573	66.21488598	0.00000112
13		Jul-Sep	571	499		Residual	14	664865.9345	47490.4		
14		Oct-Dec	1432	1523		Total	15	3809438.938			
15	4	Jan-Mar	524	409							
16		Apr-Jun	673	265			Coefficients	Standard Error	t Stat	P-value	
17		Jul-Sep	621	571		Intercept	-6.148643	103.709405	-0.05929	0.953561309	
18		Oct-Dec	1469	1432		X Variable 1	0.952053	0.116999339	8.13725	0.00000112	
19	5	Jan-Mar	286	524							
20		Apr-Jun	455	673							
21		Jul-Sep	460	621							
22		Oct-Dec	1680	1469							
23											

Whereas the *p value* for the seasonal term is extremely significant, 0.0000049, that for the trend term is equally insignificant, 0.878628. Accordingly, we will eliminate the lagged one column and re-run the regression. The results are as shown in Figure 14.37.

The elimination of the trend component improved the significance of the *p values*. The fourth-order autoregression model given in this figure is

$$Y_t = -6.1486 + 0.9521\ Y_{t-4}$$

The R^2 value of 0.8255 and the *p value* of 0.00000112 suggests the model is statistically significant.

Winters' Exponential Smoothing Forecasting Method

Winters' model

Exponential smoothing with a seasonal and/or trend component.

Winters' exponential smoothing forecasting method, generally called **Winters' model,** extends Holt's model by including a provision for a seasonal component in addition to a trend component. The formulas for using Winters' model are:

Smoothed Values:	$E_t = \alpha(X_t/S_{t-L}) + (1-\alpha)(E_{t-1} + T_{t-1})$
Trend Term Update:	$T_t = \beta(E_t - E_{t-1}) + (1-\beta)T_{t-1}$
Seasonality Update:	$S_t = \gamma(X_t/E_t) + (1-\gamma)S_{t-L}$
Forecast for Next Period:	$F_{t+1} = (E_t + T_t)S_{t-L+1}$
for k periods in the future:	$F_{t+k} = (E_t + kT_t)S_{t-L+k}$

Figure 14.38 — Winters' model computations for eggnog sales

	Year	Quarter	Year	X_t	E_t	T_t	S_t	F_t	Error	Abs. Error
1			WINTERS' MODEL		$\alpha = 0.1$		$\beta = 0.01$		$\gamma = 0.6$	
3	1	Jan-Mar	1	473	473.0	0.00	1.000			
4		Apr-Jun	2	599	485.6	0.13	1.000	473.0	126.0	126.0
5		Jul-Sep	3	854	522.6	0.49	1.000	485.7	368.3	368.3
6		Oct-Dec	4	1632	633.9	1.60	1.000	523.0	1109.0	1109.0
7	2	Jan-Mar	5	211	593.1	1.18	0.613	635.5	-424.5	424.5
8		Apr-Jun	6	313	566.1	0.90	0.732	594.3	-281.3	281.3
9		Jul-Sep	7	499	560.2	0.83	0.934	567.0	-68.0	68.0
10		Oct-Dec	8	1523	657.3	1.79	1.790	561.1	961.9	961.9
11	3	Jan-Mar	9	409	659.8	1.80	0.617	404.3	4.7	4.7
12		Apr-Jun	10	265	631.7	1.50	0.544	484.1	-219.1	219.1
13		Jul-Sep	11	571	631.0	1.48	0.917	591.6	-20.6	20.6
14		Oct-Dec	12	1432	649.2	1.64	2.040	1132.3	299.7	299.7
15	4	Jan-Mar	13	524	670.6	1.84	0.716	401.8	122.2	122.2
16		Apr-Jun	14	673	728.8	2.41	0.772	366.1	306.9	306.9
17		Jul-Sep	15	621	725.9	2.35	0.880	670.4	-49.4	49.4
18		Oct-Dec	16	1469	727.4	2.34	2.028	1485.3	-16.3	16.3
19	5	Jan-Mar	17	286	696.7	2.01	0.533	522.3	-236.3	236.3
20		Apr-Jun	18	455	687.8	1.91	0.706	539.3	-84.3	84.3
21		Jul-Sep	19	460	673.0	1.74	0.762	607.0	-147.0	147.0
22		Oct-Dec	20	1680	690.2	1.89	2.272	1368.1	311.9	311.9
23	6		1	21				368.6		
24			2	22				489.7		
25			3	23				530.3		
26			4	24				1584.9		
27							MSE =	158027.8	MAD =	271.4

Compare these formulas with those used in Holt's model. The equation for the smoothed value is the same as with Holt's method except that in place of the actual value, there is the quotient of the actual over the seasonal. This adjusts the actual value for seasonality. The trend equation is the same. The new equation is the seasonality equation:

$$S_t = \gamma(X_t/E_t) + (1 - \gamma)S_{t-L}$$

Winters' model introduces a third weight, γ, to go along with α and β. γ is a weight between 0 and 1 that is placed on the present seasonality effect, which is measured by the ratio of the current actual value to the current smoothed value. The old seasonal value, S_{t-L}, is weighted by $1 - \gamma$. The forecast is similar to Holt's except that the sum of the smoothed value and the trend value is multiplied by the seasonal effect to produce the next forecast. Shown in Figure 14.38 are the results of applying Winters' method to the eggnog sales data of Figure 14.33.

In order to initialize this model, we have set the initial smoothed value equal to the first quarter's sales value, 473. The initial trend is set to zero, and the four initial seasonal factors are set equal to one. In addition, the first year's values for E_t and F_t use the seasonal factors from the current year instead of those for the prior year since there are no prior values. The computations for E_t and F_t, and also those for S_t, after the first year use seasonal factors from the corresponding quarter of the prior year.

Figure 14.39   Modified Winters' model computations for eggnog sales

	A	B	C	D	E	F	G	H	I	J
	Year	Quarter	Year	X_t	E_t	T_t	S_t	F_t	Error	Abs. Error
					WINTERS' MODEL		$\alpha = 0.1$		$\beta = 0.01$	$\gamma = 0.6$
3	1	Jan-Mar	1	473	473.0	0.00	0.500			
4		Apr-Jun	2	599	511.3	0.38	0.700	331.1	267.9	267.9
5		Jul-Sep	3	854	567.2	0.94	0.800	409.3	444.7	444.7
6		Oct-Dec	4	1632	582.3	1.08	2.300	1306.8	325.2	325.2
7	2	Jan-Mar	5	211	567.3	0.92	0.423	291.7	-80.7	80.7
8		Apr-Jun	6	313	556.1	0.80	0.618	397.7	-84.7	84.7
9		Jul-Sep	7	499	563.6	0.86	0.851	445.5	53.5	53.5
10		Oct-Dec	8	1523	574.2	0.96	2.511	1298.2	224.8	224.8
11	3	Jan-Mar	9	409	614.3	1.35	0.569	243.4	165.6	165.6
12		Apr-Jun	10	265	597.0	1.17	0.513	380.3	-115.3	115.3
13		Jul-Sep	11	571	605.4	1.24	0.906	509.2	61.8	61.8
14		Oct-Dec	12	1432	603.0	1.20	2.429	1523.6	-91.6	91.6
15	4	Jan-Mar	13	524	635.9	1.52	0.722	343.6	180.4	180.4
16		Apr-Jun	14	673	704.8	2.19	0.778	327.3	345.7	345.7
17		Jul-Sep	15	621	704.8	2.17	0.891	640.8	-19.8	19.8
18		Oct-Dec	16	1469	696.7	2.07	2.237	1717.5	-248.5	248.5
19	5	Jan-Mar	17	286	668.5	1.77	0.545	504.5	-218.5	218.5
20		Apr-Jun	18	455	661.7	1.68	0.724	521.7	-66.7	66.7
21		Jul-Sep	19	460	648.7	1.53	0.782	591.3	-131.3	131.3
22		Oct-Dec	20	1680	660.3	1.63	2.421	1454.4	225.6	225.6
23	6		21					361.0		
24			22					480.3		
25			23					520.2		
26			24					1614.6		
27							MSE =	43671.9	MAD =	176.4

The Winters' model results are forecasts of 368.6, 489.7, 530.3, and 1584.9 for the four quarters of year 6. The final smoothed sales value is 690.2, the final trend is 1.89. The final four-quarter seasonal factors are 0.533, 0.706, 0.762, and 2.272. These indicate that most of the sales, over 50 per cent, occur in the fourth quarter.

Computed at the bottom of the worksheet are values for both MSE and MAD. The formulas used are those previously presented for prior examples. Six input items at the top of the worksheet can be changed in order to search for a better fitting model. The six inputs include the values for the three smoothing constants, α, β and γ. In addition, these include the three initial values of 473 for the initial smoothed value, 0.0 for the initial trend, and 1.0 for the four initial seasonal effects. For example, changing the four initial seasonal factors to value near the final four values of Figure 14.38 improves both the MSE and MAD as shown in Figure 14.39.

For those readers who wish to use this Excel worksheet without entering the formulas, it is given on the CD-ROM enclosed with this book. It is in the folder called *Chapter 14 Templates* and is entitled *Winters' Model*.

Time Series Decomposition Forecasting Method

The three prior seasonal forecasting methods explicitly consider the seasonal variation for forecasting. Another approach is to first deseasonalize the time series data, forecast the **deseasonalized time series,** and then adjust these deseasonalized forecasts with the estimated seasonal effects. This approach is called **time series decomposition** and is usually based on the multiplicative model as given in Section 14.2.

$$Y_t = T_t \cdot S_t \cdot C_t \cdot I_t$$

However, due to the unlikelihood of having sufficient stable time series data necessary to discern a multi-year cyclical component, the model we use is

$$Y_t = T_t \cdot S_t \cdot I_t$$

This is consistent with our approach throughout this chapter.

The general approach of time series decomposition consists of the following five steps.

1. Determine the seasonality of the time series by computing a seasonal index, S_t, for each season (each quarter, each month, and so on).
2. Divide each time series data value by the appropriate seasonal index to deseasonalize it.
3. Identify a trend model appropriate for the deseasonalized time series.
4. Forecast deseasonalized values with the trend model.
5. Multiply the deseasonalized forecasts times the appropriate seasonal index to compute the final seasonalized forecasts.

The first of these five steps, determining the seasonal indexes, requires the most computations. A number of methods have been developed for computing them, but they are all based on the ratio-to-moving-averages approach demonstrated in our presentation on the CD-ROM.

A seasonal index is a number that represents how much above or below the average a time series value is. For example, for quarterly data, suppose the four indexes for the four seasons (quarters) are 0.4, 1.6, 0.8, and 1.2. These values indicate that the first quarter values are 60% less than the average, the second quarter is 60% greater, the third is 20% less, and the fourth is 20% above the average. Seasonal indexes are essentially the same as seasonal factors computed for Winters' model (for example, see Column G of figures 14.38 and 14.39).

Once the seasonal indexes are computed, each time series value is divided by its corresponding index to eliminate the seasonal effect. For example, suppose a second quarter sales time series value is 567,158, and the second quarter seasonal index is 1.6. The corresponding *deseasonalized* or *seasonally adjusted* value is 354,474 (= 567,158/1.6).

In Steps 4 and 5 of the decomposition procedure, an appropriate trend model is identified and then used to forecast the deseasonalized time series. Suppose the deseasonalized forecast for the second quarter of next year is 423,548. The final seasonalized forecast for the second quarter of next year is 677,676.8 (= 1.6 × 423,548).

Deseasonalized time series
A time series in which the seasonal effect has been removed. Also called a seasonally adjusted time series.

Time series decomposition
Breaking down a time series into its components.

DEMONSTRATION PROBLEM 14.6

Given here are the quarterly data for shipments of household appliances in the United States for 5 recent years as published by the U.S. Department of Commerce. Using these data, perform Step 3 of the forecasting procedure of Section 14.3 for the seasonal forecasting methods. Also, perform Step 4 for the next four quarters.

YEAR	QUARTER	SHIPMENTS
1	1	4009
	2	4321
	3	4224
	4	3944
2	1	4123
	2	4522
	3	4657
	4	4030
3	1	4493
	2	4806
	3	4551
	4	4485
4	1	4595
	2	4799
	3	4417
	4	4258
5	1	4245
	2	4900
	3	4585
	4	4533

SOLUTION

Seasonal Multiple Regression

The seasonal regression model results are given below.

	A	B	C	D	E	F	G	H	I	J	K	L	M	N
1	Year	Quarter	Shipments	Time	Q(1)	Q(2)	Q(3)		SUMMARY OUTPUT					
2	1	1	4009	1	1	0	0							
3		2	4321	2	0	1	0		*Regression Statistics*					
4		3	4224	3	0	0	1		Multiple R	0.84271				
5		4	3944	4	0	0	0		R Square	0.71016				
6	2	1	4123	5	1	0	0		Adjusted R Square	0.63287				
7		2	4522	6	0	1	0		Standard Error	166.073				
8		3	4657	7	0	0	1		Observations	20				
9		4	4030	8	0	0	0							
10	3	1	4493	9	1	0	0		ANOVA					
11		2	4806	10	0	1	0			*df*	*SS*	*MS*	*F*	*Significance F*
12		3	4551	11	0	0	1		Regression	4	1013668.775	253417	9.18831	0.00058519
13		4	4485	12	0	0	0		Residual	15	413705.775	27580.4		
14	4	1	4595	13	1	0	0		Total	19	1427374.55			
15		2	4799	14	0	1	0							
16		3	4417	15	0	0	1				*Coefficients*	*Standard Error*	*t Stat*	*P-value*
17		4	4258	16	0	0	0		Intercept	3929.98	108.2666321	36.299	4.9E-16	
18	5	1	4245	17	1	0	0		Time	26.6687	6.564628821	4.06249	0.00102	
19		2	4900	18	0	1	0		Q(1)	123.006	106.8644149	1.15105	0.26773	
20		3	4585	19	0	0	1		Q(2)	472.938	105.8514592	4.46794	0.00045	
21		4	4533	20	0	0	0		Q(3)	263.469	105.2390058	2.50353	0.02434	
22														

As you will note, the *p value* for the overall relationship, 0.00058519, indicates a significant relationship. In addition, the *p value* for the trend term, 0.00102, suggests there is a significant trend component. However, one of the three seasonal factors is not significant. However, it is necessary to either use all the seasonal factors or none of them so the resulting model is

$$\hat{Y} = 3929.98 + 26.6687t + 123.006Q_1 + 472.938Q_2 + 263.469Q_3$$

It can be used to compute forecasts for the historical data and in order to measure how well the model fits the time series. The following figure provides three measures of the overall error, MSE, MAD, and ME or forecast bias. In addition it presents forecasts for the next four quarters.

	A	B	C	D	E	F	G	H	I	J	K	L
1	Year	Quarter	Shipments	Time	Q(1)	Q(2)	Q(3)	Forecast	Error	Abs. Error		
2	1	1	4009	1	1	0	0	4080	-71	71		
3		2	4321	2	0	1	0	4456	-135	135		
4		3	4224	3	0	0	1	4273	-49	49		
5		4	3944	4	0	0	0	4037	-93	93		
6	2	1	4123	5	1	0	0	4186	-63	63		
7		2	4522	6	0	1	0	4563	-41	41		
8		3	4657	7	0	0	1	4380	277	277		
9		4	4030	8	0	0	0	4143	-113	113		
10	3	1	4493	9	1	0	0	4293	200	200		
11		2	4806	10	0	1	0	4670	136	136		
12		3	4551	11	0	0	1	4487	64	64		
13		4	4485	12	0	0	0	4250	235	235		
14	4	1	4595	13	1	0	0	4400	195	195		
15		2	4799	14	0	1	0	4776	23	23		
16		3	4417	15	0	0	1	4593	-176	176		
17		4	4258	16	0	0	0	4357	-99	99		
18	5	1	4245	17	1	0	0	4506	-261	261		
19		2	4900	18	0	1	0	4883	17	17		
20		3	4585	19	0	0	1	4700	-115	115		
21		4	4533	20	0	0	0	4463	70	70		
22	6	1		21	1	0	0	4613	0.00	121.7	= MAD	
23		2		22	0	1	0	4990	Bias =			
24		3		23	0	0	1	4807				
25		4		24	0	0	0	4570				
26						MSE =		20685.3				
27												

Seasonal Autoregression

The seasonal autoregression model will include a trend and a quarterly seasonal effect. Accordingly, the two independent variables will be the dependent variable lagged by one time period for trend and lagged by four time periods for seasonal. The results follow.

	A	B	C	D	E	F	G	H	I	J	K	L
1	Year	Quarter	Shipments	Lagged 1	Lagged 4		SUMMARY OUTPUT					
2	1	1	4009									
3		2	4321	4009				Regression Statistics				
4		3	4224	4321			Multiple R	0.60014				
5		4	3944	4224			R Square	0.36016				
6	2	1	4123	3944	4009		Adjusted R Squa	0.26173				
7		2	4522	4123	4321		Standard Error	208.339				
8		3	4657	4522	4224		Observations	16				
9		4	4030	4657	3944							
10	3	1	4493	4030	4123		ANOVA					
11		2	4806	4493	4522			df	SS	MS	F	Significance F
12		3	4551	4806	4657		Regression	2	317625	158812	3.65885	0.054884364
13		4	4485	4551	4030		Residual	13	564266	43405.1		
14	4	1	4595	4485	4493		Total	15	881891			
15		2	4799	4595	4806							
16		3	4417	4799	4551			Coefficients	ndard Er	t Stat	P-value	
17		4	4258	4417	4485		Intercept	2098.69	1097.4	1.91246	0.0781	
18	5	1	4245	4258	4595		X Variable 1	0.01672	0.1987	0.08418	0.9342	
19		2	4900	4245	4799		X Variable 2	0.53003	0.2036	2.60312	0.02188	
20		3	4585	4900	4417							
21		4	4533	4585	4258							
22												

The *p value* for the trend, 0.9342, is not significant. Accordingly, the model is re-run with just the seasonal variable with the following results.

	A	B	C	D	E	F	G	H	I	J	K
1	Year	Quarter	Shipments	Lagged 4		SUMMARY OUTPUT					
2	1	1	4009								
3		2	4321				Regression Statistics				
4		3	4224			Multiple R	0.59985				
5		4	3944			R Square	0.35981				
6	2	1	4123	4009		Adjusted R Square	0.31409				
7		2	4522	4321		Standard Error	200.815				
8		3	4657	4224		Observations	16				
9		4	4030	3944							
10	3	1	4493	4123		ANOVA					
11		2	4806	4522			df	SS	MS	F	Significance F
12		3	4551	4657		Regression	1	317317.362	317317	7.86867	0.014038265
13		4	4485	4030		Residual	14	564573.575	40326.7		
14	4	1	4595	4493		Total	15	881890.938			
15		2	4799	4806							
16		3	4417	4551			Coefficients	Standard Error	t Stat	P-value	
17		4	4258	4485		Intercept	2155.12	837.414913	2.57354	0.02209	
18	5	1	4245	4595		X Variable 1	0.53417	0.19042827	2.80511	0.01404	
19		2	4900	4799							
20		3	4585	4417							
21		4	4533	4258							
22											

Both the *p value* for the seasonal term and for the overall relationship are improved. The resulting fourth-order autoregression model is

$$Y_t = 2155.12 + 0.53417Y_{t-4}$$

The *p value* of 0.0140 suggests a significant relationship. The following provides the measures of overall and error and forecasts for the next year.

	A	B	C	D	E	F	G	H
1	Year	Quarter	Shipments	Lagged 4	Forecast	Error	Abs. Error	
2	1	1	4009					
3		2	4321					
4		3	4224					
5		4	3944					
6	2	1	4123	4009	4297	-174	174	
7		2	4522	4321	4463	59	59	
8		3	4657	4224	4411	246	246	
9		4	4030	3944	4262	-232	232	
10	3	1	4493	4123	4358	135	135	
11		2	4806	4522	4571	235	235	
12		3	4551	4657	4643	-92	92	
13		4	4485	4030	4308	177	177	
14	4	1	4595	4493	4555	40	40	
15		2	4799	4806	4722	77	77	
16		3	4417	4551	4586	-169	169	
17		4	4258	4485	4551	-293	293	
18	5	1	4245	4595	4610	-365	365	
19		2	4900	4799	4719	181	181	
20		3	4585	4417	4515	70	70	
21		4	4533	4258	4430	103	103	
22	6	1		4245	4423	0.00	165.5	= MAD
23		2		4900	4773	Bias =		
24		3		4585	4604			
25		4		4533	4577			
26				MSE =	35285.8			
27								

Winters' Exponential Smoothing

For Winters' model, we set the initial values for E_t, T_t, and the first year's S_ts as shown in the following results. Experimentation with values for α, β, and γ resulted in identifying $\alpha = 0.3$, $\beta = 0.1$, and $\gamma = 0.3$ as having low values for the three error measures.

	A	B	C	D	E	F	G	H	I	J	K	L	M
1			WINTERS' MODEL		$\alpha = 0.3$		$\beta = 0.1$		$\gamma = 0.3$				
2	Year	Quarter	Year	X_t	E_t	T_t	S_t	F_t	Error	Abs.Error			
3	1	1	1	4009	4009.0	0.00	1.000						
4		2	2	4321	4102.6	9.36	1.000	4009.0	312.0	312.0			
5		3	3	4224	4145.6	12.72	1.000	4112.0	112.0	112.0			
6		4	4	3944	4094.0	6.29	1.000	4158.3	-214.3	214.3			
7	2	1	5	4123	4107.1	6.97	1.001	4100.3	22.7	22.7			
8		2	6	4522	4236.5	19.21	1.020	4114.1	407.9	407.9			
9		3	7	4657	4376.1	31.25	1.019	4255.7	401.3	401.3			
10		4	8	4030	4294.1	19.93	0.982	4407.3	-377.3	377.3			
11	3	1	9	4493	4366.2	25.14	1.010	4319.1	173.9	173.9			
12		2	10	4806	4487.1	34.73	1.035	4480.1	325.9	325.9			
13		3	11	4551	4504.8	33.02	1.017	4609.0	-58.0	58.0			
14		4	12	4485	4547.3	33.96	0.983	4454.1	30.9	30.9			
15	4	1	13	4595	4572.4	33.08	1.008	4624.9	-29.9	29.9			
16		2	14	4799	4614.2	33.95	1.037	4768.8	30.2	30.2			
17		3	15	4417	4557.2	24.86	1.002	4725.1	-308.1	308.1			
18		4	16	4258	4507.0	17.35	0.972	4504.1	-246.1	246.1			
19	5	1	17	4245	4430.2	7.94	0.993	4561.2	-316.2	316.2			
20		2	18	4900	4524.5	16.57	1.051	4601.7	298.3	298.3			
21		3	19	4585	4551.0	17.57	1.004	4551.8	33.2	33.2			
22		4	20	4533	4597.8	20.49	0.976	4438.4	94.6	94.6			
23	6	1	21					4586.7	36.5				
24		2	22					4873.9	Bias =				
25		3	23					4677.4					
26		4	24					4566.6					
27							MSE =	59118.2	MAD =	199.6			

Sheet1 / Sheet2 / **Sheet4** / Sheet5 / Sheet3

Setting the initial values for T_t and the first year's S_ts about equal to the final values from the above results, yields the following.

	A	B	C	D	E	F	G	H	I	J	K	L	M
1			WINTERS' MODEL		$\alpha = 0.3$		$\beta = 0.1$		$\gamma = 0.3$				
2	Year	Quarter	Year	X_t	E_t	T_t	S_t	F_t	Error	Abs.Error			
3	1	1	1	4009	4009.0	20.00	0.980						
4		2	2	4321	4054.9	22.59	1.050	4230.5	90.6	90.6			
5		3	3	4224	4121.4	26.98	1.000	4077.5	146.5	146.5			
6		4	4	3944	4123.7	24.51	0.970	4024.0	-80.0	80.0			
7	2	1	5	4123	4165.9	26.28	0.983	4065.2	57.8	57.8			
8		2	6	4522	4226.5	29.71	1.056	4401.8	120.2	120.2			
9		3	7	4657	4376.5	41.74	1.019	4256.2	400.8	400.8			
10		4	8	4030	4339.1	33.83	0.958	4285.6	-255.6	255.6			
11	3	1	9	4493	4432.4	39.78	0.992	4298.2	194.8	194.8			
12		2	10	4806	4495.9	42.15	1.060	4722.5	83.5	83.5			
13		3	11	4551	4516.2	39.96	1.016	4625.3	-74.3	74.3			
14		4	12	4485	4594.3	43.78	0.963	4363.1	121.9	121.9			
15	4	1	13	4595	4636.1	43.58	0.992	4601.7	-6.7	6.7			
16		2	14	4799	4634.1	39.03	1.053	4959.9	-160.9	160.9			
17		3	15	4417	4575.7	29.28	1.001	4746.9	-329.9	329.9			
18		4	16	4258	4549.7	23.75	0.955	4435.6	-177.6	177.6			
19	5	1	17	4245	4485.4	14.95	0.978	4536.1	-291.1	291.1			
20		2	18	4900	4546.8	19.59	1.060	4737.0	163.0	163.0			
21		3	19	4585	4571.1	20.06	1.001	4569.3	15.7	15.7			
22		4	20	4533	4637.8	24.72	0.962	4384.6	148.4	148.4			
23	6	1	21					4560.9	8.8				
24		2	22					4969.0	Bias =				
25		3	23					4718.4					
26		4	24					4555.4					
27							MSE =	33948.0	MAD =	153.6			

Sheet1 / Sheet2 / **Sheet4** / Sheet5 / Sheet3

Time Series Decomposition

The time series decomposition results, computed elsewhere, are comparable to those for the above three methods. For example, decomposition resulted in a trend term of 24.494 whereas our seasonal multiple regression model yielded 26.68 and Winters' model yielded 24.72. The decomposition approach identified seasonal indexes of 0.985, 1.059, 1.005, and 0.951. Winters' model resulted in seasonal factors of 0.978, 1.060, 1.001, and 0.962.

The values for MSE, MAD and forecast bias are 21,218.01, 125.2, and 0.48, respectively. The forecasts for the next four quarters are 4682, 4707, 4731, and 4756.

Forecasts for Next Four Quarters

The seasonal multiple regression model had the lowest values for all three error measures. Accordingly, the forecasts for the *best* model are 4613, 4990, 4807, and 4570 for the next year. The following figure shows the actual time series data indicated with the diamond-shaped symbols. The solid line shows the forecasts.

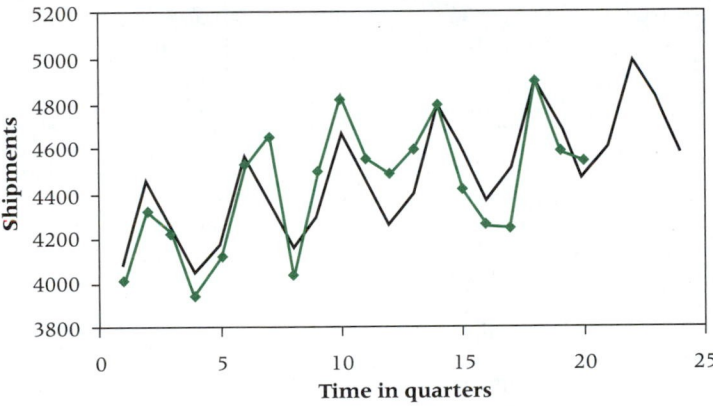

14.21 The U.S. Department of Agriculture publishes statistics on the production of various types of food commodities by month. Shown here are the production figures on broccoli for January 1996 through December 1997. Use these data to analyze the effects of seasonality and trend.

14.6
Problems

MONTH	BROCCOLI (1 MILLION POUNDS)	MONTH	BROCCOLI (1 MILLION POUNDS)
January (1996)	132.5	January (1997)	104.9
February	164.8	February	99.3
March	141.2	March	102.0
April	133.8	April	122.4
May	138.4	May	112.1
June	150.9	June	108.4
July	146.6	July	119.0
August	146.9	August	119.0
September	138.7	September	114.9
October	128.0	October	106.0
November	112.4	November	111.7
December	121.0	December	112.3

14.22 The U.S. Department of Commerce publishes census information on manufacturing. Included in these figures are monthly shipment data for the paperboard container and box industry shown here for 6 years. The shipment figures are given in millions of dollars. Use the data to analyze the effects of seasonality and trend.

MONTH	SHIPMENTS		MONTH	SHIPMENTS
January (year 1)	1891		January (year 4)	2336
February	1986		February	2474
March	1987		March	2546
April	1987		April	2566
May	2000		May	2473
June	2082		June	2572
July	1878		July	2336
August	2074		August	2518
September	2086		September	2454
October	2045		October	2559
November	1945		November	2384
December	1861		December	2305

MONTH	SHIPMENTS		MONTH	SHIPMENTS
January (year 2)	1936		January (year 5)	2389
February	2104		February	2463
March	2126		March	2522
April	2131		April	2417
May	2163		May	2468
June	2346		June	2492
July	2109		July	2304
August	2211		August	2511
September	2268		September	2494
October	2285		October	2530
November	2107		November	2381
December	2077		December	2211

MONTH	SHIPMENTS		MONTH	SHIPMENTS
January (year 3)	2183		January (year 6)	2377
February	2230		February	2381
March	2222		March	2268
April	2319		April	2407
May	2369		May	2367
June	2529		June	2446
July	2267		July	2341
August	2457		August	2491
September	2524		September	2452
October	2502		October	2561
November	2314		November	2377
December	2277		December	2277

14.23 Use Winters' model to exponentially smooth the data given in Problem 14.7. Use $\alpha = .5$, $\beta = .3$, and $\gamma = .7$. Measure the error using MAD.

Summary

Forecasting for business and industry is a necessary activity for the efficient functioning of organizations; however, forecasting is not an exact science. The purpose of this chapter is to provide a somewhat structured approach to forecasting. Although such an approach will not result in *perfect* predictions regarding the future, it usually will result in predictions that are more accurate than *guesstimates*.

Forecasting techniques are classified as qualitative and quantitative. Qualitative methods rely on *expert* opinion to establish a forecast of future conditions. Such an approach is necessary when historical data for the process being predicted is either scarce or nonexistent. These approaches provide a structure for combining multiple opinions into a consistent forecast.

Quantitative methods rely on *historical* data for forecasting. These methods are divided into two categories: time series methods and causal methods. Time series methods are the primary topic of this chapter. These are techniques that attempt to forecast future values for a variable based solely on past values for the variable. Causal methods, on the other hand, attempt to find relationships between the variable to be predicted and other variables thought to influence the forecast variable. A summary of the characteristics of the qualitative techniques and the two categories of quantitative techniques is presented in Table 14.2.

A time series represents data values for a variable recorded over time at regular intervals. The analysis of a time series begins with a time series plot. This is a graph of the variable values on the vertical axis and time on the horizontal axis. Such a presentation oftentimes provides a visual identification of the make-up of the time series. In particular, the time series data may appear to be made up of a trend component, a seasonal component, a cyclical component, and an irregular component. Whereas the trend is a long range overall increase or decrease in the time series values, the seasonal variation is change within a one-year period that is repeated each year. Cyclical variation is also a repeating pattern that is completed in a period of two or more years. Oftentimes these are business or economic cycles that have an effect on the forecast variable. The irregular variation represents all the changes in the time series values that are not attributable to the trend, seasonal, or cyclical components.

In an attempt to help explain the values of a time series, two models are oftentimes used. One of these is called the additive model because it assumes the four components are added together to provide the value for the forecast variable. The second model is termed the multiplicative model. It combines the four components by multiplying the four together. It assumes the trend provides the basic value for the time series and that the other three components' contributions are deviations from the trend.

Our approach to forecasting consists of four steps: (1) identification of the time series form, (2) selection of the methods suitable for the time series form, (3) evaluation of the suitable methods to identify the *best* method, and (4) forecasting future values for the variable with the *best* method.

A time series that appears to include only an irregular component is called a stationary time series. The values for it appear to be random variations about a constant average value. A number of forecasting techniques are designed to be effectively forecast stationary time series data. Those presented within this chapter include the naïve approach, moving averages, weighted moving averages, and simple exponential smoothing.

Forecasting methods suitable for non-stationary time series data divide into two classes. First are those suitable for time series data that appear to include a trend component in addition to an irregular component. Second are those for time series data that include a seasonal, a trend, and an irregular component, or that include just a seasonal and an irregular component. The trend forecasting methods considered within this chapter include linear trend projection, non-linear trend projection, an autoregression model for trend, and Holt's form of exponential smoothing. On the other hand, the seasonal methods include a multiple regression model with additive seasonal components, autoregression model for seasonal and trend or for only seasonal, Winters' form of exponential smoothing, and time series decomposition. Figure 14.17 summarizes the use of these twelve forecasting approaches.

Key Terms

additive model	mean square error (MSE)
autocorrelation	moving average
autoregression	multiplicative model
averaging models	naïve forecasting method
causal techniques	non-linear trend projection
cyclical component	non-stationary time series
damping factor	percentage error (PE)
deseasonalized time series	quadratic regression model
Durbin-Watson test	qualitative techniques
error of an individual forecast	quantitative techniques
exponential smoothing	seasonal autoregression
first-differences approach	seasonal component
folded annual time series plot	seasonal multiple regression
forecast bias	simple average
forecasting	simple average model
forecasting error	stationary time series
Holt's model	time series
irregular component	time series decomposition
irregular variations	time series plot
linear trend projection	time series techniques
mean absolute deviation (MAD)	trend autoregression
mean absolute percentage error (MAPE)	trend component
mean error (ME)	weighted moving average
mean percentage error (MPE)	Winters' model

SUPPLEMENTARY PROBLEMS

14.24 Following are the average yields of long-term new corporate bonds over a several-month period published by the Office of Market Finance of the U.S. Department of the Treasury.

MONTH	YIELD	MONTH	YIELD
1	10.08	13	7.91
2	10.05	14	7.73
3	9.24	15	7.39
4	9.23	16	7.48
5	9.69	17	7.52
6	9.55	18	7.48
7	9.37	19	7.35
8	8.55	20	7.04
9	8.36	21	6.88
10	8.59	22	6.88
11	7.99	23	7.17
12	8.12	24	7.22

a. Explore trends in these data by using regression trend analysis. How strong are the models? Is the quadratic model significantly stronger than the linear trend model?

b. Use a 4-month moving average to forecast values for each of the ensuing months.

c. Use simple exponential smoothing to forecast values for each of the ensuing months. Let $\alpha = .3$ and then let $\alpha = .7$. Which weight produces better forecasts?

d. Compute MAD for the forecasts obtained in (b) and (c) and compare the results.

14.25 Following are data on the quantity (million pounds) of the U.S. domestic fishing catch for human food from 1980 through 1997. The data are published by the U.S. National Oceanic and Atmosphere Administration.

a. Use a 3-year moving average to forecast the quantity of fish for the years 1983 through 1997 for these data. Compute the error of each forecast and then determine the mean absolute deviation of error for the forecast.

b. Use exponential smoothing and $\alpha = .2$ to forecast the data from 1983 through 1997. Let the forecast for 1981 equal the actual value for 1980. Compute the error of each forecast and then determine the mean absolute deviation of error for the forecast.

c. Compare the results obtained in (a) and (b) by using MAD. Which technique seems to perform better? Why?

YEAR	QUANTITY	YEAR	QUANTITY
1980	3,654	1989	6,204
1981	3,547	1990	7,041
1982	3,285	1991	7,031
1983	3,238	1992	7,618
1984	3,320	1993	8,214
1985	3,294	1994	7,936
1986	3,393	1995	7,667
1987	3,946	1996	7,474
1988	4,588	1997	7,248

14.26 The U.S. Department of Commerce publishes a series of census documents referred to as *Current Industrial Reports.* Included in these documents are the *Manufacturers' Shipments, Inventories, and Orders: 1991–1995.* Displayed here is a portion of these data representing the shipments of chemicals and allied products from January 1991 through December 1995. Use seasonal multiple regression to forecast shipments for the next year.

TIME PERIOD	CHEMICALS AND ALLIED PRODUCTS ($BILLION)	TIME PERIOD	CHEMICALS AND ALLIED PRODUCTS ($BILLION)
January (1991)	23.701	July	24.821
February	24.189	August	25.560
March	24.200	September	27.218
April	24.971	October	25.650
May	24.560	November	25.589
June	24.992	December	25.370
July	22.566	January (1994)	25.316
August	24.037	February	26.435
September	25.047	March	29.346
October	24.115	April	28.983
November	23.034	May	28.424
December	22.590	June	30.149
January (1992)	23.347	July	26.746
February	24.122	August	28.966
March	25.282	September	30.783
April	25.426	October	28.594
May	25.185	November	28.762
June	26.486	December	29.018
July	24.088	January (1995)	28.931
August	24.672	February	30.456
September	26.072	March	32.372
October	24.328	April	30.905
November	23.826	May	30.743
December	24.373	June	32.794
January (1993)	24.207	July	29.342
February	25.772	August	30.765
March	27.591	September	31.637
April	26.958	October	30.206
May	25.920	November	30.842
June	28.460	December	31.090

14.27 The National Cable Television Association publishes data on the cable television market. Shown here are "the number of basic cable subscribers" and "as percentage of household with TVs" from the years 1976 to 1997. Develop a regression model to predict the number of basic cable subscribers from the variable "as percentage of households with TVs" using these data. Use this model to predict the number of basic cable subscribers if the value of the variable "as percentage of households with TVs" is 55%. Discuss the strength of the regression model.

YEAR	BASIC CABLE SUBSCRIBERS	AS PERCENTAGE OF HOUSEHOLDS WITH TVs
1976	10,787,970	15.1
1977	12,168,450	16.6
1978	13,391,910	17.9
1979	14,814,380	19.4
1980	17,671,490	22.6
1981	23,219,200	28.3
1982	29,340,570	35.0
1983	34,113,790	40.5
1984	37,290,870	43.7
1985	39,872,520	46.2
1986	42,237,140	48.1
1987	44,970,880	50.5
1988	48,636,520	53.8
1989	52,564,470	57.1
1990	54,871,330	59.0
1991	55,786,390	60.6
1992	57,211,600	61.5
1993	58,834,440	62.5
1994	60,483,600	63.4
1995	62,956,470	65.7
1996	64,654,180	66.7
1997	65,929,420	67.3

14.28 The U.S. Bureau of Labor Statistics releases consumer price indexes (CPIs) for selected items in the publication *Monthly Labor Review.* Shown here are the CPIs for apparel and upkeep for the years 1983 through 1998. Use the data to answer the following questions.

   a. Compute a 4-year moving average to forecast the CPIs from 1987 through 1998.
   b. Compute a 4-year weighted moving average to forecast the CPIs from 1987 through 1998. Weight the most recent year by 4, the next most recent year by 3, the next year by 2, and the last year of the four by 1.
   c. Determine the errors for (a) and (b). Compute MSE for (a) and (b). Compare the MSE values and comment on the effectiveness of the moving average versus the weighted moving average for these data.

YEAR	APPAREL AND UPKEEP
1983	100.2
1984	102.1
1985	105.0
1986	105.9
1987	110.6
1988	115.4
1989	118.6
1990	124.1
1991	128.7
1992	131.9
1993	133.7
1994	133.4
1995	132.0
1996	131.7
1997	132.9
1998	133.0

14.29 Use the data and the regression model developed in Problem 14.27 to compute a Durbin-Watson test to determine whether significant autocorrelation is present. Let $\alpha = .05$.

14.30 In the *Survey of Current Business,* the U.S. Department of Commerce publishes data on farm commodity prices. Given are the cotton prices from November of year 1 through February of year 4. The prices are indexes with a base of 100 from the period of 1910 through 1914. Use these data to develop autoregression models for a 1-month lag and a 4-month lag. Compare the results of these two models. Which model seems to yield better predictions?

TIME PERIOD	COTTON PRICES	TIME PERIOD	COTTON PRICES
November (year 1)	552	January (year 3)	571
December	519	February	573
January (year 2)	505	March	582
February	512	April	587
March	541	May	592
April	549	June	570
May	552	July	560
June	526	August	565
July	531	September	547
August	545	October	529
September	549	November	514
October	570	December	469
November	576	January (year 4)	436
December	568	February	419

14.31 The Board of Governors of the Federal Reserve System publishes data on mortgage debt outstanding by type of property and holder. Below are the amounts of residential nonfarm debt (in billions of dollars) held by savings institutions in the United States over a 10-year period. Use these data to develop an autoregression model with a 1-period lag. Discuss the strength of the model.

YEAR	DEBT
1	529
2	554
3	559
4	602
5	672
6	669
7	600
8	538
9	490
10	470

14.32 The U.S. Department of Commerce publishes data on industrial machinery and equipment. Shown here are the shipments (in billions of dollars) of industrial machinery and equipment from the first quarter of year 1 through the fourth quarter of year 6. Use these data to determine the seasonal indexes for the data through time-series decomposition methods. Use seasonal multiple regression to forecast.

TIME PERIOD	INDUSTRIAL MACHINERY AND EQUIPMENT SHIPMENTS
1st quarter (year 1)	54.019
2nd quarter	56.495
3rd quarter	50.169
4th quarter	52.891
1st quarter (year 2)	51.915
2nd quarter	55.101
3rd quarter	53.419
4th quarter	57.236
1st quarter (year 3)	57.063
2nd quarter	62.488
3rd quarter	60.373
4th quarter	63.334
1st quarter (year 4)	62.723
2nd quarter	68.380
3rd quarter	63.256
4th quarter	66.446
1st quarter (year 5)	65.445
2nd quarter	68.011
3rd quarter	63.245
4th quarter	66.872
1st quarter (year 6)	59.714
2nd quarter	63.590
3rd quarter	58.088
4th quarter	61.443

14.33 The data shown here, from the Investment Company Institute, show that the equity fund assets of mutual funds have been growing since 1981. At the same time, the assets of mutual funds in taxable money markets have been increasing since 1980. Use these data to develop a regression model to forecast the equity fund assets by the taxable money market assets. All figures are given in billion-dollar units. Conduct a Durbin-Watson test on the data and the regression model to determine whether significant autocorrelation is present. Let $\alpha = .01$.

YEAR	EQUITY FUNDS	TAXABLE MONEY MARKETS
1980	44.4	74.5
1981	41.2	181.9
1982	53.7	206.6
1983	77.0	162.5
1984	83.1	209.7
1985	116.9	207.5
1986	161.5	228.3
1987	180.7	254.7
1988	194.8	272.3
1989	249.0	358.7
1990	245.8	414.7
1991	411.6	452.6
1992	522.8	451.4
1993	749.0	461.9
1994	866.4	500.4
1995	1269.0	629.7
1996	1750.9	761.8
1997	2399.3	898.1

14.34 The purchasing-power value figures for the minimum wage in 1997 dollars for the years 1980 through 1997 are shown here. Use these data and exponential smoothing to develop forecasts for the years 1981 through 1997. Try $\alpha = .1$, $.5$, and $.8$, and compare the results using MAPE. Discuss your findings. Select the value of alpha that worked best and use your exponential smoothing results to predict the figure for 1998.

YEAR	PURCHASING POWER	YEAR	PURCHASING POWER
1980	$6.04	1989	$4.34
1981	5.92	1990	4.67
1982	5.57	1991	5.01
1983	5.40	1992	4.86
1984	5.17	1993	4.72
1985	5.00	1994	4.60
1986	4.91	1995	4.48
1987	4.73	1996	4.86
1988	4.55	1997	5.15

14.35 Shown here is the Excel output for a regression analysis to predict the number of business bankruptcy filings over a 16-year period by the number of consumer bankruptcy filings. How strong is the model? Note the residuals. Compute a Durbin-Watson statistic from the data and discuss the presence of autocorrelation in this model.

Regression Statistics	
Multiple R	0.53
R Square	0.28
Adjusted R Square	0.23
Standard Error	8179.8
Observations	16

ANOVA

	df	SS	MS	F	Significance F
Regression	1	364069877.4	3.64E+08	5.44	0
Residual	14	936737379.6	66909813		
Total	15	1300807257			

	Coefficients	Standard Error	t Stat	P-value
Intercept	75532.436	4980.088	15.17	4.4E-10
Consumer Bankruptcies	−0.016	0.007	−2.33	0.035

Observation	Predicted Business Bankruptcies	Residuals
1	70638.58	−1338.58
2	71024.28	−8588.28
3	71054.61	−7050.61
4	70161.99	1115.01
5	68462.72	12772.28
6	67733.25	14712.75
7	66882.45	−3029.45
8	65834.05	−2599.05
9	64230.61	622.39
10	61801.70	9747.30
11	61354.16	9288.84
12	62738.76	−434.76
13	63249.36	−10875.36
14	61767.01	−9808.01
15	57826.69	−4277.69
16	54283.80	−256.80

ANALYZING THE DATABASES

1. Use the agricultural time-series database and the variable Green Beans to forecast the number of green beans for period 169 by using the following techniques.
   a. Five-period moving average
   b. Simple exponential smoothing with $\alpha = .6$
   c. Time-series linear trend model
   d. Seasonal multiple regression

2. Conduct a seasonal multiple regression on Carrots in the agricultural database. What are the seasonal indexes? These data actually represent 14 years of 12-month data. Do the seasonal indexes indicate the presence of some seasonal effects? Run one autoregression model to predict Carrots by a 1-month lag and another by a 12-month lag. Compare the two models. Because vegetables are somewhat seasonal, is the 12-month lag model significant?

3. Use the energy database to forecast 1999 U.S. Coal Production by using simple exponential smoothing of previous U.S. Coal Production data. Let $\alpha = .2$ and .8. Compare the forecast with the actual figure. Which of the three models produces the figure with the least error?

4. Use the international labor database to develop a regression model to predict the Unemployment Rate for Germany by the Unemployment Rate of Italy. Test for autocorrelation and discuss its presence or absence in this regression analysis.

5. Use seasonal multiple regression and the stock market database to explore trends and cycles in the Composite index. Is there any seasonability in this variable of stock market data?

CASE

DEBOURGH MANUFACTURING COMPANY

The DeBourgh Manufacturing Company was founded in 1909 as a metal-fabricating company in Minnesota. In the 1980s, the company ran into hard times, as did the rest of the metal-fabricating industry. Among the problems that DeBourgh faced were declining sales, deteriorating labor relations, and increasing costs. Labor unions had resisted cost-cutting measures. Losses were piling up in the heavy job-shop fabrication division, which was the largest of the company's three divisions. A division that made pedestrian steel bridges closed in 1990. The remaining company division, producer of All-American lockers, had to move to a lower-cost environment.

In 1990, with the company's survival at stake, the firm made a risky decision and moved everything from its high-cost location in Minnesota to a lower-cost area in La Junta, Colorado. Eighty semitrailer trucks were used to move equipment and inventory 1000 miles at a cost of $1.2 million. The company was relocated to a building in La Junta that had stood vacant for three years. Only 10 of the Minnesota workers transferred with the company, which quickly hired and trained 80 more workers in La Junta. By moving to La Junta, the company was able to go nonunion.

DeBourgh also faced a financial crisis. A bank that had been loaning the company money for 35 years would no longer do so. In addition, a costly severance package was worked out with Minnesota owners to keep production going during the move. An internal stock-purchase "earnout" was arranged between company president Steven C. Berg and his three aunts, who were the other principal owners.

The roof of the building that was to be the new home of DeBourgh Manufacturing in La Junta was badly in need of repair. During the first few weeks of production, heavy rains fell on the area and production was all but halted. However, DeBourgh was able to overcome these obstacles. One year later, locker sales achieved record-high sales levels each month. The company is now more profitable than ever with sales of over $8,000,000 in 1998. Much credit has been given to the positive spirit of teamwork fostered among its approximately 100 employees. Emphasis has shifted to employee involvement in decision making, quality, teamwork, employee participation in compensation action, and shared profits. In addition, DeBourgh became a more socially responsible company by doing more for the town in which it is located and by using paints that are more environmentally friendly.

DISCUSSION

1. After its move in 1990 to La Junta, Colorado, and its new initiatives, the DeBourgh Manufacturing Company began an upward climb of record sales. Suppose the figures shown here are the DeBourgh monthly sales figures from January 1991 through December 1999 (in $1000s). Are there any trends in the data? Does DeBourgh have a seasonal component to its sales? Shown after the sales figures is Excel output for a seasonal multiple regression analysis of the sales figures using a 12-month seasonality. Examine the data, the output, and any additional analysis you feel is helpful, and write a short report on DeBourgh sales. Include a discussion of the general direction of sales, any seasonal tendencies, and any cycles that might be occurring.

MONTH	1991	1992	1993	1994	1995	1996	1997	1998	1999
January	139.7	165.1	177.8	228.6	266.7	431.8	381.0	431.8	495.3
February	114.3	177.8	203.2	254.0	317.5	457.2	406.4	444.5	533.4
March	101.6	177.8	228.6	266.7	368.3	457.2	431.8	495.3	635.0
April	152.4	203.2	279.4	342.9	431.8	482.6	457.2	533.4	673.1
May	215.9	241.3	317.5	355.6	457.2	533.4	495.3	558.8	749.3
June	228.6	279.4	330.2	406.4	571.5	622.3	584.2	647.7	812.8
July	215.9	292.1	368.3	444.5	546.1	660.4	609.6	673.1	800.1
August	190.5	317.5	355.6	431.8	482.6	520.7	558.8	660.4	736.6
September	177.8	203.2	241.3	330.2	431.8	508.0	508.0	609.6	685.8
October	139.7	177.8	215.9	330.2	406.4	482.6	495.3	584.2	635.0
November	139.7	165.1	215.9	304.8	393.7	457.2	444.5	520.7	622.3
December	152.4	177.8	203.2	292.1	406.4	431.8	419.1	482.6	622.3

	A	B	C	D	E	F	G	H	I	J	K	L	M	N	O	P
1			**DeBourgh Manufacturing Company Sales**													
2																
3	Year	Month	Sales	Time	M1	M2	M3	M4	M5	M6	M7	M8	M9	M10	M11	
4	1991	Jan.	139.7	1	1	0	0	0	0	0	0	0	0	0	0	
5		Feb.	114.3	2	0	1	0	0	0	0	0	0	0	0	0	
6		Mar	101.6	3	0	0	1	0	0	0	0	0	0	0	0	
7		Apr	152.4	4	0	0	0	1	0	0	0	0	0	0	0	
8		May	215.9	5	0	0	0	0	1	0	0	0	0	0	0	
9		Jun	228.6	6	0	0	0	0	0	1	0	0	0	0	0	
10		Jul	215.9	7	0	0	0	0	0	0	1	0	0	0	0	
11		Aug	190.5	8	0	0	0	0	0	0	0	1	0	0	0	
12		Sep	177.8	9	0	0	0	0	0	0	0	0	1	0	0	
13		Oct	139.7	10	0	0	0	0	0	0	0	0	0	1	0	
14		Nov	139.7	11	0	0	0	0	0	0	0	0	0	0	1	
15		Dec	152.4	12	0	0	0	0	0	0	0	0	0	0	0	
16	1992	Jan	165.1	13	1	0	0	0	0	0	0	0	0	0	0	
17		Feb	177.8	14	0	1	0	0	0	0	0	0	0	0	0	
18		Mar	177.8	15	0	0	1	0	0	0	0	0	0	0	0	
19		Apr	203.2	16	0	0	0	1	0	0	0	0	0	0	0	
20		May	241.3	17	0	0	0	0	1	0	0	0	0	0	0	
21		Jun	279.4	18	0	0	0	0	0	1	0	0	0	0	0	
22		Jul	292.1	19	0	0	0	0	0	0	1	0	0	0	0	
23		Aug	317.5	20	0	0	0	0	0	0	0	1	0	0	0	
24		Sep	203.2	21	0	0	0	0	0	0	0	0	1	0	0	
25		Oct	177.8	22	0	0	0	0	0	0	0	0	0	1	0	
26		Nov	165.1	23	0	0	0	0	0	0	0	0	0	0	1	
27		Dec	177.8	24	0	0	0	0	0	0	0	0	0	0	0	
28	1993	Jan	177.8	25	1	0	0	0	0	0	0	0	0	0	0	
29		Feb	203.2	26	0	1	0	0	0	0	0	0	0	0	0	

	A	B	C	D	E	F
114	SUMMARY OUTPUT					
115						
116	*Regression Statistics*					
117	Multiple R	0.97792				
118	R Square	0.95633				
119	Adjusted R Square	0.95081				
120	Standard Error	38.6384				
121	Observations	108				
122						
123	ANOVA					
124		*df*	*SS*	*MS*	*F*	*Significance F*
125	Regression	12	3105740.503	258811.7086	173.3590613	0.00000
126	Residual	95	141827.6733	1492.922877		
127	Total	107	3247568.176			
128						
129		*Coefficients*	*Standard Error*	*t Stat*	*P-value*	
130	Intercept	50.8	14.75527	3.44284	0.00086	
131	Time	5.05648	0.12000	42.13832	0.00000	
132	M1	3.41019	18.26206	0.18674	0.85227	
133	M2	19.5204	18.25378	1.06939	0.28760	
134	M3	42.6861	18.24629	2.33944	0.02141	
135	M4	81.3741	18.23958	4.46140	0.00002	
136	M5	117.24	18.23366	6.42986	0.00000	
137	M6	174.272	18.22852	9.56041	0.00000	
138	M7	183.327	18.22418	10.05954	0.00000	
139	M8	138.759	18.22062	7.61551	0.00000	
140	M9	71.6139	18.21785	3.93097	0.00016	
141	M10	41.1574	18.21588	2.25942	0.02614	
142	M11	13.5231	18.21469	0.74243	0.45966	

	B	C	D	E	F	G	H	I	J	K	L	M	N	O	P	Q	R	S
100	Jan	495.3	97	1	0	0	0	0	0	0	0	0	0	0	544.689	-49.389	49.389	
101	Feb	533.4	98	0	1	0	0	0	0	0	0	0	0	0	565.856	-32.456	32.456	
102	Mar	635	99	0	0	1	0	0	0	0	0	0	0	0	594.078	40.922	40.922	
103	Apr	673.1	100	0	0	0	1	0	0	0	0	0	0	0	637.822	35.278	35.278	
104	May	749.3	101	0	0	0	0	1	0	0	0	0	0	0	678.744	70.556	70.556	
105	Jun	812.8	102	0	0	0	0	0	1	0	0	0	0	0	740.833	71.967	71.967	
106	Jul	800.1	103	0	0	0	0	0	0	1	0	0	0	0	754.944	45.156	45.156	
107	Aug	736.6	104	0	0	0	0	0	0	0	1	0	0	0	715.433	21.167	21.167	
108	Sep	685.8	105	0	0	0	0	0	0	0	0	1	0	0	653.344	32.456	32.456	
109	Oct	635	106	0	0	0	0	0	0	0	0	0	1	0	627.944	7.056	7.056	
110	Nov	622.3	107	0	0	0	0	0	0	0	0	0	0	1	605.367	16.933	16.933	
111	Dec	622.3	108	0	0	0	0	0	0	0	0	0	0	0	596.900	25.400	25.400	
112															Forecasts			
113	Jan		109	1	0	0	0	0	0	0	0	0	0	0	**605.367**	0.00	30.21 = MAD	
114	Feb		110	0	1	0	0	0	0	0	0	0	0	0	**626.533**	Bias =		
115	Mar		111	0	0	1	0	0	0	0	0	0	0	0	**654.756**			
116	Apr		112	0	0	0	1	0	0	0	0	0	0	0	**698.500**			
117	May		113	0	0	0	0	1	0	0	0	0	0	0	**739.422**			
118	Jun		114	0	0	0	0	0	1	0	0	0	0	0	**801.511**			
119	Jul		115	0	0	0	0	0	0	1	0	0	0	0	**815.622**			
120	Aug		116	0	0	0	0	0	0	0	1	0	0	0	**776.111**			
121	Sep		117	0	0	0	0	0	0	0	0	1	0	0	**714.022**			
122	Oct		118	0	0	0	0	0	0	0	0	0	1	0	**688.622**			
123	Nov		119	0	0	0	0	0	0	0	0	0	0	1	**666.044**			
124	Dec		120	0	0	0	0	0	0	0	0	0	0	0	**657.578**			
125														MSE =	1313.22			
126																		

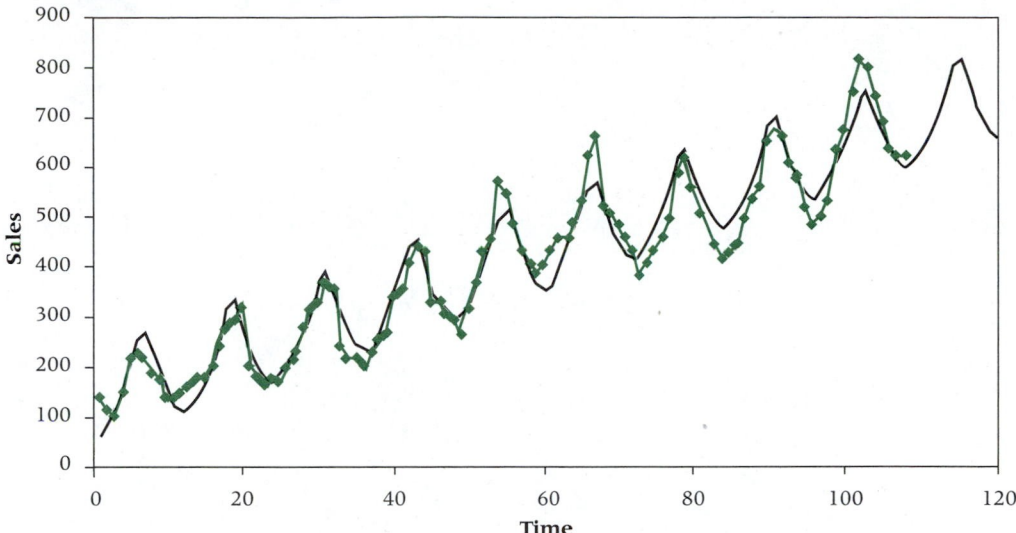

2. Suppose DeBourgh accountants have been able to compute a per-unit cost of lockers for each year since 1988, as reported here. Use techniques in this chapter to analyze the data. Forecast the per-unit labor costs through the year 2000. Use smoothing techniques, moving averages, trend analysis, and any others that seem appropriate. Calculate the error of the forecasts and determine which forecasting method seems to do the best job of minimizing error. Study the data and explain the behavior of the per-unit labor cost since 1988. Think about the company history and objectives since 1988.

YEAR	PER-UNIT LABOR COST
1988	$80.15
1989	85.29
1990	85.75
1991	64.23
1992	63.70
1993	62.54
1994	60.19
1995	59.84
1996	57.29
1997	58.74
1998	55.01
1999	56.20

ADAPTED FROM: "DeBourgh Manufacturing Company: A Move That Saved a Company," *Real-World Lessons for America's Small Businesses: Insights from the Blue Chip Enterprise Initiative.* Published by *Nation's Business* magazine on behalf of Connecticut Mutual Life Insurance Company and the U.S. Chamber of Commerce in association with the Blue Chip Enterprise Initiative, 1992. See also DeBourgh's homepage, http://www.debourgh.com, and the website containing Colorado Springs top business stories located at http://www.csbj.com/1998/981113/top_stor.htm.

15
Statistical Quality Control

Learning Objectives

This chapter presents basic concepts in quality control, with a particular emphasis on statistical quality control techniques, thereby enabling you to:

1. Understand the concepts of quality, quality control, and total quality management.

2. Understand the importance of statistical quality control in total quality management.

3. Learn about process analysis and some process analysis tools, including Pareto charts, fishbone diagrams, and control charts.

4. Learn how to construct $\bar{X}$ charts, R charts, P charts, and c charts.

What is quality? Quality means different things to different people. If you asked commuters whether or not their automobiles have quality, the response would vary according to each individual's perspective. One person's view of a quality automobile is one that goes 75,000 miles without needing any major repair work. Other people perceive automobile quality as comfortable seats and extra electronic gadgetry. These people look for "bells and whistles" along with form-fitting, cushy seats in a quality car. Still other automobile consumers define automobile quality as the presence of numerous safety features.

In this chapter, we examine various definitions of quality and discuss some of the main concepts of quality and quality control. We explore some techniques for analyzing processes. Finally, we learn how to construct and interpret control charts.

15.1
Introduction to Quality Control

Quality
When a product delivers what is stipulated in its specifications.

Transcendent quality
A view of quality that implies that a product has an innate excellence, uncompromising standards, and high achievement.

Product quality
A view of quality in which quality is measurable in the product based on the fact that there are perceived differences in products and quality products possess more attributes.

User quality
A view of quality in which the quality of the product is determined by the user.

Manufacturing quality
A view of quality in which the emphasis is on the manufacturer's ability to target consistently the requirements for the product with little variability.

There are almost as many definitions of quality as there are people and products. Assume, however, that in the marketplace buyer and seller agree on the specifications of a product. One definition of **quality,** then, is *when a product delivers what is stipulated for it in its specifications.*

From this point of view, quality is present when the producer delivers what has been specified in the product description, as agreed upon by both buyer and seller.

Most automobile purchasers agree that a Lexus or a Cadillac has quality. If a buyer agrees to purchase a Saturn with few extra accessories, and if the Saturn is delivered to the buyer in the condition specified, the Saturn is a quality car even if it is not a Lexus or a Cadillac.

There are other definitions of quality. Philip B. Crosby, author of *Quality Is Free* and *Quality Without Tears,* and a well-known expert on quality, has said that "quality is conformance to requirements."[*] The product requirements must be met by the producer to assure quality. This notion of quality is similar to the one based on specifications. Armand V. Feigenbaum, a well-known quality authority, says in his book *Total Quality Control* that "quality is a customer determination" as opposed to management's determination or a designer's determination.[†] He says that this determination is based on the customer's experience with the product or service and that it is always a moving target.

Another authority on quality control is David A. Garvin, author of *Managing Quality.* Garvin advocates defining quality as having at least five dimensions: transcendent, product, user, manufacturing, and value.[‡] **Transcendent quality** *implies that a product has an "innate excellence."* It has *"uncompromising standards and high achievement."* Garvin says that this definition offers little practical guidance to business people. **Product quality** means that quality is measurable in the product. *There are perceived differences in products, and quality products have more attributes.* For example, a baseball glove with more stitching has more quality. A hard-disk-drive personal computer with more memory has more quality. Tires with more tread have more quality.

User quality means that *the quality of a product is determined by the consumer.* This dimension of quality is closest to the definitions of quality based on customer requirements and product specifications. **Manufacturing quality** derives mainly from engineering and manufacturing practices. Once the specifications are determined, *quality is measured by the manufacturer's ability to target the requirements consistently with little variability.* The fifth dimension of quality according to Garvin is **value quality,** which *has to do with price and costs*—that is, did the customer get his or her money's worth?

---

[*]Philip B. Crosby. *Quality Without Tears.* New York, McGraw-Hill Book Company, 1984.

[†]Armand V. Feigenbaum. *Total Quality Control,* 3d ed. New York, McGraw-Hill Book Company, 1991.

[‡]David A. Garvin. *Managing Quality.* New York, The Free Press, a division of Macmillan, Inc., 1988.

What Is Quality Control?

How does a company know whether it is producing a quality product? One way is to practice quality control. **Quality control** (sometimes referred to as quality assurance) is *the collection of strategies, techniques, and actions taken by an organization to assure itself that it is producing a quality product.*

The process begins at the product planning and design phase, where attributes of the product are determined and specified. Each attribute of the product is a potential contributor to overall product quality. For quality control to be possible, measurable attributes and specifications must be established against which the actual attributes of the product can be compared.

Quality control can be undertaken in two distinct ways: after-process control and in-process control. **After-process quality control** involves *inspecting the attributes of a finished product to determine whether the product is acceptable, is in need of rework, or is to be rejected and scrapped.* The after-process quality control method was the leading quality control technique for American manufacturers for several decades until the 1980s. The emphasis in the after-process method is on *weeding out* defective products before they reach the consumer. The problem with this method is that it does not generate information that can correct in-process problems or raw-materials problems. Two main outcomes of the after-process methodology are (1) reporting the number of defects produced during a specific period of time and (2) screening defective products from consumers. Because American companies dominated world markets in many areas for several decades during and after World War II, their managers had little interest in changing from the after-process method.

However, as Japan, other Asian nations, and Western European countries began to compete strongly with the United States in the world market in the late 1970s and 1980s, U.S. companies began to reexamine quality control methods. As a result, many U.S. companies, following the example of Japanese and European manufacturers, developed quality control programs based on in-process control. **In-process quality control** techniques *measure product attributes at various intervals throughout the manufacturing process* in an effort to pinpoint problem areas. This information enables quality control personnel in conjunction with production personnel to make corrections in operations as products are being made. This intervention in turn opens the door to opportunities for improving the process and the product.

Total Quality Management

W. Edwards Deming, who has been referred to as the "father of the quality movement," advocated that the achievement of quality is an organic phenomenon that begins with top managers' commitment and extends all the way to suppliers on one side and consumers on the other. Deming believed that quality control is a long-term total company effort. The effort called for by Deming is **total quality management (TQM).** Total quality management involves all members of the organization—from the CEO to the line worker—in improving quality. In addition, the goals and objectives of the organization come under the purview of quality control and can be measured in quality terms. Suppliers, raw materials, worker training, and opportunity for workers to make improvements all are part of total quality management. The antithesis of total quality management is when a company gives a quality control department total responsibility for improving product quality.

Deming presented a cause-and-effect explanation of the impact of total quality management on a company. This has become known as the Deming chain reaction.* The chain

---

*W. Edwards Deming. *Out of the Crisis.* Cambridge, MA, Massachusetts Institute of Technology Center for Advanced Engineering Study, 1986.

Value quality
A view of quality having to do with price and costs and whether the consumer got his or her money's worth.

Quality control
The collection of strategies, techniques, and actions taken by an organization to ensure the production of quality products.

After-process quality control
A type of quality control in which product attributes are measured by inspection after the manufacturing process is completed to determine whether the product is acceptable.

In-process quality control
A quality control method in which product attributes are measured at various intervals throughout the manufacturing process.

Total quality management (TQM)
A program that occurs when all members of an organization are involved in improving quality; all goals and objectives of the organization come under the purview of quality control and are measured in quality terms.

reaction begins with improving quality. Improving quality will decrease costs because of less reworking, fewer mistakes, fewer delays and snags, and better use of machine time and materials. From the reduced costs comes an improvement in productivity because

$$\text{Productivity} = \frac{\text{Output}}{\text{Input}}.$$

A reduction of costs generates more output for less input and, hence, increases productivity. As productivity improves, a company is more able to capture the market with better quality and lower prices. This enables a company to stay in business and provide more jobs.

Deming listed 14 points which, if followed, can lead to improved total quality management.*

1. Create constancy of purpose for improvement of product and service.
2. Adopt the new philosophy.
3. Cease dependence on mass inspection.
4. End the practice of awarding business on price tag alone.
5. Improve constantly and forever the system of production and service.
6. Institute training.
7. Institute leadership.
8. Drive out fear.
9. Break down barriers between staff areas.
10. Eliminate slogans.
11. Eliminate numerical quotas.
12. Remove barriers to pride of workmanship.
13. Institute a vigorous program of education and retraining.
14. Take action to accomplish the transformation.

The first point indicates the need to seek constant improvement in process, innovation, design, and technique. The second point suggests that to truly make changes, a new, positive point of view must be taken; in other words, the viewpoint that poor quality is acceptable must be changed. The third point is a call for change from after-process inspection to in-process inspection. Deming pointed out that after-process inspection has nothing to do with improving the product or the service. The fourth point indicates that a company should be careful in awarding contracts to suppliers and vendors: Purchasers should look more for quality and reliability in a supplier than for just low price. Deming called for long-term supplier relationships in which the company and supplier agree on quality standards.

Point 5 is a statement that quality is not a one-time activity. Management and labor should be constantly on the lookout for ways to improve the product. Institute training, the sixth point, implies that training is an essential element in total quality management. Workers need to learn how to do their jobs correctly and learn techniques that will result in higher quality. Point 7, institute leadership, is a call for a new management based on showing, doing, and supporting rather than ordering and punishing. The eighth point results in establishing a "safe" work environment, where workers feel free to share ideas and make suggestions without the threat of punitive measures. Point 9, breaking down barriers, means reducing competition and conflicts between departments and groups. It is a call for more of a team approach—the notion that "we're all in this together."

---

*Mary Walton. *The Deming Management Method*. New York, Perigee Books, published by The Putnam Publishing Group, 1986.

Deming did not believe that slogans help effect quality products. Point 10 addresses the fact that a quality control program is not a movement of slogans. Point 11 is a statement that quotas do not help companies make quality products. In fact, pressure to make quotas can result in inefficiencies, errors, and lack of quality. Point 12 says that managers must find ways to make it easier for workers to produce quality products; faulty equipment and poor-quality supplies do not allow workers to take pride in what is produced. Point 13 is a call for total reeducation and training within a company about new methods and understandings that can result in a higher quality product. Point 14 implies that rhetoric is not the answer; there must be a call for action in order to institute change and promote higher quality.

Some Important Quality Concepts

Of the several widely used techniques in quality control, five in particular warrant discussion: benchmarking, just-in-time inventory systems, reengineering, six sigma, and team building.

BENCHMARKING One practice that has been increasingly used by American companies to improve quality is benchmarking. **Benchmarking** is *a method in which a company attempts to develop and establish total quality management from product to process by examining and emulating the best practices and techniques used in its industry or other industries.* The ultimate objective of benchmarking is to use a positive, proactive process to make changes that will effect superior performance. The process of benchmarking involves studying competitors and learning from the best in the industry.

One of the American pioneers in what is called "competitive benchmarking" was Xerox. Xerox was struggling to hold on to its market share against foreign competition. At one point, other companies could sell a machine for what it cost Xerox to make a machine. Xerox set out to find out why. The company instituted a benchmarking process in which the internal workings and features of competing machines were studied in depth. Xerox attempted to emulate and learn from the best of these features in developing its own products. In time, benchmarking was so successful within the company that top managers included benchmarking as a major corporate effort.*

JUST-IN-TIME INVENTORY SYSTEMS Another technique used to improve quality control is the just-in-time system for inventory, which focuses on raw materials, subparts, and suppliers. Ideally, a **just-in-time inventory system** means that *no extra raw materials or inventory of parts for production are stored.* Necessary supplies and parts needed for production arrive "just in time." The advantage of this system is that holding costs, personnel, and space needed to manage inventory are reduced. Even within the production process, as subparts are assembled and merged, the just-in-time philosophy can be applied to smooth the process and eliminate bottlenecks.

A production facility is unlikely to become 100% just-in-time. One of the residual effects of installing a just-in-time system throughout the production process is that, as the inventory "fat" is trimmed from the production process, the pressure on the system to produce often discloses problems that had gone undetected. For example, one subpart being made on two machines may not be produced in enough quantity to supply the next step. Installation of the just-in-time system shows that this station is a bottleneck. The company might choose to add another machine to produce more subparts, change the production schedule, or develop another strategy. As the bottleneck is loosened and the

Benchmarking
A quality control method in which a company attempts to develop and establish total quality management from product to process by examining and emulating the best practices and techniques used in their industry or other industries.

Just-in-time inventory system
An inventory system in which little or no extra raw materials or parts for production are stored.

---

*Robert C. Camp. *Benchmarking.* Milwaukee, WI, Quality Press, ASQC, 1989.

problem is corrected, other areas of weakness may emerge. Thus, the residual effect of a just-in-time inventory system can be the opportunity for production managers to work their way methodically through a maze of previously unidentified problems that otherwise would not normally be recognized.

A just-in-time inventory system can change the relationship between supplier and producer. Most companies using this system have fewer suppliers than they did before installing the system. The tendency is for manufacturers to give suppliers longer contracts under the just-in-time system. However, the suppliers are expected to produce raw materials and subparts to a specified quality and to deliver the goods as near to just in time as possible. Just-in-time suppliers may even build production or warehouse facilities next to the producer's. In the just-in-time system, the suppliers become part of total quality management.

Reengineering
A radical approach to total quality management in which the core business process of a company is redesigned.

REENGINEERING   A more radical approach to improving quality is reengineering. Whereas total quality approaches like Deming's 14 points call for continuous improvement, **reengineering** is *the complete redesigning of the core business process in a company*. It involves innovation and is often a complete departure from the company's usual way of doing business.

Reengineering is not a fine-tuning of the present process nor is it mere downsizing of a company. Reengineering starts with a blank sheet of paper and an idea about where the company would like to be in the future. Without considering the present limitations or constraints of the company, the reengineering process works backward from where the company wants to be in the future and then attempts to determine what it would take to get there. From this information, the company cuts or adds, reshapes, or redesigns itself to achieve the new goal. In other words, the reengineering approach involves determining what the company would be like if it could start from scratch and then redesigning the process to make it work that way.

Reengineering affects almost every functional area of the company, including information systems, financial reporting systems, the manufacturing environment, suppliers, shipping, and maintenance. Reengineering is usually painful and difficult for a company. Companies that have been most successful in implementing reengineering are those facing big shifts in the nature of competition and needing to make major changes to stay in business.

Some recommendations to consider in implementing reengineering in a company are to (1) get the strategy straight first, (2) lead from the top, (3) create a sense of urgency, (4) design from the outside in, (5) manage the firm's consultant, and (6) combine top-down and bottom-up initiatives. Getting the strategy straight is crucial because the strategy drives the changes. The company must determine what business it wants to be in and how to make money in it. The company's strategy determines its operations.

Reengineering involves cross-functional operation. It must be led by top managers who have the authority and the leadership capacity to oversee, direct, and facilitate the implementation of tough decisions. Because internal political pressure or the satisfaction of making small gains can bog down the process, a sense of urgency must be created for reengineering to be successful and to sustain the energy needed to complete the changes.

The focus of reengineering is outside the company; the process begins with the customer. Current operations may have some merit, but time is spent determining the need of the marketplace and how to meet that need.

Although the leadership for reengineering must come from the top, employees must buy into the changes through participation. Often, line workers have useful ideas and suggestions that can make the company more profitable and productive. Hence, there must be both top-down and bottom-up participation in the process.*

---

*THIS SECTION ADAPTED FROM: Thomas A. Stewart, "Reengineering: The Hot New Managing Tool," *Fortune*, August 23, 1993, 40–48. Copyright © 1993 Time Inc. All rights reserved.

SIX SIGMA   Currently, a very popular approach to total quality management is six sigma. Six sigma is both a methodology and a measurement. Originally developed in the electronics industry, **six sigma** *measures the capability of a process to perform defect-free work, where a defect is defined as anything that results in customer dissatisfaction.* Six sigma is derived from a previous quality scheme in which a process was considered to be producing quality results if $\pm 3\sigma$ or 99.74% of the products or attributes were within specification. (Note: The standard normal distribution table, Table A.5, produces an area of .4987 for a Z score of 3. Doubling that and converting to a percentage yields 99.74%, which is the portion of a normal distribution that falls within $\mu \pm 3\sigma$.) Six sigma methodology requires that $\pm 6\sigma$ of the product be within specification. The probability associated with a Z score of 6 is .49999983. Doubling this and converting to a percentage yields 99.99966%. Thus the goal of six sigma methodology is to have 99.99966% of the product or attributes be within specification, or no more than .00034% = .0000034 out of specification. This means that no more than 3.4 of the product or attributes per million can be defective. Essentially, this calls for the process to approach a defect-free status.

Why six sigma? There are several reasons for the adoption of a six sigma philosophy. First, in some industries the three sigma philosophy is simply unacceptable. For example, the three sigma goal of having 99.74% of the product or attribute be in specification in the prescription drug industry implies that it is acceptable to have .26% incorrectly filled prescriptions, or 2600 out of every million prescriptions filled. In the airline industry, the three sigma goal implies that it is acceptable to have 2600 unsatisfactory landings by commercial aircraft out of every million landings. In contrast, a six sigma approach would require that there be no more than 3.4 incorrectly filled prescriptions or 3.4 unsatisfactory landings per million, with a goal of approaching zero.

A second reason for adopting a six sigma approach is that it forces companies who adopt it to work much harder and more quickly to discover and reduce sources of variation in processes. It "raises the bar" of the quality goals of a firm, causing the company to place even more emphasis on continuous quality improvement. A third reason is that six sigma dedication to quality may be required to attain world-class status and be a top competitor in the international market.

Pande, Neuman, and Cavanagh suggest that there are six themes of six sigma. Theme one is a *genuine focus on the customer*. With six sigma, the customer is the top priority, and "the measures of Six Sigma performance begin with the customer." Theme two is *data- and fact-driven management*. Under a six sigma philosophy, decision-making is based on data and information instead of opinions and assumptions. Theme three is a *process focus, management, and improvement*. Under six sigma, breakthroughs are made on the process. Thus, there is a focus on measurement, improving the efficiency, and designing goods and services in light of the process or operation used to produce the product or service. Theme four is *proactive management* meaning that management acts in advance of events rather than after events. This includes defining and reviewing goals, setting clear priorities, and focusing on prevention. Theme five is *boundaryless collaboration,* which means breaking down barriers and improving teamwork. Such collaboration is encouraged in all directions in the organization as well as with customers and suppliers. Theme six is a *drive for perfection; tolerance for failure.* While it seems that these two are contradictory, under six sigma, they are complementary. In order to move toward perfection, risks have to be taken and management needs to be able to accept setbacks that can come with such risk taking.*

The first step in implementing six sigma is to determine the number of opportunities or chances to perform an operation successfully for each component. More complex parts

Six sigma
A total quality management approach that measures the capability of a process to perform defect-free work, where a defect is defined as anything that results in customer dissatisfaction.

---

* Pande, Peter S.; Robert P. Neuman; and Roland R. Cavanagh. *The Six Sigma Way.* New York: McGraw-Hill. 2000.

or products contain larger opportunity counts. The occurrence of defects is characterized by the defects pet unit (dpu). The dpu is then normalized by the total number of opportunities yielding defects per million opportunities, which is then related to the level of sigma using the normal distribution.*

Implementing a six sigma approach to quality will likely require a company to commit additional resources to quality improvement. One such resource is an in-house person who receives special training in six sigma and is referred to as a "black belt." Six sigma black belts are in-company experts who teach and implement six sigma ideas and tools within the company. Arising from various disciplines within the firm, they serve as change agents who lead improvement projects and interact with management on improvement plans.

TEAM BUILDING   In the past, the traditional business approach to decision making in the United States was to let managers decide what is best for the company and act upon that decision. In recent years, the U.S. business culture has been changing to include more team building approaches to decision making. **Team building** occurs *when a group of employees are organized as an entity to undertake management tasks and perform other functions such as organizing, developing, and overseeing projects.*

The result of team building is that more and more workers are taking over managerial responsibilities. There are fewer lines of demarcation between worker and manager and between union and nonunion. Workers are invited to work on a par with managers to remove obstacles that prevent a company from delivering a quality product. The old "us and them" point of view is being replaced by a cooperative relationship between managers and workers in reaching common goals under team building.

People on teams often represent different business functions such as design, production, marketing, and finance. Working together, team members are able to incorporate input from a variety of viewpoints; their decisions result in more comprehensive thinking, more effective time utilization, and fewer instances of mistaken direction due to failure to consider all factors. For example, teams are organized to design, develop, and produce new products; to oversee the building of new plants; and to reorganize and restructure office layouts for more efficient operation, among other things.

One particular type of team that was introduced to American companies by the Japanese is the quality circle. A **quality circle** is *a small group of workers,* usually from the same department or work area, and their supervisor, *who meet regularly to consider quality issues.* The size of the group ranges from 4 to 15 members, and they meet as often as once a week.† The meetings are usually on company time and members of the circle are compensated. The supervisor may be the leader of the circle, but the members of the group determine the agenda and reach their own conclusions.

Because a great amount of brainstorming occurs in quality circles, employees need to feel that they will not be penalized for ideas shared. Each person is to be respected as an integral, important part of the team. Because employees may hold negative attitudes toward such groups—remembering similar meetings in the past—a few meetings may be necessary before a quality circle becomes effective. Eventually, if the participants realize that the quality circle is a serious, useful endeavor, they may begin to offer information about specific items in the production process that need attention. It is essential, however, that top managers support such efforts and provide the impetus for these groups. The quality circle movement has failed more often than not because of lack of support from top managers.

Team building
When a group of employees are organized as an entity to undertake management tasks and perform other functions such as organizing, developing, and overseeing projects.

Quality circle
A small group of workers consisting of supervisors and 4 to 15 employees who meet frequently and regularly to consider quality issues in their department or area of the business.

---

*THIS SECTION ADAPTED FROM: "Six-Sigma Analysis: A Route to Quality and Affordability," University of Delaware Center for Composite Materials' website at http://www.ccm.udel.edu/publications/tech_briefs/114.html and "Program Background," American Society of Quality (ASQ)'s website at http://www.asq.org/products/sigma/.

†Philip C. Thompson. *Quality Circles: How to Make Them Work in America.* New York, AMACOM, 1982.

Quality circles and other teams tend to elevate workers from a level of "just installing their part" to a level of "analyst." The team concept has potential for generating communication and cooperation in a company. The concepts of team building and quality circles have been introduced in various types of businesses, ranging from hospitals to chemical production facilities.

Much of what transpires in the business world involves processes. A **process** is *"a series of actions, changes, or functions that bring about a result."*[*] Processes usually involve the manufacturing, production, assembling, or development of some output from a given input. Generally, in a meaningful system, value is added to the input as part of the process. In the area of production, processes are often the main focus of decision makers. Production processes abound in the chemical, steel, automotive, appliance, computer, furniture, and clothing manufacture industries, as well as many others. Production layouts vary, but it is not difficult to picture an assembly line with its raw materials, parts, and supplies being processed into a finished product that becomes worth more than the sum of the parts and materials that went into it. However, processes are not limited to the area of production. Virtually all other areas of business involve processes. The processing of a check from the moment it is used for a purchase, through the financial institution, and back to the user is one example. The hiring of new employees by a human resources department involves a process that might begin with a job description and end with the training of a new employee. Many different processes occur within healthcare facilities. One process involves the flow of a patient from check-in at a hospital through an operation to recovery and release. Meanwhile, the dietary and foods department prepares food and delivers it to various points in the hospital as part of another process. The patient's paperwork follows still another process, and central supply processes medical supplies from the vendor to the floor supervisor.

Identifying and understanding the role of bottlenecks in a process are key to quality improvement. Finding ways to smooth the flow and shorten the cycle can result in greater productivity and higher quality. For these and other reasons, process analysis is an important component of total quality management. Several diagnostic techniques are available as process analysis tools. Among the more prominent ones are flowcharts, Pareto analysis, cause-and-effect (fishbone) diagrams, and control charts.

15.2
Process Analysis

Process
A series of actions, changes, or functions that bring about a result.

Flowcharts

Very commonly, particularly in nonmanufacturing settings, no one maps out the complete flow of sequential stages of various processes in a business. For example, one NASA subcontractor was responsible for processing the paperwork for change items on space projects. Change requests would begin at NASA and be sent to the subcontractor's building. The requests would be processed there and returned to NASA in about 14 days. Exactly what happened to the paperwork during the two-week period? As part of a quality effort, NASA asked the contractor to study the process. No one had taken a hard look at where the paperwork went, how long it sat on various people's desks, and how many different people handled it. The contractor soon became involved in process analysis.

One of the first activities that should take place in process analysis is the flowcharting of the process from beginning to end. A **flowchart** is *a schematic representation of all the activities and interactions that occur in a process.* It includes decision points, activities, input/output, start/stop, and a flowline. Figure 15.1 displays some of the symbols used in flowcharting.

Flowchart
A schematic representation of all the activities and interactions that occur in a process.

---

[*]*The American Heritage Dictionary of the English Language,* 3d ed. Boston: Houghton Mifflin Company, 1992.

Figure 15.1

Flowchart symbols

SOURCE: G. A. Silver and J. B. Silver. *Introduction to Systems Analysis,* Englewood Cliffs, NJ, Prentice-Hall, 1976, 142–147.

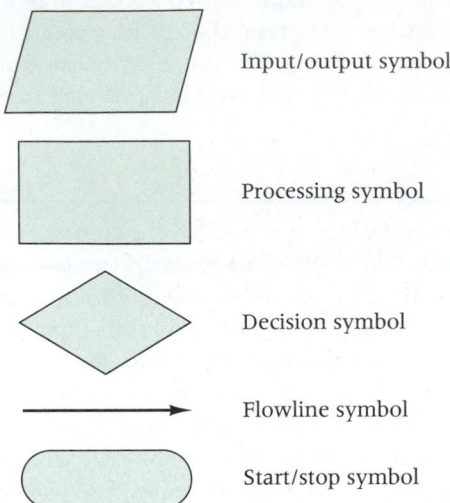

Input/output symbol

Processing symbol

Decision symbol

Flowline symbol

Start/stop symbol

The parallelogram represents input into the process or output from the process. In the case of the dietary/foods department at the hospital, the input includes uncooked food, utensils, plates, containers, and liquids. The output is the prepared meal delivered to the patient's room. The processing symbol is a rectangle that represents an activity. For the dietary/foods department, that activity could include cooking carrots or loading food carts. The decision symbol, a diamond, is used at points in the process where decisions are made that can result in different pathways. In some hospitals, the dietary/foods department supports a hospital cafeteria as well as patient meals. At some point in the process, the decision must be made as to whether the food is destined for a patient room or the cafeteria. The cafeteria food may follow a general menu whereas patient food may have to be individualized for particular health conditions. The arrow is the flowline symbol designating to the flowchart user the sequence of activities of the process. The flowline in the hospital food example would follow the pathway of the food from raw ingredients (vegetables, meat, flour, etc.) to the delivered product in patient rooms or in the cafeteria. The elongated oral represents the starting and stopping points in the process.

As an example, suppose we want to flowchart the process of obtaining a home improvement loan of $10,000 from a bank. The process begins with the customer entering the bank. The flow takes the customer to a receptionist, who poses a decision dilemma. For what purpose has the customer come to the bank? Is it to get information, to cash a check, to deposit money, to buy a money order, to get a loan, or to invest money? Because we are charting the loan process, we follow the flowline to the loan department. The customer arrives in the loan department and is met by another receptionist who asks what type and size of loan the person needs. For small personal loans, the customer is given a form to submit for loan consideration with no need to see a loan officer. For larger loans, such as the home improvement loan, the customer is given a form to fill out and is assigned to see a loan officer. The small personal loans are evaluated and the customer is given a response immediately. If the answer is yes, word of the decision is conveyed to a teller who cuts a check for the customer. For larger loans, the customer is interviewed by a loan officer, who then makes a decision. If the answer is yes, a contract is drawn up and signed. The customer is then sent to a teller who has the check for the loan. Figure 15.2 is a possible flowchart for this scenario.

Figure 15.2   Flowchart of loan process

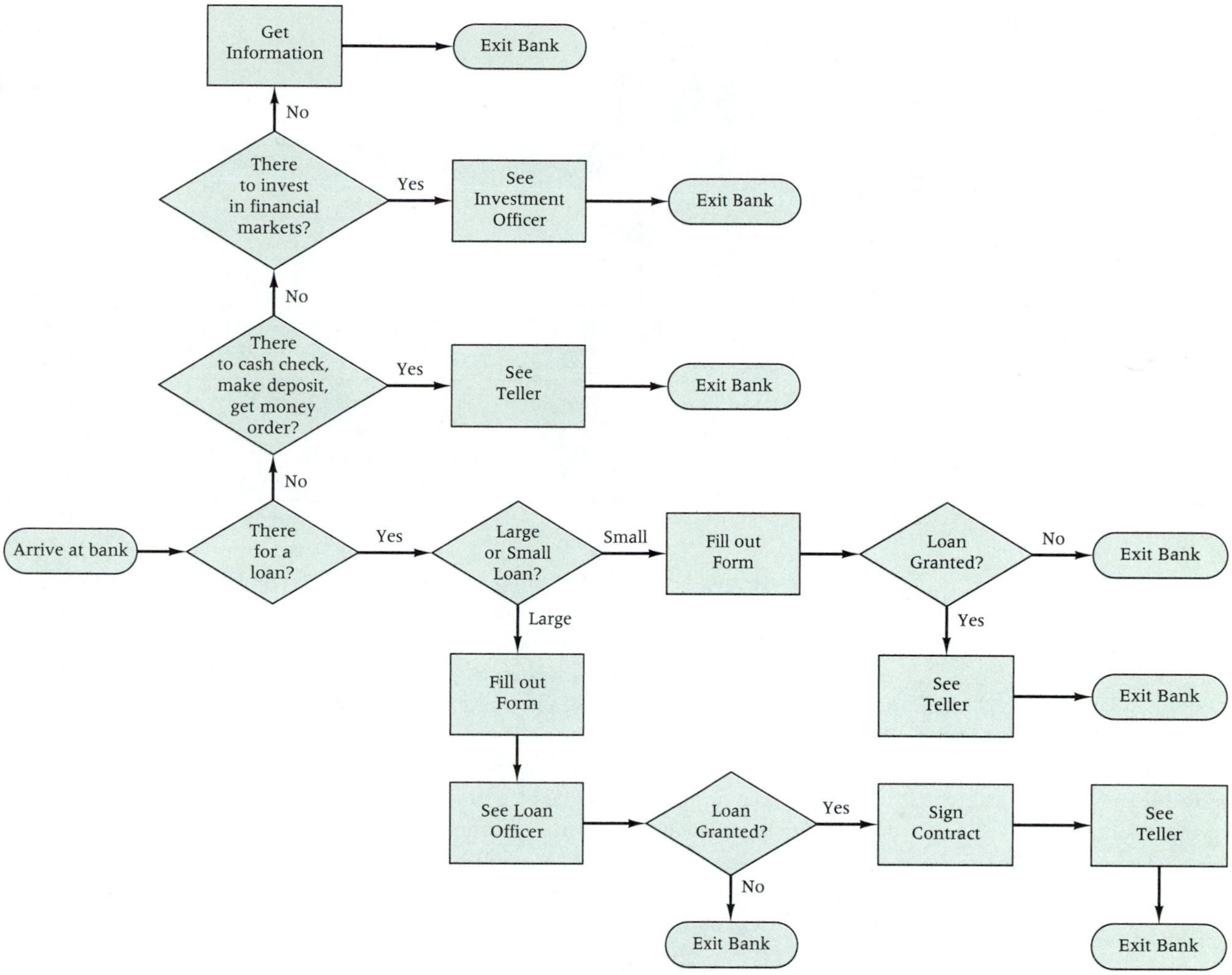

Pareto Analysis

Once the process has been mapped by such techniques as the flowchart, procedures for identifying bottlenecks and problem causes can begin. One technique for displaying problem causes is Pareto analysis. **Pareto analysis** is *a quantitative tallying of the number and types of defects that occur with a product or service.* Analysts use this tally to produce *a vertical bar chart that displays the most common types of defects, ranked in order of occurrence from left to right.* The bar chart is called a **Pareto chart.**

Pareto charts were named after an Italian economist, Vilfredo Pareto, who observed more than 100 years ago that most of Italy's wealth was controlled by a few families. That is, a few key families were the major drivers behind the Italian economy. Quality expert J. M. Juran applied this notion to the quality field by observing that often poor quality can be addressed by attacking a few major causes that result in most of the problems. A Pareto chart enables quality control decision makers to separate the most important defects from trivial defects, which helps them to set priorities for needed quality-improvement work.

Pareto analysis
A quantitative tallying of the number and types of defects that occur with a product or service, often recorded in a Pareto chart.

Pareto chart
A vertical bar chart in which the number and types of defects for a product or service are graphed in order of magnitude from greatest to least.

Figure 15.3

Excel Pareto chart for electric motor problem

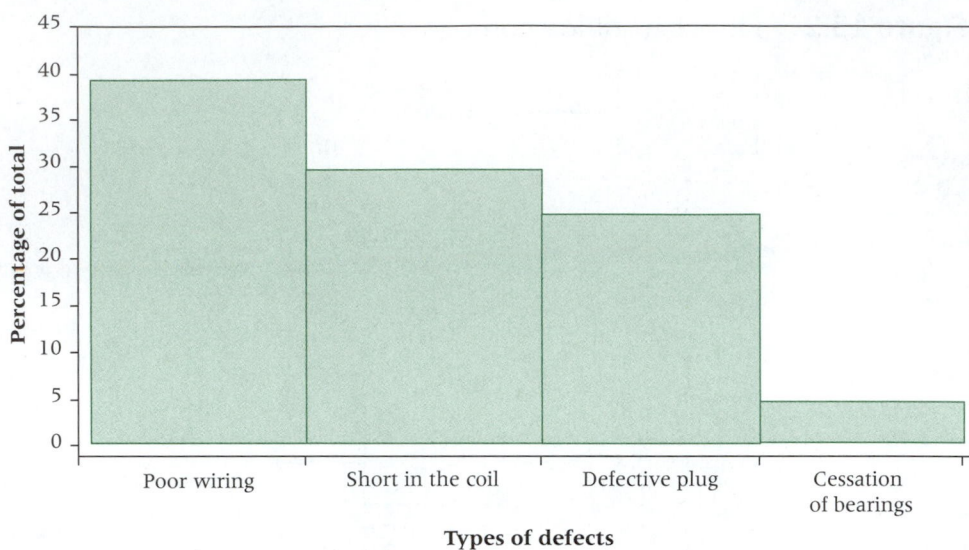

Cause-and-effect diagram
A tool for displaying possible causes for a quality problem and the interrelationships among the causes.

Fishbone diagram
A display of possible causes of a quality problem and the interrelationships among the causes. The problem is diagrammed along the main line of the "fish" and possible causes are diagrammed as line segments angled off in such a way as to give the appearance of a fish skeleton.

Ishikawa diagram
A tool developed by Kaoru Ishikawa as a way to display possible causes of a quality problem and the interrelationships of the causes.

Suppose the number of electric motors being rejected by inspectors for a company has been increasing. Company officials examine the records of several hundred of the motors in which at least one defect was found to determine which defects occurred more frequently. They find that 40% of the defects involved poor wiring, 30% involved a short in the coil, 25% involved a defective plug, and 5% involved cessation of bearings. Figure 15.3 is an Excel-produced Pareto chart constructed from this information. It shows that the main three problems with defective motors—poor wiring, a short in the coil, and a defective plug—account for 95% of the problems. From the Pareto chart, decision makers can formulate a logical plan for reducing the number of defects.

Company officials and workers would probably begin to improve quality by examining the segments of the production process that involve the wiring. Next, they would study the construction of the coil, then examine the plugs used and the plug-supplier process.

Cause-and-Effect (Fishbone) Diagrams

Another tool for identifying problem causes is the **cause-and-effect diagram,** sometimes referred to as **fishbone diagram** or **Ishikawa diagram.** This diagram was developed by Kaoru Ishikawa in the 1940s as a way to *display possible causes of a problem and the interrelationships among the causes.* The causes can be arrived at through brainstorming, investigating, surveying, observing, and other information-gathering techniques.

The name "fishbone diagram" comes from the shape of the diagram, which looks like a fish skeleton with the problem at the head of the fish and possible causes flaring out on both sides of the main "bone." Subcauses can be included along each "fishbone."

Suppose officials at the company producing the electric motor want to construct a fishbone diagram for the poor wiring problem shown as the major problem in Figure 15.3. Some of the possible causes of poor wiring might be raw materials, equipment, workers, or methods. Some possible raw material causes might be vendor problems (and their source of materials), transportation damage, or damage during storage (inventory). Possible causes of equipment failure might be out-of-date equipment, equipment that is out of adjustment, poor maintenance of equipment, or lack of effective tools. Poor wiring might also be the result of worker error, which can include lack of training or improper training, poor attitude, or excessive absenteeism that results in lack of consistency. Methods causes

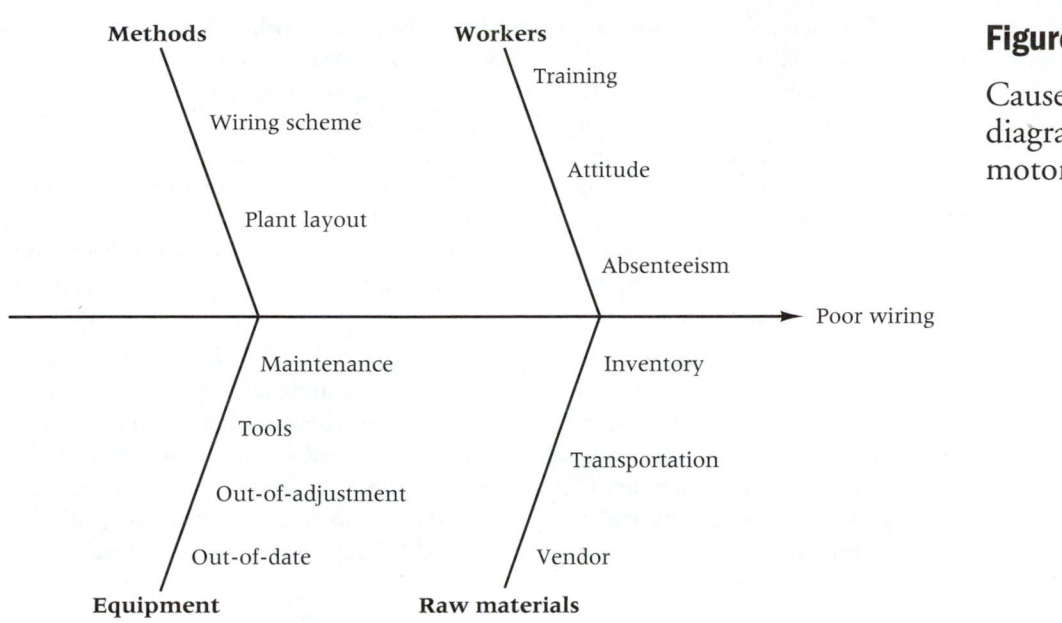

Figure 15.4

Cause-and-effect diagram for electric motor problems

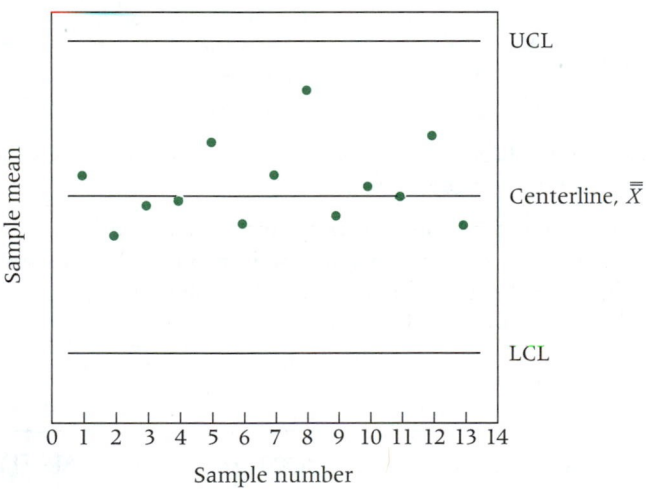

Figure 15.5

$\overline{X}$ control chart

can include poor wiring schemes and inefficient plant layouts. Figure 15.4 is a fishbone diagram of this problem and its possible causes.

Control Charts

A fourth diagnostic technique that has worldwide acceptance is the control chart. According to Armand V. Feigenbaum, a renowned expert on control charts, a **control chart** is *a graphical method for evaluating whether a process is or is not in a "state of statistical control."*[*] There are several kinds of control charts. Figure 15.5 is an $\overline{X}$ control chart. In the next section, we will explore control charts in more detail.

Control chart
A graphical method for evaluating whether a process is or is not in a state of statistical control.

---

*Armand V. Feigenbaum. *Total Quality Control.* New York, McGraw-Hill Book Company, 1991.

15.2
Problems

15.1 For each of the following scenarios, sketch the process with activities and decision points. Construct a flowchart by using the symbols depicted in Figure 15.1.

   a. A customer enters the office of an insurance agent wanting to purchase auto insurance and leaves the office with a paid policy in hand.

   b. A truckload of men's shirts enters a warehouse at the main distribution center for a men's retail clothing company that has four stores in that area. The shirts are on the racks inside the stores ready for sale.

   c. A couple enters a restaurant to enjoy dinner out. An hour and a half later, they leave, satisfied, having paid their bill. Construct the flowchart from the restaurant's point of view.

15.2 An airline company uses a central telephone bank and a semi-automated telephone process to take reservations. It has been receiving an unusually high number of customer complaints about its reservation system. The company conducted a survey of customers, asking them whether they had encountered any of the following problems in making reservations: busy signal, disconnection, poor connection, too long a wait to talk to someone, could not get through to an agent, connected with the wrong person. Suppose a survey of 744 complaining customers resulted in the following frequency tally.

NUMBER OF COMPLAINTS	COMPLAINT
184	Too long a wait
10	Transferred to the wrong person
85	Could not get through to an agent
37	Got disconnected
420	Busy signal
8	Poor connection

Construct a Pareto diagram from this information to display the various problems encountered in making reservations.

15.3 A bank has just sent out a monthly reconciliation statement to a customer with an error in the person's monthly income. Brainstorm to determine some possible causes of this error. Try to think of some possible causes for the causes. Construct a fishbone diagram to display your results.

15.3
Control Charts

Control charts have been in existence for nearly 70 years. Walter A. Shewhart is credited with developing control charts at Bell Laboratories in the 1920s. Shewhart and others, including one of his understudies at Bell Laboratories, W. Edwards Deming, were able to apply control charts to industrial processes as a tool to assist in controlling variation. The use of control charts in the United States failed to gain momentum after World War II because the success of American manufacturers in the world market reduced the apparent need for such a tool. As the Japanese and other international manufacturers became more competitive by using such tools, the control chart increased in popularity in the United States.

Control charts are easy to use and understand. Often it is the line workers who record and plot product measurements on the charts. In more automated settings, sensors record chart values and send them into an information system, which compiles the charts. Control charts are used mainly to monitor product variation. The charts enable operators, technicians, and managers to see when a process gets out of control, which in turn improves quality and increases productivity.

Variation

If there were no variations between manufactured items, there would be no reason for control charts. However, variation occurs for virtually any product or service. Variation can occur among units within a lot and can occur between lots. Among the reasons for product variation are differences in raw materials, differences in workers, differences in machines, changes in the environment, and wear and tear of machinery. Small variations can be caused by unnoticeable events, such as a passing truck that creates vibrations or dust that affects machine operation. Variations need to be measured, recorded, and studied so that out-of-control conditions can be identified and corrections can be made in the process.

Types of Control Charts

The two general types of control charts are (1) control charts for measurements and (2) control charts for attribute compliance items. In this section, we discuss two types of control charts for measurements, $\bar{X}$ charts and R charts. We also discuss two types of control charts for attribute compliance, P charts and c charts.

Each control chart has a **centerline,** an **upper control limit (UCL),** and a **lower control limit (LCL).** Data are recorded on the control chart, and the chart is examined for disturbing patterns or for data points that indicate a process is out of control. Once a process is determined to be out of control, measures can be taken to correct the problem causing the deviation.

$\bar{X}$ CHART   An $\bar{X}$ **chart** is *a graph of sample means computed for a series of small random samples over a period of time.* The means are average measurements of some product characteristic. For example, the measurement could be the volume of fluid in a liter of rubbing alcohol, the thickness of a piece of sheet metal, or the size of a hole in a plastic part. These sample means are plotted on a graph that contains a centerline and upper and lower control limits (UCL and LCL).

$\bar{X}$ charts can be made from standards or without standards.* Companies sometimes have smoothed their process to the point where they have standard centerlines and control limits for a product. These standards are usually used when a company is producing products that have been made for some time and in situations where managers have little interest in monitoring the overall measure of location for the product. In this text, we will study only situations in which no standard is given. It is fairly common to compute $\bar{X}$ charts when there are no standards—especially if a company is producing a new product, is closely monitoring proposed standards, or expects a change in the process. Many firms want to monitor the standards, so they recompute the standards for each chart. In the no-standards situation, the standards (such as mean and standard deviation) are estimated by using the sample data.

The centerline for an $\bar{X}$ chart is the average of the sample means, $\bar{\bar{X}}$. The $\bar{X}$ chart has an upper control limit (UCL) that is three standard deviations of means above the centerline $(+3\sigma_{\bar{x}})$. The lower boundary of the $\bar{X}$ chart, called the lower control limit (LCL), is three standard deviations of means below the centerline $(-3\sigma_{\bar{x}})$. Recall the empirical rule presented in Chapter 3 stating that if data are normally distributed, approximately 99.7% of all values will be within three standard deviations of the mean. Because the shape of the sampling distribution of $\bar{X}$ is normal for large sample sizes regardless of the population shape, the empirical rule applies. However, because small samples are often used, an

Centerline
The middle horizontal line of a control chart, often determined either by a product or service specification or by computing an expected value from sample information.

Upper control limit (UCL)
The top-end line of a control chart, usually situated approximately three standard deviations of the statistic above the centerline; data points above this line indicate quality control problems.

Lower control limit (LCL)
The bottom-end line of a control chart, usually situated approximately three standard deviations of the statistic below the centerline; data points below this line indicate quality control problems.

$\bar{X}$ chart
A quality control chart for measurements that graphs the sample means computed for a series of small random samples over a period of time.

---

*Armand V. Feigenbaum. *Total Quality Control.* New York, McGraw-Hill Book Company, 1991.

approximation of the three standard deviations of means is used to determine UCL and LCL. This approximation can be made using either sample ranges or sample standard deviations. For small sample sizes ($n \leq 15$ is acceptable, but $n \leq 10$ is preferred), a weighted value of the average range is a good approximation of the three-standard-deviation distance to UCL and LCL. The range is very easy to compute (difference of extreme values), which is particularly useful when a wide array of nontechnical workers are involved in control chart computations. When sample sizes are larger, a weighted average of the sample standard deviations ($\overline{S}$) is a good estimate of the three standard deviations of means. The drawback to using the sample standard deviation is that it must always be computed, whereas the sample range can often be determined at a glance. Most control charts are constructed with small sample sizes; therefore, the range is more widely used in constructing control charts.

Table A.10 contains the weights applied to the average sample range or the average sample standard deviation to compute upper and lower control limits. The value of A_2 is used for ranges and the value of A_3 is used for standard deviations. The following steps are used to produce an $\overline{X}$ chart.

1. Decide on the quality to be measured.
2. Determine a sample size.
3. Gather 20 to 30 samples.
4. Compute the sample average, $\overline{X}$, for each sample.
5. Compute the sample range, R, for each sample.
6. Determine the average sample mean for all samples, $\overline{\overline{X}}$, as

$$\overline{\overline{X}} = \frac{\Sigma \overline{X}}{k},$$

   where k is the number of samples.
7. Determine the average sample range for all samples, R, as

$$\overline{R} = \frac{\Sigma R}{k},$$

   or determine the average sample standard deviation for all samples, $\overline{S}$, as

$$\overline{S} = \frac{\Sigma S}{k}.$$

8. Using the size of the samples, n_i, determine the value of A_2 if using the range and A_3 if using standard deviations.
9. Construct the centerline, the upper control limit, and the lower control limit. For ranges:

$$\overline{\overline{X}} \text{ is the centerline.}$$
$$\overline{\overline{X}} + A_2\overline{R} \text{ is the UCL.}$$
$$\overline{\overline{X}} - A_2\overline{R} \text{ is the LCL.}$$

   For standard deviations:

$$\overline{\overline{X}} \text{ is the centerline.}$$
$$\overline{\overline{X}} + A_3\overline{S} \text{ is the UCL.}$$
$$\overline{\overline{X}} - A_3\overline{S} \text{ is the LCL.}$$

A manufacturing facility is producing bearings. The diameter specified for the bearings is 5 mm. Every 10 minutes, six bearings are sampled and their diameters are measured and recorded. Twenty of these samples of six bearings are gathered. Use the resulting data and construct an $\bar{X}$ chart.

SAMPLE 1	SAMPLE 2	SAMPLE 3	SAMPLE 4	SAMPLE 5
5.13	4.96	5.21	5.02	5.12
4.92	4.98	4.87	5.09	5.08
5.01	4.95	5.02	4.99	5.09
4.88	4.96	5.08	5.02	5.13
5.05	5.01	5.12	5.03	5.06
4.97	4.89	5.04	5.01	5.13

SAMPLE 6	SAMPLE 7	SAMPLE 8	SAMPLE 9	SAMPLE 10
4.98	4.99	4.96	4.96	5.03
5.02	5.00	5.01	5.00	4.99
4.97	5.00	5.02	4.91	4.96
4.99	5.02	5.05	4.87	5.14
4.98	5.01	5.04	4.96	5.11
4.99	5.01	5.02	5.01	5.04

SAMPLE 11	SAMPLE 12	SAMPLE 13	SAMPLE 14	SAMPLE 15
4.91	4.97	5.09	4.96	4.99
4.93	4.91	4.96	4.99	4.97
5.04	5.02	5.05	4.82	5.01
5.00	4.93	5.12	5.03	4.98
4.90	4.95	5.06	5.00	4.96
4.82	4.96	5.01	4.96	5.02

SAMPLE 16	SAMPLE 17	SAMPLE 18	SAMPLE 19	SAMPLE 20
5.01	5.05	4.96	4.90	5.04
5.04	4.97	4.93	4.85	5.03
5.09	5.04	4.97	5.02	4.97
5.07	5.03	5.01	5.01	4.99
5.12	5.09	4.98	4.88	5.05
5.13	5.01	4.92	4.86	5.06

SOLUTION

Compute the value of $\bar{X}$ for each sample and average these values, obtaining $\bar{\bar{X}}$.

$$\bar{\bar{X}} = \frac{\bar{X}_1 + \bar{X}_2 + \bar{X}_3 + \cdots + \bar{X}_{20}}{20}$$

$$= \frac{4.9933 + 4.9583 + 5.0566 + \cdots + 5.0233}{20}$$

$$= \frac{100.043}{20} = 5.00215 \text{ (the centerline)}$$

Compute the values of R and average them, obtaining $\bar{R}$.

$$\bar{R} = \frac{R_1 + R_2 + R_3 + \cdots + R_{20}}{20}$$

$$= \frac{.25 + .12 + .34 + \cdots + .09}{20} = \frac{2.72}{20} = .136$$

Determine the value of A_2 by using $n_i = 6$ (size of the sample) from Table A.10: $A_2 = .483$.

The UCL is

$$\bar{\bar{X}} + A_2\bar{R} = 5.00215 + (.483)(.136) = 5.00215 + .06569 = 5.06784.$$

The LCL is

$$\bar{\bar{X}} - A_2\bar{R} = 5.00215 + (.483)(.136) = 5.00215 - .06569 = 4.93646.$$

Using the standard deviation instead of the range,

$$\bar{S} = \frac{S_1 + S_2 + S_3 + \cdots + S_{20}}{20}$$
$$= \frac{.0905 + .0397 + .1136 + \cdots + .0356}{20}$$
$$= .0494$$

Determine the value of A_3 by using $n_i = 6$ (sample size) from Table A.10: $A_3 = 1.287$.

The UCL is

$$\bar{\bar{X}} + A_3\bar{S} = 5.00215 + (1.287)(.0494) = 5.00215 + .06358 = 5.06573.$$

The LCL is

$$\bar{\bar{X}} - A_3\bar{S} = 5.00215 - (1.287)(.0494) = 5.00215 - .06358 = 4.93857.$$

The following graph depicts the $\bar{X}$ control chart using the range (rather than the standard deviation) as the measure of dispersion to compute LCL and UCL. Observe that if the standard deviation is used instead of the range to compute LCL and UCL, there is little, if any, perceptible difference in LCL and UCL by the two methods because of the precision (or lack thereof) of this chart.

Note that the sample means for samples 5 and 16 are above the UCL and the sample means for samples 11 and 19 are below the LCL. This result indicates that these four samples are out of control and alerts the production supervisor or worker to initiate further investigation of bearings produced during these periods. All other samples are within the control limits.

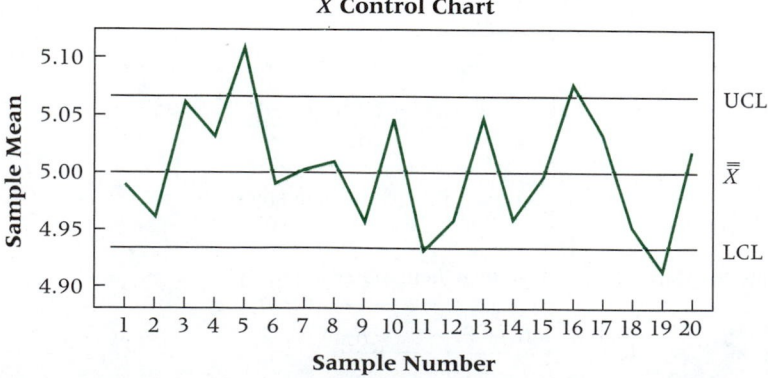

Shown next is the FAST ⧄ STAT output for this problem. Note that the FAST ⧄ STAT output is nearly identical to the control chart just shown.

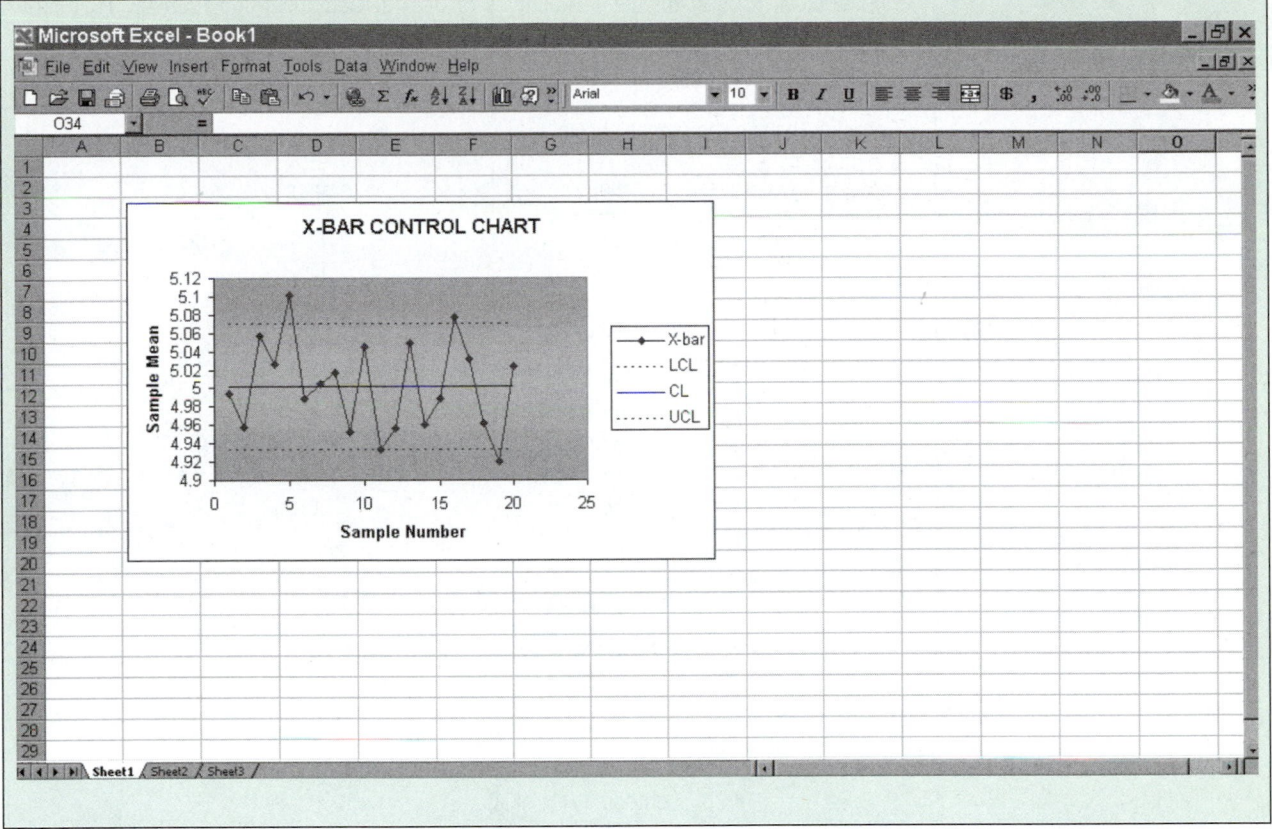

R CHARTS   An **R chart** is *a plot of the sample ranges* and often is used in conjunction with an $\overline{X}$ chart. Whereas $\overline{X}$ charts are used to plot the location values, $\overline{X}$, for each sample, R charts are used to plot the variation of each sample as measured by the sample range. The centerline of an R chart is the average range, $\overline{R}$. Lower control limits (LCLs) are determined by $D_3\overline{R}$, where D_3 is a weight applied to $\overline{R}$, reflecting sample size. The value of D_3 can be obtained from Table A.10. Upper control limits (UCLs) are determined by $D_4\overline{R}$, where D_4 is a value obtained from Table A.10, which also reflects sample size. The following steps lead to an R chart.

R chart

A plot of sample ranges used in quality control.

1. Decide on the quality to be measured.
2. Determine a sample size.
3. Gather 20 to 30 samples.
4. Compute the sample range, R, for each sample.
5. Determine the average sample range for all samples, $\overline{R}$, as

$$\overline{R} = \frac{\Sigma R}{k},$$

where k = the number of samples.
6. Using the size of the samples, n_i, find the values of D_3 and D_4 in Table A.10.
7. Construct the centerline and control limits.

$$\text{Centerline} = \overline{R}$$
$$\text{UCL} = D_4\overline{R}$$
$$\text{LCL} = D_3\overline{R}$$

Construct an *R* chart for the 20 samples of data in Demonstration Problem 15.1 on bearings.

SOLUTION

Compute the sample ranges shown.

SAMPLE	RANGE	SAMPLE	RANGE
1	.25	11	.22
2	.12	12	.11
3	.34	13	.16
4	.10	14	.21
5	.07	15	.06
6	.05	16	.12
7	.03	17	.12
8	.09	18	.09
9	.14	19	.17
10	.18	20	.09

Compute $\bar{R}$.

$$\bar{R} = \frac{.25 + .12 + .34 + \cdots + .09}{20} = \frac{2.72}{20} = .136$$

For $n_i = 6$, $D_3 = 0$ and $D_4 = 2.004$ (from Table A.10).

$$\text{Centerline } \bar{R} = .136$$
$$\text{LCL} = D_3\bar{R} = (0)(.136) = 0$$
$$\text{UCL} = D_4\bar{R} = (2.004)(.136) = .2725$$

The resulting *R* chart for these data is shown next, followed by the FAST ⟍ STAT output. Note that the range for sample 3 is out of control (beyond the UCL). The range of values in sample 3 appears to be unacceptable. Further investigation of the population from which this sample was drawn is warranted.

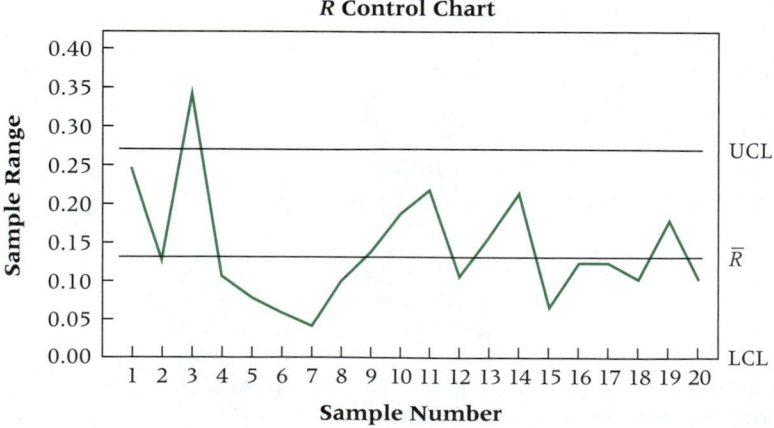

R Control Chart

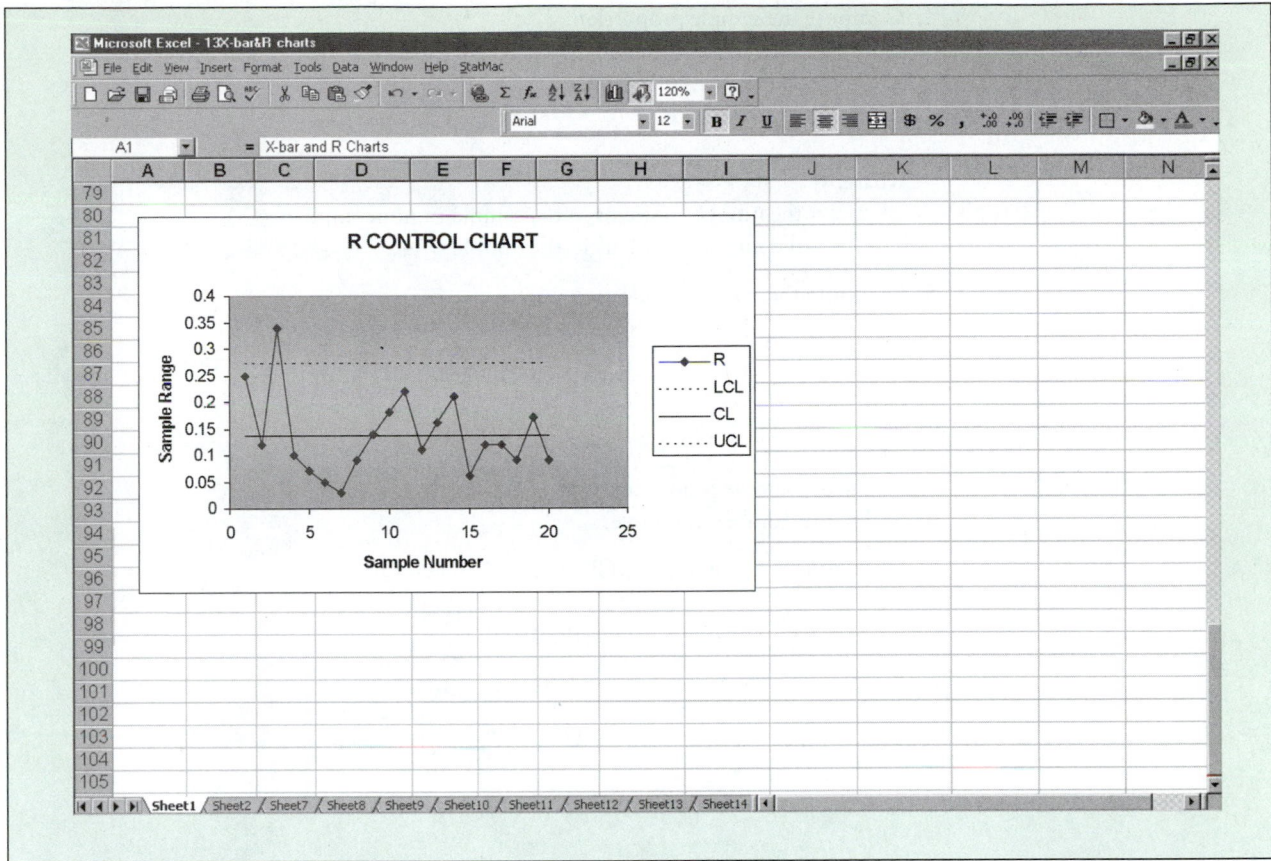

P CHARTS When product attributes are measurable, $\overline{X}$ charts and R charts can be formulated from the data. Sometimes, however, product inspection yields no measurement—only a yes-or-no type of conclusion based on whether the item complies with the specifications. For this type of data, no measure is available from which to average or determine the range. However, attribute compliance can be depicted graphically by a P chart. A **P chart** *graphs the proportion of sample items in noncompliance for multiple samples.*

For example, suppose a company producing electric motors samples 40 motors three times a week for a month. For each group of 40 motors, it determines the proportion of the sample group that does not comply with the specifications. It then plots these sample proportions, $\hat{p}$, on a P chart to identify trends or samples with unacceptably high proportions of nonconformance. Other P chart applications include determining whether a gallon of paint has been manufactured with acceptable texture, a pane of glass contains cracks, or a tire has a defective tread.

Like the $\overline{X}$ chart and the R chart, a P chart contains a centerline. The centerline is the average of the sample proportions. Upper and lower control limits are computed from the average of the sample proportions plus or minus three standard deviations of proportions. The following are the steps for constructing a P chart.

1. Decide on the quality to be measured.
2. Determine a sample size.
3. Gather 20 to 30 samples.

P chart

A quality control chart for attribute compliance that graphs the proportion of sample items in noncompliance with specifications for multiple samples.

4. Compute the sample proportion.

$$\hat{p} = \frac{n_{non}}{n}$$

where:
    n_{non} = the number of items in the sample in noncompliance
    n = the number of items in the sample

5. Compute the average proportion.

$$P = \frac{\Sigma \hat{p}}{k}$$

where:
    $\hat{p}$ = the sample proportion
    k = the number of samples

6. Determine the centerline, UCL, and LCL, when $Q = 1 - P$.

$$\text{Centerline} = P$$

$$\text{UCL} = P + 3\sqrt{(P \cdot Q)/n}$$

$$\text{LCL} = P - 3\sqrt{(P \cdot Q)/n}$$

DEMONSTRATION PROBLEM 15.3	A company produces bond paper and, at regular intervals, samples of 50 sheets of paper are inspected. Suppose 20 random samples of 50 sheets of paper each are taken during a certain period of time, with the following numbers of sheets in noncompliance per sample. Construct a P chart from these data.

SAMPLE	n	NUMBER OUT OF COMPLIANCE
1	50	4
2	50	3
3	50	1
4	50	0
5	50	5
6	50	2
7	50	3
8	50	1
9	50	4
10	50	2
11	50	2
12	50	6
13	50	0
14	50	2
15	50	1
16	50	6
17	50	2
18	50	3
19	50	1
20	50	5

SOLUTION

From the data, $n = 50$. The values of $\hat{p}$ follow.

SAMPLE	$\hat{p}$ (OUT OF COMPLIANCE)
1	$4/50 = .08$
2	$3/50 = .06$
3	$1/50 = .02$
4	$0/50 = .00$
5	$5/50 = .10$
6	$2/50 = .04$
7	$3/50 = .06$
8	$1/50 = .02$
9	$4/50 = .08$
10	$2/50 = .04$
11	$2/50 = .04$
12	$6/50 = .12$
13	$0/50 = .00$
14	$2/50 = .04$
15	$1/50 = .02$
16	$6/50 = .12$
17	$2/50 = .04$
18	$3/50 = .06$
19	$1/50 = .02$
20	$5/50 = .10$

The value of P is obtained by averaging these $\hat{p}$ values.

$$P = \frac{\hat{p}_1 + \hat{p}_2 + \hat{p}_3 + \cdots + \hat{p}_{20}}{20}$$

$$= \frac{.08 + .06 + .02 + \cdots + .10}{20} = \frac{1.06}{20} = .053$$

The centerline is $P = .053$.

The UCL is

$$P + 3\sqrt{\frac{P \cdot Q}{n}} = .053 + 3\sqrt{\frac{(.053)(.947)}{50}} = .053 + .095 = .148.$$

The LCL is

$$P - 3\sqrt{\frac{P \cdot Q}{n}} = .053 - 3\sqrt{\frac{(.053)(.947)}{50}} = -.042.$$

To have $-.042$ item in noncompliance is impossible, so the lower control limit is 0.

Following is the P chart for this problem. Note that all 20 proportions are within the quality control limits.

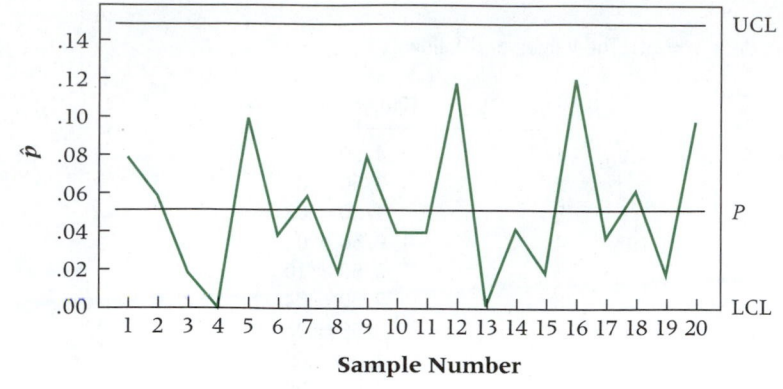

Shown below is the FAST ⦿ STAT output for this *P* chart. Note that the computer output is essentially the same as the graph just shown.

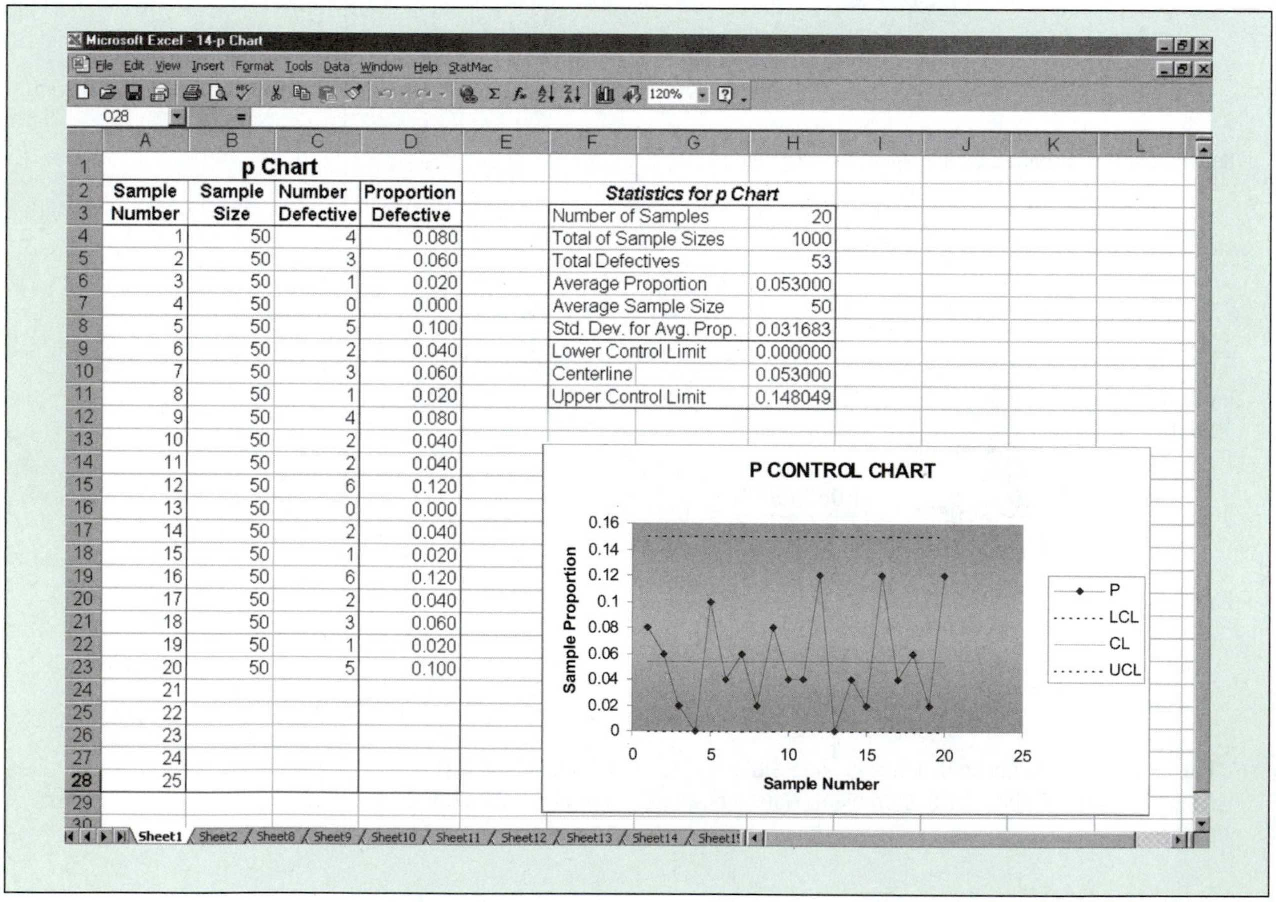

Microsoft Excel - 14-p Chart

File Edit View Insert Format Tools Data Window Help StatMac

O28

p Chart

Sample Number	Sample Size	Number Defective	Proportion Defective
1	50	4	0.080
2	50	3	0.060
3	50	1	0.020
4	50	0	0.000
5	50	5	0.100
6	50	2	0.040
7	50	3	0.060
8	50	1	0.020
9	50	4	0.080
10	50	2	0.040
11	50	2	0.040
12	50	6	0.120
13	50	0	0.000
14	50	2	0.040
15	50	1	0.020
16	50	6	0.120
17	50	2	0.040
18	50	3	0.060
19	50	1	0.020
20	50	5	0.100
21			
22			
23			
24			
25			

Statistics for p Chart

Number of Samples	20
Total of Sample Sizes	1000
Total Defectives	53
Average Proportion	0.053000
Average Sample Size	50
Std. Dev. for Avg. Prop.	0.031683
Lower Control Limit	0.000000
Centerline	0.053000
Upper Control Limit	0.148049

P CONTROL CHART

Sheet1 Sheet2 Sheet8 Sheet9 Sheet10 Sheet11 Sheet12 Sheet13 Sheet14 Sheet15

c CHARTS The *c* chart is less widely used than the $\overline{X}$, the *R*, or the *P* chart. Like the *P* chart, the *c* chart attempts to formulate information about defective items. However, whereas the *P* chart is a control chart that displays the proportion of items in a sample that are out of compliance with specifications, a ***c* chart** *displays the number of nonconformances per item or unit.* Examples of nonconformances could be paint flaws, scratches, openings drilled too large or too small, or shorts in electric wires. The *c* chart allows for multiple nonconforming features per item or unit. For example, if an item is a radio, there can be scratches (multiple) in the paint, poor soldering, bad wiring, broken dials, burned-out light bulbs, and broken antennae. A unit need not be an item such as a computer chip. It can be a bolt of cloth, 4 feet of wire, or a 2 × 4 board. The requirement is that the unit remain consistent throughout the test or experiment.

In computing a *c* chart, a *c* value is determined for each item or unit by tallying the total nonconformances for the item or unit. The centerline is computed by averaging the *c* values for all items or units. Because in theory nonconformances per item or unit are rare, the Poisson distribution is used as the basis for the *c* chart. The long-run average for the Poisson distribution is λ, and the analogous long-run average for a *c* chart is $\bar{c}$ (the average of the *c* values for the items or units studies), which is used as the centerline value. Upper control limits (UCL) and lower control limits (LCL) are computed by adding or subtracting three standard deviations of the mean, $\bar{c}$, from the centerline value, $\bar{c}$. The standard deviation of a Poisson distribution is the square root of λ; likewise, the standard deviation of $\bar{c}$ is the square root of $\bar{c}$. The UCL is thus determined by $\bar{c} + 3\sqrt{\bar{c}}$ and the LCL is given by $\bar{c} - 3\sqrt{\bar{c}}$. The following steps are used for constructing a *c* chart.

1. Decide on nonconformances to be evaluated.
2. Determine the number of items of units to be studied. (This number should be at least 25.)
3. Gather items or units.
4. Determine the value of *c* for each item or unit by summing the number of nonconformances in the item or unit.
5. Calculate the value of $\bar{c}$.

$$\bar{c} = \frac{c_1 + c_2 + c_3 + \cdots + c_i}{i}$$

where:
   i = number of items
   c_i = number of nonconformances per item

6. Determine the centerline, UCL, and LCL.

$$\text{Centerline} = \bar{c}$$
$$\text{UCL} = \bar{c} + 3\sqrt{\bar{c}}$$
$$\text{LCL} = \bar{c} - 3\sqrt{\bar{c}}$$

c chart
A quality control chart for attribute compliance that displays the number of nonconformances per item or unit.

A manufacturer produces gauges to measure oil pressure. As part of the company's statistical process control, 25 gauges are randomly selected and tested for nonconformances. The results are shown here. Use these data to construct a c chart that displays the nonconformances per item.

ITEM NUMBER	NUMBER OF NONCONFORMANCES	ITEM NUMBER	NUMBER OF NONCONFORMANCES
1	2	14	2
2	0	15	1
3	3	16	4
4	1	17	0
5	2	18	2
6	5	19	3
7	3	20	2
8	2	21	1
9	0	22	3
10	0	23	2
11	4	24	0
12	3	25	3
13	2		

SOLUTION

Determine the centerline, UCL, and LCL.

Centerline:

$$\bar{c} = \frac{(2+0+3+1+2+5+3+2+0+0+4+3+2+2+1+4+0+2+3+2+1+3+2+0+3)}{25}$$

$$= \frac{50}{25} = 2.0$$

UCL:

$$\bar{c} + 3\sqrt{\bar{c}} = 2.0 + 3\sqrt{2.0} = 2.0 + 4.2 = 6.2$$

LCL:

$$\bar{c} - 3\sqrt{\bar{c}} = 2.0 - 3\sqrt{2.0} = 2.0 - 4.2 = -2.2$$

The lower control limit cannot be less than zero; thus, the LCL is 0. The graph of the control chart, followed by the FAST ⬊ STAT c chart, is shown next. Note that none of the points are beyond the control limits and there is a healthy deviation of points both above and below the centerline. This indicates a process that is relatively in control, with an average of two nonconformances per item.

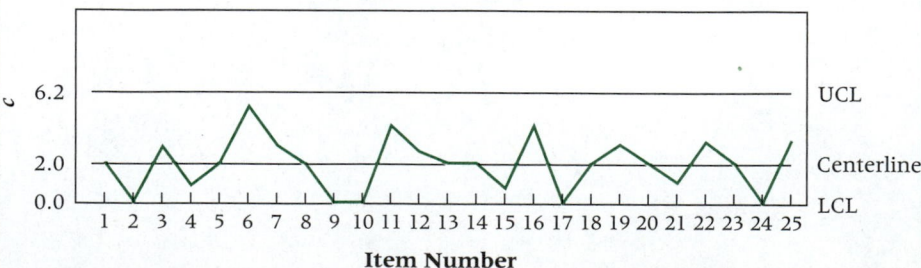

Item Number

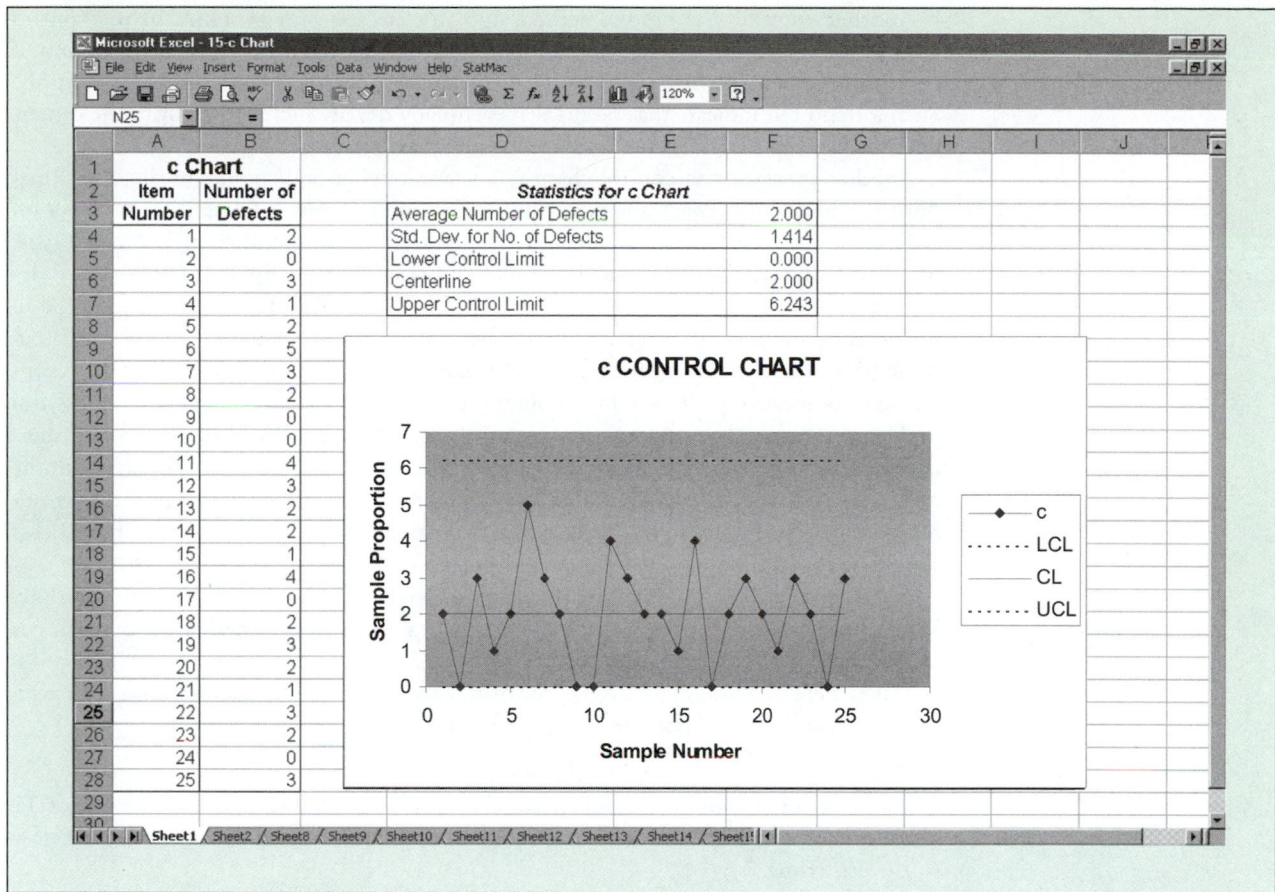

Interpreting Control Charts

How can control charts be used to monitor processes? When is a process out of control? There are several things to look for in evaluating the points plotted on a control chart. Obviously, one concern is points that are outside the control limits. Control chart outer limits (UCL and LCL) are established at three standard deviations above and below the centerline. The empirical rule discussed in Chapter 3 and the Z table value for $Z = 3$ indicate that approximately 99.7% of all values should be within three standard deviations of the mean of the statistic. Applying this rule to control charts suggests that fewer than .3% of all points should be beyond the upper and lower control limits by chance. Thus, one of the more elementary items a control chart observer looks for is points outside LCL and UCL. If the system is "in control," virtually no data points should be outside these limits. Workers responsible for process control should investigate samples in which sample members are outside the LCL and UCL. In the case of the c chart, items that are above the UCL line contain an inordinate number of nonconformances in relation to the average. The occurrence of points beyond the control limits should be a flag calling for further investigation.

Several other criteria can be used to determine whether a control chart is plotting a process that is out of control. In general, there *should* be random fluctuation above and below the centerline within the UCL and LCL. However, a process can be out of control if too many consecutive points are above or below the centerline. Eight or more consecutive points on one side of the centerline are considered too many.*

---

*James R. Evans and William M. Lindsay. *The Management and Control of Quality.* 4th Edition. Cincinnati: South-Western College Publishing, 1999.

Another criterion for process control operators to look for is trends in the control charts. At any point in the process, is a trend emerging in the data? As a rule of thumb, if six or more points are increasing or are decreasing, the process may be out of control.[*] Such a trend can indicate that points will eventually deviate increasingly from the centerline (the gap between the centerline and the points will increase).

Another concern with control charts is an overabundance of points in the outer one-third of the region between the centerline and the outer limits (LCL and UCL). By a rationalization similar to that imposed on LCL and UCL, the empirical rule and the table of Z values show that approximately 95% of all points should be within two standard deviations of the centerline. With this in mind, fewer than 5% of the points should be in the outer one-third of the region between the centerline and the outer control limits (because 95% should be within two-thirds of the region). A rule to follow is that if two out of three consecutive points are in the outer one-third of the chart, there may be a control problem. Likewise, because approximately 68% of all values should be within one standard deviation of the mean (empirical rule, Z table for $Z = 1$), only 32% should be in the outer two-thirds of the control chart above and below the centerline. As a rule, if four out of five successive points are in the outer two-thirds of the control chart, the process should be further investigated.[†]

Another consideration in evaluating control charts is the location of the centerline. With each successive batch of samples, it is important to observe whether the centerline is shifting away from specifications.

Listed here is a summary of the control chart abnormalities for which a statistical process controller should be determined.

1. Points are above UCL and/or below LCL.
2. Eight or more consecutive points are above or below the centerline. Ten out of 11 points are above or below the centerline. Twelve out of 14 points are above or below the centerline.
3. A trend of six or more consecutive points (increasing or decreasing) is present.
4. Two out of three consecutive values are in the outer one-third.
5. Four out of five consecutive values are in the outer two-thirds.
6. The centerline shifts from chart to chart.

Figure 15.6 contains several control charts, each of which has one of these types of problems. The chart in (a) contains points above and below the outer control limits. The one in (b) has eight consecutive points on one side of the centerline. The chart in (c) has seven consecutive increasing points. In (d), at least two out of three consecutive points are in the outer one-third of the control chart. In (e), at least four out of five consecutive points are in the outer two-thirds of the chart.

In investigating control chart abnormalities, several possible causes may be found. Some of them are listed here.[‡]

1. Changes in the physical environment
2. Worker fatigue
3. Worn tools
4. Changes in operators or machines

---

[*]Richard E. DeVor, Tsong-how Chang, and John W. Sutherland. *Statistical Quality Design and Control.* New York, Macmillan, Inc., 1992.

[†]DeVor, Chang, and Sutherland; Evans and Lindsay.

[‡]Eugene L. Grant and Richard S. Leavenworth. *Statistical Quality Control,* 5th ed. New York, McGraw-Hill Book Company, 1980.

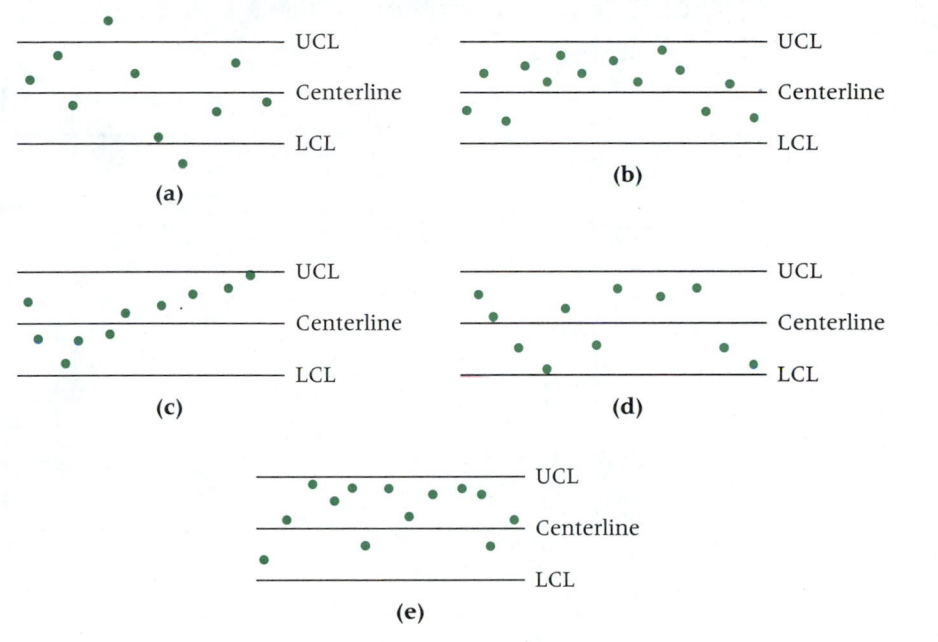

Figure 15.6

Control charts
with problems

5. Maintenance
6. Changes in worker skills
7. Changes in materials
8. Process modification

The statistical process control person should beware that control chart abnormalities can arise because of measurement errors or incorrect calculation of control limits. Judgment should be exercised so as not to overcontrol the process by readjusting to every oddity that appears to be out of the ordinary on a control chart.

Analysis Using Excel

Using the **Chart Wizard,** it is possible to construct a Pareto chart using Excel. Excel has no direct means for constructing control charts. However, **FAST ＼ STAT** can be used to construct each of the four types of control charts presented in this section.

To construct a Pareto chart with Excel, enter the "problems" in one column of the spreadsheet and the percentages of each in another column. The **Data-Sort-Descending** sequence of commands from the Excel menu bar can be used to order the percentages from highest to lowest. In using the **Sort** command, be careful to rearrange the "problems" so that each problem corresponds with its respective percentage. Then use **Chart Wizard** to create a bar chart of the percentages. At Step 1 of the **Chart Wizard,** select the **Custom Types** tab. From the list of custom types, select **Line—Column on 2 Axes.** Finish the Pareto chart using the **Chart Wizard** in the usual way.

To construct X-bar and R control charts using **FAST ＼ STAT** enter the data from the samples into an Excel worksheet. Place in each row the results for one sample. The macro allows up to 100 samples to be entered. The number of items in each sample can be up to eight items. Accordingly, the data values can fill a table of up to 100 rows and eight columns. Figure 15.7 presents the dialog box for the X-bar and R charts. Note that in addition to the location of the sample data, the dialog box requires an A_2 value from

Figure 15.7

Dialog box for *X*-bar chart and *R* chart

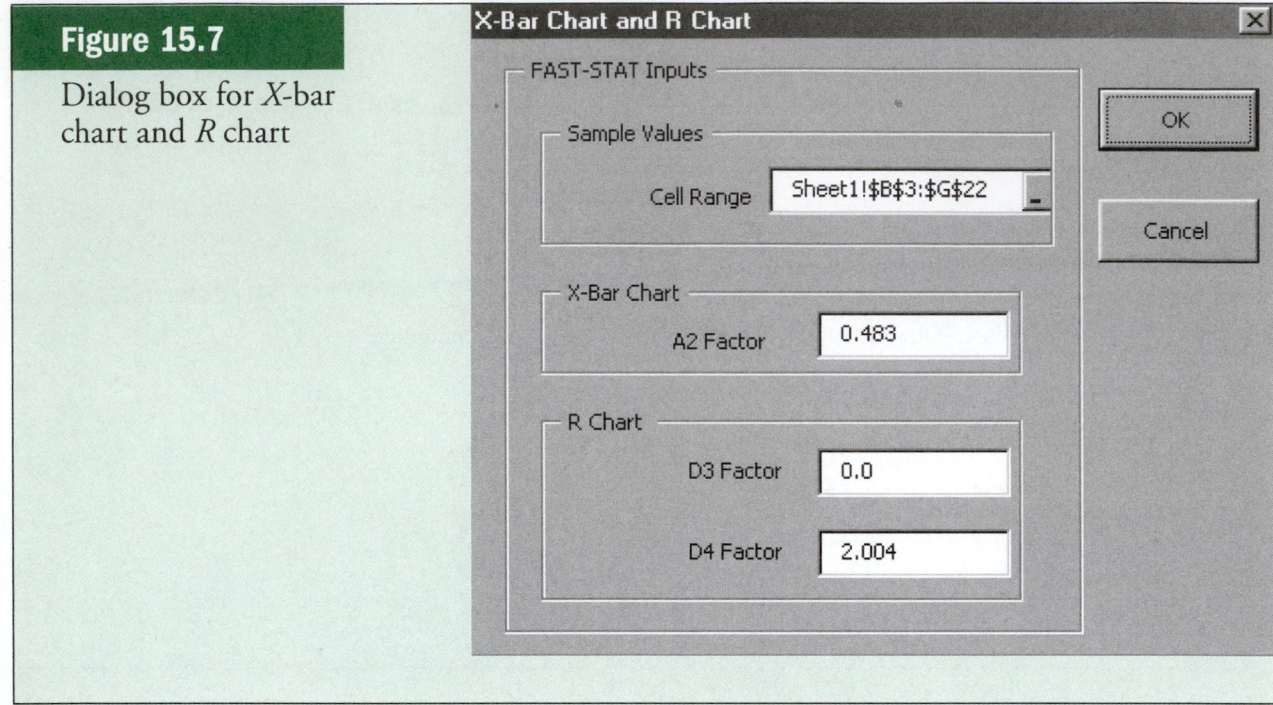

Figure 15.8

Dialog boxes for *P* chart and *c* chart

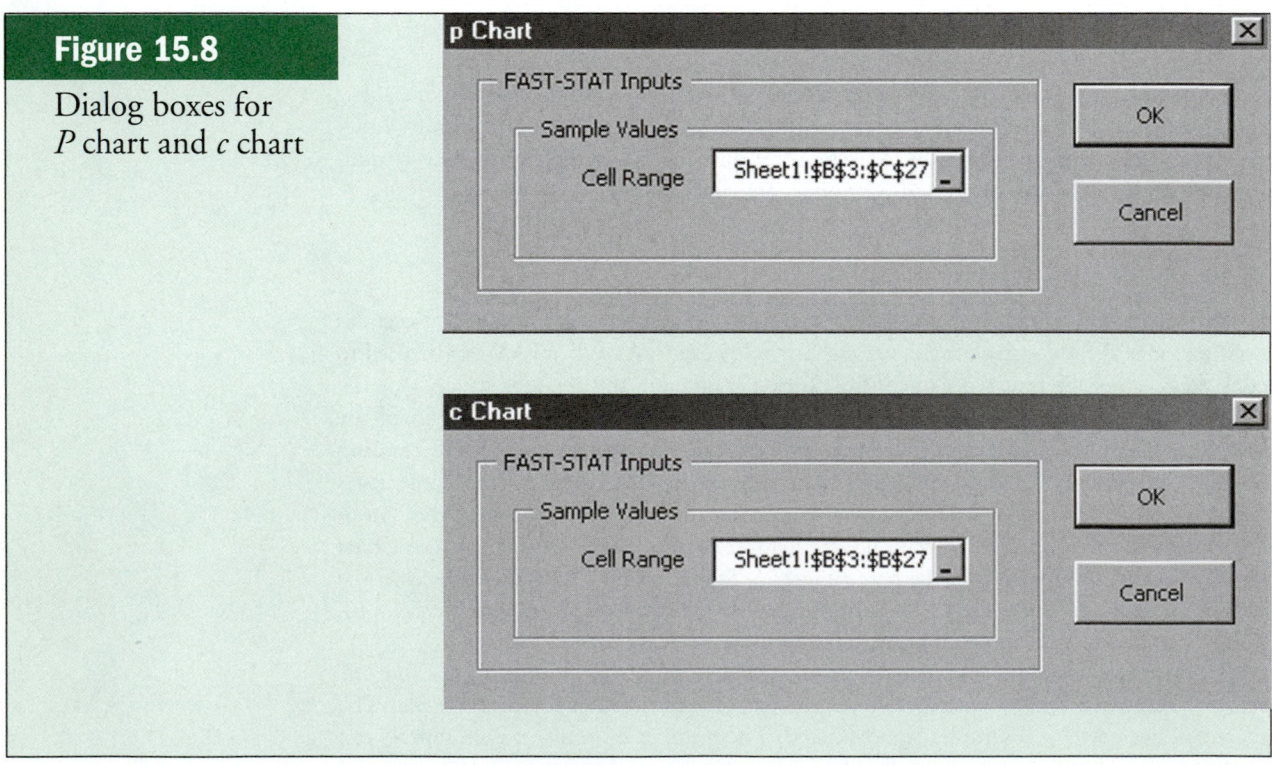

Table A.10 for the *X*-bar chart, and the D_3 and D_4 values from Table A.10 for an *R* chart. With the completion of the dialog box and a click on the **OK** button, both an *X*-bar chart and an *R* chart will be constructed.

The dialog boxes for both the *P* chart and *c* chart are shown in Figure 15.8. Each of these requires only the location of the data. For the *P* chart, the data consist of two columns of data. The first column gives the sample size for each sample and the second the number of nonconformances in each sample. On the other hand, the *c* chart only has one column of data. It is the number of nonconformances for each sample.

The output from **FAST** ⧵ **STAT** for control charts consists of each of the four control chart graphs displayed in Demonstration Problems 15.1–15.4 respectively. In addition, for the *X*-bar chart and the *R* chart the output shown in Figure 15.9 is given. This output includes the sample values, the sample means, the sample ranges, the average of the sample means, the average of the sample ranges, the value of the centerline, the value of the upper control limit (UCL), and the value of the lower control limit (LCL).

The *P* chart has a similar output as shown in Figure 15.10 and also the *c* chart as shown in Figure 15.11.

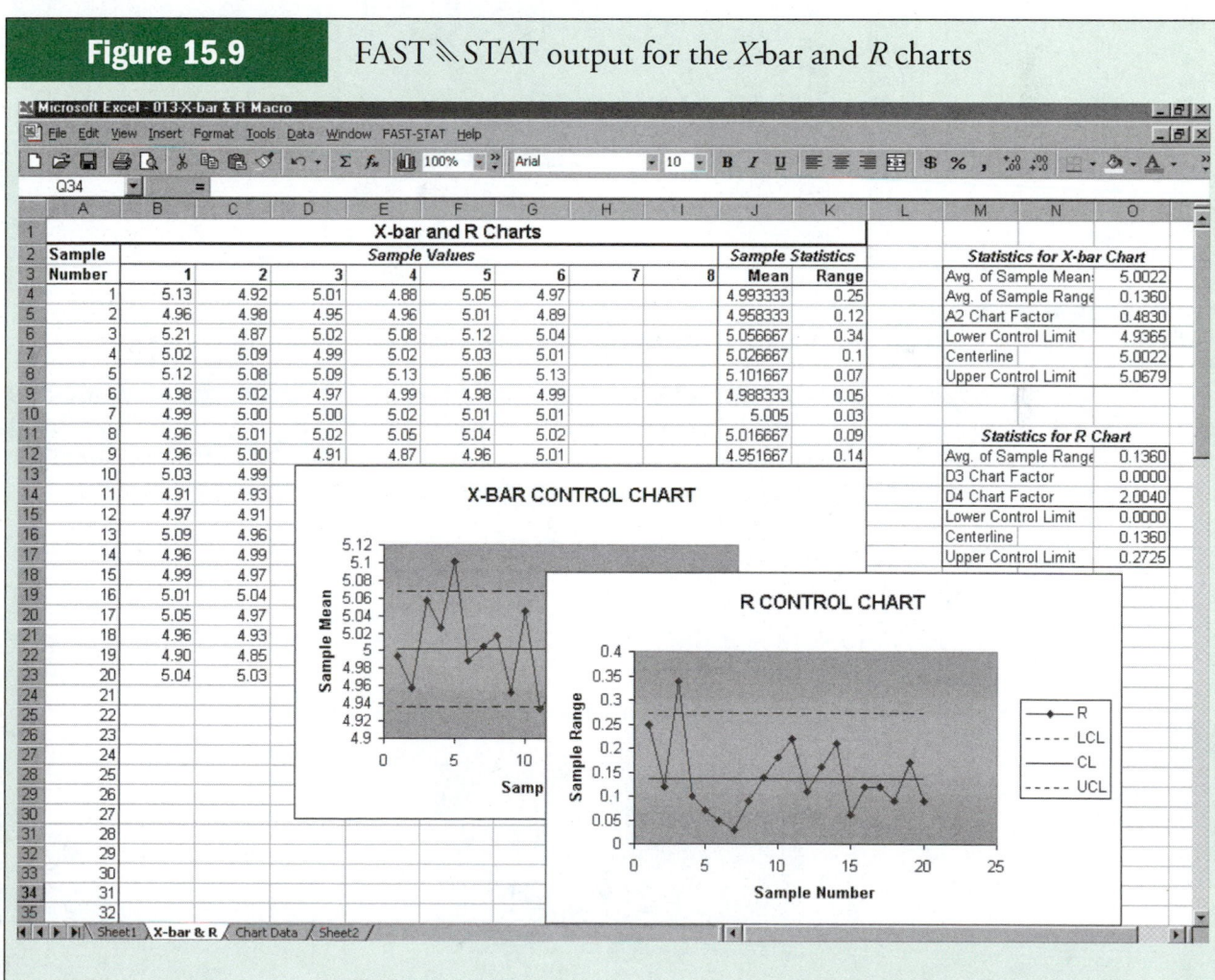

Figure 15.9 FAST ⧵ STAT output for the *X*-bar and *R* charts

Figure 15.10 FAST\STAT output for the *P* chart

Microsoft Excel - 014-p Chart macro

p Chart

Sample Number	Sample Size	No. Out of Conforma	Proportion Out of Conf.
1	100	4	0.04
2	100	2	0.02
3	100	2	0.02
4	100	4	0.04
5	100	2	0.02
6	100	0	0
7	100	2	0.02
8	100	2	0.02
9	100	4	0.04
10	100	2	0.02
11	100	6	0.06
12	100	0	0
13	100	4	0.04
14	100	2	0.02
15	100	4	0.04
16	100	6	0.06
17	100	4	0.04
18	100	4	0.04
19	100	6	0.06
20	100	4	0.04
21	100	8	0.08
22	100	2	0.02
23	100	4	0.04
24	100	0	0
25	100	0	0
26			
27			
28			
29			

Statistics for p Chart

Number of Samples	25
Total of Sample Sizes	2500
Total Out of Conforma	78
Average Proportion	0.0312
Average Sample Size	100
Std.Dev.for Avg.Prop.	0.0174
Lower Control Limit	0.0000
Centerline	0.0312
Upper Control Limit	0.0834

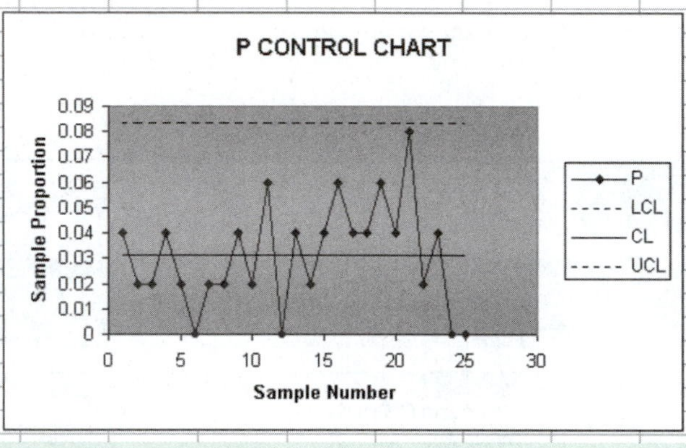

Figure 15.11 FAST\STAT output for the *c* chart

Microsoft Excel - 015-c Chart macro

c Chart

Item Number	Number of Nonconformances
1	0
2	1
3	3
4	0
5	2
6	0
7	0
8	0
9	0
10	0
11	0
12	1
13	1
14	1
15	0
16	3
17	1
18	1
19	1
20	1
21	2
22	0
23	3
24	0
25	1

Statistics for c Chart

Average Number of No	0.8800
Std.Dev.for No. of Nor	0.9381
Lower Control Limit	0.0000
Centerline	0.8800
Upper Control Limit	3.6942

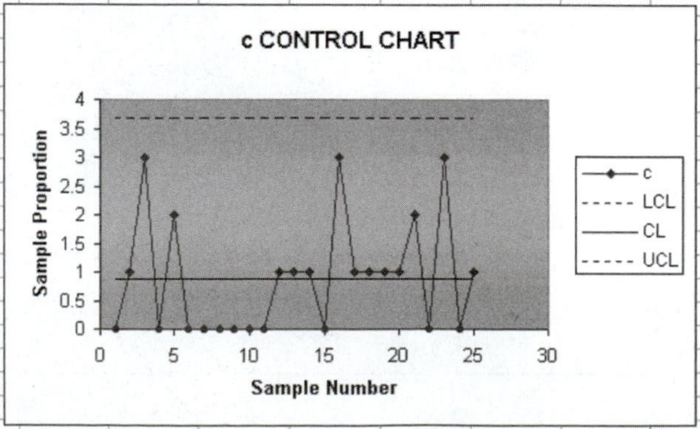

15.4 A food-processing company makes potato chips, pretzels, and cheese chips. Although its products are packaged and sold by weight, the company has been taking sample bags of cheese chips and counting the number of chips in each bag. Shown here is the number of chips per bag for five samples of seven bags of chips. Use these data to construct an $\bar{X}$ chart and an R chart. Discuss the results.

SAMPLE 1	SAMPLE 2	SAMPLE 3	SAMPLE 4	SAMPLE 5
25	22	30	32	25
23	21	23	26	23
29	24	22	27	29
31	25	26	28	27
26	23	28	25	27
28	26	27	25	26
27	29	21	31	24

15.5 A toy-manufacturing company has been given a large order for small plastic whistles that will be given away by a large fast-food hamburger chain with its kiddie meal. Seven random samples of four whistles have been taken. The weight of each whistle has been ascertained in grams. The data are shown here. Use these data to construct an $\bar{X}$ chart and an R chart. Are there any managerial decisions that should be made on the basis of these findings?

SAMPLE 1	SAMPLE 2	SAMPLE 3	SAMPLE 4	SAMPLE 5	SAMPLE 6	SAMPLE 7
4.1	3.6	4.0	4.6	3.9	5.1	4.6
5.2	4.3	4.8	4.8	3.8	4.7	4.4
3.9	3.9	5.1	4.7	4.6	4.8	4.0
5.0	4.6	5.3	4.7	4.9	4.3	4.5

15.6 A machine operator at a pencil-manufacturing facility gathered 10 different random samples of 100 pencils. The operator's inspection was to determine whether the pencils were in compliance or out of compliance with specifications. The results of this inspection are shown here. Use these data to construct a P chart. Comment on the results of this chart.

SAMPLE	SIZE	NUMBER OUT OF COMPLIANCE	SAMPLE	SIZE	NUMBER OUT OF COMPLIANCE
1	100	2	6	100	5
2	100	7	7	100	2
3	100	4	8	100	0
4	100	3	9	100	1
5	100	3	10	100	6

15.7 A large manufacturer makes valves. Currently it is producing a particular valve for use in industrial engines. As a part of a quality control effort, the company engineers randomly sample seven groups of 40 valves and inspect them to determine whether they are in or out of compliance. Results are shown here. Use the information to construct a P chart. Comment on the chart.

SAMPLE	SIZE	NUMBER OUT OF COMPLIANCE
1	40	1
2	40	0
3	40	1
4	40	3
5	40	2
6	40	5
7	40	2

15.8   A firm in the upper Midwest manufactures light bulbs. Before the bulbs are released for shipment, a sample of bulbs is selected for inspection. Inspectors look for nonconformances such as scratches, weak or broken filaments, incorrectly bored turns, insufficient outside contacts, and others. A sample of 35 60-watt bulbs has just been inspected, and the results are shown here. Use these data to construct a *c* chart. Discuss the findings.

BULB NUMBER	NUMBER OF NONCONFORMANCES	BULB NUMBER	NUMBER OF NONCONFORMANCES
1	0	19	2
2	1	20	0
3	0	21	0
4	0	22	1
5	3	23	0
6	0	24	0
7	1	25	0
8	0	26	2
9	0	27	0
10	0	28	0
11	2	29	1
12	0	30	0
13	0	31	0
14	2	32	0
15	0	33	0
16	1	34	3
17	3	35	0
18	0		

15.9   A soft-drink bottling company has just run a long line of 12-ounce soft-drink cans filled with cola. A sample of 32 cans is selected by inspectors looking for nonconforming items. Among the things the inspectors look for are paint defects on the can, improper seal, incorrect volume, leaking contents, incorrect mixture of carbonation and syrup in the soft drink, and out-of-spec syrup mixture. The results of this inspection are given here. Construct a *c* chart from the data and comment on the results.

CAN NUMBER	NUMBER OF NONCONFORMANCES	CAN NUMBER	NUMBER OF NONCONFORMANCES
1	2	17	3
2	1	18	1
3	1	19	2
4	0	20	0
5	2	21	0
6	1	22	1
7	2	23	4
8	0	24	0
9	1	25	2
10	3	26	1
11	1	27	1
12	4	28	3
13	2	29	0
14	1	30	1
15	0	31	2
16	1	32	0

15.10 Examine the three control charts shown. Discuss any and all control problems that may be apparent from these control charts.

a.

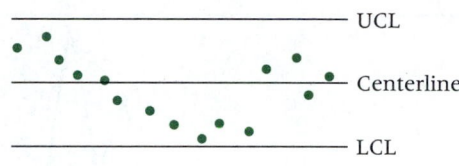

b.

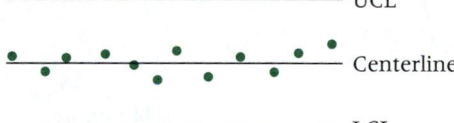

c.

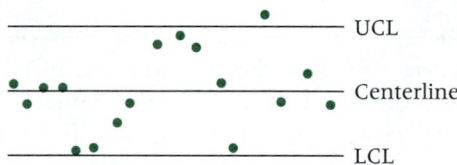

15.11 Study each of the following FAST ⟍ STAT control charts and determine whether any of them indicate problems in the processes. Comment on each chart.

a.

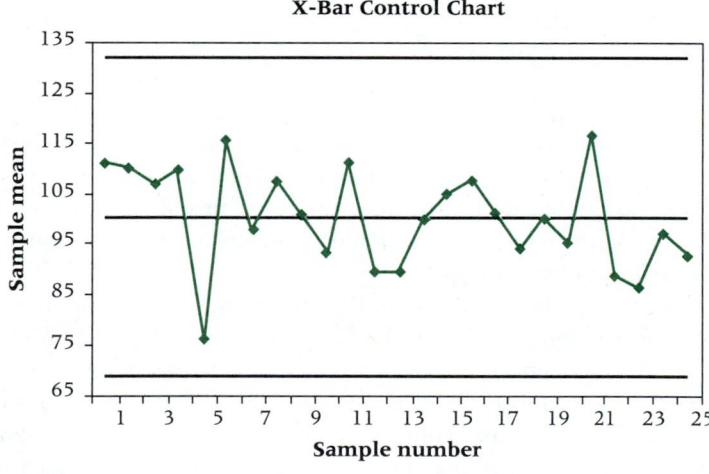

b.

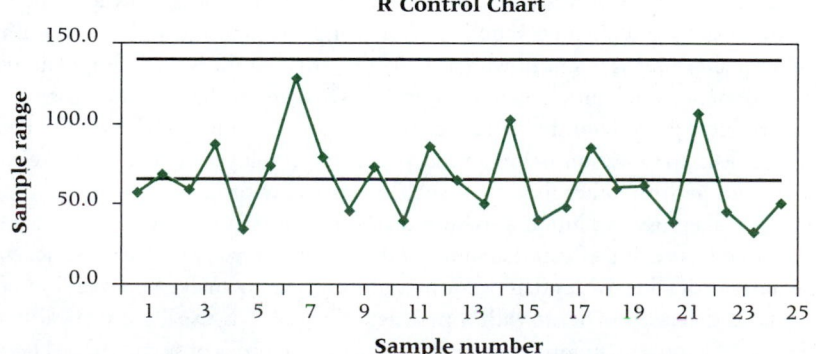

c.

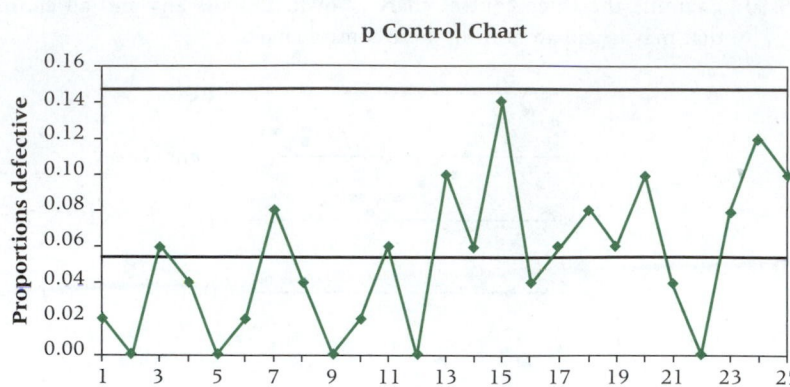

p Control Chart

Summary

Quality means many different things to different people. One definition is that a quality product is one that delivers to the customer those attributes that have been agreed upon by both buyer and seller. Leading quality experts such as Philip B. Crosby, Armand V. Feigenbaum, and David A. Garvin suggest various divergent views on the notion of quality.

Quality control is the collection of strategies, techniques, and actions that an organization can use to ensure the production of a quality product. For decades, U.S. companies used after-process quality control, which essentially consisted of inspectors determining whether a product complied with its specifications. During the 1980s, U.S. companies joined Western European and Asian businesses in instituting in-process quality control, which enables the producer to determine weaknesses and flaws during the production process.

Total quality management is when all members of an organization—from the CEO to the line worker—are involved in improving quality. One of the main proponents of total quality management was W. Edwards Deming. Deming was known for his cause-and-effect explanation of total quality management in a company, which is sometimes referred to as the Deming chain reaction. In addition, Deming presented 14 points that can lead to improved total quality management.

Five important quality concepts are benchmarking, just-in-time inventory systems, reengineering, six sigma, and team building. Benchmarking is a technique through which a company attempts to develop product and process excellence by examining and emulating the best practices and techniques used in the industry. Just-in-time inventory systems are inventory systems that focus on raw materials, subparts, and suppliers. Just in time is a philosophy of coordination and cooperation between supplier and manufacturer such that a part or raw material arrives just as it is needed. This approach saves on inventory and also serves as a catalyst for discovering bottlenecks and inefficiencies. It changes the manufacturer–supplier relationship. Reengineering is a radical approach to total quality management in which the core business process is redesigned. Six sigma is a methodology and a measurement. A goal of six sigma is that no more than 3.4 attributes or products per million be defective. It is essentially a philosophy of zero defects. Team building is the creation of organized groups of employees that undertake management tasks and perform other functions, such as overseeing projects.

Four diagnostic techniques used in analyzing processes are flowcharts, Pareto analysis, fishbone (cause-and-effect) diagrams, and control charts. Flowcharts are schematic representations of all activities that occur in a process. Pareto analysis is a method of examining types of defects that occur with a product. The result is usually a vertical bar chart that depicts the most common types of defects ranked in order of occurrence. The fishbone diagram displays potential causes of quality problems. The diagram is shaped like a fish skeleton, with the head being the problem and the skeletal bones being the potential causes. A control chart is a graphic method of evaluating whether a process is or is not in a state of statistical control.

Control charts are used to monitor product variation, thus enabling operators, technicians, and managers to see when a process gets out of control. The $\bar{X}$ chart and the R chart are two types of control charts for measurements. The $\bar{X}$ chart is a graph of sample means computed on a series of small random samples over time. The R chart is plot of sample ranges. The $\bar{X}$ chart plots the measure of location, whereas the R chart plots a measure of variability. The P chart and the c chart are two types of control charts for nonconformance. The P chart graphs the proportions of sample items that are in noncompliance. The c chart displays the number of nonconformances per item for a series of sampled items. All four types of control chart are plotted around a centerline and upper and lower control limits. The control limits are located three standard deviations from the centerline.

Key Terms

after-process quality control
benchmarking
c chart
cause-and-effect diagram
centerline
control chart
fishbone diagram
flowchart
in-process quality control
Ishikawa diagram
just-in-time inventory system
lower control limit (LCL)
manufacturing quality
P chart
Pareto analysis
Pareto chart

process
product quality
quality
quality circle
quality control
R chart
reengineering
six sigma
team building
total quality management (TQM)
transcendent quality
upper control limit (UCL)
user quality
value quality
$\bar{X}$ chart

SUPPLEMENTARY PROBLEMS

15.12 Create a flowchart from the following sequence of activities: Begin. Flow to activity A. Flow to decision B. If yes, flow to activity C. If no, flow to activity D. From C flow to activity E and to activity F. From F, flow to decision G. If yes, flow to decision H. If no at G, stop. At H, if yes, flow to activity I and on to activity J and then stop. If no at H, flow to activity J and stop. At D, flow to activity K, flow to L, and flow to decision M. If yes at M, stop. If no at M, flow to activity N, then stop.

15.13 An examination of rejects shows that there are at least 10 problems. A frequency tally of the problems follows. Construct a Pareto chart for these data.

PROBLEM	FREQUENCY
1	673
2	29
3	108
4	379
5	73
6	564
7	12
8	402
9	54
10	202

15.14 A brainstorm session on possible causes of a problem resulted in five possible causes: A, B, C, D, and E. Cause A has three possible subcauses, cause B has four, cause C has two, cause D has five, and cause E has three. Construct a fishbone diagram for this problem and its possible causes.

15.15 A bottled-water company has been randomly inspecting bottles of water to determine whether they are acceptable for delivery and sale. The inspectors are looking at water quality, bottle condition, and seal tightness. A series of 10 random samples of 50 bottles each has been taken. Some bottles have been rejected. Use the following information on the number of bottles from each batch that were rejected as being out of compliance to construct a P chart.

SAMPLE	n	NUMBER OUT OF COMPLIANCE
1	50	3
2	50	11
3	50	7
4	50	2
5	50	5
6	50	8
7	50	0
8	50	9
9	50	1
10	50	6

15.16 A fruit juice company sells a glass container filled with 24 ounces of cranapple juice. Inspectors are concerned about the consistency of volume of fill in these containers. Every 2 hours for 3 days of production, a sample of five containers is randomly selected and the volume of fill is measured. The results follow.

SAMPLE 1	SAMPLE 2	SAMPLE 3	SAMPLE 4
24.05	24.01	24.03	23.98
24.01	24.02	23.95	24.00
24.02	24.10	24.00	24.01
23.99	24.03	24.01	24.01
24.04	24.08	23.99	24.00

SAMPLE 5	SAMPLE 6	SAMPLE 7	SAMPLE 8
23.97	24.02	24.01	24.08
23.99	24.05	24.00	24.03
24.02	24.01	24.00	24.00
24.01	24.00	23.97	24.05
24.00	24.01	24.02	24.01

SAMPLE 9	SAMPLE 10	SAMPLE 11	SAMPLE 12
24.00	24.00	24.01	24.00
24.02	24.01	23.99	24.05
24.03	24.00	24.02	24.04
24.01	24.00	24.03	24.02
24.01	24.00	24.01	24.00

Use this information to construct $\overline{X}$ and R charts and comment on any samples that are out of compliance.

15.17 A metal-manufacturing company produces sheet metal. Statistical quality control technicians randomly select sheets to be inspected for blemishes and size problems. The number of nonconformances per sheet is tallied. Shown here are the results of testing 36 sheets of metal. Use the data to construct a c chart. What is the centerline? What is the meaning of the centerline value?

SHEET NUMBER	NUMBER OF NONCON- FORMANCES	SHEET NUMBER	NUMBER OF NONCON- FORMANCES
1	4	19	1
2	2	20	3
3	1	21	4
4	1	22	0
5	3	23	2
6	0	24	3
7	4	25	0
8	5	26	0
9	2	27	4
10	1	28	2
11	2	29	5
12	0	30	3
13	5	31	1
14	4	32	2
15	1	33	0
16	2	34	4
17	1	35	2
18	0	36	3

15.18 A manufacturing company produces cylindrical tubes for engines that are specified to be 1.20 cm thick. As part of the company's statistical quality control effort, random samples of four tubes are taken each hour. The tubes are measured to determine whether they are within thickness tolerances. Shown here are the thickness data in centimeters for nine samples of tubes. Use these data to develop an $\overline{X}$ chart and an R chart. Comment on whether or not the process appears to be in control at this point.

SAMPLE 1	SAMPLE 2	SAMPLE 3	SAMPLE 4	SAMPLE 5
1.22	1.20	1.21	1.16	1.24
1.19	1.20	1.18	1.17	1.20
1.20	1.22	1.17	1.20	1.21
1.23	1.20	1.20	1.16	1.18

SAMPLE 6	SAMPLE 7	SAMPLE 8	SAMPLE 9
1.19	1.24	1.17	1.22
1.21	1.17	1.23	1.17
1.21	1.18	1.22	1.16
1.20	1.19	1.16	1.19

15.19 A manufacturer produces digital watches. Every 2 hours a sample of six watches is selected randomly to be tested. Each watch is run for exactly 15 minutes and is timed by an accurate, precise timing device. Because of the variation among watches, they do not all run the same. Shown here are the data from eight different samples given in minutes. Use these data to construct $\bar{X}$ and R charts. Observe the results and comment on whether or not the process is in control.

SAMPLE 1	SAMPLE 2	SAMPLE 3	SAMPLE 4
15.01	15.03	14.96	15.00
14.99	14.96	14.97	15.01
14.99	15.01	14.96	14.97
15.00	15.02	14.99	15.01
14.98	14.97	15.01	14.99
14.99	15.01	14.98	14.96

SAMPLE 5	SAMPLE 6	SAMPLE 7	SAMPLE 8
15.02	15.02	15.03	14.96
15.03	15.01	15.04	14.99
14.99	14.97	15.03	15.02
15.01	15.00	15.00	15.01
15.02	15.01	15.01	14.98
15.01	14.99	14.99	15.02

15.20 A company produces outdoor home thermometers. For a variety of reasons, a thermometer can be tested and found to be out of compliance with company specification. The company takes samples of thermometers on a regular basis and tests each one to determine whether it meets company standards. Shown here are data from 12 different random samples of 75 thermometers. Use these data to construct a P chart. Comment on the pattern of points in the chart.

SAMPLE	n	NUMBER OUT OF COMPLIANCE
1	75	9
2	75	3
3	75	0
4	75	2
5	75	7
6	75	14
7	75	11
8	75	8
9	75	5
10	75	4
11	75	0
12	75	7

15.21 A plastics company makes thousands of plastic bottles for another company that manufactures saline solution for users of soft contact lenses. The plastics company randomly inspects a sample of its bottles as part of its quality control program. Inspectors look for blemishes on the bottle, size and thickness, closability, leaks, labeling problems, and so on. Shown here are the results of tests completed on 25 bottles. Use these data to construct a c chart. Observe the results and comment on the chart.

BOTTLE NUMBER	NUMBER OF NONCON-FORMANCES	BOTTLE NUMBER	NUMBER OF NONCON-FORMANCES
1	1	14	0
2	0	15	0
3	1	16	0
4	0	17	1
5	0	18	0
6	2	19	0
7	1	20	1
8	1	21	0
9	0	22	1
10	1	23	2
11	0	24	0
12	2	25	1
13	1		

15.22 A bathtub manufacturer closely inspects several tubs on every shift for nonconformances such as leaks, lack of symmetry, unstable base, drain malfunctions, and so on. The following list gives the number of nonconformances per tub for 40 tubs. Use these data to construct a c chart of nonconformances for bathtubs. Comment on the results of this chart.

TUB	NUMBER OF NONCONFORMANCES	TUB	NUMBER OF NONCONFORMANCES
1	3	21	1
2	2	22	0
3	3	23	2
4	1	24	1
5	4	25	3
6	2	26	2
7	2	27	2
8	1	28	1
9	4	29	0
10	2	30	4
11	3	31	3
12	0	32	2
13	2	33	2
14	5	34	1
15	1	35	1
16	3	36	1
17	4	37	3
18	3	38	0
19	2	39	1
20	0	40	4

15.23 A glass manufacturer produces hand mirrors. Each mirror is supposed to meet company standards for such things as glass thickness, reflectability, size of handle, quality of glass, color of handle, and so on. To control for these features, the company quality people randomly sample 40 mirrors every shift and determine how many of the mirrors are out of compliance on at least one feature. Shown here are the data for 15 such samples. Use the data to construct a P chart. Observe the results and comment on the control of the process as indicated by the chart.

SAMPLE	n	NUMBER OUT OF COMPLIANCE
1	40	2
2	40	0
3	40	6
4	40	3
5	40	1
6	40	1
7	40	5
8	40	0
9	40	4
10	40	3
11	40	2
12	40	2
13	40	6
14	40	1
15	40	0

15.24 Study the FAST ＼ STAT $\overline{X}$ chart on the fill of a product that is supposed to contain 12 ounces. Does the process appear to be out of control? Why or why not?

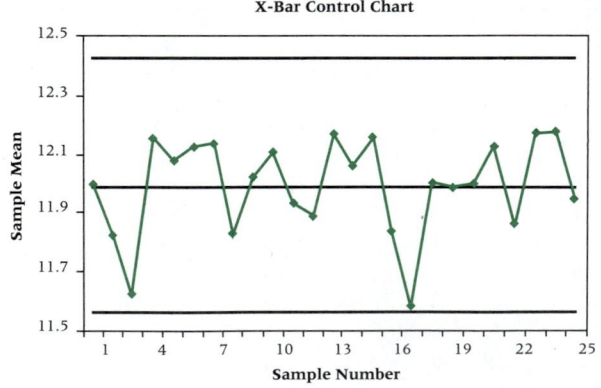

15.25 Study the FAST ＼ STAT R chart for the product and data used in Problem 15.24. Comment on the state of the production process for this item.

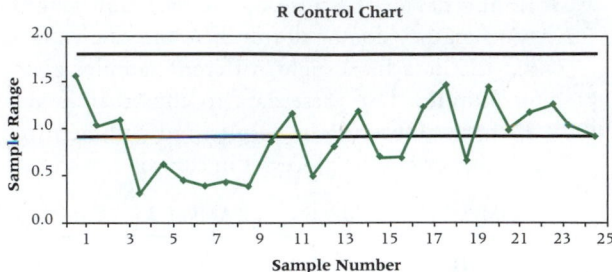

15.26 Study the FAST ＼ STAT P chart for a manufactured item. The chart represents the results of testing 30 items at a time for compliance. Twenty-five different samples were taken for this chart. Discuss the results and the implications for the production process.

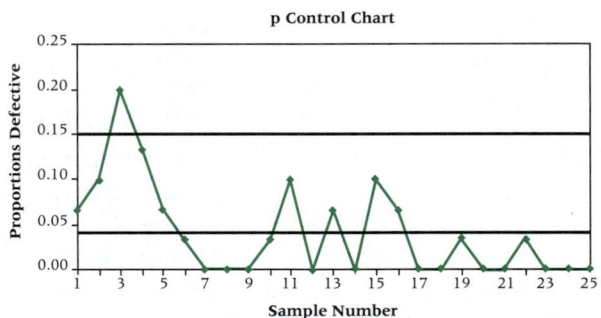

15.27 Study the FAST ＼ STAT c chart for nonconformances for a part produced in a manufacturing process. Comment on the results.

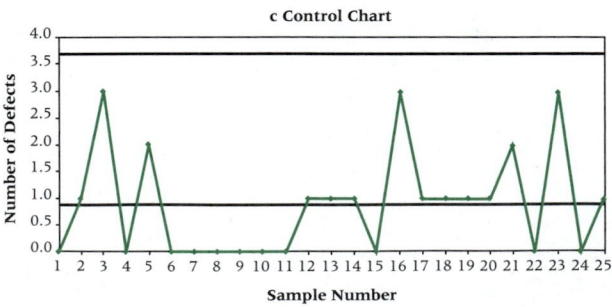

ANALYZING THE DATABASES

1. A dairy company in the manufacturing database has been testing its quart milk container fills for volume in four-container samples. Shown here are the results of 10 such samples and the volume measurements in quarts. Use the information to construct both an $\overline{X}$ and an R chart for the data. Discuss the results. What are the centerlines, LCL, and UCL for each of these charts?

SAMPLE NUMBER	MEASUREMENTS			
1	.98	1.01	1.05	1.03
2	1.02	.94	.97	1.02
3	1.11	1.02	.93	1.01
4	.95	.98	1.02	.96
5	1.03	1.01	1.01	.95
6	1.04	.93	.91	.96
7	.94	1.12	1.10	1.03
8	1.03	.92	.98	1.03
9	1.01	1.01	.99	1.00
10	1.05	.96	1.00	1.04

2. A hospital in the hospital database takes weekly samples of patient account statements for 12 weeks with each sample containing 40 accounts. Auditors analyze the account statements, looking for nonconforming statements. Shown here are the results of the 12 samples. Use these data to construct a P chart for proportion of nonconforming statements. What is the centerline? What are UCL and LCL? Comment on the control chart.

SAMPLE	NUMBER OF NONCONFORMING STATEMENTS
1	1
2	0
3	6
4	3
5	0
6	2
7	8
8	3
9	5
10	2
11	2
12	1

CASE

ROBOTRON

Robotron Corporation of Southfield, Michigan, has been a manufacturer of bonding products for the automobile industry for more than two decades. For several years, Robotron felt it produced and delivered a high-quality product because it rarely received complaints from customers. However, early in the 1980s General Motors gave Robotron an order for induction bonding machines to cure adhesive in auto door joints. Actually, the machines were shipped to a Michigan plant of Sanyo Manufacturing that GM was using as a door builder.

The Japanese firm was very unhappy with the quality of the machines. Robotron President Leonard Brzozowski went to the Sanyo plant to investigate the situation in person and learned that the Japanese had a much higher quality standard than the usual American customers. Tolerances were much smaller and inspection was more demanding. Brzozowski said that he realized for the first time that the philosophy, engineering, management, and shop practices of Robotron did not qualify the company for world competition. Brzozowski said that this was the most embarrassing time of his professional career. What should Robotron do about this situation?

Brzozowski began by having a group of hourly employees driven to the Sanyo plant. There they met the customer and heard firsthand the many complaints about their product. The workers could see the difference in quality between their machines and those of Sanyo. The plant visit was extremely effective. On the way home, the Robotron workers started discussing things they could do to improve quality.

The company took several steps to begin the process of quality improvement. It established new inspection procedures, bought more accurate inspection tools, changed internal control procedures, and developed baselines against which to measure progress. Teams were organized and sent out to customers six months after a purchase to determine customer satisfaction. A hotline was established for customers to call to report product dissatisfaction.

For one month, engineers assembled machinery in the shop under the direction of hourly employees. This gave the engineers a new awareness of the importance of accurate, clear drawings; designing smaller, lighter weight details; and minimizing the number of machined surfaces.

Robotron's effort has paid off handsomely. Claims under warranty dropped 40% in three years, during which time orders rose at a compound annual rate of 13.5%. The company cut costs and streamlined procedures and processes. Sales increased and new markets have been opened.

Early in 1998, Robotron merged with ELOTHERM, a European company, so that Robotron could more easily enjoy a presence in the European market and at the same time provide ELOTHERM opportunities in North America. Robotron's bonding business will expand into induction heating systems, heat treating, tube welding, and electrical discharge machines. In 1997, Robotron received ISO-9001 certification. The company maintains a quality system that takes corrective action in response to customer complaints, employee suggestions, or supplier defects. Robotron currently employs 140 people in Michigan.

DISCUSSION

1. As a part of quality improvement, it is highly likely that Robotron analyzed its manufacturing processes. Suppose that as Robotron improved quality, the company wanted to examine other processes including the flow of work orders from the time they are received until they are filled. Following is a verbal sketch of some of the activities that might take place in such a flow. Using this as a start, add some of your own ideas and draw a flowchart for the process of work orders.

    Work order flow: Received at mail room. Sent to order processing office. Examined by order processing clerk who decides whether the item is a standard item or a custom item. If it is a standard item, the order is sent to the warehouse. If the item is available, the item is shipped and the order is sent to the billing department. If the item is not available, the order is sent to the plant where it is received by a manufacturing clerk. The clerk checks to determine whether such an item is being manufactured. If so, the order is sent to the end of the assembly line where it will be tagged with one such item. If not, the order is sent to the beginning of the assembly line and flows along the assembly line with the item as it is being made. In either case, when the part comes off the assembly line, the order is attached to the item and they are sent to shipping. They shipping clerk then ships the item and sends the order to billing. If the ordered item is a customized part, the order is sent straight from the order processing clerk to manufacturing where it follows the same procedures as enumerated above for standard items that have not been manufactured yet.

2. Virtually all quality manufacturers use some type of control chart to monitor performance. Suppose the FAST ⟍ STAT control charts shown here are for two different parts produced by Robotron during a particular period. Part 173 is specified to weigh 183 grams. Part 248 contains an opening that is specified to be 35 mm in diameter. Study these charts and report to Robotron what you found. Is there any reason for concern? Is everything in control?

X-Bar Control Chart

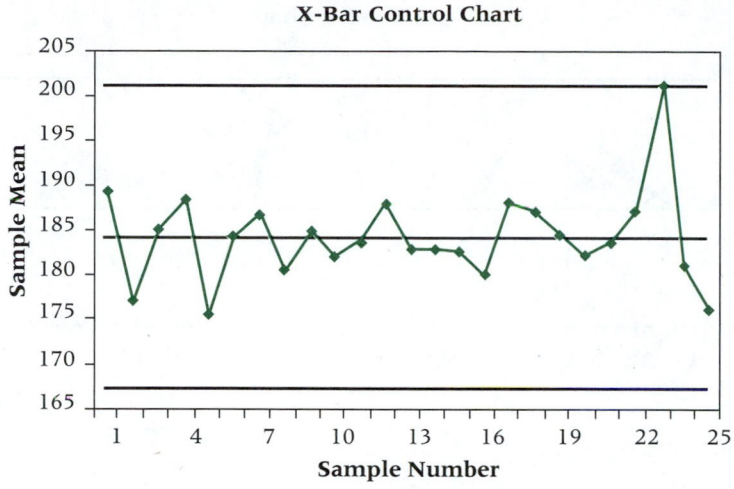

R Control Chart

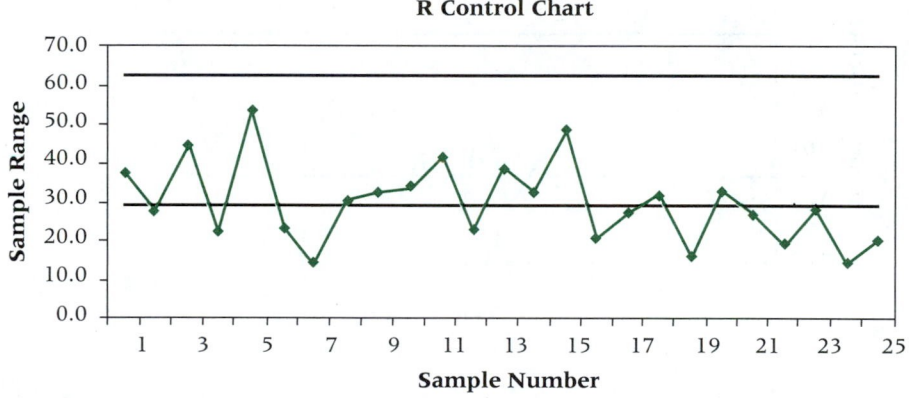

X-Bar Control Chart

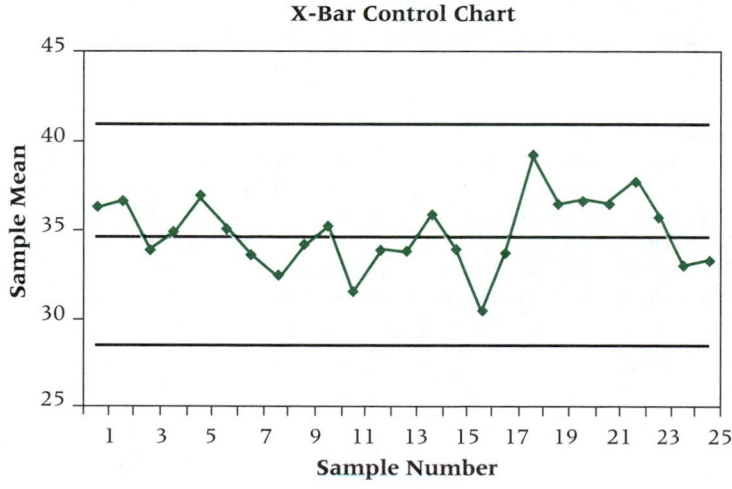

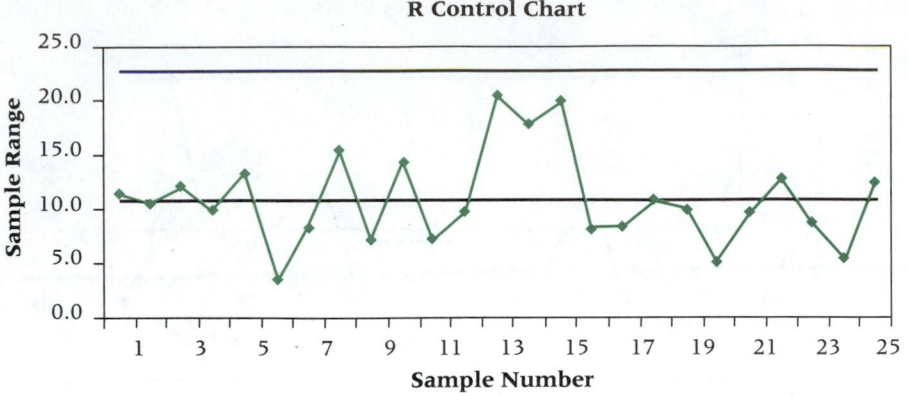

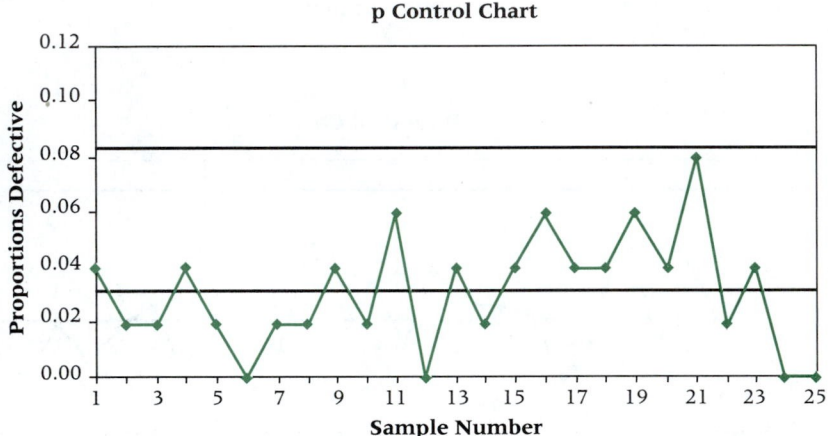

3. Suppose Robotron also keeps *P* charts on nonconformance. The FAST ⦚ STAT chart shown here represents the proportion of nonconforming items for a given part over 25 samples. Study the chart and write a brief report to Robotron about what you learned from the chart. Think about overall performance, out-of-control samples, samples that have outstanding performance, and any general trends that you see.

ADAPTED FROM: "ROBOTRON: Qualifying For Global Competition," *Real-World Lessons for America's Small Businesses: Insights from the Blue Chip Enterprise Initiative.* Published by *Nation's Business* magazine on behalf of Connecticut Mutual Life Insurance Company and the U.S. Chamber of Commerce in association with The Blue Chip Enterprise Initiative 1994. See also Robotron's Internet websites at http://www.robotron.com/ aboutus/iso9001.html; http://www.robotron.com/products; http://www.robotron.com//aboutus/merger.html; http://www.robotron.com/aboutus/jobs.html

A

Tables

Table A.1
Random Numbers

12651	61646	11769	75109	86996	97669	25757	32535	07122	76763
81769	74436	02630	72310	45049	18029	07469	42341	98173	79260
36737	98863	77240	76251	00654	64688	09343	70278	67331	98729
82861	54371	76610	94934	72748	44124	05610	53750	95938	01485
21325	15732	24127	37431	09723	63529	73977	95218	96074	42138
74146	47887	62463	23045	41490	07954	22597	60012	98866	90959
90759	64410	54179	66075	61051	75385	51378	08360	95946	95547
55683	98078	02238	91540	21219	17720	87817	41705	95785	12563
79686	17969	76061	83748	55920	83612	41540	86492	06447	60568
70333	00201	86201	69716	78185	62154	77930	67663	29529	75116
14042	53536	07779	04157	41172	36473	42123	43929	50533	33437
59911	08256	06596	48416	69770	68797	56080	14223	59199	30162
62368	62623	62742	14891	39247	52242	98832	69533	91174	57979
57529	97751	54976	48957	74599	08759	78494	52785	68526	64618
15469	90574	78033	66885	13936	42117	71831	22961	94225	31816
18625	23674	53850	32827	81647	80820	00420	63555	74489	80141
74626	68394	88562	70745	23701	45630	65891	58220	35442	60414
11119	16519	27384	90199	79210	76965	99546	30323	31664	22845
41101	17336	48951	53674	17880	45260	08575	49321	36191	17095
32123	91576	84221	78902	82010	30847	62329	63898	23268	74283
26091	68409	69704	82267	14751	13151	93115	01437	56945	89661
67680	79790	48462	59278	44185	29616	76531	19589	83139	28454
15184	19260	14073	07026	25264	08388	27182	22557	61501	67481
58010	45039	57181	10238	36874	28546	37444	80824	63981	39942
56425	53996	86245	32623	78858	08143	60377	42925	42815	11159
82630	84066	13592	60642	17904	99718	63432	88642	37858	25431
14927	40909	23900	48761	44860	92467	31742	87142	03607	32059
23740	22505	07489	85986	74420	21744	97711	36648	35620	97949
32990	97446	03711	63824	07953	85965	87089	11687	92414	67257
05310	24058	91946	78437	34365	82469	12430	84754	19354	72745
21839	39937	27534	88913	49055	19218	47712	67677	51889	70926
08833	42549	93981	94051	28382	83725	72643	64233	97252	17133
58336	11139	47479	00931	91560	95372	97642	33856	54825	55680
62032	91144	75478	47431	52726	30289	42411	91886	51818	78292
45171	30557	53116	04118	58301	24375	65609	85810	18620	49198
91611	62656	60128	35609	63698	78356	50682	22505	01692	36291
55472	63819	86314	49174	93582	73604	78614	78849	23096	72825
18573	09729	74091	53994	10970	86557	65661	41854	26037	53296
60866	02955	90288	82136	83644	94455	06560	78029	98768	71296
45043	55608	82767	60890	74646	79485	13619	98868	40857	19415
17831	09737	79473	75945	28394	79334	70577	38048	03607	06932
40137	03981	07585	18128	11178	32601	27994	05641	22600	86064
77776	31343	14576	97706	16039	47517	43300	59080	80392	63189
69605	44104	40103	95635	05635	81673	68657	09559	23510	95875
19916	52934	26499	09821	97331	80993	61299	36979	73599	35055
02606	58552	07678	56619	65325	30705	99582	53390	46357	13244
65183	73160	87131	35530	47946	09854	18080	02321	05809	04893
10740	98914	44916	11322	89717	88189	30143	52687	19420	60061
98642	89822	71691	51573	83666	61642	46683	33761	47542	23551
60139	25601	93663	25547	02654	94829	48672	28736	84994	13071

Table A.2
Binomial Probability
Distribution

					$n = 1$				
					Probability				
X	.1	.2	.3	.4	.5	.6	.7	.8	.9
0	.900	.800	.700	.600	.500	.400	.300	.200	.100
1	.100	.200	.300	.400	.500	.600	.700	.800	.900

					$n = 2$				
					Probability				
X	.1	.2	.3	.4	.5	.6	.7	.8	.9
0	.810	.640	.490	.360	.250	.160	.090	.040	.010
1	.180	.320	.420	.480	.500	.480	.420	.320	.180
2	.010	.040	.090	.160	.250	.360	.490	.640	.810

					$n = 3$				
					Probability				
X	.1	.2	.3	.4	.5	.6	.7	.8	.9
0	.729	.512	.343	.216	.125	.064	.027	.008	.001
1	.243	.384	.441	.432	.375	.288	.189	.096	.027
2	.027	.096	.189	.288	.375	.432	.441	.384	.243
3	.001	.008	.027	.064	.125	.216	.343	.512	.729

					$n = 4$				
					Probability				
X	.1	.2	.3	.4	.5	.6	.7	.8	.9
0	.656	.410	.240	.130	.063	.026	.008	.002	.000
1	.292	.410	.412	.346	.250	.154	.076	.026	.004
2	.049	.154	.265	.346	.375	.346	.265	.154	.049
3	.004	.026	.076	.154	.250	.346	.412	.410	.292
4	.000	.002	.008	.026	.063	.130	.240	.410	.656

					$n = 5$				
					Probability				
X	.1	.2	.3	.4	.5	.6	.7	.8	.9
0	.590	.328	.168	.078	.031	.010	.002	.000	.000
1	.328	.410	.360	.259	.156	.077	.028	.006	.000
2	.073	.205	.309	.346	.313	.230	.132	.051	.008
3	.008	.051	.132	.230	.313	.346	.309	.205	.073
4	.000	.006	.028	.077	.156	.259	.360	.410	.328
5	.000	.000	.002	.010	.031	.078	.168	.328	.590

					$n = 6$				
					Probability				
X	.1	.2	.3	.4	.5	.6	.7	.8	.9
0	.531	.262	.118	.047	.016	.004	.001	.000	.000
1	.354	.393	.303	.187	.094	.037	.010	.002	.000
2	.098	.246	.324	.311	.234	.138	.060	.015	.001
3	.015	.082	.185	.276	.313	.276	.185	.082	.015
4	.001	.015	.060	.138	.234	.311	.324	.246	.098
5	.000	.002	.010	.037	.094	.187	.303	.393	.354
6	.000	.000	.001	.004	.016	.047	.118	.262	.531

Continued

					$n = 7$				
					Probability				
X	.1	.2	.3	.4	.5	.6	.7	.8	.9
0	.478	.210	.082	.028	.008	.002	.000	.000	.000
1	.372	.367	.247	.131	.055	.017	.004	.000	.000
2	.124	.275	.318	.261	.164	.077	.025	.004	.000
3	.023	.115	.227	.290	.273	.194	.097	.029	.003
4	.003	.029	.097	.194	.273	.290	.227	.115	.023
5	.000	.004	.025	.077	.164	.261	.318	.275	.124
6	.000	.000	.004	.017	.055	.131	.247	.367	.372
7	.000	.000	.000	.002	.008	.028	.082	.210	.478

					$n = 8$				
					Probability				
X	.1	.2	.3	.4	.5	.6	.7	.8	.9
0	.430	.168	.058	.017	.004	.001	.000	.000	.000
1	.383	.336	.198	.090	.031	.008	.001	.000	.000
2	.149	.294	.296	.209	.109	.041	.010	.001	.000
3	.033	.147	.254	.279	.219	.124	.047	.009	.000
4	.005	.046	.136	.232	.273	.232	.136	.046	.005
5	.000	.009	.047	.124	.219	.279	.254	.147	.033
6	.000	.001	.010	.041	.109	.209	.296	.294	.149
7	.000	.000	.001	.008	.031	.090	.198	.336	.383
8	.000	.000	.000	.001	.004	.017	.058	.168	.430

					$n = 9$				
					Probability				
X	.1	.2	.3	.4	.5	.6	.7	.8	.9
0	.387	.134	.040	.010	.002	.000	.000	.000	.000
1	.387	.302	.156	.060	.018	.004	.000	.000	.000
2	.172	.302	.267	.161	.070	.021	.004	.000	.000
3	.045	.176	.267	.251	.164	.074	.021	.003	.000
4	.007	.066	.172	.251	.246	.167	.074	.017	.001
5	.001	.017	.074	.167	.246	.251	.172	.066	.007
6	.000	.003	.021	.074	.164	.251	.267	.176	.045
7	.000	.000	.004	.021	.070	.161	.267	.302	.172
8	.000	.000	.000	.004	.018	.060	.156	.302	.387
9	.000	.000	.000	.000	.002	.010	.040	.134	.387

					$n = 10$				
					Probability				
X	.1	.2	.3	.4	.5	.6	.7	.8	.9
0	.349	.107	.028	.006	.001	.000	.000	.000	.000
1	.387	.268	.121	.040	.010	.002	.000	.000	.000
2	.194	.302	.233	.121	.044	.011	.001	.000	.000
3	.057	.201	.267	.215	.117	.042	.009	.001	.000
4	.011	.088	.200	.251	.205	.111	.037	.006	.000
5	.001	.026	.103	.201	.246	.201	.103	.026	.001
6	.000	.006	.037	.111	.205	.251	.200	.088	.011
7	.000	.001	.009	.042	.117	.215	.267	.201	.057
8	.000	.000	.001	.011	.044	.121	.233	.302	.194
9	.000	.000	.000	.002	.010	.040	.121	.268	.387
10	.000	.000	.000	.000	.001	.006	.028	.107	.349

					$n = 11$				
					Probability				
X	.1	.2	.3	.4	.5	.6	.7	.8	.9
0	.314	.086	.020	.004	.000	.000	.000	.000	.000
1	.384	.236	.093	.027	.005	.001	.000	.000	.000
2	.213	.295	.200	.089	.027	.005	.001	.000	.000
3	.071	.221	.257	.177	.081	.023	.004	.000	.000
4	.016	.111	.220	.236	.161	.070	.017	.002	.000
5	.002	.039	.132	.221	.226	.147	.057	.010	.000
6	.000	.010	.057	.147	.226	.221	.132	.039	.002
7	.000	.002	.017	.070	.161	.236	.220	.111	.016
8	.000	.000	.004	.023	.081	.177	.257	.221	.071
9	.000	.000	.001	.005	.027	.089	.200	.295	.213
10	.000	.000	.000	.001	.005	.027	.093	.236	.384
11	.000	.000	.000	.000	.000	.004	.020	.086	.314

					$n = 12$				
					Probability				
X	.1	.2	.3	.4	.5	.6	.7	.8	.9
0	.282	.069	.014	.002	.000	.000	.000	.000	.000
1	.377	.206	.071	.017	.003	.000	.000	.000	.000
2	.230	.283	.168	.064	.016	.002	.000	.000	.000
3	.085	.236	.240	.142	.054	.012	.001	.000	.000
4	.021	.133	.231	.213	.121	.042	.008	.001	.000
5	.004	.053	.158	.227	.193	.101	.029	.003	.000
6	.000	.016	.079	.177	.226	.177	.079	.016	.000
7	.000	.003	.029	.101	.193	.227	.158	.053	.004
8	.000	.001	.008	.042	.121	.213	.231	.133	.021
9	.000	.000	.001	.012	.054	.142	.240	.236	.085
10	.000	.000	.000	.002	.016	.064	.168	.283	.230
11	.000	.000	.000	.000	.003	.017	.071	.206	.377
12	.000	.000	.000	.000	.000	.002	.014	.069	.282

					$n = 13$				
					Probability				
X	.1	.2	.3	.4	.5	.6	.7	.8	.9
0	.254	.055	.010	.001	.000	.000	.000	.000	.000
1	.367	.179	.054	.011	.002	.000	.000	.000	.000
2	.245	.268	.139	.045	.010	.001	.000	.000	.000
3	.100	.246	.218	.111	.035	.006	.001	.000	.000
4	.028	.154	.234	.184	.087	.024	.003	.000	.000
5	.006	.069	.180	.221	.157	.066	.014	.001	.000
6	.001	.023	.103	.197	.209	.131	.044	.006	.000
7	.000	.006	.044	.131	.209	.197	.103	.023	.001
8	.000	.001	.014	.066	.157	.221	.180	.069	.006
9	.000	.000	.003	.024	.087	.184	.234	.154	.028
10	.000	.000	.001	.006	.035	.111	.218	.246	.100
11	.000	.000	.000	.001	.010	.045	.139	.268	.245
12	.000	.000	.000	.000	.002	.011	.054	.179	.367
13	.000	.000	.000	.000	.000	.001	.010	.055	.254

Continued

					n = 14				
					Probability				
X	.1	.2	.3	.4	.5	.6	.7	.8	.9
0	.229	.044	.007	.001	.000	.000	.000	.000	.000
1	.356	.154	.041	.007	.001	.000	.000	.000	.000
2	.257	.250	.113	.032	.006	.001	.000	.000	.000
3	.114	.250	.194	.085	.022	.003	.000	.000	.000
4	.035	.172	.229	.155	.061	.014	.001	.000	.000
5	.008	.086	.196	.207	.122	.041	.007	.000	.000
6	.001	.032	.126	.207	.183	.092	.023	.002	.000
7	.000	.009	.062	.157	.209	.157	.062	.009	.000
8	.000	.002	.023	.092	.183	.207	.126	.032	.001
9	.000	.000	.007	.041	.122	.207	.196	.086	.008
10	.000	.000	.001	.014	.061	.155	.229	.172	.035
11	.000	.000	.000	.003	.022	.085	.194	.250	.114
12	.000	.000	.000	.001	.006	.032	.113	.250	.257
13	.000	.000	.000	.000	.001	.007	.041	.154	.356
14	.000	.000	.000	.000	.000	.001	.007	.044	.229

					n = 15				
					Probability				
X	.1	.2	.3	.4	.5	.6	.7	.8	.9
0	.206	.035	.005	.000	.000	.000	.000	.000	.000
1	.343	.132	.031	.005	.000	.000	.000	.000	.000
2	.267	.231	.092	.022	.003	.000	.000	.000	.000
3	.129	.250	.170	.063	.014	.002	.000	.000	.000
4	.043	.188	.219	.127	.042	.007	.001	.000	.000
5	.010	.103	.206	.186	.092	.024	.003	.000	.000
6	.002	.043	.147	.207	.153	.061	.012	.001	.000
7	.000	.014	.081	.177	.196	.118	.035	.003	.000
8	.000	.003	.035	.118	.196	.177	.081	.014	.000
9	.000	.001	.012	.061	.153	.207	.147	.043	.002
10	.000	.000	.003	.024	.092	.186	.206	.103	.010
11	.000	.000	.001	.007	.042	.127	.219	.188	.043
12	.000	.000	.000	.002	.014	.063	.170	.250	.129
13	.000	.000	.000	.000	.003	.022	.092	.231	.267
14	.000	.000	.000	.000	.000	.005	.031	.132	.343
15	.000	.000	.000	.000	.000	.000	.005	.035	.206

					n = 16				
					Probability				
X	.1	.2	.3	.4	.5	.6	.7	.8	.9
0	.185	.028	.003	.000	.000	.000	.000	.000	.000
1	.329	.113	.023	.003	.000	.000	.000	.000	.000
2	.275	.211	.073	.015	.002	.000	.000	.000	.000
3	.142	.246	.146	.047	.009	.001	.000	.000	.000
4	.051	.200	.204	.101	.028	.004	.000	.000	.000
5	.014	.120	.210	.162	.067	.014	.001	.000	.000
6	.003	.055	.165	.198	.122	.039	.006	.000	.000
7	.000	.020	.101	.189	.175	.084	.019	.001	.000
8	.000	.006	.049	.142	.196	.142	.049	.006	.000
9	.000	.001	.019	.084	.175	.189	.101	.020	.000
10	.000	.000	.006	.039	.122	.198	.165	.055	.003
11	.000	.000	.001	.014	.067	.162	.210	.120	.014
12	.000	.000	.000	.004	.028	.101	.204	.200	.051
13	.000	.000	.000	.001	.009	.047	.146	.246	.142
14	.000	.000	.000	.000	.002	.015	.073	.211	.275
15	.000	.000	.000	.000	.000	.003	.023	.113	.329
16	.000	.000	.000	.000	.000	.000	.003	.028	.185

				$n = 17$					
				Probability					
X	.1	.2	.3	.4	.5	.6	.7	.8	.9
0	.167	.023	.002	.000	.000	.000	.000	.000	.000
1	.315	.096	.017	.002	.000	.000	.000	.000	.000
2	.280	.191	.058	.010	.001	.000	.000	.000	.000
3	.156	.239	.125	.034	.005	.000	.000	.000	.000
4	.060	.209	.187	.080	.018	.002	.000	.000	.000
5	.017	.136	.208	.138	.047	.008	.001	.000	.000
6	.004	.068	.178	.184	.094	.024	.003	.000	.000
7	.001	.027	.120	.193	.148	.057	.009	.000	.000
8	.000	.008	.064	.161	.185	.107	.028	.002	.000
9	.000	.002	.028	.107	.185	.161	.064	.008	.000
10	.000	.000	.009	.057	.148	.193	.120	.027	.001
11	.000	.000	.003	.024	.094	.184	.178	.068	.004
12	.000	.000	.001	.008	.047	.138	.208	.136	.017
13	.000	.000	.000	.002	.018	.080	.187	.209	.060
14	.000	.000	.000	.000	.005	.034	.125	.239	.156
15	.000	.000	.000	.000	.001	.010	.058	.191	.280
16	.000	.000	.000	.000	.000	.002	.017	.096	.315
17	.000	.000	.000	.000	.000	.000	.002	.023	.167

				$n = 18$					
				Probability					
X	.1	.2	.3	.4	.5	.6	.7	.8	.9
0	.150	.018	.002	.000	.000	.000	.000	.000	.000
1	.300	.081	.013	.001	.000	.000	.000	.000	.000
2	.284	.172	.046	.007	.001	.000	.000	.000	.000
3	.168	.230	.105	.025	.003	.000	.000	.000	.000
4	.070	.215	.168	.061	.012	.001	.000	.000	.000
5	.022	.151	.202	.115	.033	.004	.000	.000	.000
6	.005	.082	.187	.166	.071	.015	.001	.000	.000
7	.001	.035	.138	.189	.121	.037	.005	.000	.000
8	.000	.012	.081	.173	.167	.077	.015	.001	.000
9	.000	.003	.039	.128	.185	.128	.039	.003	.000
10	.000	.001	.015	.077	.167	.173	.081	.012	.000
11	.000	.000	.005	.037	.121	.189	.138	.035	.001
12	.000	.000	.001	.015	.071	.166	.187	.082	.005
13	.000	.000	.000	.004	.033	.115	.202	.151	.022
14	.000	.000	.000	.001	.012	.061	.168	.215	.070
15	.000	.000	.000	.000	.003	.025	.105	.230	.168
16	.000	.000	.000	.000	.001	.007	.046	.172	.284
17	.000	.000	.000	.000	.000	.001	.013	.081	.300
18	.000	.000	.000	.000	.000	.000	.002	.018	.150

Continued

Table A.2
Continued

					n = 19				
					Probability				
X	.1	.2	.3	.4	.5	.6	.7	.8	.9
0	.135	.014	.001	.000	.000	.000	.000	.000	.000
1	.285	.068	.009	.001	.000	.000	.000	.000	.000
2	.285	.154	.036	.005	.000	.000	.000	.000	.000
3	.180	.218	.087	.017	.002	.000	.000	.000	.000
4	.080	.218	.149	.047	.007	.001	.000	.000	.000
5	.027	.164	.192	.093	.022	.002	.000	.000	.000
6	.007	.095	.192	.145	.052	.008	.001	.000	.000
7	.001	.044	.153	.180	.096	.024	.002	.000	.000
8	.000	.017	.098	.180	.144	.053	.008	.000	.000
9	.000	.005	.051	.146	.176	.098	.022	.001	.000
10	.000	.001	.022	.098	.176	.146	.051	.005	.000
11	.000	.000	.008	.053	.144	.180	.098	.017	.000
12	.000	.000	.002	.024	.096	.180	.153	.044	.001
13	.000	.000	.001	.008	.052	.145	.192	.095	.007
14	.000	.000	.000	.002	.022	.093	.192	.164	.027
15	.000	.000	.000	.001	.007	.047	.149	.218	.080
16	.000	.000	.000	.000	.002	.017	.087	.218	.180
17	.000	.000	.000	.000	.000	.005	.036	.154	.285
18	.000	.000	.000	.000	.000	.001	.009	.068	.285
19	.000	.000	.000	.000	.000	.000	.001	.014	.135

					n = 20				
					Probability				
X	.1	.2	.3	.4	.5	.6	.7	.8	.9
0	.122	.012	.001	.000	.000	.000	.000	.000	.000
1	.270	.058	.007	.000	.000	.000	.000	.000	.000
2	.285	.137	.028	.003	.000	.000	.000	.000	.000
3	.190	.205	.072	.012	.001	.000	.000	.000	.000
4	.090	.218	.130	.035	.005	.000	.000	.000	.000
5	.032	.175	.179	.075	.015	.001	.000	.000	.000
6	.009	.109	.192	.124	.037	.005	.000	.000	.000
7	.002	.055	.164	.166	.074	.015	.001	.000	.000
8	.000	.022	.114	.180	.120	.035	.004	.000	.000
9	.000	.007	.065	.160	.160	.071	.012	.000	.000
10	.000	.002	.031	.117	.176	.117	.031	.002	.000
11	.000	.000	.012	.071	.160	.160	.065	.007	.000
12	.000	.000	.004	.035	.120	.180	.114	.022	.000
13	.000	.000	.001	.015	.074	.166	.164	.055	.002
14	.000	.000	.000	.005	.037	.124	.192	.109	.009
15	.000	.000	.000	.001	.015	.075	.179	.175	.032
16	.000	.000	.000	.000	.005	.035	.130	.218	.090
17	.000	.000	.000	.000	.001	.012	.072	.205	.190
18	.000	.000	.000	.000	.000	.003	.028	.137	.285
19	.000	.000	.000	.000	.000	.000	.007	.058	.270
20	.000	.000	.000	.000	.000	.000	.001	.012	.122

				$n = 25$					
				Probability					
X	.1	.2	.3	.4	.5	.6	.7	.8	.9
0	.072	.004	.000	.000	.000	.000	.000	.000	.000
1	.199	.024	.001	.000	.000	.000	.000	.000	.000
2	.266	.071	.007	.000	.000	.000	.000	.000	.000
3	.226	.136	.024	.002	.000	.000	.000	.000	.000
4	.138	.187	.057	.007	.000	.000	.000	.000	.000
5	.065	.196	.103	.020	.002	.000	.000	.000	.000
6	.024	.163	.147	.044	.005	.000	.000	.000	.000
7	.007	.111	.171	.080	.014	.001	.000	.000	.000
8	.002	.062	.165	.120	.032	.003	.000	.000	.000
9	.000	.029	.134	.151	.061	.009	.000	.000	.000
10	.000	.012	.092	.161	.097	.021	.001	.000	.000
11	.000	.004	.054	.147	.133	.043	.004	.000	.000
12	.000	.001	.027	.114	.155	.076	.011	.000	.000
13	.000	.000	.011	.076	.155	.114	.027	.001	.000
14	.000	.000	.004	.043	.133	.147	.054	.004	.000
15	.000	.000	.001	.021	.097	.161	.092	.012	.000
16	.000	.000	.000	.009	.061	.151	.134	.029	.000
17	.000	.000	.000	.003	.032	.120	.165	.062	.002
18	.000	.000	.000	.001	.014	.080	.171	.111	.007
19	.000	.000	.000	.000	.005	.044	.147	.163	.024
20	.000	.000	.000	.000	.002	.020	.103	.196	.065
21	.000	.000	.000	.000	.000	.007	.057	.187	.138
22	.000	.000	.000	.000	.000	.002	.024	.136	.226
23	.000	.000	.000	.000	.000	.000	.007	.071	.266
24	.000	.000	.000	.000	.000	.000	.001	.024	.199
25	.000	.000	.000	.000	.000	.000	.000	.004	.072

Table A.3
Poisson Probabilities $\dfrac{\lambda^X e^{-\lambda}}{X!}$

	λ									
X	0.005	0.01	0.02	0.03	0.04	0.05	0.06	0.07	0.08	0.09
0	.9950	.9900	.9802	.9704	.9608	.9512	.9418	.9324	.9231	.9139
1	.0050	.0099	.0192	.0291	.0384	.0476	.0565	.0653	.0738	.0823
2	.0000	.0000	.0002	.0004	.0008	.0012	.0017	.0023	.0030	.0037
3	.0000	.0000	.0000	.0000	.0000	.0000	.0000	.0001	.0001	.0001

X	0.1	0.2	0.3	0.4	0.5	0.6	0.7	0.8	0.9	1.0
0	.9048	.8187	.7408	.6703	.6065	.5488	.4966	.4493	.4066	.3679
1	.0905	.1637	.2222	.2681	.3033	.3293	.3476	.3595	.3659	.3679
2	.0045	.0164	.0333	.0536	.0758	.0988	.1217	.1438	.1647	.1839
3	.0002	.0011	.0033	.0072	.0126	.0198	.0284	.0383	.0494	.0613
4	.0000	.0001	.0002	.0007	.0016	.0030	.0050	.0077	.0111	.0153
5	.0000	.0000	.0000	.0001	.0002	.0004	.0007	.0012	.0020	.0031
6	.0000	.0000	.0000	.0000	.0000	.0000	.0001	.0002	.0003	.0005
7	.0000	.0000	.0000	.0000	.0000	.0000	.0000	.0000	.0000	.0001

X	1.1	1.2	1.3	1.4	1.5	1.6	1.7	1.8	1.9	2.0
0	.3329	.3012	.2725	.2466	.2231	.2019	.1827	.1653	.1496	.1353
1	.3662	.3614	.3543	.3452	.3347	.3230	.3106	.2975	.2842	.2707
2	.2014	.2169	.2303	.2417	.2510	.2584	.2640	.2678	.2700	.2707
3	.0738	.0867	.0998	.1128	.1255	.1378	.1496	.1607	.1710	.1804
4	.0203	.0260	.0324	.0395	.0471	.0551	.0636	.0723	.0812	.0902
5	.0045	.0062	.0084	.0111	.0141	.0176	.0216	.0260	.0309	.0361
6	.0008	.0012	.0018	.0026	.0035	.0047	.0061	.0078	.0098	.0120
7	.0001	.0002	.0003	.0005	.0008	.0011	.0015	.0020	.0027	.0034
8	.0000	.0000	.0001	.0001	.0001	.0002	.0003	.0005	.0006	.0009
9	.0000	.0000	.0000	.0000	.0000	.0000	.0001	.0001	.0001	.0002

X	2.1	2.2	2.3	2.4	2.5	2.6	2.7	2.8	2.9	3.0
0	.1225	.1108	.1003	.0907	.0821	.0743	.0672	.0608	.0550	.0498
1	.2572	.2438	.2306	.2177	.2052	.1931	.1815	.1703	.1596	.1494
2	.2700	.2681	.2652	.2613	.2565	.2510	.2450	.2384	.2314	.2240
3	.1890	.1966	.2033	.2090	.2138	.2176	.2205	.2225	.2237	.2240
4	.0992	.1082	.1169	.1254	.1336	.1414	.1488	.1557	.1622	.1680
5	.0417	.0476	.0538	.0602	.0668	.0735	.0804	.0872	.0940	.1008
6	.0146	.0174	.0206	.0241	.0278	.0319	.0362	.0407	.0455	.0504
7	.0044	.0055	.0068	.0083	.0099	.0118	.0139	.0163	.0188	.0216
8	.0011	.0015	.0019	.0025	.0031	.0038	.0047	.0057	.0068	.0081
9	.0003	.0004	.0005	.0007	.0009	.0011	.0014	.0018	.0022	.0027
10	.0001	.0001	.0001	.0002	.0002	.0003	.0004	.0005	.0006	.0008
11	.0000	.0000	.0000	.0000	.0000	.0001	.0001	.0001	.0002	.0002
12	.0000	.0000	.0000	.0000	.0000	.0000	.0000	.0000	.0000	.0001

					λ					
X	3.1	3.2	3.3	3.4	3.5	3.6	3.7	3.8	3.9	4.0
0	.0450	.0408	.0369	.0334	.0302	.0273	.0247	.0224	.0202	.0183
1	.1397	.1304	.1217	.1135	.1057	.0984	.0915	.0850	.0789	.0733
2	.2165	.2087	.2008	.1929	.1850	.1771	.1692	.1615	.1539	.1465
3	.2237	.2226	.2209	.2186	.2158	.2125	.2087	.2046	.2001	.1954
4	.1734	.1781	.1823	.1858	.1888	.1912	.1931	.1944	.1951	.1954
5	.1075	.1140	.1203	.1264	.1322	.1377	.1429	.1477	.1522	.1563
6	.0555	.0608	.0662	.0716	.0771	.0826	.0881	.0936	.0989	.1042
7	.0246	.0278	.0312	.0348	.0385	.0425	.0466	.0508	.0551	.0595
8	.0095	.0111	.0129	.0148	.0169	.0191	.0215	.0241	.0269	.0298
9	.0033	.0040	.0047	.0056	.0066	.0076	.0089	.0102	.0116	.0132
10	.0010	.0013	.0016	.0019	.0023	.0028	.0033	.0039	.0045	.0053
11	.0003	.0004	.0005	.0006	.0007	.0009	.0011	.0013	.0016	.0019
12	.0001	.0001	.0001	.0002	.0002	.0003	.0003	.0004	.0005	.0006
13	.0000	.0000	.0000	.0000	.0001	.0001	.0001	.0001	.0002	.0002
14	.0000	.0000	.0000	.0000	.0000	.0000	.0000	.0000	.0000	.0001

X	4.1	4.2	4.3	4.4	4.5	4.6	4.7	4.8	4.9	5.0
0	.0166	.0150	.0136	.0123	.0111	.0101	.0091	.0082	.0074	.0067
1	.0679	.0630	.0583	.0540	.0500	.0462	.0427	.0395	.0365	.0337
2	.1393	.1323	.1254	.1188	.1125	.1063	.1005	.0948	.0894	.0842
3	.1904	.1852	.1798	.1743	.1687	.1631	.1574	.1517	.1460	.1404
4	.1951	.1944	.1933	.1917	.1898	.1875	.1849	.1820	.1789	.1755
5	.1600	.1633	.1662	.1687	.1708	.1725	.1738	.1747	.1753	.1755
6	.1093	.1143	.1191	.1237	.1281	.1323	.1362	.1398	.1432	.1462
7	.0640	.0686	.0732	.0778	.0824	.0869	.0914	.0959	.1002	.1044
8	.0328	.0360	.0393	.0428	.0463	.0500	.0537	.0575	.0614	.0653
9	.0150	.0168	.0188	.0209	.0232	.0255	.0280	.0307	.0334	.0363
10	.0061	.0071	.0081	.0092	.0104	.0118	.0132	.0147	.0164	.0181
11	.0023	.0027	.0032	.0037	.0043	.0049	.0056	.0064	.0073	.0082
12	.0008	.0009	.0011	.0014	.0016	.0019	.0022	.0026	.0030	.0034
13	.0002	.0003	.0004	.0005	.0006	.0007	.0008	.0009	.0011	.0013
14	.0001	.0001	.0001	.0001	.0002	.0002	.0003	.0003	.0004	.0005
15	.0000	.0000	.0000	.0000	.0001	.0001	.0001	.0001	.0001	.0002

Continued

Table A.3

Continued

X	5.1	5.2	5.3	5.4	5.5	5.6	5.7	5.8	5.9	6.0
0	.0061	.0055	.0050	.0045	.0041	.0037	.0033	.0030	.0027	.0025
1	.0311	.0287	.0265	.0244	.0225	.0207	.0191	.0176	.0162	.0149
2	.0793	.0746	.0701	.0659	.0618	.0580	.0544	.0509	.0477	.0446
3	.1348	.1293	.1239	.1185	.1133	.1082	.1033	.0985	.0938	.0892
4	.1719	.1681	.1641	.1600	.1558	.1515	.1472	.1428	.1383	.1339
5	.1753	.1748	.1740	.1728	.1714	.1697	.1678	.1656	.1632	.1606
6	.1490	.1515	.1537	.1555	.1571	.1584	.1594	.1601	.1605	.1606
7	.1086	.1125	.1163	.1200	.1234	.1267	.1298	.1326	.1353	.1377
8	.0692	.0731	.0771	.0810	.0849	.0887	.0925	.0962	.0998	.1033
9	.0392	.0423	.0454	.0486	.0519	.0552	.0586	.0620	.0654	.0688
10	.0200	.0220	.0241	.0262	.0285	.0309	.0334	.0359	.0386	.0413
11	.0093	.0104	.0116	.0129	.0143	.0157	.0173	.0190	.0207	.0225
12	.0039	.0045	.0051	.0058	.0065	.0073	.0082	.0092	.0102	.0113
13	.0015	.0018	.0021	.0024	.0028	.0032	.0036	.0041	.0046	.0052
14	.0006	.0007	.0008	.0009	.0011	.0013	.0015	.0017	.0019	.0022
15	.0002	.0002	.0003	.0003	.0004	.0005	.0006	.0007	.0008	.0009
16	.0001	.0001	.0001	.0001	.0001	.0002	.0002	.0002	.0003	.0003
17	.0000	.0000	.0000	.0000	.0000	.0001	.0001	.0001	.0001	.0001

X	6.1	6.2	6.3	6.4	6.5	6.6	6.7	6.8	6.9	7.0
0	.0022	.0020	.0018	.0017	.0015	.0014	.0012	.0011	.0010	.0009
1	.0137	.0126	.0116	.0106	.0098	.0090	.0082	.0076	.0070	.0064
2	.0417	.0390	.0364	.0340	.0318	.0296	.0276	.0258	.0240	.0223
3	.0848	.0806	.0765	.0726	.0688	.0652	.0617	.0584	.0552	.0521
4	.1294	.1269	.1205	.1162	.1118	.1076	.1034	.0992	.0952	.0912
5	.1579	.1549	.1519	.1487	.1454	.1420	.1385	.1349	.1314	.1277
6	.1605	.1601	.1595	.1586	.1575	.1562	.1546	.1529	.1511	.1490
7	.1399	.1418	.1435	.1450	.1462	.1472	.1480	.1486	.1489	.1490
8	.1066	.1099	.1130	.1160	.1188	.1215	.1240	.1263	.1284	.1304
9	.0723	.0757	.0791	.0825	.0858	.0891	.0923	.0954	.0985	.1014
10	.0441	.0469	.0498	.0528	.0558	.0588	.0618	.0649	.0679	.0710
11	.0245	.0265	.0285	.0307	.0330	.0353	.0377	.0401	.0426	.0452
12	.0124	.0137	.0150	.0164	.0179	.0194	.0210	.0227	.0245	.0264
13	.0058	.0065	.0073	.0081	.0089	.0098	.0108	.0119	.0130	.0142
14	.0025	.0029	.0033	.0037	.0041	.0046	.0052	.0058	.0064	.0071
15	.0010	.0012	.0014	.0016	.0018	.0020	.0023	.0026	.0029	.0033
16	.0004	.0005	.0005	.0006	.0007	.0008	.0010	.0011	.0013	.0014
17	.0001	.0002	.0002	.0002	.0003	.0003	.0004	.0004	.0005	.0006
18	.0000	.0001	.0001	.0001	.0001	.0001	.0001	.0002	.0002	.0002
19	.0000	.0000	.0000	.0000	.0000	.0000	.0000	.0001	.0001	.0001

					λ					
X	7.1	7.2	7.3	7.4	7.5	7.6	7.7	7.8	7.9	8.0
0	.0008	.0007	.0007	.0006	.0006	.0005	.0005	.0004	.0004	.0003
1	.0059	.0054	.0049	.0045	.0041	.0038	.0035	.0032	.0029	.0027
2	.0208	.0194	.0180	.0167	.0156	.0145	.0134	.0125	.0116	.0107
3	.0492	.0464	.0438	.0413	.0389	.0366	.0345	.0324	.0305	.0286
4	.0874	.0836	.0799	.0764	.0729	.0696	.0663	.0632	.0602	.0573
5	.1241	.1204	.1167	.1130	.1094	.1057	.1021	.0986	.0951	.0916
6	.1468	.1445	.1420	.1394	.1367	.1339	.1311	.1282	.1252	.1221
7	.1489	.1486	.1481	.1474	.1465	.1454	.1442	.1428	.1413	.1396
8	.1321	.1337	.1351	.1363	.1373	.1382	.1388	.1392	.1395	.1396
9	.1042	.1070	.1096	.1121	.1144	.1167	.1187	.1207	.1224	.1241
10	.0740	.0770	.0800	.0829	.0858	.0887	.0914	.0941	.0967	.0993
11	.0478	.0504	.0532	.0558	.0585	.0613	.0640	.0667	.0695	.0722
12	.0283	.0303	.0323	.0344	.0366	.0388	.0411	.0434	.0457	.0481
13	.0154	.0168	.0181	.0196	.0211	.0227	.0243	.0260	.0278	.0296
14	.0078	.0086	.0095	.0104	.0113	.0123	.0134	.0145	.0157	.0169
15	.0037	.0041	.0046	.0051	.0057	.0062	.0069	.0075	.0083	.0090
16	.0016	.0019	.0021	.0024	.0026	.0030	.0033	.0037	.0041	.0045
17	.0007	.0008	.0009	.0010	.0012	.0013	.0015	.0017	.0019	.0021
18	.0003	.0003	.0004	.0004	.0005	.0006	.0006	.0007	.0008	.0009
19	.0001	.0001	.0001	.0002	.0002	.0002	.0003	.0003	.0003	.0004
20	.0000	.0000	.0001	.0001	.0001	.0001	.0001	.0001	.0001	.0002
21	.0000	.0000	.0000	.0000	.0000	.0000	.0000	.0000	.0001	.0001

X	8.1	8.2	8.3	8.4	8.5	8.6	8.7	8.8	8.9	9.0
0	.0003	.0003	.0002	.0002	.0002	.0002	.0002	.0002	.0001	.0001
1	.0025	.0023	.0021	.0019	.0017	.0016	.0014	.0013	.0012	.0011
2	.0100	.0092	.0086	.0079	.0074	.0068	.0063	.0058	.0054	.0050
3	.0269	.0252	.0237	.0222	.0208	.0195	.0183	.0171	.0160	.0150
4	.0544	.0517	.0491	.0466	.0443	.0420	.0398	.0377	.0357	.0337
5	.0882	.0849	.0816	.0784	.0752	.0722	.0692	.0663	.0635	.0607
6	.1191	.1160	.1128	.1097	.1066	.1034	.1003	.0972	.0941	.0911
7	.1378	.1358	.1338	.1317	.1294	.1271	.1247	.1222	.1197	.1171
8	.1395	.1392	.1388	.1382	.1375	.1366	.1356	.1344	.1332	.1318
9	.1256	.1269	.1280	.1290	.1299	.1306	.1311	.1315	.1317	.1318
10	.1017	.1040	.1063	.1084	.1104	.1123	.1140	.1157	.1172	.1180
11	.0749	.0776	.0802	.0828	.0853	.0878	.0902	.0925	.0948	.0970
12	.0505	.0530	.0555	.0579	.0604	.0629	.0654	.0679	.0703	.0728
13	.0315	.0334	.0354	.0374	.0395	.0416	.0438	.0459	.0481	.0504
14	.0182	.0196	.0210	.0225	.0240	.0256	.0272	.0289	.0306	.0324
15	.0098	.0107	.0116	.0126	.0136	.0147	.0158	.0169	.0182	.0194
16	.0050	.0055	.0060	.0066	.0072	.0079	.0086	.0093	.0101	.0109
17	.0024	.0020	.0029	.0033	.0036	.0040	.0044	.0048	.0053	.0058
18	.0011	.0012	.0014	.0015	.0017	.0019	.0021	.0024	.0026	.0029
19	.0005	.0005	.0006	.0007	.0008	.0009	.0010	.0011	.0012	.0014
20	.0002	.0002	.0002	.0003	.0003	.0004	.0004	.0005	.0005	.0006
21	.0001	.0001	.0001	.0001	.0001	.0002	.0002	.0002	.0002	.0003
22	.0000	.0000	.0000	.0000	.0001	.0001	.0001	.0001	.0001	.0001

Continued

Table A.3

Continued

X	λ									
	9.1	9.2	9.3	9.4	9.5	9.6	9.7	9.8	9.9	10.0
0	.0001	.0001	.0001	.0001	.0001	.0001	.0001	.0001	.0001	.0000
1	.0010	.0009	.0009	.0008	.0007	.0007	.0006	.0005	.0005	.0005
2	.0046	.0043	.0040	.0037	.0034	.0031	.0029	.0027	.0025	.0023
3	.0140	.0131	.0123	.0115	.0107	.0100	.0093	.0087	.0081	.0076
4	.0319	.0302	.0285	.0269	.0254	.0240	.0226	.0213	.0201	.0189
5	.0581	.0555	.0530	.0506	.0483	.0460	.0439	.0418	.0398	.0378
6	.0881	.0851	.0822	.0793	.0764	.0736	.0709	.0682	.0656	.0631
7	.1145	.1118	.1091	.1064	.1037	.1010	.0982	.0955	.0928	.0901
8	.1302	.1286	.1269	.1251	.1232	.1212	.1191	.1170	.1148	.1126
9	.1317	.1315	.1311	.1306	.1300	.1293	.1284	.1274	.1263	.1251
10	.1198	.1210	.1219	.1228	.1235	.1241	.1245	.1249	.1250	.1251
11	.0991	.1012	.1031	.1049	.1067	.1083	.1098	.1112	.1125	.1137
12	.0752	.0776	.0799	.0822	.0844	.0866	.0888	.0908	.0928	.0948
13	.0526	.0549	.0572	.0594	.0617	.0640	.0662	.0685	.0707	.0729
14	.0342	.0361	.0380	.0399	.0419	.0439	.0459	.0479	.0500	.0521
15	.0208	.0221	.0235	.0250	.0265	.0281	.0297	.0313	.0330	.0347
16	.0118	.0127	.0137	.0147	.0157	.0168	.0180	.0192	.0204	.0217
17	.0063	.0069	.0075	.0081	.0088	.0095	.0103	.0111	.0119	.0128
18	.0032	.0035	.0039	.0042	.0046	.0051	.0055	.0060	.0065	.0071
19	.0015	.0017	.0019	.0021	.0023	.0026	.0028	.0031	.0034	.0037
20	.0007	.0008	.0009	.0010	.0011	.0012	.0014	.0015	.0017	.0019
21	.0003	.0003	.0004	.0004	.0005	.0006	.0006	.0007	.0008	.0009
22	.0001	.0001	.0002	.0002	.0002	.0002	.0003	.0003	.0004	.0004
23	.0000	.0001	.0001	.0001	.0001	.0001	.0001	.0001	.0002	.0002
24	.0000	.0000	.0000	.0000	.0000	.0000	.0000	.0001	.0001	.0001

X	e^{-x}	X	e^{-x}	X	e^{-x}	X	e^{-x}
0.0	1.0000	3.0	0.0498	6.0	0.00248	9.0	0.00012
0.1	0.9048	3.1	0.0450	6.1	0.00224	9.1	0.00011
0.2	0.8187	3.2	0.0408	6.2	0.00203	9.2	0.00010
0.3	0.7408	3.3	0.0369	6.3	0.00184	9.3	0.00009
0.4	0.6703	3.4	0.0334	6.4	0.00166	9.4	0.00008
0.5	0.6065	3.5	0.0302	6.5	0.00150	9.5	0.00007
0.6	0.5488	3.6	0.0273	0.6	0.00136	9.6	0.00007
0.7	0.4966	3.7	0.0247	6.7	0.00123	9.7	0.00006
0.8	0.4493	3.8	0.0224	6.8	0.00111	9.8	0.00006
0.9	0.4066	3.9	0.0202	6.9	0.00101	9.9	0.00005
1.0	0.3679	4.0	0.0183	7.0	0.00091	10.0	0.00005
1.1	0.3329	4.1	0.0166	7.1	0.00083		
1.2	0.3012	4.2	0.0150	7.2	0.00075		
1.3	0.2725	4.3	0.0136	7.3	0.00068		
1.4	0.2466	4.4	0.0123	7.4	0.00061		
1.5	0.2231	4.5	0.0111	7.5	0.00055		
1.6	0.2019	4.6	0.0101	7.6	0.00050		
1.7	0.1827	4.7	0.0091	7.7	0.00045		
1.8	0.1653	4.8	0.0082	7.8	0.00041		
1.9	0.1496	4.9	0.0074	7.9	0.00037		
2.0	0.1353	5.0	0.0067	8.0	0.00034		
2.1	0.1225	5.1	0.0061	8.1	0.00030		
2.2	0.1108	5.2	0.0055	8.2	0.00027		
2.3	0.1003	5.3	0.0050	8.3	0.00025		
2.4	0.0907	5.4	0.0045	8.4	0.00022		
2.5	0.0821	5.5	0.0041	8.5	0.00020		
2.6	0.0743	5.6	0.0037	8.6	0.00018		
2.7	0.0672	5.7	0.0033	8.7	0.00017		
2.8	0.0608	5.8	0.0030	8.8	0.00015		
2.9	0.0550	5.9	0.0027	8.9	0.00014		

Table A.4
The e^{-x} Table

Table A.5
Areas of the Standard Normal Distribution

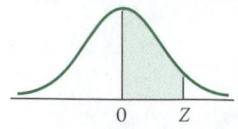

The entries in this table are the probabilities that a standard normal random variable is between 0 and Z (the shaded area).

SECOND DECIMAL PLACE IN Z

Z	0.00	0.01	0.02	0.03	0.04	0.05	0.06	0.07	0.08	0.09
0.0	.0000	.0040	.0080	.0120	.0160	.0199	.0239	.0279	.0319	.0359
0.1	.0398	.0438	.0478	.0517	.0557	.0596	.0636	.0675	.0714	.0753
0.2	.0793	.0832	.0871	.0910	.0948	.0987	.1026	.1064	.1103	.1141
0.3	.1179	.1217	.1255	.1293	.1331	.1368	.1406	.1443	.1480	.1517
0.4	.1554	.1591	.1628	.1664	.1700	.1736	.1772	.1808	.1844	.1879
0.5	.1915	.1950	.1985	.2019	.2054	.2088	.2123	.2157	.2190	.2224
0.6	.2257	.2291	.2324	.2357	.2389	.2422	.2454	.2486	.2517	.2549
0.7	.2580	.2611	.2642	.2673	.2704	.2734	.2764	.2794	.2823	.2852
0.8	.2881	.2910	.2939	.2967	.2995	.3023	.3051	.3078	.3106	.3133
0.9	.3159	.3186	.3212	.3238	.3264	.3289	.3315	.3340	.3365	.3389
1.0	.3413	.3438	.3461	.3485	.3508	.3531	.3554	.3577	.3599	.3621
1.1	.3643	.3665	.3686	.3708	.3729	.3749	.3770	.3790	.3810	.3830
1.2	.3849	.3869	.3888	.3907	.3925	.3944	.3962	.3980	.3997	.4015
1.3	.4032	.4049	.4066	.4082	.4099	.4115	.4131	.4147	.4162	.4177
1.4	.4192	.4207	.4222	.4236	.4251	.4265	.4279	.4292	.4306	.4319
1.5	.4332	.4345	.4357	.4370	.4382	.4394	.4406	.4418	.4429	.4441
1.6	.4452	.4463	.4474	.4484	.4495	.4505	.4515	.4525	.4535	.4545
1.7	.4554	.4564	.4573	.4582	.4591	.4599	.4608	.4616	.4625	.4633
1.8	.4641	.4649	.4656	.4664	.4671	.4678	.4686	.4693	.4699	.4706
1.9	.4713	.4719	.4726	.4732	.4738	.4744	.4750	.4756	.4761	.4767
2.0	.4772	.4778	.4783	.4788	.4793	.4798	.4803	.4808	.4812	.4817
2.1	.4821	.4826	.4830	.4834	.4838	.4842	.4846	.4850	.4854	.4857
2.2	.4861	.4864	.4868	.4871	.4875	.4878	.4881	.4884	.4887	.4890
2.3	.4893	.4896	.4898	.4901	.4904	.4906	.4909	.4911	.4913	.4916
2.4	.4918	.4920	.4922	.4925	.4927	.4929	.4931	.4932	.4934	.4936
2.5	.4938	.4940	.4941	.4943	.4945	.4946	.4948	.4949	.4951	.4952
2.6	.4953	.4955	.4956	.4957	.4959	.4960	.4961	.4962	.4963	.4964
2.7	.4965	.4966	.4967	.4968	.4969	.4970	.4971	.4972	.4973	.4974
2.8	.4974	.4975	.4976	.4977	.4977	.4978	.4979	.4979	.4980	.4981
2.9	.4981	.4982	.4982	.4983	.4984	.4984	.4985	.4985	.4986	.4986
3.0	.4987	.4987	.4987	.4988	.4988	.4989	.4989	.4989	.4990	.4990
3.1	.4990	.4991	.4991	.4991	.4992	.4992	.4992	.4992	.4993	.4993
3.2	.4993	.4993	.4994	.4994	.4994	.4994	.4994	.4995	.4995	.4995
3.3	.4995	.4995	.4995	.4996	.4996	.4996	.4996	.4996	.4996	.4997
3.4	.4997	.4997	.4997	.4997	.4997	.4997	.4997	.4997	.4997	.4998
3.5	.4998									
4.0	.49997									
4.5	.499997									
5.0	.4999997									
6.0	.499999983									

Table A.6
Critical Values of t

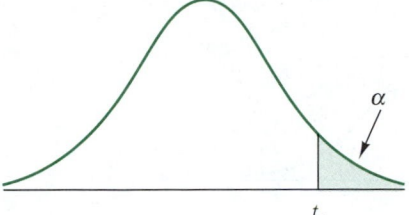

DEGREES OF FREEDOM	$t_{.100}$	$t_{.050}$	$t_{.025}$	$t_{.010}$	$t_{.005}$
1	3.078	6.314	12.706	31.821	63.657
2	1.886	2.920	4.303	6.965	9.925
3	1.638	2.353	3.182	4.541	5.841
4	1.533	2.132	2.776	3.747	4.604
5	1.476	2.015	2.571	3.365	4.032
6	1.440	1.943	2.447	3.143	3.707
7	1.415	1.895	2.365	2.998	3.499
8	1.397	1.860	2.306	2.896	3.355
9	1.383	1.833	2.262	2.821	3.250
10	1.372	1.812	2.228	2.764	3.169
11	1.363	1.796	2.201	2.718	3.106
12	1.356	1.782	2.179	2.681	3.055
13	1.350	1.771	2.160	2.650	3.012
14	1.345	1.761	2.145	2.624	2.977
15	1.341	1.753	2.131	2.602	2.947
16	1.337	1.746	2.120	2.583	2.921
17	1.333	1.740	2.110	2.567	2.898
18	1.330	1.734	2.101	2.552	2.878
19	1.328	1.729	2.093	2.539	2.861
20	1.325	1.725	2.086	2.528	2.845
21	1.323	1.721	2.080	2.518	2.831
22	1.321	1.717	2.074	2.508	2.819
23	1.319	1.714	2.069	2.500	2.808
24	1.318	1.711	2.064	2.492	2.797
25	1.316	1.708	2.060	2.485	2.787
26	1.315	1.706	2.056	2.479	2.779
27	1.314	1.703	2.052	2.473	2.771
28	1.313	1.701	2.048	2.467	2.763
29	1.311	1.699	2.045	2.462	2.756
30	1.310	1.697	2.042	2.457	2.750
40	1.303	1.684	2.021	2.423	2.704
60	1.296	1.671	2.000	2.390	2.660
120	1.289	1.658	1.980	2.358	2.617
∞	1.282	1.645	1.960	2.326	2.576

Table A.7

Percentage Points of the F Distribution

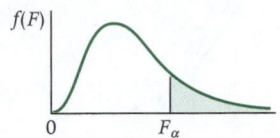

v_2	v_1 — $\alpha = .10$ Numerator Degrees of Freedom								
	1	*2*	*3*	*4*	*5*	*6*	*7*	*8*	*9*
1	39.86	49.50	53.59	55.83	57.24	58.20	58.91	59.44	59.86
2	8.53	9.00	9.16	9.24	9.29	9.33	9.35	9.37	9.38
3	5.54	5.46	5.39	5.34	5.31	5.28	5.27	5.25	5.24
4	4.54	4.32	4.19	4.11	4.05	4.01	3.98	3.95	3.94
5	4.06	3.78	3.62	3.52	3.45	3.40	3.37	3.34	3.32
6	3.78	3.46	3.29	3.18	3.11	3.05	3.01	2.98	2.96
7	3.59	3.26	3.07	2.96	2.88	2.83	2.78	2.75	2.72
8	3.46	3.11	2.92	2.81	2.73	2.67	2.62	2.59	2.56
9	3.36	3.01	2.81	2.69	2.61	2.55	2.51	2.47	2.44
10	3.29	2.92	2.73	2.61	2.52	2.46	2.41	2.38	2.35
11	3.23	2.86	2.66	2.54	2.45	2.39	2.34	2.30	2.27
12	3.18	2.81	2.61	2.48	2.39	2.33	2.28	2.24	2.21
13	3.14	2.76	2.56	2.43	2.35	2.28	2.23	2.20	2.16
14	3.10	2.73	2.52	2.39	2.31	2.24	2.19	2.15	2.12
15	3.07	2.70	2.49	2.36	2.27	2.21	2.16	2.12	2.09
16	3.05	2.67	2.46	2.33	2.24	2.18	2.13	2.09	2.06
17	3.03	2.64	2.44	2.31	2.22	2.15	2.10	2.06	2.03
18	3.01	2.62	2.42	2.29	2.20	2.13	2.08	2.04	2.00
19	2.99	2.61	2.40	2.27	2.18	2.11	2.06	2.02	1.98
20	2.97	2.59	2.38	2.25	2.16	2.09	2.04	2.00	1.96
21	2.96	2.57	2.36	2.23	2.14	2.08	2.02	1.98	1.95
22	2.95	2.56	2.35	2.22	2.13	2.06	2.01	1.97	1.93
23	2.94	2.55	2.34	2.21	2.11	2.05	1.99	1.95	1.92
24	2.93	2.54	2.33	2.19	2.10	2.04	1.98	1.94	1.91
25	2.92	2.53	2.32	2.18	2.09	2.02	1.97	1.93	1.89
26	2.91	2.52	2.31	2.17	2.08	2.01	1.96	1.92	1.88
27	2.90	2.51	2.30	2.17	2.07	2.00	1.95	1.91	1.87
28	2.89	2.50	2.29	2.16	2.06	2.00	1.94	1.90	1.87
29	2.89	2.50	2.28	2.15	2.06	1.99	1.93	1.89	1.86
30	2.88	2.49	2.28	2.14	2.05	1.98	1.93	1.88	1.85
40	2.84	2.44	2.23	2.09	2.00	1.93	1.87	1.83	1.79
60	2.79	2.39	2.18	2.04	1.95	1.87	1.82	1.77	1.74
120	2.75	2.35	2.13	1.99	1.90	1.82	1.77	1.72	1.68
∞	2.71	2.30	2.08	1.94	1.85	1.77	1.72	1.67	1.63

Denominator Degrees of Freedom

				$\alpha = .10$						v_1	
			Numerator Degrees of Freedom								
10	*12*	*15*	*20*	*24*	*30*	*40*	*60*	*120*	*∞*		*v_2*
60.19	60.71	61.22	61.74	62.00	62.26	62.53	62.79	63.06	63.33	1	
9.39	9.41	9.42	9.44	9.45	9.46	9.47	9.47	9.48	9.49	2	
5.23	5.22	5.20	5.18	5.18	5.17	5.16	5.15	5.14	5.13	3	
3.92	3.90	3.87	3.84	3.83	3.82	3.80	3.79	3.78	3.76	4	
3.30	3.27	3.24	3.21	3.19	3.17	3.16	3.14	3.12	3.10	5	
2.94	2.90	2.87	2.84	2.82	2.80	2.78	2.76	2.74	2.72	6	
2.70	2.67	2.63	2.59	2.58	2.56	2.54	2.51	2.49	2.47	7	
2.54	2.50	2.46	2.42	2.40	2.38	2.36	2.34	2.32	2.29	8	
2.42	2.38	2.34	2.30	2.28	2.25	2.23	2.21	2.18	2.16	9	
2.32	2.28	2.24	2.20	2.18	2.16	2.13	2.11	2.08	2.06	10	
2.25	2.21	2.17	2.12	2.10	2.08	2.05	2.03	2.00	1.97	11	
2.19	2.15	2.10	2.06	2.04	2.01	1.99	1.96	1.93	1.90	12	
2.14	2.10	2.05	2.01	1.98	1.96	1.93	1.90	1.88	1.85	13	
2.10	2.05	2.01	1.96	1.94	1.91	1.89	1.86	1.83	1.80	14	
2.06	2.02	1.97	1.92	1.90	1.87	1.85	1.82	1.79	1.76	15	
2.03	1.99	1.94	1.89	1.87	1.84	1.81	1.78	1.75	1.72	16	
2.00	1.96	1.91	1.86	1.84	1.81	1.78	1.75	1.72	1.69	17	
1.98	1.93	1.89	1.84	1.81	1.78	1.75	1.72	1.69	1.66	18	
1.96	1.91	1.86	1.81	1.79	1.76	1.73	1.70	1.67	1.63	19	
1.94	1.89	1.84	1.79	1.77	1.74	1.71	1.68	1.64	1.61	20	
1.92	1.87	1.83	1.78	1.75	1.72	1.69	1.66	1.62	1.59	21	
1.90	1.86	1.81	1.76	1.73	1.70	1.67	1.64	1.60	1.57	22	
1.89	1.84	1.80	1.74	1.72	1.69	1.66	1.62	1.59	1.55	23	
1.88	1.83	1.78	1.73	1.70	1.67	1.64	1.61	1.57	1.53	24	
1.87	1.82	1.77	1.72	1.69	1.66	1.63	1.59	1.56	1.52	25	
1.86	1.81	1.76	1.71	1.68	1.65	1.61	1.58	1.54	1.50	26	
1.85	1.80	1.75	1.70	1.67	1.64	1.60	1.57	1.53	1.49	27	
1.84	1.79	1.74	1.69	1.66	1.63	1.59	1.56	1.52	1.48	28	
1.83	1.78	1.73	1.68	1.65	1.62	1.58	1.55	1.51	1.47	29	
1.82	1.77	1.72	1.67	1.64	1.61	1.57	1.54	1.50	1.46	30	
1.76	1.71	1.66	1.61	1.57	1.54	1.51	1.47	1.42	1.38	40	
1.71	1.66	1.60	1.54	1.51	1.48	1.44	1.40	1.35	1.29	60	
1.65	1.60	1.55	1.48	1.45	1.41	1.37	1.32	1.26	1.19	120	
1.60	1.55	1.49	1.42	1.38	1.34	1.30	1.24	1.17	1.00	∞	

Denominator Degrees of Freedom

Continued

Table A.7
Percentage Points of the F Distribution, *continued*

v_2	v_1 1	2	3	4	5	6	7	8	9
				$\alpha = .05$					
				Numerator Degrees of Freedom					
1	161.40	199.50	215.70	224.60	230.20	234.00	236.80	238.90	240.50
2	18.51	19.00	19.16	19.25	19.30	19.33	19.35	19.37	19.38
3	10.13	9.55	9.28	9.12	9.01	8.94	8.89	8.85	8.81
4	7.71	6.94	6.59	6.39	6.26	6.16	6.09	6.04	6.00
5	6.61	5.79	5.41	5.19	5.05	4.95	4.88	4.82	4.77
6	5.99	5.14	4.76	4.53	4.39	4.28	4.21	4.15	4.10
7	5.59	4.74	4.35	4.12	3.97	3.87	3.79	3.73	3.68
8	5.32	4.46	4.07	3.84	3.69	3.58	3.50	3.44	3.39
9	5.12	4.26	3.86	3.63	3.48	3.37	3.29	3.23	3.18
10	4.96	4.10	3.71	3.48	3.33	3.22	3.14	3.07	3.02
11	4.84	3.98	3.59	3.36	3.20	3.09	3.01	2.95	2.90
12	4.75	3.89	3.49	3.26	3.11	3.00	2.91	2.85	2.80
13	4.67	3.81	3.41	3.18	3.03	2.92	2.83	2.77	2.71
14	4.60	3.74	3.34	3.11	2.96	2.85	2.76	2.70	2.65
15	4.54	3.68	3.29	3.06	2.90	2.79	2.71	2.64	2.59
16	4.49	3.63	3.24	3.01	2.85	2.74	2.66	2.59	2.54
17	4.45	3.59	3.20	2.96	2.81	2.70	2.61	2.55	2.49
18	4.41	3.55	3.16	2.93	2.77	2.66	2.58	2.51	2.46
19	4.38	3.52	3.13	2.90	2.74	2.63	2.54	2.48	2.42
20	4.35	3.49	3.10	2.87	2.71	2.60	2.51	2.45	2.39
21	4.32	3.47	3.07	2.84	2.68	2.57	2.49	2.42	2.37
22	4.30	3.44	3.05	2.82	2.66	2.55	2.46	2.40	2.34
23	4.28	3.42	3.03	2.80	2.64	2.53	2.44	2.37	2.32
24	4.26	3.40	3.01	2.78	2.62	2.51	2.42	2.36	2.30
25	4.24	3.39	2.99	2.76	2.60	2.49	2.40	2.34	2.28
26	4.23	3.37	2.98	2.74	2.59	2.47	2.39	2.32	2.27
27	4.21	3.35	2.96	2.73	2.57	2.46	2.37	2.31	2.25
28	4.20	3.34	2.95	2.71	2.56	2.45	2.36	2.29	2.24
29	4.18	3.33	2.93	2.70	2.55	2.43	2.35	2.28	2.22
30	4.17	3.32	2.92	2.69	2.53	2.42	2.33	2.27	2.21
40	4.08	3.23	2.84	2.61	2.45	2.34	2.25	2.18	2.12
60	4.00	3.15	2.76	2.53	2.37	2.25	2.17	2.10	2.04
120	3.92	3.07	2.68	2.45	2.29	2.17	2.09	2.02	1.96
∞	3.84	3.00	2.60	2.37	2.21	2.10	2.01	1.94	1.88

Denominator Degrees of Freedom

α = .05										v_1
Numerator Degrees of Freedom										
10	12	15	20	24	30	40	60	120	∞	v_2
241.90	243.90	245.90	248.00	249.10	250.10	251.10	252.20	253.30	254.30	1
19.40	19.41	19.43	19.45	19.45	19.46	19.47	19.48	19.49	19.50	2
8.79	8.74	8.70	8.66	8.64	8.62	8.59	8.57	8.55	8.53	3
5.96	5.91	5.86	5.80	5.77	5.75	5.72	5.69	5.66	5.63	4
4.74	4.68	4.62	4.56	4.53	4.50	4.46	4.43	4.40	4.36	5
4.06	4.00	3.94	3.87	3.84	3.81	3.77	3.74	3.70	3.67	6
3.64	3.57	3.51	3.44	3.41	3.38	3.34	3.30	3.27	3.23	7
3.35	3.28	3.22	3.15	3.12	3.08	3.04	3.01	2.97	2.93	8
3.14	3.07	3.01	2.94	2.90	2.86	2.83	2.79	2.75	2.71	9
2.98	2.91	2.85	2.77	2.74	2.70	2.66	2.62	2.58	2.54	10
2.85	2.79	2.72	2.65	2.61	2.57	2.53	2.49	2.45	2.40	11
2.75	2.69	2.62	2.54	2.51	2.47	2.43	2.38	2.34	2.30	12
2.67	2.60	2.53	2.46	2.42	2.38	2.34	2.30	2.25	2.21	13
2.60	2.53	2.46	2.39	2.35	2.31	2.27	2.22	2.18	2.13	14
2.54	2.48	2.40	2.33	2.29	2.25	2.20	2.16	2.11	2.07	15
2.49	2.42	2.35	2.28	2.24	2.19	2.15	2.11	2.06	2.01	16
2.45	2.38	2.31	2.23	2.19	2.15	2.10	2.06	2.01	1.96	17
2.41	2.34	2.27	2.19	2.15	2.11	2.06	2.02	1.97	1.92	18
2.38	2.31	2.23	2.16	2.11	2.07	2.03	1.98	1.93	1.88	19
2.35	2.28	2.20	2.12	2.08	2.04	1.99	1.95	1.90	1.84	20
2.32	2.25	2.18	2.10	2.05	2.01	1.96	1.92	1.87	1.81	21
2.30	2.23	2.15	2.07	2.03	1.98	1.94	1.89	1.84	1.78	22
2.27	2.20	2.13	2.05	2.01	1.96	1.91	1.86	1.81	1.76	23
2.25	2.18	2.11	2.03	1.98	1.94	1.89	1.84	1.79	1.73	24
2.24	2.16	2.09	2.01	1.96	1.92	1.87	1.82	1.77	1.71	25
2.22	2.15	2.07	1.99	1.95	1.90	1.85	1.80	1.75	1.69	26
2.20	2.13	2.06	1.97	1.93	1.88	1.84	1.79	1.73	1.67	27
2.19	2.12	2.04	1.96	1.91	1.87	1.82	1.77	1.71	1.65	28
2.18	2.10	2.03	1.94	1.90	1.85	1.81	1.75	1.70	1.64	29
2.16	2.09	2.01	1.93	1.89	1.84	1.79	1.74	1.68	1.62	30
2.08	2.00	1.92	1.84	1.79	1.74	1.69	1.64	1.58	1.51	40
1.99	1.92	1.84	1.75	1.70	1.65	1.59	1.53	1.47	1.39	60
1.91	1.83	1.75	1.66	1.61	1.55	1.50	1.43	1.35	1.25	120
1.83	1.75	1.67	1.57	1.52	1.46	1.39	1.32	1.22	1.00	∞

Denominator Degrees of Freedom

Continued

Table A.7
Percentage Points of the F Distribution, *continued*

v_2	v_1				$\alpha = .025$				
				Numerator Degrees of Freedom					
	1	2	3	4	5	6	7	8	9
1	647.80	799.50	864.20	899.60	921.80	937.10	948.20	956.70	963.30
2	38.51	39.00	39.17	39.25	39.30	39.33	39.36	39.37	39.39
3	17.44	16.04	15.44	15.10	14.88	14.73	14.62	14.54	14.47
4	12.22	10.65	9.98	9.60	9.36	9.20	9.07	8.98	8.90
5	10.01	8.43	7.76	7.39	7.15	6.98	6.85	6.76	6.68
6	8.81	7.26	6.60	6.23	5.99	5.82	5.70	5.60	5.52
7	8.07	6.54	5.89	5.52	5.29	5.12	4.99	4.90	4.82
8	7.57	6.06	5.42	5.05	4.82	4.65	4.53	4.43	4.36
9	7.21	5.71	5.08	4.72	4.48	4.32	4.20	4.10	4.03
10	6.94	5.46	4.83	4.47	4.24	4.07	3.95	3.85	3.78
11	6.72	5.26	4.63	4.28	4.04	3.88	3.76	3.66	3.59
12	6.55	5.10	4.47	4.12	3.89	3.73	3.61	3.51	3.44
13	6.41	4.97	4.35	4.00	3.77	3.60	3.48	3.39	3.31
14	6.30	4.86	4.24	3.89	3.66	3.50	3.38	3.29	3.21
15	6.20	4.77	4.15	3.80	3.58	3.41	3.29	3.20	3.12
16	6.12	4.69	4.08	3.73	3.50	3.34	3.22	3.12	3.05
17	6.04	4.62	4.01	3.66	3.44	3.28	3.16	3.06	2.98
18	5.98	4.56	3.95	3.61	3.38	3.22	3.10	3.01	2.93
19	5.92	4.51	3.90	3.56	3.33	3.17	3.05	2.96	2.88
20	5.87	4.46	3.86	3.51	3.29	3.13	3.01	2.91	2.84
21	5.83	4.42	3.82	3.48	3.25	3.09	2.97	2.87	2.80
22	5.79	4.38	3.78	3.44	3.22	3.05	2.93	2.84	2.76
23	5.75	4.35	3.75	3.41	3.18	3.02	2.90	2.81	2.73
24	5.72	4.32	3.72	3.38	3.15	2.99	2.87	2.78	2.70
25	5.69	4.29	3.69	3.35	3.13	2.97	2.85	2.75	2.68
26	5.66	4.27	3.67	3.33	3.10	2.94	2.82	2.73	2.65
27	5.63	4.24	3.65	3.31	3.08	2.92	2.80	2.71	2.63
28	5.61	4.22	3.63	3.29	3.06	2.90	2.78	2.69	2.61
29	5.59	4.20	3.61	3.27	3.04	2.88	2.76	2.67	2.59
30	5.57	4.18	3.59	3.25	3.03	2.87	2.75	2.65	2.57
40	5.42	4.05	3.46	3.13	2.90	2.74	2.62	2.53	2.45
60	5.29	3.93	3.34	3.01	2.79	2.63	2.51	2.41	2.33
120	5.15	3.80	3.23	2.89	2.67	2.52	2.39	2.30	2.22
∞	5.02	3.69	3.12	2.79	2.57	2.41	2.29	2.19	2.11

Denominator Degrees of Freedom

$\alpha = .025$										v_1	
Numerator Degrees of Freedom											v_2
10	12	15	20	24	30	40	60	120	∞		
968.60	976.70	984.90	993.10	997.20	1001.00	1006.00	1010.00	1014.00	1018.00	1	
39.40	39.41	39.43	39.45	39.46	39.46	39.47	39.48	39.49	39.50	2	
14.42	14.34	14.25	14.17	14.12	14.08	14.04	13.99	13.95	13.90	3	
8.84	8.75	8.66	8.56	8.51	8.46	8.41	8.36	8.31	8.26	4	
6.62	6.52	6.43	6.33	6.28	6.23	6.18	6.12	6.07	6.02	5	
5.46	5.37	5.27	5.17	5.12	5.07	5.01	4.96	4.90	4.85	6	
4.76	4.67	4.57	4.47	4.42	4.36	4.31	4.25	4.20	4.14	7	
4.30	4.20	4.10	4.00	3.95	3.89	3.84	3.78	3.73	3.67	8	
3.96	3.87	3.77	3.67	3.61	3.56	3.51	3.45	3.39	3.33	9	
3.72	3.62	3.52	3.42	3.37	3.31	3.26	3.20	3.14	3.08	10	
3.53	3.43	3.33	3.23	3.17	3.12	3.06	3.00	2.94	2.88	11	
3.37	3.28	3.18	3.07	3.02	2.96	2.91	2.85	2.79	2.72	12	
3.25	3.15	3.05	2.95	2.89	2.84	2.78	2.72	2.66	2.60	13	
3.15	3.05	2.95	2.84	2.79	2.73	2.67	2.61	2.55	2.49	14	
3.06	2.96	2.86	2.76	2.70	2.64	2.59	2.52	2.46	2.40	15	
2.99	2.89	2.79	2.68	2.63	2.57	2.51	2.45	2.38	2.32	16	
2.92	2.82	2.72	2.62	2.56	2.50	2.44	2.38	2.32	2.25	17	
2.87	2.77	2.67	2.56	2.50	2.44	2.38	2.32	2.26	2.19	18	
2.82	2.72	2.62	2.51	2.45	2.39	2.33	2.27	2.20	2.13	19	
2.77	2.68	2.57	2.46	2.41	2.35	2.29	2.22	2.16	2.09	20	
2.73	2.64	2.53	2.42	2.37	2.31	2.25	2.18	2.11	2.04	21	
2.70	2.60	2.50	2.39	2.33	2.27	2.21	2.14	2.08	2.00	22	
2.67	2.57	2.47	2.36	2.30	2.24	2.18	2.11	2.04	1.97	23	
2.64	2.54	2.44	2.33	2.27	2.21	2.15	2.08	2.01	1.94	24	
2.61	2.51	2.41	2.30	2.24	2.18	2.12	2.05	1.98	1.91	25	
2.59	2.49	2.39	2.28	2.22	2.16	2.09	2.03	1.95	1.88	26	
2.57	2.47	2.36	2.25	2.19	2.13	2.07	2.00	1.93	1.85	27	
2.55	2.45	2.34	2.23	2.17	2.11	2.05	1.98	1.91	1.83	28	
2.53	2.43	2.32	2.21	2.15	2.09	2.03	1.96	1.89	1.81	29	
2.51	2.41	2.31	2.20	2.14	2.07	2.01	1.94	1.87	1.79	30	
2.39	2.29	2.18	2.07	2.01	1.94	1.88	1.80	1.72	1.64	40	
2.27	2.17	2.06	1.94	1.88	1.82	1.74	1.67	1.58	1.48	60	
2.16	2.05	1.94	1.82	1.76	1.69	1.61	1.53	1.43	1.31	120	
2.05	1.94	1.83	1.71	1.64	1.57	1.48	1.39	1.27	1.00	∞	

Denominator Degrees of Freedom

Continued

Table A.7
Percentage Points of the F Distribution, *continued*

v_2	$\alpha = .01$ Numerator Degrees of Freedom								
	1	2	3	4	5	6	7	8	9
1	4,052.00	4,999.50	5,403.00	5,625.00	5,764.00	5,859.00	5,928.00	5,982.00	6,022.00
2	98.50	99.00	99.17	99.25	99.30	99.33	99.36	99.37	99.39
3	34.12	30.82	29.46	28.71	28.24	27.91	27.67	27.49	27.35
4	21.20	18.00	16.69	15.98	15.52	15.21	14.98	14.80	14.66
5	16.26	13.27	12.06	11.39	10.97	10.67	10.46	10.29	10.16
6	13.75	10.92	9.78	9.15	8.75	8.47	8.26	8.10	7.98
7	12.25	9.55	8.45	7.85	7.46	7.19	6.99	6.84	6.72
8	11.26	8.65	7.59	7.01	6.63	6.37	6.18	6.03	5.91
9	10.56	8.02	6.99	6.42	6.06	5.80	5.61	5.47	5.35
10	10.04	7.56	6.55	5.99	5.64	5.39	5.20	5.06	4.94
11	9.65	7.21	6.22	5.67	5.32	5.07	4.89	4.74	4.63
12	9.33	6.93	5.95	5.41	5.06	4.82	4.64	4.50	4.39
13	9.07	6.70	5.74	5.21	4.86	4.62	4.44	4.30	4.19
14	8.86	6.51	5.56	5.04	4.69	4.46	4.28	4.14	4.03
15	8.68	6.36	5.42	4.89	4.56	4.32	4.14	4.00	3.89
16	8.53	6.23	5.29	4.77	4.44	4.20	4.03	3.89	3.78
17	8.40	6.11	5.18	4.67	4.34	4.10	3.93	3.79	3.68
18	8.29	6.01	5.09	4.58	4.25	4.01	3.84	3.71	3.60
19	8.18	5.93	5.01	4.50	4.17	3.94	3.77	3.63	3.52
20	8.10	5.85	4.94	4.43	4.10	3.87	3.70	3.56	3.46
21	8.02	5.78	4.87	4.37	4.04	3.81	3.64	3.51	3.40
22	7.95	5.72	4.82	4.31	3.99	3.76	3.59	3.45	3.35
23	7.88	5.66	4.76	4.26	3.94	3.71	3.54	3.41	3.30
24	7.82	5.61	4.72	4.22	3.90	3.67	3.50	3.36	3.26
25	7.77	5.57	4.68	4.18	3.85	3.63	3.46	3.32	3.22
26	7.72	5.53	4.64	4.14	3.82	3.59	3.42	3.29	3.18
27	7.68	5.49	4.60	4.11	3.78	3.56	3.39	3.26	3.15
28	7.64	5.45	4.57	4.07	3.75	3.53	3.36	3.23	3.12
29	7.60	5.42	4.54	4.04	3.73	3.50	3.33	3.20	3.09
30	7.56	5.39	4.51	4.02	3.70	3.47	3.30	3.17	3.07
40	7.31	5.18	4.31	3.83	3.51	3.29	3.12	2.99	2.89
60	7.08	4.98	4.13	3.65	3.34	3.12	2.95	2.82	2.72
120	6.85	4.79	3.95	3.48	3.17	2.96	2.79	2.66	2.56
∞	6.63	4.61	3.78	3.32	3.02	2.80	2.64	2.51	2.41

Denominator Degrees of Freedom

$\alpha = .01$											v_1
Numerator Degrees of Freedom											
10	12	15	20	24	30	40	60	120	∞		v_2
6,056.00	6,106.00	6,157.00	6,209.00	6,235.00	6,261.00	6,287.00	6,313.00	6,339.00	6,366.00	1	
99.40	99.42	99.43	99.45	99.46	99.47	99.47	99.48	99.49	99.50	2	
27.23	27.05	26.87	26.69	26.60	26.50	26.41	26.32	26.22	26.13	3	
14.55	14.37	14.20	14.02	13.93	13.84	13.75	13.65	13.56	13.46	4	
10.05	9.89	9.72	9.55	9.47	9.38	9.29	9.20	9.11	9.02	5	
7.87	7.72	7.56	7.40	7.31	7.23	7.14	7.06	6.97	6.88	6	
6.62	6.47	6.31	6.16	6.07	5.99	5.91	5.82	5.74	5.65	7	
5.81	5.67	5.52	5.36	5.28	5.20	5.12	5.03	4.95	4.86	8	
5.26	5.11	4.96	4.81	4.73	4.65	4.57	4.48	4.40	4.31	9	
4.85	4.71	4.56	4.41	4.33	4.25	4.17	4.08	4.00	3.91	10	
4.54	4.40	4.25	4.10	4.02	3.94	3.86	3.78	3.69	3.60	11	
4.30	4.16	4.01	3.86	3.78	3.70	3.62	3.54	3.45	3.36	12	
4.10	3.96	3.82	3.66	3.59	3.51	3.43	3.34	3.25	3.17	13	
3.94	3.80	3.66	3.51	3.43	3.35	3.27	3.18	3.09	3.00	14	
3.80	3.67	3.52	3.37	3.29	3.21	3.13	3.05	2.96	2.87	15	
3.69	3.55	3.41	3.26	3.18	3.10	3.02	2.93	2.84	2.75	16	
3.59	3.46	3.31	3.16	3.08	3.00	2.92	2.83	2.75	2.65	17	
3.51	3.37	3.23	3.08	3.00	2.92	2.84	2.75	2.66	2.57	18	
3.43	3.30	3.15	3.00	2.92	2.84	2.76	2.67	2.58	2.49	19	
3.37	3.23	3.09	2.94	2.86	2.78	2.69	2.61	2.52	2.42	20	
3.31	3.17	3.03	2.88	2.80	2.72	2.64	2.55	2.46	2.36	21	
3.26	3.12	2.98	2.83	2.75	2.67	2.58	2.50	2.40	2.31	22	
3.21	3.07	2.93	2.78	2.70	2.62	2.54	2.45	2.35	2.26	23	
3.17	3.03	2.89	2.74	2.66	2.58	2.49	2.40	2.31	2.21	24	
3.13	2.99	2.85	2.70	2.62	2.54	2.45	2.36	2.27	2.17	25	
3.09	2.96	2.81	2.66	2.58	2.50	2.42	2.33	2.23	2.13	26	
3.06	2.93	2.78	2.63	2.55	2.47	2.38	2.29	2.20	2.10	27	
3.03	2.90	2.75	2.60	2.52	2.44	2.35	2.26	2.17	2.06	28	
3.00	2.87	2.73	2.57	2.49	2.41	2.33	2.23	2.14	2.03	29	
2.98	2.84	2.70	2.55	2.47	2.39	2.30	2.21	2.11	2.01	30	
2.80	2.66	2.52	2.37	2.29	2.20	2.11	2.02	1.92	1.80	40	
2.63	2.50	2.35	2.20	2.12	2.03	1.94	1.84	1.73	1.60	60	
2.47	2.34	2.19	2.03	1.95	1.86	1.76	1.66	1.53	1.38	120	
2.32	2.18	2.04	1.88	1.79	1.70	1.59	1.47	1.32	1.00	∞	

Denominator Degrees of Freedom

Continued

Table A.7
Percentage Points of the F Distribution, *continued*

	v_1					$\alpha = .005$				
					Numerator Degrees of Freedom					
v_2		1	2	3	4	5	6	7	8	9
	1	16211.00	20000.00	21615.00	22500.00	23056.00	23437.00	23715.00	23925.00	24091.00
	2	198.50	199.00	199.20	199.20	199.30	199.30	199.40	199.40	199.40
	3	55.55	49.80	47.47	46.19	45.39	44.84	44.43	44.13	43.88
	4	31.33	26.28	24.26	23.15	22.46	21.97	21.62	21.35	21.14
	5	22.78	18.31	16.53	15.56	14.94	14.51	14.20	13.96	13.77
	6	18.63	14.54	12.92	12.03	11.46	11.07	10.79	10.57	10.39
	7	16.24	12.40	10.88	10.05	9.52	9.16	8.89	8.68	8.51
	8	14.69	11.04	9.60	8.81	8.30	7.95	7.69	7.50	7.34
	9	13.61	10.11	8.72	7.96	7.47	7.13	6.88	6.69	6.54
	10	12.83	9.43	8.08	7.34	6.87	6.54	6.30	6.12	5.97
	11	12.23	8.91	7.60	6.88	6.42	6.10	5.86	5.68	5.54
	12	11.75	8.51	7.23	6.52	6.07	5.76	5.52	5.35	5.20
	13	11.37	8.19	6.93	6.23	5.79	5.48	5.25	5.08	4.94
	14	11.06	7.92	6.68	6.00	5.56	5.26	5.03	4.86	4.72
	15	10.80	7.70	6.48	5.80	5.37	5.07	4.85	4.67	4.54
	16	10.58	7.51	6.30	5.64	5.21	4.91	4.69	4.52	4.38
	17	10.38	7.35	6.16	5.50	5.07	4.78	4.56	4.39	4.25
	18	10.22	7.21	6.03	5.37	4.96	4.66	4.44	4.28	4.14
	19	10.07	7.09	5.92	5.27	4.85	4.56	4.34	4.18	4.04
	20	9.94	6.99	5.82	5.17	4.76	4.47	4.26	4.09	3.96
	21	9.83	6.89	5.73	5.09	4.68	4.39	4.18	4.01	3.88
	22	9.73	6.81	5.65	5.02	4.61	4.32	4.11	3.94	3.81
	23	9.63	6.73	5.58	4.95	4.54	4.26	4.05	3.88	3.75
	24	9.55	6.66	5.52	4.89	4.49	4.20	3.99	3.83	3.69
	25	9.48	6.60	5.46	4.84	4.43	4.15	3.94	3.78	3.64
	26	9.41	6.54	5.41	4.79	4.38	4.10	3.89	3.73	3.60
	27	9.34	6.49	5.36	4.74	4.34	4.06	3.85	3.69	3.56
	28	9.28	6.44	5.32	4.70	4.30	4.02	3.81	3.65	3.52
	29	9.23	6.40	5.28	4.66	4.26	3.98	3.77	3.61	3.48
	30	9.18	6.35	5.24	4.62	4.23	3.95	3.74	3.58	3.45
	40	8.83	6.07	4.98	4.37	3.99	3.71	3.51	3.35	3.22
	60	8.49	5.79	4.73	4.14	3.76	3.49	3.29	3.13	3.01
	120	8.18	5.54	4.50	3.92	3.55	3.28	3.09	2.93	2.81
	∞	7.88	5.30	4.28	3.72	3.35	3.09	2.90	2.74	2.62

Denominator Degrees of Freedom

$\alpha = .005$										v_1
Numerator Degrees of Freedom										
10	12	15	20	24	30	40	60	120	∞	v_2
24224.00	24426.00	24630.00	24836.00	24940.00	25044.00	25148.00	25253.00	25359.00	25465.00	1
199.40	199.40	199.40	199.40	199.50	199.50	199.50	199.50	199.50	199.50	2
43.69	43.39	43.08	42.78	42.62	42.47	42.31	42.15	41.99	41.83	3
20.97	20.70	20.44	20.17	20.03	19.89	19.75	19.61	19.47	19.32	4
13.62	13.38	13.15	12.90	12.78	12.66	12.53	12.40	12.27	12.14	5
10.25	10.03	9.81	9.59	9.47	9.36	9.24	9.12	9.00	8.88	6
8.38	8.18	7.97	7.75	7.65	7.53	7.42	7.31	7.19	7.08	7
7.21	7.01	6.81	6.61	6.50	6.40	6.29	6.18	6.06	5.95	8
6.42	6.23	6.03	5.83	5.73	5.62	5.52	5.41	5.30	5.19	9
5.85	5.66	5.47	5.27	5.17	5.07	4.97	4.86	4.75	4.64	10
5.42	5.24	5.05	4.86	4.76	4.65	4.55	4.44	4.34	4.23	11
5.09	4.91	4.72	4.53	4.43	4.33	4.23	4.12	4.01	3.90	12
4.82	4.64	4.46	4.27	4.17	4.07	3.97	3.87	3.76	3.65	13
4.60	4.43	4.25	4.06	3.96	3.86	3.76	3.66	3.55	3.44	14
4.42	4.25	4.07	3.88	3.79	3.69	3.58	3.48	3.37	3.26	15
4.27	4.10	3.92	3.73	3.64	3.54	3.44	3.33	3.22	3.11	16
4.14	3.97	3.79	3.61	3.51	3.41	3.31	3.21	3.10	2.98	17
4.03	3.86	3.68	3.50	3.40	3.30	3.20	3.10	2.99	2.87	18
3.93	3.76	3.59	3.40	3.31	3.21	3.11	3.00	2.89	2.78	19
3.85	3.68	3.50	3.32	3.22	3.12	3.02	2.92	2.81	2.69	20
3.77	3.60	3.43	3.24	3.15	3.05	2.95	2.84	2.73	2.61	21
3.70	3.54	3.36	3.18	3.08	2.98	2.88	2.77	2.66	2.55	22
3.64	3.47	3.30	3.12	3.02	2.92	2.82	2.71	2.60	2.48	23
3.59	3.42	3.25	3.06	2.97	2.87	2.77	2.66	2.55	2.43	24
3.54	3.37	3.20	3.01	2.92	2.82	2.72	2.61	2.50	2.38	25
3.49	3.33	3.15	2.97	2.87	2.77	2.67	2.56	2.45	2.33	26
3.45	3.28	3.11	2.93	2.83	2.73	2.63	2.52	2.41	2.29	27
3.41	3.25	3.07	2.89	2.79	2.69	2.59	2.48	2.37	2.25	28
3.38	3.21	3.04	2.86	2.76	2.66	2.56	2.45	2.33	2.21	29
3.34	3.18	3.01	2.82	2.73	2.63	2.52	2.42	2.30	2.18	30
3.12	2.95	2.78	2.60	2.50	2.40	2.30	2.18	2.06	1.93	40
2.90	2.74	2.57	2.39	2.29	2.19	2.08	1.96	1.83	1.69	60
2.71	2.54	2.37	2.19	2.09	1.98	1.87	1.75	1.61	1.43	120
2.52	2.36	2.19	2.00	1.90	1.79	1.67	1.53	1.36	1.00	∞

Denominator Degrees of Freedom

Table A.8
The Chi-Square Table

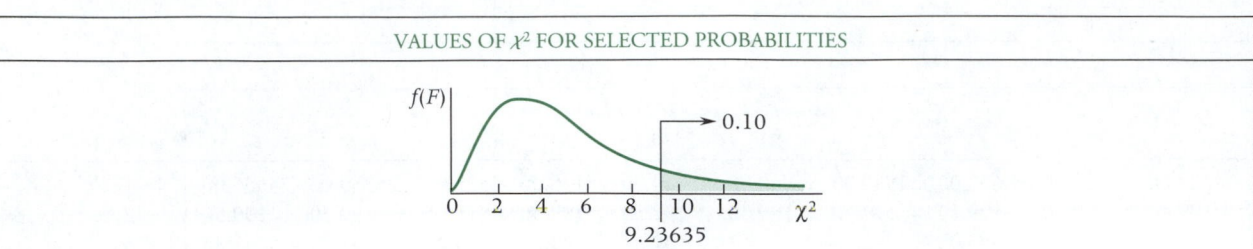

VALUES OF χ^2 FOR SELECTED PROBABILITIES

Example: df (Number of degrees of freedom) = 5, the tail above $\chi^2 = 9.23635$ represents 0.10 or 10% of the area under the curve.

AREA IN UPPER TAIL

Degrees of Freedom	.995	.99	.975	.95	.90	.10	.05	.025	.01	.005
1	$392,704 \times 10^{-10}$	$157,088 \times 10^{-9}$	$982,069 \times 10^{-9}$	$393,214 \times 10^{-8}$	.0157908	2.70554	3.84146	5.02389	6.63490	7.87944
2	.0100251	.0201007	.0506356	.102587	.210720	4.60517	5.99147	7.37776	9.21034	10.5966
3	.0717212	.114832	.215795	.351846	.584375	6.25139	7.81473	9.34840	11.3449	12.8381
4	.206990	.297110	.484419	.710721	1.063623	7.77944	9.48773	11.1433	13.2767	14.8602
5	.411740	.554300	.831211	1.145476	1.61031	9.23635	11.0705	12.8325	15.0863	16.7496
6	.675727	.872085	1.237347	1.63539	2.20413	10.6446	12.5916	14.4494	16.8119	18.5476
7	.989265	1.239043	1.68987	2.16735	2.83311	12.0170	14.0671	16.0128	18.4753	20.2777
8	1.344419	1.646482	2.17973	2.73264	3.48954	13.3616	15.5073	17.5346	20.0902	21.9550
9	1.734926	2.087912	2.70039	3.32511	4.16816	14.6837	16.9190	19.0228	21.6660	23.5893
10	2.15585	2.55821	3.24697	3.94030	4.86518	15.9871	18.3070	20.4831	23.2093	25.1882
11	2.60321	3.05347	3.81575	4.57481	5.57779	17.2750	19.6751	21.9200	24.7250	26.7569
12	3.07382	3.57056	4.40379	5.22603	6.30380	18.5494	21.0261	23.3367	26.2170	28.2995
13	3.56503	4.10691	5.00874	5.89186	7.04150	19.8119	22.3621	24.7356	27.6883	29.8194
14	4.07468	4.66043	5.62872	6.57063	7.78953	21.0642	23.6848	26.1190	29.1413	31.3193
15	4.60094	5.22935	6.26214	7.26094	8.54675	22.3072	24.9958	27.4884	30.5779	32.8013
16	5.14224	5.81221	6.90766	7.96164	9.31223	23.5418	26.2962	28.8454	31.9999	34.2672
17	5.69724	6.40776	7.56418	8.67176	10.0852	24.7690	27.5871	30.1910	33.4087	35.7185
18	6.26481	7.01491	8.23075	9.39046	10.8649	25.9894	28.8693	31.5264	34.8053	37.1564
19	6.84398	7.63273	8.90655	10.1170	11.6509	27.2036	30.1435	32.8523	36.1908	38.5822
20	7.43386	8.26040	9.59083	10.8508	12.4426	28.4120	31.4104	34.1696	37.5662	39.9968
21	8.03366	8.89720	10.28293	11.5913	13.2396	29.6151	32.6705	35.4789	38.9321	41.4010
22	8.64272	9.54249	10.9823	12.3380	14.0415	30.8133	33.9244	36.7807	40.2894	42.7958
23	9.26042	10.19567	11.6885	13.0905	14.8479	32.0069	35.1725	38.0757	41.6384	44.1813
24	9.88623	10.8564	12.4011	13.8484	15.6587	33.1963	36.4151	39.3641	42.9798	45.5585
25	10.5197	11.5240	13.1197	14.6114	16.4734	34.3816	37.6525	40.6465	44.3141	46.9278
26	11.1603	12.1981	13.8439	15.3791	17.2919	35.5631	38.8852	41.9232	45.6417	48.2899
27	11.8076	12.8786	14.5733	16.1513	18.1138	36.7412	40.1133	43.1944	46.9630	49.6449
28	12.4613	13.5648	15.3079	16.9279	18.9392	37.9159	41.3372	44.4607	48.2782	50.9933
29	13.1211	14.2565	16.0471	17.7083	19.7677	39.0875	42.5569	45.7222	49.5879	52.3356
30	13.7867	14.9535	16.7908	18.4926	20.5992	40.2560	43.7729	46.9792	50.8922	53.6720
40	20.7065	22.1643	24.4331	26.5093	29.0505	51.8050	55.7585	59.3417	63.6907	66.7659
50	27.9907	29.7067	32.3574	34.7642	37.6886	63.1671	67.5048	71.4202	76.1539	79.4900
60	35.5346	37.4848	40.4817	43.1879	46.4589	74.3970	79.0819	83.2976	88.3794	91.9517
70	43.2752	45.4418	48.7576	51.7393	55.3290	85.5271	90.5312	95.0231	100.425	104.215
80	51.1720	53.5400	57.1532	60.3915	64.2778	96.5782	101.879	106.629	112.329	116.321
90	59.1963	61.7541	65.6466	69.1260	73.2912	107.567	113.145	118.136	124.116	128.299
100	67.3276	70.0648	74.2219	77.9295	82.3581	118.498	124.342	129.561	135.807	140.169

This table is reprinted by permission of Biometrika Trustees from Table 8, Percentage Points of the χ^2 Distribution, by E. S. Pearson and H. O. Hartley, *Biometrika Tables for Statisticians,* Vol. 1, 3d edition, 1966. Copyright © 1966, *Biometrika* Trustees. Used with permission.

Entries in the table give the critical values for a one-tailed Durbin-Watson test for autocorrelation. For a two-tailed test, the level of significance is doubled.

SIGNIFICANT POINTS OF d_L AND d_U: $\alpha = .05$
NUMBER OF INDEPENDENT VARIABLES

k	1		2		3		4		5	
n	d_L	d_U	d_L	d_U	d_L	d_U	d_L	d_U	d_L	d_U
15	1.08	1.36	0.95	1.54	0.82	1.75	0.69	1.97	0.56	2.21
16	1.10	1.37	0.98	1.54	0.86	1.73	0.74	1.93	0.62	2.15
17	1.13	1.38	1.02	1.54	0.90	1.71	0.78	1.90	0.67	2.10
18	1.16	1.39	1.05	1.53	0.93	1.69	0.82	1.87	0.71	2.06
19	1.18	1.40	1.08	1.53	0.97	1.68	0.86	1.85	0.75	2.02
20	1.20	1.41	1.10	1.54	1.00	1.68	0.90	1.83	0.79	1.99
21	1.22	1.42	1.13	1.54	1.03	1.67	0.93	1.81	0.83	1.96
22	1.24	1.43	1.15	1.54	1.05	1.66	0.96	1.80	0.86	1.94
23	1.26	1.44	1.17	1.54	1.08	1.66	0.99	1.79	0.90	1.92
24	1.27	1.45	1.19	1.55	1.10	1.66	1.01	1.78	0.93	1.90
25	1.29	1.45	1.21	1.55	1.12	1.66	1.04	1.77	0.95	1.89
26	1.30	1.46	1.22	1.55	1.14	1.65	1.06	1.76	0.98	1.88
27	1.32	1.47	1.24	1.56	1.16	1.65	1.08	1.76	1.01	1.86
28	1.33	1.48	1.26	1.56	1.18	1.65	1.10	1.75	1.03	1.85
29	1.34	1.48	1.27	1.56	1.20	1.65	1.12	1.74	1.05	1.84
30	1.35	1.49	1.28	1.57	1.21	1.65	1.14	1.74	1.07	1.83
31	1.36	1.50	1.30	1.57	1.23	1.65	1.16	1.74	1.09	1.83
32	1.37	1.50	1.31	1.57	1.24	1.65	1.18	1.73	1.11	1.82
33	1.38	1.51	1.32	1.58	1.26	1.65	1.19	1.73	1.13	1.81
34	1.39	1.51	1.33	1.58	1.27	1.65	1.21	1.73	1.15	1.81
35	1.40	1.52	1.34	1.58	1.28	1.65	1.22	1.73	1.16	1.80
36	1.41	1.52	1.35	1.59	1.29	1.65	1.24	1.73	1.18	1.80
37	1.42	1.53	1.36	1.59	1.31	1.66	1.25	1.72	1.19	1.80
38	1.43	1.54	1.37	1.59	1.32	1.66	1.26	1.72	1.21	1.79
39	1.43	1.54	1.38	1.60	1.33	1.66	1.27	1.72	1.22	1.79
40	1.44	1.54	1.39	1.60	1.34	1.66	1.29	1.72	1.23	1.79
45	1.48	1.57	1.43	1.62	1.38	1.67	1.34	1.72	1.29	1.78
50	1.50	1.59	1.46	1.63	1.42	1.67	1.38	1.72	1.34	1.77
55	1.53	1.60	1.49	1.64	1.45	1.68	1.41	1.72	1.38	1.77
60	1.55	1.62	1.51	1.65	1.48	1.69	1.44	1.73	1.41	1.77
65	1.57	1.63	1.54	1.66	1.50	1.70	1.47	1.73	1.44	1.77
70	1.58	1.64	1.55	1.67	1.52	1.70	1.49	1.74	1.46	1.77
75	1.60	1.65	1.57	1.68	1.54	1.71	1.51	1.74	1.49	1.77
80	1.61	1.66	1.59	1.69	1.56	1.72	1.53	1.74	1.51	1.77
85	1.62	1.67	1.60	1.70	1.57	1.72	1.55	1.75	1.52	1.77
90	1.63	1.68	1.61	1.70	1.59	1.73	1.57	1.75	1.54	1.78
95	1.64	1.69	1.62	1.71	1.60	1.73	1.58	1.75	1.56	1.78
100	1.65	1.69	1.63	1.72	1.61	1.74	1.59	1.76	1.57	1.78

This table is reprinted by permission of *Biometrika* Trustees from J. Durbin and G. S. Watson, "Testing for Serial Correlation in Least Square Regression II," *Biometrika,* vol. 38, 1951, pp. 159–78.

Continued

Table A.9
Continued

	SIGNIFICANT POINTS OF d_L AND d_U: $\alpha = .01$ NUMBER OF INDEPENDENT VARIABLES								
k	1		2		3		4		5	
n	d_L	d_U	d_L	d_U	d_L	d_U	d_L	d_U	d_L	d_U
15	0.81	1.07	0.70	1.25	0.59	1.46	0.49	1.70	0.39	1.96
16	0.84	1.09	0.74	1.25	0.63	1.44	0.53	1.66	0.44	1.90
17	0.87	1.10	0.77	1.25	0.67	1.43	0.57	1.63	0.48	1.85
18	0.90	1.12	0.80	1.26	0.71	1.42	0.61	1.60	0.52	1.80
19	0.93	1.13	0.83	1.26	0.74	1.41	0.65	1.58	0.56	1.77
20	0.95	1.15	0.86	1.27	0.77	1.41	0.68	1.57	0.60	1.74
21	0.97	1.16	0.89	1.27	0.80	1.41	0.72	1.55	0.63	1.71
22	1.00	1.17	0.91	1.28	0.83	1.40	0.75	1.54	0.66	1.69
23	1.02	1.19	0.94	1.29	0.86	1.40	0.77	1.53	0.70	1.67
24	1.04	1.20	0.96	1.30	0.88	1.41	0.80	1.53	0.72	1.66
25	1.05	1.21	0.98	1.30	0.90	1.41	0.83	1.52	0.75	1.65
26	1.07	1.22	1.00	1.31	0.93	1.41	0.85	1.52	0.78	1.64
27	1.09	1.23	1.02	1.32	0.95	1.41	0.88	1.51	0.81	1.63
28	1.10	1.24	1.04	1.32	0.97	1.41	0.90	1.51	0.83	1.62
29	1.12	1.25	1.05	1.33	0.99	1.42	0.92	1.51	0.85	1.61
30	1.13	1.26	1.07	1.34	1.01	1.42	0.94	1.51	0.88	1.61
31	1.15	1.27	1.08	1.34	1.02	1.42	0.96	1.51	0.90	1.60
32	1.16	1.28	1.10	1.35	1.04	1.43	0.98	1.51	0.92	1.60
33	1.17	1.29	1.11	1.36	1.05	1.43	1.00	1.51	0.94	1.59
34	1.18	1.30	1.13	1.36	1.07	1.43	1.01	1.51	0.95	1.59
35	1.19	1.31	1.14	1.37	1.08	1.44	1.03	1.51	0.97	1.59
36	1.21	1.32	1.15	1.38	1.10	1.44	1.04	1.51	0.99	1.59
37	1.22	1.32	1.16	1.38	1.11	1.45	1.06	1.51	1.00	1.59
38	1.23	1.33	1.18	1.39	1.12	1.45	1.07	1.52	1.02	1.58
39	1.24	1.34	1.19	1.39	1.14	1.45	1.09	1.52	1.03	1.58
40	1.25	1.34	1.20	1.40	1.15	1.46	1.10	1.52	1.05	1.58
45	1.29	1.38	1.24	1.42	1.20	1.48	1.16	1.53	1.11	1.58
50	1.32	1.40	1.28	1.45	1.24	1.49	1.20	1.54	1.16	1.59
55	1.36	1.43	1.32	1.47	1.28	1.51	1.25	1.55	1.21	1.59
60	1.38	1.45	1.35	1.48	1.32	1.52	1.28	1.56	1.25	1.60
65	1.41	1.47	1.38	1.50	1.35	1.53	1.31	1.57	1.28	1.61
70	1.43	1.49	1.40	1.52	1.37	1.55	1.34	1.58	1.31	1.61
75	1.45	1.50	1.42	1.53	1.39	1.56	1.37	1.59	1.34	1.62
80	1.47	1.52	1.44	1.54	1.42	1.57	1.39	1.60	1.36	1.62
85	1.48	1.53	1.46	1.55	1.43	1.58	1.41	1.60	1.39	1.63
90	1.50	1.54	1.47	1.56	1.45	1.59	1.43	1.61	1.41	1.64
95	1.51	1.55	1.49	1.57	1.47	1.60	1.45	1.62	1.42	1.64
100	1.52	1.56	1.50	1.58	1.48	1.60	1.46	1.63	1.44	1.65

Table A.10
Factors for Control Charts

NUMBER OF ITEMS IN SAMPLE	AVERAGES		RANGES		
	Factors for Control Limits		Factors for Central Line	Factors for Control Limits	
n	A_2	A_3	d_2	D_3	D_4
2	1.880	2.659	1.128	0	3.267
3	1.023	1.954	1.693	0	2.575
4	0.729	1.628	2.059	0	2.282
5	0.577	1.427	2.326	0	2.115
6	0.483	1.287	2.534	0	2.004
7	0.419	1.182	2.704	0.076	1.924
8	0.373	1.099	2.847	0.136	1.864
9	0.337	1.032	2.970	0.184	1.816
10	0.308	0.975	3.078	0.223	1.777
11	0.285	0.927	3.173	0.256	1.744
12	0.266	0.886	3.258	0.284	1.716
13	0.249	0.850	3.336	0.308	1.692
14	0.235	0.817	3.407	0.329	1.671
15	0.223	0.789	3.472	0.348	1.652

Adapted from American Society for Testing and Materials, *Manual on Quality Control of Materials,* 1951, Table B2, p. 115. For a more detailed table and explanation, see Acheson J. Duncan, *Quality Control and Industrial Statistics,* 3d ed. (Homewood, Ill.: Richard D. Irwin, 1974), Table M, p. 927.

B
Answers to Selected Odd-Numbered Quantitative Problems

Chapter 1

1.1
 a. ratio
 b. ratio
 c. ordinal
 d. nominal
 e. ratio
 f. ratio
 g. nominal
 h. ratio

1.3
 a. 900 electric contractors
 b. 35 electric contractors
 c. average score for 35 participants
 d. average score for all 900 electric contractors

Chapter 2

No answers given

Chapter 3

3.1 4

3.3 294

3.5 −1

3.7 114, 127.5, 143.5

3.9 624, 489, 751

3.11
 a. 8
 b. 2.041
 c. 6.204
 d. 2.491
 e. 4
 f. 0.69, −0.92, −0.11, 1.89, −1.32, −0.52, 0.29

3.13
 a. 4.598
 b. 4.598

3.15 58,631.359; 242.139

3.17
 $\sigma_1 = 14.2215$
 $\sigma_2 = 14.5344$
 $CV_1 = 21.71\%$
 $CV_2 = 10.20\%$

3.19
 a. 623
 b. MAD = 119.37
 c. $\sigma^2 = 29,728.23$
 d. $\sigma = 172.42$
 e. $Q_1 = 2.75$
    $Q_3 = 323$
    IQR= 161
 f. $Z = -0.70$
 g. CV = 58.98%

3.21 3, 2.67

3.23 1.07, 0.56, −0.84, −1.87

3.25 mean = 51, median = 54, mode = 59. The distribution is skewed to the left.

3.27 0.451. Slight skewness to the right.

3.29 $Q_1 = 66$, $Q_3 = 90$, Median = 74, $IQR = 24$, inner fences: 30 and 126, outer fences: −6 and 162. No extreme outliers. 21 is a mild outlier. Positively skewed.

3.31 27.5, 47.5, 20, 62

3.33 $\mu = 928,256$, $\sigma = 430,872$

3.35
 a. $\overline{X} = 5.277$, Median = 4.44, No Mode
 b. $Q_3 = 7.13$, $Q_1 = 4.06$, $IQR = 3.07$, MAD = 1.5798, $S^2 = 3.598$, S = 1.897
 c. $S_k = 1.324$
 d. Box: $Q_1 = 4.06$, $Q_2 = 4.44$, $Q_3 = 7.13$
 Inner fences: −0.545, 11.735
 Outer fences: −5.15, 16.34

3.37 $\sigma = 2.167$, between 4.1 and 10.9

3.39
 a. $\mu = 7,655$; $\sigma = 7,472.53$
 b. $\mu = 4,925$; $\sigma = 2,626.19$
 c. $CV_1 = 97.62\%$, $CV_2 = 53.32\%$

3.41
 a. $Q_1 = 47.4$, $Q_3 = 80.75$, Median = 53.65, IQR = 33.35
 Inner fences: −2.625, 130.775
 Outer fences: −52.65, 180.80
 b. & c. Three extreme outliers at upper end. Positively skewed.

Chapter 4

4.1 15, .60

4.3 {4, 8, 10, 14, 16, 18, 20, 22, 26, 28, 30}

4.5 20, combinations, .60

4.7 38,760

4.9
 a. .7167
 b. .5000
 c. .65
 d. .5167

4.11 not solvable

4.13
 a. .86
 b. .31
 c. .14

4.15
 a. .2807
 b. .0526
 c. .0000
 d. .0000

4.17
 a. .0122
 b. .0144

4.19
 a. .57
 b. .3225
 c. .4775
 d. .5225
 e. .6775
 f. .0475

4.21
 a. .039
 b. .571
 c. .129

4.23
 a. .2286
 b. .2297
 c. .3231
 d. .0000

4.25 not independent

4.27
 a. .4054
 b. .3261
 c. .4074
 d. .32

4.29 **a.** .03
  b. .2875
  c. .3354
  d. .9759
4.31 **a.** .45
  b. .95
  c. .4743, .4269, .0988
  d. .2748, .4533, .2719
4.33 .65, .859, .6205
4.35 **a.** .0897
  b. .0000
  c. .2821
  d. .0000
  e. .3636
  f. .3810
  g. .4615
  h. .2051
4.37 **a.** .91
  b. .09
  c. .3462
  d. .13
4.39 **a.** .042
  b. .034
  c. .2625
  d. .1976
  e. .525
4.41 **a.** .43
  b. .189
  c. .6143
  d. .699
4.43 **a.** .312
  b. .572
  c. .9176
  d. .22
  e. .9533
4.45 **a.** .20
  b. .6429
  c. .40
  d. .60
  e. .40
  f. .33
4.47 .329, .0623
4.49 .8456, .1149, .0396

Chapter 5

5.1 **a.** .0036
  b. .1147
  c. .3822
  d. .5838
5.5 .142, .508, .1376, .1898, 11.44
5.7 **a.** .1032
  b. .0000
  c. .0352
  d. .3480
5.9 **a.** .0538
  b. .1539

  c. .4142
  d. .0672
  e. .0244
  f. .3702
5.11 **a.** 6.3, 2.51
  b. 1.3, 1.14
  c. 8.9, 2.98
  d. 0.6, .775
5.13 **a.** .0037
  b. .1584
  c. .0008
5.15 **a.** .5488
  b. .3293
  c. .1220
  d. .8913
  e. .1912
5.17 **a.** .5091
  b. .2937
  c. .4167
  d. .0014
5.19 **a.** .0529
  b. .0294
  c. .4235
5.21 **a.** .30
  b. .0238
  c. .2381
5.23 .0474
5.25 **a.** .1378
  b. .4493
  c. .0236
5.27 .9463
5.29 **a.** .382, 7
  b. 5, .0023
  c. .0117
5.31 .174
5.33 .5488, .0232, .3012
5.35 **a.** .0002
  b. .0595
  d. .2330
5.37 **a.** .0907
  b. .0358
  c. .1517
  d. .8781
5.39 **a.** .2189
  b. .7504
  c. .0922
5.41 **a.** .0215
  b. .1317
  c. .7907

Chapter 6

6.1 **a.** 1/40
  b. 220, 11.547
  c. .25
  d. .3750
  e. .6250

6.3	2.97, 0.10, .2941
6.5	**a.** 981.5, .000294, .2353, .0000, .2353
6.7	**a.** .8944
	b. .0122
	c. .2144
6.9	**a.** .1788
	b. .0329
	c. .1476
6.11	**a.** 188.25
	b. 244.65
	c. 163.81
	d. 206.11
6.13	22.2
6.15	**a.** $P(X \leq 16.5 \mid \mu = 21$ and $\sigma = 2.51)$
	b. $P(10.5 \leq X \leq 20.5 \mid \mu = 12.5$ and $\sigma = 2.5)$
	c. $P(21.5 \leq X \leq 22.5 \mid \mu = 24$ and $\sigma = 3.10)$
	d. $P(X > 14.51 \mid \mu = 7.2$ and $\sigma = 1.99)$
6.17	**a.** .1170, .120
	b. .4090, .415
	c. .1985, .196
	d. .0009, .001
6.19	.0495
6.21	**a.** .1314
	b. .6767
	c. .0132
	d. .0916
6.23	**a.** 0.31, 0.31
	b. 1.43, 1.43
	c. 0.91, 0.91
	d. 0.17, 0.17
6.25	**a.** .0735
	b. .0000
6.27	**a.** 53.4
	b. .1065
	c. .1703
6.29	2 years 9.52%
6.31	3.915
6.33	**a.** .0498
	b. .3834
	c. .2212
6.35	2498.96 million
6.37	**a.** 8.33
	b. .4980
	c. .4512
	d. .0007
6.39	.2643, .0071, .2660
6.41	$\lambda = 6.67$/hour, .4262
6.43	.0693, .7357, .0004
6.45	1.585, 2.745
6.47	**a.** 60.75
	b. 33
	c. .0455
	d. .0733
6.49	.0262, 40, .2941
6.51	.4493, .1353
6.53	13.64, .0123, .0002

Chapter 7

7.7	825
7.13	**a.** .0548
	b. .7881
	c. .0082
	d. .8575
	e. .1664
7.15	11.11
7.17	**a.** .9772
	b. .2385
	c. .1469
	d. .1230
7.19	.0000
7.21	**a.** .1894
	b. .0559
	c. .0000
	d. 16.4964
7.23	**a.** .1492
	b. .9404
	c. .1985
	d. .1445
	e. .0000
7.25	.26
7.27	**a.** 1230
	b. .0409
	c. 1.000
7.29	**a.** .1020
	b. .7568
	c. .2981
7.31	55, 45, 90, 25, 35
7.37	**a.** .3156
	b. .00003
	c. .1736
7.41	**a.** .0021
	b. .9265
	c. .0281
7.43	**a.** .0314
	b. .2420
	c. .2250
	d. .1469
	e. .0000
7.45	**a.** .8534
	b. .0256
	c. .0007
7.49	.6819, .0110, .0023
7.51	.9147

Chapter 8

8.1	**a.** $24.11 \leq \mu \leq 25.89$
	b. $113.17 \leq \mu \leq 126.03$
	c. $3.136 \leq \mu \leq 3.702$
	d. $54.55 \leq \mu \leq 58.85$
8.3	$45.92 \leq \mu \leq 48.08$
8.5	5.3, $5.13 \leq \mu \leq 5.47$
8.7	$2.853 \leq \mu \leq 3.759$
8.9	$23.036 \leq \mu \leq 26.030$

8.11	$314.45 \leq \mu \leq 323.89$
8.13	$15.631 \leq \mu \leq 16.545$, 16.088
8.15	$2.26886 \leq \mu \leq 2.45346$, 2.36116, .0923
8.17	$36.77 \leq \mu \leq 62.83$
8.19	**a.** $.316 \leq P \leq .704$
	b. $.777 \leq P \leq .863$
	c. $.456 \leq P \leq .504$
	d. $.246 \leq P \leq .394$
8.21	$.38 \leq P \leq .56$
	$.364 \leq P \leq .576$
	$.33 \leq P \leq .61$
8.23	$.4287 \leq P \leq .5113$
	$.2488 \leq P \leq .3112$
8.25	**a.** .266
	b. $.246 \leq P \leq .286$
8.27	$.5935 \leq P \leq .6665$
8.29	**a.** 2522
	b. 601
	c. 268
8.31	106
8.33	1083
8.35	97
8.37	722
8.39	$36.231 \leq \mu \leq 38.281$
8.41	$.542 \leq P \leq .596$
8.43	$5.893 \leq \mu \leq 7.541$
8.45	$.726 \leq P \leq .814$
8.47	$34.11 \leq \mu \leq 53.29$
8.49	$-0.20 \leq \mu \leq 5.16$
8.51	543
8.53	$5.181 \leq \mu \leq 5.226$

Chapter 9

9.1	$Z = -2.39$, reject
9.3	$Z = 1.97$, fail to reject
9.5	$Z = -6.39$, reject
9.7	$Z = -2.17$, reject
9.9	$t = 1.66$, fail to reject
9.11	$t = 1.96$, fail to reject
9.13	$t = 0.73$, fail to reject
9.15	$Z = -1.89$, fail to reject
9.17	$Z = 1.23$, fail to reject
9.19	$Z = -3.12$, reject
9.21	$Z = 1.34$, fail to reject
9.23	$Z = -3.72$, reject
9.25	$t = -5.70$, reject
9.27	$Z = 1.53$, fail to reject
9.29	$Z = 2.05$, fail to reject, .01
9.31	$t = 4.50$, reject
9.33	$t = -2.25$, reject

Chapter 10

10.1	**a.** $Z = -0.53$, fail to reject
	b. $(\overline{X}_1 - \overline{X}_2)_c = 1.963$, $(\overline{X}_1 - \overline{X}_2) = -0.818$, fail to reject
	c. .2981

10.3	$Z = -5.39$, reject
10.5	$Z = 7.43$, reject
10.7	$t = -1.91$, reject
10.9	$t = -4.66$, reject
10.11	$t = 1.21$, fail to reject
10.13	$t = 3.31$, reject
10.15	$t = 0.875$, fail to reject
10.17	$t = 3.61$, reject
10.19	**a.** $Z = 0.75$, fail to reject
	b. $Z = 4.83$, reject
10.21	$Z = -2.37$, fail to reject
10.23	$Z = -0.67$, fail to reject
10.25	$F = 1.81$, fail to reject
10.27	$F = 1.21$, fail to reject
10.29	$F = 1.53$, fail to reject
10.31	$Z = 0.16$, fail to reject
10.33	$t = -2.34$, fail to reject
10.35	$F = 1.64$, fail to reject
10.37	$F = 5.12$, reject
10.39	$Z = 8.49$, reject
10.41	$t = 9.00$, reject
10.43	$Z = 2.64$, reject
10.45	$F = 1.95$, fail to reject
10.47	$t = 4.78$, reject
10.49	$Z = 5.96$, reject

Chapter 11

11.1	$F = 11.07$, reject
11.3	$F = 13.00$, reject
11.5	$F = 92.67$, reject
11.7	$F = 11.03$, reject
11.9	$F_{treatment} = 1.48$, fail to reject
	$F_{blocks} = 44.10$, reject
11.11	$F_{treatment} = 15.37$, reject
	$F_{blocks} = 16.38$, reject
11.13	$F_R = 38.37$, reject; $F_C = 4.42$, fail to reject;
	$F_I = 0.63$, fail to reject
11.15	$F_R = 14.77$, reject; $F_C = 25.61$, reject; $F_I = 0.86$, fail to reject
11.17	$F_R = 0.14$, fail to reject; $F_C = 7.47$, reject;
	$F_I = 0.38$, fail to reject
11.19	$\chi^2 = 198.48$, reject
11.21	$\chi^2 = 10.27$, fail to reject
11.23	$\chi^2 = 3.8095$, fail to reject
11.25	$\chi^2 = 34.97$, reject
11.27	$\chi^2 = 6.43$, reject
11.29	$\chi^2 = 3.93$, fail to reject
11.31	$F = 16.19$, reject
11.33	$F_R = 38.21$, reject; $F_C = 0.23$, fail to reject;
	$F_I = 1.30$, fail to reject
11.35	$\chi^2 = 1.65$, fail to reject
11.37	$F = 7.38$, reject
11.39	$F_B = 103.70$, reject; $F_A = 3.02$, fail to reject
11.41	$F = 0.46$, fail to reject
11.43	$\chi^2 = 3.09$, fail to reject
11.45	$F_{treatment} = 13.64$, reject
	$F_{blocks} = 2.50$, reject

11.47 Two-way ANOVA with 3 columns and 4 rows. 3 observations per cell. $F_R = 4.30$, *p-value* = .014, significant at $\alpha = .05$. The null hypothesis is rejected for rows. $F_C = 0.53$, *p-value* = .594 is not significant. Fail to reject the null hypothesis for columns. $F_I = 0.99$, *p-value* = .453 is not significant. Fail to reject the null hypothesis for interaction effects.

Chapter 12

12.3 $\hat{Y} = 16.5 + 0.162X$

12.5 $\hat{Y} = -46.29 + 15.24X$

12.7 -0.496825; 162503.1; $\hat{Y} = 162503.1 - 0.496825X$

12.9 $\hat{Y} = -2.31307 + 0.05557X$

12.11 $\hat{Y} = 50.50556 - 1.64599X$, $S_e = 4.3914$, $r^2 = .909$; Residuals: 4.7244, -0.9836, -0.3996, -6.7537, 2.7683, 0.6442

12.13 $S_e = 9794.72$; $r^2 = .617$; Residuals: $10,706.94$, $-13,228.65$, $-11,510.27$, $4,597.34$, $11,907.69$, $5,151.98$, $-8,395.59$, $2,220.33$, $-7,026.95$, $5,577.19$

12.15 $S_e = 0.2735$, $r^2 = .63$ Residuals: -0.49834, -0.25720, -0.25277, 0.29850, 0.13065, -0.03720, 0.08276, 0.09381, 0.21158, 0.22820

12.17 Error terms non-independent, non-constant error variance, non-linear regression

12.19 $\hat{Y} = -599.3674 + 19.2204X$; $S_e = 13.539$; $r^2 = .688$

12.21 $t = -3.59$, reject

12.23 $t = 3.70$, reject

12.25 $t = -1.32$, fail to reject

12.27 $441.555 \le E(Y_{40}) \le 685.045$

12.31 $r = -0.927$

12.33 $r = .975$

12.35 $r = -.3453$, $r^2 = .1192$

12.37 $r = .975$, $r = .985$, $r = .957$

12.39 **a.** $\hat{Y} = -11.335 + 0.355X$
    b. Predicted Values: 7.48, 5.35, 3.22, 6.415, 9.255, 10.675, 4.64, 9.965
      Residuals: -2.48, -0.35, 3.78, -2.415, 0.745, 1.325, -1.64, 1.035
    c. SSE = 32.4649
    d. $S_e = 2.3261$
    e. $r^2 = .608$
    f. $r = .780$
    g. $t = 3.05$, reject

12.41 **a.** $20.92 \le E(Y_{60}) \le 6.8$
    b. $20.994 \le Y \le 37.688$

12.43 $r = .909$, $r^2 = .826$

12.45 $r = .636$

12.47 $S_e = 21.13$, $r^2 = .906$

12.49 $r = .90$

12.51 $\hat{Y} = -54.35604 + 2.40107X$, $\hat{Y}_{100} = 185.751$, $S_e = 17.886$, $r^2 = .91$, $t = 7.80$, reject

Chapter 13

13.1 $\hat{Y} = 25.03 - 0.0497X_1 + 1.928X_2$, 28.586

13.3 $\hat{Y} = 121.62 - 0.174X_1 + 6.02X_2 + 0.00026X_3 + 0.0041X_4$, 4

13.5 per capita consumption $= -538 + 0.23368$ paper consumption $+ 18.09$ fish consumption $- 0.2116$ gasoline consumption

13.7 $8, \hat{Y} = 56.594 + 0.208X_1 - 0.368X_2 - 0.219X_3 + 0.351X_4 + 0.260X_5 - 0.131X_6 + 0.044X_7 - 0.048X_8$, $S_e = 10.258$, $R^2 = .440$, adj. $R^2 = .268$

13.9 per capita consumption $= -538 + 0.23368$ paper consumption $+ 18.09$ fish consumption $- 0.2116$ gasoline consumption; $F = 24.63$, $p = .002$; $t_1 = 5.31$, $p = .003$; $t_2 = 0.98$, $p = .373$; $t_3 = -0.93$, $p = .397$; $S_e = 2085$; $R^2 = .937$; adj. $R^2 = .899$

13.11 $\hat{Y} = 3.981 + 0.07322X_1 - 0.03232X_2 - 0.003886X_3$, $F = 100.47$ significant at $\alpha = .01$, $t = 3.50$ for X_1 significant at $\alpha = .01$, $S_e = 0.2331$, $R^2 = .965$, adj. $R^2 = .955$

13.13 3 predictors, 15 observations, $\hat{Y} = 657.053 + 5.710 X_1 - 0.417 X_2 - 3.471 X_3$, $R^2 = .842$, adjusted $R^2 = .630$, $S_e = 109.43$, $F = 8.96$ with $p = .0027$, X_1 significant at $\alpha = .01$, X_3 significant at $\alpha = .05$

13.15 3, $R^2 = .806$, X_3 and X_4 Significant predictors

13.17 $\hat{Y} = 7.066 - 0.0855$ Hours $+ 9.614$ Probability $+ 10.507$ French Quarter, $F = 6.80$ significant at $\alpha = .01$, $t = 3.97$ for French Quarter (dummy variable) significant at $\alpha = .01$, $S_e = 4.02$, $R^2 = .671$, adj. $R^2 = .573$

13.19 Simple Model: $\hat{Y} = -147.27 + 27.128X$, $F = 229$ with $p = .000$, $S_e = 27.27$, $R^2 = .97$, adj. $R^2 = .966$ Quadratic Model: $\hat{Y} = -22.01 + 3.385X$, $0.9373X^2$, $F = 578.76$ with $p = .000$, $S_e = 12.3$, $R^2 = .995$, adj. $R^2 = .993$, for X: $t = 0.75$, for X^2: $t = 5.33$.

13.21 $\log Y = 0.5797 + 0.82096 X$, $F = 68.83$ with $p = .000$, $S_e = 0.1261$, $R^2 = .852$

13.23 $\log$ liabilities $= 3.1256 + 0.012846$ failures, $F = 19.98$ with $p = .001$, $S_e = 0.2862$, $R^2 = .666$

13.25 $C_p = 2.01$ for X_2, X_3

13.27 $C_p = 3.97$ for dividend, income

13.29

	Y	X_1	X_2
X_1	$-.653$		
X_2	$-.891$	$.650$	
X_3	$.821$	$-.615$	$-.688$

13.31 $r_{\text{div., gain/loss}} = -.522$, $r_{\text{div., income}} = .682$, $r_{\text{gain/loss, income}} = .092$

13.33 $\hat{Y} = 137.27 + 0.0025X_1 + 29.206X_2$, $F = 10.89$ with *p*-value of .005, $S_e = 9.401$, $R^2 = .731$, adj. $R^2 = .664$. For X_1, $t = 0.01$ with *p*-value of .99. For X_2, $t = 4.47$ with *p*-value of .002.

13.35 Model with X_1, X_1^2 is strongest, $C_p = 4.07$

13.37 Grain $= -4.675 + 0.4732$ Oilseed $+ 1.18$ Livestock, $R^2 = .901$, adj. $R^2 = .877$, $S_e = 1.761$, $F = 36.55$ with *p*-value of .000, $t_{\text{oilseed}} = 3.74$ with *p*-value of .006, $t_{\text{livestock}} = 3.78$ with *p*-value of .005.

13.39 Employment $= 71.03 + 0.4620$ Naval Vessels $+ 0.02082$ Commercial, $F = 1.22$ with *p*-value of .386 (not significant), $R^2 = .379$, adj. $R^2 = .068$, $t_{\text{naval vessels}} = 0.67$ with *p*-value of .541, $t_{\text{commercial}} = 1.07$ with *p*-value of .345.

13.41 The regression model is: Corn = −2718.36 + 6.2445 Soybeans − 0.7682 Wheat, R^2 = .803, Adj. R^2 = .746, S_e = 862.38, F = 14.25 with p-value = .0034. t-ratio for soybeans = 4.20 with p-value = .004 and t-ratio for wheat = −0.75 with p-value = .476.

13.43 R^2 did not decrease. Dropping out variable X_3 did not impact the model significantly.

Chapter 14

14.1 ME = −0.033, MAD= 1.367, MSE = 2.27

14.3 ME = 5.375, MAD = 5.375, MSE = 23.65, MPE = 20.01, MAPE = 20.01

14.5 **a.** Forecasts: 44.75, 52.75, 61.50, 64.75, 70.50, 81.00
 Errors: 14.25, 13.25, 9.50, 21.25, 30.50, 16.00
 b. Forecasts: 53.25, 56.375, 62.875, 67.25, 76.375, 89.125
 Errors: 5.75, 9.625, 8.125, 18.75, 24.625, 7.875
 c. Difference in Errors: 8.5, 3.626, 1.375, 2.5, 5.875, 8.125

14.7 Forecasts (α = .3): 9, 8.7, 8.8, 9.1, 9.7, 9.9, 9.8
 Errors: −1.1, 0.3, 1.0, 1.9, 0.6, −0.4, −0.7
 Forecasts (α = .7): 8.6, 8.1, 8.7, 9.5, 10.6, 10.4, 9.8
 Errors: −0.7, 0.9, 1.1, 1.5, −0.3, −0.9, −0.7
 Forecasts (3-mo. avg.): 8.5, 8.4, 8.9, 9.9, 10.4, 9.6
 Errors: 0.5, 1.4, 1.1, 0.4, −0.9, −0.5

14.9 Forecasts (α = .2): 332, 404.4, 427.1, 386.1, 350.7, 315, 325.2, 362.6, 423.5, 453, 477.4, 554.9
 Errors: 362, 113.6, 205.1, 177.1, 178.7, 51, 186.8, 304.4, 147.5, 122, 387.6, 54.1
 Forecasts (α = .9): 332, 657.8, 532, 253, 213.4, 176.1, 347, 495.5, 649.9, 578.9, 575.4, 836
 Errors: 362, 139.8, 310, 44, 41.4, 189.9, 165, 171.5, 78.9, 3.9, 289.6, 227
 MAD for α = .2, 190.8; MAD for α = .9, 168.6

14.11 Members = 167,986−75.998Year; R^2 = .846; S_e = 140.9; F = 66.17, reject

14.13 Food = 0.7542 + 0.6788 shelter, R^2 = .622, D = 1.11, reject

14.15 Assets = 1379 + 136.68 failures, R^2 = .379, D = 2.49, fail to reject

14.17 1 lag model: $\hat{Y}$ = 158 + 0.589 X, R^2 = .353
 2 lag model: $\hat{Y}$ = 401−0.065 X, R^2 = .05

14.19 for β = .3, MAD = 1.06; for β = .7, MAD = 1.32

14.21 Trend: R^2 = .572, F = 29.38
 Seasonality: R^2 = .158, F = 0.205

14.23 MAD = 0.84

14.25 **a.** $\text{MAD}_{\text{moving average}}$ = 722.44
 b. $\text{MAD}_{\alpha=.2}$ = 1235.13
 c. The three-year moving average produced a smaller MAD (722.44) than did exponential smoothing with α = .2 (MAD = 1235.13). Using MAD as the criterion, the three-year moving average was a better forecasting tool than the exponential smoothing with α = .2.

14.27 $\hat{Y}$ = −5997939 + 1032214 X, $\hat{Y}$ (55) = 50773831, R^2 = .996, F = 4829.65 (p = .000), S_e = 1243142

14.29 D = .16, reject

14.31 $\hat{Y}$ = 81 + 0.849 X, R^2 = .558, S_e = 50.18, F = 8.83 (p = .021)

14.33 D = 0.48, reject, there is significant autocorrelation

14.35 D = 0.98, reject, there is significant autocorrelation

Chapter 15

15.5 $\bar{\bar{X}}$ = 4.51, UCL = 5.17, LCL = 3.85
 $\bar{R}$ = 0.90, UCL = 2.05, LCL = 0

15.7 P = .05, UCL =.153, LCL = .000

15.9 $\bar{c}$ = 1.34375, UCL = 4.82136, LCL = .000

15.11 $\bar{X}$ chart, in control; R chart, in control; P chart, inordinate number in outer 1/3 but at lower end

15.15 P = .1040, LCL = 0.000, UCL = .2335

15.17 centerline = 2.13889, UCL = 6.52637, LCL = 0.00

15.19 $\bar{X}$ chart: centerline = 14.99854, UCL = 15.02269, LCL = 14.97439
 R chart: centerline = .05, UCL = .1002, LCL = 0.0000

15.21 centerline = 0.64, UCL = 3.04, LCL = 0.00

15.23 P = 0.06, LCL = 0.000, UCL = .1726

15.25 Out-of-control; four of five in outer 2/3; ten of eleven below centerline

15.27 In control

Glossary

a posteriori After the experiment; pairwise comparisons made by the researcher *after* determining that there is a significant overall F value from ANOVA; also called *post hoc*.

Additive model Value of time series variable equals the sum of the four component effects.

Adjusted R^2 A modified value of R^2 in which the degrees of freedom are taken into account, thereby allowing the researcher to determine whether the value of R^2 is inflated for a particular multiple regression model.

After-process quality control A type of quality control in which product attributes are measured by inspection after the manufacturing process is completed to determine whether the product is acceptable.

All possible regressions A multiple regression search procedure in which all possible multiple linear regression models are determined from the data using all variables.

alpha (α) The probability of committing a Type I error; also called the level of significance.

Alternative hypothesis The hypothesis that complements the null hypothesis; usually it is the hypothesis that the researcher is interested in proving.

Analysis of variance (ANOVA) A technique for statistically analyzing the data from a completely randomized design; uses the F test to determine whether there is a significant difference in two or more independent groups.

Arithmetic mean The average of a group of numbers.

Autocorrelation A problem that arises in regression analysis when the data occur over time and the error terms are correlated; also called serial correlation.

Autoregression A multiple regression forecasting technique in which the independent variables are time-lagged versions of the dependent variable.

Averaging models Forecasting models in which the forecast is the average of several preceding time periods.

Backward elimination A step-by-step multiple regression search procedure that begins with a full model containing all predictors. A search is made to determine if there are any nonsignificant independent variables in the model. If there are no nonsignificant predictors, then the backward process ends with the full model. If there are nonsignificant predictors, then the predictor with the smallest absolute value of t is eliminated and a new model is developed with the remaining variables. This procedure continues until only variables with significant t values remain in the model.

Bayes' rule An extension of the conditional law of probabilities discovered by Thomas Bayes that can be used to revise probabilities.

Benchmarking A quality control method in which a company attempts to develop and establish total quality management from product to process by examining and emulating the best practices and techniques used in their industry.

beta (β) The probability of committing a Type II error.

Bimodal Data sets that have two modes.

Binomial distribution Widely known discrete distribution in which there are only two possibilities on any one trial.

Blocking Variable A variable that the researcher wants to control but is not the treatment variable of interest.

Bounds The error portion of the confidence interval that is added and/or subtracted from the point estimate to form the confidence interval.

Box and whisker plot A diagram that utilizes the upper and lower quartiles along with the median and the two most extreme values to depict a distribution graphically; sometimes called a box plot.

c chart A quality control chart for attribute compliance that displays the number of nonconformances per item or unit.

Causal techniques Forecasting techniques based on historic data for the variable being forecast and one or more explanatory variables.

Cause-and-effect diagram A tool for displaying possible causes for a quality problem and the interrelationships among the causes; also called a fishbone diagram or an Ishikawa diagram.

Census A process of gathering data from the whole population for a given measurement of interest.

Centerline The middle horizontal line of a control chart, often determined either by a product or service specification or by computing an expected value from sample information.

Central limit theorem A theorem that states that regardless of the shape of a population, the distributions of sample means and proportions are normal if sample sizes are large.

Chi-square distribution A continuous distribution determined by the sum of the squares of k independent random variables.

Chi-square goodness-of-fit test A statistical test used to analyze probabilities of multinomial distribution trials along a single dimension; compares expected, or theoretical, frequencies of categories from a population distribution to the observed, or actual, frequencies from a distribution.

Chi-square test of independence A statistical test used to analyze the frequencies of two variables with multiple categories to determine whether the two variables are independent.

Classical method of assigning probabilities Probabilities assigned based on rules and laws.

Classification variable   The independent variable of an experimental design that was present prior to the experiment and is not the result of the researcher's manipulations or control.

Classifications   The subcategories of the independent variable used by the researcher in the experimental design; also called levels.

Class mark   Another name for class midpoint; the midpoint of each class interval in grouped data.

Class midpoint   For any given class interval of a frequency distribution, the value halfway across the class interval; the average of the two class endpoints.

Cluster (or area) sampling   A type of random sampling in which the population is divided into nonoverlapping areas or clusters and elements are randomly sampled from the areas or clusters.

Coefficient of determination (r^2)   The proportion of variability of the dependent variable accounted for or explained by the independent variable in a regression model.

Coefficient of multiple determination (R^2)   The proportion of variation of the dependent variable accounted for by the independent variables in the regression model.

Coefficient of skewness   A measure of the degree of skewness that exists in a distribution of numbers; compares the mean and the median in light of the magnitude of the standard deviation.

Coefficient of variation (CV)   The ratio of the standard deviation to the mean, expressed as a percentage.

Collectively exhaustive events   A list containing all possible elementary events for an experiment.

Combinations   Used to determine the number of possible ways n things can happen from N total possibilities when sampling without replacement.

Complement of a union   The only possible case other than the union of sets X and Y; the probability that neither X nor Y is in the outcome.

Complementary events   Two events, one of which comprises all the elementary events of an experiment that are not in the other event.

Completely randomized design   An experimental design wherein there is one treatment or independent variable with two or more treatment levels and one dependent variable. This design is analyzed by analysis of variance.

Concomitant variables   Variables that are not being controlled by the researcher in the experiment but can have an effect on the outcome of the treatment being studied; also called confounding variables.

Conditional probability   The probability of the occurrence of one event given that another event has occurred.

Confounding variables   Variables that are not being controlled by the researcher in the experiment but can have an effect on the outcome of the treatment being studied; also called concomitant variables.

Contingency analysis   Another name for the chi-square test of independence.

Contingency table   A two-way table that contains the frequencies of responses to two questions; also called a raw values matrix.

Continuous data   Numeric data that take on values at every point over a given interval.

Continuous distributions   Distributions constructed from continuous random variables.

Continuous random variables   Variables that take on values at every point over a given interval.

Convenience sampling   A nonrandom sampling technique in which items for the sample are selected for the convenience of the researcher.

Correction for continuity   A correction made when a binomial distribution problem is approximated by the normal distribution because a discrete distribution problem is being approximated by a continuous distribution.

Correlation   A measure of the degree of relatedness of two or more variables.

Critical value   The value that divides the nonrejection region from the rejection region.

Critical value method   A method of testing hypotheses in which the sample statistic is compared to a critical value in order to reach a conclusion about rejecting or failing to reject the null hypothesis.

Cumulative frequency   A running total of frequencies through the classes of a frequency distribution.

Cyclical component   Periodic increases and decreases that occur over more than a one-year period.

Damping factor   One minus the exponential smoothing constant.

Degrees of freedom   A mathematical adjustment made to the size of the sample; used along with α to locate values in statistical tables.

Dependent samples   Two or more samples selected in such a way as to be dependent or related; each item or person in one sample has a corresponding matched or related item in the other samples. Also called related samples.

Dependent variable   In regression analysis, the variable that is being predicted.

Descriptive statistics   Statistics that have been gathered on a group to describe or reach conclusions about that same group.

Deseasonalized time series   A time series in which the seasonal effect has been removed. Also called a seasonally adjusted time series.

Deterministic model   Mathematical models that produce an "exact" output for a given input.

Deviation from the mean   The difference between a number and the average of the set of numbers of which the number is a part.

Discrete Data   Numeric data in which the set of all possible values is at most a finite or a countably infinite number of possible values.

Discrete distributions   Distributions constructed from discrete random variables.

Discrete random variables   Random variables in which the set of all possible values is at most a finite or a countably infinite number of possible values.

Disproportionate stratified random sampling   A type of stratified random sampling in which the proportions of items selected from the strata for the final sample do not reflect the proportions of the strata in the population.

Dummy variable   Another name for a qualitative or indicator variable; usually coded as 0 or 1 and represents whether or not a given item or person possesses a certain characteristic.

Durbin-Watson test   A statistical test for determining whether significant autocorrelation is present in a time-series regression model.

Elementary events   Events that cannot be decomposed or broken down into other events.

Empirical rule   A guideline that states the approximate percentage of values that fall within a given number of standard deviations of a mean of a set of data that are normally distributed.

Error of an individual forecast   The difference between the actual value and the forecast of that value.

Error of estimation   The difference between the statistic computed to estimate a parameter and the parameter.

Event   An outcome of an experiment.

Experiment   A process that produces outcomes.

Experimental design   A plan and a structure to test hypotheses in which the researcher either controls or manipulates one or more variables.

Exponential distribution   A continuous distribution closely related to the Poisson distribution that describes the times between random occurrences.

Exponential smoothing   A forecasting technique in which a weighting system is used to determine the importance of previous time periods in the forecast.

F distribution   A distribution based on the ratio of two random variances; used in testing two variances and in analysis of variance.

F value   The ratio of two sample variances, used to reach statistical conclusions regarding the null hypothesis; in ANOVA, the ratio of the treatment variance to the error variance.

Factorial design   An experimental design in which two or more independent variables are studied simultaneously and every level of each treatment is studied under the conditions of every level of all other treatments. Also called a factorial experiment.

Factors   Another name for the independent variables of an experimental design.

Finite correction factor   A statistical adjustment made to the Z formula for sample means; adjusts for the fact that a population is finite and the size is known.

First-differences approach   A method of transforming data in an attempt to reduce or remove autocorrelation from a time-series regression model; results in each data value being subtracted from each succeeding time period data value, producing a new, transformed value.

Fishbone diagram   A display of possible causes of a quality problem and the interrelationships among the causes. The problem is diagrammed along the main line of the "fish" and possible causes are diagrammed as line segments angled off in such a way as to give the appearance of a fish skeleton. Also called an Ishikawa diagram or a cause-and-effect diagram.

Flowchart   A schematic representation of all the activities and interactions that occur in a process.

Folded annual time series plot   A multi-year time series plot with only one year on the horizontal axis.

Forecasting   The art or science of predicting the future.

Forecast bias   Tendency of forecasting method to over or under predict. Also called average or mean error (ME).

Forecasting error   A single measure of the overall error of a forecast for an entire set of data.

Forward selection   A multiple regression search procedure that is essentially the same as stepwise regression analysis except that once a variable is entered into the process, it is never deleted.

Frame   A list, map, directory, or some other source that is being used to represent the population in the process of sampling.

Frequency distribution   A summary of data presented in the form of class intervals and frequencies.

Frequency polygon   A graph constructed by plotting a dot for the frequencies at the class midpoints and connecting the dots.

General linear regression model   Regression models that take the form of $Y = \beta_0 + \beta_1 X_1 + \beta_2 X_2 + \ldots + \beta_k X_k + \epsilon$, where the parameters, β_i, are linear.

Grouped data   Data that have been organized into a frequency distribution.

Heteroscedasticity   The condition that occurs when the error variances produced by a regression model are not constant.

Histogram   A type of vertical bar chart constructed by graphing line segments for the frequencies of classes across the class intervals and connecting each to the X axis to form a series of rectangles.

Holt's Model   Exponential smoothing with a trend component.

Homoscedasticity   The condition that occurs when the error variances produced by a regression model are constant.

Hypergeometric distribution   A distribution of probabilities of the occurrence of X items in a sample of n when there are A of that same item in a population of N.

Hypothesis testing   A process of testing hypotheses about parameters by setting up null and alternative hypotheses, gathering sample data, computing statistics from the samples, and using statistical techniques to reach conclusions about the hypotheses.

Independent events Events such that the occurrence or non-occurrence of one has no effect on the occurrence of the others.

Independent samples Two or more samples in which the selected items are related only by chance.

Independent variable In regression analysis, the predictor variable.

Indicator variable Another name for a dummy or qualitative variable; usually coded as 0 or 1 and represents whether or not a given item or person possesses a certain characteristic.

Inferential statistics Statistics that have been gathered from a sample and used to reach conclusions about the population from which the sample was taken.

In-process quality control A quality control method in which product attributes are measured at various intervals throughout the manufacturing process.

Interaction When the effects of one treatment in an experimental design vary according to the levels of treatment of the other effect(s).

Interquartile range The range of values between the first and the third quartile.

Intersection The portion of the population that contains elements that lie in both or all groups of interest.

Interval estimate A range of values within which it is estimated with some confidence the population parameter lies.

Interval level data Next to highest level of data. These data have all the properties of ordinal level data, but in addition, intervals between consecutive numbers have meaning.

Irregular component Changes in the time series not attributable to the other three components.

Ishikawa diagram A tool developed by Kaoru Ishikawa as a way to display possible causes of a quality problem and the interrelationships of the causes; also called a fishbone diagram or a cause-and-effect diagram.

Joint probability The probability of the intersection occurring, or the probability of two or more events happening at once.

Judgment sampling A nonrandom sampling technique in which items selected for the sample are chosen by the judgment of the researcher.

Just-in-time inventory system An inventory system in which little or no extra raw materials or parts for production are stored.

Kurtosis The amount of peakedness of a distribution.

Lambda (λ) Denotes the long-run average of a Poisson distribution.

Least squares analysis The process by which a regression model is developed based on calculus techniques that attempt to produce a minimum sum of the squared error values.

Leptokurtic Distributions that are high and thin.

Level of significance The probability of committing a Type I error; also known as alpha.

Levels The subcategories of the independent variable used by the researcher in the experimental design; also called classifications.

Linear trend projection Forecasting by fitting a linear equation to a time series.

Lower control limit (LCL) The bottom-end line of a control chart, usually situated approximately three standard deviations of the statistic below the centerline; data points below this line indicate quality control problems.

Manufacturing quality A view of quality in which the emphasis is on the manufacturer's ability to target consistently the requirements for the product with little variability.

Marginal probability A probability computed by dividing a subtotal of the population by the total of the population.

Matched pairs data Data or measurements gathered from pairs of items or persons that are matched on some characteristic or from a before-and-after design and then separated into different samples; also called paired data or related measures.

Matched-pairs test A *t* test to test the differences in two related or matched samples; sometimes called the *t* test for related measures or the correlated *t* test.

Mean absolute deviation (MAD) The average of the absolute values of the deviations around the mean for a set of numbers.

Mean absolute percentage error (MAPE) The average of the absolute values of the percentage errors of a forecast.

Mean error (ME) The average of all the errors of forecast for a group of data.

Mean percentage error (MPE) The average of the percentage errors of a forecast.

Mean square error (MSE) The average of all errors squared of a forecast for a group of data.

Measures of central tendency One type of measure that is used to yield information about the center of a group of numbers.

Measures of shape Tools that can be used to describe the shape of a distribution of data.

Measures of variability Statistics that describe the spread or dispersion of a set of data.

Median The middle value in an ordered array of numbers.

Mesokurtic Distributions that are normal in shape—that is, not too high or too flat.

Metric data Interval and ratio level data; also called quantitative data.

mn counting rule A rule used in probability to count the number of ways two operations can occur if the first operation has *m* possibilities and the second operation has *n* possibilities.

Mode The most frequently occurring value in a set of data.

Moving average When an average of data from previous time periods is used to forecast the value for ensuing time periods and this average is modified at each new time period by including more recent values not in the previous average and dropping out values from the more distant time periods that were in the average. It is continually updated at each new time period.

Multicollinearity   A problematic condition that occurs when two or more of the independent variables of a multiple regression model are highly correlated.

Multimodal   Data sets that contain more than two modes.

Multiple comparisons   Statistical techniques used to compare pairs of treatment means when the analysis of variance yields an overall significant difference in the treatment means.

Multiple regression   Regression analysis with one dependent variable and two or more independent variables or at least one nonlinear independent variable.

Multiplicative model   Value of time series variable equals the product of the four component effects.

Mutually exclusive events   Events such that the occurrence of one precludes the occurrence of the other.

Naïve forecasting method   Forecasting based on the assumption that the more recent time periods of data represent the best forecast for future values.

Nominal level data   The lowest level of data measurement; used only to classify or categorize.

Nonlinear regression model   Multiple regression models in which the models are nonlinear, such as polynomial models, logarithmic models, and exponential models.

Non-linear trend projection   Forecasting by fitting a non-linear equation to a time series.

Nonrandom sampling   Sampling in which not every unit of the population has the same probability of being selected into the sample.

Nonrandom sampling techniques   Sampling techniques used to select elements from the population by any mechanism that does not involve a random selection process.

Nonrejection region   Any portion of a distribution that is not in the rejection region. If the observed statistic falls in this region, the decision is to fail to reject the null hypothesis.

Nonsampling errors   All errors other than sampling errors.

Normal distribution   A widely known and much-used continuous distribution that fits the measurements of many human characteristics and many machine-produced items.

Null hypothesis   The hypothesis that assumes the status quo—that the old theory, method, or standard is still true; the complement of the alternative hypothesis.

Observed significance level   Another name for the p-value method of testing hypotheses.

Ogive   A cumulative frequency polygon; plotted by graphing a dot at each class endpoint for the cumulative or decumulative frequency value and connecting the dots.

One-tailed test   A statistical test wherein the researcher is interested only in testing one side of the distribution.

One-way analysis of variance   The process used to analyze a completely randomized experimental design. This process involves computing a ratio of the variance between treatment levels of the independent variable to the error variance. This ratio is an F value, which is then used to determine whether there are any significant differences between the means of the treatment levels.

Ordinal level data   Next-higher level of data from nominal level data; can be used to order or rank items, objects, or people.

Outliers   Data points that lie apart from the rest of the points.

P chart   A quality control chart for attribute compliance that graphs the proportion of sample items in noncompliance with specifications for multiple samples.

p-value method   A method of testing hypotheses in which there is no preset level of α. The probability of getting a test statistic at least as extreme as the observed test statistic is computed under the assumption that the null hypothesis is true. This probability is called the p value, and it is the smallest value of α for which the null hypothesis can be rejected.

Paired data   Data gathered from pairs of items or persons that are matched on some characteristic or from a before-and-after design and then separated into different samples; also called matched pairs data or related measures.

Parameter   A descriptive measure of the population.

Pareto analysis   A quantitative tallying of the number and types of defects that occur with a product or service, often recorded in a Pareto chart.

Pareto chart   A vertical bar chart in which the number and types of defects for a product or service are graphed in order of magnitude from greatest to least.

Partial regression coefficient   The coefficient of an independent variable in a multiple regression model that represents the increase that will occur in the value of the dependent variable from a 1-unit increase in the independent variable if all other variables are held constant.

Pearson product-moment correlation coefficient (r)   A correlation measure used to determine the degree of relatedness of two variables that are at least of interval level.

Percentage error (PE)   The ratio of the error of a forecast to the actual value being forecast, multiplied by 100.

Pie chart   A circular depiction of data where the area of the whole pie represents 100% of the data being studied and slices represent a percentage breakdown of the sublevels.

Platykurtic   Distributions that are flat and spread out.

Point estimate   An estimate of a population parameter constructed from a statistic taken from a sample.

Poisson distribution   A discrete distribution that is constructed from the probability of occurrence of rare events over an interval; focuses only on the number of discrete occurrences over some interval or continuum.

Population   A collection of persons, objects, or items of interest.

Power   The probability of rejecting a false null hypothesis.

Probabilistic model A model that includes an error term that allows for various values of output to occur for a given value of input.

Probability matrix A two-dimensional table that displays the marginal and intersection probabilities of a given problem.

Process A series of actions, changes, or functions that bring about a result.

Product quality A view of quality in which quality is measurable in the product based on the fact that there are perceived differences in products and quality products possess more attributes.

Proportionate stratified random sampling A type of stratified random sampling in which the proportions of the items selected for the sample from the strata reflect the proportions of the strata in the population.

Quadratic regression model A multiple regression model in which the predictors are a variable and the square of the variable.

Qualitative data Data of the nominal or ordinal level that classifies by a nonnumeric or numeric label or category.

Qualitative techniques Forecasting techniques based on historic data.

Qualitative variable Another name for a dummy or indicator variable; represents whether or not a given item or person possesses a certain characteristic and is usually coded as 0 or 1.

Quality When a product delivers what is stipulated in its specifications.

Quality circle A small group of workers consisting of supervisors and six to 10 employees who meet frequently and regularly to consider quality issues in their department or area of the business.

Quality control The collection of strategies, techniques, and actions taken by an organization to ensure the production of quality products.

Quantitative data Data of the interval or ratio level that measures on a naturally occurring numeric scale.

Quantitative techniques Forecasting techniques based on historic data.

Quartiles Measures of central tendency that divide a group of data into four subgroups or parts.

Quota sampling A nonrandom sampling technique in which the population is stratified on some characteristic and then elements selected for the sample are chosen by nonrandom processes.

R chart A plot of sample ranges used in quality control.

R^2 The coefficient of multiple determination; a value that ranges from 0 to 1 and represents the proportion of the dependent variable in a multiple regression model that is accounted for by the independent variables.

Random sampling Sampling in which every unit of the population has the same probability of being selected for the sample.

Random variable A variable that contains the outcomes of a chance experiment.

Randomized block design An experimental design in which there is one independent variable of interest and a second variable, known as a blocking variable, that is used to control for confounding or concomitant variables.

Range The difference between the largest and the smallest values in a set of numbers.

Ratio level data Highest level of data measurement; contains the same properties as interval level data, with the additional property that zero has meaning and represents the absence of the phenomenon being measured.

Rectangular distribution A relatively simple continuous distribution in which the same height is obtained over a range of values; also referred to as the uniform distribution.

Reengineering A radical approach to total quality management in which the core business processes of a company is redesigned.

Regression The process of constructing a mathematical model or function that can be used to predict or determine one variable by any other variable.

Rejection region If a computed statistic lies in this portion of a distribution, the null hypothesis will be rejected.

Related measures Another name for matched pairs or paired data in which measurements are taken from pairs of items or persons matched on some characteristic or from a before-and-after design and then separated into different samples.

Related samples Another name for dependent samples, where each item in one sample has a corresponding matched or related item in the other sample.

Relative frequency The proportion of the total frequencies that fall into any given class interval in a frequency distribution.

Relative frequency of occurrence Assigning probability based on cumulated historical data.

Repeated measures design A randomized block design in which each block level is an individual item or person, and that person or item is measured across all treatments.

Residual The difference between the actual Y value and the Y value predicted by the regression model; the error of the regression model in predicting each value of the dependent variable.

Residual plot A type of graph in which the residuals for a particular regression model are plotted along with their associated values of X.

Response plane A plane fit in a three-dimensional space and that represents the response surface defined by a multiple regression model with two independent first-order variables.

Response surface The surface defined by a multiple regression model.

Response variable The dependent variable in a multiple regression model; the variable that the researcher is trying to predict.

Robust Describes a statistical technique that is relatively insensitive to minor violations in one or more of its underlying assumptions.

Sample   A portion of the whole.

Sample proportion   The quotient of the frequency at which a given characteristic occurs in a sample and the number of items in the sample.

Sample-size estimation   An estimate of the size of sample necessary to fulfill the requirements of a particular level of confidence and to be within a specified amount of error.

Sample space   A complete roster or listing of all elementary events for an experiment.

Sampling error   Error that occurs when the sample is not representative of the population.

Scatter plot   A plot or graph of the pairs of data from a simple regression analysis.

Search procedures   Processes whereby more than one multiple regression model is developed for a given database, and the models are compared and sorted by different criteria, depending on the given procedure.

Seasonal autoregression   Autoregression model with an independent variable lagged by season period.

Seasonal component   Periodic increases and decreases that occur within one year.

Seasonal multiple regression   Forecasting by using dummy variables to represent seasons.

Set notation   The use of braces to group numbers that have some specified characteristic.

Simple average   The arithmetic mean or average for the values of a given number of time periods of data.

Simple average model   A forecasting averaging model in which the forecast for the next time period is the average of values for a given number of previous time periods.

Simple random sampling   The most elementary of the random sampling techniques; involves numbering each item in the population and using a list or roster of random numbers to select items for the sample.

Simple regression   Bivariate, linear regression.

Six sigma   A total quality management approach that measures the capability of a process to perform defect-free work, where a defect is defined as anything that results in customer dissatisfaction.

Skewness   The lack of symmetry of a distribution of values.

Snowball sampling   A nonrandom sampling technique in which survey subjects who fit a desired profile are selected based on referral from other survey respondents who also fit the desired profile.

Standard deviation   The square root of the variance.

Standard error of the estimate (S_e)   A standard deviation of the error of a regression model.

Standard error of the mean   The standard deviation of the distribution of sample means.

Standard error of the proportion   The standard deviation of the distribution of sample proportions.

Standardized normal distribution Z distribution; a distribution of Z scores produced for values from a normal distribution with a mean of 0 and a standard deviation of 1.

Stationary time series   A time series that does not include a trend, seasonal, or cyclical component—only an irregular component.

Statistic   A descriptive measure of a sample.

Statistics   A science dealing with the collection, analysis, interpretation, and presentation of numerical data.

Stepwise regression   A step-by-step multiple regression search procedure that begins by developing a regression model with a single predictor variable and adds and deletes predictors one step at a time, examining the fit of the model at each step until there are no more significant predictors remaining outside the model.

Stratified random sampling   A type of random sampling in which the population is divided into various nonoverlapping strata and then items are randomly selected into the sample from each stratum.

Subjective probability   A probability assigned based on the intuition or reasoning of the person determining the probability.

Sum of squares of error (SSE)   The sum of the residuals squared for a regression model.

Sum of squares of X   The sum of the squared deviations about the mean of a set of values.

Systematic sampling   A random sampling technique in which every kth item or person is selected from the population.

t distribution   A distribution that describes the sample data in small samples when the standard deviation is unknown and the population is normally distributed.

t value   The computed value of t used to reach statistical conclusions regarding the null hypothesis in small-sample analysis.

Team building   When a group of employees are organized as an entity to undertake management tasks and perform other functions such as organizing, developing, and overseeing projects.

Time series   Data values for a variable recorded over a period of time at regular intervals.

Time series decomposition   Breaking down a time series into its components.

Time series plot   A graph of the values for a variable versus time.

Time series techniques   Forecasting techniques based on historic data only for the variable being forecast.

Total quality management (TQM)   A program that occurs when all members of an organization are involved in improving quality; all goals and objectives of the organization come under the purview of quality control and are measured in quality terms.

Transcendent quality   A view of quality that implies that a product has an innate excellence, uncompromising standards, and high achievement.

Treatment variable   The independent variable of an experimental design that the researcher either controls or modifies.

Trend autoregression   Autoregression model with a one-period lagged independent variable.

Trend component   Long-term upward or downward change in a time series.

Two-stage sampling   Cluster sampling done in two stages: A first round of samples is taken and then a second round is taken from within the first samples.

Two-tailed test   A statistical test wherein the researcher is interested in testing both sides of the distribution.

Two-way analysis of variance (two-way ANOVA)   The process used to statistically test the effects of variables in factorial designs with two independent variables.

Type I error   An error committed by rejecting a true null hypothesis.

Type II error   An error committed by failing to reject a false null hypothesis.

Ungrouped data   Raw data, or data that have not been summarized in any way.

Uniform distribution   A relatively simple continuous distribution in which the same height is obtained over a range of values; also called the rectangular distribution.

Union   A new set of elements formed by combining the elements of two or more other sets.

Union probability   The probability of one event occurring or the other event occurring or both occurring.

Upper control limit (UCL)   The top-end line of a control chart, usually situated approximately three standard deviations of the statistic above the centerline; data points above this line indicate quality control problems.

User quality   A view of quality in which the quality of the product is determined by the user.

Value quality   A view of quality having to do with price and costs and whether the consumer got his or her money's worth.

Variance   The average of the squared deviations about the arithmetic mean for a set of numbers.

Variance inflation factor   A statistic computed using the R^2 value of a regression model developed by predicting one independent variable of a regression analysis by other independent variables; used to determine whether there is multicollinearity among the variables.

Weighted moving average   A moving average in which different weights are applied to the data values from different time periods.

Winters' model   Exponential smoothing with a seasonal and/or trend component.

$\bar{X}$ chart   A quality control chart for measurements that graphs the sample means computed for a series of small random samples over a period of time.

Z distribution   A distribution of Z scores; a normal distribution with a mean of 0 and a standard deviation of 1.

Z score   The number of standard deviations a value (X) is above or below the mean of a set of numbers when the data are normally distributed.

Index